MW01253136

The Most Successful Way of JEE Preparation

DPP

Daily Practice Problems

JEE Main & Advanced

Electromagnetic Induction & Alternating Current

with Magnetic Effect of Current & Magnetism

PHYSICS Vol-7

The Most Successful Way of JEE Preparation

DPP

Daily Practice Problems

JEE Main & Advanced

Electromagnetic Induction & Alternating Current

with Magnetic Effect of Current & Magnetism

PHYSICS Vol-7

Nitesh Bharti

arihant
ARIHANT PRAKASHAN (SERIES), MEERUT

DPP

Daily Practice Problems

arihant

ARIHANT PRAKASHAN (SERIES), MEERUT

Administrative & Production Offices

Corporate Office 'Ramchhaya' 4577/15, Agarwal Road, Darya Ganj, New Delhi -110002
Tele: 011- 47630600, 43518550; Fax: 011- 23280316

Head Office Kalindi, TP Nagar, Meerut (UP) - 250002
Tel: 0121-2401479, 2512970, 4004199; Fax: 0121-2401648

Sales & Support Offices

Agra, Ahmedabad, Bengaluru, Bareilly, Chennai, Delhi, Guwahati, Hyderabad, Jaipur, Jhansi, Kolkata, Lucknow, Meerut, Nagpur & Pune

ISBN : 978-93-13193-37-1

Price : ₹230

Published by Arihant Publications (I) Ltd.

Production Team

Publishing Managers Mahendra Singh Rawat & Keshav Mohan
Project Head Sorabh Chaudhary
Cover Designer Shanu Mahad
Inner Designer Mazher Chaudhary
Page Layouting Sunil Sharma, Hemant
Proof Reading Ashish

For further information about the products from Arihant, log on to www.arihantbooks.com or email to info@arihantbooks.com

/arihantpub /@arihantpub Arihant Publications /arihantpub

ABOUT DPP

Daily Practice Problems

Indian Institutes of Technology better known as IITs have always been the dream destination of each science group students who want to pursue their career as Engineers. These are the institutes which have got very high value not even in India but in the whole world. To get the admission in such institutes one need to face entrance Examinations; JEE Main & JEE Advanced, out of which JEE Main is considered as the preliminary exam & JEE Advanced is the final exam to get into the IITs. Approximately per year 12 lakh students enroll themselves for JEE Main and only the top 2.25 lakh in JEE Main merit can attempt for JEE Advanced, the fact clearly express the degree of toughness of JEE Main & JEE Advanced.

To have success in such an exam one needs to be very planned & systematic in its preparation. A very popular wise saying 'Rome was not built in a day' suggests that to achieve a big target like IITs we need to work continuously. For continuous work, one need to have proper planning that how much work he/she has to do on regular basis, i.e., on daily basis. Each student preparing for JEE has many books but unfortunately none of them gives the students a plan for daily studies, which is the most effective success key for JEE.

Keeping the daily practice needs of the students in mind we have come up with DPP Series (Daily Practice Problems) having approximately 26 books. The primary focus of this series is on achieving success through practice & proper planning. For that we have divided the whole syllabus of JEE Main & JEE Advanced; Physics, Chemistry & Mathematics, into different volumes (around 8 volumes of Mathematics & 9 volumes each of Physics & Chemistry), each volume has Daily Practice Problems Sheets having specific questions on various Topics of the individual chapter of the volume ensuring the complete Practice & Assessment of the topic.

DPP

Daily Practice Problems

In this way we have tried to give you an effective daily planning for JEE keeping in mind the daily hours that a student can devote to his practice, you are just required to practice one sheet of each subject on daily basis. This habit of daily practice will keep you away from exam stress and thoughts like; how much I have prepared/am I preparing properly or not/ do I need to study more on daily basis or what I am doing is ok, etc., which regularly tease your mind and can give you panic at any stage of your preparation.

Some special features that we have tried to incorporate in our Daily Practice Problems (DPP) series are discussed below

1. Micro Level coverage of all the topics of a Chapter/ Unit given in JEE Main & Advanced Syllabus.
2. Each DPP has various types of questions; Subjective Questions, Single/ Multiple Correct Options, Statements type, Comprehension based, Integer Value Answer type, Matching, etc., making you practice & ready for all the question formats of JEE.
3. New & innovative questions in each DPP challenging & improving your problem solving skills.
4. Along with Topical Coverage, Revisal DPPs for JEE Main & JEE Advanced with each Chapter, covering multidisciplinary questions proving comprehensive practice.
5. JEE Main & JEE Advanced Archive (collection of last 13 years' exams questions) with each chapter.
6. Complete Solutions for each DPP.

We hope that DPP series will prove to be very beneficial for the students preparing for JEE, by giving them daily practice in a systematic & planned manner. The suggestions/criticism from the students, teachers & parents about DPP series will be welcomed whole heartedly and we promise to update this series accordingly from time to time.

Daily Practice Problems

DAILY TASKS...

1 MAGNETIC EFFECT OF CURRENT

2 MAGNETISM

DPP

Daily Practice Problems

❸ ELECTROMAGNETIC INDUCTION

❹ ALTERNATING CURRENT

1. Magnetic Effect of Current

DPP

DAILY PRACTICE PROBLEMS

DPP-1 Magnetic Force and Motion of Charged Particle in Magnetic Field

Subjective Questions

Working Space

Direction (*Q. Nos. 1-4*) *These questions are subjective in nature, need to be solved completely on notebook.*

1 i. A charged particle is projected in the magnetic field $\mathbf{B} = (3\hat{\mathbf{i}} + 4\hat{\mathbf{j}}) \times 10^{-2}$ T Acceleration of the particle is found to be

$$\mathbf{a} = (x\hat{\mathbf{i}} + 2\hat{\mathbf{j}})\ \text{m/s}^2$$

Find the value of x.

ii. A proton is moving in the region of magnetic field. When its velocity is $\mathbf{v} = (2\hat{\mathbf{i}} + 3\hat{\mathbf{j}}) \times 10^6$ m/s, it experiences a force of $\mathbf{F} = -(1.28 \times 10^{-13})\hat{\mathbf{k}}$ N. When its velocity is along Z-axis, it experiences a force along X-axis. Find the magnetic field.

iii. A particle of mass m and charge q is projected from origin with velocity $\mathbf{v} = v_0\hat{\mathbf{i}} - v_0\hat{\mathbf{k}}$ in a uniform magnetic field $\mathbf{B} = -B_0\hat{\mathbf{k}}$. Find the velocity and position of particle at time t second.

2 A charged particle of charge q and mass m enters a uniform magnetic field **B** at an angle of θ shown in the figure. Find the

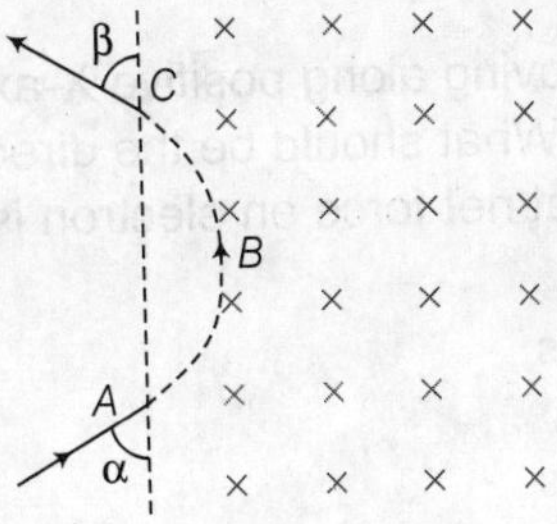

i. length of path travelled by particle in the region of magnetic field.
ii. time spent by the particle in magnetic field.
iii. distance AC.
iv. relation between angle α and β.
v. ratio of KE of particle at point A and at C.
vi. change in momentum of particle at A and C.

3 A particle of mass 1.6×10^{-27} kg and charge 1.6×10^{-19} C enters a region of a uniform magnetic field of strength 1 T along the direction as shown in figure. The speed of the particle is 10^{-7} ms^{-1}. The magnetic field is directed along the inward normal to the plane of the paper. The particle leaves the region of the field at the point F. Based on the above facts, answer the following :

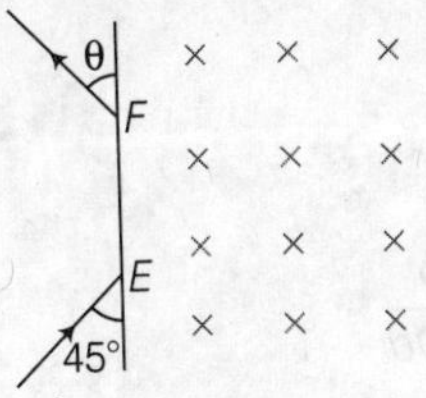

i. the value of θ.
ii. the radius of the circular path and the length EF.

4 A proton and an α-particle are projected with the same kinetic energy at right angles to a uniform magnetic field. The radius of their circular paths are r_p and r_α, respectively. Find the ratio of $\frac{r_p}{r_\alpha}$.

Working Space

Only One Option Correct Type

Direction *(Q. Nos. 5-20) This section contains 16 multiple choice questions. Each question has four choices (a), (b), (c) and (d), out of which ONLY ONE is correct.*

5 A particle with a specific charge s is fired with a speed v towards a wall at a distance d perpendicular to the wall. What minimum magnetic field must exist in this region for the particle not to hit the wall?

a. $\frac{v}{sd}$

b. $\frac{2v}{sd}$

c. $\frac{v}{2sd}$

d. $\frac{v}{4sd}$

6 Three ions H^+, He^+ and O^{+2} having same kinetic energy passing through a region in which there is a uniform magnetic field perpendicular to their velocity, then

a. H^+ will be least deflected

b. He^+ and O^{+2} will be deflected equally

c. O^{+2} will be deflected most

d. all will be deflected equally

7 A charged particle moves with a velocity $\mathbf{v} = a\hat{\mathbf{i}} + d\,\hat{\mathbf{j}}$ in a magnetic field $\mathbf{B} = A\hat{\mathbf{i}} + D\hat{\mathbf{j}}$. The force acting on the particle has magnitude of

a. $F = 0$, if $aD = dA$

b. $F = 0$, if $aD = -dA$

c. $F = 0$, if $aA = -dD$

d. $F \propto (a^2 + b^2)^{1/2} \times (A^2 + D^2)^{1/2}$

8 An electron is moving along positive X-axis. A uniform electric field exists towards negative Y-axis. What should be the direction of magnetic field of suitable magnitude so that net force on electron is zero?

a. Positive Z-axis

b. Negative Z-axis

c. Positive Y-axis

d. Negative Y-axis

9 A particle with a charge Q moving with a momentum p enters a uniform magnetic field normally. The magnetic field has magnitude B and is confined to a region of width d, where $d < \frac{p}{BQ}$. The particle is deflected by an angle θ in crossing the field.

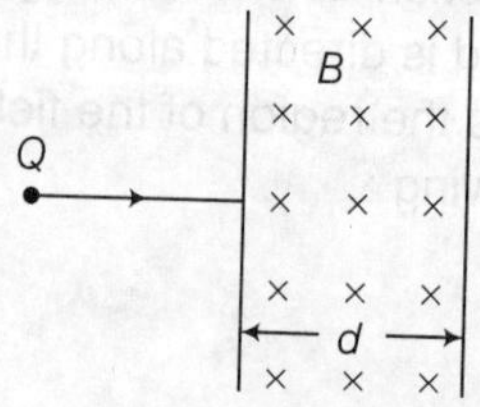

a. $\sin\theta = \frac{BQd}{p}$

b. $\sin\theta = \frac{p}{BQd}$

c. $\sin\theta = \frac{Bp}{Qd}$

d. $\sin\theta = \frac{pd}{BQ}$

10 A uniform magnetic field is directed out of the page. A charged particle moving in the plane of the page follows a clockwise spiral of decreasing radius as shown in the figure. A reasonable explanation is that a

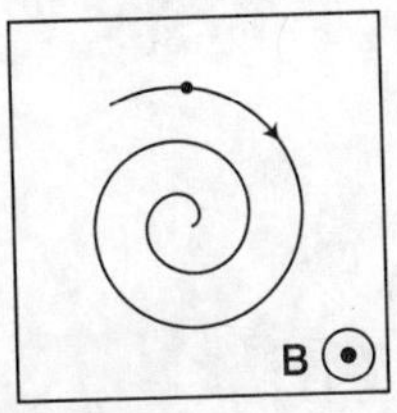

a. charge is positive and slowing down
b. charge is negative and slowing down
c. charge is positive and speeding up
d. charge is negative and speeding up

11 For the given figure, a charged sphere of mass *m* and charge *q* starts sliding from rest on a vertical fixed circular track of radius *R* from the position shown below. There exists a uniform and constant horizontal magnetic field of induction *B*.

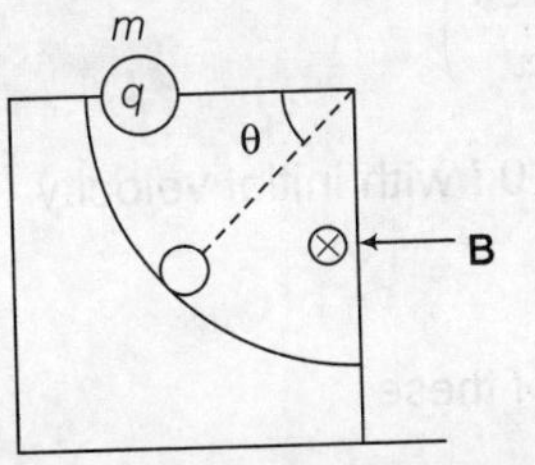

The maximum force exerted by the track on the sphere is
a. mg
b. $3mg - qB\sqrt{2gR}$
c. $3mg + qB\sqrt{2gR}$
d. $mg - qB\sqrt{2gR}$

12 A charged particle begins to move from the origin in a region which has a uniform magnetic field in the *X*-direction and a uniform electric field in the *Y*-direction. Its speed is *v* when it reaches to the point (x, y, z), then *v* will depend
a. only on *x*
b. only on *y*
c. on both *x* and *y* but not *z*
d. on *x*, *y* and *z*

13 Two charged particles having charges q_1 and q_2 and masses m_1 and m_2 are projected with same velocity in a region of a uniform magnetic field. They follow the trajectory as shown in the figure below.

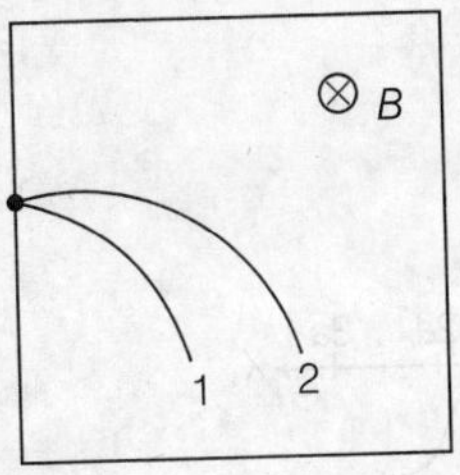

From this, we can conclude that
a. $q_1 > q_2$
b. $q_1 < q_2$
c. $m_1 < m_2$
d. None of these

Working Space

14 A particle with charge $+Q$ and mass m enters a magnetic field of magnitude B existing only to the right of boundary YZ. The direction of motion of the particle is perpendicular to the direction of **B**. Let $T = 2\pi \dfrac{m}{QB}$.

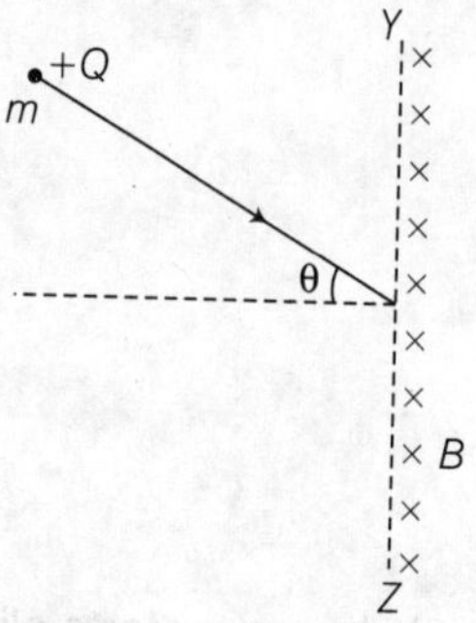

The time spent by the particle in the field will be

a. $T\theta$

b. $2T\theta$

c. $T\left(\dfrac{\pi + 2\theta}{2\pi}\right)$

d. $T\left(\dfrac{\pi - 2\theta}{2\pi}\right)$

15 A charged particle moves in a magnetic field $\mathbf{B} = 10\,\hat{\mathbf{i}}$ with initial velocity $\mathbf{u} = 5\hat{\mathbf{i}} + 4\hat{\mathbf{j}}$. The path of the particle will be

a. straight line

b. circle

c. helical path

d. None of these

16 A particle of specific charge α starts moving from the origin under the action of an electric field $\mathbf{E} = E_0\hat{\mathbf{i}}$ and magnetic field $\mathbf{B} = B_0\,\hat{\mathbf{k}}$. Its velocity at $(x_0, y_0, 0)$ is $(4\hat{\mathbf{i}} + 3\hat{\mathbf{j}})$. The value of x_0 is

a. $\dfrac{13}{2} \cdot \dfrac{\alpha E_0}{B_0}$

b. $\dfrac{16\alpha B_0}{E_0}$

c. $\dfrac{25}{2\alpha E_0}$

d. $\dfrac{5\alpha}{2B_0}$

17 A magnetic field $\mathbf{B} = B_0\hat{\mathbf{j}}$ exists in the region $a < x < 2a$ and $\mathbf{B} = -B_0\hat{\mathbf{j}}$ in the region $2a < x < 3a$, where B_0 is a positive constant. A positive point charge moving with a velocity $\mathbf{v} = v_0\hat{\mathbf{i}}$, where v_0 is a positive constant enters the magnetic field at $x = a$. The trajectory of the charge in this region can be like

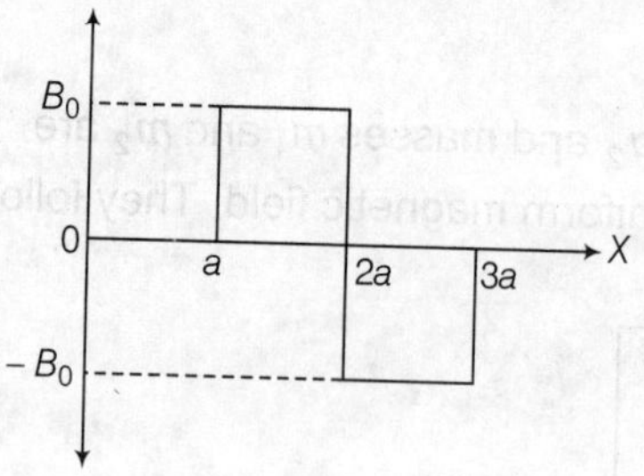

a.
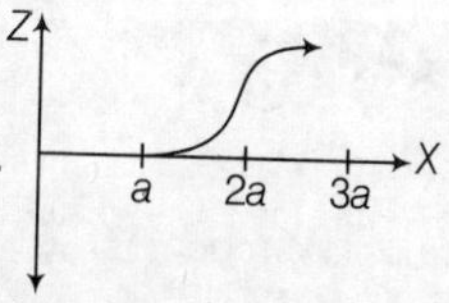

b.
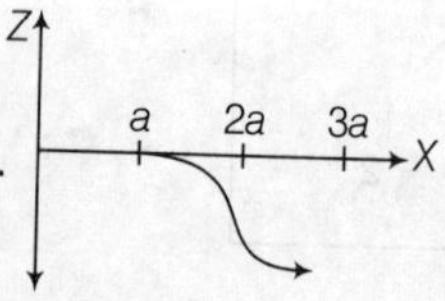

c.
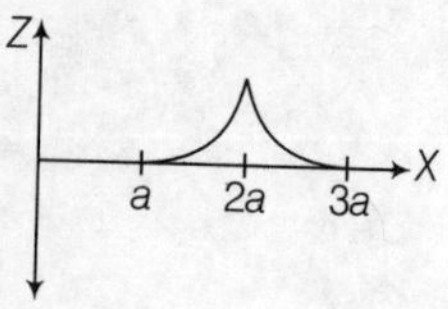

d.
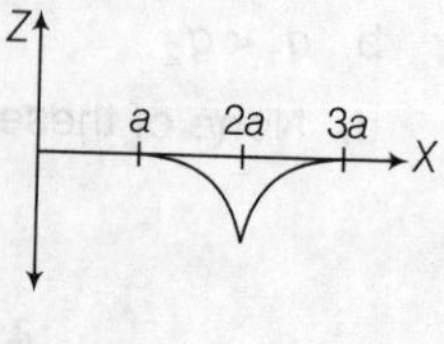

18 A charged particle of specific charge α is released from origin at $t = 0$ with velocity $\mathbf{v} = v_0(\hat{\mathbf{i}} + \hat{\mathbf{j}})$ in a uniform magnetic field $\mathbf{B} = B_0\hat{\mathbf{i}}$. The coordinates of particle at time $t = \dfrac{\pi}{B_0\alpha}$ are

a. $\left(\dfrac{v_0}{2B_0\alpha}, \dfrac{\sqrt{2}v_0}{\alpha B_0}, \dfrac{-v_0}{B_0\alpha}\right)$
b. $\left(\dfrac{-v_0}{2B_0\alpha}, 0, 0\right)$
c. $\left(0, \dfrac{2v_0}{B_0\alpha}, \dfrac{v_0\pi}{2B_0\alpha}\right)$
d. $\left(\dfrac{v_0\pi}{B_0\alpha}, 0, \dfrac{-2v_0}{B_0\alpha}\right)$

19 A particle of charge per unit mass α is released with a velocity $\mathbf{v} = v_0\hat{\mathbf{i}}$ in a uniform magnetic field $\mathbf{B} = -B_0\hat{\mathbf{k}}$. If the particle passes through $(0, y, 0)$, then y is equal to

a. $\dfrac{-2v_0}{B_0\alpha}$
b. $\dfrac{v_0}{B_0\alpha}$
c. $\dfrac{2v_0}{B_0\alpha}$
d. $\dfrac{-v_0}{B_0\alpha}$

20 An electron moving with a speed u along the positive X-axis at $y = 0$ enter a uniform magnetic field which exists to the right of Y-axis. The electron exists from the region after sometime with the speed v at coordinate y, then

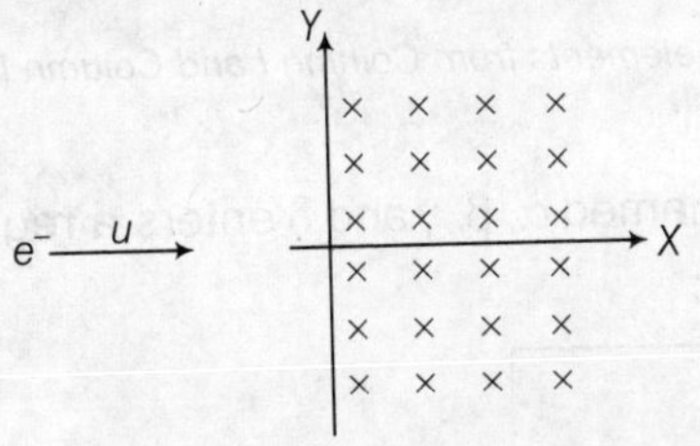

a. $v > u, y < 0$
b. $v = u, y > 0$
c. $v > u, y > 0$
d. $v = u, y < 0$

One or More than One Options Correct Type

Direction *(Q. Nos. 21-23) This section contains 3 multiple choice questions. Each question has four choices (a), (b), (c) and (d), out of which ONE or MORE THAN ONE are correct.*

21 A charged particle with velocity $\mathbf{v} = x\hat{\mathbf{i}} + y\hat{\mathbf{j}}$ moves in a magnetic field $\mathbf{B} = y\hat{\mathbf{i}} + x\hat{\mathbf{j}}$. The force acting on the particle has magnitude F. Which one of the following statement(s) is/are correct?

a. If $x = y$, no force will act on charged particle
b. If $x > y$, $F \propto (x^2 - y^2)$
c. If $x > y$, force will act along Z-axis
d. If $y > x$, force will act along X-axis

22 A particle of mass m and charge q moving with velocity v enters Region II normal to the boundary as shown in the figure. Region II having a uniform magnetic field B is perpendicular to the plane of the paper. The length of the Region II is l. Choose the correct choice(s).

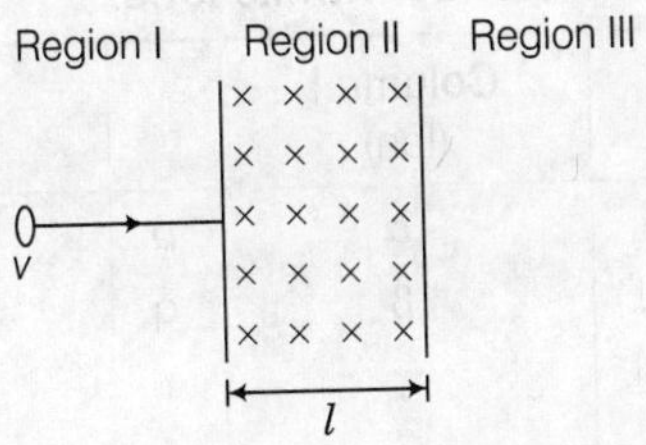

a. The particle enters Region III only if its velocity $v > \dfrac{qlB}{m}$
b. The particle enters Region III only if its velocity $v < \dfrac{qlB}{m}$
c. Path length of the particle in Region II is maximum when velocity $v = \dfrac{qlB}{m}$
d. Time spent in Region II is same for any velocity v as long as the particle returns to Region I

23 For a positively charged particle moving in a XY-plane initially along the X-axis, there is a sudden change in its path due to the presence of electric and/or magnetic fields beyond P. The curved path is shown in the XY-plane and is found to be non-circular.

Working Space

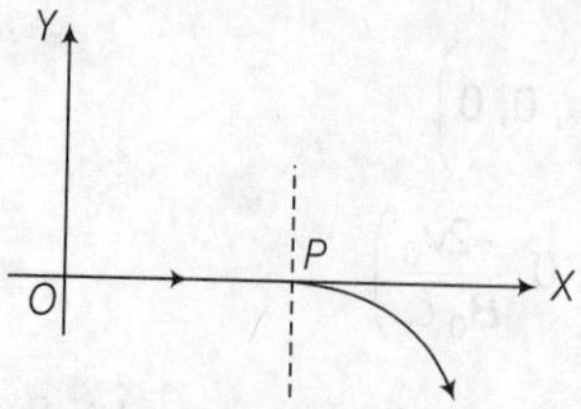

Which one of the following is correct for the given situation?

a. $\mathbf{E} = 0, \mathbf{B} = b\hat{\mathbf{j}} + c\hat{\mathbf{k}}$

b. $\mathbf{E} = a\hat{\mathbf{i}}, \mathbf{B} = c\hat{\mathbf{k}} + a\hat{\mathbf{i}}$

c. $\mathbf{F}_{\text{net}} = qa\hat{\mathbf{i}} - qvc\hat{\mathbf{j}} + qvb\hat{\mathbf{k}}$

d. $\mathbf{F}_{\text{net}} = qa\hat{\mathbf{i}} - qvc\hat{\mathbf{j}}$

Matching List Type

Direction *(Q. No. 24) Choices for the correct combination of elements from Column I and Column II are given as options (a), (b), (c) and (d), out of which one is correct.*

24 A beam consisting of 4 different types of ions named α, β, γ and δ enters a region of magnetic field as shown in the figure below :

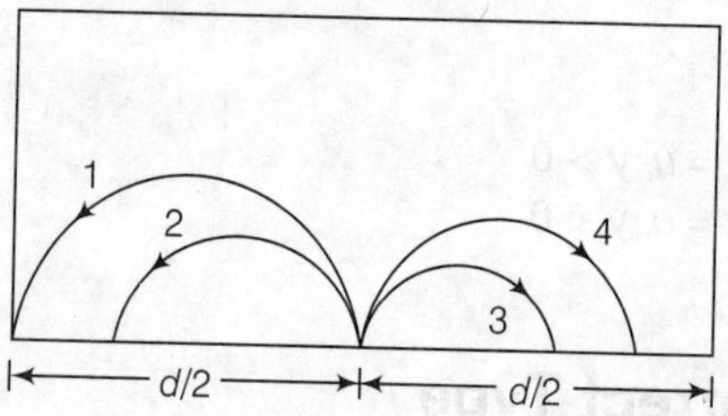

Magnetic field is perpendicular to the velocity of beam. All ions in the beam travels with same speed.

Mass and charge of ions are given as

Ion	Mass	Charge
α	$2m$	e
β	$4m$	e^-
γ	$2m$	e^-
δ	$3m$	e

Now, match the trajectories with its ions.

	Column I (Ion)		Column II (Path)
i.	α	p.	1
ii.	β	q.	2
iii.	γ	r.	3
iv	δ	s.	4

Codes

	i	ii	iii	iv
a.	p	r	s	q
b.	q	s	r	p
c.	r	p	q	s
d.	s	q	p	r

Working Space

Comprehension Type

Direction (Q. Nos. 25 and 26) *This section contains a paragraph, which describing theory, experiments, data, etc. Two questions related to the paragraph have been given. Each question has only one correct answer among the four given options (a), (b), (c) and (d).*

Passage

Magnetic force on a charged particle is given by $\mathbf{F}_m = q(\mathbf{v} \times \mathbf{B})$ and electrostatic force $\mathbf{F}_e = q\mathbf{F}$.

A particle having charge $q = 1\,\text{C}$ and mass $m = 1\,\text{kg}$ is released from the rest at origin. There are electric and magnetic fields given by

$\mathbf{E} = (10\hat{\mathbf{i}})$ N/C for $x \le 1.8$ m and

$\mathbf{B} = (-5\hat{\mathbf{k}})$ T for $1.8\text{ m} \le x \le 2.4$ m

A screen is placed parallel to YZ-plane at $x = 3.0$ m. Neglect gravity forces.

25 y-coordinate of particle, where it collides with the screen is

a. $\frac{0.6(\sqrt{3}-1)}{\sqrt{3}}$ m

b. $\frac{0.6(\sqrt{3}+1)}{\sqrt{3}}$ m

c. $1.2(\sqrt{3}+1)$ m

d. $\frac{1.2(\sqrt{3}-1)}{\sqrt{3}}$ m

26 Time after which the particle will collide the screen is

a. $\frac{1}{5}\left(3+\frac{\pi}{6}+\frac{1}{\sqrt{3}}\right)$ s

b. $\frac{1}{5}\left(6+\frac{\pi}{3}+\sqrt{3}\right)$ s

c. $\frac{1}{3}\left(5+\frac{\pi}{6}+\frac{1}{\sqrt{3}}\right)$ s

d. $\frac{1}{3}\left(6+\frac{\pi}{18}+\sqrt{3}\right)$ s

Statement Type

Direction (Q. Nos. 27 and 28) *This section is based on Statement I and Statement II. Select the correct answer from the codes given below.*

a. Both Statement I and Statement II are correct and Statement II is the correct explanation of Statement I
b. Both Statement I and Statement II are correct but Statement II is not the correct explanation of Statement I
c. Statement I is correct but Statement II is incorrect
d. Statement II is correct but Statement I is incorrect

27 **Statement I** A charged particle is moving in a circle with constant speed in a uniform magnetic field. If we increase the speed of particle to twice, then its acceleration becomes four times.

Statement II In circular path, acceleration of particle is given by $\frac{v^2}{r}$.

28 **Statement I** A magnetic field produces a force on a moving charge but not on a static charge.

Statement II A moving charge produces a magnetic field.

DPP-2 Biot-Savart's Law and Its Applications

Subjective Questions

Direction (Q. Nos. 1-4) *These questions are subjective in nature, need to be solved completely on notebook.*

Working Space

1 Determine magnetic field at point P in the following cases :

i.

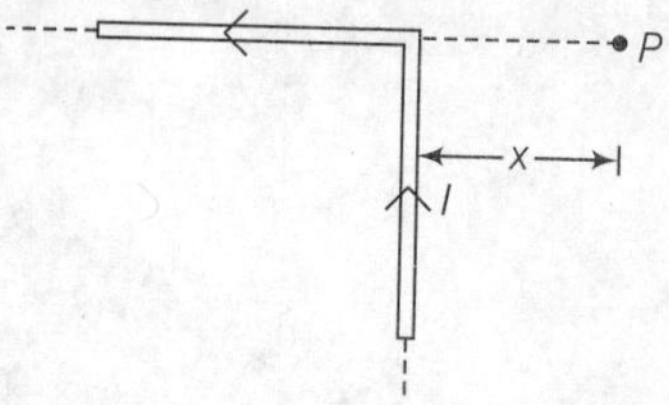

ii.

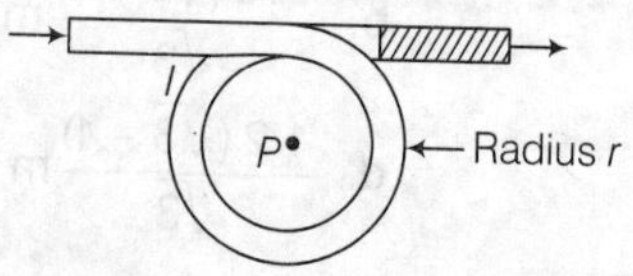

iii.

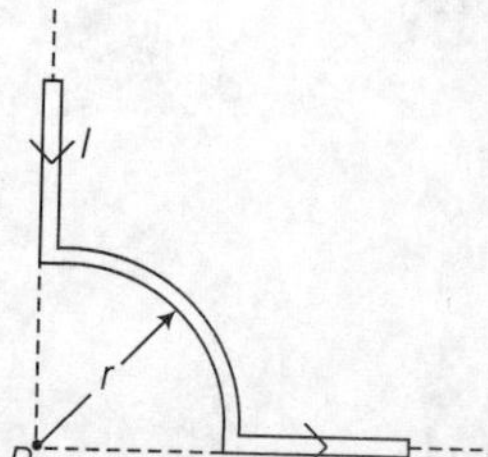

iv.

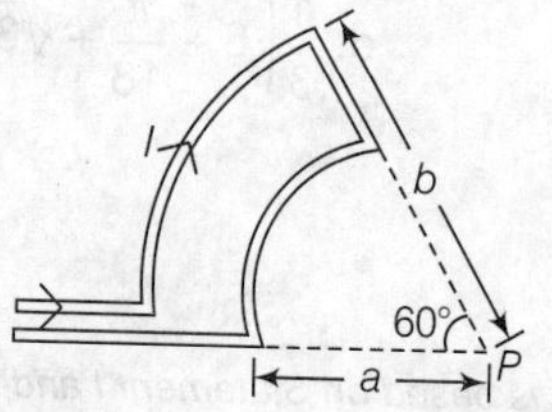

2 Two long straight wires a and b are 2.0 m apart perpendicular to the plane of the paper as shown in figure. The wire a carries a current of 9.6 A directed into the plane of the figure.

The magnetic field at the point P at a distance of 10/11 m from the wire b is zero. Find

i. the magnitude and direction of the current in b.

ii. magnitude of the magnetic field B at the points.

iii. force per unit length on the wire b.

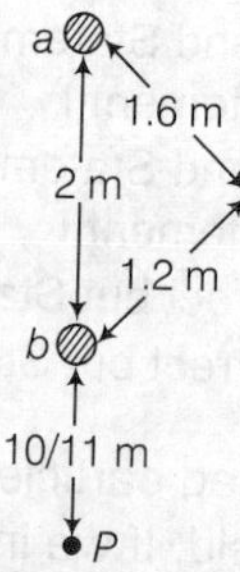

3 A battery is connected between two points A and B on the circumference of a uniform conducting ring of radius r and resistance R as shown in the figure.

One of the arcs AB of the ring subtends angle θ at the centre. Show that the magnetic field at the centre of the coil is zero and independent of θ.

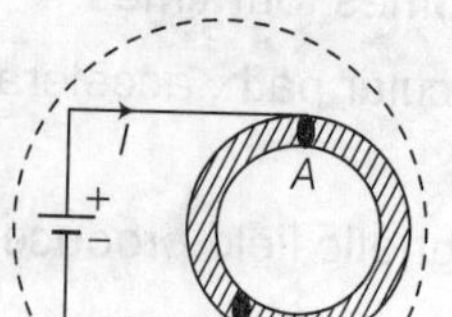

4 A pair of stationary and infinitely long bent wire is placed in the XY-plane as shown in figure. Each wire carries current of 10 A. The segments l and m are along the X-axis. The segments P and Q are parallel to the Y-axis such that $OS = OR = 0.02$ m. Find the magnitude and direction of the magnetic induction at the origin O.

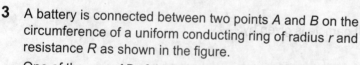

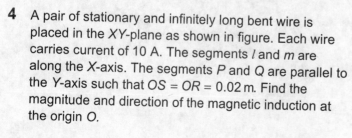

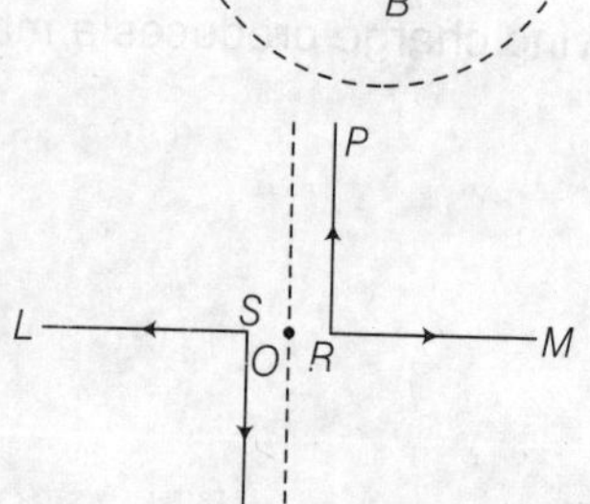

Working Space

Only One Option Correct Type

Direction *(Q. Nos. 5-15) This section contains 11 multiple choice questions. Each question has four choices (a), (b), (c) and (d), out of which ONLY ONE is correct.*

5 A current of i ampere is flowing through each of the bent wires as shown in the figure. The magnitude of magnetic field at O is

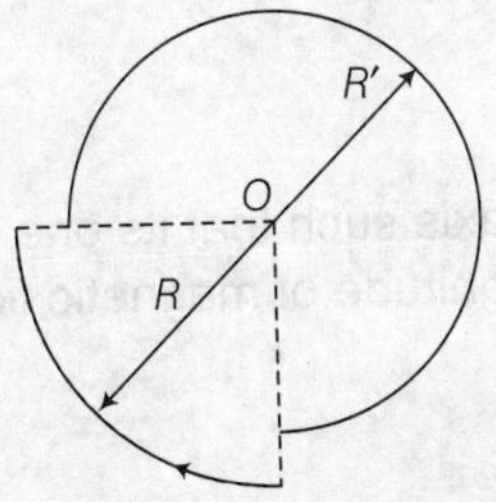

a. $\frac{\mu_0 i}{4}\left(\frac{1}{R}+\frac{2}{R'}\right)$

b. $\frac{\mu_0 i}{4}\left(\frac{1}{R}+\frac{3}{R'}\right)$

c. $\frac{\mu_0 i}{8}\left(\frac{1}{R}+\frac{3}{2R'}\right)$

d. $\frac{\mu_0 i}{8}\left(\frac{1}{R}+\frac{3}{R'}\right)$

6 Find the magnetic field at point P due to arrangement as shown in the figure.

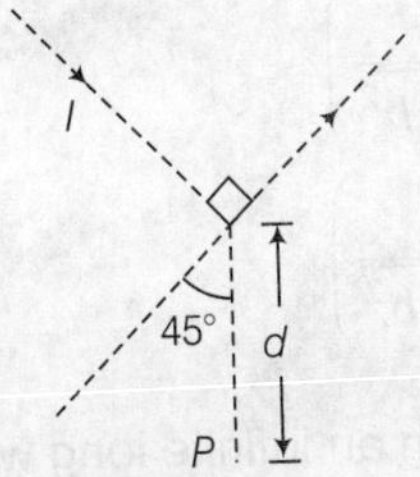

a. $\frac{\mu_0 I}{\sqrt{2}\,\pi d}\left(1-\frac{1}{\sqrt{2}}\right)$

b. $\frac{2\mu_0 I}{\sqrt{2}\,\pi d}$

c. $\frac{\mu_0 I}{\sqrt{2}\,\pi d}$

d. $\frac{\mu_0 I}{\sqrt{2}\,\pi d}\left(1+\frac{1}{\sqrt{2}}\right)$

7 Two long and parallel straight wires A and B carrying currents of 8.0 A and 5.0 A in the same directions are separated by a distance of 4 cm. Estimate the force on a 10 cm section of wire A.

a. 1.5×10^{-5} N

b. 2×10^{-5} N

c. 4×10^{-5} N

d. 3.2×10^{-5} N

8 Equal current i flows in two segments of a circular loop in the direction as shown in the figure. Radius of the loop is r. The magnitude of magnetic field induction at the centre of the loop is

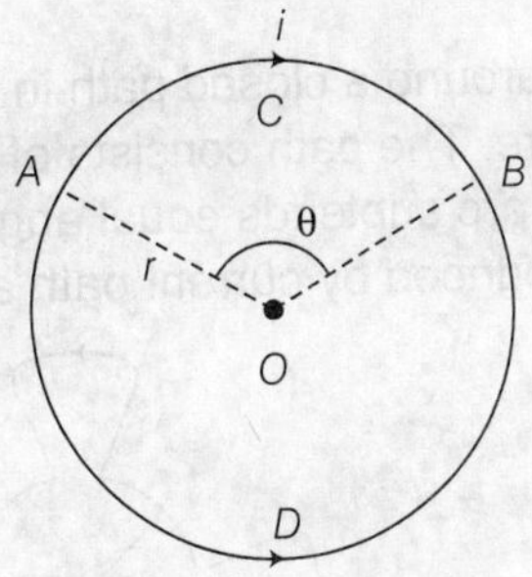

a. zero

b. $\frac{\mu_0 i\theta}{3\pi r}$

c. $\frac{\mu_0}{2\pi}\cdot\frac{i}{r}(\pi-\theta)$

d. $\frac{\mu_0}{2\pi}\cdot\frac{i}{r}(2\pi-\theta)$

9 Two wires PQ and QR carry equal currents i as shown in the figure. One end of both the wires extends to infinity $\angle PQR=\theta$. The magnitude of the magnetic field at O on the bisector angle of these two wires at a distance r from the point Q is

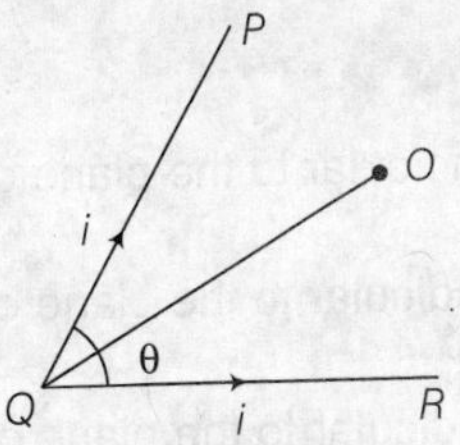

a. $\frac{\mu_0}{4\pi}\cdot\frac{i}{r}\sin\frac{\theta}{2}$

b. $\frac{\mu_0}{4\pi}\cdot\frac{i}{r}\cot\theta$

c. $\frac{\mu_0}{4\pi}\cdot\frac{i}{r}\tan\frac{\theta}{2}$

d. $\frac{\mu_0}{2\pi}\cdot\frac{i}{r}\cdot\frac{(1+\cos\theta/2)}{\sin\theta/2}$

Working Space

10 For a square loop $PQRS$ made with a wire of cross-section current i enters from point P and leaves from point S. The magnitude of magnetic field induction at the centre O of the square is

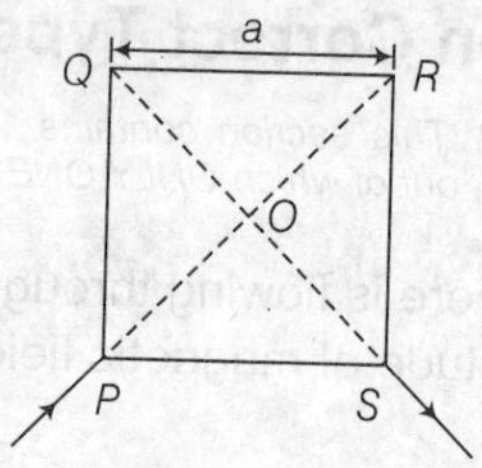

a. $\frac{\mu_0}{4\pi} \cdot \frac{2\sqrt{2}i}{a}$

b. $\frac{\mu_0}{4\pi} \cdot \frac{4\sqrt{2}i}{a}$

c. $\frac{\mu_0}{4} \cdot \frac{2\sqrt{2}i}{a}$

d. zero

11 An infinitely long wire carrying current i is along Y-axis such that its one end is at point $(0, b)$ while the wire extends up to ∞. The magnitude of magnetic field strength at point $P\ (a, 0)$ is

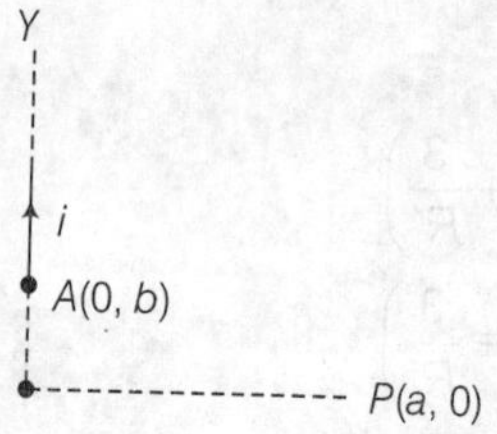

a. $\frac{\mu_0 i}{4\pi a}\left(1+\frac{b}{\sqrt{a^2+b^2}}\right)$

b. $\frac{\mu_0 i}{4\pi a}\left(1-\frac{b}{\sqrt{a^2+b^2}}\right)$

c. $\frac{\mu_0 i}{4\pi a}\left(1-\frac{a}{\sqrt{a^2+b^2}}\right)$

d. $\frac{\mu_0 i}{4\pi a}\left(1+\frac{a}{\sqrt{a^2+b^2}}\right)$

12 A current I flows in an infinite long wire with its cross-section in the form of a semicircular ring of radius R. The magnitude of the magnetic induction along its axis is **[AIEEE 2011]**

a. $\frac{\mu_0 I}{2\pi^2 R}$

b. $\frac{\mu_0 I}{2\pi R}$

c. $\frac{\mu_0 I}{4\pi R}$

d. $\frac{\mu_0 I}{\pi^2 R}$

13 A current I flows around a closed path in the horizontal plane of the circle as shown in the figure. The path consists of eight arcs with alternating radii r and $2r$. Each segment of arc subtends equal angle at the common centre P. The magnetic field produced by current path at point P is

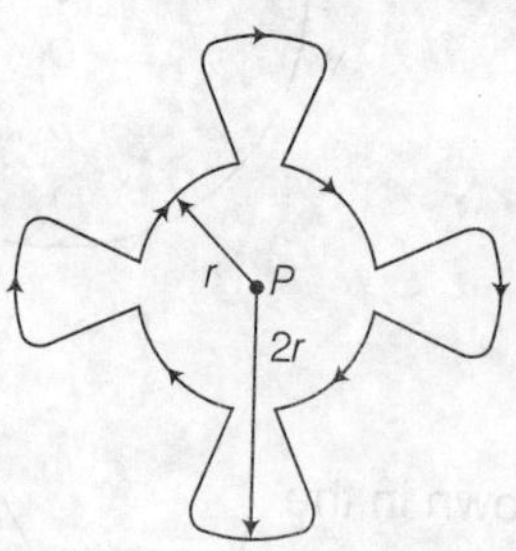

a. $\frac{3}{8} \cdot \frac{\mu_0 I}{r}$; perpendicular to the plane of the paper and directed inwards

b. $\frac{3}{8} \cdot \frac{\mu_0 I}{r}$; perpendicular to the plane of the paper and outwards

c. $\frac{1}{8} \cdot \frac{\mu_0 I}{r}$; perpendicular to the plane of the paper and inwards

d. $\frac{1}{8} \cdot \frac{\mu_0 I}{r}$; perpendicular to the plane of the paper and outwards

14 An equilateral triangle of side l is formed from a piece of a wire of a uniform resistance. The current i is fed as shown in the figure. The magnitude of the magnetic field at its centre O is

a. $\frac{\sqrt{3}\mu_0 i}{2\pi l}$ b. $\frac{3\sqrt{3}\mu_0 i}{2\pi l}$ c. $\frac{\mu_0 i}{2\pi l}$ d. zero

15. A wire loop $PQRSP$ formed by two joining semicircular wires of radii R_1 and R_2 carries a current I as shown in the figure. The magnitude of the magnetic induction at the centre C is

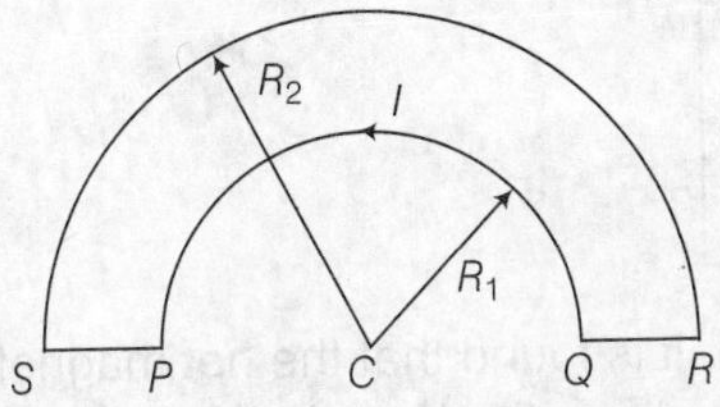

a. $\frac{\mu_0 I}{4}\left(\frac{1}{R_1}+\frac{1}{R_2}\right)$ b. $\frac{\mu_0 I}{4\pi}\left(\frac{1}{R_1}+\frac{1}{R_2}\right)$

c. $\frac{\mu_0 I}{4\pi}\left(\frac{1}{R_1}+\frac{1}{R_2}\right)$ d. $\frac{\mu_0 I}{4}\left(\frac{1}{R_1}-\frac{1}{R_2}\right)$

Comprehension Type

Direction *(Q. Nos. 16-19) This section contains 2 paragraphs, each describing theory, experiments, data, etc. Four questions related to the paragraphs have been given. Each question has only one correct answer among the four given options (a), (b), (c) and (d).*

Passage I

A current loop $ABCD$ is held fixed on the plane of the paper as shown in the figure.

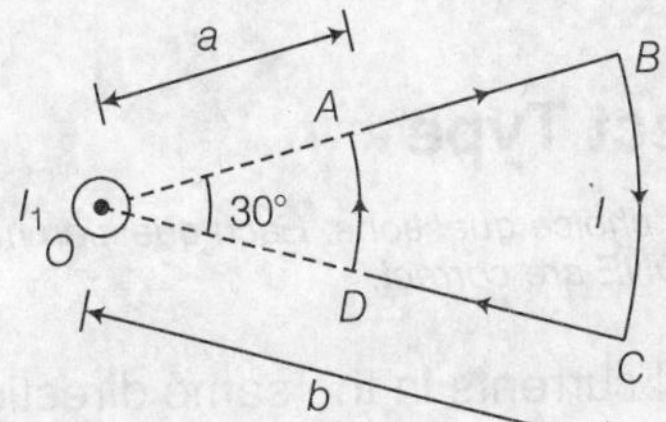

The arcs BC (radius $= b$) and DA (radius $= a$) of the loop are joined by two straight wires AB and CD. A steady current I is flowing in the loop. Angle made by AB and CD at the origin O is 30°. Another straight thin wire with steady current I_1 flowing out of the plane of the paper is kept at the origin.

16 The magnitude of the magnetic field (B) due to loop $ABCD$ at the origin (O) is **[AIEEE 2009]**

a. zero b. $\frac{\mu_0 I(b-a)}{24ab}$

c. $\frac{\mu_0 I}{4\pi}\left[\frac{b-a}{ab}\right]$ d. $\frac{\mu_0 I}{4\pi}\left[2(b-a)+\frac{\pi}{3}(a+b)\right]$

17 Due to the presence of the current I_1 at the origin, **[AIEEE 2009]**

a. the forces on AB and DC are zero
b. the forces on AD and BC are zero
c. the magnitude of the net force on the loop is given by $\frac{\mu_0 II_1}{4\pi}\left[2(b-a)+\frac{\pi}{3}(a-b)\right]$
d. The magnitude of the net force on the loop is given by $\frac{\mu_0 II_1}{24ab}(b-a)$

Passage II

The figure shows a circular loop of radius a with two long parallel wires (numbered 1 and 2) all in the plane of the paper. The distance of each wire from the centre of the loop is d. The loop and the wires are carrying the same current I. The current in the loop is in the counter-clockwise direction if seen from above

[JEE Advanced 2014]

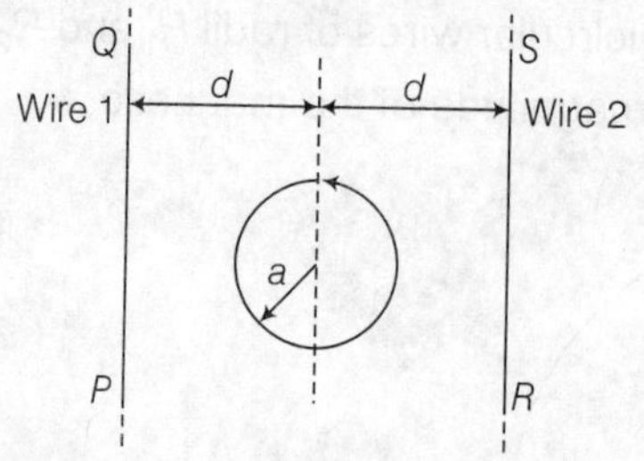

18 When $d \approx a$ but wires are not touching the loop, it is found that the net magnetic field on the axis of the loop is zero at a height h above the loop. In that case,

a. current in wire 1 and wire 2 is the direction PQ and RS, respectively and $h \approx a$
b. current in wire 1 and wire 2 is the direction PQ and SR, respectively and $h \approx a$
c. current in wire 1 and wire 2 is the direction PQ and SR, respectively and $h \approx 1.2a$
d. current in wire 1 and wire 2 is the direction PQ and RS, respectively and $h \approx 1.2a$

19 Consider $d >> a$, and the loop is rotated about is diameter parallel to the wires by 30° from the position shown in the figure. If the currents in the wires are in the opposite directions, then torque on the loop at its new position will be (assume that the net field due to the wires is constant over the loop)

a. $\frac{\mu_0 I^2 a^2}{d}$
b. $\frac{\mu_0 I^2 a^2}{2d}$
c. $\frac{\sqrt{3}\mu_0 I^2 a^2}{d}$
d. $\frac{\sqrt{3}\mu_0 I^2 a^2}{2d}$

One or More than One Options Correct Type

Direction *(Q. Nos. 20 and 21) This section contains 2 multiple choice questions. Each question has four choices (a), (b), (c) and (d), out of which ONE or MORE THAN ONE are correct.*

20 Two long thin, parallel conductors carrying equal currents in the same direction are fixed parallel to the X-axis, one passing through $y = a$ and the other through $y = -a$. The resultant magnetic field due to the two conductors at any point is 1. Which of the following are correct?

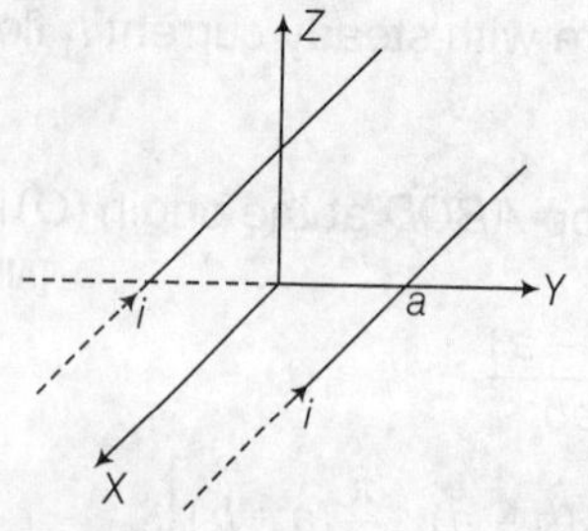

a. $B = 0$, for all points on X-axis
b. At all points on the Y-axis, excluding the origin, B has only a z-component
c. At all points on Z-axis, excluding the origin, B has only a y-component
d. B can't have an x-component

Working Space

21 Which of the following statement(s) is(are) correct in the given figure? **[IIT JEE 2006]**

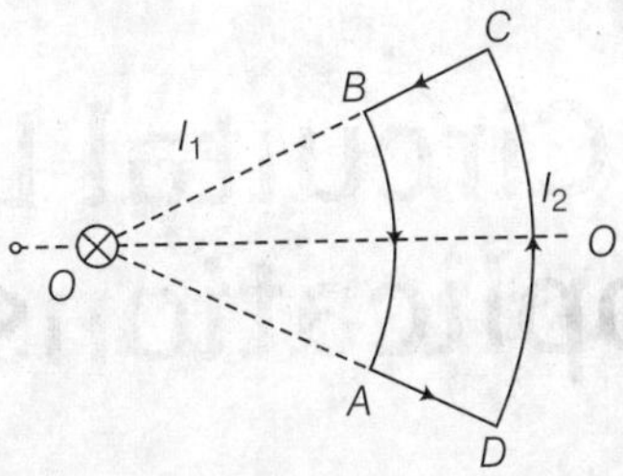

Infinitely long wire kept perpendicular to the paper carrying current inwards, then

a. net force on the loop is zero
b. net torque on the loop is zero
c. loop will rotate clockwise about axis OO' when seen from O
d. loop will rotate anti-clockwise about OO' when seen from O

One Integer Value Correct Type

Direction *(Q. No. 22) This section contains 1 question. When worked out will result in an integer from 0 to 9 (both inclusive).*

22 A steady current I goes through a wire loop PQR having shape of a right angled triangle with $PQ = 3x$, $PR = 4x$ and $QR = 5x$. If the magnitude of the magnetic field at P due to this loop is $k\left(\frac{\mu_0 I}{48\pi x}\right)$, then find the value of k. **[IIT JEE 2009]**

Matching List Type

Direction *(Q. No. 23) Choices for the correct combination of elements from Column I and Column II are given as options (a), (b), (c) and (d), out of which one is correct.*

23 Consider a closed loop in the form of a trapezium carrying current I. Match the Column I with Column II regarding the magnitude of magnetic field at point P. Mark the correct option from the given codes.

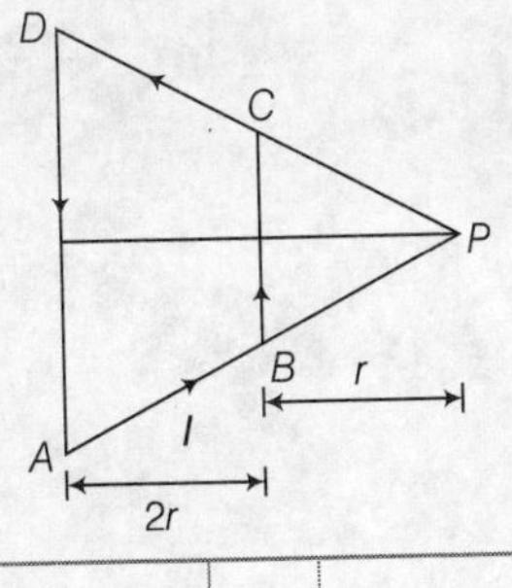

	Column I		Column II
i.	Magnetic field due to AB	p.	is greater than that due to DA
ii.	Magnetic field due to BC	q.	is greater than that due to CD
iii.	Magnetic field due to DA	r.	is not equal to zero
iv.	Magnetic field due to complete figure	s.	is zero

Codes

	i	ii	iii	iv
a.	q,r	p,s	p	r
b.	p,q,r	p,s	r	q
c.	r,s	p,s	q	q,s
d.	r,p	p,s	p,q	r,q

DPP-3 Ampere's Circuital Law and Its Applications

Subjective Questions

Direction (Q. Nos. 1-4) *These questions are subjective in nature, need to be solved completely on notebook.*

Working Space

1 A co-axial cable used in a particular application carries a current of 1 A in inner conductor and a current of 3 A in outer conductor.

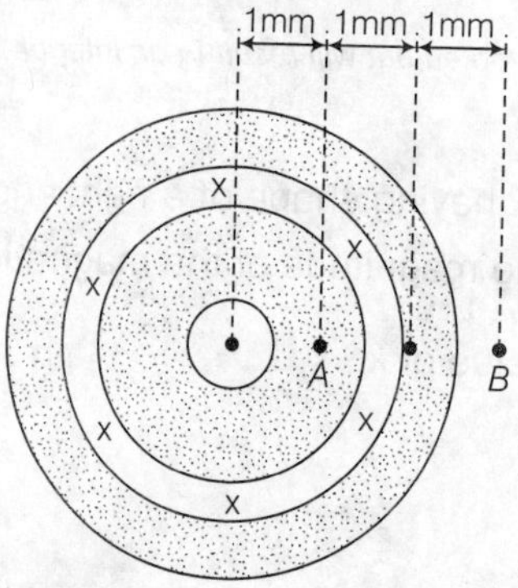

The central conductor is surrounded by a rubber layer which is surrounded by a outer conductor which is surrounded by a another layer. Determine magnitude and direction of magnetic field at points *A* and *B*.

2 For the figure shown, a long straight wire of circular cross-section (radius *a*) carrying steady current *I*. The current *I* is uniformly distributed across this cross-section.

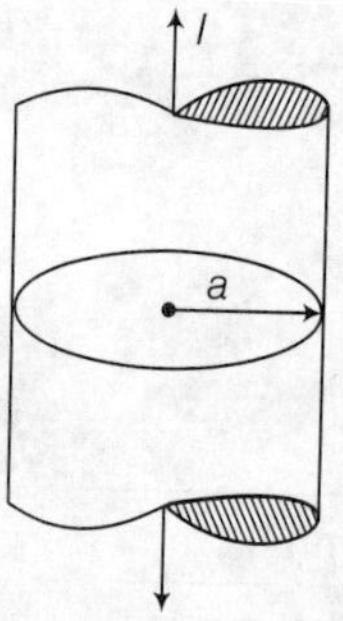

Calculate the magnetic field in the region when

i. $r < a$
ii. $r > a$

3 A toroid has a core of inner radius 20 cm and outer radius 22 cm around which 4200 turns of a wire are wound. If the current in the wire is 10 A, then what is the magnetic field

i. inside the core of toroid?
ii. outside the toroid?
iii. in the empty space surrounded by toroid?

Working Space

4 If the current density in a linear conductor of radius a varies with r according to relation $J = kr^2$, where k is a constant and r is the distance of a point from the axis of conductor. Find the magnetic field induction at a point distance r from the axis when

i. $r < a$

ii. $r > a$

Only One Option Correct Type

***Direction** (Q. Nos. 5-14) This section contains 10 multiple choice questions. Each question has four choices (a), (b), (c) and (d), out of which ONLY ONE is correct.*

5 A hollow cylinder having infinite length and carrying a uniform current per unit length λ along the circumference as shown in the figure. Magnetic field inside the cylinder is

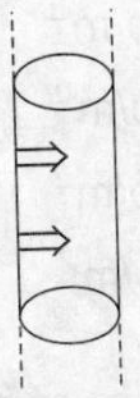

a. $\frac{\mu_0 \lambda}{2}$

b. $\mu_0 \lambda$

c. $2\mu_0 \lambda$

d. None of the above

6 Figure shows an amperian path $ABCDA$. Part ABC is in vertical plane. $PSTU$, while CDA is in horizontal plane $PQRS$.

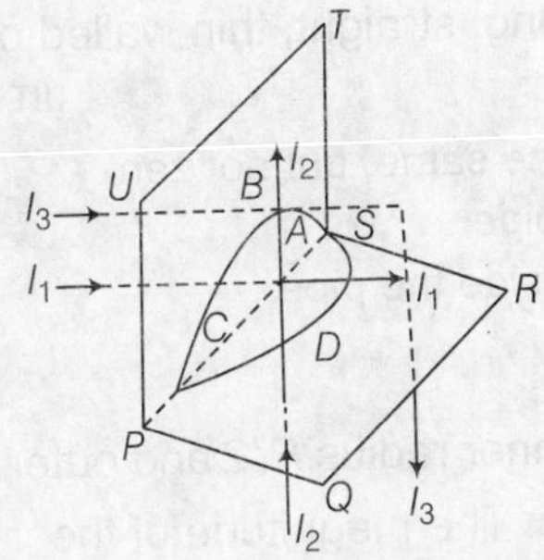

The direction of circulation along the path is shown by an arrow near points B and D. $\oint \mathbf{B} \cdot d\mathbf{l}$ for this path according to Ampere's law will be

a. $(I_1 - I_2 + I_3)\mu_0$

b. $(-I_1 + I_2)\mu_0$

c. $I_3 \mu_0$

d. $(I_1 + I_2)\mu_0$

7 A co-axial cable is made up of two conductors. Inner conductor is solid and is of radius R_1 and the outer conductor is hollow of inner radius R_2 and outer radius R_3. The space between conductor is filled with air. The inner and outer conductors are carrying equal and opposite currents. Then, the variation of magnetic field with distance from the axis is best plotted as

a.
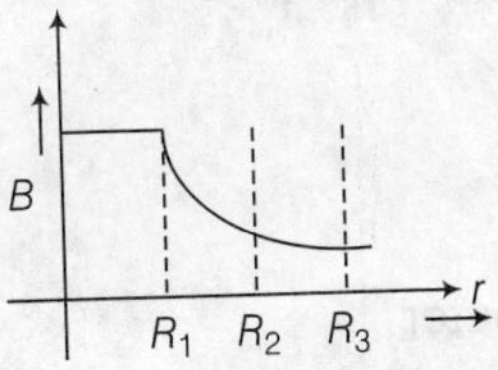

b.
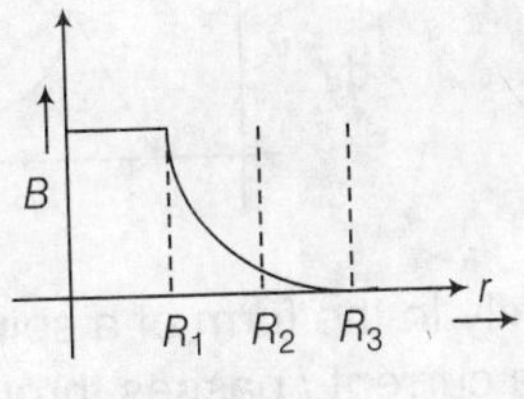

c.
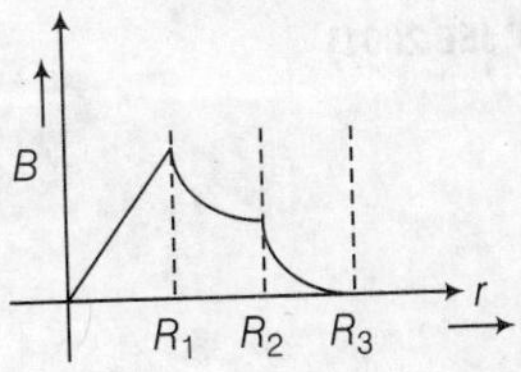

d.
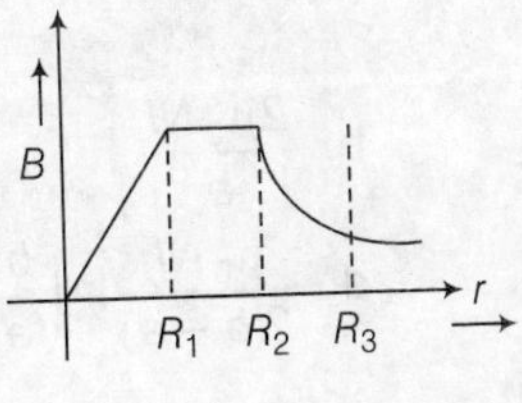

Working Space

8 A closely wound solenoid 80 cm long has 5 layers of windings of 400 turns each. The diameter of the solenoid is 1.8 cm. If the current carried is 8.0 A, then estimate the magnitude of B inside the solenoid near its centre.

a. 1.5×10^{-2} T, opposite to the axis of solenoid

b. 2.5×10^{-2} T, along the axis of solenoid

c. 3.5×10^{-2} T, along the axis of solenoid

d. 1.5×10^{-2} T, opposite to the axis of solenoid

9. A long solenoid has 200 turns/cm and carries a current i. The magnetic field at its centre is 6.28×10^{-2} Wb. Another long solenoid has 100 turns/cm and it carries a current $i/3$. The value of magnetic field at its centre is **[AIEEE 2009]**

a. 1.05×10^{-4} Wb/m^2

b. 1.05×10^{-2} Wb/m^2

c. 1.05×10^{-4} Wb/m^2

d. 1.05×10^{-3} Wb/m^2

10. A long straight wire of radius a carries a steady current i. The current is uniformly distributed across its cross-section. The ratio of the magnetic field at $a/2$ and $2a$ is **[AIEEE 2007]**

a. 1/2 b. 1/4

c. 4 d. 1

11 A current I flows along the length of an infinitely long, straight, thin walled pipe. Then, **[IIT JEE 1993]**

a. the magnetic field at all points inside the pipe is the same, but not zero

b. the magnetic field is zero only on the axis of the pipe

c. the magnetic field is different at different points inside the pipe

d. magnetic field at any point inside the pipe is zero

12 An infinitely long hollow conducting cylinder with inner radius $R/2$ and outer radius R carries a uniform current density along its length. The magnitude of the magnetic field, $|\mathbf{B}|$ as a function of radial distance r from the axis is best represented by **[IIT JEE 2012]**

a.
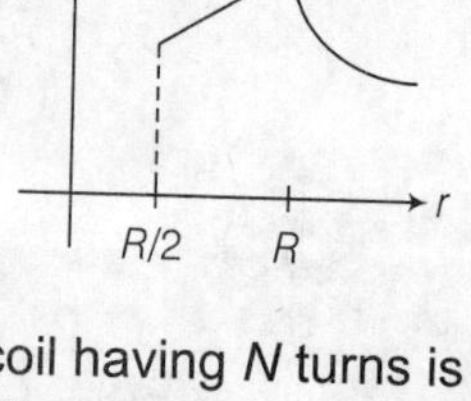

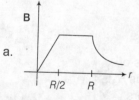

b.
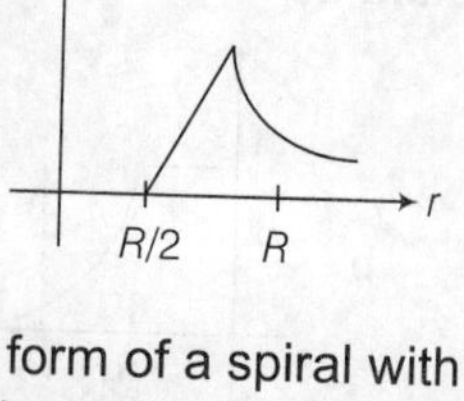

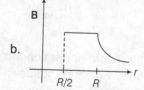

c.
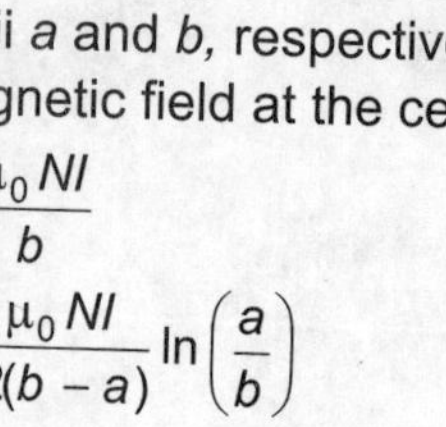

d.
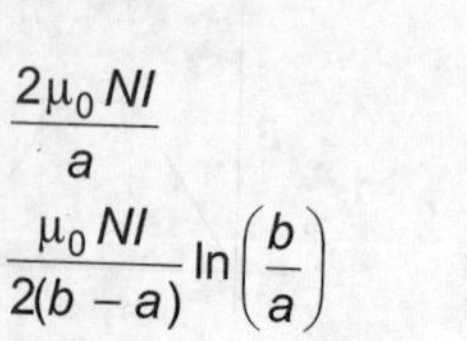

13 A coil having N turns is wound tightly in the form of a spiral with inner and outer radii a and b, respectively. When a current I passes through the coil, then magnetic field at the centre is **[IIT JEE 2001]**

a. $\dfrac{\mu_0 NI}{b}$ b. $\dfrac{2\mu_0 NI}{a}$

c. $\dfrac{\mu_0 NI}{2(b-a)} \ln\left(\dfrac{a}{b}\right)$ d. $\dfrac{\mu_0 NI}{2(b-a)} \ln\left(\dfrac{b}{a}\right)$

14 A long insulated copper wire is closely wound as a spiral of N turns. The spiral has inner radius a and outer radius b. The spiral lies in the XY-plane and a steady current I flows through the wire. The z-component of the magnetic field at the centre of the spiral is **[JEE Advanced 2014]**

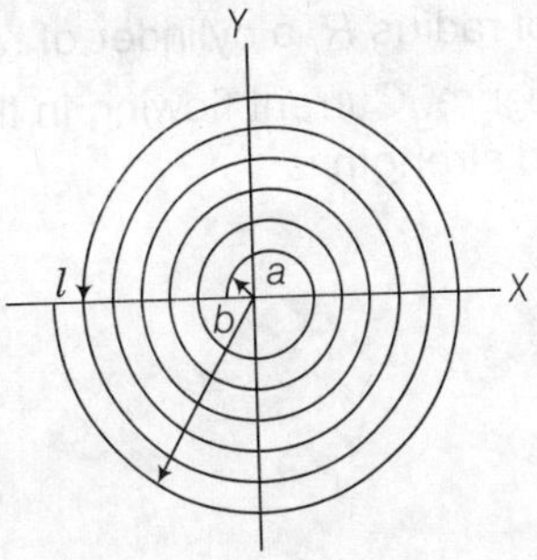

a. $\frac{\mu_0 NI}{2(b-a)} \ln\left(\frac{b}{a}\right)$

b. $\frac{\mu_0 NI}{2(b-a)} \ln\left(\frac{b+a}{b-a}\right)$

c $\frac{\mu_0 NI}{2b} \ln\left(\frac{b}{a}\right)$

d. $\frac{\mu_0 NI}{2b} \ln\left(\frac{b+a}{b-a}\right)$

Comprehension Type

Direction *(Q. Nos. 15-17) This section contains a paragraph, which describing theory, experiments, data, etc. Three questions related to the paragraph have been given. Each question has only one correct answer among the four given options (a), (b), (c) and (d).*

Passage

Curves in the graph shown as function of radial distance r, the magnitude B of the magnetic field inside and outside four long wires m, n, o and p carrying currents that are uniformly distributed across the cross-section of the wires. Overlapping portions of the plots are indicated by double labels.

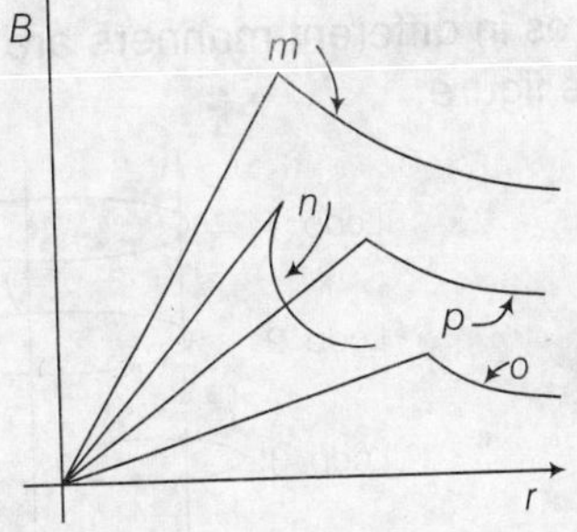

15 Which wire has the greatest radius?

a. m
b. n
c. o
d. p

16 Which wire has the greatest magnitude of the magnetic field on the surface?

a. m
b. n
c. o
d. p

17 The current density in the wire m is

a. greater than in wire n
b. less than in wire n
c. equal to that in wire n
d. not comparable due to lack of information

One or More than One Options Correct Type

Direction *(Q. Nos. 18-20) This section contains 3 multiple choice questions. Each question has four choices (a), (b), (c) and (d), out of which ONE or MORE THAN ONE are correct.*

18 A long thick conducting cylinder of radius R carries a uniformly distributed current over its cross-section. Which of the following statement(s) is/are correct?

a. The magnetic field strength is maximum on the surface
b. The magnetic field strength is zero on the surface
c. The strength of magnetic field inside the cylinder will vary as inversely proportional to r
d. The energy density of magnetic field outside the cylinder varies as inversely proportional to $\frac{1}{r^2}$

Working Space

19 From a cylinder of radius R, a cylinder of radius $R/2$ is removed as shown in the figure. Current flowing in the remaining cylinder is I. Magnetic field strength is

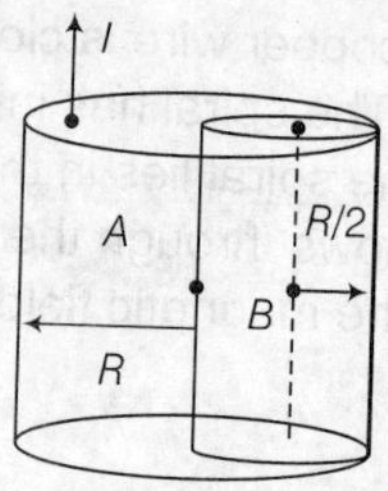

a. zero at point A
b. zero at point B
c. $\frac{\mu_0 I}{3\pi R}$ at point A
d. $\frac{\mu_0 I}{3\pi R}$ at point B

20 Two identical current carrying co-axial loops, carry current I in an opposite sense. A simple amperian loop passes through both of them once. Calling the loop as C,

a. $\int_C \mathbf{B} \cdot d\mathbf{l} = m\, 2\mu_0 I$
b. the value of $\int_C \mathbf{B} \cdot d\mathbf{l}$ is independent of sense of C
c. there may be a point on C, where **B** and dl are perpendicular
d. **B** vanishes everywhere on C

Matching List Type

Direction *(Q. No. 21) Choices for the correct combination of elements from Column I and Column II are given as options (a), (b), (c) and (d), out of which one is correct.*

21 Three wires are carrying same constant current i in different directions. Four loops enclosing the wires in different manners are shown in the figure. The direction of $d\mathbf{l}$ is shown in the figure.

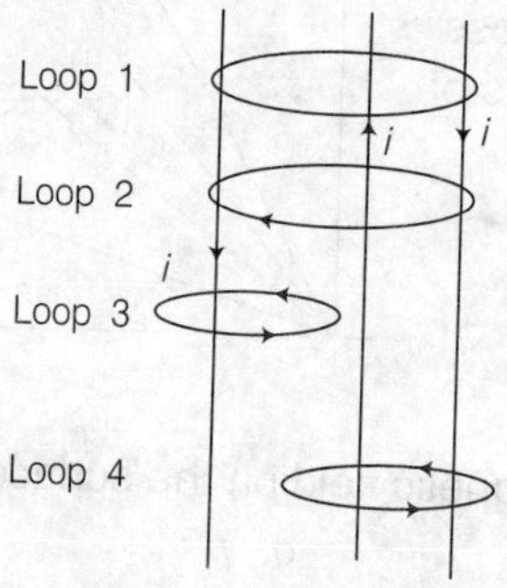

Then, match the Column I with Column II and select the correct option from the codes given below.

	Column I		Column II
i.	Along closed loop 1	p.	$\oint \mathbf{B} \cdot d\mathbf{l} = \mu_0 i$
ii.	Along closed loop 2	q.	$\oint \mathbf{B} \cdot d\mathbf{l} = -\mu_0 i$
iii.	Along closed loop 3	r.	$\oint \mathbf{B} \cdot d\mathbf{l} = 0$
iv.	Along closed loop 4	s.	net work done by the magnetic force to move a unit charge along the loop is zero.
		t.	$\oint \mathbf{B} \cdot d\mathbf{l} = \mu_0 (2i)$

Codes

	i	ii	iii	iv
a.	q,s	p,s	r,s	r,s
b.	p,q	p,s	q,s	s,p
c.	p	q	r,s	p,s
d.	q,r	t,s	p,q	s,t

One Integer Value Correct Type

Direction (Q. Nos. 22-24) *This section contains 3 questions. When worked out will result in an integer from 0 to 9 (both inclusive).*

22 Three identical long solenoids P, Q and R are connected to each other as shown in the figure. If the magnetic field at the centre of P is 2.0 T, then what would be the field (in T) at the centre of Q? Assume that the field due to any solenoid is confined within the volume of that solenoid only.

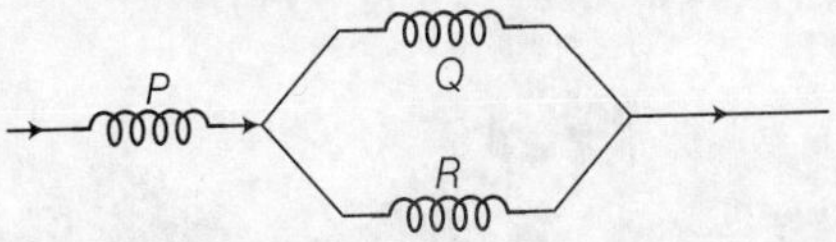

23 A packed bundle of 100 long straight, insulated wires forms a cylinder of radius $R = 0.5$ cm. If each wire carries a current of 2 A, then magnetic force per unit length on a wire located at 2 mm from the centre of the bundle is $704 \times N \times 10^{-6}$ N. Find the value of N.

24 A long conducting cylinder of radius R carries a current I. The current density J is a function of radius r as $J = br$, where b is a constant. Magnetic field at a distance $r > R$ measured from the axis is $B = \dfrac{\mu_0 bR^3}{Nr_2}$, then find the value of N.

Statement Type

Direction (Q. Nos. 25 and 26) *This section is based on Statement I and Statement II. Select the correct answer from the codes given below.*

a. Both Statement I and Statement II are correct and Statement II is the correct explanation of Statement I
b. Both Statement I and Statement II are correct but Statement II is not the correct explanation of Statement I
c. Statement I is correct but Statement II is incorrect
d. Statement II is correct but Statement I is incorrect

25 **Statement I** If we consider two wires carrying currents I_1 and I_2, then for the Amperian loop shown.

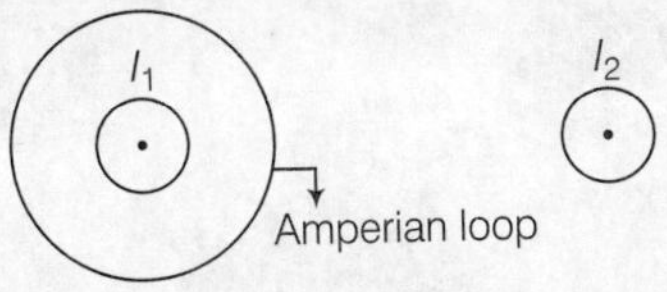

Then, $\oint \mathbf{B} \cdot dl = \mu_0 I_1$, even when the second wire is present or not.

Statement II At some points magnetic fields of both wires oppose each other and add up at some points and so, in sum over complete path these effects balance each other.

26 **Statement I** Inside a thick conductor, magnetic field increases linearly with the distance from the axis of conductor.

Statement II Current enclosed increases linearly with the distance from the axis of conductor.

DPP-4 Magnetic Force on a Current Carrying Wire

Subjective Questions

Direction *(Q. Nos. 1-4) These questions are subjective in nature, need to be solved completely on notebook.*

Working Space

1 For the given figure, cube has side length of 40 cm. Four segments of copper wire *ab*, *bc*, *cd* and *da* form a closed loop that carries a current of $I = 5$ A in the sense as shown in the figure.

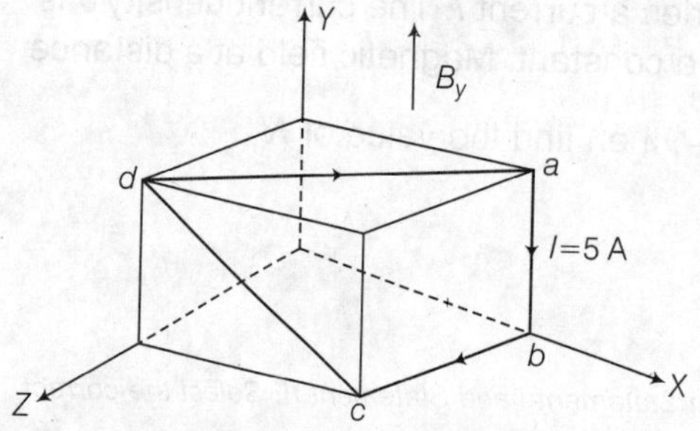

A uniform magnetic field $\mathbf{B} = 0.02\hat{\mathbf{j}}$ T exists in the region. Determine the magnitude and direction of force on each wire segment.

2 The free electrons in a conducting wire are in a constant thermal motion. If such a wire carrying no current is placed in a magnetic field, is there a magnetic force on each free electron? Is there a magnetic force on the wire?

3 Find the magnetic force on a current carrying wire joining two fixed points *a* and *b* in a uniform magnetic field **B** perpendicular to the plane of paper as shown in each case below.

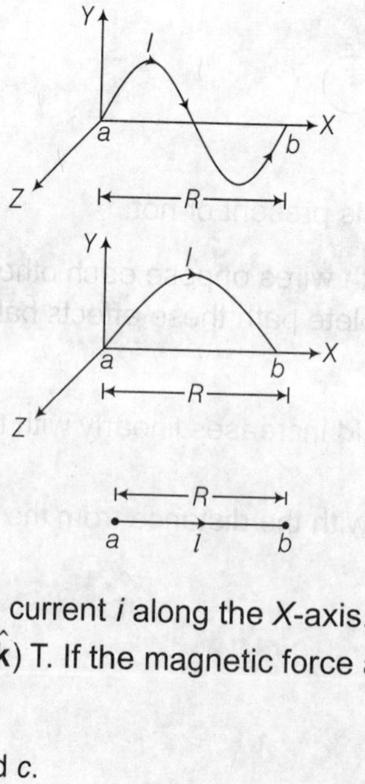

4 A wire of length *l* carries a current *i* along the *X*-axis. A magnetic field exists which is given as $\mathbf{B} = B_0(\hat{\mathbf{i}} + \hat{\mathbf{j}} + \hat{\mathbf{k}})$ T. If the magnetic force acting on the wire is given as $\mathbf{F} = a\hat{\mathbf{i}} + b\hat{\mathbf{j}} + c\hat{\mathbf{k}}$, then

i. find the value of *a, b* and *c*.

ii. also, find the magnitude of the force (**F**) acting on the wire.

Only One Option Correct Type

Direction (Q. Nos. 5-16) *This section contains 12 multiple choice questions. Each question has four choices (a), (b), (c) and (d), out of which ONLY ONE is correct.*

5 A conductor *ABCDEF* with each side of length *L* is bent as shown in the figure. It is carrying a current *I* in a uniform magnetic induction *B* parallel to the positive *Y*-direction. The force experienced by the wire is

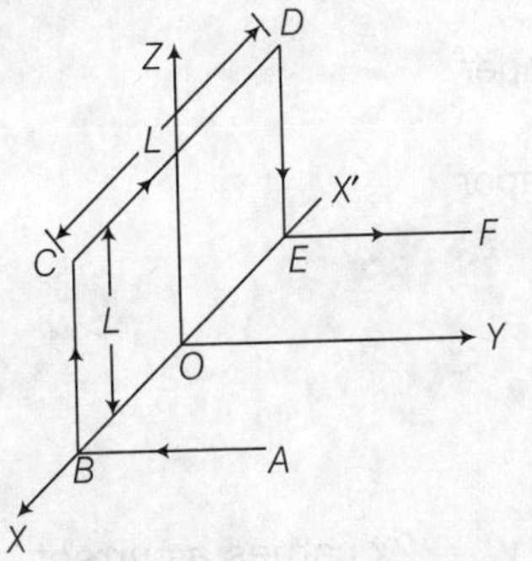

a. *BIL* in the positive *Y*-direction
b. *BIL* in the negative *Z*-direction
c. 3*BIL*
d. zero

6 A closed loop lying in the *XY*-plane carries a current. If a uniform magnetic field **B** is present in the region, the force acting on the loop will be zero, if **B** is in

a. the *X*-direction
b. the *Y*-direction
c. the *Z*-direction
d. any of the above directions

7 A circular loop having mass *m* is kept above the ground (*XZ*-plane) at some height. The coil carries a current *i* in the direction shown in the figure. In which direction, a uniform magnetic field **B** be applied so that the magnetic force balances the weight of the coil?

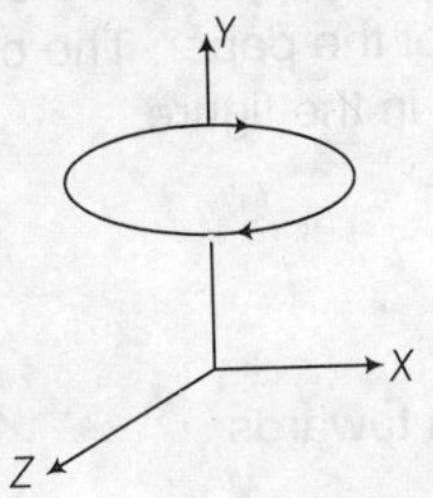

a. Positive *X*-direction
b. Negative *X*-direction
c. Positive *Z*-direction
d. None of these

8 A square loop *ABCD* carrying a current *i* is placed near and coplanar with a long straight conductor *XY* carrying a current *I*.

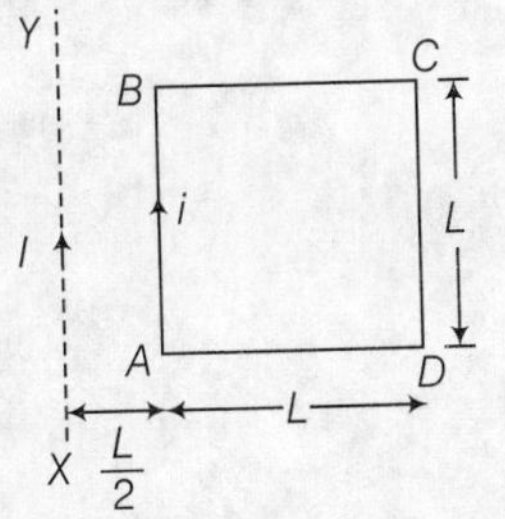

Mark the correct option.

a. There is no net force on the loop
b. The loop will be attracted by the conductor only if the current in the loop flows clockwise
c. The loop will be attracted by the conductor only if the current in the loop flows anti-clockwise
d. The loop will always be attracted by the conductor

9 A straight rod of mass m and length L is suspended from the identical spring as shown in the figure. The spring stretched by a distance x_0 due to weight of the wire. The circuit has total resistance $R\ \Omega$. When the magnetic field perpendicular to the plane of paper is switched ON, then springs are observed to extend further by the same distance.

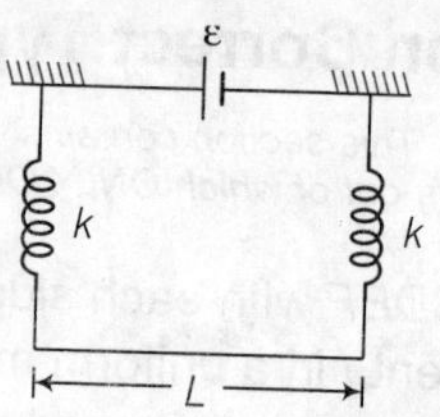

The magnetic field strength is

a. $\frac{mgR}{\varepsilon L}$; directed outward from the plane of the paper

b. $\frac{mgR}{2\varepsilon x_0}$; directed outward from the plane of the paper

c. $\frac{mgR}{\varepsilon L}$; directed into the plane of the paper

d. $\frac{mgR}{\varepsilon x_0}$; directed into the plane of the paper

10 A conducting wire bent in the form of a parabola $y^2 = 2x$ carries a current $i = 2$ A as shown in figure. The wire is placed in a uniform magnetic field $\mathbf{B} = -4\hat{\mathbf{k}}$ T.

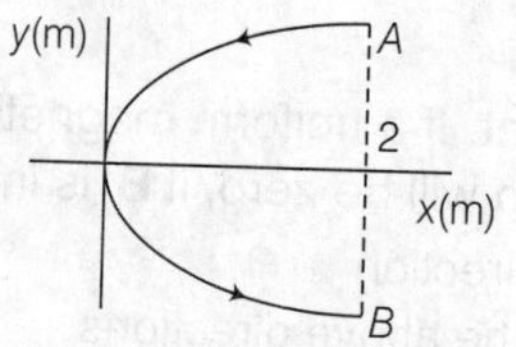

The magnetic force on the wire is

a. $-16\hat{i}$

b. $32\hat{i}$

c. $-32\hat{i}$

d. $16\hat{i}$

11 A straight current carrying conductor is placed in such a way that the current in the conductor flows in the direction out of the plane of the paper. The conductor is placed between two poles of two magnets as shown in the figure.

The conductor will experience a force in the direction towards

a. P

b. Q

c. R

d. S

12 A circular current carrying wire having radius R is placed in XY-plane with its centre at origin O. There is a non-uniform magnetic field $\mathbf{B} = \frac{B_0 x}{2R}\mathbf{k}$ (here B_0 is a positive constant) is existing in the region.

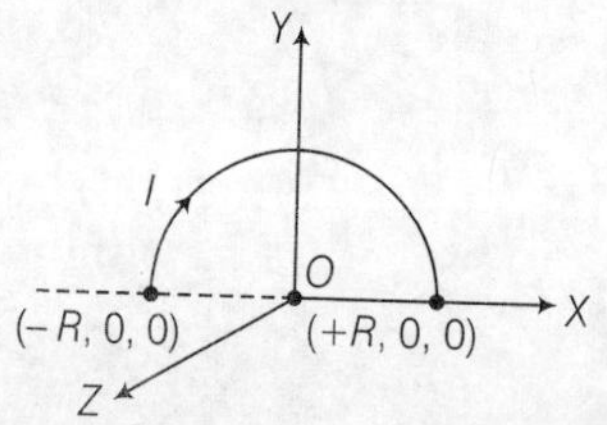

The magnetic force acting on semicircular wire will be along

a. $-X$-axis

b. $+Y$-axis

c. $-Y$-axis

d. $+X$-axis

13 A metal ring of radius $r = 0.5$ m with its plane normal to a uniform magnetic field B of induction 0.2 T carries current $I = 100$ A. The tension (in newton) developed in the ring is

a. 100

b. 50

c. 25

d. 10

14 For the given figure, *X* and *Y* are two long straight parallel conductors each carrying current 2 A. Force on each wire is *F* newton. When current in each is changed to 1 A and reversed in direction, then the force on each is now

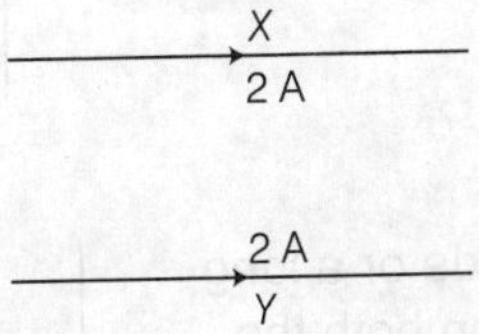

a. $\frac{F}{4}$ and unchanged in direction
b. $\frac{F}{2}$ and reversed in direction
c. $\frac{F}{2}$ and unchanged in direction
d. $\frac{F}{4}$ and reversed in direction

15 A conducting rod of length l and mass m is moving down a smooth inclined plane of inclination θ with constant velocity v. A current I is flowing in the conductor in a direction perpendicular to paper inward. A vertically upward magnetic field **B** exists in space. Then, magnitude of magnetic field **B** is

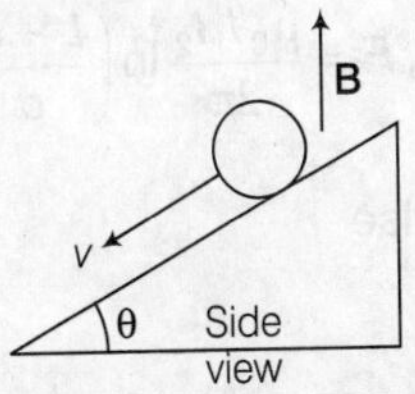

a. $\frac{mg}{Il}\sin\theta$
b. $\frac{mg}{Il}\tan\theta$
c. $\frac{mg\cos\theta}{Il}$
d. $\frac{mg}{Il\sin\theta}$

16 A uniform conducting wire *ABC* has a mass of 10 g. A current of 2 A flows through it. The wire is kept in a uniform magnetic field $B = 2$ T perpendicular to plane *ABC*. The acceleration of the wire is

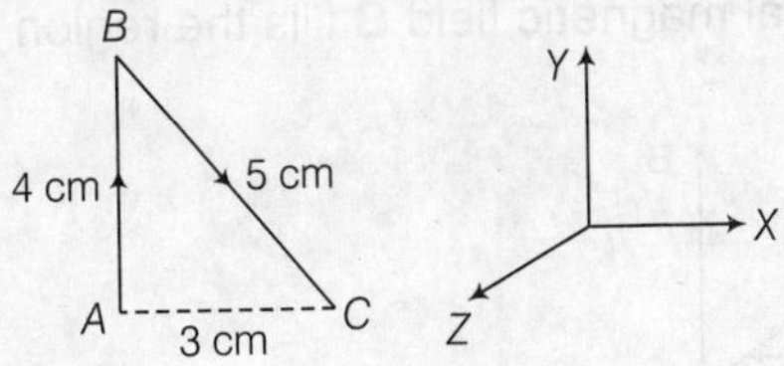

a. zero
b. 12 m/s^2 along *Y*-axis
c. 1.2×10^{-3} m/s^2 along *Y*-axis
d. 12 m/s^2 along *X*-axis

One or More than One Options Correct Type

Direction *(Q. Nos. 17-19) This section contains 3 multiple choice questions. Each question has four choices (a), (b), (c) and (d), out of which ONE or MORE THAN ONE are correct.*

17 Four wires of a cube of side *a* carry equal currents *i* in the directions shown in figure. A uniform magnetic field $\mathbf{B} = B_0\hat{\mathbf{j}}$ exists in space, then

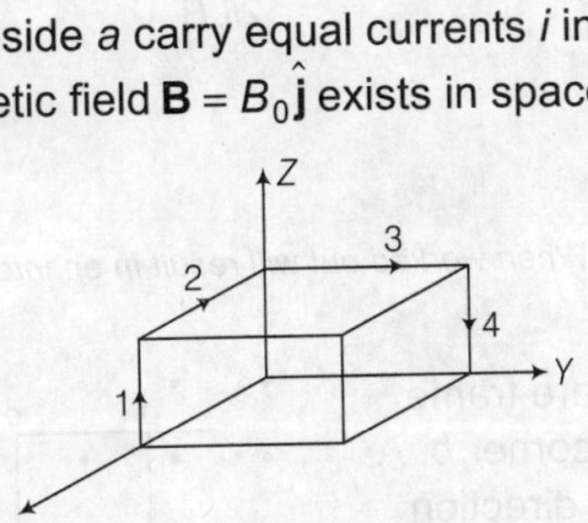

a. force on wire 1 is iaB_0 in negative *X*-direction
b. force on wire 2 is iaB_0 in negative *Z*-direction
c. force on wire 3 is zero
d. force on wire 4 is iaB_0 in positive *Z*-direction

18 Four infinitely long wires carrying equal currents are placed parallel and equidistant as shown in the figure. Then, magnitude of forces

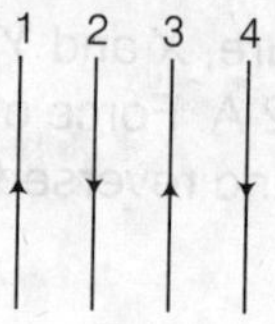

a. on 1 and 4 are equal
b. on 2 and 3 are equal
c. on 2 is maximum
d. on 4 is minimum

19 A wire B of finite length L is kept on the right hand side or a long wire A as shown in the figure. Direction of currents on both the wires are shown in figure.

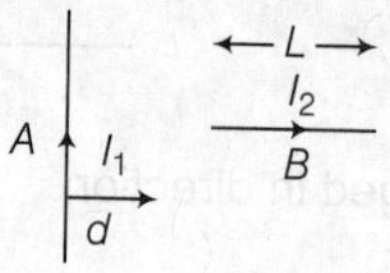

Suppose F is the force on wire B and τ the torque on it. Then,

a. F is upward, $F = \frac{\mu_0 I_1 I_2}{2\pi} \ln\left(\frac{L+d}{d}\right)$

b. F is downward, $F = \frac{\mu_0 I_1 I_2}{2\pi} \ln\left(\frac{L+d}{d}\right)$

c. τ is clockwise
d. τ is anti-clockwise

Comprehension Type

Direction *(Q. Nos. 20 and 21) This section contains a paragraph, which describing theory, experiments, data, etc. Two questions related to the paragraph have been given. Each question has only one correct answer among the four given options (a), (b), (c) and (d).*

Passage

A conducting bar with mass m and length L slides over horizontal rails that are connected to a voltage source. The voltage source maintains a constant current I in the rails, bar and a constant uniform, vertical magnetic field **B** fills the region between rails as shown in the figure.

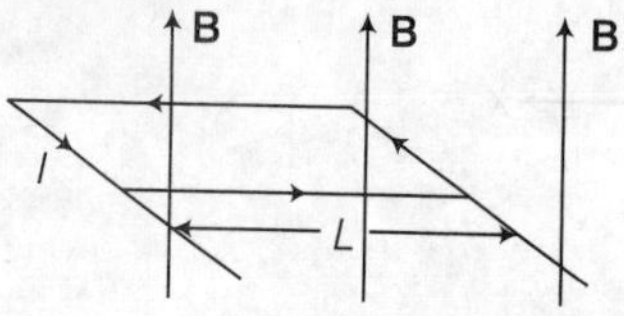

20 Find the magnitude and direction of the net force on the conducting bar

a. ILB, to the right
b. ILB, to the left
c. $2ILB$, to the right
d. $2ILB$, to the left

21 If the bar has mass m, find the distance d that the bar must move along the rails from rest to attain speed v.

a. $\frac{3v^2m}{2ILB}$
b. $\frac{5v^2m}{2ILB}$
c. $\frac{v^2m}{ILB}$
d. $\frac{v^2m}{2ILB}$

One Integer Value Correct Type

Direction *(Q. Nos. 22-24) This section contains 3 questions. When worked out will result in an integer from 0 to 9 (both inclusive).*

22 A current of 2 A enters at the corner d of a square frame $abcd$ of side 20 cm and leaves at the opposite corner b. A magnetic field $B = 0.1\,T$ exists in the space in a direction perpendicular to the plane of the frame as shown in the figure.

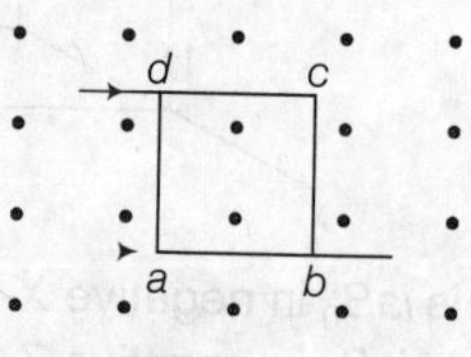

The magnitude of the magnetic forces on the four sides of the frame is $x \times 10^{-2}$ N. Find the value of x.

Working Space

23 A stiff metal rod is kept over two knife edges of length $L = 1$ m as shown in the figure.

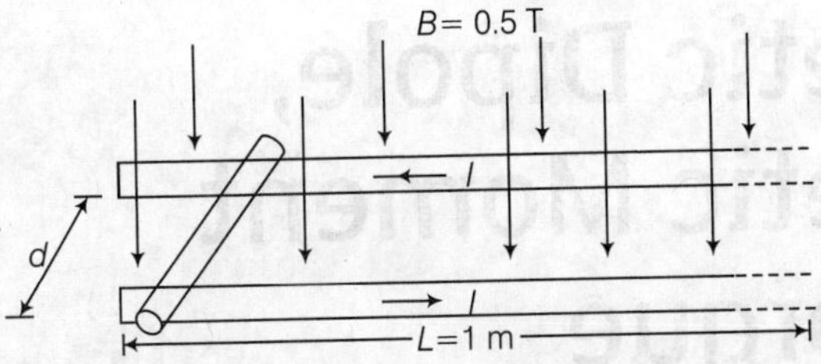

The rod carries a current of 16 A (in the direction shown) and rolls along the rails without slipping due to the magnetic force produced by a uniform magnetic field of 0.5 T perpendicular and pointing downwards. If rod starts from rest and its speed when it leaves the rails is $\frac{k}{\sqrt{5}}$ ms^{-1}, then find the value of k.

24 Let a wire is placed in a magnetic field that varies with distance from origin as

$$B = B_0\left(1+\frac{x}{a}\right)\hat{\mathbf{k}}$$

Ends of wire are at $(a, 0)$ and $(2a, 0)$ and it carries a current i. If force on wire is

$$F = B_0 i\left(\frac{ka}{2}\right)(-\hat{\mathbf{j}}).$$

Then, find the value of k.

Statement Type

Direction *(Q. Nos. 25 and 26) This section is based on Statement I and Statement II. Select the correct answer from the codes given below.*

a. Both Statement I and Statement II are correct and Statement II is the correct explanation of Statement I
b. Both Statement I and Statement II are correct but Statement II is not the correct explanation of Statement I
c. Statement I is correct but Statement II is incorrect
d. Statement II is correct but Statement I is incorrect

25 **Statement I** A battery supplies a current of 3 A to a load. Force acting per unit length of conductors 1.8 cm apart is 10^{-4} Nm^{-1} repulsive.
Statement II Magnetic force between two current carrying conductors is repulsive when they carry current in opposite directions.

26 **Statement I** Two long wires carrying currents I_1 and I_2 are arranged as shown in the figure. Wire carrying I_1 current is along the X-axis and wire carrying I_2 current is along a line parallel to Y-axis given by $x = 0$ and $z = d$. Force exerted at O, is $\frac{F}{l} = \frac{\mu_0 I_1 I_2}{2\pi d}$.

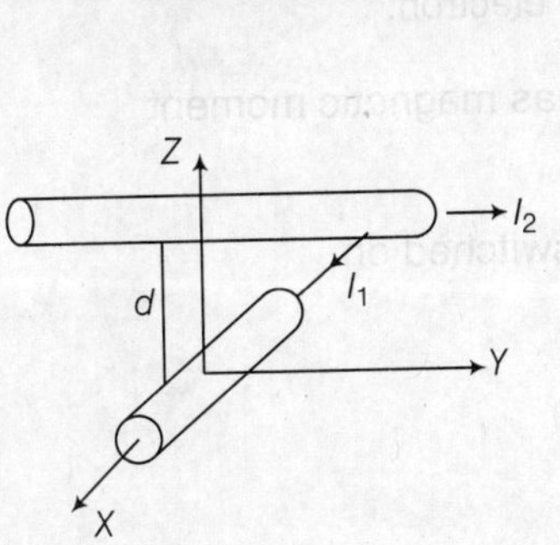

Statement II For two parallel vectors, $\mathbf{A} \times \mathbf{B} = 0$.

DPP-5 Magnetic Dipole, Magnetic Moment and Torque

Subjective Questions

***Direction** (Q. Nos. 1-4) These questions are subjective in nature, need to be solved completely on notebook.*

Working Space

1 A uniform conducting wire of length $12a$ and resistance R is wound up as a coil in the shape of

i. an equilateral triangle of side a.

ii. a square of side a.

iii. a regular hexagon of side a.

The coil is connected to a voltage source of V_o volts. Find magnetic dipole moment of coil in each case.

2 Consider three different orientation of the current carrying circular loop placed in a uniform magnetic field as shown in the figure below.

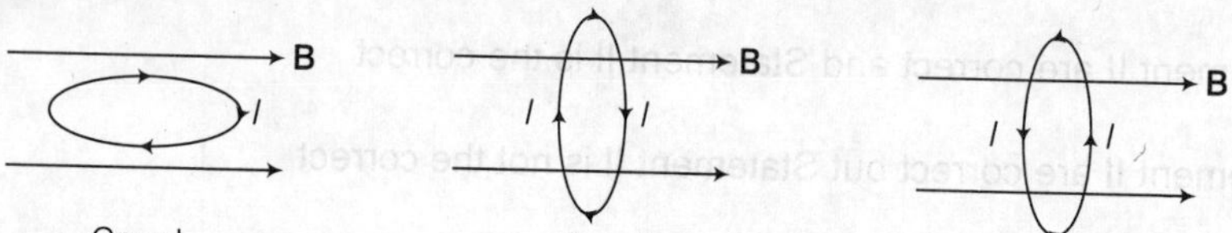

Case I (Plane of loop parallel to **B**)

Case II (Plane of loop normal to **B**)

Case III (Plane of loop normal to **B**)

In which of the above cases, potential energy is

i. maximum.

ii. minimum.

iii. zero.

3 An electron is orbiting in a circular orbit of radius r with a speed v.

i. If the orbiting electron behaves as a current loop of current i, then find the value of i.

ii. Also, find the magnetic moment of the orbiting electron.

4 A current carrying ring with its centre at origin has magnetic moment $\mathbf{M} = (4\hat{\mathbf{i}} - 3\hat{\mathbf{j}})\,\text{A-m}^2$.

At time $t = 0$, a magnetic field $\mathbf{B} = (3\hat{\mathbf{i}} + 4\hat{\mathbf{j}})$ T is switched on.

i. Find the initial torque acting on the loop.

ii. The potential energy of the loop at $t = 0$.

Working Space

Only One Option Correct Type

Direction *(Q. Nos. 5-13) This section contains 9 multiple choice questions. Each question has four choices (a), (b), (c) and (d), out of which ONLY ONE is correct.*

5 A loop carrying current I lies in the XY-plane as shown in the figure. The unit vector $\hat{\mathbf{k}}$ is coming out of the plane of the paper. The magnetic moment of the current loop is

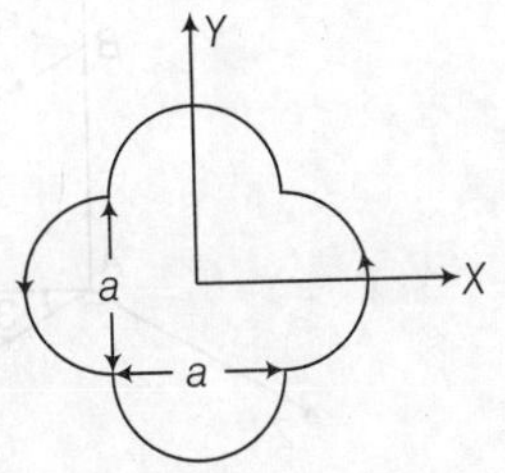

a. $a^2 I \hat{\mathbf{k}}$
b. $\left(\frac{\pi}{2}+1\right) a^2 I \hat{\mathbf{k}}$
c. $-\left(\frac{\pi}{2}+1\right) a^2 I \hat{\mathbf{k}}$
d. $(2\pi + 1) a^2 I \hat{\mathbf{k}}$

6 The figure represents four positions of a current carrying coil in a magnetic field directed towards right. $\hat{\mathbf{n}}$ represents the direction of area vector of the coil. The correct order of potential energy is

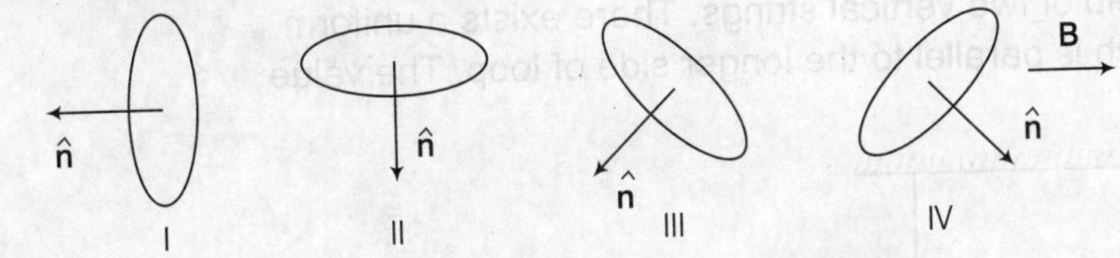

a. I > III > II > IV
b. I < III < II < IV
c. IV < I < II < III
d. III > II > IV > I

7 A conducting ring of mass 2 kg and radius = 0.5 m is placed on a smooth horizontal plane. The ring carries a current i = 4 A. A horizontal magnetic field B = 10 T is switched on at time t = 0 as shown in figure. The initial angular acceleration of the ring will be

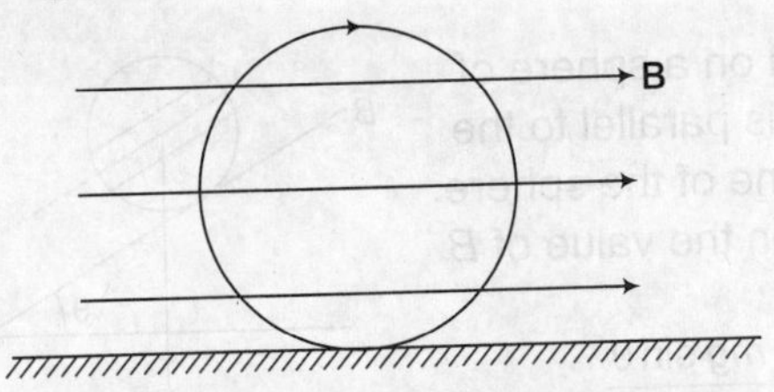

a. 40π rads^{-2}
b. 20π rads^{-2}
c. 5π rads^{-2}
d. 15π rads^{-2}

8 The magnetic moment of a circular orbit of radius r carrying a charge q and rotating with velocity v is given by

a. $\frac{qvr}{2\pi}$
b. $\frac{qvr}{2}$
c. $qv\pi r$
d. $qv\pi r^2$

9 A rectangular coil PQ has $2n$ turns, an area $2a$ and carries a current $2I$ (as shown). The plane of the coil is at 60° to a horizontal uniform magnetic field of flux density B. The torque on the coil due to magnetic force is

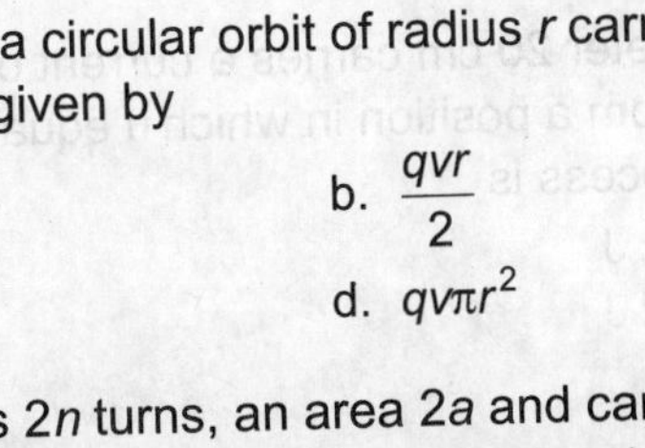

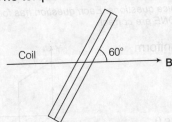

a. $BnaI \sin 60°$
b. $8BnaI \cos 60°$
c. $4naI \sin 60°$
d. None of these

Working Space

10 Figure shows a square current carrying loop *ABCD* of side 10 cm and current $i = 10\,A$. The magnetic moment **M** of the loop is

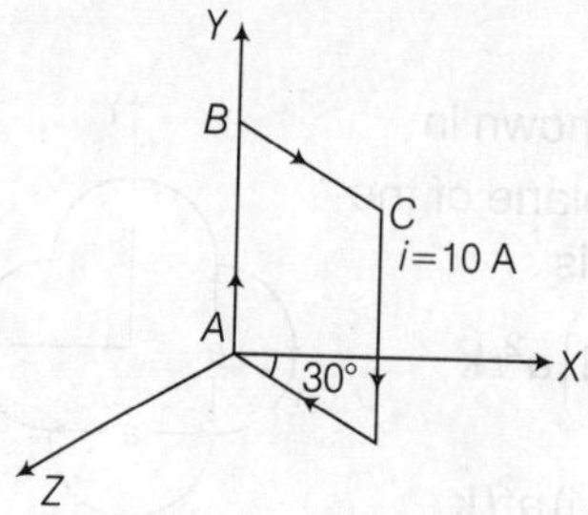

a. $0.05(\hat{\mathbf{i}} - \sqrt{3}\,\hat{\mathbf{k}})$ A-m^2
b. $0.05(\hat{\mathbf{j}} + \hat{\mathbf{k}})$ A-m^2
c. $0.05(\sqrt{3}\,\hat{\mathbf{i}} + \hat{\mathbf{k}})$ A-m^2
d. $(\hat{\mathbf{i}} + \hat{\mathbf{k}})$ A-m^2

11 A uniform conducting rectangular loop of side *l*, *b* and mass *m* carrying current *I* is hanging horizontally with the help of two vertical strings. There exists a uniform horizontal magnetic field *B* which is parallel to the longer side of loop. The value of tension which is least, is

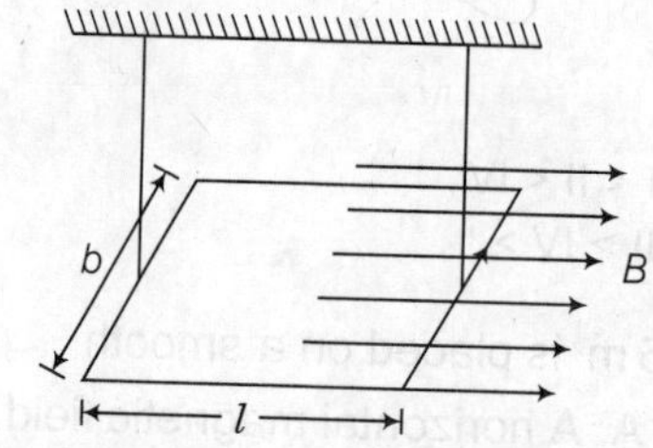

a. $\dfrac{mg - Bb}{2}$
b. $\dfrac{mg + Bb}{2}$
c. $\dfrac{mg - 2IBb}{2}$
d. $\dfrac{mg + 2Bb}{2}$

12 For the figure, a coil of single turn is wound on a sphere of radius *r* and mass *m*. The plane of the coil is parallel to the inclined plane and lies in the equatorial plane of the sphere. If the sphere is in rotational equilibrium, then the value of *B* is (current in the coil is *I*)

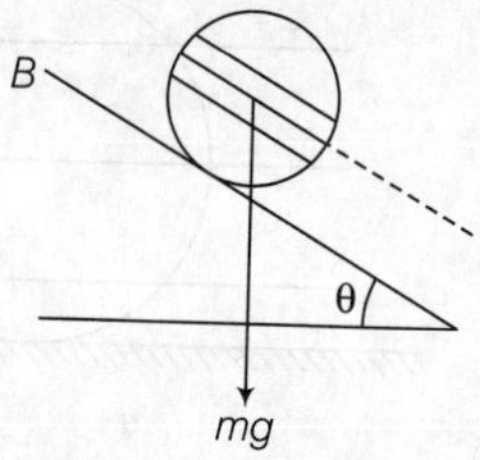

a. $\dfrac{mg}{\pi Ir}$
b. $\dfrac{mg\sin\theta}{\pi I}$
c. $\dfrac{mg\sin\theta}{\pi Ir}$
d. None of these

13 A circular coil of 100 turns and effective diameter 20 cm carries a current of 0.5 A. It is to be turned in a magnetic field $B = 2$ T from a position in which θ equal zero to θ equals 180°. The work required in the process is

a. π J
b. 2π J
c. 4π J
d. 8π J

One or More than One Options Correct Type

Direction *(Q. Nos. 14-16) This section contains 3 multiple choice questions. Each question has four choices (a), (b), (c) and (d), out of which ONE or MORE THAN ONE are correct.*

14 A rectangular loop $(a \times b)$ carries a current *i*. A uniform magnetic field $\mathbf{B} = B_0\hat{\mathbf{i}}$ exists in space. Then,

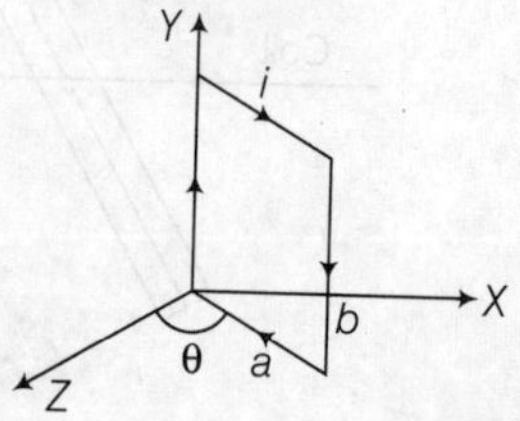

a. torque on the loop is $iabB_0\sin\theta$
b. torque on the loop is in negative *Y*-direction
c. if allowed to move the loop turn so as to increase θ
d. if allowed to move the loop turn so as to decrease θ

15 The rectangular coil having 100 turns is turned in a uniform magnetic field of $\frac{0.05}{\sqrt{2}}\hat{\mathbf{j}}$ as shown in the figure. Then,

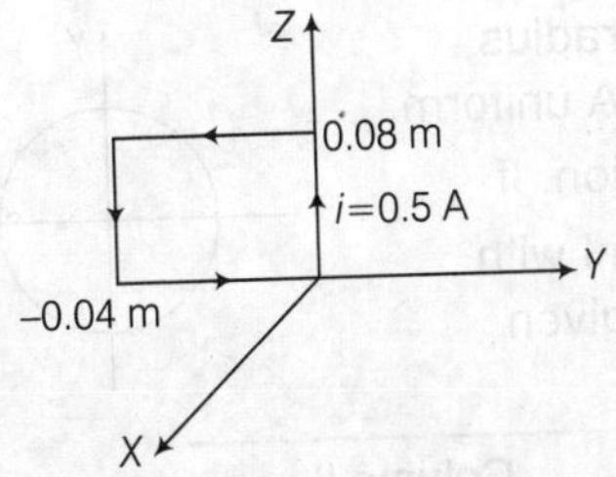

a. the magnetic dipole moment of the current carrying coil is $1.6 \times 10^{-2}\,\hat{\mathbf{i}}$ (A-m^2)

b. the torque acting on the coil is 5.66×10^{-5} (N-m) $\hat{\mathbf{k}}$

c. the torque acting on the coil is 0.566×10^{-5} (N-m) $\hat{\mathbf{k}}$

d. the magnetic dipole moment of coil is $16 \times 10^{-2}\,\hat{\mathbf{i}}$ (A-m^2)

16 A current carrying ring with its centre at origin and moment of inertia 2×10^{-2} kg-m^2 about an axis passing through the centre and perpendicular to the plane has magnetic moment $\mathbf{M} = (3\hat{\mathbf{i}} - 4\hat{\mathbf{j}})$ A-m^2. At time $t = 0$ a magnetic field $\mathbf{B} = (4\hat{\mathbf{i}} + 3\hat{\mathbf{j}})$ T is switched on. Then,

a. angular acceleration of the ring at time $t = 0$ is 2500 rad s^{-2}

b. maximum angular velocity of the ring is $50\sqrt{2}$ rad s^{-1}

c. angular acceleration of the ring at time $t = 0$ is 5000 rad s^{-2}

d. maximum angular velocity of the ring is $25\sqrt{2}$ rad s^{-1}

Comprehension Type

***Direction** (Q. Nos. 17 and 18) This section contains a paragraph, which describing theory, experiments, data, etc. Two questions related to the paragraph have been given. Each question has only one correct answer among the four given options (a), (b), (c) and (d).*

Passage

A uniform, constant magnetic field **B** is directed at an angle of 45° to the axis in *XY*-plane. *PQRS* is a rigid square wire frame carrying a steady current I_0 with its centre at origin *O*. At time $t = 0$, the frame is at rest in the position shown in the figure with its side parallel to *X* and *Y*-axis. Each side on frame is of mass *M* and length *L*.

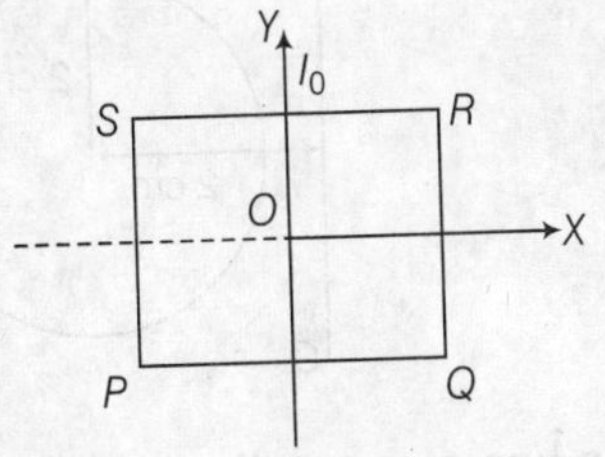

17 The torque τ acting on the frame due to magnetic field and axis about which system will rotate is

a. $\frac{I_0 L^2 B}{\sqrt{2}}(\hat{\mathbf{i}} + \hat{\mathbf{j}})$ and *SQ*

b. $\frac{I_0 L^2 B}{\sqrt{2}}(-\hat{\mathbf{i}} + \hat{\mathbf{j}})$ and *SQ*

c. $\frac{I_0 L^2 B}{2}(-\hat{\mathbf{i}} - \hat{\mathbf{j}})$ and *PR*

d. $\frac{I_0 L^2 B}{2}(\hat{\mathbf{i}} + \hat{\mathbf{j}})$ and *PR*

18 Find the angle by which the frame rotates under the action of this torque in a short interval of time Δt. (Δt is so short that any variation in the torque during this interval may be neglected)

a. $\frac{3I_0 B}{4M}(\Delta t)^2$

b. $\frac{4I_0 B}{M}(\Delta t)^2$

c. $\frac{2I_0 B}{3M}(\Delta t)^2$

d. $\frac{4I_0 B}{3M}(\Delta t)^2$

Matching List Type

Direction *(Q. No. 19) Choices for the correct combination of elements from Column I and Column II are given as options (a), (b), (c) and (d), out of which one is correct.*

Working Space

19 A circular current carrying loop of 100 turns and radius 10 cm is placed in XY-plane as shown in figure. A uniform magnetic field $\mathbf{B} = (-\hat{\mathbf{i}} + \hat{\mathbf{k}})$ T is present in the region. If current in the loop is 5 A, then match the Column I with Column II and select the correct option from the given codes.

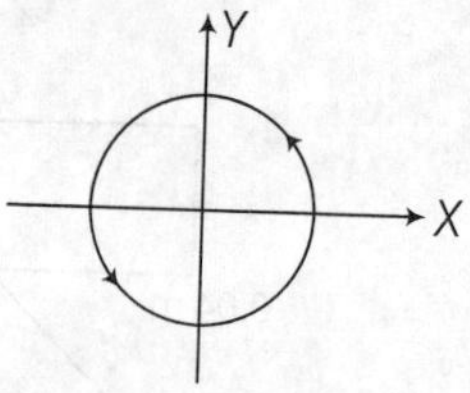

	Column I		Column II
i.	Magnitude and direction of moment (in A - m^2) of the loop are	p.	zero
ii.	Magnitude and direction of torque (in N-m) on the loop are	q.	5π
iii.	Magnitude and direction of net force (in N-m) on the current loop are	r.	along positive Z-axis
		s.	along negative Y-axis

Codes

	i	ii	iii			i	ii	iii
a.	p	q,r	s		b.	q,r	q,s	p
c.	s,q	p	r		d.	r	p	q,s

One Integer Value Correct Type

Direction *(Q. Nos. 20-22) This section contains 3 questions. When worked out will result in an integer from 0 to 9 (both inclusive).*

20 A current carrying small circular loop is placed at a distance 2 cm from an infinitely long current carrying conductor as shown in the figure. Wire and loop lies in same plane. The loop has 100 turns, its effective radius is 2 cm. The torque acting on the loop in the given situation is $x\mu_0$. Find the value of x.

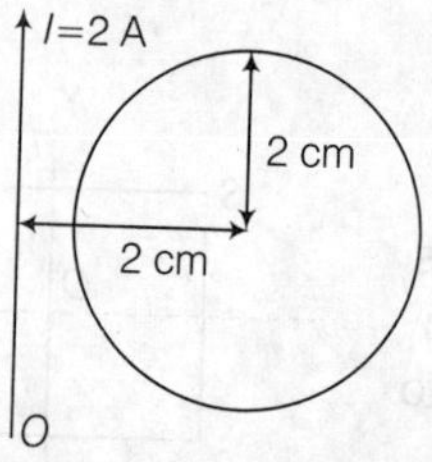

21 A compass needle has a magnetic moment of 10 mA-m^2. At its location, the earth's magnetic field is 55 μT North at 48° below horizontal. Difference in maximum and minimum potential energy of needle field system is 10^{-k} J. Find the value of k.

22 A non-conducting sphere has mass of 100 g and radius 20 cm. A flat compact coil of wire with 5 turns is wrapped tightly around it with each turns concentric with the sphere. This sphere is placed on an inclined plane such that plane of coil is parallel to the inclined plane. A uniform magnetic field of 0.5 T exists in the region in vertically upwards direction. If a current of $\frac{k}{\pi}$ A is required to rest the sphere in equilibrium, then find the value of k.

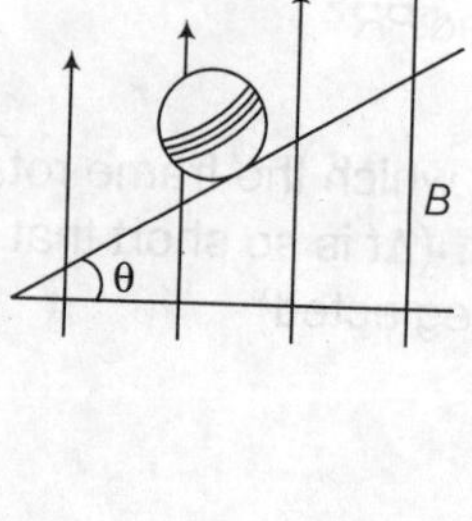

DPP-6 Moving Coil Galvanometer and Its Conversion

Subjective Questions

Working Space

Direction *(Q. Nos. 1-4) These questions are subjective in nature, need to be solved completely on notebook.*

1 The figure given below shows a section of moving coil galvanometer having rectangular coil of n turns. The area vector of the coil is **A** and the magnetic field at the site of the coil is **B**.

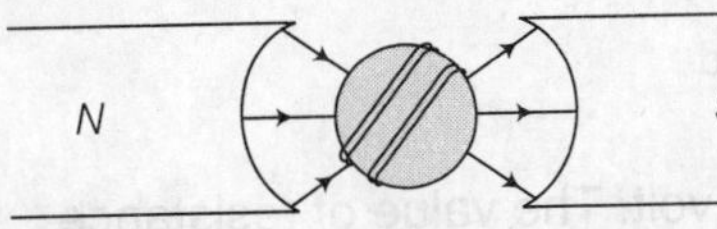

With reference to the above figure, find

i. magnitude and direction of torque acting on the coil when current i is passed through it.

ii. value of the deflecting torque considering the pole pieces are made cylindrical.

iii. the restoring torque acting on the coil if the deflection of the coil is θ and the torsional constant of suspension strip is k.

2 A moving coil galvanometer has n divisions on its scale and N number of turns of the coil.

i. If the current sensitivity is I_s, then find the expression for the maximum current it can measure.

ii. If $I_s = 20\mu$A/div and $n = 30$, then find the full scale deflection of the galvanometer if it is used as ammeter.

iii. What change will be observed in voltage sensitivity if the number of turns N of the coil is made large?

3 A galvanometer as voltmeter reads 5 V at full scale deflection and is graded according to its resistance per volt at full scale deflection as 5000 ΩV^{-1}. How will you convert it into a voltmeter that reads 20 V at full scale deflection? Will it still be graded as 5000 ΩV^{-1}? Will you prefer this voltmeter to one that is graded as 2000 ΩV^{-1}?

4 A moving coil galvanometer experiences torque $= ki$, where i is current. If N coils of area A each and moment of inertia I is kept in magnetic field B. **[IIT JEE 2005]**

i. Find the value of k in terms of given parameters.

ii. If current i deflection is $\pi/2$, then find out torsional constant of spring.

iii. If a charge Q passes suddenly through the galvanometer, then find out maximum angle of deflection.

Only One Option Correct Type

Working Space

Direction *(Q. Nos. 5-18) This section contains 14 multiple choice questions. Each question has four options (a), (b), (c) and (d), out of which ONLY ONE is correct.*

5 A candidate connects a moving coil ammeter *A*, a moving coil voltmeter *V* and a resistance *R* as shown in the figure. If the voltmeter reads 20 V and ammeter reads 4 A, then *R* is

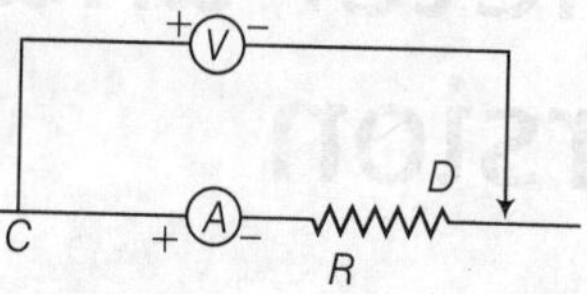

a. equal to 5 Ω
b. greater than 5 Ω
c. less than 5 Ω
d. None of the above

6 If a shunt of $\frac{1}{10}$th of the coil resistance is applied to a moving coil galvanometer, then its sensitivity becomes

a. 10 fold
b. 11 fold
c. $\frac{1}{10}$ fold
d. $\frac{1}{11}$ fold

7 A voltmeter has a resistance of *G* Ω and range *V* volt. The value of resistance used in series to convert it into voltmeter of range *nV* volt is

a. nG
b. $(n-1)G$
c. G/n
d. $G/n-1$

8 An ammeter reads up to 1 A. Its internal resistance is 0.81 Ω. To increase the range to 10 A, the value of the required shunt is

a. 0.03 Ω
b. 0.3 Ω
c. 0.9 Ω
d. 0.09 Ω

9 A moving coil galvanometer has 150 equal divisions. Its current sensitivity is 10 div/mA and voltage sensitivity is 2 div/mV. In order that each division read 1 V, the resistance (in ohm) needed to be connected is series with the coil is **[AIEEE 2005]**

a. 99995
b. 9995
c. 10^3
d. 10^5

10 In an ammeter, 0.5% of main current passes through a galvanometer. If the resistance of galvanometer is *G*, then the resistance of ammeter will be

a. $G/200$
b. $G/199$
c. $199\,G$
d. $200\,G$

11 The scale of a galvanometer of resistance 100 Ω contains 25 divisions. It gives a deflection of one division on passing a current of 4×10^{-4} A. The resistance (in ohm) to be added to it so that it may become a voltmeter of range 2.5 V is

a. 100
b. 150
c. 250
d. 300

12 A current of 200 μA deflects the coil of a moving coil galvanometer through 30°. What should be the current to cause the rotation through $\pi/10$ rad? What is the sensitivity of the galvanometer?

a. $100\,\mu A$, $0.15°\,\mu A^{-1}$
b. $120\,\mu A$, $0.15°\,\mu A^{-1}$
c. $150\,\mu A$, $0.15°\,\mu A^{-1}$
d. $200\,\mu A$, $0.15°\,\mu A^{-1}$

13 The deflection produced in a galvanometer is reduced to 45 divisions from 55 when a shunt of 8 Ω is used. Calculate the resistance of the galvanometer.

a. 72 Ω
b. 36 Ω
c. 32 Ω
d. 48 Ω

14 To increase the current sensitivity of a moving coil galvanometer by 50%, its resistance is increased so that its new resistance is twice of its initial resistance. Its voltage sensitivity changes are

a. increased by a factor of 2
b. decreased by a factor of 1/2
c. increased by a factor of 4
d. decreased by a factor of 1/4

15 A galvanometer with a coil resistance 12 Ω shows full scale deflection for a current of 2.5 mA. The ratio of net resistance of an ammeter of range 0 to 7.5 A and voltmeter of range 0 to 10 V is

a. 10^{-12}
b. 10^{-7}
c. 10^{-6}
d. 10^{-8}

16 A moving coil galvanometer has 100 turns and each turn has an area of 2 cm^2. The magnetic field produced by the magnet is 0.01 T. The deflection in the galvanometer coil is 0.05 rad when a current of 10 mA is passed through it. Find the torsional constant of the spiral spring.

a. 3×10^{-4} N-m rad^{-1}
b. 4×10^{-5} N-m rad^{-1}
c. 5×10^{-6} N-m rad^{-1}
d. 7×10^{-7} N-m rad^{-1}

17 A galvanometer of resistance 100 Ω gives a full scale deflection for a current of 10^{-5} A. To convert it into an ammeter capable of measuring up to 1 A, we should connect a resistance of

a. 1 Ω in parallel
b. 10^{-3} Ω in parallel
c. 10^{5} Ω in series
d. 100 Ω in series

18 A moving coil galvanometer of resistance 100 Ω is used as an ammeter using a resistance 0.1 Ω. The maximum deflection current in the galvanometer is 100 μA. Find the minimum current in the circuit so that the ammeter shows maximum deflection.

a. 100.1 mA
b. 1.0001 mA
c. 10.01 mA
d. 1.001 mA

One or More than One Options Correct Type

***Direction** (Q. Nos. 19 and 20) This section contains 2 multiple choice questions. Each question has four options (a), (b), (c) and (d), out of which ONE or MORE THAN ONE are correct.*

19 The galvanometer cannot be such used as an ammeter to measure the value of current in a given circuit. The following reasons are

a. galvanometer gives full scale deflection for a small current
b. galvanometer has a large resistance
c. a linear scale cannot be designed so that $I \propto \phi$
d. a galvanometer can give inaccurate values

Working Space

20 A microammeter has a resistance of 100 Ω and a full scale range of 50 μA. It can be used as a voltmeter or as a higher range ammeter provided a resistance combination. Choose the correct option(s).

a. 50 V range with 10 kΩ resistance in series
b. 10 V range with 200 kΩ resistance in series
c. 5 mA range with 1 Ω resistance in parallel
d. 10 mA range with 1 Ω resistance in parallel

Matching List Type

***Direction** (Q. No. 21) Choices for the correct combination of elements from Column I and Column II are given as options (a), (b), (c) and (d), out of which one is correct.*

21 DC measuring instruments are based on magnetic effect of current, whereas AC measuring instruments are based on heating effect of current. Now, match entries of Column I with entries of Column II with their correct proportionality.

	Column I		Column II
i.	Deflection of AC ammeter	p.	V
ii.	Deflection of DC ammeter	q.	V^2
iii.	Deflection of AC voltmeter	r.	I
iv.	Deflection of DC voltmeter	s.	I^2

Codes

	i	ii	iii	iv
a.	q	p	s	r
b.	s	p	r	q
c.	p	q	r	s
d.	s	r	q	p

Comprehension Type

***Direction** (Q. Nos. 22-24) This section contains a paragraph, which describing theory, experiments, data, etc. Three questions related to the paragraph have been given. Each question has only one correct answer among the four given options (a), (b), (c) and (d).*

Passage

A moving coil galvanometer essentially consists of a rectangular coil in a magnetic field. When a current flows through the coil, a torque is produced and this rotates the coil. Magnitude of this torque is

$$\tau = BINA\cos\theta$$

22 If a uniform magnetic field is used in the galvanometer, then scale of galvanometer looks like

a. 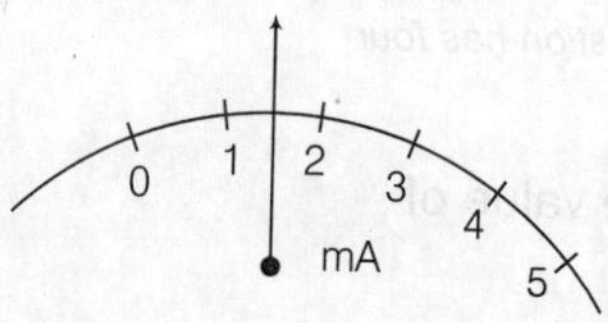

b. 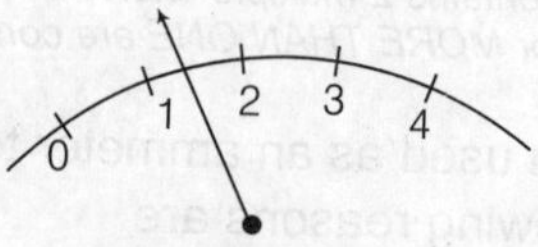

c. 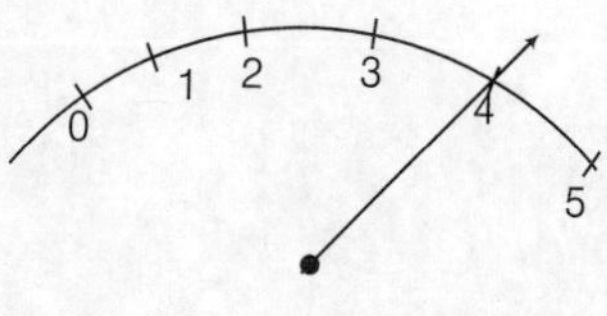

d. 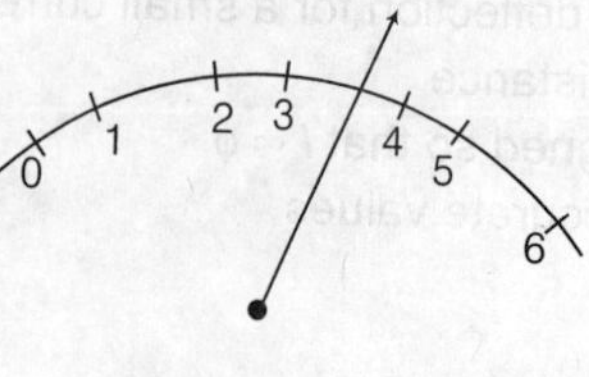

Working Space

23 A rectangular coil of N turns pivoted about a vertical axis carrying a current I is placed in the region of a uniform horizontal magnetic field B. When plane of coil makes an angle θ with magnetic field, torque required to per cent it rotating is τ_1. When coil rotates through 90°, then torque required will be τ_2. Magnitude of magnetic field B will be

a. $\dfrac{\sqrt{\tau_1^2 + \tau_2^2}}{NIA}$ b. $\dfrac{2\sqrt{\tau_1^2 + \tau_2^2}}{NIA}$

c. $\dfrac{\tau_1 + \tau_2}{2(NIA)}$ d. $\dfrac{\tau_1 + \tau_2}{NIA}$

24 In a practical galvanometer, the field is made radial instead of uniform because

a. when field is radial, torque is maximum
b. when field is radial, torque is uniform
c. when field is radial, a linear scale is possible
d. a non-linear scale is difficult to read and calibrate

One Integer Value Correct Type

Direction *(Q. Nos. 25 and 26) This section contains 2 questions. When worked out will result in an integer from 0 to 9 (both inclusive).*

25 A voltmeter reads 5 V at full scale deflection and is graded according to its resistance per volt of full scale as 5000 Ω/V. This has been converted into a voltmeter that reads 20 V at full scale reflection. Now, grading of instrument is $N \times 10^3$ Ω/V. Find the value of N.

26 For the given circuit shown,

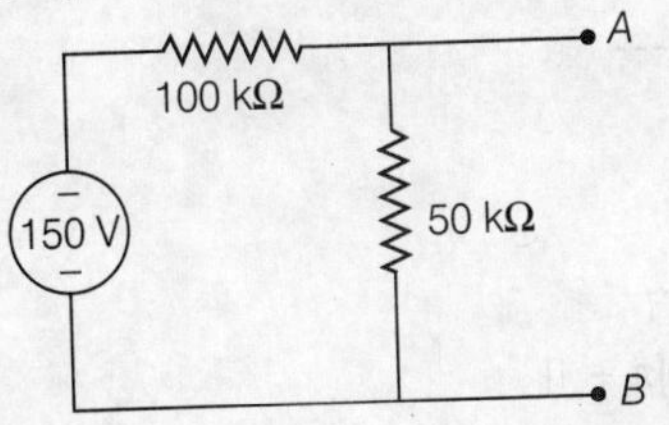

If a voltmeter with sensitivity of 1k Ω/V is connected between points A and B, then percentage error in the reading is found $0.1 \times k$. Find the value of k.

SET - 1

Revisal Problems for JEE Main

Only One Option Correct Type

Direction *(Q. Nos. 1-27) This section contains 27 multiple choice questions. Each question has four choices (a), (b), (c) and (d), out of which ONLY ONE is correct.*

Working Space

1 For the magnetic field due to a small element of a current carrying conductor at a point to be maximum, the angle between the element and the line joining the element to point P must be

a. 0°
b. 90°
c. 180°
d. 45°

2 A current of 30 A is flowing in a vertical straight wire. If the horizontal component of earth's magnetic field is 2×10^{-5} T, then the position of null-point will be

a. 0.9 m
b. 0.3 mm
c. 0.3 cm
d. 0.3 m

3 A length L of wire carrying current I is bent into a circle of one turn. The field at the centre of the coil is B_1. A similar wire of length L carrying current I is bent into a square of one turn. The field at its centre is B_2. Then,

a. $B_1 > B_2$
b. $B_1 < B_2$
c. $B_1 = B_2$
d. Nothing can be predicted

4 An electric current is flowing in a very long pin as shown in the figure. The value of magnetic flux density at point O will be

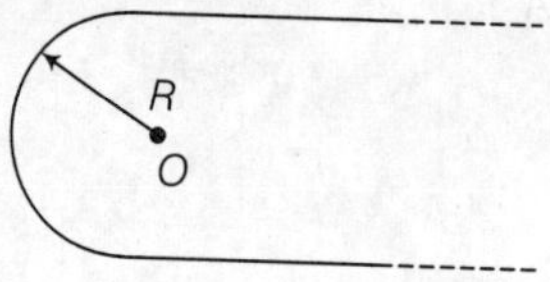

a. $\frac{\mu_0 i}{4\pi R}[\pi + 2]$
b. $\frac{\mu_0 i}{4\pi R}[\pi + 1]$
c. $\frac{\mu_0 i}{4\pi R}[\pi - 2]$
d. $\frac{\mu_0 i}{4\pi R}[\pi - 1]$

5 A current i is flowing in a conductor shaped as shown in the figure. The radius of curved part is r and length of straight portions is very large. The value of magnetic field at the centre will be

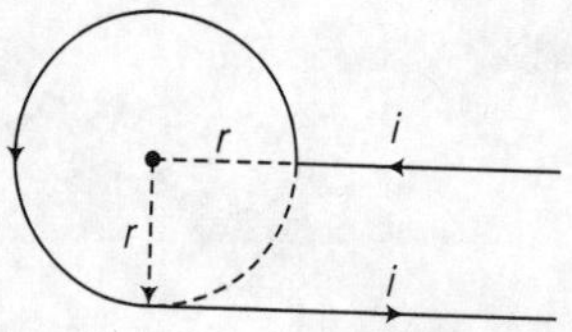

a. $\frac{\mu_0 i}{4\pi r}\left[\frac{3\pi}{2} + 1\right]$
b. $\frac{\mu_0 i}{4\pi r}\left[\frac{3\pi}{2} - 1\right]$
c. $\frac{\mu_0 i}{4\pi r}\left[\frac{\pi}{2} + 1\right]$
d. $\frac{\mu_0 i}{4\pi r}\left[\frac{\pi}{2} - 1\right]$

6 A circular arc of wire of radius of curvature r subtends an angle of $\frac{\pi}{4}$ rad at its centre. If i current is flowing in it, then the magnetic induction at its centre will be

a. $\frac{\mu_0 i}{8r}$
b. $\frac{\mu_0 i}{4r}$
c. $\frac{\mu_0 i}{16r}$
d. 0

7 A 6.28 m long wire is turned into a coil of diameter 0.2 m and current of 1 A is passed in it. The magnetic induction at its centre will be

a. 6.28×10^{-5} T
b. 0
c. 6.28 T
d. 6.28×10^{-3} T

8 A long straight wire is turned into a loop of radius 10 cm (as shown in figure). If a current of 8 A is passed through the loop, then the value of the magnetic field B at the centre C of the loop will be (Wb/m^2)

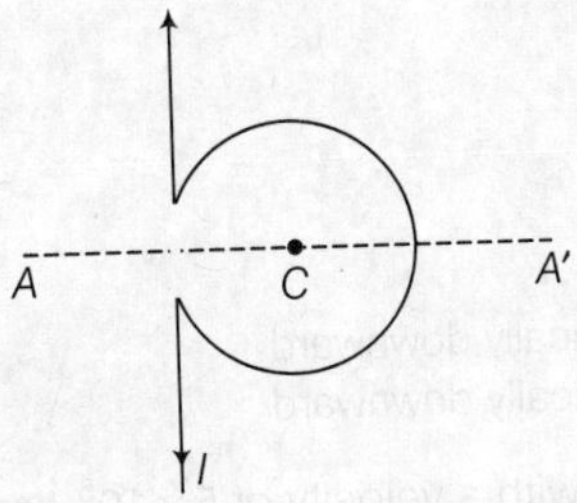

a. 3.424×10^{-5}, vertically upwards
b. 3.424×10^{-5}, vertically downwards
c. 4.24×10^{-5}, vertically upwards
d. 4.24×10^{-5}, vertically downwards

9 The magnetic dipole moment of a coil is 5.4×10^{-6} J/T and it is in stable equilibrium position in an external magnetic field whose strength is 0.80 T. Then, the work done in rotating the coil by 180° is

a. 4.32 μJ
b. 2.16 μJ
c. 8.6 μJ
d. None of these

10 4 A current is passing through a coil of radius 5 cm and 100 turns. The magnetic moment of the coil is

a. 3.14 A-m^2
b. 3.14 A-cm^2
c. 314 A-m^2
d. 0.0314 A-cm^2

11 You are given a closed circuit with radii a and b as shown in figure carrying current i. The magnetic dipole moment of the circuit is

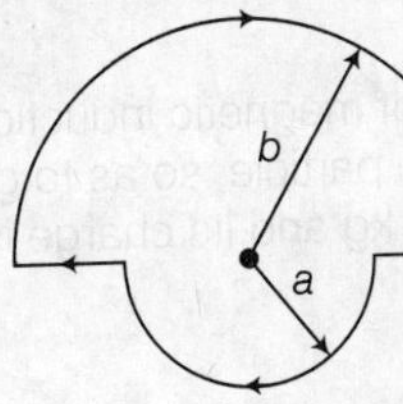

a. $\pi (a^2 + b^2) i$
b. $\frac{1}{2} \pi (a^2 + b^2) i$
c. $\pi (a^2 - b^2) i$
d. $\frac{1}{2} \pi (a^2 - b^2) i$

12 A proton, a deuteron and an α-particle are accelerated through same potential difference and then they enter a normal uniform magnetic field. The ratio of their kinetic energies will be

a. 2 : 1 : 3
b. 1 : 1 : 2
c. 1 : 1 : 1
d. 1 : 2 : 4

13 A proton of energy 8 eV is moving in a circular path in a uniform magnetic field. The energy of an α-particle moving in the same magnetic field and along the same path will be

a. 4 eV
b. 2 eV
c. 8 eV
d. 6 eV

14 If an α-particle moving with velocity v enters a perpendicular to a magnetic field, then the magnetic force acting on it will be

a. 1 eVB
b. 2 eVB
c. 0
d. 4 eVB

Working Space

15 A potential difference of 600 V is applied across the plates of a parallel plate condenser placed in a magnetic field. The separation between the plates is 3 mm. An electron projected vertically upward parallel to the plates with a velocity of 2×10^6 m/s moves undeflected between the plates. The magnitude and direction of the magnetic field in the region between the condenser plates will be (in Wb/m^2) (Given, charge of electron $= -1.6 \times 10^{-19}$ C)

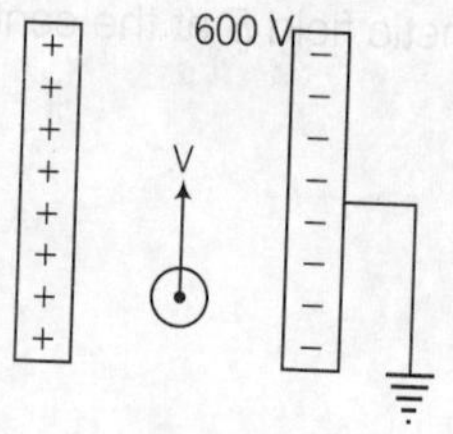

a. 0.1 vertically downward
b. 0.2 vertically downward
c. 0.3 vertically upward
d. 0.4 vertically downward

16 An α-particle is moving in a magnetic field of $(3\hat{i} + 2\hat{j})$ T with a velocity of $5 \times 10^5\ \hat{i}$ m/s. The magnetic force acting on the particle will be

a. 3.2×10^{-13} dyne
b. 3.2×10^{13} N
c. 0
d. 3.2×10^{-13} N

17 A beam of protons enters a uniform magnetic field of 0.3 T with a velocity of 4×10^5 m/s at an angle of 60° to the field. The radius of the helical path taken by the beam and the pitch of the helix (which is the distance travelled by a proton parallel to the magnetic field during one period of rotation) will be respectively (Given, mass of the proton $= 1.7 \times 10^{-27}$ kg)

a. 1.226×10^{-2} m, 4.45×10^{-3} m
b. 1.226×10^{-2} m, 4.45×10^{-2} m
c. 1.226×10^{-3} m, 4.45×10^{-3} m
d. 1.226×10^{-4} m, 4.45×10^{-4} m

18 A proton is to circulate the earth along the equator with a speed of 1.0×10^7 m/s. The minimum magnetic field which should be created at the equator for this purpose. (Given, the mass of proton $= 1.7 \times 10^{-27}$ kg and radius of earth $= 6.37 \times 10^6$ m), will be (in Wb/m^2)

a. 1.6×10^{-19}
b. 1.67×10^{-8}
c. 1.0×10^{-7}
d. 2×10^{-7}

19 An α-particle is describing a circle of radius 0.45 m in a field of magnetic induction 1.2 Wb/m^2. The potential difference required to accelerate the particle, so as to give this much energy to it (Given, the mass of α-particle is 6.8×10^{-27} kg and its charge is C), will be

a. 6×10^6 V
b. 2.3×10^{-12} V
c. 7×10^6 V
d. 3.2×10^{-12} V

20 A straight horizontal stretched of copper wire carries a current $i = 30$ A. The linear mass density of the wire is 45 g/m^3. What is the magnitude of the magnetic field needed to float the wire, that is to be balance its weight?

a. 147 G
b. 441 G
c. 14.7 G
d. 0 G

21 If an electron is moving with velocity v in an orbit of radius r in a hydrogen atom, then the equivalent magnetic moment will be

a. $\frac{\mu_0 e}{2r}$
b. $\frac{ev}{r^2}$
c. $\frac{ev \times 10^{-7}}{r^3}$
d. $\frac{evr}{2}$

22 On account of the orbital motion of an electron, its magnetic moment will be (h = Planck constant, e = electronic charge and m = mass of an electron)

a. $\frac{eh}{4\pi m}$
b. $\frac{h}{4\pi m}$
c. $\frac{eh}{2\pi}$
d. $\frac{eh}{2\pi m}$

Working Space

23 A 5 cm × 12 cm coil with number of turns 600 is placed in a magnetic field of strength 0.10 T. The maximum magnetic torque acting on it when a current of 10^{-5} A is passed through it, will be

a. 3.6×10^{-6} N-m
b. 3.6×10^{-6} dyne-cm
c. 3.6×10^{6} N-m
d. 3.6×10^{6} dyne-m

24 The effective radius of a coil of 100 turns is 0.05 m and a current of 0.1 A is flowing in it. The work required to turn this coil in an external magnetic field of 1.5 T through 180° will be, if initially the plane of the coil is normal to the magnetic field

a. 0.236 J
b. 0.236 erg
c. 236 J
d. 236 erg

25 A circular coil of 20 turns and radius 10 cm is placed in uniform magnetic field of 0.10 T normal to the plane of the coil. If the current in the coil is 5 A, then the torque acting on the coil will be

a. 31.4 N-m
b. 3.14 N-m
c. 0.314 N-m
d. zero

26 A beam of protons is moving horizontally towards you. As it approaches, it passes through a magnetic field directed downward. The beam deflects

a. to your left side
b. to your right side
c. does not deflect
d. nothing can be said

27 If a particle moves in a circular path in clockwise direction after entering into a downward vertical magnetic field, then charge on the particle is

a. positive
b. negative
c. nothing can be said
d. neutral

Statement Type

Direction (Q. Nos. 28-30) *This section is based on Statement I and Statement II. Select the correct answer from the codes given below.*

a. Both Statement I and Statement II are correct and Statement II is the correct explanation of Statement I
b. Both Statement I and Statement II are correct but Statement II is not the correct explanation of Statement I
c. Statement I is correct but Statement II is incorrect
d. Statement II is correct but Statement I is incorrect

28 **Statement I** Where there are no currents, the magnetic field cannot be both unidirectional and non-uniform.

Statement II If magnetic field is unidirectional and non-uniform, then Ampere's circuital law is violated.

29 **Statement I** It is not possible to find out exact N-S direction while sitting in your drawing room with lights on as current in wires in your home can affect a compass.

Statement II Deflection in compass needle depends on the direction and magnitude of current in wire and distance of compass and wire.

30 **Statement I** Two long insulated wires carrying equal currents cross at right angles to each other. Then, the net magnetic force on either wire is zero, but net torque is non-zero.

Statement II Force between two long conductors is

$$F = \frac{\mu_0 I_1 I_2}{2\pi d} l$$

and

$$\text{torque} = \mathbf{M} \times \mathbf{B}$$

SET - 2

Revisal Problems for JEE Advanced

Only One Option Correct Type

Direction *(Q. Nos. 1-13) This section contains 13 multiple choice questions. Each question has four choices (a), (b), (c) and (d), out of which ONLY ONE is correct.*

Working Space

1 A charged particle *A* of charge $q = 2$ C has velocity $v = 100$ m/s. When it passes through point *A* and has velocity in the direction shown in the figure. The strength of magnetic field at point *B* due to this moving charge is ($r = 2$ m)

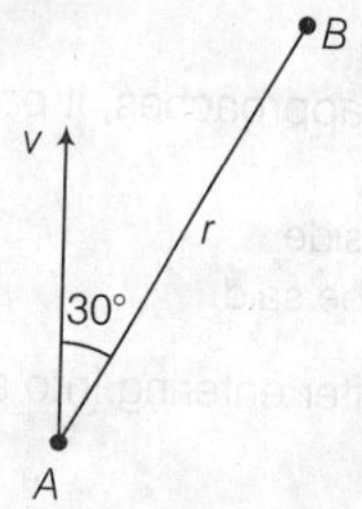

a. 2.5 μT
b. 5.0 μT
c. 2.0 μT
d. None of these

2 An electron (charge e^-, mass m) is revolving around a fixed proton in circular path of radius *r*. The magnetic field at the centre due to electron is

a. 0
b. $\dfrac{\mu_0 e^2}{8\pi r^2 \sqrt{\pi m \varepsilon_0 r}}$
c. $\dfrac{\mu_0 e}{8\pi r \sqrt{\pi m \varepsilon_0 r}}$
d. $\dfrac{\mu_0 e}{4\pi r^2 \sqrt{\pi m \varepsilon_0 r}}$

3 A point charge is moving in clockwise direction in a circle with constant speed. Consider the magnetic field produced by the charge at a point *P* (not centre of the circle) on the axis of the circle, then

a. it is constant in magnitude only
b. it is constant in direction only
c. it is constant in direction and magnitude both
d. it is not constant in magnitude and direction both

4 An α-particle is moving along a circle of radius *R* with a constant angular velocity ω. Point *A* lies in the same plane at a distance 2*R* from the centre. Point *A* records magnetic field produced by α-particle. If the minimum time interval between two successive times at which *A* records zero magnetic field is *t*, the angular speed ω in terms of *t* is

a. $\dfrac{2\pi}{t}$
b. $\dfrac{2\pi}{3t}$
c. $\dfrac{\pi}{3t}$
d. $\dfrac{\pi}{t}$

5 A particle is moving with velocity $\mathbf{v} = \hat{\mathbf{i}} - 3\hat{\mathbf{j}}$ and it produces an electric field at a point given by $\mathbf{E} = 2\hat{\mathbf{k}}$. It will produce magnetic field at that point equal to (all quantities are in SI units)

a. $\dfrac{6\hat{\mathbf{i}} - 2\hat{\mathbf{j}}}{c^2}$
b. $\dfrac{6\hat{\mathbf{i}} + 2\hat{\mathbf{j}}}{c^2}$
c. zero
d. Cannot be determined from the given data

Working Space

6 Determine the magnitude of magnetic field at the centre of the current carrying wire arrangement shown in the figure. The arrangement extends to infinity. The wires joining the successive squares are along the line passing through the centre.

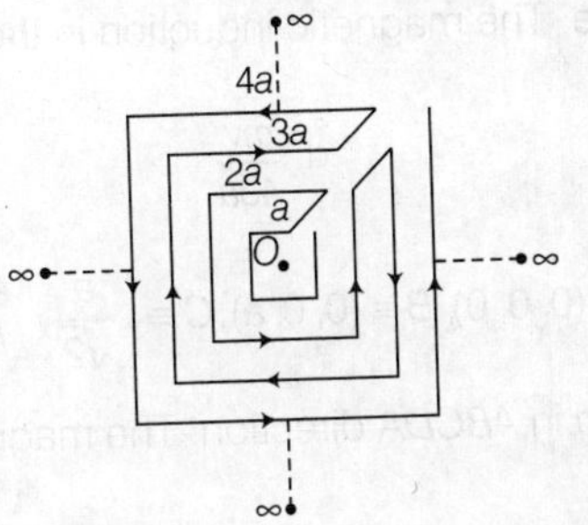

a. $\frac{\mu_0 i}{\sqrt{2}\,\pi a}$

b. 0

c. $\frac{2\sqrt{2}\,\mu_0 i}{\pi a}\ln 2$

d. None of these

7 There exists a uniform magnetic and electric field of magnitude 1 T and 1 V/m respectively, along positive *Y*-axis. A charged particle of mass 1 kg and of charge 1 C is having velocity 1 ms^{-1} along *X*-axis and is at origin at $t = 0$. Then, the coordinates of particle at time π seconds will be

a. (0, 1, 2)

b. $(0, -\pi^2/2, -2)$

c. $(2, \pi^2/2, 2)$

d. $(0, \pi^2/2, 2)$

8 Two large conducting planes carrying current perpendicular to *X*-axis and placed at $(d, 0)$ and $(2d, 0)$ as shown in figure. Current per unit width in both the planes is same and current is flowing in the outward direction. The variation of magnetic induction (taken as positive if it is in positive *Y*-direction) as function of x $(0 \le x \le 3d)$ is best represented by

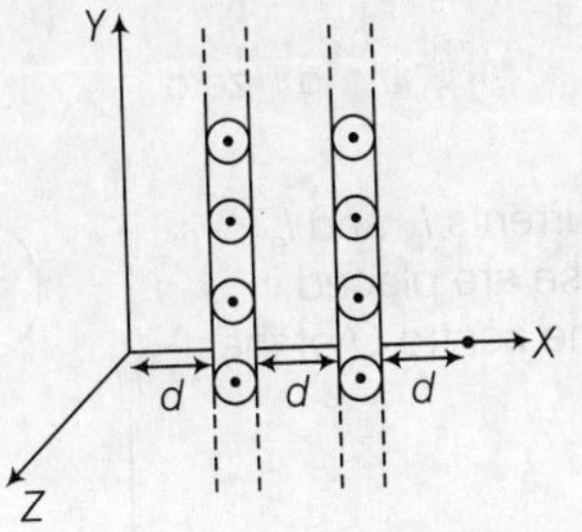

a.

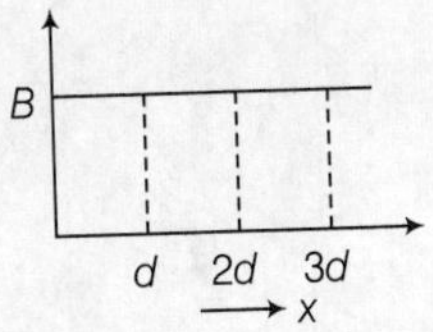

b.

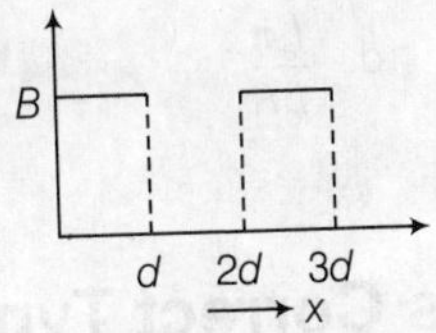

c.

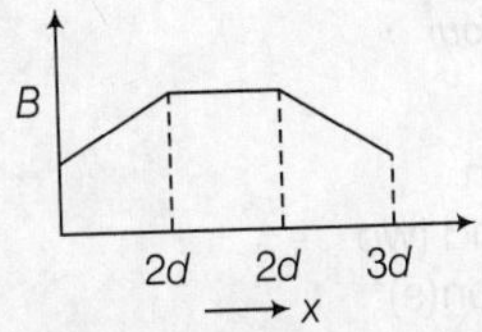

d.

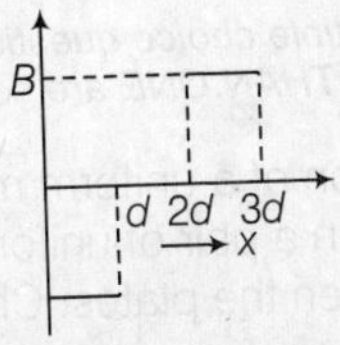

9 A particle of positive charge q and mass m enters with velocity $v\,\hat{\mathbf{j}}$ at the origin in a magnetic field $B(-\hat{\mathbf{k}})$ which is present in the whole space. The charge makes a perfectly inelastic collision with identical particle at rest but free to move at its maximum y-coordinate. After collision, the combined charge will move on trajectory (where, $r = \frac{mv}{qB}$)

a. $y = \frac{mv}{qB}(-\hat{\mathbf{i}})$

b. $(x + r)^2 + (y - r/2)^2 = r^2/4$

c. $(x - r)^2 + (y - r)^2 = r^2$

d. $(x - r)^2 + (y + r/2)^2 = r^2/4$

10 A uniform magnetic field exists in region which forms an equilateral triangle of side a. The magnetic field is perpendicular to the plane of the triangle. A charge q enters into this magnetic field perpendicularly with speed v along perpendicular bisector of one side and comes out along perpendicular bisector of other side. The magnetic induction in the triangle is

a. $\frac{mv}{qa}$ b. $\frac{2mv}{qa}$ c. $\frac{mv}{2qa}$ d. $\frac{mv}{4qa}$

11 Coordinates of four corners of a square loop are $A \equiv (0, 0, 0)$, $B \equiv (0, 0, a)$, $C \equiv \left(\frac{a}{\sqrt{2}}, \frac{a}{\sqrt{2}}, a\right)$ and $D \equiv \left(\frac{a}{\sqrt{2}}, \frac{a}{\sqrt{2}}, 0\right)$. A current I is flowing in the loop in $ABCDA$ direction. The magnetic moment of the loop would be

a. $\left(\frac{a^2}{\sqrt{2}}\hat{\mathbf{i}} + \frac{a^2}{\sqrt{2}}\hat{\mathbf{k}}\right)I$ b. $\left(\frac{a^2}{\sqrt{2}}\hat{\mathbf{j}} - \frac{a^2}{\sqrt{2}}\hat{\mathbf{i}}\right)I$

c. $\left(\frac{a^2}{\sqrt{2}}\hat{\mathbf{j}} + \frac{a^2}{\sqrt{2}}\hat{\mathbf{i}}\right)I$ d. $\left(\frac{a^2}{\sqrt{2}}\hat{\mathbf{i}} - \frac{a^2}{\sqrt{2}}\hat{\mathbf{j}}\right)I$

12 The resistances of three parts of a circular loop is shown in the figure. The magnetic field at the centre O is

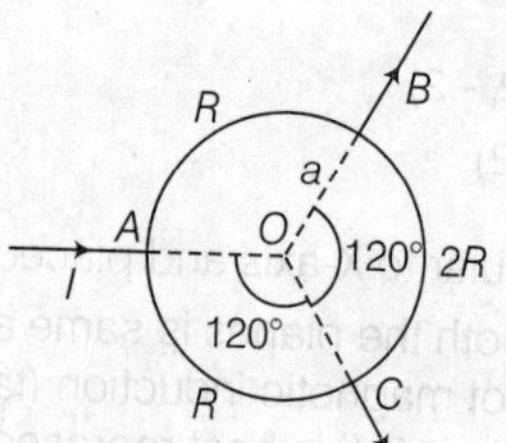

a. $\frac{\mu_0 I}{6a}$ b. $\frac{\mu_0 I}{3a}$ c. $\frac{2\mu_0 I}{3a}$ d. zero

13 Circular loop of a wire and a long straight wire carry currents I_c and I_e respectively as shown in the figure. Assuming that these are placed in the same plane, the magnetic fields will be zero and the centre O of the loop, when separation H is

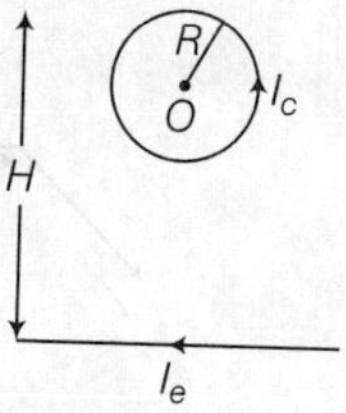

a. $\frac{I_e R}{i_c \pi}$ b. $\frac{I_c R}{I_e \pi}$

c. $\frac{I_c \pi}{I_e R}$ d. $\frac{I_e \pi}{I_c R}$

One or More than One Options Correct Type

Direction *(Q. Nos. 14-21) This section contains 8 multiple choice questions. Each question has four choices (a), (b), (c) and (d), out of which ONE or MORE THAN ONE are correct.*

14 Figure shows the path of an electron in a region of a uniform magnetic field. The path consists of two straight sections, each between a pair of uniformly charged plates and two half circles. The electric field exists only between the plates. Choose the correct option(s).

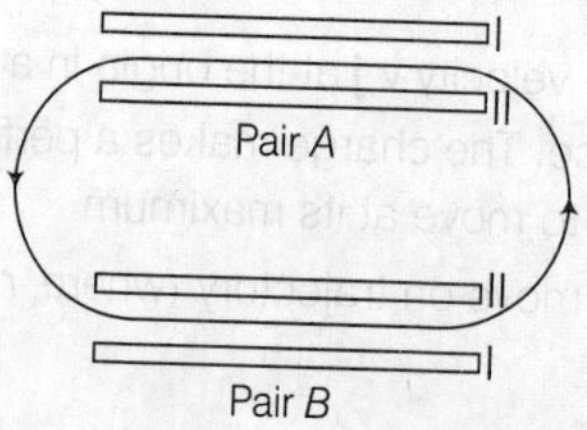

a. plate I of pair A is at higher potential than plate II of the same pair
b. plate I of pair B is at higher potential than plate II of the same pair
c. direction of the magnetic field is out of the page
d. direction of the magnetic field into the page

15 Two charged particles M and N enter a space of uniform magnetic field with velocities perpendicular to the magnetic field. The paths are as shown in the figure. The possible reasons for different paths may be

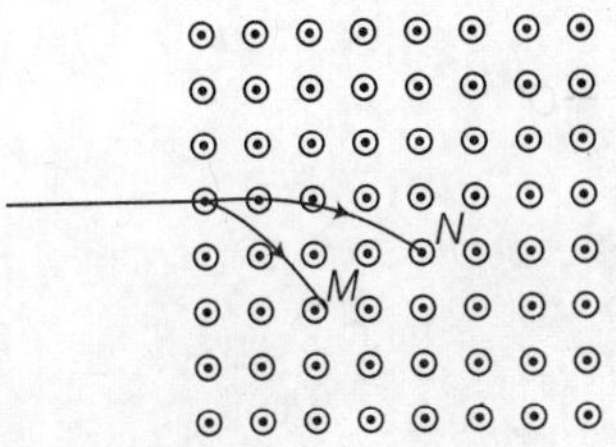

a. the charge of M is greater than that of N
b. the momentum of M is greater than that of N
c. specific charge of M is greater than that of N
d. the speed of M is less than that of N

16 Figure shows cross-section of two large parallel metal sheets carrying electric currents along their surfaces. The current in each sheet is $\frac{10}{\pi}$ Am^{-1} along the width. Consider two points A and B as shown in the figure with their positions. Choose the correct option(s).

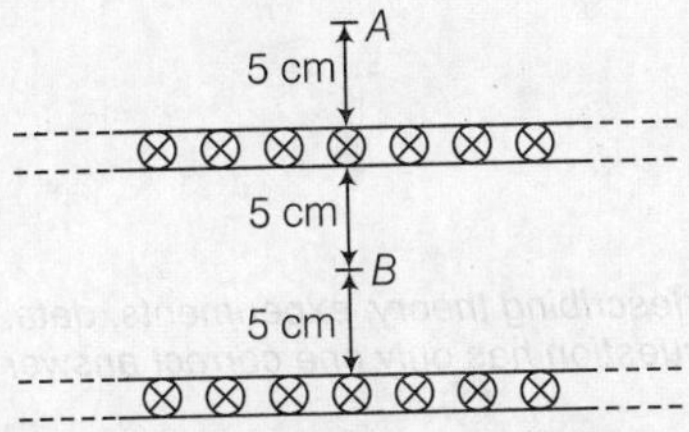

a. Magnetic field at A is $4\mu T$ along positive X-direction
b. Magnetic field at A is $4\mu T$ along negative X-direction
c. Magnetic field at B is zero
d. Magnetic field at B is $2\mu T$ along X-direction

17 A charged particle of charge q and mass m having velocity v as shown in figure in a uniform magnetic field B along negative Z-direction. Select correct alternative(s).

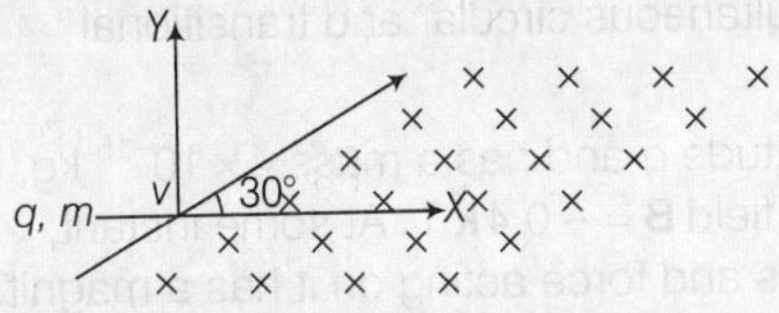

a. velocity of the particle when it comes out to the magnetic field $\mathbf{v} = v\cos 60°\ \hat{\mathbf{i}} + v\sin 60°\ \hat{\mathbf{j}}$
b. time for which the particle was in magnetic field is $\frac{\pi m}{3qB}$
c. distance travelled in magnetic field is $\frac{\pi mv}{3qB}$
d. None of the above

18 A particle of mass m and charge q moving with velocity $\mathbf{v}$ enters a region of a uniform magnetic field of induction $\mathbf{B}$. Then,
a. its path in the region of the field is always circular
b. its path in the region of the field is a circular if $\mathbf{v} \cdot \mathbf{B} = 0$
c. its path in the region of field is a straight line if $\mathbf{v} \times \mathbf{B} = 0$
d. distance travelled by the particle in time T does not depend on the angle between $\mathbf{v}$ and $\mathbf{B}$

19 A particle of charge $+q$ and mass m moving under the influence of a uniform electric field $E\hat{\mathbf{i}}$ and uniform magnetic field $B\hat{\mathbf{k}}$ follows a trajectory from P and Q as shown in figure. The velocities at P and Q are $v\hat{\mathbf{i}}$ and $-2\hat{\mathbf{j}}$. Which of the following statement(s) is/are correct?

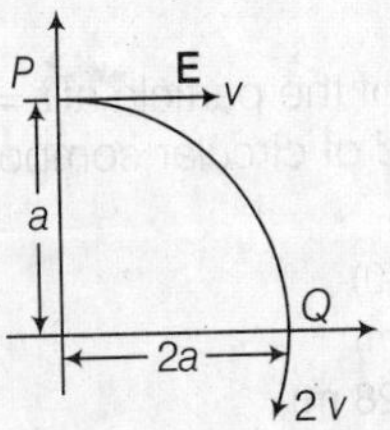

a. $E = \frac{3}{4}\left[\frac{mv^2}{qa}\right]$

b. Rate of work done by electric field at P is $\frac{3}{4}\left[\frac{mv^3}{a}\right]$

c. Rate of work done by electric field at P is zero
d. Rate of work done by both the fields at Q is zero

Working Space

20 Which of the following statement(s) is/are correct in the given figure?

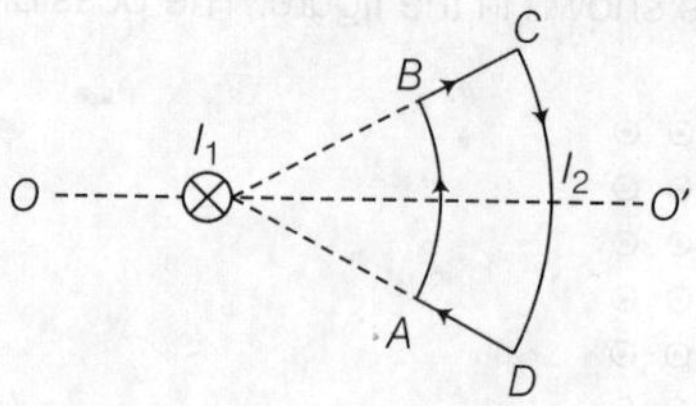

a. Net force on the loop is zero
b. Net torque on the loop is zero
c. Loop will rotate clockwise about axis OO' when seen from O
d. Loop will rotate anti-clockwise about OO' when seen from O

21 An electron and proton are moving on straight parallel paths with same velocity. They enter a semi-infinite region of a uniform magnetic field perpendicular to the velocity. Which of the following statement(s) is/are correct? **[IIT JEE 2011]**

a. They will never come out of the magnetic field region
b. They will come out travelling along parallel paths
c. They will come out at the same time
d. They will come out at different times

Comprehension Type

Direction (Q. Nos. 22-30) *This section contains 3 paragraphs, each describing theory, experiments, data, etc. Nine questions related to the paragraphs have been given. Each question has only one correct answer among the four given options (a), (b), (c) and (d).*

Passage I

As a charged particle q is moving with a velocity $\mathbf{v}$, enters a uniform magnetic field $\mathbf{B}$, it experiences a force $\mathbf{F} = q(\mathbf{v} \times \mathbf{B})$. For $\theta = 0°$ or $180°$, θ being the angle between $\mathbf{v}$ and $\mathbf{B}$, force experienced is zero and the particle passes undeflected. For $\theta = 90°$, the particle moves along a circular arc and the magnetic force (qvB) provides the necessary centripetal force (mv^2/r). For other values of θ ($\theta \neq 0°$, $180°$, $90°$), the charged particle moves along a helical path which is the resultant motion of simultaneous circular and transitional motions.

Suppose a particle, that carries a charge of magnitude q and has a mass 4×10^{-15} kg, is moving in a region containing a uniform magnetic field $\mathbf{B} = -0.4\hat{\mathbf{k}}$ T. At some instant, velocity of the particle is $\mathbf{v} = (8\hat{\mathbf{i}} - 6\hat{\mathbf{j}} + 4\hat{\mathbf{k}}) \times 10^6$ m/s and force acting on it has a magnitude 1.6 N.

22 Motion of charged particle will be along a helical path with

a. a translational component along X-direction and a circular component in the YZ-plane
b. a translational component along Y-direction and a circular component in the XZ-plane
c. a translational component along Z-axis and a circular component in the XY-plane
d. direction of translational component and plane of circular component are uncertain

23 Angular frequency of rotation of particle also called the cyclotron frequency is

a. 8×10^5 rad s^{-1}
b. 12.5×10^4 rad s^{-1}
c. 6.2×10^6 rad s^{-1}
d. 4×10^7 rad s^{-1}

24 If the coordinates of the particle at $t = 0$ are (2 m, 1 m, 0), coordinates at a time $t = 3T$, where T is the time period of circular component of motion will be (Given, $\pi = 3.14$)

a. (2 m, 1 m, 400 m)
b. (0.142 m, 130 m, 0)
c. (2 m, 1 m, 1.884 m)
d. (142 m, 130 m, 628 m)

Working Space

Passage II

A particle of mass m and charge q is accelerated by a potential difference V volt and made to enter a magnetic field region at an angle θ with the field. At the same moment, another particle of same mass and charge is projected in the direction of the field from the same point magnetic field of induction is B.

25 What would be the speed of second particle so that both particles meet again and again after a regular interval of time, which should be minimum?

a. $\sqrt{\frac{qv}{m}} \cos\theta$ b. $\sqrt{\frac{2qv}{m}} \cos\theta$

c. $\sqrt{\frac{qv}{m}} \sin\theta$ d. $\sqrt{\frac{qv}{m}} \cos\theta$

26 Find the time interval after which they meet.

a. $\frac{2\pi m}{qB}$ b. $\frac{\pi m}{2qB}$

c. $\frac{3\pi m}{2qB}$ d. $\frac{3\pi m}{2qs}$

27 Find the distance travelled by the second particle during that interval mentioned in the above problem.

a. $\sqrt{\frac{vm}{q}} \cdot \frac{2\pi}{B} \cos\theta$

b. $\sqrt{\frac{2vm}{3q}} \cdot \frac{2\pi}{B} \cos\theta$

c. $\sqrt{\frac{2vm}{q}} \cdot \frac{2\pi}{B} \cos\theta$

d. $\frac{2}{3}\sqrt{\frac{vm}{q}} \cdot \frac{\pi}{m} \cos\theta$

Passage III

Ampere's law provides us an easy way to calculate the magnetic field due to a symmetrical distribution of current. Its mathematical expression is as

$$\int \mathbf{B} \cdot d\mathbf{l} = \mu_0 I_{\text{enclosed}}$$

The quantity on the left hand side is known as line integral of magnetic field over a closed Ampere's loop.

28 Only the current inside amperian loop contributes in

a. finding magnetic field at any point on the Ampere's loop
b. line integral of magnetic field
c. Both (a) and (b)
d. Neither of them

29 If the current density in a linear conductor of radius a varies with r according to the relation $J = kr^2$, where k is a constant and r is the distance of a point from the axis of the conductor, find the magnetic field induction at a point distance r from the axis when $r < a$. Assume relative permeability of the conductor to be unity.

a. $\frac{\mu_0 k\pi a^4}{4r}$ b. $\frac{\mu_0 kr^3}{2}$

c. $\frac{\mu_0 k\pi a^4}{2r}$ d. $\frac{\mu_0 kr^3}{4}$

30 In the above question, find the magnetic field induction at a point distance r from the axis when $r > a$. Assume relative permeability of the medium surrounding the conductor to be unity.

a. $\frac{\mu_0 k\pi a^4}{4r}$ b. $\frac{\mu_0 kr^3}{2}$

c. $\frac{\mu_0 k\pi a^4}{2r}$ d. $\frac{\mu_0 kr^2}{4}$

Matching List Type

Working Space

Direction *(Q. Nos. 31-35) Choices for the correct combination of elements from Column I and Column II are given as options (a), (b), (c) and (d), out of which one is correct.*

31 A charged particle having non-zero velocity is subjected to certain conditions given in Column I. Column II gives possible trajectories of the particle. Match the conditions in Column I with the results in Column II and select the correct option from the given codes below.

	Column I		Column II
i.	In only uniform electric field	p.	The path of the charged particle may be a straight line.
ii.	In only uniform magnetic field	q.	The path of the charged particle may be a parabola.
iii.	In uniform magnetic and uniform electric field such that both are parallel	r.	The path of the charged particle may be a circle.
iv.	Subjected to a net force of constant magnitude	s.	The path of the charged particle may be a helix with uniform or non-uniform.

Codes

	i	ii	iii	iv
a.	r,s	q	p	r,s,q
b.	p,q	p,r,s	p,s	p,q,r,s
c.	p,q,r	p,q	r,s	s
d.	a,p	r,q	s,p	r

32 A square loop of a uniform conducting wire is shown as in figure. A current I (in amperes) enters the loop from one end exits the loop from opposite end as shown in figure. The length of one side of square loop is l metre. The wire has a uniform cross-section area and uniform linear mass density. In four situations of Column I, the loop is subjected to four different magnetic field. Under the conditions of Column I, match the Column I with corresponding results of Column II (B_0 in Column I is a positive non-zero constant)

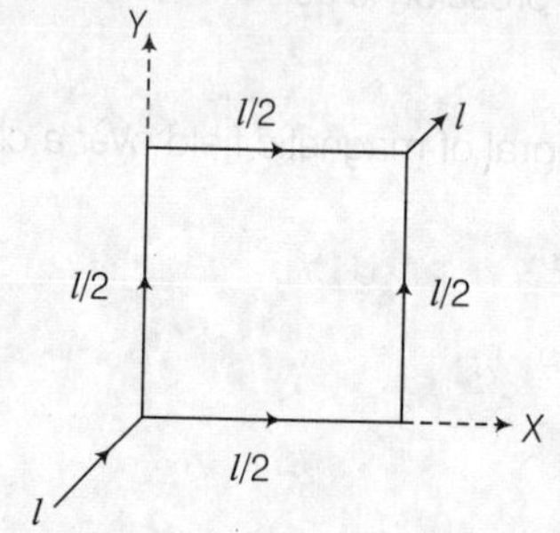

	Column I		Column II
i.	$\mathbf{B} = B_0\hat{\mathbf{i}}$ (in tesla)	p.	Magnitude of net force on loop is $\sqrt{2}B_0Il$ newton.
ii.	$\mathbf{B} = B_0\hat{\mathbf{j}}$ (in tesla)	q.	Magnitude of net force on loop is zero.
iii.	$\mathbf{B} = B_0(\hat{\mathbf{i}} + \hat{\mathbf{j}})$ (in tesla)	r.	Magnitude of net torque on loop about its centre is zero.
iv.	$\mathbf{B} = B_0\hat{\mathbf{k}}$ (in tesla)	s.	Magnitude of net force on loop is B_0Il newton.

Codes

	i	ii	iii	iv
a.	r,s	r,s	q,r	p,r
b.	q,r	r	q	p,s
c.	p	q,r	s	q,p
d.	q	r,s	s	p,q

33 Column II gives four situations in which three or four semi-infinite current carrying wires are placed in *XY*-plane as shown in the figure. The magnitude of the direction of current is shown in each figure. Column I gives statements regarding the *x* and *y*-components of magnetic field at a point *P* whose coordinates are $(0, 0, d)$. Match the statements in Column I with the corresponding figures in Column II.

	Column I		Column II
i.	The *x*-component of magnetic field at point *P* is zero in	p.	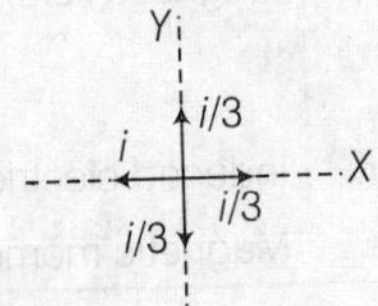
ii.	The *z*-component of magnetic field at point *P* is zero in	q.	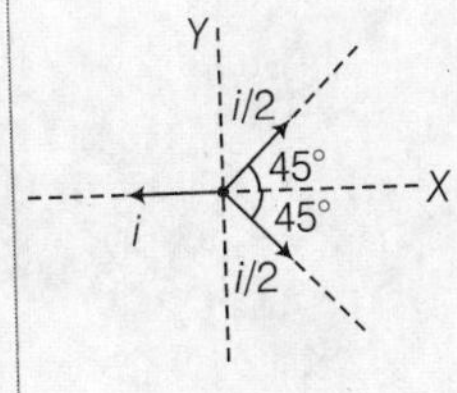
iii.	The magnitude of magnetic field at point *P* is $\frac{\mu_0 i}{4\pi d}$ in	r.	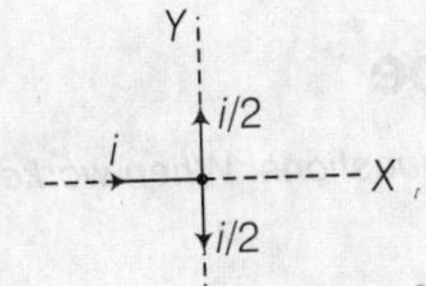
iv.	The magnitude of magnetic field at point *P* is less than $\frac{\mu_0 i}{2\pi d}$ in	s.	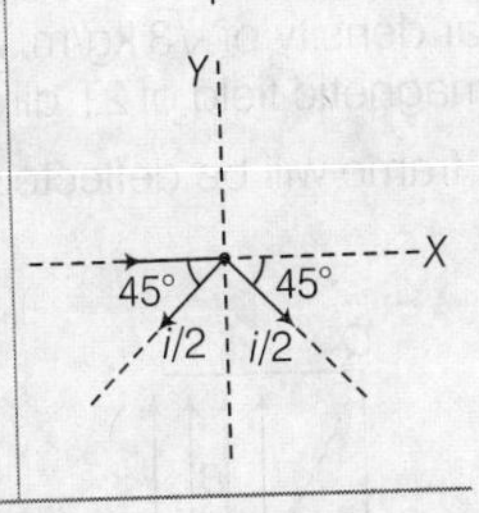

Codes

	i	ii	iii	iv
a.	p,q,r	p,q,r,s	r	p,q,r,s
b.	p,q	q,r	s,p	p,q,r
c.	q	r	s,p,q	q,r,s
d.	p,q	r,s	r,p,q	p,q,r

34 A circular current carrying loop having 100 turns and radius 10 cm is placed in *XY*-plane as shown in figure. A uniform magnetic field $\mathbf{B} = (-\hat{\mathbf{i}} + \hat{\mathbf{k}})$ tesla is present in the region. If current in the loop is 5 A, then match the Column I with Column II and select the correct option from the given codes below. **[IIT JEE 2006]**

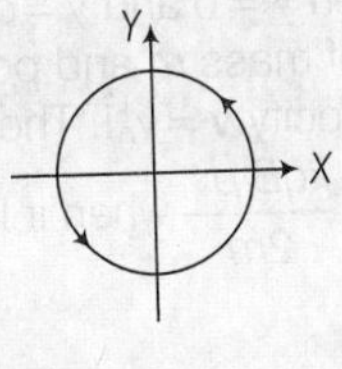

	Column I		Column II
i.	Magnitude and direction of moment (in A-m) of the loop are	p.	Zero
ii.	Magnitude and direction of torque (in N-m) on the loop are	q.	5π
iii.	Magnitude and direction of net force (in N) on the current loop are	r.	Along positive *Z*-axis
iv.	Direction of magnetic field of loop at the centre is	s.	Along negative *Y*-axis

Codes

	i	ii	iii	iv
a.	q,r	q	p	r
b.	q,r	q,s	p	r
c.	q,p	r,s	p,r	q,s
d.	s,p	s	r	q

35 Some laws/processes are given in Column I. Match these with the physical phenomena given in Column II. **[IIT JEE 2006]**

Working Space

	Column I		Column II
i.	Dielectric ring uniformly charged	p.	Time independent electrostatic field out of system
ii.	Dielectric ring uniformly charged rotating with angular velocity ω	q.	Magnetic field
iii.	Constant current in ring i_0	r.	Induced electric field
iv.	$i = i_0 \cos \omega t$	s.	Magnetic moment

Codes

	i	ii	iii	iv
a.	p	p,q,s	q,s	q,r
b.	p,q	q,s	r	q
c.	p,q	q,r	s,r	r
d.	s	r	p,q	p,q,r

One Integer Value Correct Type

Direction *(Q. Nos. 36-39) This section contains 4 questions. When worked out will result in an integer from 0 to 9 (both inclusive).*

36 As shown in the figure, three sided frame is pivoted at P and Q and hangs vertically. Its sides are of same length and have a linear density of $\sqrt{3}$ kg/m. A current of $10\sqrt{3}$ A is sent through the frame, which is in a uniform magnetic field of 2T directed upwards as shown in the figure. Then, angle through which the frame will be deflected in equilibrium is $\frac{\pi}{k}$, Find the value of k. (Given, $g = 10\ \text{m/s}^2$)

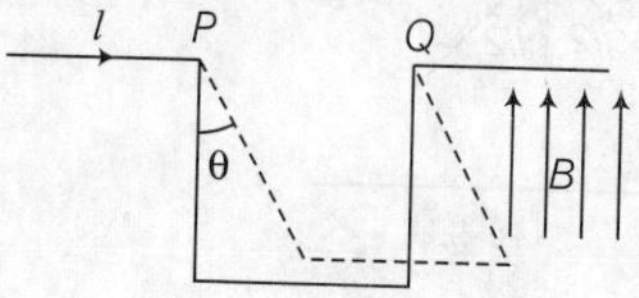

37 A non-uniform magnetic field $\mathbf{B} = B_0\left(1+\frac{y}{d}\right)\hat{\mathbf{k}}$ is present in region of space in between $y = 0$ and $y = d$. The lines are shown in the figure. A particle of mass m and positive charge q is moving. Given, an initial velocity $\mathbf{v} = v_0\hat{\mathbf{i}}$. The x-component of velocity of the particle is $v_x = v_0 - \frac{kqB_0d}{2m}$ when it leaves the field, then find the value of k.

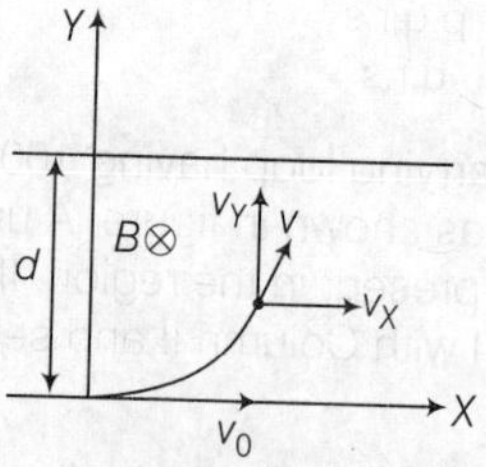

38 A cylindrical cavity of figure a exists inside a cylinder of diameter $2a$ as shown in the figure. Both the cylinder and the cavity are infinitely long. A uniform current density J flows along the length. If the magnitude of the magnetic field at the point P is given by $\frac{N}{12}\mu_0 aJ$, then the value of N is **[IIT JEE 2012]**

39 Two parallel wires in the plane of the paper are distance X_0 apart. A point charge is moving with speed u between the wires in the same plane at a distance X_1 from one of the wires. When the wires carry current of magnitude I in the same direction, the radius of curvature of the path of the point charge is R_1. In contrast, if the current I in the two wires have directions opposite to each other, the radius of curvature of the path is R_2. If $\frac{X_0}{X_1} = 3$, then find the value of $\frac{R_1}{R_2}$.

JEE Main & AIEEE Archive

(Compilation of Last 13 Years' Questions)

1 A conductor lies along the Z-axis at $-1.5 \le z < 1.5$ m and carries a fixed current of 10 A in $-\hat{\mathbf{k}}$ direction (see figure). For a field $\mathbf{B} = 3.0 \times 10^{-4} e^{-0.2x}\,\hat{\mathbf{j}}$ T, find the power required to move the conductor at constant speed to $x = 2$ m, $y = -0$ in 5×10^{-3} s. Assuming parallel motion along the X-axis. **(JEE Main 2014)**

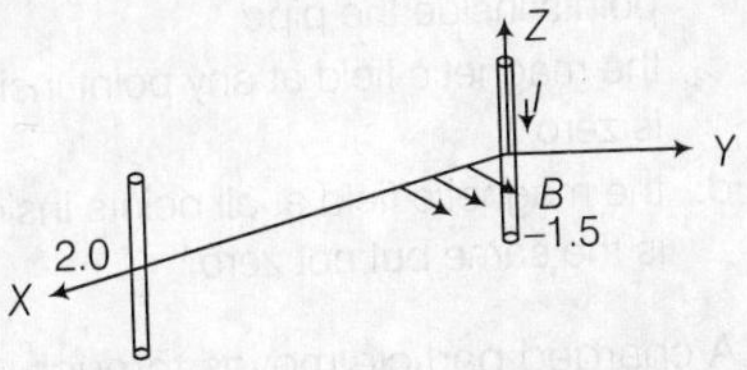

a. 1.57 W
b. 2.97 W
c. 14.85 W
d. 29.7 W

2 Proton, deuteron and α-particles of same kinetic energy are moving in circular trajectories in a constant magnetic field. The radii of proton, deuteron and α-particle are respectively r_p, r_d and r_α. Which one of the following relations is correct? **(AIEEE 2012)**

a. $r_\alpha = r_p = r_d$
b. $r_\alpha = r_p < r_d$
c. $r_\alpha > r_d > r_p$
d. $r_\alpha = r_d > r_p$

3 A charge Q is uniformly distributed over the surface of non-conducting disc of radius R. The disc rotates about an axis perpendicular to its plane and passing through its centre with an angular velocity ω. As a result of this rotation, a magnetic field of induction B is obtained at the centre of the disc.
If we keep both the amount of charge placed on the disc and its angular velocity to be constant and vary the radius of the disc, then the variation of the magnetic induction at the centre of the disc will be represented by the figure. **(AIEEE 2012)**

a.
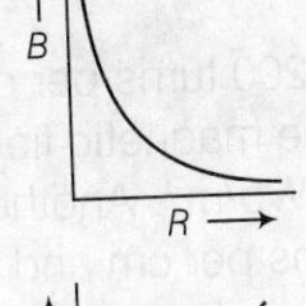

b.
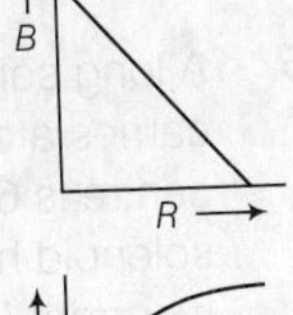

c.
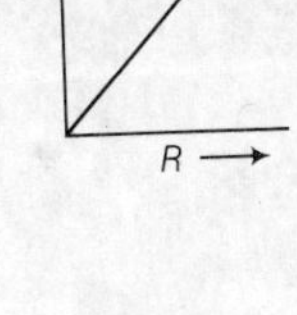

d.
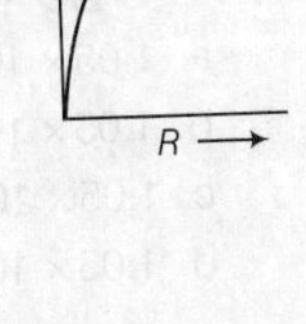

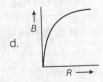

4 A thin circular disc of radius R is uniformly charged with density $\sigma > 0$ per unit area. The disc rotates about its axis with a uniform angular speed ω. The magnetic moment of the disc is **(AIEEE 2011)**

a. $2\pi R^4 \sigma\omega$
b. $\pi R^4 \sigma\omega$
c. $\dfrac{\pi R^4}{2}\sigma\omega$
d. $\dfrac{\pi R^4}{4}\sigma\omega$

5 A current I flows in an infinitely long wire with cross-section in the form of a semicircular ring of radius R. The magnitude of the magnetic induction along its axis is **(AIEEE 2011)**

a. $\dfrac{\mu_0 I}{2\pi^2 R}$
b. $\dfrac{\mu_0 I}{2\pi R}$
c. $\dfrac{\mu_0 I}{4\pi R}$
d. $\dfrac{\mu_0 I}{\pi^2 R}$

6 Two long parallel wires are at a distance $2d$ apart. They carry steady equal current flowing out of the plane of the paper as shown in the figure. The variation of the magnetic field along the line xx' is given by **(AIEEE 2010)**

a.
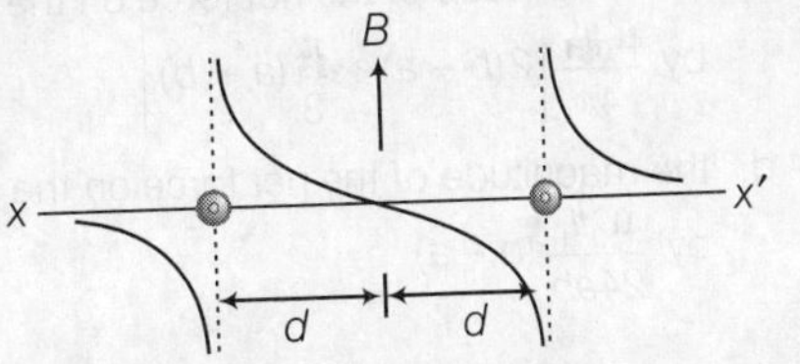

b.
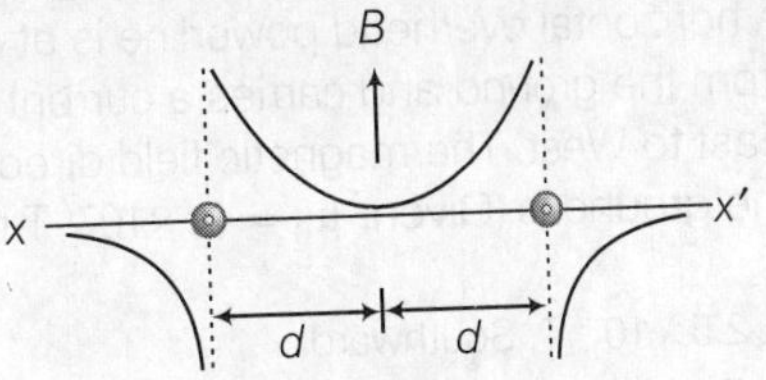

c.
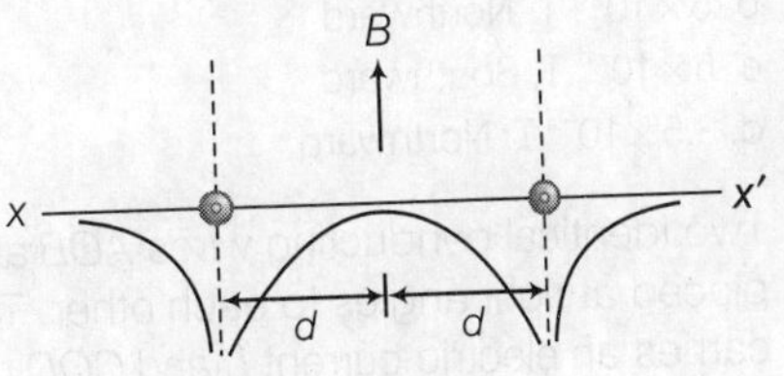

d.
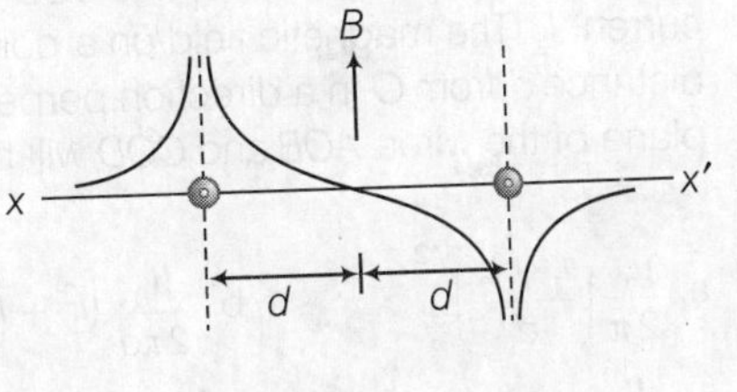

7 The magnitude of the magnetic field B. Due to loop $ABCD$ at the origin O is **(AIEEE 2009)**

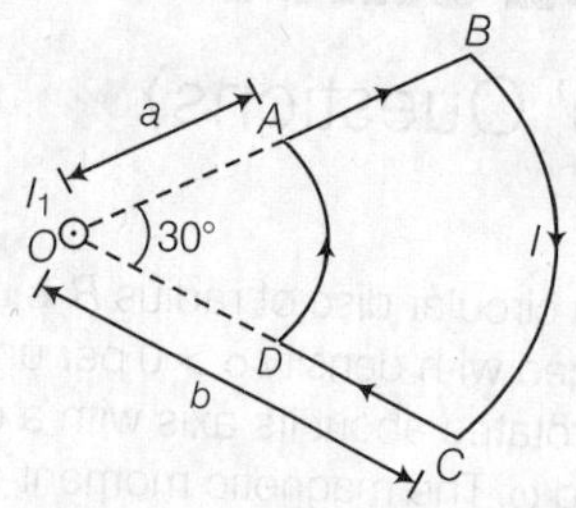

a. zero

b. $\frac{\mu_0 I(b-a)}{24ab}$

c. $\frac{\mu_0 I}{4\pi}\left[\frac{b-a}{ab}\right]$

d. $\frac{\mu_0 I}{4\pi}\left[2(b-a)+\frac{\pi}{3}(a+b)\right]$

8 Due to the presence of the current I_1 at the origin in the figure, **(AIEEE 2009)**

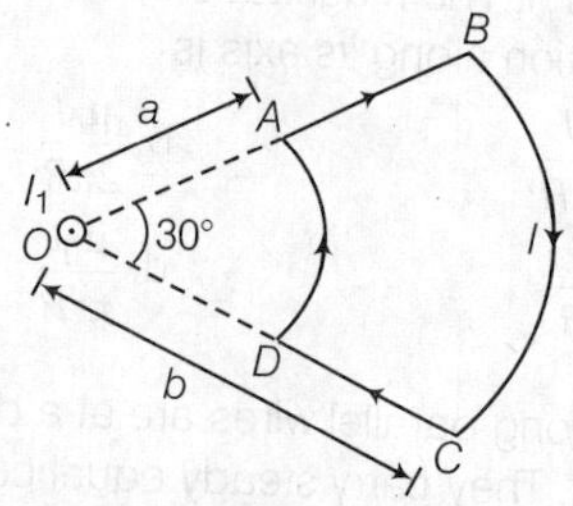

a. the forces on AB and DC are zero

b. the forces on AD and BC are zero

c. the magnitude of the net force on the loop is given by $\frac{\mu_0 I I_1}{4\pi}\left[2(b-a)+\frac{\pi}{3}(a+b)\right]$

d. the magnitude of the net force on the loop is given by $\frac{\mu_0 I I_1}{24ab}(b-a)$

9 A horizontal overhead powerline is at a height of 4 m from the ground and carries a current of 100 A from East to West. The magnetic field directly below it on the ground is (Given, $\mu_0 = 4\pi\times10^{-7}\ \text{T mA}^{-1}$) **(AIEEE 2008)**

a. 2.5×10^{-7} T, Southward

b. 5×10^{-6} T, Northward

c. 5×10^{-6} T, Southward

d. 2.5×10^{-7} T, Northward

10 Two identical conducting wires AOB and COD are placed at right angles to each other. The wire AOB carries an electric current I_1 and COD carries a current I_2. The magnetic field on a point lying at a distance d from O in a direction perpendicular to the plane of the wires AOB and COD will be given by **(AIEEE 2007)**

a. $\frac{\mu_0}{2\pi}\left(\frac{I_1+I_2}{d}\right)^{1/2}$

b. $\frac{\mu_0}{2\pi d}(I_1^2+I_2^2)^{1/2}$

c. $\frac{\mu_0}{2\pi d}(I_1+I_2)$

d. $\frac{\mu_0}{2\pi d}(I_1^2+I_2^2)$

11 A long straight wire of radius a carries a steady current I. The current is uniformly distributed across its cross-section. The ratio of the magnetic field at $\frac{a}{2}$ and $2a$ is **(AIEEE 2007)**

a. 1/4

b. 4

c. 1

d. 1/2

12 A current I flows along the length of an infinitely long, straight, thin walled pipe. Then, **(AIEEE 2007)**

a. the magnetic field is zero only on the axis of the pipe

b. the magnetic field is different at different points inside the pipe

c. the magnetic field at any point inside the pipe is zero

d. the magnetic field at all points inside the pipe is the same but not zero

13 A charged particle moves through a magnetic field perpendicular to its direction. Then, **(AIEEE 2007)**

a. the momentum changes but the kinetic energy is constant

b. both momentum and kinetic energy of the particle are not constant

c. both momentum and kinetic energy of the particle are constant

d. kinetic energy changes but the momentum is constant

14 In a region, steady and uniform electric and magnetic fields are present. These two fields are parallel to each other. A charged particle is released from rest in this region. The path of the particle will be a **(AIEEE 2006)**

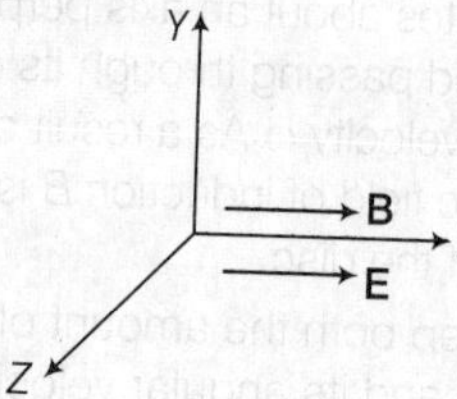

a. helix

b. straight line

c. ellipse

d. circle

15. A long solenoid has 200 turns per cm and carries a current I. The magnetic field at its centre is $6.28\times10^{-2}\ \text{Wb/m}^2$. Another long solenoid has 100 turns per cm and it carries a current $I/3$. The value of the magnetic field at its centre is **(AIEEE 2006)**

a. $1.05\times10^{-2}\ \text{Wb/m}^2$

b. $1.05\times10^{-5}\ \text{Wb/m}^2$

c. $1.05\times10^{-3}\ \text{Wb/m}^2$

d. $1.05\times10^{-4}\ \text{Wb/m}^2$

16 Two thin, long, parallel wires, separated by a distance d carry a current of I ampere in the same direction. They will **(AIEEE 2005)**

a. attract each other with a force of $\frac{\mu_0 I^2}{(2\pi d)}$

b. repel each other with a force of $\frac{\mu_0 I^2}{(2\pi d)}$

c. attract each other with a force of $\frac{\mu_0 I^2}{(2\pi d^2)}$

d. repel each other with a force of $\frac{\mu_0 I^2}{(2\pi d^2)}$

17 Two concentric coils each of radius equal to 2π cm are placed at right angles to each other. 3 A and 4 A are the currents flowing in each coil, respectively. The magnetic induction (in Wb/m^2) at the centre of the coils will be (Given, $\mu_0 = 4\pi \times 10^{-7}$ Wb/A-m) **(AIEEE 2005)**

a. 12×10^{-5}

b. 10^{-5}

c. 5×10^{-5}

d. 7×10^{-5}

18 A magnetic needle is kept in a non-uniform magnetic field. It experiences **(AIEEE 2005)**

a. a torque but not a force

b. neither a force nor a torque

c. a force and a torque

d. a force but not a torque

19 A charged particle of mass m and charge q travels on a circular path of radius r that is perpendicular to a magnetic field B. The time taken by the particle to complete one revolution is **(AIEEE 2005)**

a. $\frac{2\pi mq}{B}$ b. $\frac{2\pi q^2 B}{m}$

c. $\frac{2\pi qB}{m}$ d. $\frac{2\pi m}{qB}$

20 A moving coil galvanometer has 150 equal divisions. Its current sensitivity is 10 divisions per milliampere and voltage sensitivity is 2 divisions per millivolt. In order that each division reads 1 V, the resistance (in Ω) needed to be connected in series with the coil will be **(AIEEE 2005)**

a. 10^3 b. 10^5

c. 99995 d. 9995

21 A current I ampere flows along an infinitely long straight thin walled tube, then the magnetic induction at any point inside the tube is **(AIEEE 2004)**

a. infinite

b. zero

c. $\frac{\mu_0}{4\pi} \cdot \frac{2I}{r}$ T

d. $\frac{2I}{r}$ T

22 A long wire carries a steady current. It is bent into a circle of one turn and the magnetic field at the centre of the coil is B. It is then bent into a circular loop of n turns. The magnetic field at the centre of the coil will be **(AIEEE 2004)**

a. nB b. n^2B

c. $2nB$ d. $2n^2B$

23 The magnetic field due to a current carrying circular loop of radius 3 cm at a point on the axis at a distance of 4 cm from the centre is 54 μT. What will be its value at the centre of the loop? **(AIEEE 2004)**

a. 250 μT b. 150 μT

c. 125 μT d. 75 μT

24 Two long conductors, separated by a distance d carry currents I_1 and I_2 in the same direction. They exert a force F on each other. Now, the current in one of them is increased to two times and its direction is reversed. The distance is also increased to $3d$. The new value of the force between them is **(AIEEE 2004)**

a. $-2F$ b. $\frac{F}{3}$

c. $-\frac{2F}{3}$ d. $-\frac{F}{3}$

25 The length of a magnet is large compared to its width and breadth. The time period of its oscillation in a vibration magnetometer is 2 s. The magnet is cut along its length into three equal parts and three parts are then placed on each other with their like poles together. The time period of this combination will be **(AIEEE 2004)**

a. 2 s b. $\frac{2}{3}$ s

c. $2\sqrt{3}$ s d. $\frac{2}{\sqrt{3}}$ s

26 A magnetic needle lying parallel to a magnetic field requires W unit of work to turn it through 60°. The torque needed to maintain the needle in this position will be **(AIEEE 2003)**

a. $\sqrt{3}W$ b. W

c. $\left(\frac{\sqrt{3}}{2}\right)W$ d. $2W$

27 The magnetic lines of force inside a bar magnet

a. are from North pole to South pole of the magnet

b. do not exist

c. depend upon the area of cross-section of the bar magnet

d. are from South pole to North pole of the magnet

28 A particle of mass M and charge Q moving with velocity **v** describes a circular path of radius R when subjected to a uniform transverse magnetic field of induction B. The work done by the field when the particle completes one full circle is **(AIEEE 2003)**

a. $\left(\frac{Mv^2}{R}\right) \times 2\pi R$
b. zero
c. $BQ \times 2\pi R$
d. $BQv \times 2\pi R$

29 A particle of charge -16×10^{-18} C moving with velocity 10 ms^{-1} along the X-axis enters a region where a magnetic field of induction B is along the Y-axis and an electric field of magnitude 10^4 Vm^{-1} is along the negative Z-axis. If the charged particle continues moving along the X-axis, the magnitude of B is **(AIEEE 2003)**

a. 10^3 Wb/m^2
b. 10^5 Wb/m^2
c. 10^{16} Wb/m^2
d. 10^{-3} Wb/m^2

30 An ammeter reads up to 1 A. Its internal resistance is 0.81 Ω. To increase the range to 10 A, the value of the required shunt is **(AIEEE 2003)**

a. 0.03 Ω
b. 0.3 Ω
c. 0.9 Ω
d. 0.09 Ω

31 If in a circular coil A of radius R, current I is flowing and in another coil B of radius $2R$ a current $2I$ is flowing, then the ratio of the magnetic fields B_A and B_B produced by them will be **(AIEEE 2002)**

a. 1
b. 2
c. $\frac{1}{2}$
d. 4

32 If an electron and a proton having same momentum enter perpendicularly to a magnetic field, then **(AIEEE 2002)**

a. curved path of electron and proton will be same (ignoring the sense of revolution)
b. they will move undeflected
c. curved path of electron is more curved than that of proton
d. path of proton is more curved

33 The time period of a charged particle undergoing a circular motion in a uniform magnetic field is independent of its **(AIEEE 2002)**

a. speed
b. mass
c. charge
d. magnetic induction

34 At a specific instant, emission of radioactive compound is deflected in a magnetic field. The compound can emit **(AIEEE 2002)**

(i) electrons
(ii) protons
(iii) He^{2+}
(iv) neutrons

The emission at the instant can be

a. (i), (ii), (iii)
b. (i), (ii), (iii), (iv)
c. (iv)
d. (ii), (iii)

35 Wires 1 and 2 carrying currents I_1 and I_2 respectively are inclined at an angle θ to each other. What is the force on a small element dl of wire 2 at a distance r from wire 1 (as shown in figure) due to the magnetic field of wire 1?

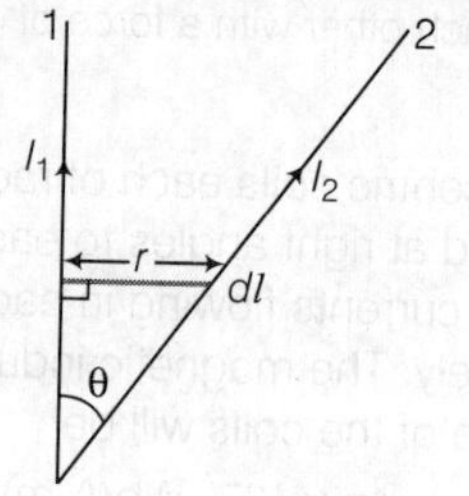

(AIEEE 2002)

a. $\frac{\mu_0}{2\pi r} I_1 I_2 \, dl \tan\theta$
b. $\frac{\mu_0}{2\pi r} I_1 I_2 \, dl \sin\theta$
c. $\frac{\mu_0}{2\pi r} I_1 I_2 \, dl \cos\theta$
d. $\frac{\mu_0}{4\pi r} I_1 I_2 \, dl \sin\theta$

36 A conducting square loop of side L and resistance R moves in its plane with a uniform velocity v perpendicular to one of its sides. A magnetic induction B constant in time and space, pointing perpendicular and into the plane at the loop exists everywhere with half the loop outside the field as shown in the figure. The induced emf is **(AIEEE 2002)**

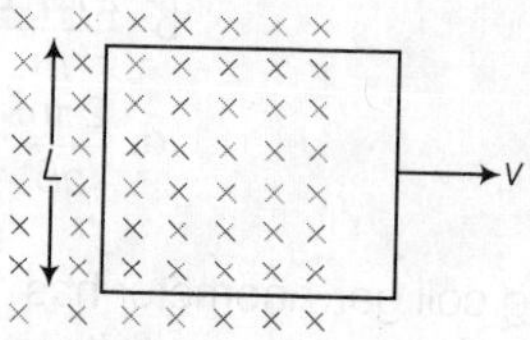

a. zero
b. RvB
c. $\frac{vBL}{R}$
d. vBL

37 If an ammeter is to be used in place of a voltmeter, then we must connect with the ammeter a **(AIEEE 2002)**

a. low resistance in parallel
b. high resistance in parallel
c. high resistance in series
d. low resistance in series

38 If a current is passed through a spring, then the spring will **(AIEEE 2002)**

a. expand
b. compress
c. remain same
d. None of the above

JEE Advanced & IIT JEE Archive

(Compilation of Last 10 Years' Questions)

Passage for Q. Nos. (1-2)

The figure shows a circular loop of radius a with two long parallel wires (numbered 1 and 2) all in the plane of the paper. The distance of each wire from the centre of the loop is d. The loop and the wires are carrying the same current I. The current in the loop is in the counter-clockwise direction if seen from above. **(2014 Adv., Comprehension Type)**

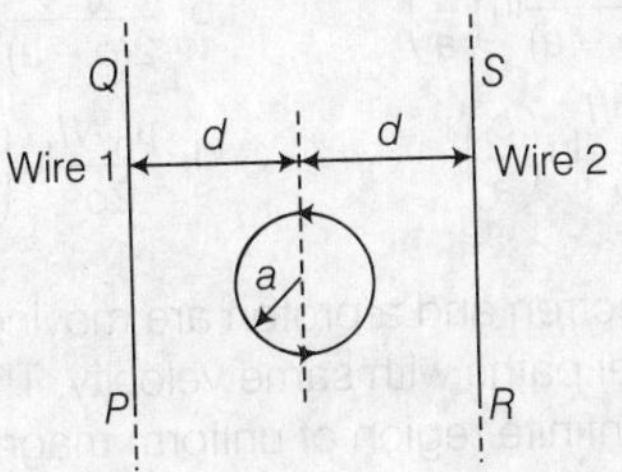

1 When $d \approx a$ but wires are not touching the loop, it is found that the net magnetic field on the axis of the loop is zero at a height h above the loop. In that case

a. current in wire 1 and wire 2 is the direction PQ and RS, respectively and $h \approx a$

b. current in wire 1 and wire 2 is the direction PQ and SR, respectively and $h \approx a$

c. current in wire 1 and wire 2 is the direction PQ and SR, respectively and $h \approx 1.2a$

d. current in wire 1 and wire 2 is the direction PQ and RS, respectively and $h \approx 1.2a$

2 Consider $d >> a$, and the loop is rotated about its diameter parallel to the wires by 30° from the position shown in the figure. If the currents in the wires are in the opposite directions, the torque on the loop at its new position will be (assume that the net field due to the wires is constant over the loop)

a. $\frac{\mu_0 I^2 a^2}{d}$ b. $\frac{\mu_0 I^2 a^2}{2d}$

c. $\frac{\sqrt{3}\mu_0 I^2 a^2}{d}$ d. $\frac{\sqrt{3}\mu_0 I^2 a^2}{2d}$

3 Two parallel wires in the plane of the paper are distance X_0 apart. A point charge is moving with speed u between the wires in the same plane at a distance X_1 from one of the wires. When the wires carry current of magnitude I in the same direction, the radius of curvature of the path of the point charge is R_1. In contrast, if the currents I in the two wires have directions opposite to each other, the radius of curvature of the path is R_2. If $\frac{X_0}{X_1} = 3$, then the value of $\frac{R_1}{R_2}$ is

(2014 Adv., Integer Type)

4 A particle of mass M and positive charge Q, moving with a constant velocity $\mathbf{u}_1 = 4\hat{\mathbf{i}}\ \text{ms}^{-1}$, enters a region of uniform static magnetic field normal to the XY-plane. The region of the magnetic field extends from $x = 0$ to $x = L$ for all values of y. After passing through this region, the particle emerges on the other side after 10 milliseconds with a velocity $\mathbf{u}_2 = 2(\sqrt{3}\hat{\mathbf{i}} + \hat{\mathbf{j}})\ \text{ms}^{-1}$. The correct statement(s) is (are)

(2013 Adv., One or More than One Options Correct Type)

a. the direction of the magnetic field is $-Z$-direction

b. the direction of the magnetic field is $+Z$-direction

c. the magnitude of the magnetic field is $\frac{50\pi M}{3Q}$ units

d. the magnitude of the magnetic field is $\frac{100\pi M}{3Q}$ units

5 A steady current I flows along an infinitely long hollow cylindrical conductor of radius R. This cylinder is placed co-axially inside an infinite solenoid of radius $2R$. The solenoid has n turns per unit length and carries a steady current I. Consider a point P at a distance r from the common axis. The correct statement(s) is (are)

(2013 Adv., One or More than One Options Correct Type)

a. In the region $0 < r < R$, the magnetic field is non-zero

b. In the region $R < r < 2R$, the magnetic field is along the common axis

c. In the region $R < r < 2R$, the magnetic field is tangential to the circle of radius r, centered on the axis

d. In the region $r > 2R$, the magnetic field is non-zero

6 Consider the motion of a positive point charge in a region where there are simultaneous uniform electric and magnetic fields $\mathbf{E} = E_0\hat{\mathbf{j}}$ and $\mathbf{B} = B_0\hat{\mathbf{j}}$. At time $t = 0$, this charge has velocity $\mathbf{v}$ in the XY-plane, making an angle θ with the X-axis. Which of the following options is(are) correct for time $t > 0$?

(2013 Adv., One or More than One Options Correct Type)

a. If $\theta = 0°$, the charge moves in a circular path in the XZ-plane

b. If $\theta = 0°$, the charge undergoes helical motion with constant pitch along the Y-axis

c. If $\theta = 0°$, the charge undergoes helical motion with its pitch increasing with time, along the Y-axis

d. If $\theta = 90°$, the charge undergoes linear but accelerated motion along the Y-axis

7 A loop carrying current I lies in the XY-plane as shown in the figure. The unit vector $\hat{\mathbf{k}}$ is coming out of the plane of the paper. The magnetic moment of the current loop is

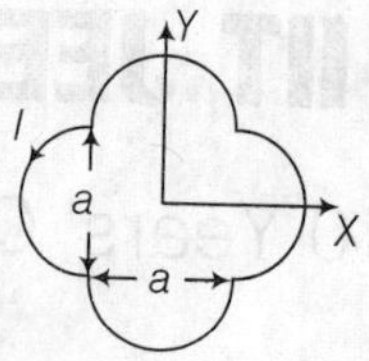

(2012, Only One Option Correct Type)

a. $a^2 I \hat{\mathbf{k}}$

b. $\left(\frac{\pi}{2}+1\right) a^2 I \hat{\mathbf{k}}$

c. $-\left(\frac{\pi}{2}+1\right) a^2 I \hat{\mathbf{k}}$

d. $(2\pi+1) a^2 I \hat{\mathbf{k}}$

8 An infinitely long hollow conducting cylinder with inner radius $R/2$ and outer radius R carries a uniform current density along its length. The magnitude of the magnetic field, $|\mathbf{B}|$ as a function of the radial distance r from the axis is best represented by

(2012, Only One Option Correct Type)

a. 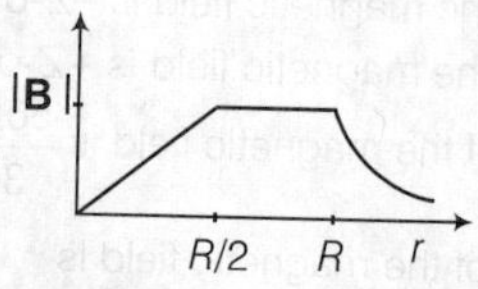

b. 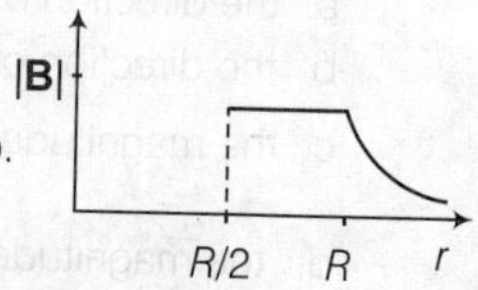

c. 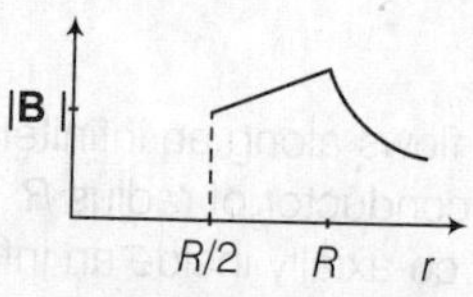

d. 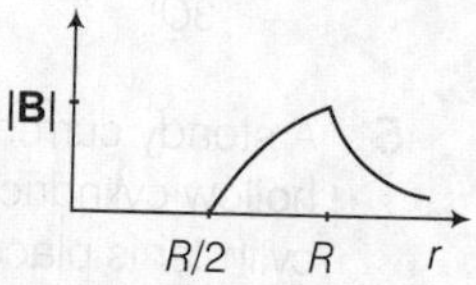

9 A cylindrical cavity of diameter a exists inside a cylinder of diameter $2a$ as shown in the figure. Both the cylinder and the cavity are infinitely long. A uniform current density J flows along the length. If the magnitude of the magnetic field at the point P is given by $\frac{N}{12}\mu_0 aJ$, then the value of N is

(2012, Integer Type)

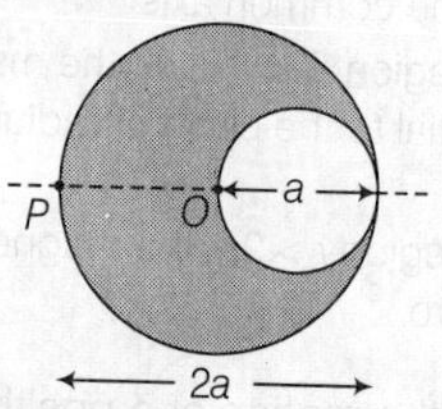

10 Which of the field patterns given in the figure is valid for electric field as well as for magnetic field?

(2011, Only One Option Correct Type)

a. 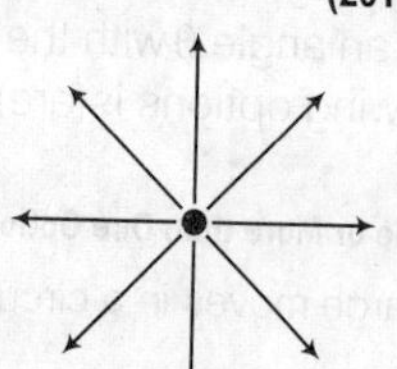

b. 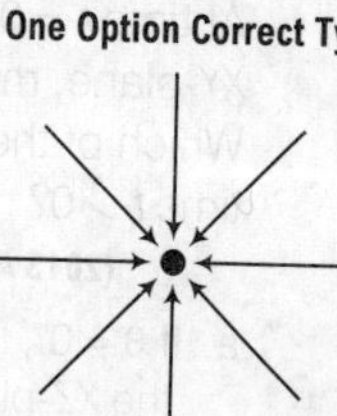

c. 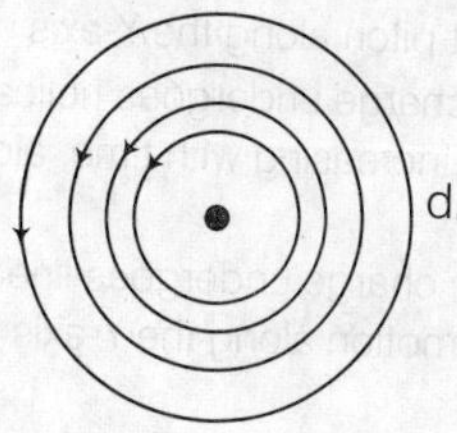

d. 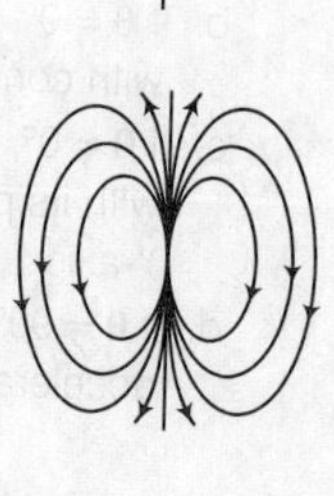

11 A long insulated copper wire is closely wound as a spiral of N turns. The spiral has inner radius a and outer radius b. The spiral lies in the XY-plane and a steady current I flows through the wire. The Z-component of the magnetic field at the centre of the spiral is

(2011, Only One Option Correct Type)

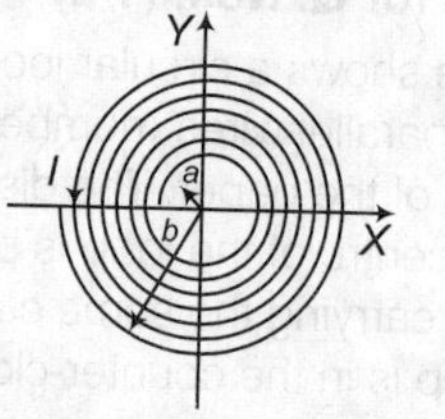

a. $\frac{\mu_0 NI}{2(b-a)}\ln\left(\frac{b}{a}\right)$

b. $\frac{\mu_0 NI}{2(b-a)}\ln\left(\frac{b+a}{b-a}\right)$

c. $\frac{\mu_0 NI}{2b}\ln\left(\frac{b}{a}\right)$

d. $\frac{\mu_0 NI}{2b}\ln\left(\frac{b+a}{b-a}\right)$

12 An electron and a proton are moving on straight parallel paths with same velocity. They enter a semi-infinite region of uniform magnetic field perpendicular to the velocity. Which of the following statement(s) is/are true?

(2011, One or More than One Options Correct Type)

a. They will never come out of the magnetic field region

b. They will come out travelling along parallel paths

c. They will come out at the same time

d. They will come out at different times

13 A steady current I goes through a wire loop PQR having shape of a right angle triangle with $PQ = 3x$, $PR = 4x$ and $QR = 5x$. If the magnitude of the magnetic field at P due to this loop is $k\left(\frac{\mu_0 I}{48\pi x}\right)$, then find the value of k.

(2009, Integer Type)

14 A particle of mass m and charge q, moving with velocity v enters Region II normal to the boundary as shown in the figure. Region II has a uniform magnetic field B perpendicular to the plane of the paper. The length of the Region II is l. Choose the correct choice(s).

(2008, One or More than One Options Correct Type)

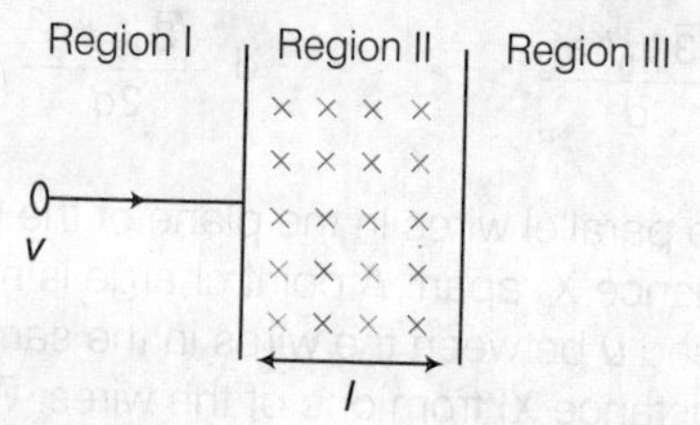

a. The particle enters Region III only if its velocity $v > \frac{qlB}{m}$

b. The particle enters Region III only if its velocity $v < \frac{qlB}{m}$

c. Path length of the particle in Region II is maximum when velocity $v = qlB/m$

d. Time spent in Region II is same for any velocity v as long as the particle returns to Region I

15. **Statement I** The sensitivity of a moving coil galvanometer is increased by placing a suitable magnetic material as a core inside the coil.

Statement II Soft iron has a high magnetic permeability and cannot be easily magnetised or demagnetised.

(2008, Statement Type)

a. Both Statement I and Statement II are correct and Statement II is the correct explanation of Statement I
b. Both Statement I and Statement II are correct but Statement II is not the correct explanation of Statement I
c. Statement I is correct but Statement II is incorrect
d. Statement II is correct but Statement I is incorrect

16 Two wires each carrying a steady current I are shown in four configurations in Column I. Some of the resulting effects are described in Column II. Match the Statements in Column I with the statements in Column II.

(2007, Matching Type)

Column I		Column II
i. Point P is situated midway between the wires.	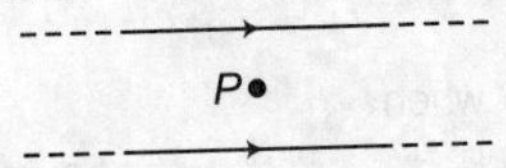	p. The magnetic fields (B) at P due to the currents in the wires are in the same direction.
ii. Point P is situated at the mid-point of the line joining the centres of the circular wires, which have same radii.	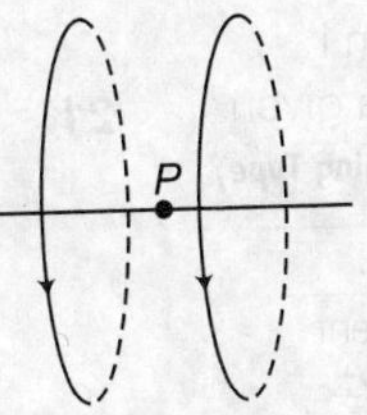	q. The magnetic fields (B) at P due to the currents in the wires are in opposite directions.
iii. Point P is situated at the mid-point of the line joining the centres of the circular wires, which have same radii.	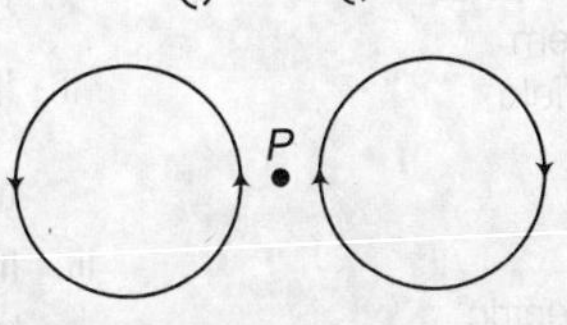	r. There is no magnetic field at P.
iv. Point P is situated at the common centre of the wires.	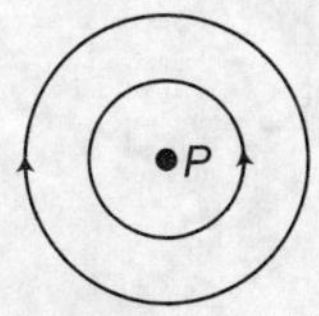	s. The wires repel each other.

Codes

	i	ii	iii	iv
a.	p	q	r,s	p,q
b.	q,r	p	q,r	q,s
c.	q,r	p,s	r	s
d.	q	q	r	p

17 A magnetic field $\mathbf{B} = B_0\hat{\mathbf{j}}$ exists in the region $a < x < 2a$ and $\mathbf{B} = -B_0\hat{\mathbf{j}}$, in the region $2a < x < 3a$,

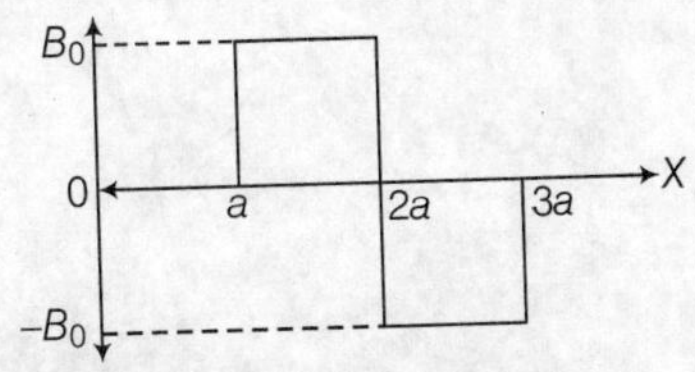

where B_0 is a positive constant. A positive point charge moving with a velocity $\mathbf{v} = v_0\hat{\mathbf{i}}$, where v_0 is a positive constant, enters the magnetic field at $x = a$. The trajectory of the charge in this region can be like

(2007, Only One Option Correct Type)

a. 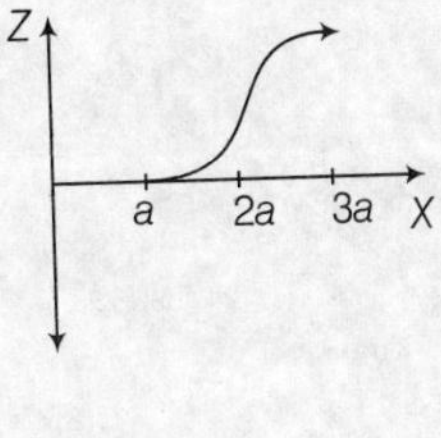
b. 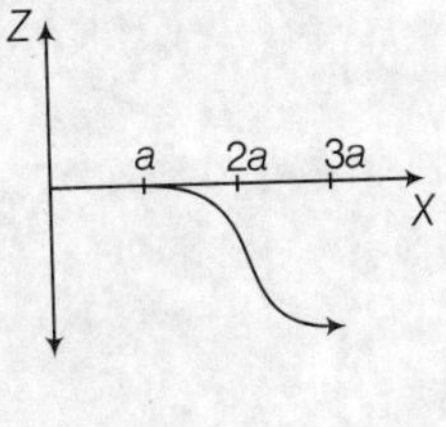
c. 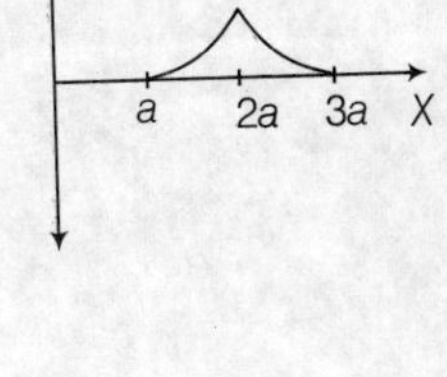
d. 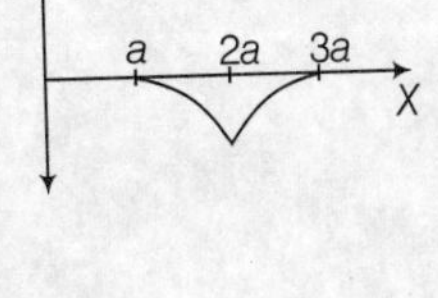

18 Which of the following statement(s) is(are) correct in the given figure?

(2006, One or More than One Options Correct Type)

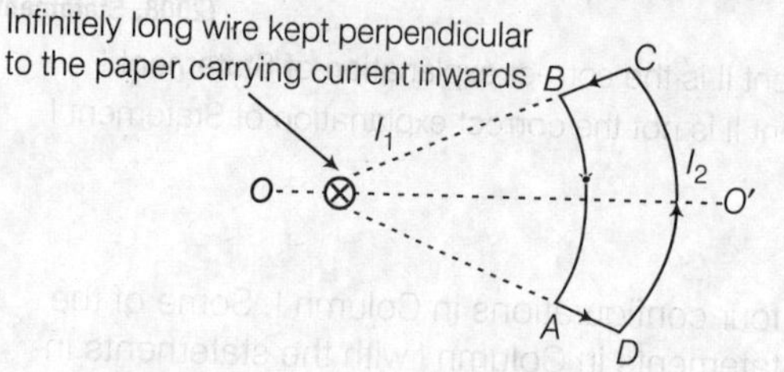

a. net force on the loop is zero
b. net torque on the loop is zero
c. loop will rotate clockwise about axis OO' when seen from O
d. loop will rotate anti-clockwise about OO' when seen from O

19 Some laws/processes are given in Column I. Match these with the physical phenomena given in Column II. **(2006, Matching Type)**

	Column I		Column II
i.	Dielectric ring uniformly charged	p.	Time independent electrostatic field out of system
ii.	Dielectric ring uniformly charged rotating with angular velocity ω	q.	Magnetic field
iii.	Constant current in ring i_0	r.	Induced electric field
iv.	$i = i_0 \cos \omega t$	s.	Magnetic moment

Codes

	i	ii	iii	iv
a.	p,s	q,r	p,r	p
b.	p	q	r	q,r
c.	p	p,q,s	q,s	q,r
d.	p	q	r,s	r,s

20 A field line is shown in the figure. This field cannot represent

(2006, One or More than One Options Correct Type)

a. magnetic field
b. electrostatic field
c. induced electric field
d. gravitational field

21. A moving coil galvanometer experiences torque $= ki$, where i is current. If N coils of area A each and moment of inertia I is kept in magnetic field B. **(2005, Subjective Type)**

i. Find k in terms of given parameters.
ii. If for current i deflection is $\frac{\pi}{2}$, find out torsional constant of spring.
iii. If a charge Q is passed suddenly through the galvanometer, find out maximum angle of deflection.

2. Magnetism

DPP
DAILY PRACTICE PROBLEMS

DPP-1 Bar Magnet and Magnetic Field Lines

Subjective Questions

Working Space

Direction *(Q. Nos. 1-4) These questions are subjective in nature, need to be solved completely on notebook.*

1 A solenoid of length l and radius r containing n number of turns and carrying current i can be replaced by a South pole and a North pole having equal pole strength separated by a distance d (i.e. like a bar magnet).

i. If $l = 10$ cm, $r = 1$ cm, $n = 200$ and $i = 10$ A, then find the magnetic field at that point on the axis at a distance of 10 cm from the centre due to North pole and South pole.

ii. Also, find the net magnetic field at that point in the above case (i).

2 i. Three identical bar magnets are rivetted at the centre in same plane.

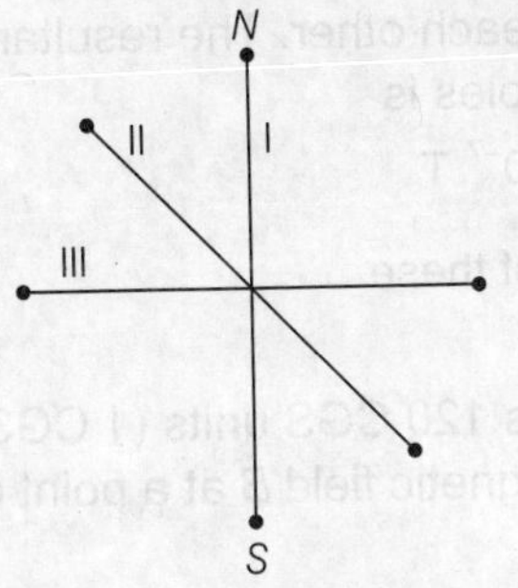

The system is placed at rest in a magnetic field. It is found that system of magnets does not show any motion.

a. Determine polarity of other two magnets.

b. What you conclude about external field?

ii. A bar magnet of magnetic moment **M** and moment of inertia I (about an axis through centre and perpendicular to plane of magnet) is cut into two equal pieces.

a. Perpendicular to its length.

b. Parallel to its length.

Find the ratio of initial and final magnetic moments and of time periods of oscillations in a uniform externally applied field.

3 Two identical short bar magnets each of magnetic moment 12.5 A-m^2 are placed at a separation of 10 cm between their centres such that their axes are perpendicular to each other. Find the magnetic field at a point mid-way between the two magnets.

4 There are two current carrying planar coils made each from identical wires of length L. C_1 is circular (radius R) and C_2 is square (side a). They are so constructed that they have same frequency of oscillation when they are placed in the same uniform **B** and carry the same current i. Find a in terms of R.

Only One Option Correct Type

Working Space

Direction *(Q. Nos. 5-19) This section contains 15 multiple choice questions. Each question has four choices (a), (b), (c) and (d), out of which ONLY ONE is correct.*

5 A magnetic needle lying parallel to a magnetic field requires W units of work to turn it through 60°. The torque required to maintain the needle in this position will be **[AIEEE 2003]**

a. $\sqrt{3}\,W$ b. $-W$

c. $\frac{\sqrt{3}}{2}W$ d. $2W$

6 A bar magnet of magnetic moment 3 A-m^2 is placed in a uniform magnetic induction field of 2×10^{-5} T. If each pole of the magnet experiences a force of 6×10^{-4} N, then the length of the magnet is

a. 0.5 m b. 0.3 m

c. 0.2 m d. 0.1 m

7 Force between two identical bar magnets whose centres are r metre apart is 4.8 N when their axes are in the same line. If the separation is increased to $2r$ metre, then the force between them is reduced to

a. 2.4 N b. 1.2 N

c. 0.6 N d. 0.3 N

8 Two identical magnetic dipoles of magnetic moment 1 A-m^2 each placed at a separation of 2 m with their axis perpendicular to each other. The resultant magnetic field at a point mid-way between the dipoles is

a. 5×10^{-7} T b. $\sqrt{5} \times 10^{-7}$ T

c. $\frac{1}{2}$T d. None of these

9 A bar magnet is 10 cm long and its pole strength is 120 CGS units (1 CGS unit of pole strength = 0.1A-m). The magnitude of the magnetic field B at a point on its axis at a distance 20 cm from it is

a. 2.4×10^{-5} T b. 3.0×10^{-5} T

c. 3.4×10^{-5} T d. 2.0×10^{-5} T

10 Let r be the distance of a point on the axis of a bar magnet from its centre. The magnetic field at such a point is proportional to

a. $\frac{1}{r}$ b. $\frac{1}{r^2}$

c. $\frac{1}{r^3}$ d. None of these

11 A magnet of magnetic moment 20 CGS unit is freely suspended in a uniform magnetic field of intensity 0.3 CGS unit. The amount of work done in deflecting it by an angle of 30° in CGS unit is

a. 6 b. $3\sqrt{3}$

c. $3(2-\sqrt{3})$ d. 3

12 The magnetic field at a point x on the axis of a smaller magnet is equal to the field at a point y in the equator of the same magnet. The ratio of distance of x and y from the centre of the magnet is

a. 2^{-3} b. $2^{-1/3}$

c. 2^3 d. $2^{1/3}$

13 A bar magnet of magnetic moment 2 A-m^2 is free to rotate about a vertical axis passing through its centre. The magnet is released to rest from East-West position. Then, the KE of the magnet as it takes North-South position is (Given, $B_H = 25\,\mu T$)

a. $25\,\mu J$
b. $50\,\mu J$
c. $100\,\mu J$
d. $12.5\,\mu J$

14 If a small magnet intensity at a distance x in the end on position is 9 gauss, what will be the intensity at a distance $\frac{x}{2}$ on broadside on position?

a. 9 gauss
b. 4 gauss
c. 36 gauss
d. 4.5 gauss

15 Two equal bar magnets are kept as shown in the figure. The direction of resultant magnetic field indicated by the arrow head at the point P is (approximately)

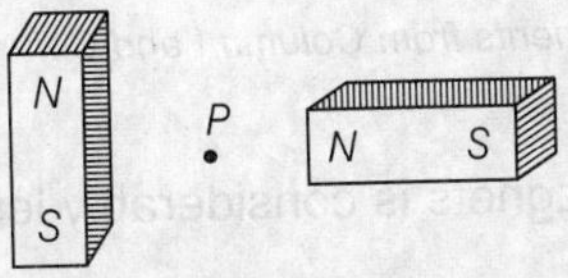

a. →
b. ↗
c. ↘
d. ↑

16 Two short magnets of equal dipole moment M are fastened perpendicularly at their centre (see figure). The magnitude of the magnetic field at a distance d from the centre on the bisector of the right angle is

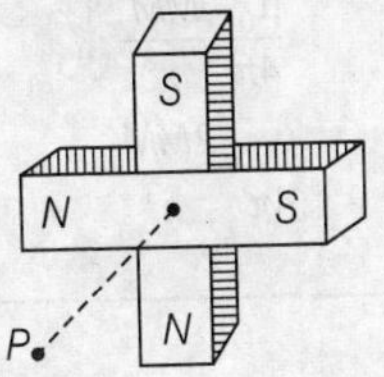

a. $\frac{\mu_0}{4\pi} \cdot \frac{M}{d^3}$
b. $\frac{\mu_0}{4\pi} \cdot \frac{M\sqrt{2}}{d^3}$
c. $\frac{\mu_0}{4\pi} \cdot \frac{2\sqrt{2}\,M}{d^3}$
d. $\frac{\mu_0}{4\pi} \cdot \frac{2M}{d^3}$

17 Two magnets of equal magnetic moment M each are placed as shown in figure. The resultant magnetic moment is

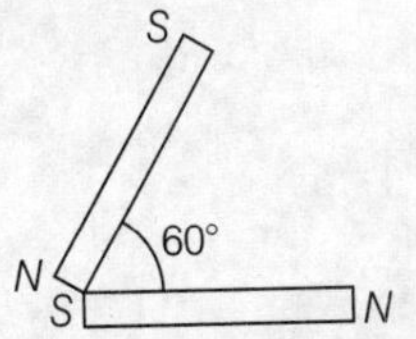

a. M
b. $\sqrt{3}\,M$
c. $\sqrt{2}\,M$
d. $M/2$

18 For the given figure, the magnetic needle has magnetic moment 6.7×10^{-2} A-m^2 and moment of inertia $I = 7.5 \times 10^{-6}$ kg-m^2. It performs 10 complete oscillations in 6.70 s. The magnitude of the magnetic field is

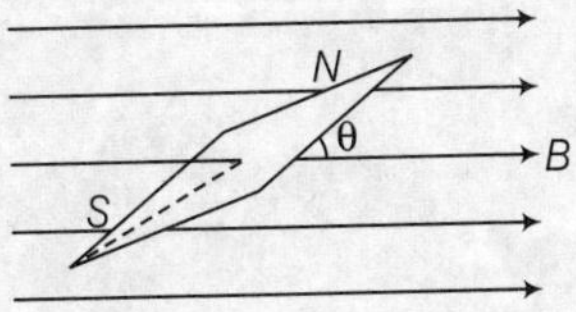

a. 1.00 T
b. 0.67 T
c. 0.01 T
d. 1.50 T

19 The magnetic field lines due to a bar magnet are correctly shown in figure. **[IITJEE 2002]**

Working Space

a. 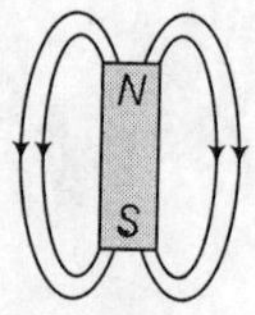

b. 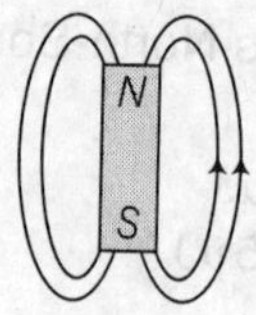

c. 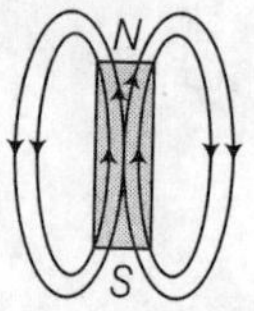

d. 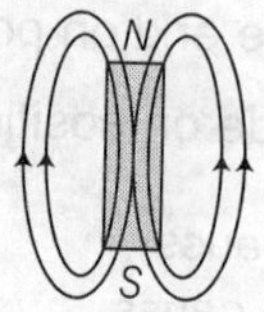

Matching List Type

***Direction** (Q. No. 20) Choices for the correct combination of elements from Column I and Column II are given as options (a), (b), (c) and (d), out of which one is correct.*

20 Match Column I with Column II, when length of magnets is considerably less than distance x.

	Column I		Column II
i.	Couple between two small magnets in the end on position.	p.	$\frac{\mu_0}{4\pi}\cdot\frac{3MM'}{x^4}$
ii.	Couple between two small magnets in the broadside on position.	q.	$\frac{\mu_0}{4\pi}\cdot\frac{6MM'}{x^4}$
iii.	Force on a magnet when it is in end on position with respect to deflecting magnet.	r.	$\frac{\mu_0}{4\pi}\cdot\frac{MM'}{x^3}$
iv.	Force on a magnet when it is in broadside on position with respect to deflecting magnet.	s.	$\frac{\mu_0}{4\pi}\cdot\frac{2MM'}{x^3}$

Codes

	i	ii	iii	iv
a.	s	r	q	p
b.	s	r	p	q
c.	q	p	r	s
d.	p	q	s	r

DPP-2 Earth's Magnetism

Subjective Questions

Working Space

Direction (Q. Nos. 1-4) *These questions are subjective in nature, need to be solved completely on notebook.*

1 Find magnitude and inclination of earth's magnetic field at

i. geomagnetic equator.

ii. at a point where geomagnetic latitude is 60°.

iii. at magnetic North pole.

Given, earth's magnetic dipole moment is 8×10^{22} A-m^2.

2 A short bar magnet of magnetic moment of $\mathbf{M} = 5.25 \times 10^{-2}$ JT^{-1} is placed with its axis perpendicular to earth's magnetic field. At what distance from centre of magnet, the resultant magnetic field is inclined at 45° with earth's field on

i. normal bisector.

ii. axis.

Given, $B_H = 0.42$ gauss.

3 Consider the dipole model for earth's magnetism.

i. Find locus of points, where $|\mathbf{B}|$ is minimum.

ii. Find locus of points with zero dip.

iii. Consider a plane formed by the dipole axis and the axis of earth. Let P is a point in magnetic equator and in S. Let Q be the point of intersection of the geographical and magnetic equators. Obtain declination and dip at P and Q.

4 A compass needle oscillates 20 times per minute at a place, where the angle of dip is 45° and 30 times per minute where the angle of dip is 30°. What is the ratio of the total magnetic field due to the earth at the two places?

Only One Option Correct Type

Direction (Q. Nos. 5-18) *This section contains 14 multiple choice questions. Each question has four choices (a), (b), (c) and (d), out of which ONLY ONE is correct.*

5 If the angles of dip at two places are 30° and 45°, respectively. Then, the ratio of B_H at the two places will be

a. $\sqrt{3}:\sqrt{2}$ b. $1:\sqrt{2}$ c. $1:\sqrt{3}$ d. $1:2$

6 At a place the earth's horizontal component of magnetic field is 0.38×10^{-4} Wb/m^2. If the angle of dip at that place is 60°, then the vertical component of the earth's field at that place will be approximately

a. 0.12×10^{-4} T b. 0.24×10^{-4} T

c. 0.40×10^{-4} T d. 0.658×10^{-4} T

7 A dip circle is at right angles to the magnetic meridian. The apparent dip is **[AIEEE 2009]**

a. 0° b. 90° c. 45° d. 60°

Working Space

8 A dip needle which is free to move in a vertical plane perpendicular to magnetic meridian will remain **[AIEEE 2007]**

a. horizontal
b. vertical
c. neither horizontal nor vertical
d. inclined

9 The true value of angle of dip at a place is 60°, the apparent dip in a plane at an angle of 30° with the magnetic meridian is **[AIEEE 2002]**

a. $\tan^{-1}\left(\frac{1}{2}\right)$
b. $\tan^{-1}(2)$
c. $\tan^{-1}\left(\frac{2}{3}\right)$
d. None of these

10 A short bar magnet is placed with its South pole towards geographical North. The neutral points are situated at a distance of 20 cm from the centre of the magnet. If $B_H = 0.3 \times 10^{-4}$ Wb/m^2, then the magnetic moment of magnet is

a. 9 A-m^2
b. 0.9 A-m^2
c. 1.2 A-m^2
d. 120 A-m^2

11 Two magnets of equal mass are joined at right angles to each other as shown in the figure. The magnet 1 has a magnetic moment 3 times that of magnet 2. This arrangement is pivoted so that it is free to rotate in the horizontal plane. In equilibrium what angle will the magnet 1 subtends with the magnetic meridian?

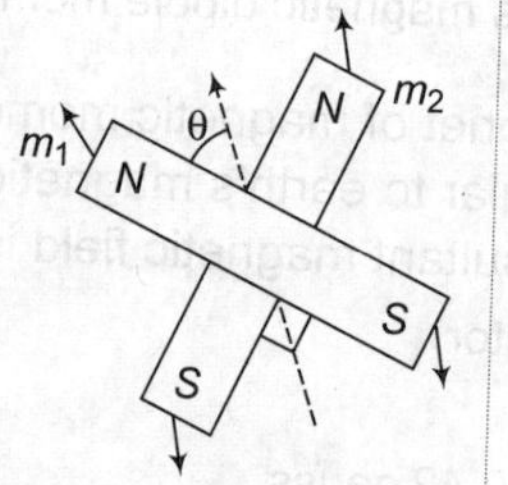

a. $\tan^{-1}\left(\frac{1}{2}\right)$
b. $\tan^{-1}\left(\frac{1}{3}\right)$
c. $\tan^{-1}(1)$
d. 0°

12 A compass needle whose magnetic moment is 60 A-m^2 pointing towards geographical North at a certain place, where the horizontal component of earth's magnetic field is 40 μWb/m^2, experiences a torque 1.2×10^{-3} N-m. What is the declination at this place?

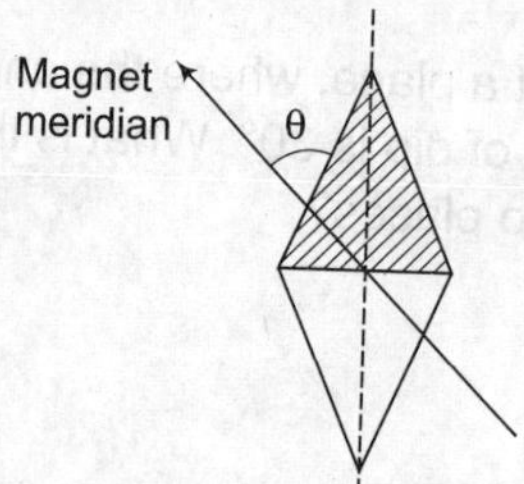

a. 30°
b. 45°
c. 60°
d. 25°

13 A thin rectangular magnet suspended freely has a period of oscillation equal to T. Now, it is broken into two equal halves (each having half of original length) and one piece is made to oscillate freely in the same field. If its time period is T', then ratio $\frac{T'}{T}$ is **[AIEEE 2003]**

a. $\frac{1}{2\sqrt{2}}$
b. $\frac{1}{2}$
c. 2
d. $\frac{1}{4}$

14 The length of a magnet is large compared to its width and breadth. The time period of oscillation in a vibration magnetometer is 2 s. The magnet is cut along its length into three equal parts and three parts are then placed on each other with their like poles together. The time period of this combination will be **[AIEEE 2004]**

a. $2\sqrt{3}$ s
b. $\frac{2}{3}$ s
c. 2 s
d. $\frac{2}{\sqrt{3}}$ s

Working Space

15 The angle of dip at a place is 60°. A magnetic needle oscillates in a horizontal plane at this place with period T. The same needle will oscillate in a vertical plane, coinciding with the magnetic meridian with a period of

a. T
b. $2T$
c. $\frac{T}{2}$
d. $\frac{T}{\sqrt{2}}$

16 A dip needle vibrates in the vertical plane perpendicular to magnetic meridian. The time period of vibration is found to be 2 s. The same needle is then allowed to vibrate in the horizontal plane and time period is again found to be 2 s. Then, the angle of dip is

a. 0°
b. 30°
c. 45°
d. 90°

17 The plane of a dip circle is set in the geographic meridian and the apparent dip is δ_1. It is then set in a vertical plane perpendicular to the geographic meridian. The apparent dip angle is δ_2. The declination θ at the place is

a. $\theta = \tan^{-1}(\tan\delta_1 \tan\delta_2)$
b. $\theta = \tan^{-1}(\tan\delta_1 + \tan\delta_2)$
c. $\theta = \tan^{-1}\left(\frac{\tan\delta_1}{\tan\delta_2}\right)$
d. $\theta = \tan^{-1}(\tan\delta_1 - \tan\delta_2)$

18 A bar magnet is oscillating in the earth's magnetic field with a period T. What happens to its period of motion if its mass is quadrupled?

a. Motion remains SHM with time period = $T/2$
b. Motion remains SHM and period remains nearly constant
c. Motion remains SHM with time period = $2T$
d. Motion remains SHM with time period = $4T$

One or More than One Options Correct Type

Direction (Q. Nos. 19-21) *This section contains 3 multiple choice questions. Each question has four choices (a), (b), (c) and (d), out of which ONE or MORE THAN ONE are correct.*

19 The angle of dip at geographical equator

a. is always zero
b. is zero at specific point
c. can be positive or negative
d. is bounded

20 Two magnets are tied together and are allowed to oscillate in earth's magnetic field. With like poles together 12 oscillations are made in one minute and with unlike poles together only 4 oscillations are made in one minute.

Now, choose the correct options.

a. Time periods are different due to change in earth's horizontal component of field
b. Time periods are different due to change in earth's vertical component of field
c. Time periods are different due to change in magnetic dipole moment of arrangement
d. Ratio of magnetic dipole moments of both magnets is 5 : 4

21 A short bar magnet is placed such that its S-pole points towards geographic North. (Given, $\mathbf{B}_H = 0.3 \times 10^{-4}\,\text{Wb/m}^2$ and $\mathbf{M} = 1.2\,\text{A-m}^2$). Then, correct options are

a. at neutral points, magnetic field of magnet is parallel to horizontal component of earth's magnetic field
b. at neutral points, magnetic field of magnet is antiparallel to horizontal component of earth's magnetic field
c. neutral points are situated at 20 cm from centre of magnet on its equatorial axis
d. neutral points are situated at 20 cm from centre of magnet on the axis of magnet

Comprehension Type

Direction (Q. Nos. 22 and 23) *This section contains a paragraph, which describing theory, experiments, data, etc. Two questions related to the paragraph have been given. Each question has only one correct answer among the four given options (a), (b), (c) and (d).*

Passage

Magnetic field of the earth is identical to magnetic field of a giant magnet held 20° West of geographic *N-S* at the centre of Earth. At equator, horizontal component of earth is 0.32 G. Vertical component can be calculated from the relation $V = H \tan \delta$, where δ is angle of dip at the place. The value of $\delta = 0°$ at equator and $\delta = 90°$ at poles.

22 At a particle place, $V = H$. The angle of dip is

a. 45°
b. 90°
c. 0°
d. None of these

23 What is the order of magnetic declination at a place on the earth?

a. 20° East
b. 10° East
c. 20° West
d. 10° West

Matching List Type

Direction (Q. No. 24) *Choices for the correct combination of elements from Column I and Column II are given as options (a), (b), (c) and (d), out of which one is correct.*

24 At many places on earth, value of magnetic elements (declination, dip and horizontal component) have same values. Lines are drawn to join their points to form magnetic maps. These are used in studying magnetism of earth and in navigation. Now, match entries of Column I with correct terms in Column II.

	Column I		Column II
i.	Isogonic lines	p.	Same B_H
ii.	Agonic lines	q.	Zero dip
iii.	Isoclinic lines	r.	Equal dip
iv.	Aclinic lines	s.	Zero declination
v.	Isodynamic lines	t.	Equal declination

Codes

	i	ii	iii	iv	v
a.	p	q	r	s	t
b.	r	s	p	q	t
c.	s	r	q	t	p
d.	t	s	t	q	p

One Integer Value Correct Type

Direction (Q. Nos. 25 and 26) *This section contains 2 questions. When worked out will result in an integer from 0 to 9 (both inclusive).*

25 If $\phi_1 = 45°$ and ϕ_2 are angles of dip observed in two vertical planes at right angle to each other and ϕ is true dip angle, then find the value of $\cot^2 \phi$.

26 If we consider earth's magnetism is due to a short bar magnet placed at centre of earth, then the angle of dip ϕ is related to magnetic latitude λ as $\tan\theta = k \tan\lambda$. Find the value of *k*.

Working Space

DPP-3 Magnetic Materials and Their Properties

Subjective Questions

Working Space

Direction (Q. Nos. 1-4) *These questions are subjective in nature, need to be solved completely on notebook.*

1 The intensity of magnetisation of a magnetic material is given by the relation

$$\mathbf{I} = \chi_B \mathbf{H}$$

With reference to the above relation, answer the following

i. Does the value of χ_B remains same for all types of magnetic material?

ii. An external magnetic field is applied to this material and the material is placed in vacuum. Find the resultant magnetic field.

2 A ferromagnetic material is formed in the shape of a ring and is placed inside a toroid having n turns per unit length. If current i is passed through the toroid such that the net magnetic field produced is B, find the expression for intensity of magnetisation of the material.

3 Find the per cent increase in the magnetic field B when the space within a current carrying toroid is filled with aluminium. The susceptibility of aluminium is 2.1×10^{-5}.

4 A bar magnet has pole strength 4.5 A-m, magnetic length 12 cm and cross-sectional area 0.9 cm^2. Find

i. intensity of magnetisation (I).

ii. magnetising intensity (H) at the centre.

iii. magnetic induction (B) at the centre of the magnet.

Only One Option Correct Type

Direction (Q. Nos. 5-22) *This section contains 18 multiple choice questions. Each question has four choices (a), (b), (c) and (d), out of which ONLY ONE is correct.*

5 For mercury at 40 K, the magnetic susceptibility χ has the value

a. $-1 \le \chi \le 0$
b. $0 < \chi \le 1$
c. $\chi >> 1$
d. $\chi << -1$

6 When a small magnet is placed above the disc of a material M (which is kept at low temperature), the magnet is levitated over the disc.
Then, disc must be of

a. diamagnetic material
b. paramagnetic material
c. ferromagnetic material
d. superconductor material

7 Curie temperature is the temperature of transition from **[AIEEE 2003]**

a. paramagnetism to diamagnetism
b. ferromagnetism to paramagnetism
c. diamagnetism to ferromagnetism
d. ferromagnetism to diamagnetism

Working Space

8 The materials suitable for making electromagnets should have **[AIEEE 2004]**

a. high retentivity and low coercivity
b. low retentivity and low coercivity
c. high retentivity and high coercivity
d. low retentivity and high coercivity

9 The material of permanent magnet has

a. high retentivity, low coercivity
b. low retentivity, high coercivity
c. low retentivity, low coercivity
d. high retentivity, high coercivity

10 Needles N_1, N_2 and N_3 are made of a ferromagnetic, a paramagnetic and a diamagnetic substance, respectively. A magnet when brought close to them, will

a. attract N_1 and N_2 strongly but repel N_3 **[AIEEE 2006]**
b. attract N_1 strongly, N_2 weakly and repel N_3 weakly
c. attract N_1 strongly, but repel N_2 and N_3 weakly
d. attract all three of them

11 For the figure, B is flux density inside a sample of unmagnetised ferromagnetic material varies with B_0, the magnetic flux density in which the sample is kept. For the sample to be suitable for making a permanent magnet

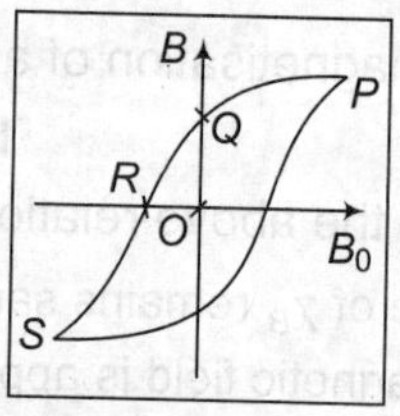

a. OQ should be large, OR should be small
b. OQ and OR should both be large
c. OQ should be small and OR should be large
d. OQ and OR should both be small

12 Relative permittivity and permeability of material are ε_r and μ_r, respectively. Which of the following values of these quantities are allowed for a diamagnetic material? **[AIEEE 2008]**

a. $\varepsilon_r = 0.5, \mu_r = 1.5$
b. $\varepsilon_r = 1.5, \mu_r = 0.5$
c. $\varepsilon_r = 0.5, \mu_r = 0.5$
d. $\varepsilon_r = 1.5, \mu_r = 1.5$

13 Select the wrong statement.

a. In a diamagnetic substance, the direction of magnetising field intensity **I** is opposite to that of magnetic intensity **H**
b. In a paramagnetic substance, the direction of magnetising field intensity **I** is along magnetic intensity **H**
c. In a ferromagnetic substance, the direction of magnetising field intensity **I** is along magnetic intensity **H**
d. In a diamagnetic substance, the direction of magnetising field intensity **I** is along magnetic intensity **H**

14 For a paramagnetic substance,

a. $\mu_r > 1, \chi > 0$
b. $\mu_r > 1, \chi < 0$
c. $\mu_r < 1, \chi > 0$
d. $\mu_r < 1, \chi < 0$

15 A magnetising field of 1600 Am^{-1} produces a magnetic flux of 2.4×10^{-5} Wb in an iron bar of cross-sectional area 0.2 cm^2. The susceptibility of an iron bar is

a. 298
b. 596
c. 1192
d. 1788

16 For iron, its density is 7500 kg/m^3 and mass 0.075 kg. If its magnetic moment is 8×10^{-7} Am^{-1}, its intensity of magnetisation is

a. 8 Am^{-1}
b. 0.8 Am^{-1}
c. 0.08 Am^{-1}
d. 0.008 Am^{-1}

17 The coercivity of a small bar magnet is 4×10^3 A/m. It is inserted inside a solenoid of 500 turns and length 1 m to demagnetise it. The amount of current to be passed through the solenoid will be

a. 2.5 A
b. 5 A
c. 8 A
d. 10 A

18 The correct variation of the intensity of magnetisation I with respect to the magnetising field H in a diamagnetic substance is described by which of the graph shown in figure below?

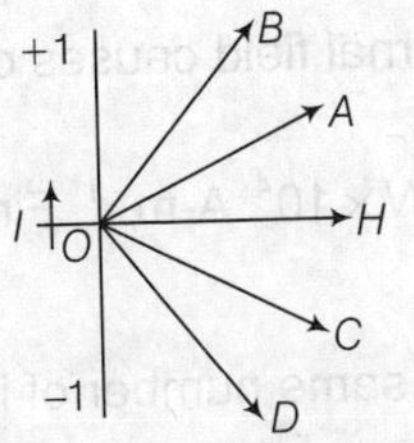

a. OD b. OC c. OB d. OA

19 A uniform magnetic field parallel to the plane of paper, existing in space initially directed from left to right. When a bar of soft iron is placed in the field parallel to it, the lines of force passing through it will be represented by figure.

a. 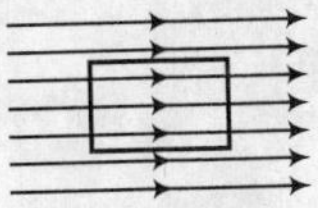

b. 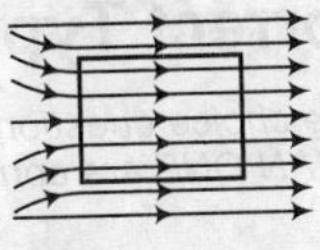

c. 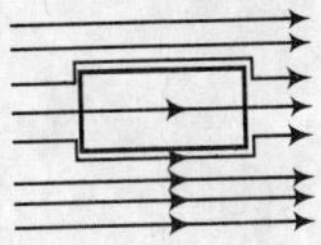

d. 

20 The variation of magnetic susceptibility (χ) with absolute temperature T for a ferromagnetic is given in figure by

a. 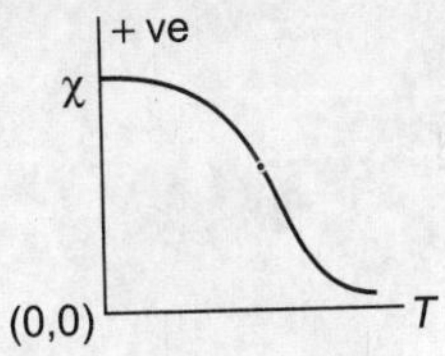

b. 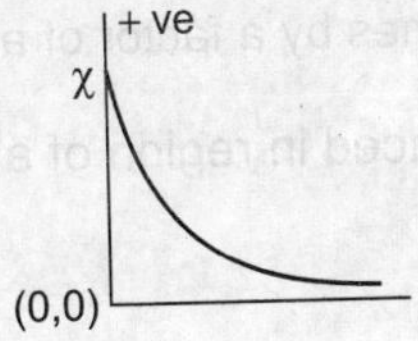

c. 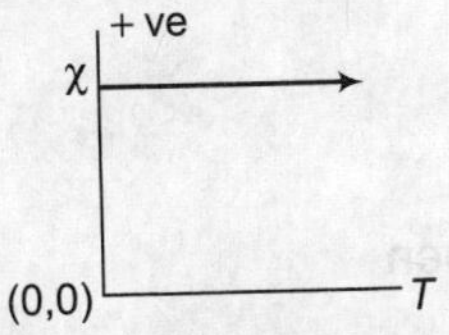

d. 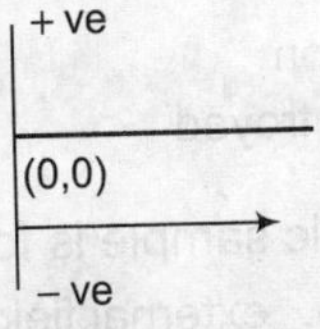

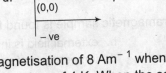

21 A paramagnetic sample shows a net magnetisation of 8 Am^{-1} when placed in an external magnetic field of 0.6 T at a temperature of 4 K. When the same sample is placed in an external magnetic field of 0.2 T at a temperature of 16 K, the magnetisation will be

a. $\frac{32}{3}\ Am^{-1}$ b. $\frac{2}{3}\ Am^{-1}$

c. 6 Am^{-1} d. 2.4 Am^{-1}

22 Nickel shows ferromagnetic property at room temperature. If the temperature is increased beyond Curie temperature, then it will show

a. paramagnetism b. anti-ferromagnetism

c. no magnetic property d. diamagnetism

One Integer Value Correct Type

Direction *(Q. Nos. 23-25) This section contains 3 questions. When worked out will result in an integer from 0 to 9 (both inclusive).*

23 The coercivity of a bar magnet is 120 A/m. It is to be demagnetised by placing it inside a solenoid of length 120 cm and number of turns 72. Find the current (in A) flowing through the solenoid.

Working Space

24 A ferromagnetic cube of side 1mm is magnetised by placing it in an external uniform magnetic field. The ferromagnetic substance has molecular mass of 55 g/mol and its density is 7.9 g/cm^3. (Assume that each atom has a dipole moment of 9.27×10^{-24} A-m^2, also assume that external field causes only 10% atom to align).

Magnetisation of sample in above condition is found $N \times 10^4$ A-m^{-1}. Find the value of N.

25 An ionised gas consists of 5×10^{21} electrons/m^3 and same number of ions. Average KE of an electron is 0.4×10^{-21} J and average KE of an ion is 7.6×10^{-21} J. When a magnetic field of 0.2 T is applied to this gas volume, magnetisation of $100N$ is produced in the gas sample. Find the value of N.

One or More than One Options Correct Type

Direction *(Q. Nos. 26-28) This section contains 3 multiple choice questions. Each question has four choices (a), (b), (c) and (d), out of which ONE or MORE THAN ONE are correct.*

26 A long solenoid has 1000 turns per metre and carries a current of 1 A. It has a soft iron core of $\mu_r = 1000$. The core is heated beyond the Curie temperature T_C. Which of the following statement(s) is/are correct?

a. The **H** field in the solenoid is (nearly) unchanged but the **B** field decreases drastically
b. The **H** and **B** fields in the solenoid are nearly unchanged
c. The magnetisation in the core reverses direction
d. The magnetisation in the core diminishes by a factor of about 10^8

27 When a ferromagnetic substance is placed in region of a strong magnetic field, then

a. domain present may increase in size
b. domain present may decrease in size
c. domain present may change in orientation
d. domain boundaries of all domains is destroyed

28 Magnetisation produced in a paramagnetic sample is found to be reduced when

a. external field is reduced
b. external field is increased
c. temperature of sample is reduced
d. temperature of sample is increased

Matching List Type

Direction *(Q. No. 29) Choices for the correct combination of elements from Column I and Column II are given as options (a), (b), (c) and (d), out of which one is correct.*

29 Match entries of Column I with entries of Column II.

	Column I		Column II
i.	Magnetic field	p.	$[AL^2]$
ii.	Magnetic moment	q.	$[ATM^{-1}]$
iii.	Ratio of magnetic moment to angular momentum	r.	$[MA^{-1}T^{-2}]$
iv.	$\sqrt{\varepsilon_0\mu_0}$	s.	$[L^{-1}T]$

Codes

	i	ii	iii	iv
a.	r	p	q	s
b.	p	q	r	s
c.	q	p	r	s
d.	r	q	p	s

Revisal Problems for JEE Main

Only One Option Correct Type

Working Space

Direction *(Q. Nos. 1-16) This section contains 16 multiple choice questions. Each question has four choices (a), (b), (c) and (d), out of which ONLY ONE is correct.*

1 The hysteresis loss for a specimen of iron weighing 12 kg is equivalent to 300 Jm^{-3} $cycle^{-1}$. Find the loss of energy per hour at 50 cycle s^{-1} (Given, density of iron is 7500 kg/m^3).

a. 86000 J
b. 86400 J
c. 95000 J
d. 50000 J

2 In a permanent magnet at room temperature,

a. magnetic moment of each molecule is zero
b. the individual molecules have non-zero magnetic moments which are all perfectly aligned
c. domains are partially aligned
d. domains are all perfectly aligned

3 A paramagnetic sample shows a net magnetisation of 8 Am^{-1} when placed in an external magnetic field of 0.6 T at a temperature of 4 K. When the same sample is placed in an external magnetic field of 0.2 T at a temperature of 16 K, the magnetisation of sample will be

a. $\frac{32}{3}$ Am^{-1}
b. $\frac{2}{3}$ Am^{-1}
c. 6 Am^{-1}
d. 2.4 Am^{-1}

4 The variation of magnetic susceptibility (χ) with temperature for a diamagnetic substance is best represented by which figure?

a.
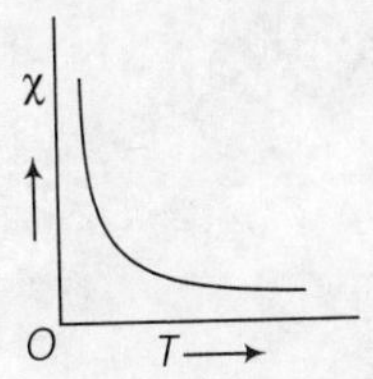

b.
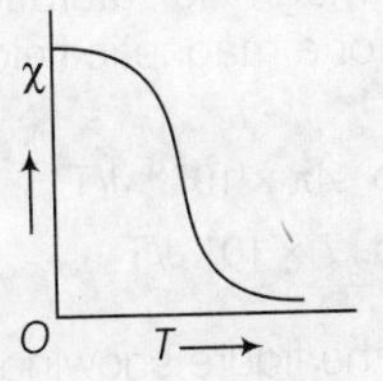

c.
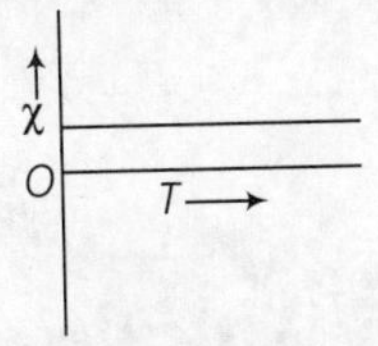

d.
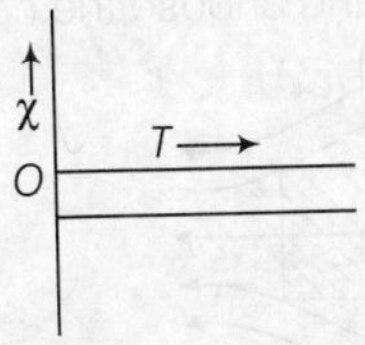

5 A short bar magnet with the North pole facing North forms a neutral point at P in the horizontal plane. If the magnet is rotated by 90° in the horizontal plane, the net magnetic induction at P is (horizontal component of the earth's magnetic field = B_H)

a. zero
b. $2B_H$
c. $\frac{\sqrt{5}}{2}B_H$
d. $\sqrt{5}\,B_H$

6 A magnetic material of volume 30 cm^3 is placed in a magnetic field of intensity 5 oersted. The magnetic moment produced due to it is 6 A/m^2. The value of magnetic induction will be

a. 0.2517 T
b. 0.025 T
c. 0.0025 T
d. 25 T

Working Space

7 The total magnetic flux in a material, which produces a pole of strength m_p when a magnetic material of cross-sectional area A is placed in a magnetic field of strength H, will be

a. $\mu_0 (AH + m_p)$
b. $\mu_0 AH$
c. $\mu_0 m_p$
d. $\mu_0 (m_p AH + A)$

8 Two bar magnets having same geometry with magnetic moments m and $2m$ are firstly placed in such a way that their similiar poles are on the same side and its period of oscillation is T_1. Now, the polarity of one of the magnets is reversed and its time period becomes T_2. Then,

a. $T_1 < T_2$
b. $T_1 > T_2$
c. $T_1 = T_2$
d. $T_2 = \infty$

9 Dimensions of B_H are

a. $[M^1L^{-1}T^{-2}A^0]$
b. $[M^0L^{-1}T^0A]$
c. $[M^0L^{-1}T^{-2}A]$
d. $[M^0L^2T^0A]$

10 The current I-H curve for a paramagnetic material is best represented by which figure?

a.
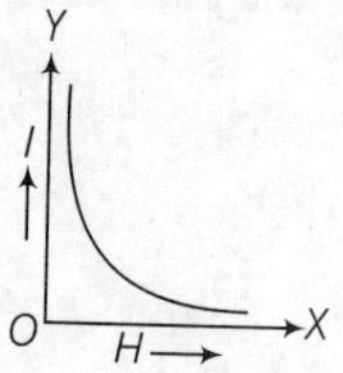

b.
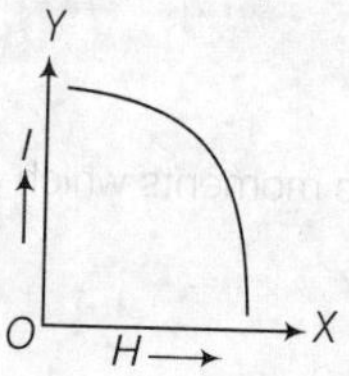

c.
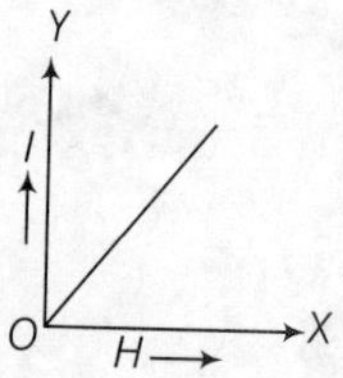

d.
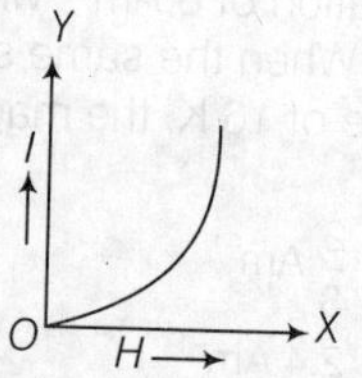

11 A sample of paramagnetic salt contains 2×10^{24} atomic dipoles, each of dipole moment 1.5×10^{-23} J/T. The sample is placed under a homogeneous magnetic field of 0.84 T and cooled a temperature of 4.2 K. The degree of magnetic saturation achieved is equal to 15%. The total dipole moment of the sample for a magnetic field of 0.98 T and a temperature of 2.8 K is

a. 3×10^{-3} J/T
b. $4\pi \times 10^{-6}$ J/T
c. 7.9 J/T
d. 7×10^4 J/T

12 Consider the given statement with respect to the figure showing a bar of diamagnetic material placed in an external magnetic field and choose the correct statement.

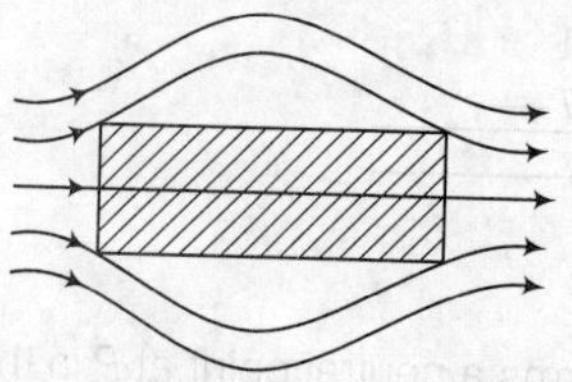

I. The field lines are repelled or expelled and the field inside the material is reduced.
II. When placed in a non-uniform magnetic field, the bar will tends to move from high to low field.
III. Reduction in the field inside the material is slight being one part in 10^5.

Correct option is

a. Only I
b. Both II and III
c. Both I and III
d. I, II and III

13 A rod of ferromagnetic material with dimensions 10 cm × 0.5 cm × 0.2 cm is placed in a magnetic field of strength 0.5×10^4 Am^{-1} as result of which a magnetic moment of 5 A-m^2 is produced in the rod. The value of magnetic induction will be

a. 0.54 T
b. 0.358 T
c. 2.519 T
d. 6.28 T

14 A Rowland ring of mean radius 15 cm has 3500 turns of wire wound on a ferromagnetic core of relative permeability 800. The magnetic field (B) in the core for a magnetising current of 1.2 A is

a. 4 T
b. 4.48 T
c. 7.9 T
d. 3 T

15 A domain in ferromagnetic iron is in the form of a cube of side length 1 μm. Estimate the number of iron atoms in the domain. The molecular mass of iron is 55 g/mol and its density is 7.9 g/cm^3.

a. 8.65×10^{-10} atoms
b. 8×10^{-13} atoms
c. 8×10^5 atoms
d. 8.65×10^{10} atoms

16 A solenoid has core of a material with relative permeability 400. The windings of the solenoid are insulated from the core and carry a current of 2 A. If the number of turns is 1000 per metre, calculate H and M.

a. 2×10^3 Am^{-1}, 8×10^5 Am^{-1}
b. 1 Am^{-1}, 1.5×10^5 Am^{-1}
c. 6.8×10^{-5} Am^{-1}, 1.2×10^{-5} Am^{-1}
d. 2.1×10^{-4} Am^{-1}, -2.6×10^{-5} Am^{-1}

Statement Type

***Direction** (Q. Nos. 17-20) This section is based on Statement I and Statement II. Select the correct answer from the codes given below.*

a. Both Statement I and Statement II are correct and Statement II is the correct explanation of Statement I
b. Both Statement I and Statement II are correct but Statement II is not the correct explanation of Statement I
c. Statement I is correct but Statement II is incorrect
d. Statement II is correct but Statement I is incorrect

17 **Statement I** When an electron makes a transition from a higher energy level to a lower energy level, its angular momentum as well as its magnetic dipole moment increases.
Statement II $\mathbf{L}_n = mv_n r_n$ and $\mathbf{M}_n = i_n A_n$.

18 **Statement I** A soft ferromagnetic core is preferred over a hard ferromagnetic material.
Statement II Area of hysteresis loop of a soft ferromagnetic material is comparatively less than the corresponding area of hard ferromagnetic material.

19 **Statement I** It is possible to achieve nearly 100% magnetisation of a paramagnetic sample.
Statement II For a ferromagnetic sample, $M = C\frac{B_0}{T}$.

20 **Statement I** Above Curie temperature, a ferromagnetic material behaves like a paramagnetic substance.
Statement II

As, $$M = \frac{CB_0}{T}$$

∴ $$M \propto \frac{1}{T}$$

SET - 2

Revisal Problems for JEE Advanced

Only One Option Correct Type

Direction *(Q. Nos. 1-10) This section contains 10 multiple choice questions. Each question has four choices (a), (b), (c) and (d), out of which ONLY ONE is correct.*

Working Space

1 A North pole of strength 50 A-m and North pole of strength 100 A-m are separated by a distance of 10 cm in air. Force between them is

a. 50×10^{-3} N
b. 5×10^{-3} N
c. 100×10^{-3} N
d. 10×10^{-3} N

2 A bar magnet with its poles 25 cm apart and of pole strength 24 A-m rests with its centre on a frictionless pivot. A force F is applied on the magnet at a distance 12 cm from the pivot. So that, it is held in equilibrium at an angle of 30° with respect to magnetic field of induction 0.25 T. The value of force F is

a. 65.62 N
b. 2.56 N
c. 6.52 N
d. 6.25 N

3 The ratio of magnetic intensities by distances x and $2x$ from the centre of magnet of length 2 cm on its axis will be

a. 4 : 1
b. 1 : 4
c. 8 : 1
d. None of these

4 Two magnets A and B are identical and these are arranged as shown in figure.

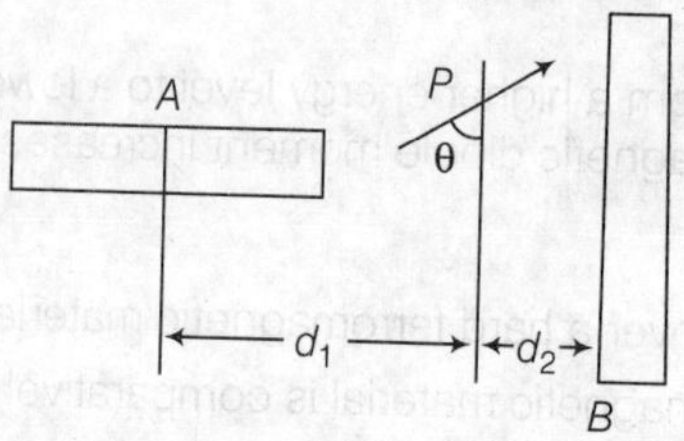

Their length is negligible in comparison to the separation between them. A magnetic needle is placed between them at a point P which gets deflected through an angle θ under the influence of magnets. The ratio of distances d_1 and d_2 will be

a. $(2 \tan \theta)^{1/3}$
b. $(2 \tan \theta)^{-1/3}$
c. $(2 \cot \theta)^{1/3}$
d. $(2 \cot \theta)^{-1/3}$

5 The period of oscillation of a freely suspended bar magnet is 4 s. If it is cut into two equal part lengthwise, then time period of each part will be

a. 4 s
b. 2 s
c. 0.5 s
d. 0.25 s

6 The length, breadth and mass of two bar magnets are same but their magnetic moments are $3M$ and $2M$ respectively. These are joined pole to pole and are suspended by a string. When oscillated in magnetic field B, time period obtained is 5 s. If the poles of either of the magnets are reversed, then time period obtained in the same magnetic field will be

a. $3\sqrt{3}$ s
b. $2\sqrt{2}$ s
c. $5\sqrt{5}$ s
d. 1s

7 A thin magnetic needle oscillates in a horizontal plane with a period T. It is broken into n equals parts. The time period of each part is

a. T
b. T/n^2
c. n^2
d. T/n

Working Space

8 A 10 cm long bar magnet of magnetic moment 1.34 A-m^2 is placed in magnetic meridian with its South pole pointing geographical South. The neutral point obtained at a distance of 15 cm from the centre on the magnet. The horizontal component of the earth's magnetic field is

a. 0.68×10^4 T
b. 0.34×10^{-4} T
c. 0.17×10^{-4} T
d. 0.92×10^4 T

9 Consider a toroidal core with a small air gap. Let magnetic field lines inside the core arc as shown in the figure.

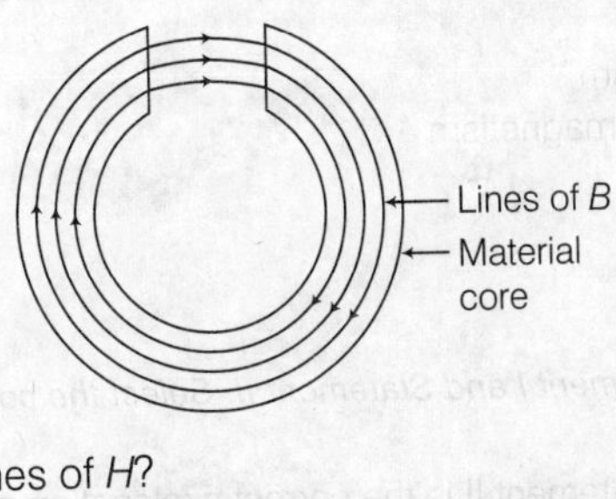

Which of these shows lines of H?

a.
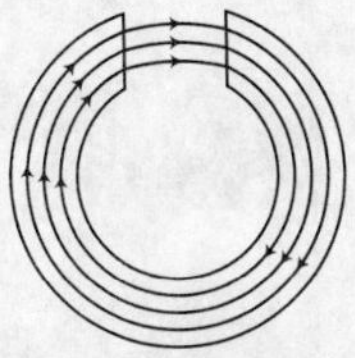

b.
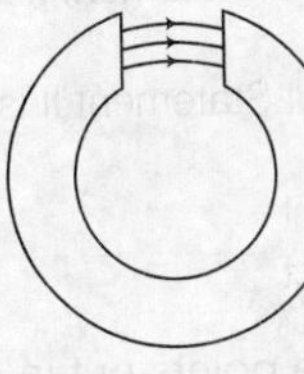

c.
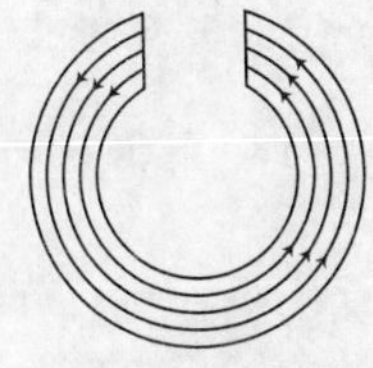

d.
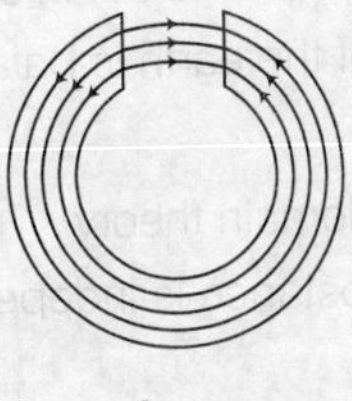

10 Magnetic moment of the earth is around 8×10^{22} A-m^2. Let this is caused only by complete magnetisation of huge iron deposit, how many kilograms of iron this would corresponds to? (Given, density of iron = 7900 kg/m^3 and number of atoms in 1m^3 = 8.63×10^{28})

a. 8.63×10^{45} kg
b. 4.31×10^{45} kg
c. 3.95×10^{20} kg
d. 9.27×10^{24} kg

One or More than One Options Correct Type

***Direction** (Q. Nos. 11-16) This section contains 6 multiple choice questions. Each question has four choices (a), (b), (c) and (d), out of which ONE or MORE THAN ONE are correct.*

11 The permanent dipole moment of the atoms of a material is not zero. The material

a. must be paramagnetic
b. must be diamagnetic
c. may be paramagnetic
d. may be ferromagnetic

12 Tick the correct options.

a. All electrons have magnetic moment
b. All protons have magnetic moment
c. All nuclei have magnetic moment
d. All atoms have magnetic moment

13 When a ferromagnetic material goes through a hysteresis loop, the magnetic susceptibility

a. has a fixed value
b. may be zero
c. may be infinity
d. may be negative

14 Magnetic dipole moment of an electron is due to its orbital motion. In this view, correct options are

a. magnetic dipole moment points opposite to angular momentum
b. ratio of magnetic dipole moment with angular momentum is a constant
c. in a multi-electron atom, spin magnetic moments points in same direction
d. total magnetic moment of an electron is the vector sum of the orbital and spin magnetic moments

15 Which of these are true regarding a ferromagnetic substance?

a. Domain's size is about 10^{-12} to $10^{-8} m^3$
b. Domains contain about 10^{17} to 10^{21} atoms
c. Above Curie temperature sample shows paramagnetic behaviour
d. Relative permeability is of order 10^3 to 10^5

Working Space

16 Which of these are true regarding diamagnetic behaviour?

a. It is not present in all substances
b. Diamagnetism arises as external field increases, speed of electron orbiting in one direction and slows down electrons orbiting in other direction
c. Copper, diamond and lead show strong diamagnetism
d. Aluminium, magnesium and platinum show weak diamagnetism

Statement Type

Direction (Q. Nos. 17 and 18) *This section is based on Statement I and Statement II. Select the correct answer from the codes given below.*

a. Both Statement I and Statement II are correct and Statement II is the correct explanation of Statement I
b. Both Statement I and Statement II are correct but Statement II is not the correct explanation of Statement I
c. Statement I is correct but Statement II is incorrect
d. Statement II is correct but Statement I is incorrect

17 **Statement I** At neutral point, a compass needle points out in any arbitrary direction.

Statement II At neutral point, magnetic field of the earth is balanced by field due to magnet.

18 **Statement I** Paramagnetism is explained by domain theory.

Statement II Susceptibility of diamagnetic substance is independent of temperature.

Matching List Type

Direction (Q. Nos. 19 and 20) *Choices for the correct combination of elements from Column I and Column II are given as options (a), (b), (c) and (d), out of which one is correct.*

19 Match the terms of Column I with the items of Column II and choose the correct option from the codes given below.

	Column I		Column II
i.	Magnetic moment	p.	Vector
ii.	Magnetic induction	q.	Scalar
iii.	Permeability	r.	$NA^{-1}m^{-1}$
iv.	Intensity of magnetisation	s.	Nm^3/Wb

Codes

	i	ii	iii	iv
a.	p,s	p,r	q	p
b.	p,r	p, s	p	q
c.	p,s	p,r	p	q
d.	p	s	q	s

20 Match the terms of Column I with the items of Column II and choose the correct option from the codes given below.

	Column I		Column II
i.	Diamagnetic	p.	$\mu >> \mu_0, \mu_r >> 1$ and $\chi >> 1$
ii.	Paramagnetic	q.	$-1 \le \chi < 0, 0 \le \mu_r < 1$ and $\mu < \mu_0$
iii.	Ferromagnetic	r.	$0 < \chi < \varepsilon, 1 < \mu_r < 1 + \varepsilon$ and $\mu > \mu_0$

Codes

	i	ii	iii
a.	p	q	r
b.	q	r	p
c.	q	p	r
d.	r	q	p

Comprehension Type

Direction *(Q. Nos. 21-25) This section contains 2 paragraphs, each describing theory, experiments, data, etc. Five questions related to the paragraphs have been given. Each question has only one correct answer among the four given options (a), (b), (c) and (d).*

Passage I

Electrical resistance of certain materials known as superconductors changes abruptly from a non-zero value to zero as their temperature is lowered below a critical temperature $T_C(0)$. An interesting property of superconductors is that their critical temperature becomes smaller than $T_C(0)$ if they are placed in a magnetic field, i.e. the critical temperature $T_C(B)$ is a function of the magnetic field strength B. The dependence of $T_C(B)$ on B is shown in the figure. **[IIT JEE 2010]**

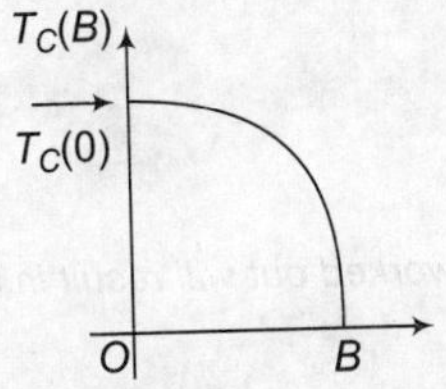

21 For the graphs given below, the resistance R of a superconductor is shown as a function of its temperature T for two different magnetic fields B_1 (solid line) and B_2 (dashed line). If B_2 is larger than B_1, which of the following graphs shows the correct variation of R with T in these fields?

a.
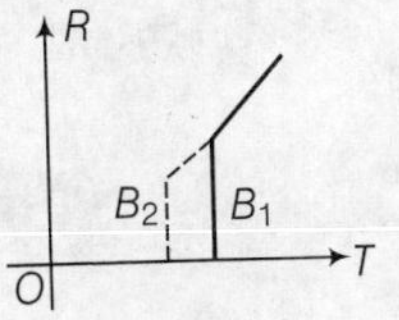

b.
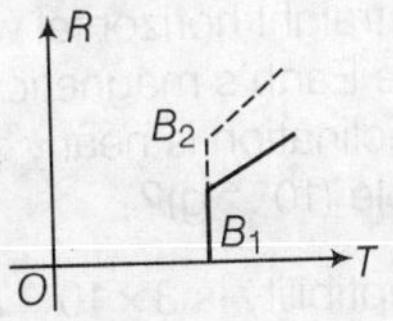

c.
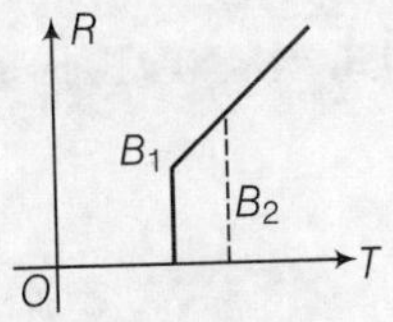

d.
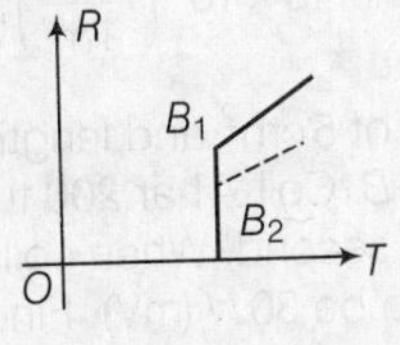

22 A superconductor has $T_C(0) = 100$ K. When a magnetic field of 7.5 T is applied, its T_C decreases to 75 K. For this material one can definitely say that when (where, T = tesla)

a. $B = 5$ T, $T_C(B) = 80$ K
b. $B = 5$ T, $75\text{ K} < T_C(B) = 100$ K
c. $B = 10$ T, $75\text{ K} < T_C(B) = 100$ K
d. $B = 10$ T, $T_C(B) = 70$ K

Passage II

Toroids and solenoids found there are in many practical applications. Consider an electromagnet as shown in the figure. The magnetic material which is almost a toroid is wound with N turns each carrying a current I.

Length of arc of magnetic material is L with an air gap of length l. Cross-section of toroid is A.

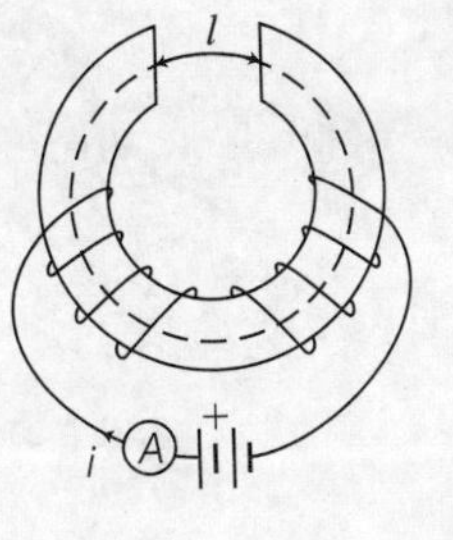

23 Magnetic field intensity in the air gap is

a. $\frac{1}{\mu_r}$ times of magnetic field intensity in the material
b. μ_r times of magnetic field intensity in the material
c. same as that of the material
d. cannot be determined due to fringing of the field

Working Space

24 Magnetic flux in the magnetic circuit is given by

a. $\frac{\mu_0 NiA}{(l+L)}$

b. $\frac{\mu_0 NiA}{\left(l+\frac{L}{\mu_r}\right)}$

c. $\frac{\mu_0 NiA}{(l\mu_r + L)}$

d. $\frac{\mu_0 \mu_r NiA}{(\mu_r l+L)}$

25 A toroid is made by bending an iron bar of length 6 cm and cross-section area of 4 cm^2 has an air gap of 1cm. It is wound with 500 turns of wire, current in wire is 20 A, for material $\mu_r = 3000$. If gap is closed by pressing the ring of toroid, then field increases by a factor of around

a. 20

b. 10

c. 40

d. 80

One Integer Value Correct Type

Direction *(Q. Nos. 26-29) This section contains 4 questions. When worked out will result in an integer from 0 to 9 (both inclusive).*

26 A 30 cm long bar magnet is placed in the magnetic meridian with its North pole pointing South. The neutral point is obtained at a distance of 40 cm from the centre of the magnet. Find the magnetic dipole moment and pole strength of the magnet. The horizontal component of the earth's magnetic field is 0.34 g.

27 A telephone cable at a place has four long straight horizontal wires carrying a current of 1.0 A in the same direction East to West. The Earth's magnetic field at the place is 0.39 g and the angle of dip is 35°. The magnetic declination is nearly zero. What is the resultant magnetic field at a point 4 cm below the cable (10^{-2} g)?

28 For a compound chromium, magnetic susceptibility is 3×10^{-4} at 300 K temperature. Curie's constant for chromium is found to be $6.45 \times 10^{N}\left(\frac{\text{A-K}}{\text{Tm}}\right)$, find the value of N.

29 A circular iron core has cross-sectional area of 5 cm^2 and length of magnetic path is 50 cm. It has wound over by two coils A and B. Coil A har 200 turns and B has 500 turns. Current in coil A changes from 0 to 15 A in π second. When coils are wound over a core with $\mu_r = 250$ induced emf in coil B is found to be $30N$ (mV). Find the value of N.

JEE Main & AIEEE Archive

(Compilation of Last 13 Years' Questions)

1 The coercivity of a small magnet where the ferromagnet gets demagnetised is 3×10^3 Am^{-1}. The current required to be passed in a solenoid of length 10 cm and number of turns 100 so that the magnet gets demagnetised when inside the solenoid is **(JEE Main 2014)**

a. 30 mA b. 60 mA
c. 3 A d. 6 A

2 Two short bar magnets of length 1 cm each, have magnetic moments 1.20 A-m^2 and 1.00 A-m ^{2} respectively. They are placed on a horizontal table parallel to each other with their *N*-poles pointing towards the South. They have a common magnetic equator and are separated by a distance of 20.0 cm. The value of the resultant horizontal magnetic induction at the mid-point *O* of the line joining their centres is close to (Horizontal component of the earth's magnetic induction is 3.6×10^{-5} Wb/m^2) **(JEE Main 2013)**

a. 3.6×10^{-5} Wb/m^2 b. 2.56×10^{-4} Wb/m^2
c. 3.50×10^{-4} Wb/m^2 d. 5.80×10^{-4} Wb/m^2

3 Relative permittivity and permeability of a material are ε_r and μ_r, respectively. Which of the following values of these quantities are allowed for a diamagnetic material? **(AIEEE 2008)**

a. $\varepsilon_r = 0.5, \mu_r = 1.5$ b. $\varepsilon_r = 1.5, \mu_r = 0.5$
c. $\varepsilon_r = 0.5, \mu_r = 0.5$ d. $\varepsilon_r = 1.5, \mu_r = 1.5$

4 Needles N_1, N_2 and N_3 are made of a ferromagnetic, a paramagnetic and a diamagnetic substance, respectively. A magnet when brought close to them will **(AIEEE 2006)**

a. attract N_1 and N_2 strongly but repel N_3
b. attract N_1 strongly, N_2 weakly and repel N_3 weakly
c. attract N_1 strongly but repel N_2 and N_3 weakly
d. attract all three of them

5 The materials suitable for making electromagnets should have **(AIEEE 2004)**

a. high retentivity and high coercivity
b. low retentivity and low coercivity
c. high retentivity and low coercivity
d. low retentivity and high coercivity

6 A thin rectangular magnet suspended freely has a period of oscillation equal to *T*. Now, it is broken into two equal halves (each having half of the original length) and one piece is made to oscillate freely in the same field. If its period of oscillation is *T*′, the ratio *T*′/*T* is **(AIEEE 2003)**

a. $\frac{1}{2\sqrt{2}}$ b. $\frac{1}{2}$
c. 2 d. $\frac{1}{4}$

JEE Advanced & IIT JEE Archive

(Compilation of Last 10 Years' Questions)

Passage for Q. Nos. (1-2)

Electrical resistance of certain materials is known as superconductors, changes abruptly from a non-zero value to zero as their temperature is lowered below a critical temperature $T_C(0)$. An interesting property of superconductors is that their critical temperature becomes smaller than $T_C(0)$ if they are placed in a magnetic field, i.e. the critical temperature $T_C(B)$ is a function of the magnetic field strength B. The dependence of $T_C(B)$ on B is shown in the figure.

(2010, Comprehension Type)

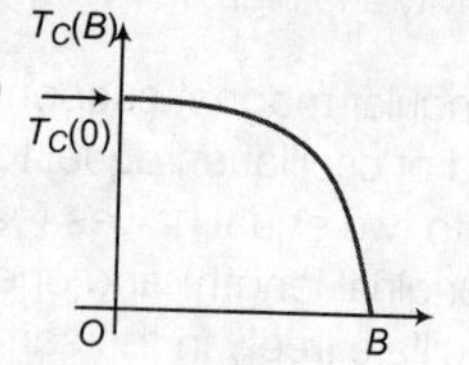

1. For the graphs given below, the resistance R of a superconductor is shown as a function of its temperature T for two different magnetic fields B_1 (solid line) and B_2 (dashed line). If B_2 is larger than B_1, which of the following graphs shows the correct variation of R with T in these fields?

a. 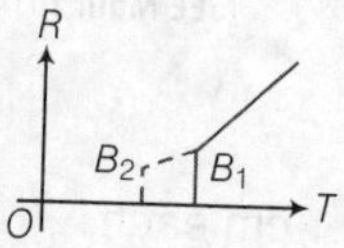

b. 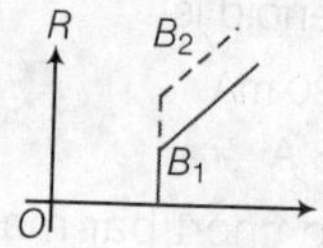

c. 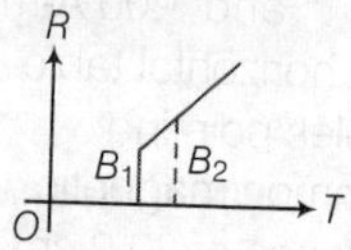

d. 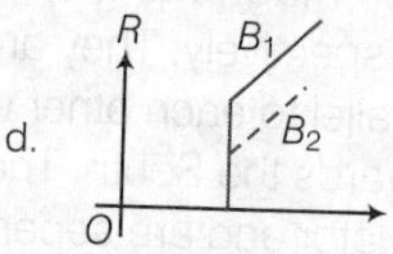

2. A superconductor has $T_C(0) = 100$ K. When a magnetic field of 7.5 T is applied, its T_C decreases to 75 K. For this material, one can definitely say that when (where, T = tesla)

a. $B = 5$ T, $T_C(B) = 80$ K

b. $B = 5$ T, $75\text{ K} < T_C(B) < 100$ K

c. $B = 10$ T, $75\text{ K} < T_C(B) < 100$ K

d. $B = 10$ T, $T_C(B) = 70$ K

3. Electromagnetic Induction

DPP
DAILY PRACTICE PROBLEMS

DPP-1 Magnetic Flux and Faraday's Law

Subjective Questions

Working Space

Direction *(Q. Nos. 1-4) These questions are subjective in nature, need to be solved completely on notebook.*

1 Determine the magnetic flux through a square of side *a*. If one side of square is parallel to and at a distance *b* from a current carrying straight wire.

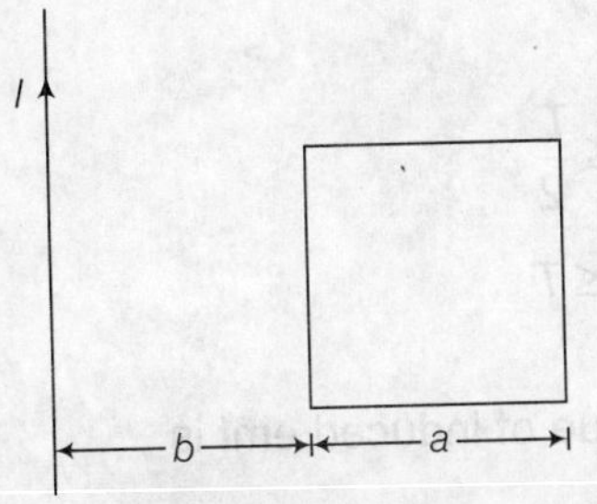

2 A power line passes over a farmer's land. The farmer's son, bright student of class XII, constructs a vertically oriented 2 m high 10 turns rectangular wire loop below the power line. Farmer's son requires an induced voltage of 120 V which can runs his laptop charger. Laptop charger requires sinusoidally varying voltage of frequency 50 Hz and peak voltage of 170 V.

Power line carries a sinusoidal voltage of frequency 60 Hz and peak value of current is 55 kA. Power line runs 7 m above farmer's land.

i. Calculate length of coil to fulfill the requirement of farmer's son.

ii. Is this technique is innovative or unethical?

3 A metal disc of radius $R = 25$ cm rotates with a constant angular velocity $\omega = 130 \text{ rad s}^{-1}$ about its axis. Find the potential difference between the centre and rim of the disc, if

i. the external magnetic field is absent.

ii. the external uniform magnetic field $B = 5.0$ mT is directed perpendicular to the disc.

4 In figure, the four rods have λ resistance per unit length. The arrangement is kept in a magnetic field of constant magnitude *B* and directed perpendicular to the plane of the figure and directing inwards. Initially, the rods as shown form a square. Now each wire starts moving with constant velocity *v* towards opposite wire.

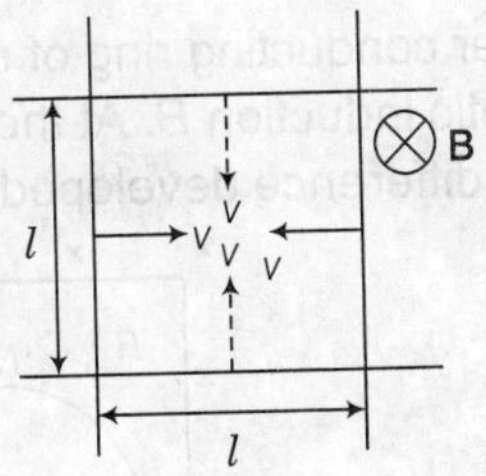

Find as a function of time

i. induced emf in the circuit.

ii. induced current in the circuit with direction.

iii. force required on each wire to keep its velocity constant.

iv. total power required to maintain constant velocity.

v. thermal power developed in the circuit.

Only One Option Correct Type

Working Space

Direction (Q. Nos. 5-18) *This section contains 14 multiple choice questions. Each question has four choices (a), (b), (c) and (d) out of which ONLY ONE is correct.*

5 One conducting U tube can slide inside another as shown in figure, maintaining electrical contacts between the tubes. The magnetic field B is perpendicular to the plane of the figure. If each tube moves towards the other at a constant speed v, then the emf induced in the circuit in terms of B, l and v, where l is the width of each tube will be

[AIEEE 2005]

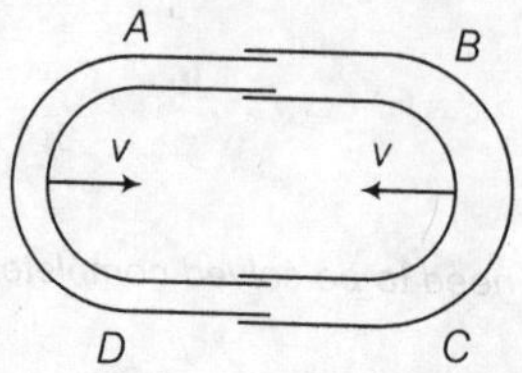

a. zero
b. $2Blv$
c. Blv
d. $-Blv$

6 The flux linked with a coil of 100 turns varies as

$$\phi = \begin{cases} \phi_m\left(1-\dfrac{4t}{T}\right); & 0 \le t \le \dfrac{T}{2} \\ \phi_m\left(\dfrac{4t}{T}-3\right); & \dfrac{T}{2} \le t \le T \end{cases}$$

where $\phi_m = 0.02$ Wb and $T = \dfrac{1}{50}$ s. Maximum value of induced emf is

a. 100 V
b. 200 V
c. 400 V
d. 600 V

7 The circuit shown in the figure, the key K is closed at $t = 0$. The current through the battery is

[AIEEE 2010]

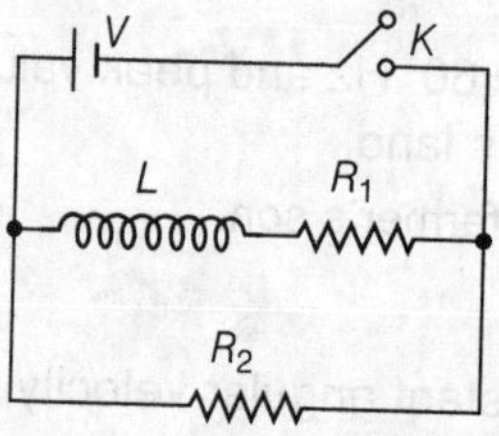

a. $\dfrac{V(R_1+R_2)}{R_1R_2}$ at $t=0$ and $\dfrac{V}{R_2}$ at $t=\infty$
b. $\dfrac{V(R_1+R_2)}{\sqrt{R_1^2+R_2^2}}$ at $t=0$ and $\dfrac{V}{R_2}$ at $t=\infty$
c. $\dfrac{V}{R_2}$ at $t=0$ and $\dfrac{V(R_1+R_2)}{R_1R_2}$ at $t=\infty$
d. $\dfrac{V}{R_2}$ at $t=0$ and $\dfrac{V(R_1+R_2)}{\sqrt{R_1^2+R_2^2}}$ at $t=\infty$

8 A thin semicircular conducting ring of radius R is falling with its plane vertical in a horizontal magnetic induction B. At the position MNQ, the speed of the ring is v and the potential difference developed across the ring is

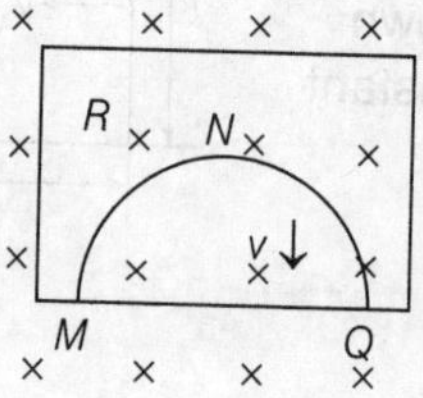

a. zero
b. $Bv\pi R^2/2$ and M is at higher potential
c. πRBv and Q is at higher potential
d. $2RBv$ and Q is at higher potential

9 If a coil is placed in a time varying magnetic field such that flux linked with the coil at any instant is $\phi(= \phi_m \sin \omega t)$. Then, the voltage induced in the coil is

a. in phase with the harmonically varying flux
b. lags behind the harmonically varying flux
c. leads the harmonically varying flux
d. depends on geometrical factors of the coil

10 A conducting wire frame is placed in a magnetic field which is directed into the paper. The magnetic field is increasing at a constant rate.

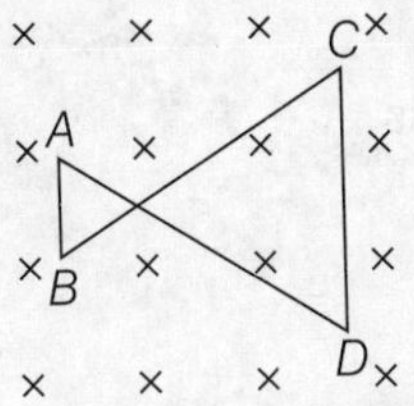

The directions of induced current in wires *AB* and *CD* are

a. *B* to *A* and *D* to *C*
b. *A* to *B* and *C* to *D*
c. *A* to *B* and *D* to *C*
d. *B* to *A* and *C* to *D*

11 As shown in the figure, a circular loop of radius *r* and resistance *R*. A variable magnetic field of induction $B = B_0 e^{-t}$ is established inside the coil. If the key (*K*) is closed, the electrical power developed right after closing the switch is equal to

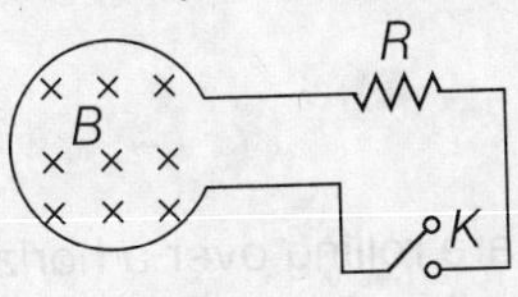

a. $\dfrac{B_0^2 \pi r^2}{R}$
b. $\dfrac{B_0 10 r^3}{R}$
c. $\dfrac{B_0^2 \pi^2 r^4 R}{5}$
d. $\dfrac{B_0^2 \pi^2 r^4}{R}$

12 A wire of length *l* velocity *v* at an angle θ in the region of a uniform magnetic field such that length of the wire always remains perpendicular to field lines.

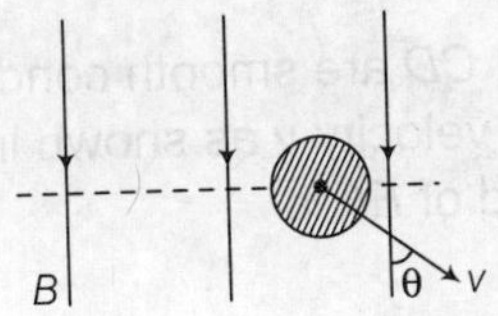

The induced emf in wire is

a. 0
b. $l(\mathbf{V} \cdot \mathbf{B})$
c. $l(\mathbf{V} \times \mathbf{B})$
d. $l(\mathbf{B} \times \mathbf{V})$

13 A metallic rod of length l is tied to a string of length $2l$ and made to rotate with angular speed ω on a horizontal table with one end of the string fixed. If there is a vertical magnetic field *B* in the region, the emf induced across the ends of the rod is **[JEE Main 2014]**

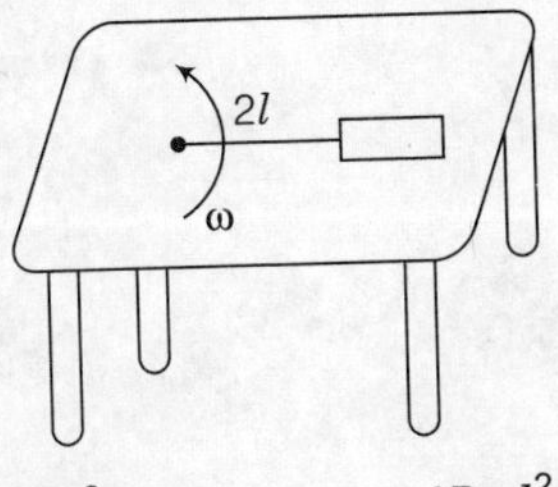

a. $\dfrac{2B\omega l^3}{2}$
b. $\dfrac{3B\omega l^3}{2}$
c. $\dfrac{4B\omega l^2}{2}$
d. $\dfrac{5B\omega l^2}{2}$

14 A semicircular piece of wire is rotated about a point P (as shown) with angular velocity ω about an axis passing through P and perpendicular to plane of wire. A magnetic field of constant magnitude B exists in the region parallel to axis of rotation.

Working Space

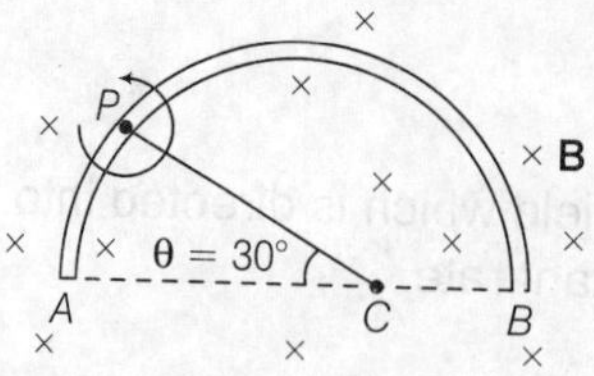

Potential difference between ends of wires (A and B) is

a. $\dfrac{B\omega R^2}{2}$
b. $\dfrac{B\omega R^2}{4}$
c. $2B\omega R^2$
d. $B\omega R^2$

15 A circular loop of radius 0.3 cm lies parallel to a much bigger circular loop of radius 20 cm. The centre of the smaller loop is on the axis of the bigger loop. The distance between their centres is 15 cm. If a current of 2.0 A flows through the bigger loop, then the flux linked with smaller loop is **[JEE Main 2014]**

a. 9.1×10^{-11} Wb
b. 6×10^{-11} Wb
c. 3.3×10^{-11} Wb
d. 6.6×10^{-9} Wb

16 Two identical conducting rings A and B of radius R are rolling over a horizontal conducting plane with same speed v but in opposite direction. A constant magnetic field B is present pointing into the plane of paper. Then, the potential difference between the highest points of the two rings is

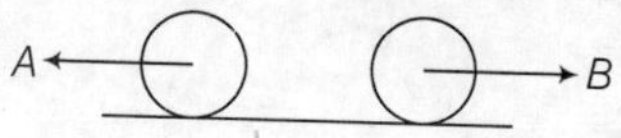

a. zero
b. $2BvR$
c. $4BvR$
d. None of these

17 PQ is an infinite current carrying conductor. AB and CD are smooth conducting rods on which a conductor EF moves with constant velocity v as shown in the figure. The force needed to maintain constant speed of EF is

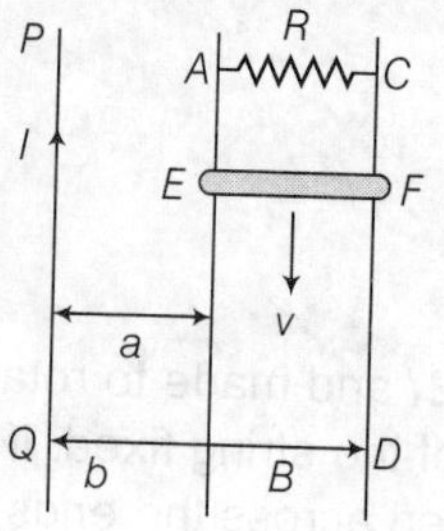

a. $\dfrac{1}{vR}\left[\dfrac{\mu_0 Iv}{2\pi}\ln\left(\dfrac{b}{a}\right)\right]^2$

b. $\left[\dfrac{\mu_0 Iv}{2\pi}\ln\left(\dfrac{b}{a}\right)\right]^2\dfrac{1}{R}$

c. $\left[\dfrac{\mu_0 Iv}{2\pi}\ln\left(\dfrac{b}{a}\right)\right]^2\dfrac{v}{R}$

d. $\dfrac{v}{R}\left[\dfrac{\mu_0 Iv}{2\pi}\ln\left(\dfrac{b}{a}\right)\right]^2$

18 The magnetic field in a region is given by $\mathbf{B} = B_0\left(1+\frac{x}{a}\right)\hat{\mathbf{k}}$. A square loop of edge length d is placed with its edge along the X- axis and Y-axis. The loop is moved with a constant velocity $\mathbf{v} = v_0\hat{\mathbf{i}}$. The emf induced in the loop at $t = 0$ is

a. $\frac{v_0 B_0 d^2}{a}$
b. $\frac{v_0 B_0 d^3}{a^2}$
c. $v_0 B_0 d$
d. zero

One or More than One Options Correct Type

Direction (Q. Nos. 19-21) *This section contains 3 multiple choice questions. Each question has four choices (a), (b), (c) and (d) out of which ONE or MORE THAN ONE are correct.*

19 A current carrying infinitely long wire is kept along the diameter of a circular wire loop, without touching it. Which of the following statement(s) is/are correct? **[IIT JEE 2012]**

a. The emf induced in the loop is zero if the current is constant
b. The emf induced in the loop is finite if the current is constant
c. The emf induced in the loop is zero if the current decreases at a steady rate
d. The emf induced in the loop is finite if the current decreases at a steady rate

20 Two metallic rings A and B, identical in shape and size but having different resistivities ρ_A and ρ_B are kept on top of two identical solenoids as shown in the figure. When current I is switched on in both the solenoids in identical manner, then rings A and B jump to heights h_A and h_B, respectively, with $h_A > h_B$.

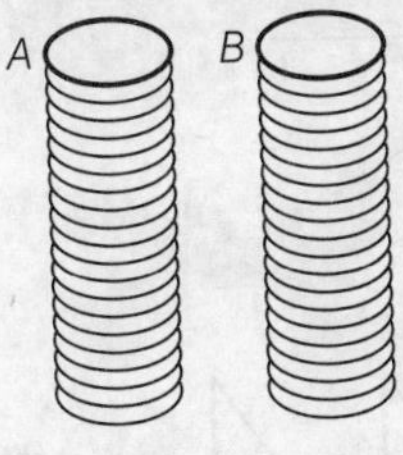

The possible relation(s) between their resistivities and their masses m_A and m_B is(are) **[IIT JEE 2009]**

a. $\rho_A > \rho_B$ and $m_A = m_B$
b. $\rho_A < \rho_B$ and $m_A = m_B$
c. $\rho_A > \rho_B$ and $m_A > m_B$
d. $\rho_A < \rho_B$ and $m_A < m_B$

21 A resistor of R ohms is connected across a square wire loop of area A. Resistance of wire loop is negligible.

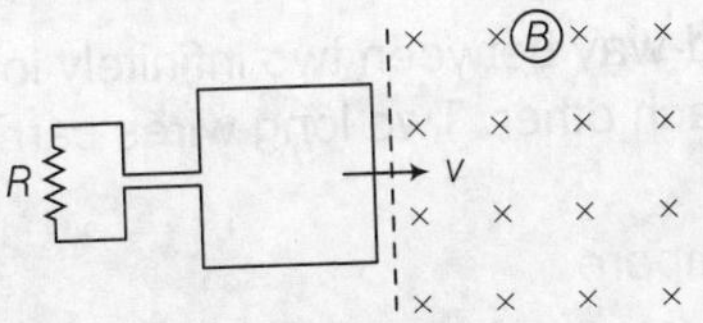

This wire loop is pulled into a region of magnetic field B. If Q is the total charge which passes through resistance R till the coil is completely pulled inside the region of magnetic field, then correct options are

a. Amount of charge which flows through R depends on velocity of loop
b. Amount of neat developed in R depends on velocity of loop
c. Amount of charge that flows through R is independent of time in which loop is pulled inside the region of magnetic field
d. Amount of charge that flows through R is $\frac{BA}{R}$

Comprehension Type

Working Space

Direction (Q. Nos. 22 and 23) *This section contains a paragraph, which describing theory, experiments, data, etc. Two questions related to the paragraph have been given. Each question has only one correct answer among the four given options (a), (b), (c) and (d).*

Passage

A wire loop enclosing a semicircle of radius R is located on the boundary of a uniform magnetic field B. At the moment $t = 0$, the loop is set into rotation with constant angular acceleration α about an axis O. The clockwise emf direction is taken to be positive.

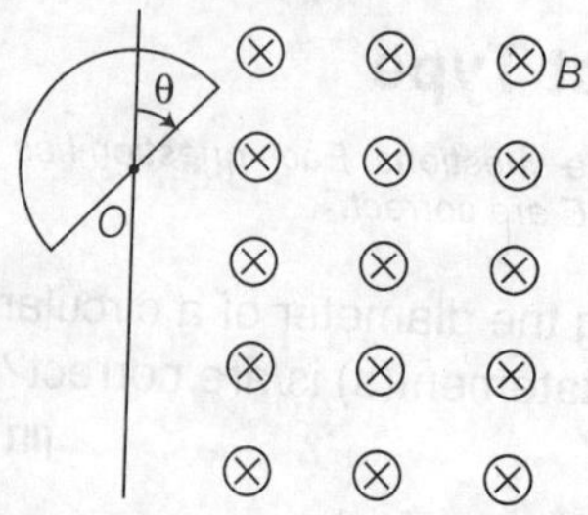

22 The variation of emf as a function of time is

a. $\frac{1}{2}BR^2\alpha t$ b. $\frac{3}{2}BR^2\alpha tA$ c. $\sqrt{3}\,BR^2\alpha t$ d. $\frac{BR^2\alpha t}{\sqrt{2}}$

23 The variation of emf as a function of time is

a. 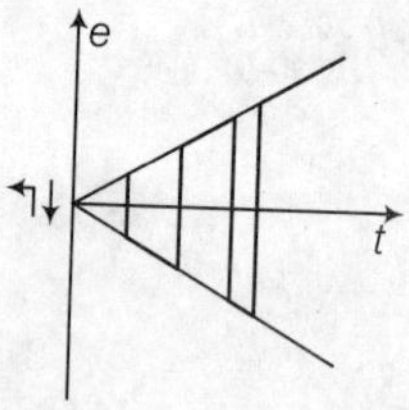

b. 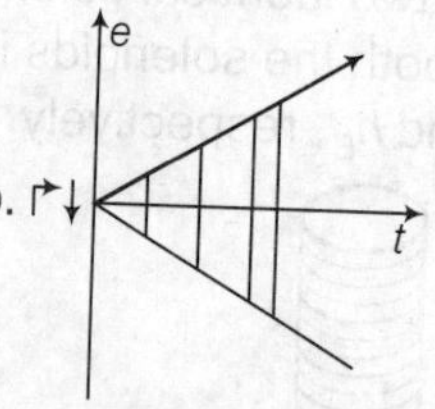

c. 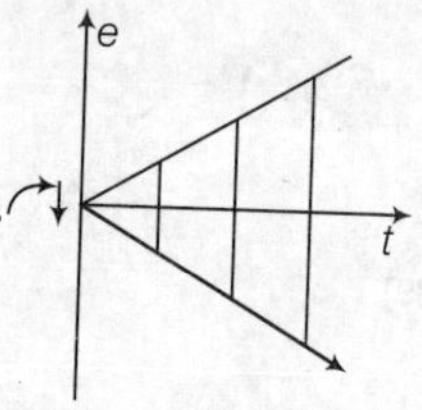

d. 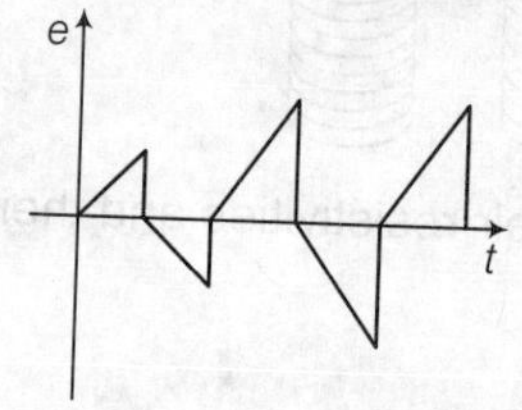

One Integer Value Correct Type

Direction (Q. Nos. 24 and 25) *This section contains 2 questions. When worked out will result in an integer from 0 to 9 (both inclusive).*

24 A square wire loop of side 6 cm is situated mid-way between two infinitely long straight conductors placed 18 cm apart from each other. Two long wires carries current in opposite directions of magnitude,

$$I = 5 \sin 120\pi t \text{ ampere}$$

Induced emf in the wire loop has a peak value of $k \times \mu_0$ (log 4) $\times 10^2$ volt

Find the value of k.

25 A light straight slider 1 m long can slide over two straight long parallel rails, vertically kept. When a uniform magnetic field of 0.6 T is applied perpendicular to the plane of rails, the slider is found moving with a constant velocity.

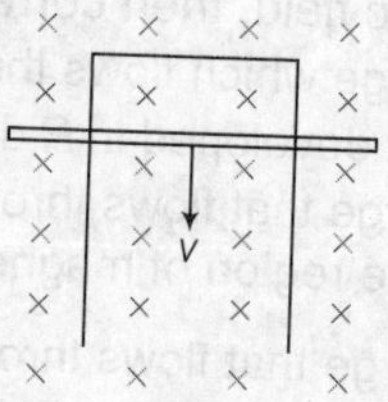

At this instant power dissipated in rails and slider due to their resistance is 1.96 W. If slider has a mass of 0.2 kg, then find the terminal velocity of the slider.

Matching List Type

Direction *(Q. No. 26) Choices for the correct combination of elements from Column I and Column II are given as options (a), (b), (c) and (d), out of which one is correct.*

26 Apply Kirchhoff's laws to the circuit given below:

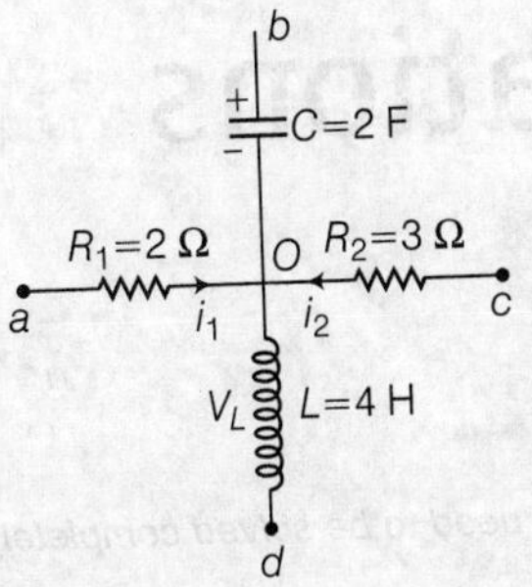

Given, $i_1 = 10e^{-2t}$ A, $i_2 = 4$A, $V_C = 3e^{-2t}$V
$C = 2$F, $L = 4$H, $R_1 = 2\Omega, R_2 = 3\Omega$.

Now, match the following columns and select the correct option from the codes given below.

	Column I		Column II
i.	Current through inductor.	p.	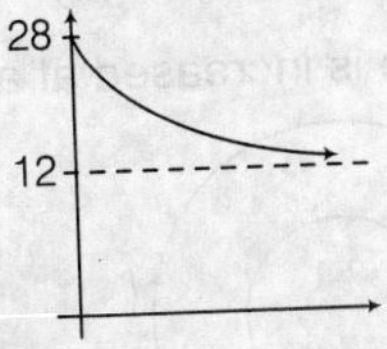
ii.	Potential drop across inductor.	q.	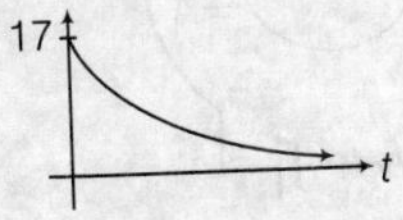
iii.	Potential difference between points *a* and *c*.	r.	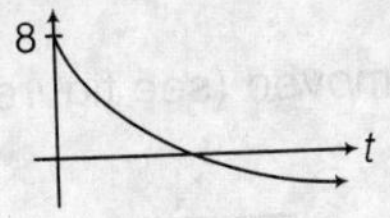
iv.	Potential difference between points *a* and *b*.	s.	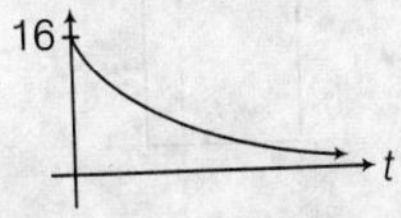
v.	Potential difference between points *c* and *d*.	t.	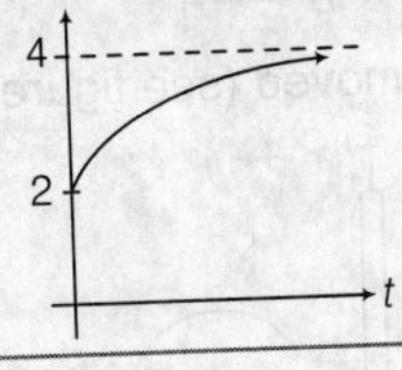

Codes

	i	ii	iii	iv	v
a.	p	q	r	s	t
b.	p	r	q	t	s
c.	t	s	r	q	p
d.	t	s	q	p	r

DPP-2 Lenz's Law and Its Applications

Subjective Questions

Direction *(Q. Nos. 1-4) These questions are subjective in nature, need to be solved completely on notebook.*

Working Space

1 Determine the direction of induced current in

i. the coil when *N*-pole of magnet is moved towards the coil.

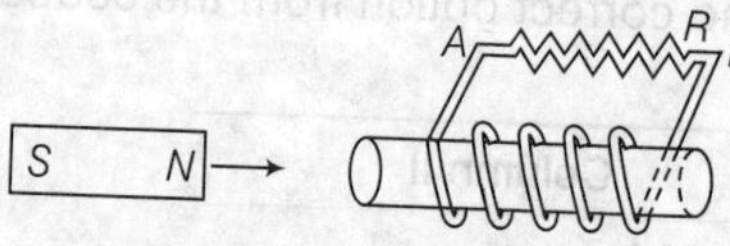

ii. the smaller loop *A*, when resistance is increased at a constant rate.

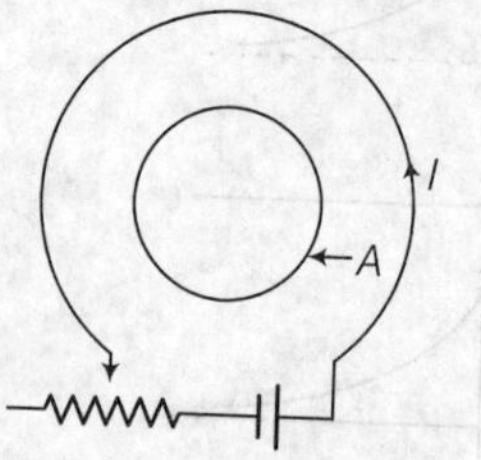

iii. the loops *A* and *B*, when both are moved (see figure) at a constant rate.

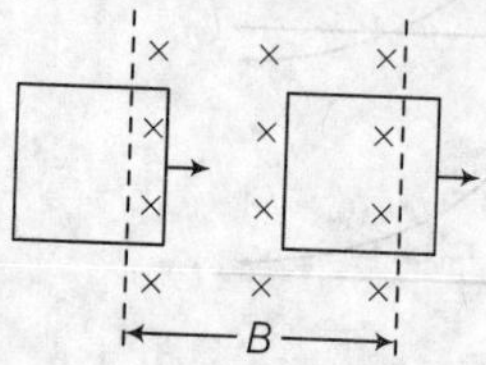

iv. the loops *A* and *B*, when both are moved (see figure) near a current carrying wire.

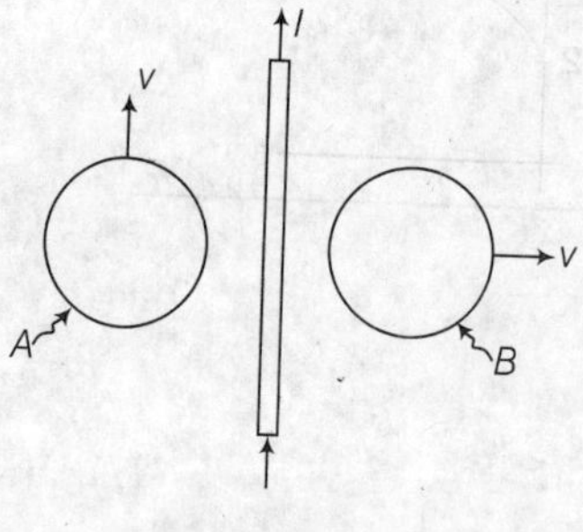

2 In figure, *P* and *S* are two coils. What shall be the direction of induced momentary current in *S* immediately

i. after the switch is closed?

ii. if the switch is opened, after it has been closed for sometime?

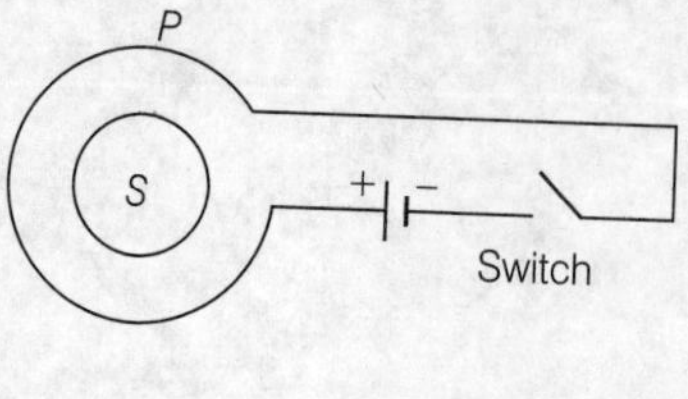

3 Three identical coils A, B and C are placed with their planes parallel to one another, figure. Coils A and C carry current as shown in the figure. Coil B and C are fixed. The coil A is moved towards B with uniform speed. Is there any induced current in B?

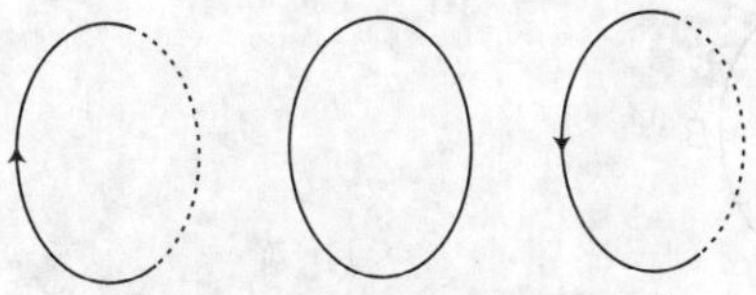

4 The magnitude of electric current is increasing from A towards B. If there is any induced current in the loop shown in the figure. What will be its direction?

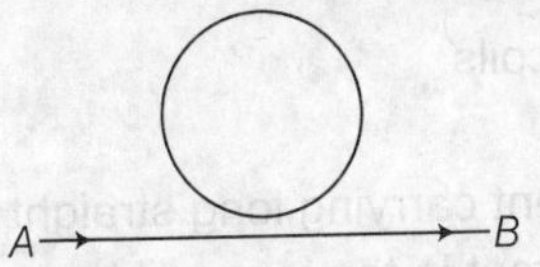

Only One Option Correct Type

***Direction** (Q. Nos. 5-13) This section contains 9 multiple choice questions. Each question has four choices (a), (b), (c) and (d) out of which ONLY ONE is correct.*

5 A conducting rod is placed on two frictionless long parallel rails kept in a region of uniform magnetic field.

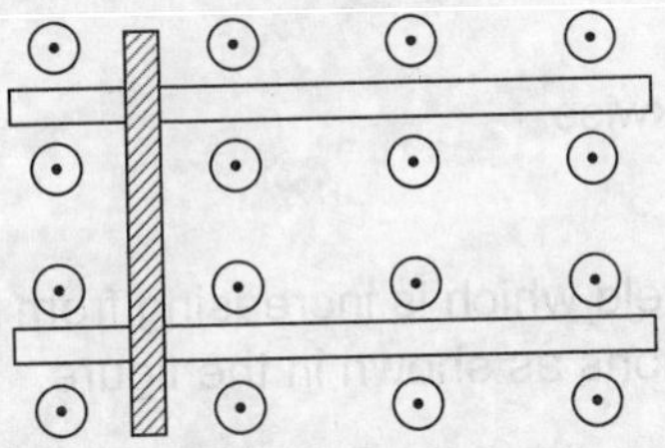

At $t = 0$, rod is given an impulse, so that it acquires a velocity $v = v_0$. After some time say t second,

a. speed of rod keeps on decreasing exponentially with time
b. speed of rod keeps on increasing exponentially with time
c. speed of rod keeps on decreasing linearly with time
d. speed of rod remains constant

6 The North pole of magnet is brought near a coil. The coil induced current in the coil as seen by an observer on the side of magnet will be

a. in the clockwise direction
b. in the anti-clockwise direction
c. initially in the clockwise and then anti-clockwise direction
d. initially in the anti-clockwise and then clockwise direction

7 Consider the situation shown in the figure. If the current I in the long straight wire XY is increased at a steady rate, then the induced emf's in loops A and B will be

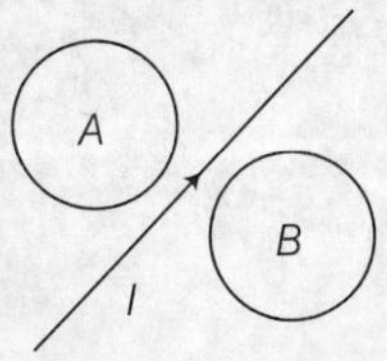

a. clockwise in A, anti-clockwise in B
b. anti-clockwise in A, clockwise in B
c. clockwise in both A and B
d. anti-clockwise in both A and B

8 i. A system *S* consists of two coils *A* and *B*. The coil *A* carries a steady current *I* while the coil *B* is suspended near by as shown in the figure. Now, if the system is heated so as to raise the temperature of two coils steadily, then

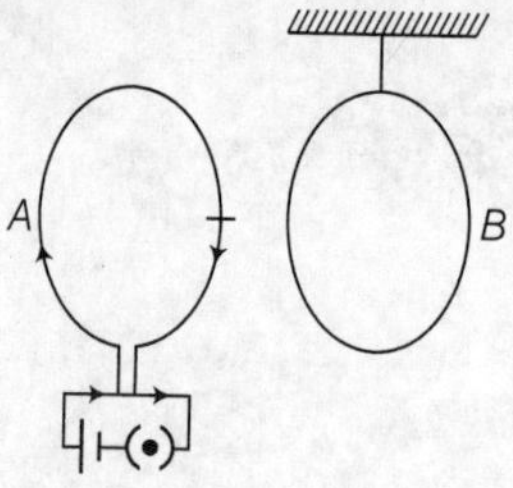

a. the two coils show attraction
b. the two coils show repulsion
c. there is no change in the position of the two coils
d. induced currents are not possible in coil *B*

ii. A square loop *PQRS* is carried away from a current carrying long straight conducting wire *CD*. The direction of induced current in the loop will be

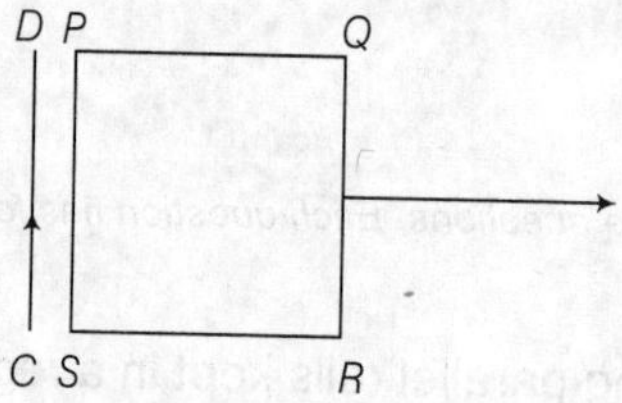

a. anti-clockwise
b. clockwise
c. sometimes clockwise sometimes anti-clockwise
d. current will not be induced

9 A metal sheet is placed in a variable magnetic field which is increasing from zero to maximum. Induced current flows in the directions as shown in the figure.

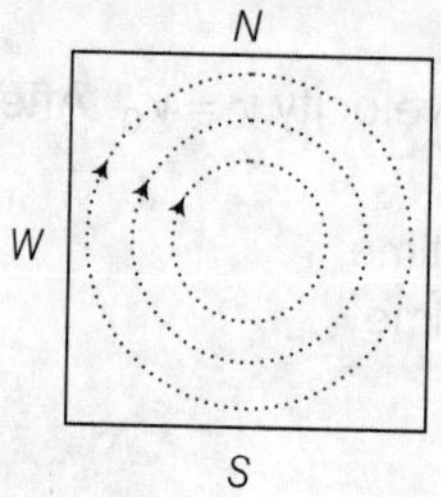

The direction of magnetic field will be
a. normal to the paper, inwards
b. normal to the paper, outwards
c. from East to West
d. from North to South

10 A copper ring having a cut such as not to form a complete loop is held horizontally and a bar magnet is dropped through the ring with its length along the axis of the ring. Then, acceleration of the falling magnet is (neglect air friction)

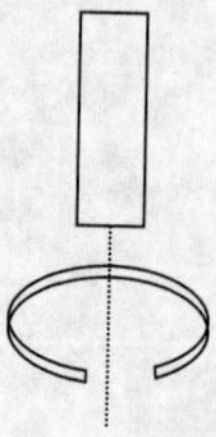

a. *g*
b. less than *g*
c. more than *g*
d. 0

Working Space

11 Two identical circular loops of metal wire are lying on a table without touching each other. Loop A carries a current which increases with time. In response, the loop B **[IIT JEE 1999]**

a. remains stationary
b. is attracted by the loop A
c. is repelled by the loop A
d. rotates about its CM, with CM fixed

12 As shown in the figure, P and Q are two coaxial conducting loops separated by some distance. When the switch S is closed, a clockwise current I_P flows in P (as seen by E) and an induced current I_Q, flows in Q. The switch remains closed for a long time. When S is opened, a current flows in Q. Then, the directions (as seen by E) are **[IIT JEE 2002]**

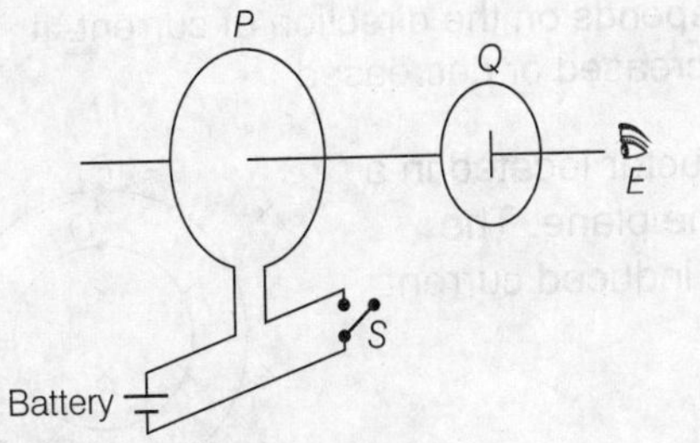

a. respectively clockwise and anti-clockwise
b. both clockwise
c. both anti-clockwise
d. respectively anti-clockwise and clockwise

13 The figure shows certain wire segments joined together to form a coplanar loop. The loop is placed in a perpendicular magnetic field in the direction going into the plane of the figure. The magnitude of the field increases with time. I_1 and I_2 are the currents in the segments ab and cd. Then, **[IIT JEE 2009]**

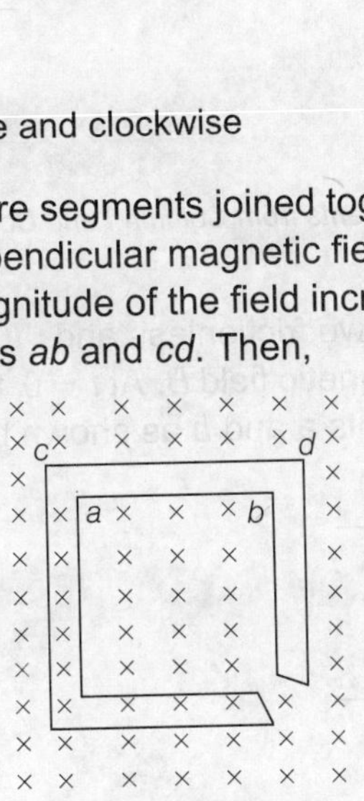

a. $I_1 > I_2$
b. $I_1 < I_2$
c. I_1 is the direction ba and I_2 is in the direction cd
d. I_1 is the direction ab and I_2 is in the direction dc

One or More than One Options Correct Type

Direction *(Q. Nos. 14-16) This section contains 3 multiple choice questions. Each question has four choices (a), (b), (c) and (d), out of which ONE or MORE THAN ONE are correct.*

14 A bar magnet is moved along the axis of copper ring placed far away from the magnet. Looking from the side of the magnet, an anti-clockwise current is found to be induced in the ring. Which of the following may be true?

a. The South pole faces the ring and the magnet moves towards it
b. The North pole faces the ring and the magnet moves towards it
c. The South pole faces the ring and the magnet moves away from it
d. The North pole faces the ring and the magnet moves away from it

Working Space

15 Two circular coils A and B are facing each other as shown in the figure. The current i through A can be altered, when

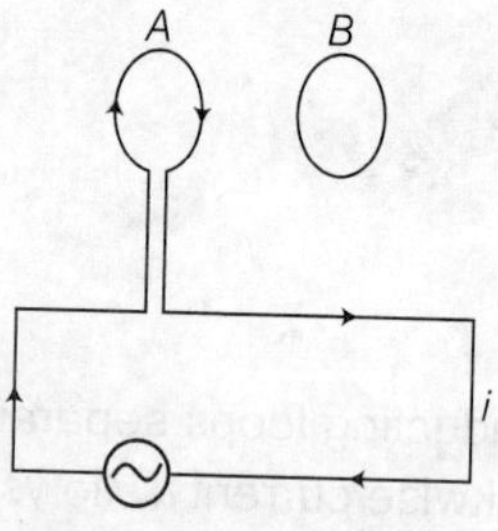

a. there will be repulsion between A and B if i is increased
b. there will be attraction between A and B if i is decreased
c. there will be neither attraction nor repulsion when i is changed
d. attraction or repulsion between A and B depends on the direction of current. It does not depend whether the current is increased or decreased

16 Figure shown, plane figure made of a conductor located in a magnetic field along the inward normal to the plane. The magnetic field starts diminishing. Then, the induced current

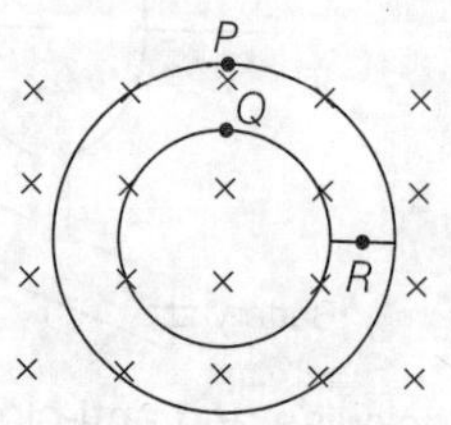

a. at point P is clockwise
b. at point Q is anti-clockwise
c. at point Q is clockwise
d. at point R is zero

Matching List Type

***Direction** (Q. No. 17) Choices for the correct combination of elements from Column I and Column II are given as options (a), (b), (c) and (d), out of which one is correct.*

17 A conducting rod (mass m, resistance R) rests on two frictionless and resistanceless rails distant l apart in a uniform magnetic field B. At $t = 0$, the rod is at rest and a source of emf E_0 is connected to points a and b as shown below.

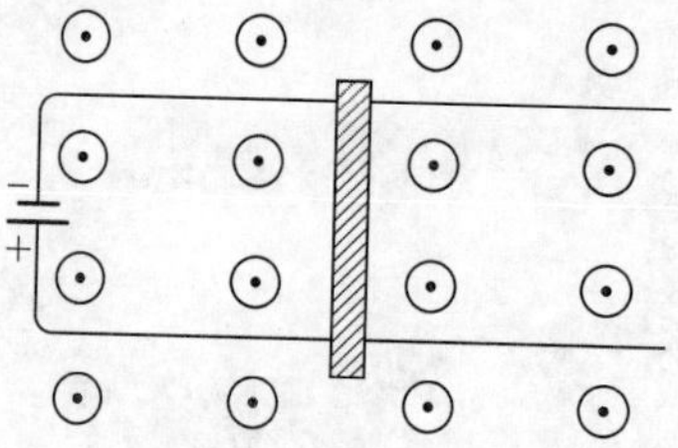

Now, match the following columns and select the correct option from the codes given below.

	Column I		Column II
i.	Speed of rod at time t when source gives a constant current I.	p.	∞
ii.	Speed of rod at time t when source gives a constant emf E_0.	q.	$\dfrac{E_0}{Bl}$
iii.	Terminal speed of rod when source gives a constant current I.	r.	$\dfrac{E_0}{Bl}\left(1-e^{-\frac{B^2l^2t}{mR}}\right)$
iv.	Terminal speed of rod when source gives a constant emf E_0.	s.	$\dfrac{E_0\, lBt}{mR}$

Codes

	i	ii	iii	iv		i	ii	iii	iv
a.	p	q	r	s	b.	q	r	s	p
c.	r	s	p	q	d.	s	r	p	q

One Integer Value Correct Type

Direction *(Q. Nos. 18 and 19) This section contains 2 questions. When worked out will result in an integer from 0 to 9 (both inclusive).*

18 Armature windings of a DC motor has a resistance of 50 Ω. The motor is connected to a 120 V line and when the motor reaches full speed against normal load, the back emf is 108 V.
The ratio of current in the circuit at $t = 0$ and at $t = \infty$ is found to be $5k$. Find the value of k.

19 A square wire frame $ABCD$ of side 1 m is placed in a region of changing magnetic field whose magnitude is decreasing at a constant rate of 1 T/s.

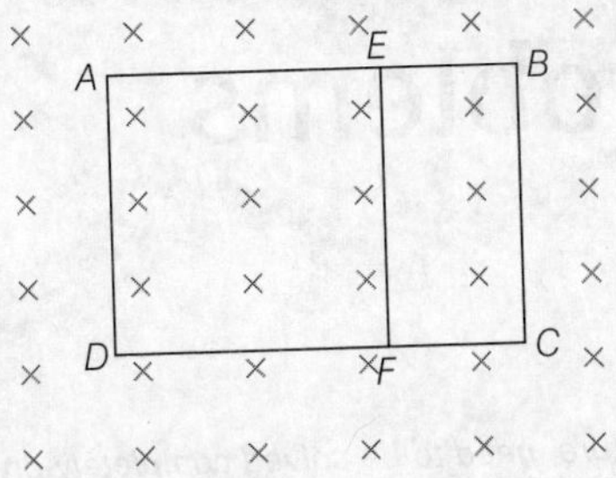

Wire segment EF is connected such that $EB = FC = 0.5$ m. Resistance per metre length of wire is 1 Ω/m. Current in segment EF is $\frac{1}{11p}$ A. Find the value of p.

Statement Type

Direction *(Q. Nos. 20 and 21) This section is based on Statement I and Statement II. Select the correct answer from the codes given below.*

a. Both Statement I and Statement II are correct and Statement II is the correct explanation of Statement I
b. Both Statement I and Statement II are correct but Statement II is not the correct explanation of Statement I
c. Statement I is correct but Statement II is incorrect
d. Statement II is correct but Statement I is incorrect

20 **Statement I** When a rod AB is moved in region of a uniform field as shown in the figure.

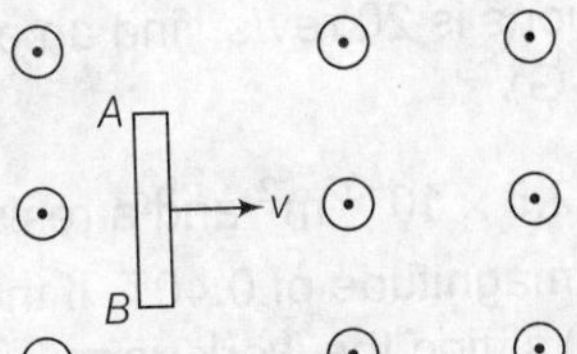

Then end A becomes negatively charged.
Statement II Force on a charged particle is given by

$$\mathbf{F} = q(\mathbf{v} \times \mathbf{B})$$

21 **Statement I** When a disc is rotated counter clockwise in a magnetic field (whose points perpendicularly into the plane of disc), rim of disc is more negative with respect to centre of disc.
Statement II Induced emf always opposes the charge which produces it.

DPP-3 Induced EMF in a Moving Rod in Uniform Magnetic Field and Circuit Problems

Subjective Questions

Direction (Q. Nos. 1-4) *These questions are subjective in nature, need to be solved completely on notebook.*

Working Space

1 A conducting rod of length l slides at constant velocity v on two parallel conducting rails, placed in a uniform and constant magnetic field B perpendicular to the plane of the rail as shown in the figure. A resistance R is connected between the ends of the rail.

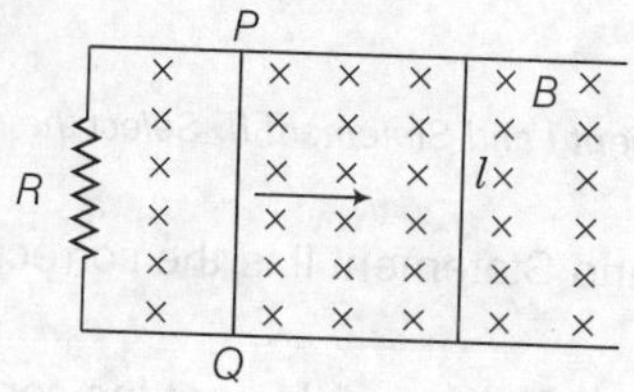

i. Identify the cause which produces changes in the magnetic flux.

ii. Identify the direction of current in the loop.

iii. Determine the emf induced in the loop.

iv. Compute the electric power dissipated in the resistor.

2 A metal rod 1.5 m long rotates about its one end in a vertical plane at right angles to the magnetic meridian. If the frequency of rotations is 20 rev/s, find an emf induced between the ends of the rods (B_H = 0.32 G).

3 In figure, a square loop has 100 turns, an area of 2.5×10^{-3} m^2 and a resistance of 100 Ω. The perpendicular magnetic field has a magnitude of 0.40T. If the loop is slowly and uniformly pulled out of the field in 1.0 s, find the work done.

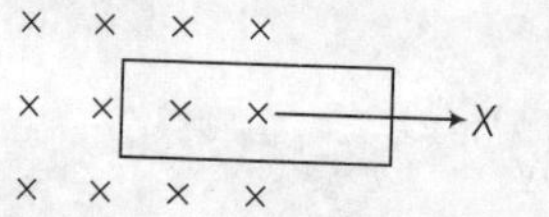

4 A wire in the form of a semicircle of radius r. One end is attached to an axis about which rotates with an angular speed ω. The axis is normal to the plane of the semicircle. The wire immersed in a uniform magnetic field B parallel to the axis. Find the induced emf between points O and P of the semicircle.

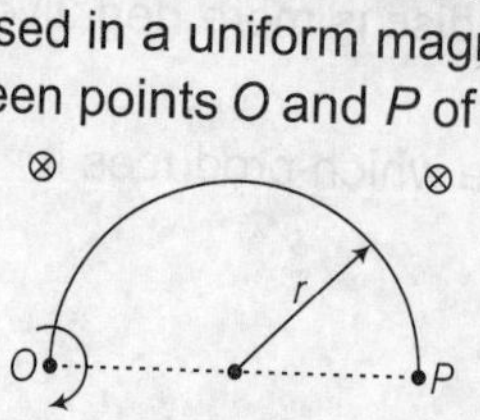

Only One Option Correct Type

Direction (Q. Nos. 5-12) *This section contains 8 multiple choice questions. Each question has four choices (a), (b), (c) and (d) out of which ONLY ONE is correct.*

5 A conducting square loop of side L and resistance R moves in its plane with a uniform velocity v perpendicular to one of its sides. A magnetic induction B, constant in time and space pointing perpendicular and into the plane of the loop exists everywhere.

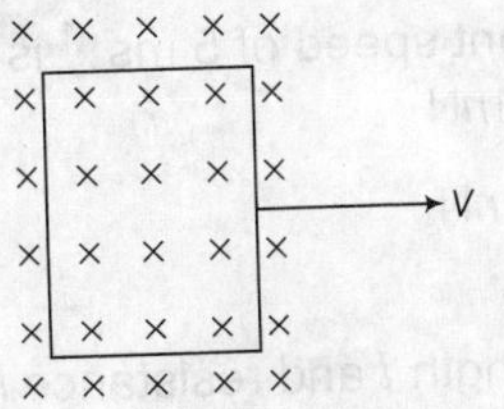

The current induced in the loop is

a. BvL/R, clockwise
b. $\frac{BvL}{R}$, anti-clockwise
c. $\frac{2BvL}{R}$, anti-clockwise
d. zero

6 A thin wire of length of 2m is perpendicular to the XY-plane. It is moved with velocity $\mathbf{v} = (2\hat{\mathbf{i}} + 3\hat{\mathbf{j}} + \hat{\mathbf{k}})\ \text{ms}^{-1}$ through a region of magnetic induction $\mathbf{B} = (\hat{\mathbf{i}} + 2\hat{\mathbf{j}})\ \text{Wb/m}^2$. Then, potential difference induced between the ends of the wire

a. 2 V
b. 4 V
c. 0 V
d. None of these

7 A long metal bar of 30 cm length is aligned along a North-South line and moves Eastward at a speed of 10 m/s. A uniform magnetic field of 4.0 T points vertically downwards. If the South end of the bar has potential of 0 V, then induced potential at the North end of the bar is

a. + 12 V
b. − 12 V
c. 0 V
d. None of these

8 A square metal of loop $PQRS$ of side 10 cm and resistance 1 Ω is moved with a constant velocity partly inside a magnetic field of 2 Wbm^{-2}, directed into the paper, as shown in the figure. This loop is connected to a network $ABCD$ of five resistors each of value 3Ω. If a steady current of 1 mA flows in the loop, then the speed of the loop is

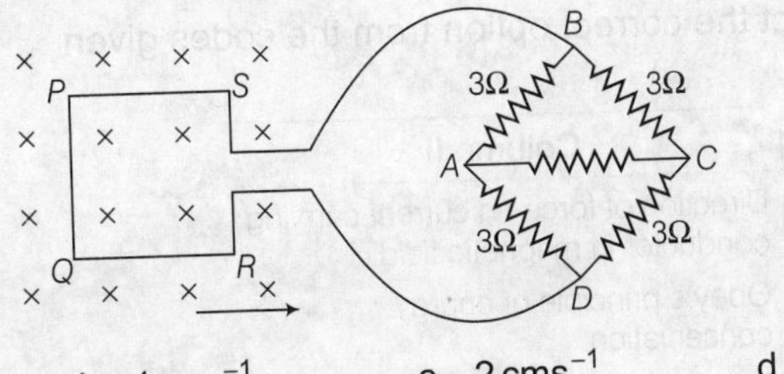

a. 0.5 cms^{-1}
b. 1 cms^{-1}
c. 2 cms^{-1}
d. 4 cms^{-1}

9 A metallic square loop $ABCD$ is moving in its own plane with velocity v in a uniform magnetic field perpendicular to its plane as shown in the figure.

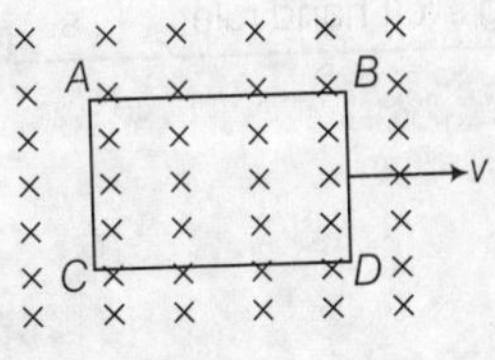

Electric field is induced

a. in AD, but not in BC
b. in BC, but not in AD
c. neither in AD nor in BC
d. in both AD and BC

Working Space

10 A rod of length b is moved at a constant speed v along a horizontal conducting rails. A magnetic field exists in the region which is provided by a long wire as shown in the figure. Rails have negligible resistance.

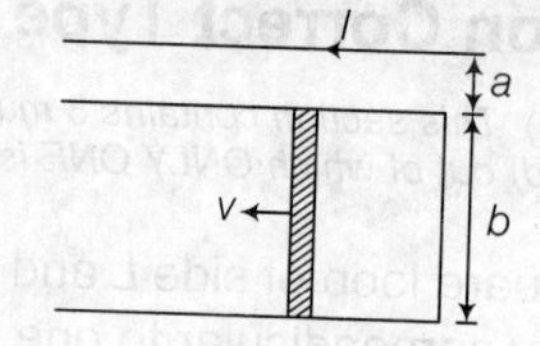

Given, $I = 100$ A, $v = 5\ \text{ms}^{-1}$

$a = 10$ mm, $b = 10$ cm

Resistance of rod = $0.4\ \Omega$

Force required to move rod on the rails with constant speed of $5\ \text{ms}^{-1}$ is

a. $12\,n$N
b. $12 \log 11\,n$N
c. $12 \log\left(\frac{11}{10}\right) n$N
d. $12 \log 2\,n$N

11 A rectangular loop has a sliding connector PQ of length l and resistance R and is moving with a speed v as shown in the figure. The set up is placed in a uniform field going into the plane of the paper.

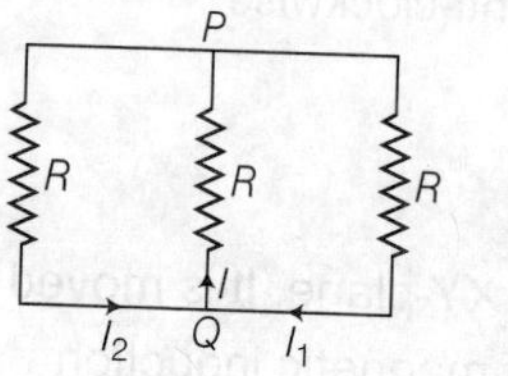

The three currents are **[AIEEE 2010]**

a. $I_1 = I_2 = \frac{Blv}{6R}; I = \frac{Blv}{3R}$
b. $I_1 = -I_2 = \frac{Blv}{R}; I = \frac{2Blv}{R}$
c. $I_1 = I_2 = \frac{Blv}{3R}; I = \frac{2Blv}{3R}$
d. $I_1 = I_2 = I = \frac{Blv}{R}$

12 An infinitely long cylinder is kept parallel to a uniform magnetic field B directed along the positive Z-axis. The direction of induced current as seen from the Z-axis will be **[IIT JEE 2005]**

a. clockwise of the positive Z-axis
b. anti-clockwise of the positive Z-axis
c. zero
d. along the magnetic field

Matching List Type

Direction *(Q. No. 13) Choices for the correct combination of elements from Column I and Column II are given as options (a), (b), (c) and (d), out of which one is correct.*

13 Match the following columns and select the correct option from the codes given below.

	Column I		Column II
i.	Faraday's law	p.	Direction of force on current carrying conductor in magnetic field.
ii.	Lenz's law	q.	Obey's principle of energy conservation
iii.	Fleming's right hand rule	r.	Direction of induced current.
iv.	Fleming's left hand rule	s.	EMI

Codes

	i	ii	iii	iv
a.	s	q	r	p
b.	p	q	r	s
c.	s	r	q	p
d.	q	r	p	s

One or More than One Options Correct Type

Direction *(Q. Nos. 14-16) This section contains 3 multiple choice questions. Each question has four choices (a), (b), (c) and (d), out of which ONE or MORE THAN ONE are correct.*

14 A field line is shown in the figure. This field cannot represents **[IIT JEE 2006]**

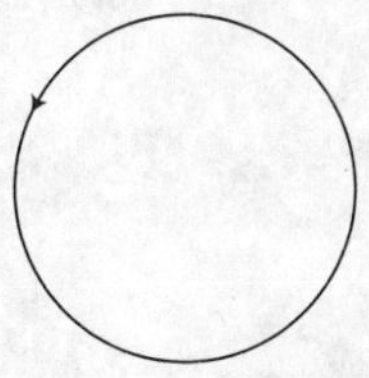

a. magnetostatic field
b. electrostatic field
c. induced electric field
d. gravitational field

15 Consider the situation shown in the figure. The wire PQ has a negligible resistance and is made to slide on the three rails with a constant speed 5 cm/s. The current in the 10 Ω resistor. Mark the correct option(s).

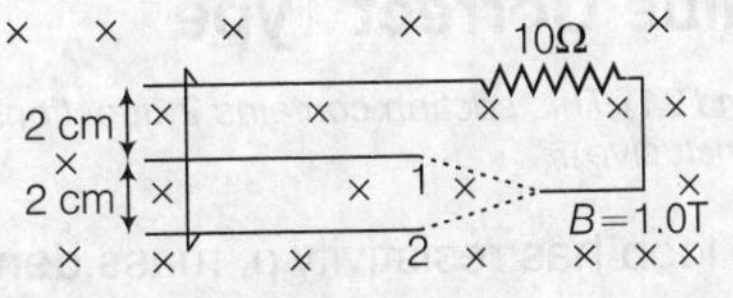

a. In position 1, $I = 0.1$ mA
b. In position 2, $I = 0.1$ mA
c. In position 1, $I = 0.2$ mA
d. In position 2, $I = 0.2$ mA

16 A conducting rod of length l is moved at constant velocity v_0 on two parallel, conducting, smooth, fixed rails, which are placed in a uniform constant magnetic field B perpendicular to the plane of rails as shown in the figure. A resistance R is connected between the two ends of the rail. Then, which of the following is/are correct?

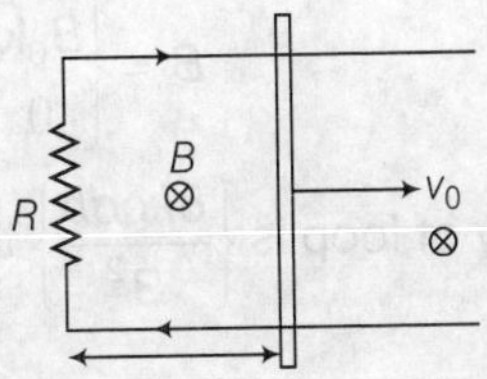

a. The thermal power dissipated in the resistor is equal to the rate of work done by an external person pulling the rod
b. If applied external force is doubled, then a part of the external power increases the velocity of the rod
c. Lenz's law is not satisfied if the rod is accelerated by external force
d. If resistance R is doubled, then power required to maintain the constant velocity v_0 becomes half

Comprehension Type

Direction *(Q. Nos. 17-19) This section contains a paragraph, which describing theory, experiments, data, etc. Three questions related to the paragraph have been given. Each question has only one correct answer among the four given options (a), (b), (c) and (d).*

Passage

In figure shown, the rod has a resistance R, the horizontal rails have negligible friction. A battery of emf E and negligible internal resistance is connected between points a and b. The rod is initially at rest.

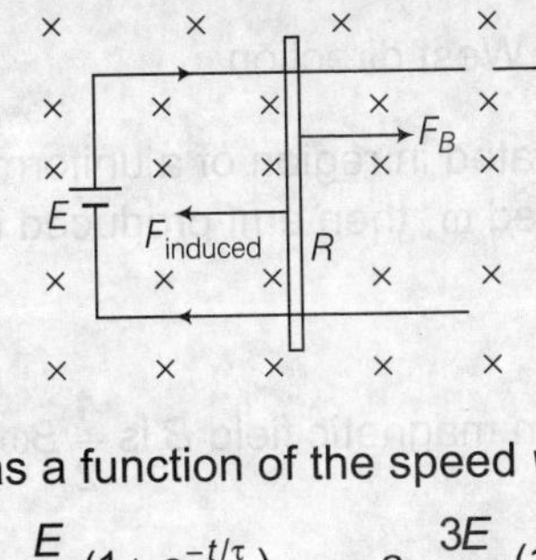

17 The force on the rod as a function of the speed v (where, $\tau = mR/B^2l^2$) is

a. $\frac{E}{Bl}(1-e^{-t/\tau})$
b. $\frac{E}{Bl}(1+e^{-t/\tau})$
c. $\frac{3E}{2Bl}(1-e^{-t/\tau})$
d. $\frac{E}{2Bl}(1-e^{-t/\tau})$

18 After sometime, the rod will approach a terminal speed. Find an expression for it.

a. $\frac{3E}{2Bl}$
b. $\frac{E}{2Bl}$
c. $\frac{E}{Bl}$
d. $\frac{2E}{Bl}$

19 The current when the rod attains its terminal speed is

a. $\frac{2E}{R}$
b. $\frac{E}{R}$
c. $\frac{3E}{2R}$
d. $\frac{E}{2R}$

One Integer Value Correct Type

Direction *(Q. Nos. 20 and 21) This section contains 2 questions. When worked out will result in an integer from 0 to 9 (both inclusive).*

20 A vertical square loop has resistivity ρ, mass density σ, diameter of wire d and side length l is initially at rest with its lower horizontal side at $y = a$. It is then pulled along negative y-direction with acceleration of b ms^{-2}. A magnetic field exists in the region given by

$$B = \begin{cases} B_0\hat{k}, & y > a \\ 0, & y \le a \end{cases}$$

If terminal velocity of loop is $\left\{\frac{8k\rho\sigma b}{B^2}\right\}$ then, find the value of k.

21 A conducting rod of length 1 m moves with a speed of 2 m/s parallel to a straight long wire carrying 4 A current. Axis of rod is kept perpendicular to the wire with its near end at a distance of 1 m from the wire. Magnitude of induced emf in rod is $N \times 10^{-7}$ log 16 volt. Find the value of N.

Statement Type

Direction *(Q. Nos. 22 and 23) This section is based on Statement I and Statement II. Select the correct answer from the codes given below.*

a. Both Statement I and Statement II are correct and Statement II is the correct explanation of Statement I
b. Both Statement I and Statement II are correct but Statement II is not the correct explanation of Statement I
c. Statement I is correct but Statement II is incorrect
d. Statement II is correct but Statement I is incorrect

22 **Statement I** Emf induced between wings tips of an aeroplane cannot be used to power a light bulb.

Statement II Aeroplane may not be moving in East to West direction.

23 **Statement I** When a metal wheel with N-spokes is rotated in region of a uniform and perpendicular magnetic field B with angular speed ω, then emf produced is $\frac{1}{2}NB\omega R^2$, where R is radius of wheel.

Statement II Induced emf in a rotating rod of length l in magnetic field B is $\frac{1}{2}B\omega l^2$.

Working Space

DPP-4 EMF Induced in a Rod or Loop in Non-Uniform Magnetic Field

Subjective Questions

Working Space

Direction (Q. Nos. 1-4) *These questions are subjective in nature, need to be solved completely on notebook.*

1 A rod of mass m and resistance r is placed on fixed, smooth conducting and resistanceless rails which are closed by a resistance R.

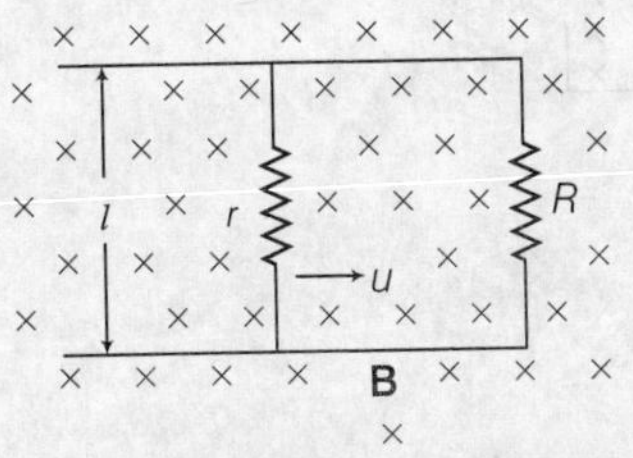

Let a uniform downward magnetic field exists in the region and let rod is projected with velocity u towards closed end of rails. Find velocity of rod, after t second of projection.

2 Two fixed long straight wires carry the same current i in opposite directions as shown in the figure below. A square loop of side b is fixed in the plane of the wires with its length parallel to one wire at a distance a.

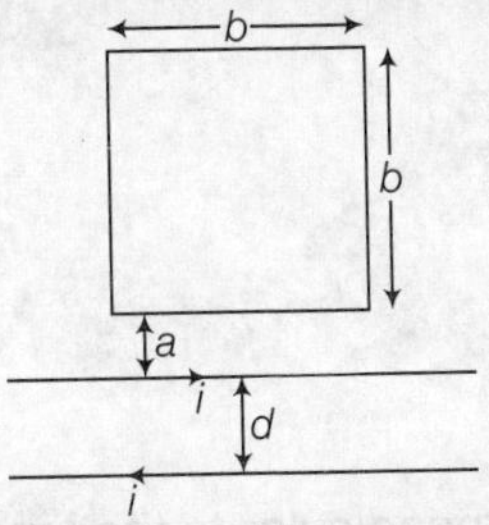

i. Calculate the induced emf in the loop if the current in both the wires is changing at the rate di/dt.

ii. What is the direction of force on the loop if di/dt is positive?

3 A current of 10 A flowing in a long straight wire situated near a rectangular circuit whose two sides of length 0.2 m are parallel to the wire. One of them is at a distance of 0.05 m and the other at a distance of 0.10 m from the wire.

i. Find the magnetic flux through the rectangular circuit.

ii. If the current decays uniformly to zero in 0.02 s. Find the emf induced in the circuit.

iii. Indicate the direction in which the induced current flows.

4 An infinite wire carries a current I. An S shaped conducting rod OB of two semicircles each of radius r is placed at an angle θ to the wire. The centre of the conductor is at a distance d from the wire. If the rod translates parallel to the wire with a velocity v. Calculate the emf induced across the ends of the rod.

Working Space

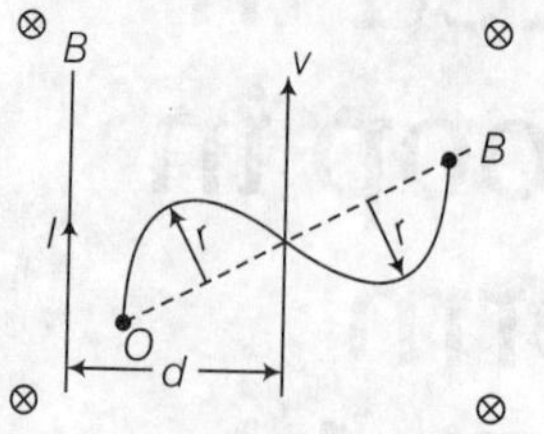

Only One Option Correct Type

Direction (Q. Nos. 5-15) *This section contains 11 multiple choice questions. Each question has four choices (a), (b), (c) and (d) out of which ONLY ONE is correct.*

5 A metallic ring connected to a rod oscillates freely like a pendulum. If now a magnetic field is applied in horizontal direction so that the pendulum now swings through the field, then pendulum will be

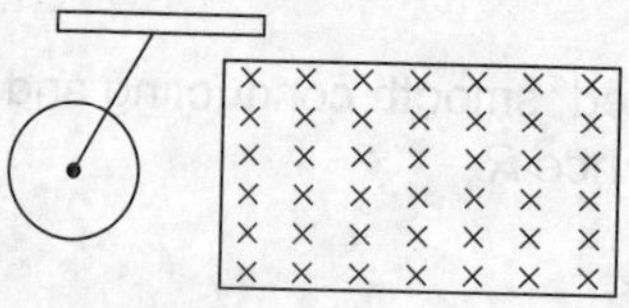

a. keep oscillating with a smaller time period
b. come to rest very soon
c. keep oscillating with a larger time period
d. keep oscillating with the new time period

6 A square loop of side a placed in the same plane as a long straight wire carrying a current i. The centre of the loop is at a distance r from wire, where $r >> a_1$ (see figure). The loop is moved away from the wire with a constant velocity v. The induced emf in the loop is

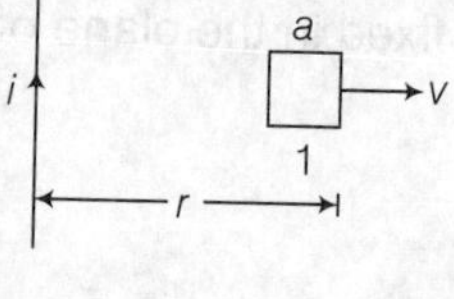

a. $\frac{\mu_0 iav}{2\pi r}$

b. $\mu_0 ia^3 v$

c. $\frac{\mu_0 iv}{2\pi}$

d. $\frac{\mu_0 ia^2 v}{2\pi r^2}$

7 A conducting rod moves with constant velocity v perpendicular to the long, straight wire carrying a current I as shown in the figure. The emf generated between the ends of the rod is

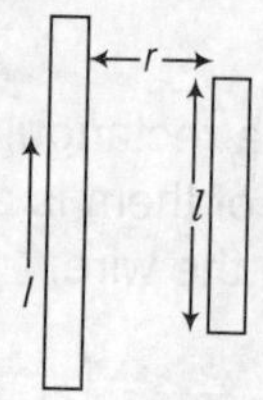

a. $\frac{\mu_0 vIl}{\pi r}$

b. $\frac{\mu_0 vIl}{2\pi r}$

c. $\frac{2\mu_0 vIl}{\pi r}$

d. $\frac{\mu_0 vIl}{4\pi r}$

8 A square loop of side 2 cm is placed in region of a magnetic field, directed out of plane of paper, with magnitude $B = 4t^2 y$, where y is measured in meters with respect to the coordinate axes as shown in the figure below:

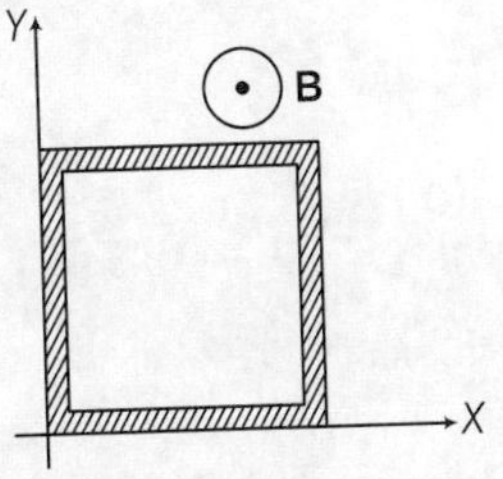

Induced emf in the loop at t = 2.5s is

a. 8×10^{-5}V anti-clockwise
b. 8×10^{-5}V clockwise
c. 4×10^{-3}V anti-clockwise
d. 4×10^{-3}V clockwise

9 The magnetic field in a region is given by $\mathbf{B} = \frac{B_0}{L} y \hat{\mathbf{k}}$, where L is a fixed length. A conducting rod of length L lies along the Y-axis between the origin and the point (0, L, 0). If the rod moves with a velocity $v = v_0 \hat{\mathbf{i}}$, then emf induced between the ends of the rod is

a. $\frac{B_0 v_0 L}{2}$
b. $\frac{2B_0 v_0 L}{3}$
c. $\frac{B_0 v_0 L}{4}$
d. $B_0 v_0 L$

10 A long straight wire is parallel to one edge as in figure. If the current in the long wire is varies in time as $I = I_0 e^{-t/\tau}$, what will be the induced emf in the loop?

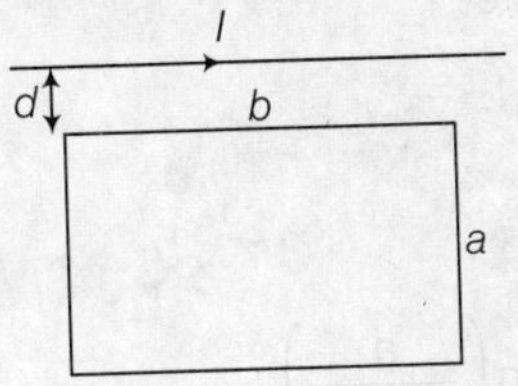

a. $\frac{\mu_0 b I_0}{\pi\tau} \ln\left(\frac{d+a}{d}\right) e^{-t/\tau}$
b. $\frac{\mu_0 b I_0}{2\pi\tau} \ln\left(\frac{d+a}{d}\right) e^{-t/\tau}$
c. $\frac{2\mu_0 b I_0}{\pi\tau} \ln\left(\frac{d+a}{d}\right) e^{-t/\tau}$
d. $\frac{\mu_0 b I_0}{\pi\tau} \ln\left(\frac{d}{d+a}\right) e^{-t/\tau}$

11 A conducting rod of length l moves with velocity v, a direction to a long wire carrying a steady current I. The axis of the rod is maintained perpendicular to the wire with near end a distance r away as shown in the figure. Find the emf induced in the rod.

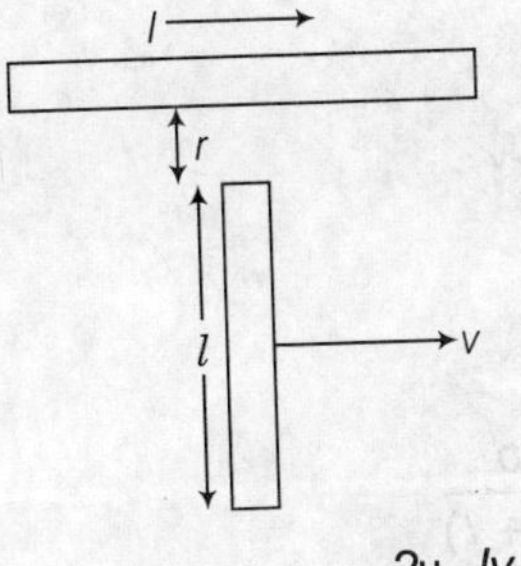

a. $\frac{\mu_0 I v}{\pi} \ln\left(\frac{r+l}{r}\right)$
b. $\frac{2\mu_0 I v}{\pi} \ln\left(\frac{r+l}{r}\right)$
c. $\frac{\mu_0 I v}{\pi} \ln\left(\frac{l}{r+l}\right)$
d. $\frac{\mu_0 I v}{2\pi} \ln\left(\frac{r+l}{r}\right)$

12 A conducting rod slides on a pair of thick metallic rails laid parallel to an infinitely long fixed wire carrying a constant current i. The centre of the rod is at a distance x from the wire. The ends of the rails are connected by resistor of resistance R.

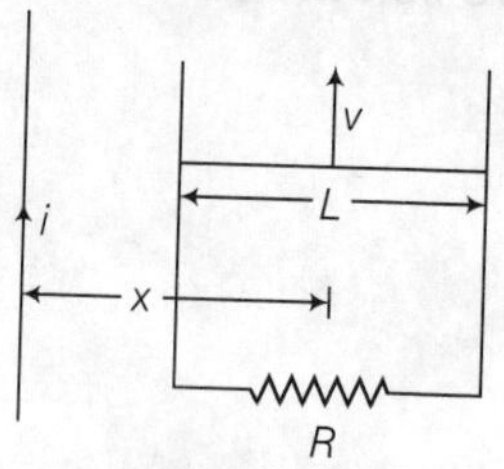

The force needed to keep the rod sliding at a constant speed v is

a. $\frac{v}{R}\left(\frac{\mu_0 i}{2\pi}\ln\frac{2x+l}{2x-l}\right)^2$
b. $\frac{v}{R}\left(\frac{\mu_0 i}{\pi}\ln\frac{2x+l}{2x-l}\right)^2$
c. $\frac{v}{R}\left(\frac{\mu_0 i}{2\pi}\ln\frac{x+l}{x-l}\right)^2$
d. $\frac{v}{R}\left(\frac{\mu_0 i}{\pi}\ln\frac{x+l}{x-l}\right)^2$

13 In previous question, what is the current in R in this situation?

a. $\frac{\mu_0 iv}{2\pi R}\ln\left(\frac{2x+l}{2x-l}\right)$
b. $\frac{\mu_0 iv}{\pi R}\ln\left(\frac{2x+l}{2x-l}\right)$
c. $\frac{\mu_0 i}{2\pi R}\ln\left(\frac{x+l}{x-l}\right)$
d. $\frac{\mu_0 i}{\pi R}\ln\left(\frac{x+l}{x-l}\right)$

14 For the situation shown in the figure, flux through the square loop is

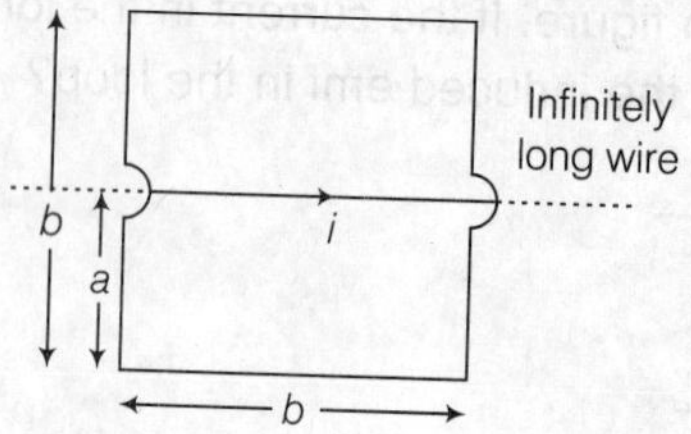

a. $\left(\frac{\mu_0 ia}{2\pi}\right)\ln\left(\frac{a}{2a-b}\right)$
b. $\left(\frac{\mu_0 ib}{2\pi}\right)\ln\left(\frac{a}{a-2b}\right)$
c. $\left(\frac{\mu_0 ib}{2\pi}\right)\ln\left(\frac{a}{b-a}\right)$
d. $\left(\frac{\mu_0 ia}{2\pi}\right)\ln\left(\frac{2a}{b-a}\right)$

15 A rectangular metallic loop of length l and width b is placed coplanarly with a long wire carrying a current i (see figure). The loop is moved perpendicularly to the wire with a speed v in the plane containing the wire and loop. The emf induced in the loop when the rear end is at a distance a from the wire is

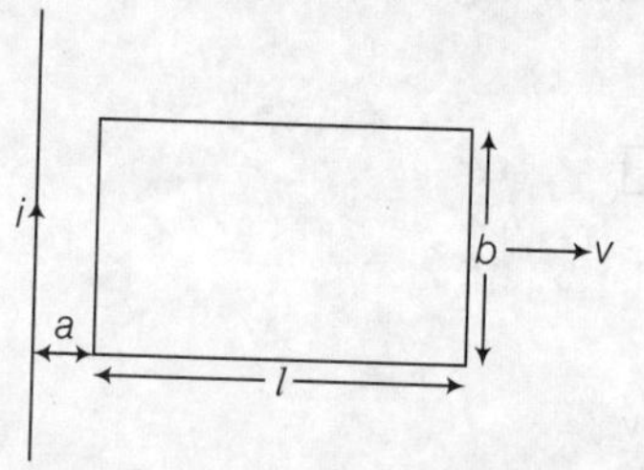

a. $\frac{\mu_0 ilvb}{2\pi a(a+2l)}$
b. $\frac{\mu_0 ilvb}{2\pi a(a+l)}$
c. $\frac{\mu_0 ilvb}{2\pi a(2a+l)}$
d. $\frac{\mu_0 iavb}{2\pi l(a+l)}$

Matching List Type

Direction (Q. No. 16) *Choices for the correct combination of elements from Column I and Column II are given as options (a), (b), (c) and (d), out of which one is correct.*

16 A square loop is placed near a long straight current carrying wire as shown in the figure.

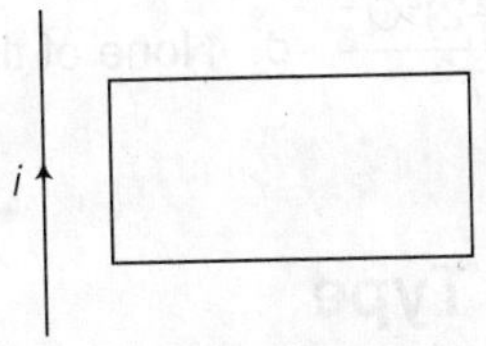

Match the following columns and select the correct option from the codes given below.

	Column I		Column II
i.	If current is increased	p.	Induced current in loop is clockwise
ii.	If current is decreased	q.	Induced current in loop is anti-clockwise
iii.	If loop is moved away from the wire	r.	Wire will attract the loop
iv.	If loop is moved toward the wire	s.	Wire will repel the loop

Codes

	i	ii	iii	iv
a.	q,s	p,r	p,r	q,s
b.	p,r	q,s	p,q	q,s
c.	p,q	r,s	s,p	q,s
d.	p,q	p,r	q,s	r,q

Comprehension Type

Direction (Q. Nos. 17-19) *This section contains a paragraph, which describing theory, experiments, data, etc. Three questions related to the paragraph have been given. Each question has only one correct answer among the four given options (a), (b), (c) and (d).*

Passage

A semicircular wire with *PQROP* has a resistance *R*. Its radius $r (OP = OQ = OR = r)$ with *POR* as the diameter at a distance *r* from a long parallel wire *AB* carrying a time current *I*, as shown in the figure. The current in wire *AB* decreases linearly from an initial value I_0 to zero in time *T*.

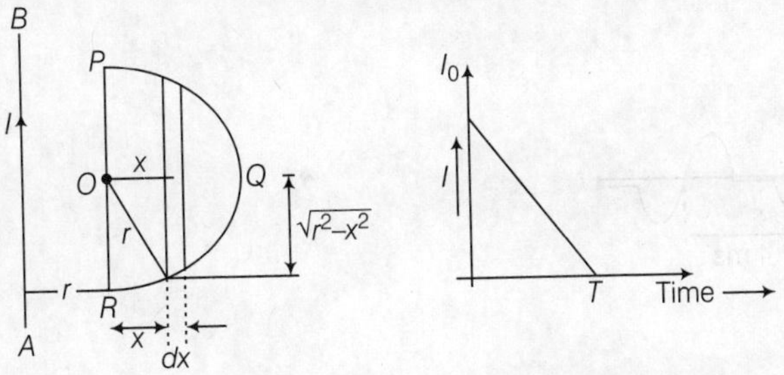

17 Find magnetic flux through *PQROP*, when current *I* in wire *AB* is

a. $\frac{\mu_0 IR}{2\pi}(\pi - 2)$

b. $\frac{2\mu_0 IR}{\pi}(\pi - 2)$

c. $\frac{\mu_0 IR}{\pi}(\pi - 2)$

d. None of these

18 Find induced emf in *PQROP* as a function time.

a. $\dfrac{\mu_0 r(\pi - 2)Q}{\pi RT^2}$ b. $\dfrac{\mu_0 r(\pi - 2)Q}{2\pi T^3}$ c. $\dfrac{2\mu_0 r(\pi - 2)^2 Q}{\pi RT^2}$ d. None of these

19 Find heat generated in *PQROP* in time *T*.

a. $\dfrac{\mu_0 r^2(\pi - 2)^2 Q^2}{\pi RT^2}$ b. $\dfrac{\mu_0^2 r^2(\pi - 2)^2 Q^2}{\pi^2 RT^3}$ c. $\dfrac{\mu_0 r^2(\pi - 2)^2 Q^2}{\pi^2 RT^2}$ d. None of these

One or More than One Options Correct Type

Direction *(Q. Nos. 20-24) This section contains 5 multiple choice questions. Each question has four choices (a), (b), (c) and (d), out of which ONE or MORE THAN ONE are correct.*

20 In the circuit shown, $C = 2\mu F$, $R = 3\Omega$ and connecting wire is of negligible resistance. The circuit is placed in an spatially uniform magnetic field pointing downwards into the plane of paper.

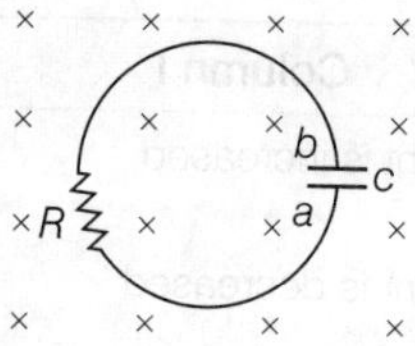

Potential difference across the capacitor is observed to increase with time as

$$V_b - V_a = V_{ba} = V_0(1 - e^{-t/\tau})$$

where, V_0 and τ are positive constants. Now, choose the correct option(s).

a. Magnetic field intensity is decreasing at a rate of $\dfrac{V_0}{\pi r^2}$, r = radius of loop

b. Magnetic field intensity is increasing at a rate of $\dfrac{V_0}{\pi r^2}$, r = radius of loop

c. Current in the circuit is $\dfrac{V_0}{R} e^{-t/\tau}$

d. Current in the circuit is $C\dfrac{V_0}{\tau} e^{-t/\tau}$

21 A small magnet is moved over a straight line and it passes through 2 coils separated by a distance of 1.5 m. If anti-clockwise current is taken positive, then choose the correct option(s).

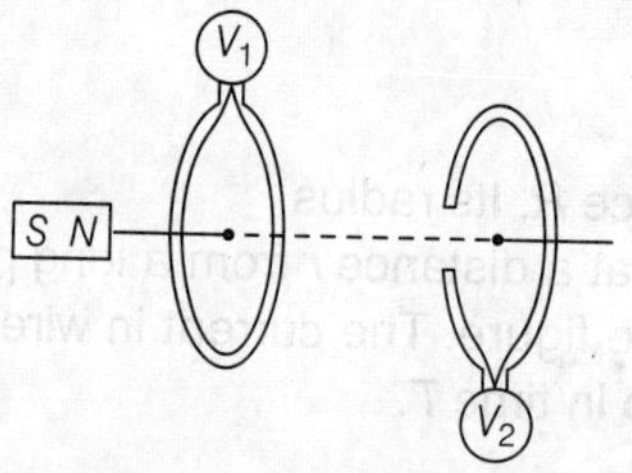

a. Voltage pulses produced are as

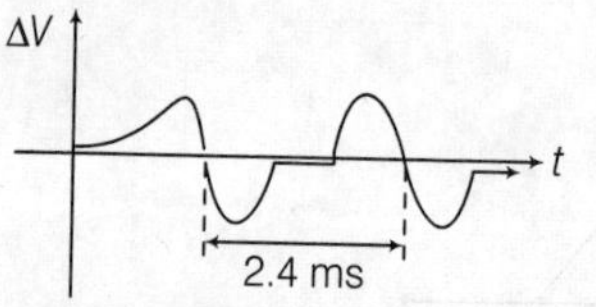

b. Voltage pulses produced are as

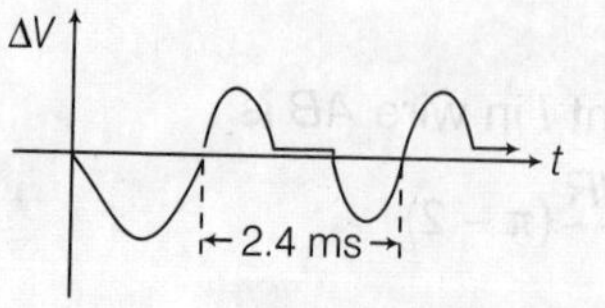

c. Speed of magnet is 625 ms^{-1}

d. Speed of magnet is $\dfrac{5}{8}$ ms^{-1}

22 A rolling axle 1.5 m long is pushed along horizontal metallic rails at a constant speed of $v = 3$ m/s. A resistor $R = 0.4\,\Omega$ is connected between rails and a magnetic field $B = 0.08$ T is applied vertically downwards as shown in the figure.

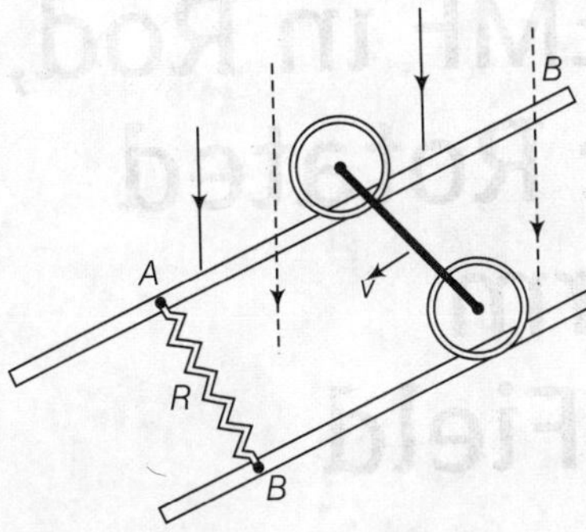

Which of these options are true?

a. Induced current in axle is 0.9 A
b. Point B is at lower potential
c. Force required to keep axle moving at constant speed is 0.018 N
d. When axle rolls beyond AB, then direction of current in resistor R is reversed

23 A circular conducting loop of radius r is kept perpendicular to magnetic field B. Field B varies with time as $B = a + bt$. Which of the following statements is/are correct?

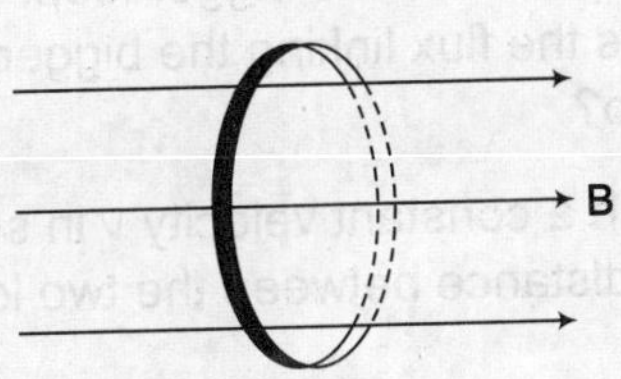

a. Flux linked with loop at $t = 0$ is zero
b. Induced emf in loop is πbr^2 volts
c. Loop is heated at rate of $\dfrac{\pi^2 b^2 r^4}{R}$
d. Face of loop facing the source of B have South polarity

24 A resistance R (Ω) and a capacitor C (F) is connected with a metal wire to form a loop and this is placed in a magnetic field that points vertically into the loop and is decreasing at a constant rate $\dfrac{dB}{dt} = -k$. Choose the correct statements.

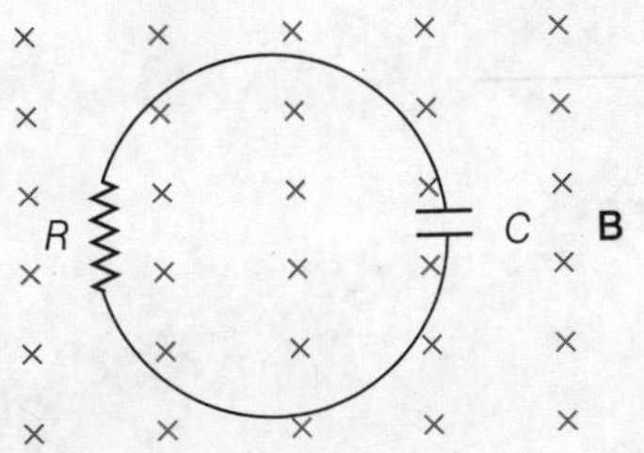

a. Induced emf is $\pi a^2 k$, where a = radius of loop
b. Charge stored in capacitor is $C\pi a^2 k$ coulombs
c. Upper plate of capacitor is negatively charged
d. Changing magnetic field induces an electric field that pushes charge

DPP-5 Induced EMF in Rod, Ring, Disc Rotated in a Uniform Magnetic Field

Subjective Questions

Direction (Q. Nos. 1-4) *These questions are subjective in nature, need to be solved completely on notebook.*

Working Space

1 A circular loop of radius 0.3 cm lies parallel to a much bigger circular loop of radius 20 cm. The centre of the small loop is on the axis of the bigger loop. The difference between their centres is 15 cm. What is the flux linking the bigger loop, if a current of 2.0 A flows through the smaller loop?

2 Loop A of radius $r << R$ moves towards loop B with a constant velocity v in such that their planes are always parallel. What is the distance between the two loops (x) when the induced emf in loop A is maximum?

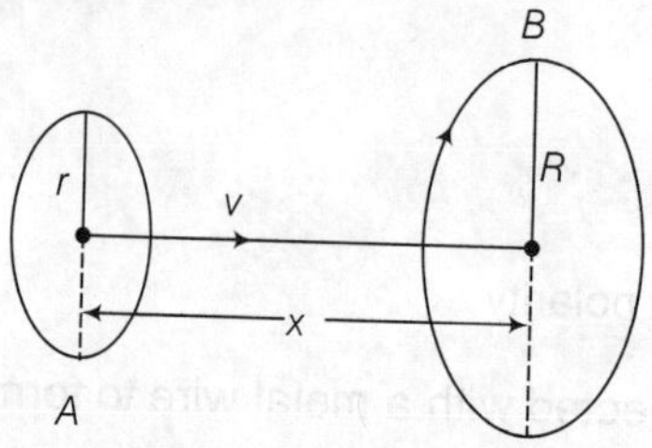

3 A copper rod moving with velocity v parallel to a long straight wire carrying current $I = 100$ A. Calculate the induced emf in the rod, where $v = 5$ m/s, $a = 1$cm, $b = 100$ cm.

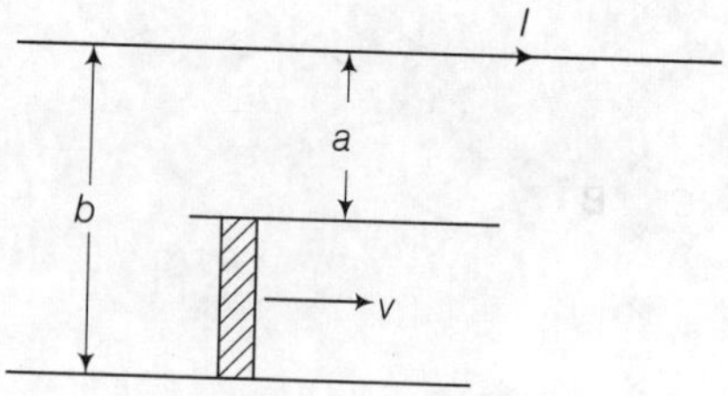

4 A stationary conducting loop in shape of letter e is placed in a uniform magnetic field directed out of plane page.

A straight conducting rod is pivoted at point O and rotates with a constant angular velocity of 2 rad/s.

Given, radius of circular portion of e shaped base $a = 50$ cm

Resistance of conductor = 5Ω / m, magnitude of field = 0.5T

i. Determine induced emf in the loop POQ.

ii. Induced current in loop POQ when $t = 0.25$ s.

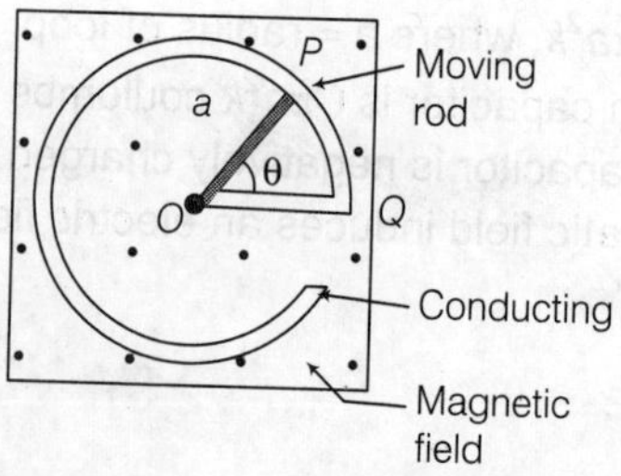

Only One Option Correct Type

Direction *(Q. Nos. 5-13) This section contains 9 multiple choice questions. Each question has four choices (a), (b), (c) and (d), out of which ONLY ONE is correct.*

5 A wheel with N spokes is rotated in a plane perpendicular to the magnetic field of earth such that an emf e is induced between axle and rim of wheel. In the same wheel, number of spokes is made $3N$ and is rotated in the same manner in the same field, then new emf is

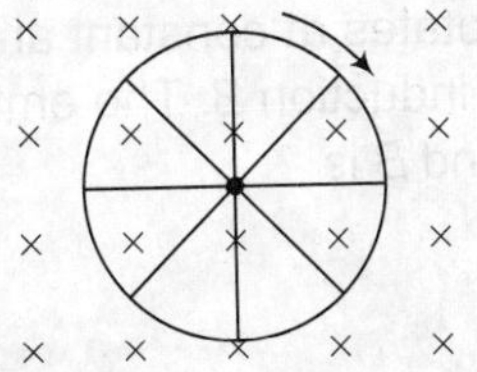

a. e b. $\frac{3}{2}e$

c. $\frac{e}{3}$ d. e

6 A rod of length 10 cm made up of conducting and non-conducting material (shaded part is non-conducting). The rod is rotated with constant angular velocity ω rad/s about point O, in a constant magnetic field 2 T as shown in the figure. The induced emf between the point A and B of rod will be

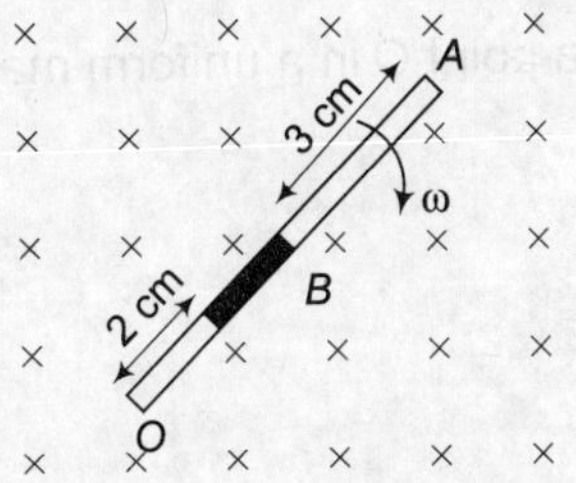

a. 0.029 V b. 0.1 V

c. 0.051 V d. 0.064 V

7 A metal conductor of length 1 m rotates vertically about one of its ends at constant angular velocity 5 rad/s. If the horizontal component of earth's magnetic field is 0.2×10^{-4} T, then the emf developed between the two ends of the conductor is **[AIEEE 2004]**

a. 5 mV b. 5 μV

c. 50 μV d. 50 mV

8 In given arrangement, slider, PQ has given an initial speed of v_0.

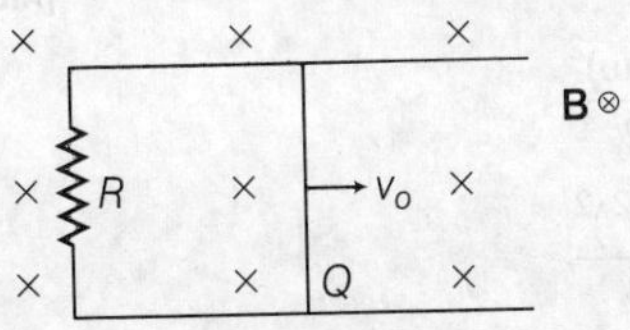

Then, graph of log v *versus* time t is given by

a. 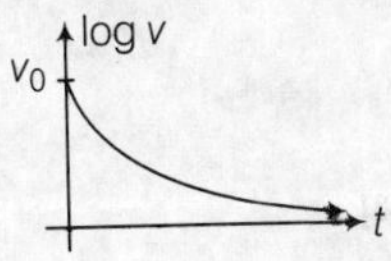

b. 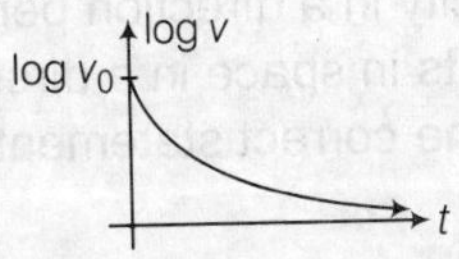

c. 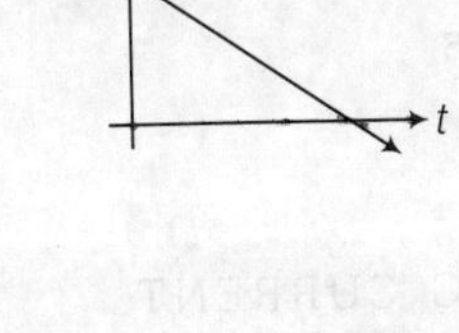

d. 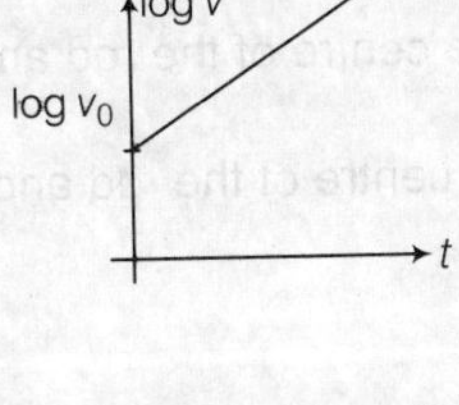

9 A rod rotates with a small but uniform angular velocity ω about its perpendicular bisector. A uniform magnetic field B exists parallel to the axis of rotation. The potential difference between centre of the rod and an end is

a. zero
b. $\frac{1}{8}B\omega l^2$
c. $\frac{1}{2}B\omega l^2$
d. $B\omega l^2$

10 A copper rod AB of length L, pivoted at one end A, rotates at constant angular velocity ω, at right angle to uniform magnetic field of induction B. The emf developed between the mid-point C of the rod and end B is

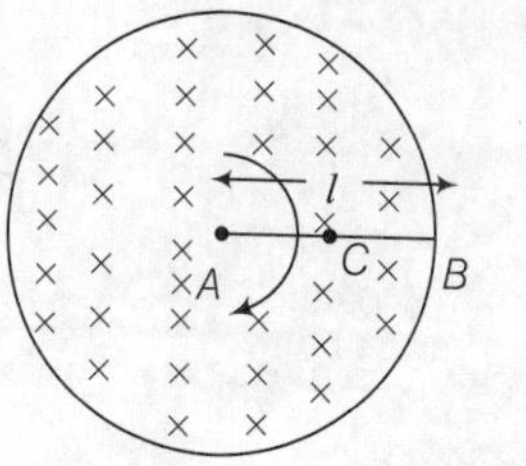

a. $\frac{B\omega l^2}{2}$
b. $\frac{B\omega l^2}{4}$
c. $\frac{3B\omega l^2}{4}$
d. $\frac{3B\omega l^2}{8}$

11 A conducting rod AC of length $4l$ is rotated about a point O in a uniform magnetic field **B** directed into the plane of the paper
$AO = l$ and $OC = 3l$. Then,

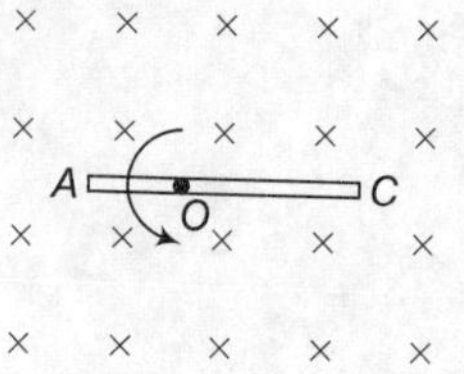

a. $V_A - V_O = \frac{B\omega l^2}{2}$
b. $V_O - V_C = \frac{7}{2}B\omega l^2$
c. $V_A - V_C = 4B\omega l^2$
d. $V_C - V_O = \frac{9}{2}B\omega l^2$

12 In a uniform magnetic field of induction B, a wire in the form of a circle of radius r rotates about the diameter of the circle with an angular frequency ω. The axis of rotation is perpendicular to the field. If the total resistance of the circle is R, the mean power dissipated per period of rotation is **[AIEEE 2004]**

a. $\frac{(B\pi r\omega)^2}{2R}$
b. $\frac{(B\pi r^2\omega)^2}{2R}$
c. $\frac{B\pi r^2\omega}{2R}$
d. $\frac{(B\pi r\omega^2)^2}{8R}$

13 A metal rod moves at a constant velocity in a direction perpendicular to its length. A constant uniform magnetic field exists in space in a direction perpendicular to the rod as well as its velocity. Select the correct statement(s) from the following. **[JEE Main 1998]**

a. The entire rod is at the same electric potential
b. There is an electric field in the rod
c. The electric potential is highest at the centre of the rod and decrease towards its ends
d. The electric potential is lowest at the centre of the rod and increases towards its ends

Comprehension Type

Direction *(Q. Nos. 14 and 15) This section contains a paragraph, which describing theory, experiments, data, etc. Two questions related to the paragraph have been given. Each question has only one correct answer among the four given options (a), (b), (c) and (d).*

Passage

A conducting ring of radius a is rotated about a point O on its periphery as shown in the figure. In a plane perpendicular to uniform magnetic field B which exists everywhere. The rotational velocity is ω.

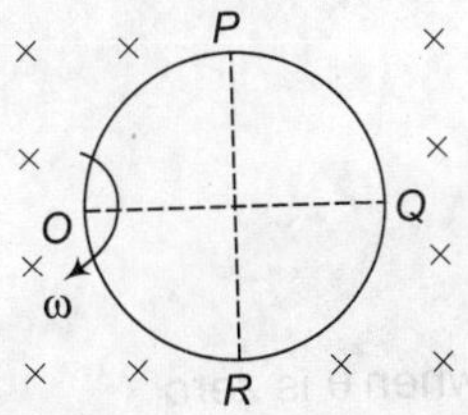

14 Select the correct statement(s) related to the magnitude of PD.

a. $V_P - V_O = \dfrac{B\omega R^2}{2}$ b. $V_P - V_Q = \dfrac{B\omega R^2}{2}$

c. $V_Q - V_O = 2B\omega R^2$ d. $V_P - V_R = 2B\omega R^2$

15 Select the correct statement(s) related to the induced current in the ring.

a. Current flows from $Q \to P \to O \to R \to Q$

b. Current flows from $Q \to R \to O \to P \to Q$

c. Current flows from $Q \to P \to O$ and from $Q \to R \to O$

d. No current flows

One Integer Value Correct Type

Direction *(Q. Nos. 16 and 17) This section contains 2 questions. When worked out will result in an integer from 0 to 9 (both inclusive).*

16 A circular coil of wire consists of exactly 100 turns with a total resistance of 0.20 Ω. The area of the coils is 100 cm^2. The coil is kept in a uniform magnetic field B as shown in the figure. The magnetic field is increased at a constant rate of 2 T/s. Find the induced current in the coil is A.

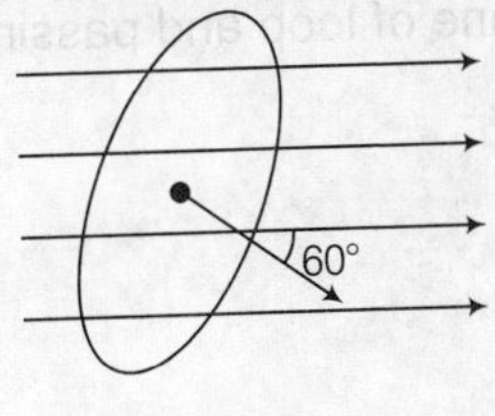

17 A square loop of side 2 cm with a capacitor of capacity 1 μF is located centrally between two current carrying wires separated by a distance of 6 cm as shown in the figure:

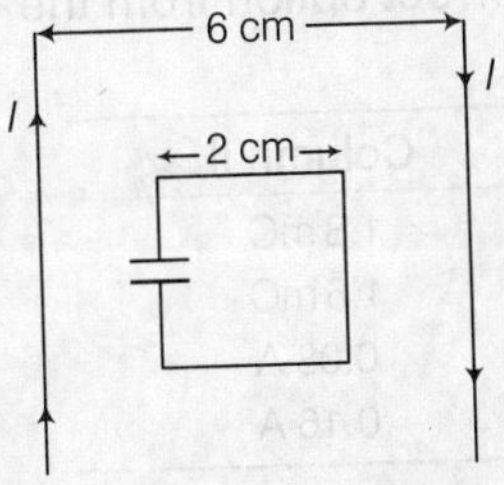

When current in wires is $I = 5\sin 50\pi t$ amperes, then maximum value of the current in square loop is $10^{-k} \cdot \pi^2 \cdot \log 2$ amperes. Find the value of $k - 2$.

One or More than One Options Correct Type

Working Space

Direction *(Q. Nos. 18 and 19) This section contains 2 multiple choice questions. Each question has four choices (a), (b), (c) and (d), out of which ONE or MORE THAN ONE are correct.*

18 A uniform circular loop of radius a and resistance R is placed perpendicular to a uniform magnetic field B. One half of the loop is rotated about the diameter with angular velocity ω as shown in the figure. Then, the current in the loop is

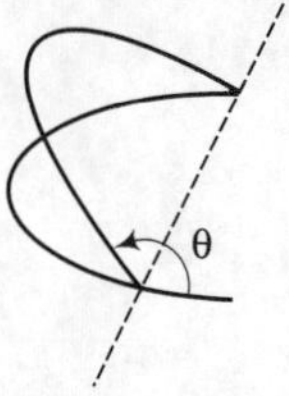

a. zero, when θ is zero

b. $\frac{\pi a^2 B\omega^2}{2R}$, when θ is zero

c. zero, when $\theta = \frac{\pi}{2}$

d. $\frac{\pi a^2 B\omega}{2R}$, when $\theta = \frac{\pi}{2}$

19 In the given figure, R is a fixed conducting ring of negligible resistance and radius a. PQ is a uniform rod of resistance r. It is hinged at the centre of the ring and rotated about this point in clockwise direction with a uniform angular velocity ω. There is a uniform magnetic field of strength B pointing inward and r is a stationary resistance. Then,

a. current through r is zero

b. current through r is $\frac{2B\omega a^2}{5r}$

c. direction of current in external resistance r is from centre of circumference

d. direction of current in external resistance r is from circumference to centre

Matching List Type

Direction *(Q. No. 20) Choices for the correct combination of elements from Column I and Column II are given as options (a), (b), (c) and (d), out of which one is correct.*

20 A square loop $ABCD$ whose area 20 cm^2 and resistance J Ω is rotated in a uniform field of $2T$ through 180° about an axis in plane of loop and passing through its centre.

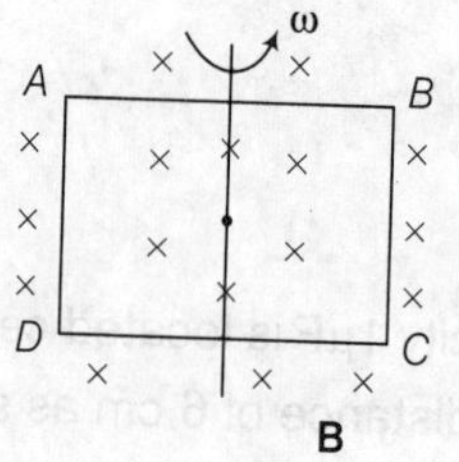

Now, match the following columns and select the correct option from the codes given below.

	Column I		Column II
i.	Current in loop at $t = 0.01$ s	p.	1.8 mC
ii.	Current in loop at $t = 0.02$ s	q.	1.6 mC
iii.	Magnitude of free charge at $t = 0.01$ s	r.	0.08 A
iv.	Magnitude of free charge at 0.02 s	s.	0.16 A

Codes

	i	ii	iii	iv
a.	s	r	r	p
b.	s	r	q	q
c.	s	r	p	q
d.	r	s	p	q

Statement Type

Direction *(Q. Nos. 21 and 22) This section is based on Statement I and Statement II. Select the correct answer from the codes given below.*

a. Both Statement I and Statement II are correct and Statement II is the correct explanation of Statement I
b. Both Statement I and Statement II are correct but Statement II is not the correct explanation of Statement I
c. Statement I is correct but Statement II is incorrect
d. Statement II is correct but Statement I is incorrect

21 **Statement I** Plane of a square loop with edge length $a = 0.2$ m is perpendicular to the earth's magnetic field $B = 15\,\mu$T. Total resistance of loop and connecting wires is $0.5\,\Omega$. Capacity of capacitor is $2\,\mu$F. If loop is suddenly collapsed by horizontal forces (see figure), then total charge stored in capacitor is $1.2\,\mu$C.

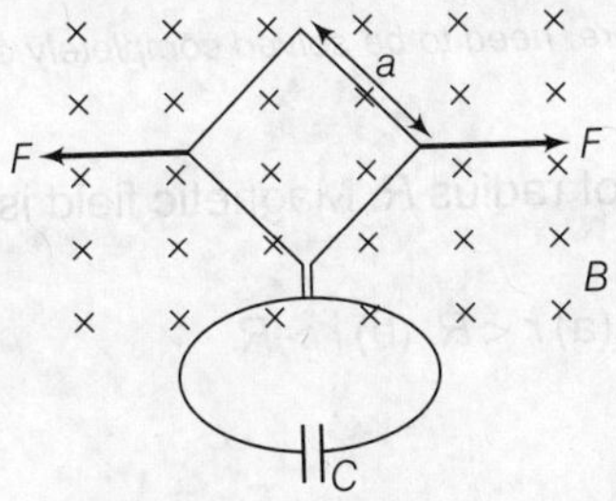

Statement II Capacity of capacitor is $C = \dfrac{Q}{E}$,

where, E = induced emf generated due to collapse of loop

and Q = charge stored by capacitor

22 **Statement I** A rectangular coil is rotated in the region of uniform magnetic field B. To keep rotating the coil, a uniform external torque must be applied to the coil.

Statement II Due to induced current, coil has its own dipole moment μ and so it experiences an opposing torque,

$$\tau = |\mu \times \mathbf{B}|$$

DPP-6 Loop in a Time Varying Magnetic Field and Induced EMF

Subjective Questions

Direction (*Q. Nos.* 1-4) *These questions are subjective in nature, need to be solved completely on notebook.*

Working Space

1 There exits magnetic field in a cylindrical region of radius R. Magnetic field is changing at the rate of αt.

i. Deduce expression for induced electric field for (a) $r < R$, (b) $r > R$.

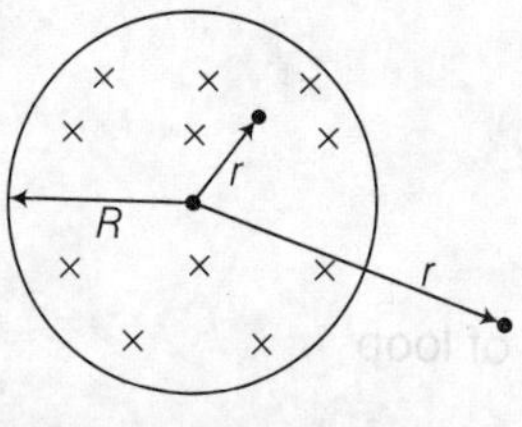

ii. Find out the direction of induced electric field.

2 A uniform magnetic field of induction B fills a cylindrical volume of radius R. A rod AB of length $2\,l$ is placed as shown in the figure. If B is changing at the rate $\frac{dB}{dt}$ find the emf that is produced by the change of magnetic field and that acts between the ends of the rod.

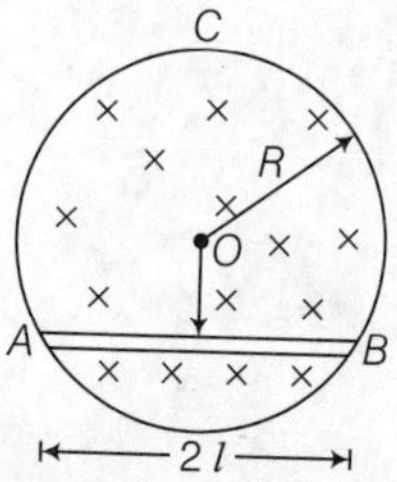

3 A rectangular frame $ABCD$, made of a uniform metal wire has a straight connection between E and F made of the same wire as shown in the figure. $AEFD$ is a square of side 1m and $EB = FC = 0.5$m. The entire circuit is placed in a steadily increasing uniform magnetic field directed into the plane of the paper and normal to it. The rate of change of the magnetic field is 1 T/s. The resistance per unit length (in Ω/m). Find the magnitude and directions of the current in segments AE, BE and EF.

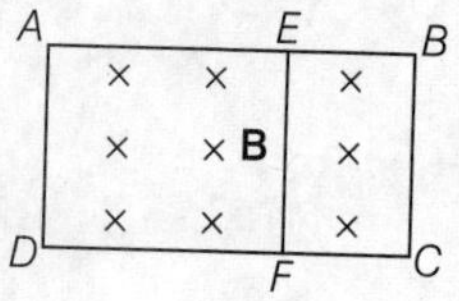

4 A thin non-conducting ring of mass m carrying a charge q can rotate freely about its axis. At $t = 0$, the ring was at rest and no magnetic field was present. Then, suddenly a magnetic field B was set perpendicular to the plane. Find the angular velocity acquired by the ring.

Only One Option Correct Type

Direction (Q. Nos. 5-13) *This section contains 9 multiple choice questions. Each question has four choices (a), (b), (c) and (d), out of which ONLY ONE is correct.*

5 Figure shows a uniform magnetic field B confined in a cylindrical volume and is increasing at a constant rate of $\frac{dB}{dt}$ Ts^{-1} having instantaneous acceleration experienced by an electron placed at P is

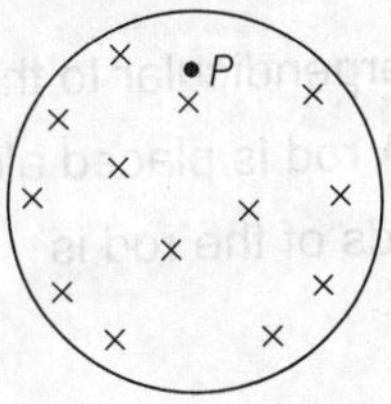

a. zero

b. $\frac{1}{2}\frac{eR}{m}\cdot\frac{dB}{dt}$ towards right

c. $\frac{eR}{m}\cdot\frac{dB}{dt}$ towards left

d. $\frac{1}{2}\frac{eR}{m}\cdot\frac{dB}{dt}$ towards left

6 The magnetic field in a certain cylindrical region is changing with time according to the law $B = 16 - 4t^2$ T. The induced electric field at point P at time $t = 2$ is

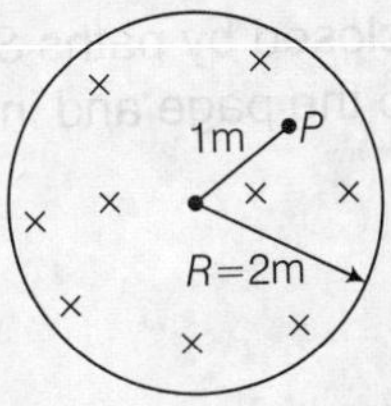

a. $-8\,V/m$

b. $+6\,V/m$

c. $-4\,V/m$

d. None of these

7 A triangular wire frame (each side = 2m) is placed in a region of time variant magnetic field having $\frac{dB}{dt} = \sqrt{3}$ T/s. The magnetic field is perpendicular to the plane of the triangle. The base of the triangle AB has a resistance 1Ω while the other two sides have resistance 2Ω each. The magnitude of potential difference between the points A and B will be

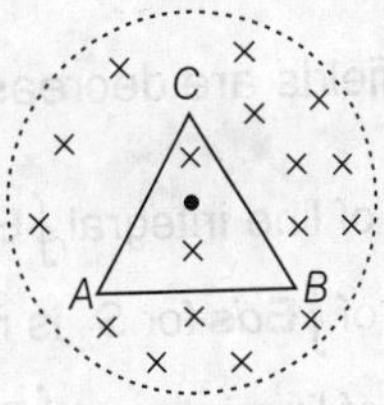

a. 0.4 V

b. 0.6 V

c. 1.2 V

d. None of these

8 In a cylindrical region, uniform magnetic field which is perpendicular to the plane of the figure is increasing with time and a conducting rod PQ is placed in the region. Then,

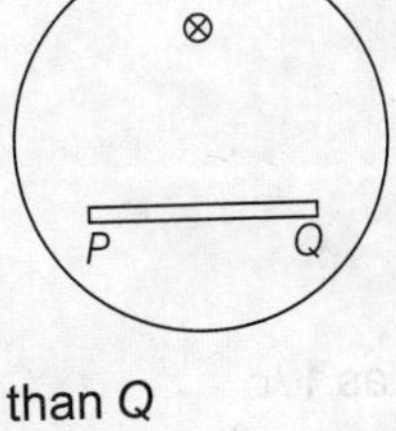

a. P will be at higher potential than Q

b. Q will be at higher potential than P

c. both P and Q will be equipotential

d. no emf will be developed

Working Space

9 A flexible wire loop in the shape of a circle has a radius that grows linearly with time. There is a magnetic field perpendicular to the plane of the loop that has a magnitude inversely proportional to the distance from the centre of the loop, $B(r) \propto \frac{1}{r}$. How does the emf ε vary with time?

a. $\varepsilon \propto t^2$
b. $\varepsilon \propto t$
c. $\varepsilon \propto \sqrt{t}$
d. ε is constant

10 A cylindrical region of uniform magnetic field exists perpendicular to the plane of paper which is increasing at a constant rate $\frac{dB}{dt} = \alpha$. A rod is placed along the diameter of the cylinder. Induced emf between the ends of the rod is

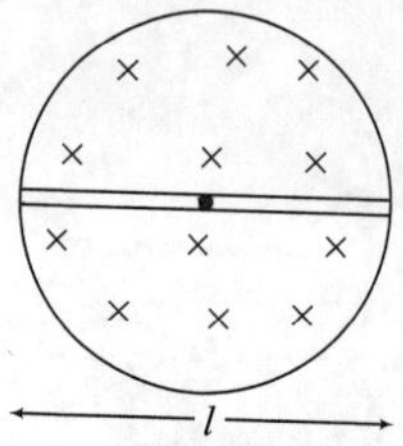

a. $\frac{\pi \alpha l^2}{2}$
b. $\pi l^2 \alpha$
c. $\frac{\pi \alpha l^2}{4}$
d. zero

11 Two circular regions of radii 20 cm and 30 cm are enclosed by paths S_1 and S_2. In region I, there is a uniform magnetic field of 50mT into the page and in region II, there is a magnetic field of 75 mT out of the page.

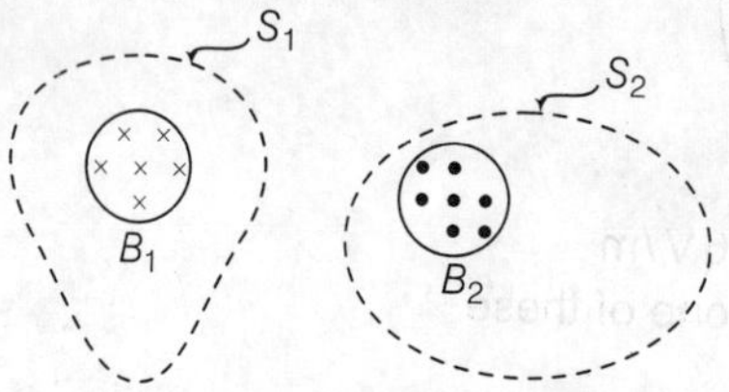

If both magnetic fields are decreasing at the rate of 8.5mT/s, then choose the correct option.

a. Absolute value of line integral $\oint \mathbf{E} d\mathbf{s}$ for S_1 is more than that of S_2
b. Absolute value of $\oint \mathbf{E} d\mathbf{s}$ for S_2 is more than that of S_1.
c. Absolute value of line integral $\oint \mathbf{E} d\mathbf{s}$ for S_1 and S_2 is same
d. Data is insufficient to reach any conclusion

12 The uniform but time varying magnetic field $B(t)$ exists in a circular region of radius a and is directed into one of the papers as shown in the figure. The magnitude of the induced electric field at point P at a distance r from the centre of the circular region.

[JEE Main 2000]

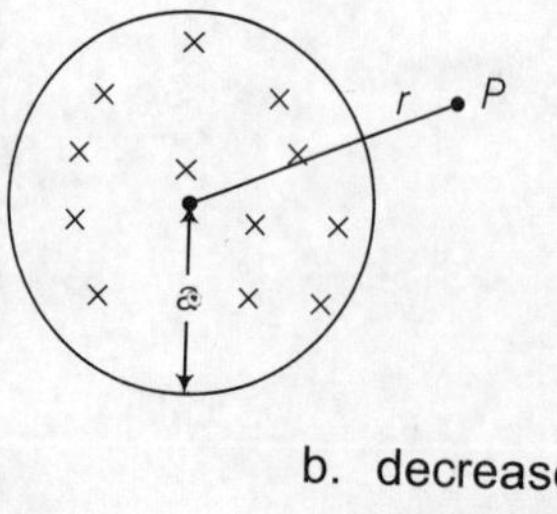

a. is zero
b. decreases as $1/r$
c. increases as r
d. decreases as $1/r^2$

13 A thin flexible wire of length L is connected to two adjacent fixed points and carries a current I in the clockwise direction, as shown in the figure. When the system is put in a uniform magnetic field of strength B going into the plane of paper, the wire takes the shape of a circle. The tension in the wire is **[JEE Main 2010]**

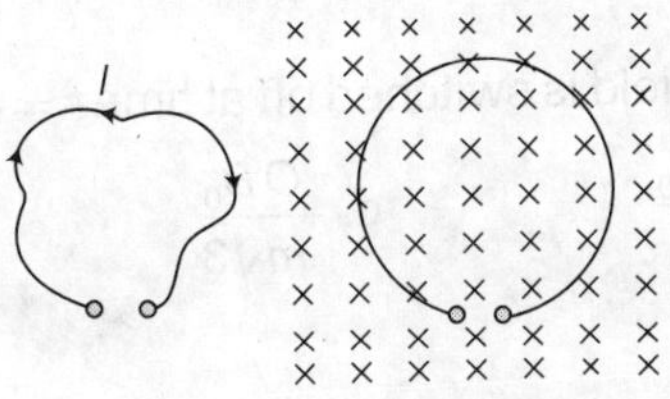

a. IBL

b. $\frac{IBL}{\pi}$

c. $\frac{IBL}{2\pi}$

d. $\frac{IBL}{4\pi}$

One or More than One Options Correct Type

Direction *(Q. Nos. 14 and 15) This section contains 2 multiple choice questions. Each question has four choices (a), (b), (c) and (d), out of which ONE or MORE THAN ONE are correct.*

14 A circular loop of radius 1 m is placed in a varying magnetic field $B = 6t$ T. Resistance per unit length is 1Ω. Choose the correct option(s).

a. Emf induced in the loop is 6π V

b. Electric field is in the tangential direction

c. Current in the loop is 3 A

d. Induced electric field is conservative in nature

15 A circular loop of radius r having N turns of a wire, is placed in a uniform and constant magnetic field B. The normal of the loop makes an angle θ with the magnetic field. Its normal rotates with the angular velocity ω such that the angle θ is constant. Choose the statement from the following:

a. Emf in the loop is $\frac{NB\omega r^2}{2} \times \cos\theta$

b. Emf induced in the loop is zero

c. Emf must be induced as the loop crosses magnetic lines

d. Emf must not be induced as flux does not change with time

Comprehension Type

Direction *(Q. Nos. 16-18) This section contains a paragraph, which describing theory, experiments, data, etc. Three questions related to the paragraph have been given. Each question has only one correct answer among the four given options (a), (b), (c) and (d).*

Passage

A horizontal surface is rough. The plane is parallel to the surface and the ring is placed on it. A charge Q uniformly distributed on this ring of radius R and mass m. At $t = 0$, a uniform magnetic field $B = \frac{B_0 t^2 \sqrt{3}}{2}$ is switched on and coefficient of friction is $\mu = \frac{QB_0 R}{2mg}$.

16 Find a force which is tangential to the ring.

a. $B_0 Rt$

b. $\frac{QB_0 Rt}{2}$

c. $QB_0 Rt\sqrt{3}$

d. $\frac{QB_0 Rt\sqrt{3}}{2}$

17 Find the time at which ring will start rotating an axis perpendicular to its plane and passing its centre (in seconds).

a. $\frac{\sqrt{3}}{2}$ b. $\frac{1}{2}$ c. $\sqrt{3}$ d. $\frac{1}{\sqrt{3}}$

Working Space

18 Find the angular velocity just after the magnetic field is switched off at time $t = \sqrt{3}$ s.

a. $\frac{2QB_0}{m\sqrt{3}}$ b. $\frac{2QB_0}{m}$ c. $\frac{QB_0}{2m}$ d. $\frac{QB_0}{m\sqrt{3}}$

Matching List Type

Direction *(Q. Nos. 19 and 20) Choices for the correct combination of elements from Column I and Column II are given as options (a), (b), (c) and (d), out of which one is correct.*

19 The magnetic field perpendicular to the plane of a conducting ring changes at the rate of $\frac{dB}{dt}$, where B is magnetic field strength and t is time. The radius of the ring is r. Match the following columns and select the correct option from the codes given below.

	Column I		Column II
i.	Magnetic flux is proportional to	p.	dB/dt
ii.	Emf induced in the ring is proportional to	q.	Parallel to the length of element of the ring
iii.	All points on the ring are at	r.	Same potential
iv.	Electric intensity at any point of the ring is	s.	None of these

Codes

	i	ii	iii	iv		i	ii	iii	iv
a.	s	p	r	q	b.	p	q	r	s
c.	p	r	s	q	d.	q	p	r	s

20 Magnetic flux in a circular coil of resistance 10Ω changes with time as shown in the figure.

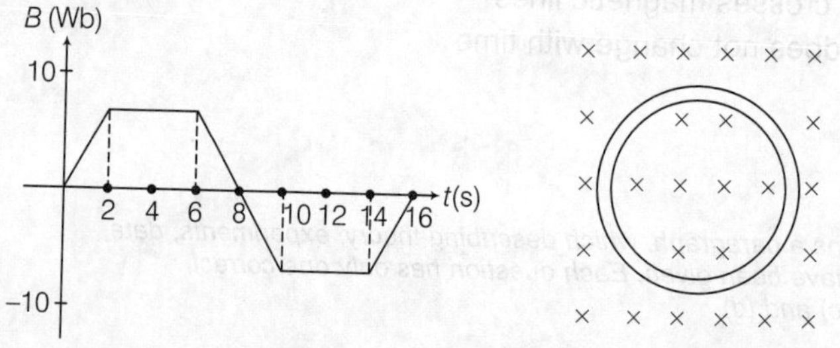

Now, match the following columns and select the correct option from the codes given below.

	Column I		Column II
i.	At 1 s induced current is	p.	Clockwise
ii.	At 5 s induced current is	q.	Anti-clockwise
iii.	At 9 s induced current is	r.	Zero
iv.	At 15 s induced current	s.	2A

Codes

	i	ii	iii	iv		i	ii	iii	iv
a.	q	r	p	q	b.	p	q	r	s
c.	q	r	p	s	d.	q	r	s	p

Working Space

One Integer Value Correct Type

Direction *(Q. Nos. 21 and 22) This section contains 2 questions. When worked out will result in an integer from 0 to 9 (both inclusive).*

21 The magnetic field of a cylindrical magnet that has a pole face radius 2.8 cm can be varied sinusoidally between the minimum value of 16.8 T and the maximum value 17.2 T at a frequency of $50/\pi$ Hz. Cross-section of the magnetic field as shown in the figure. At a radial distance of 2 cm from the axis. Find the amplitude of the electric field (in unit of $\times 10^2$ mN/C) induced by magnetic field variation.

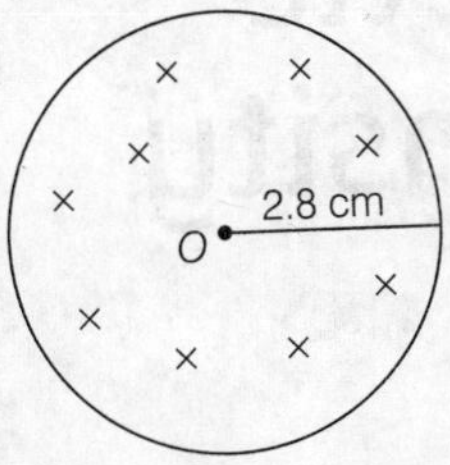

22 In three-dimensional region of space, a uniformly decreasing magnetic field points in Z-direction.

If magnetic field decays at a rate of −1.0 T/s, what is the force (in nN) acting on a charge of $1\mu C$ placed at point (10, 0) cm.

Statement Type

Direction *(Q. Nos. 23 and 24) This section is based on Statement I and Statement II. Select the correct answer from the codes given below.*

a. Both Statement I and Statement II are correct and Statement II is the correct explanation of Statement I
b. Both Statement I and Statement II are correct but Statement II is not the correct explanation of Statement I
c. Statement I is correct but Statement II is incorrect
d. Statement II is correct but Statement I is incorrect

23 **Statement I** Magnetic flux linked with a metal ring is

$$\phi_B = 3(at^3 - bt^2)\ \text{Tm}^2$$

where, $a = 2\ \text{s}^{-3}$ and $b = 6\ \text{s}^{-2}$

Resistance of ring is $3\ \Omega$.

Maximum current induced in the ring during time interval $t = 0$ to $t = 2$ s is 6 A.

Statement II Maxima of a function $y = f(x)$ occurs when $\frac{dy}{dx} = 0$.

24 **Statement I** A general form of Faraday's law of induction is

$$E = \oint \mathbf{E} \cdot d\mathbf{s} = -\frac{d\phi_B}{dt}$$

Statement II Induced current set up a magnetic field that opposes the change which produces it.

DPP-7 Self-Induction, Self-Inductance and Magnetic Energy Density

Subjective Questions

Direction (Q. Nos. 1-4) *These questions are subjective in nature, need to be solved completely on notebook.*

Working Space

1 A tightly wound solenoid is used to make another tightly wound solenoid whose diameter is 2.5 times of its initial diameter. By what factor the inductance change.

2 A toroid has a rectangular core with cross-sectional dimensions as shown in the figure.

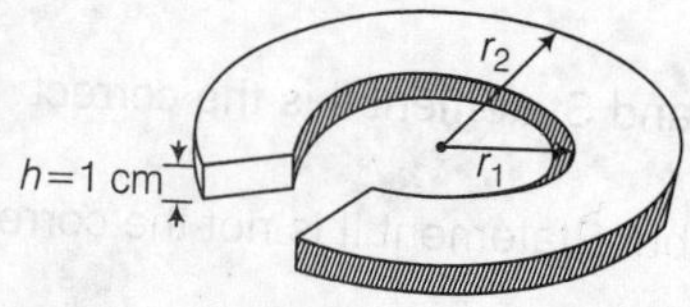

Given, $h = 1\text{cm}, r_1 = 5\text{ cm}, r_2 = 10\text{ cm}$,
total number of turns of wire wound over core (of relative permeability 900) is 5000. Find self-inductance of the toroid.

3 In the circuit shown in the figure, S_1 and S_2 are switches. S_2 remains closed for a long time and S_1 is opened. Now S_1 is closed. Just after S_1 is closed, find the potential difference V across R and di/dt (with sign) in terms of ε and L.

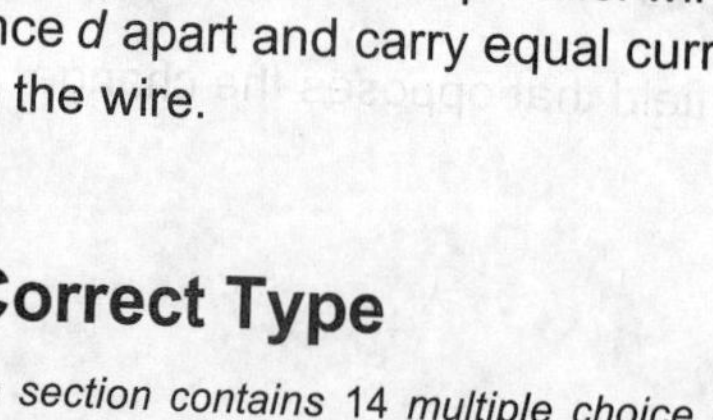

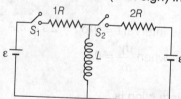

4 Find the inductance per unit length of two parallel wires each of radius a whose centres are at a distance d apart and carry equal currents in opposite directions. Neglect the flux within the wire.

Only One Option Correct Type

Direction (Q. Nos. 5-18) *This section contains 14 multiple choice questions. Each question has four choices (a), (b), (c) and (d), out of which ONLY ONE is correct.*

5 Two inductance L_1 and L_2 are placed far apart and in parallel. Their combined inductance is

[IIT JEE 1994]

a. $L_1 + L_2$ b. $(L_1 + L_2) \times \frac{L_1}{L_2}$ c. $(L_1 - L_2) \times \frac{L_2}{L_1}$ d. $\frac{L_1 L_2}{L_1 + L_2}$

6 When the current changes from +2A to – 2A in 0.05 s, an emf of 8 V is induced in a coil. The coefficient of self-induction of the coil is

a. 0.1 H
b. 0.2 H
c. 0.4 H
d. 0.8 H

7 The inductance between A and D is **[AIEEE 2002]**

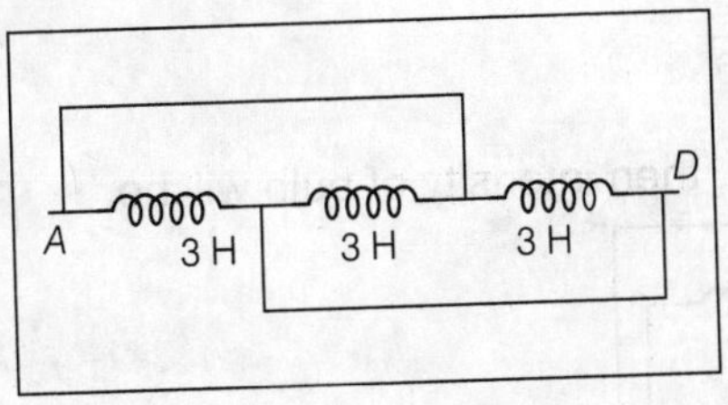

a. 3.66H
b. 9H
c. 0.66H
d. 1 H

8 In a circular coil, when number of turns is doubled, then its inductance becomes **[AIEEE 2002]**

a. 4 times
b. 2 times
c. 8 times
d. no change

9 The SI unit of inductance, the Henry cannot be written as **[IIT JEE 1998]**

a. weber ampere^{-1}

b. volt second ampere^{-1}

c. joule ampere^{-2}

d. ohm second^{-1}

10 A current of 2 A is increasing at a rate of 4 A/s through a coil of inductance 2H. The energy stored in the inductor per unit time is

a. 2 J/s
b. 1J/s
c. 16 J/s
d. 4 J/s

11 Two identical inductors carry currents that vary with time according to linear laws (as shown in figure). In which of two inductors, is the self induced emf is greater?

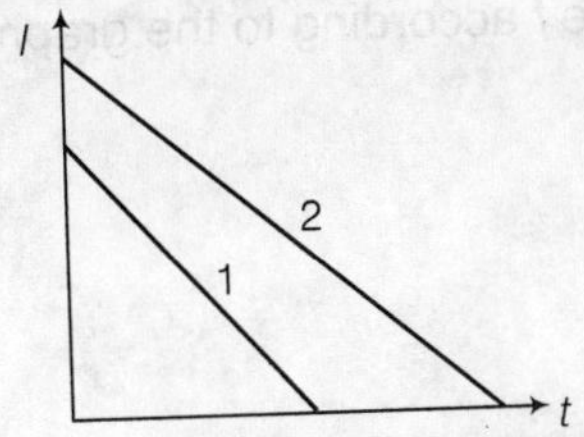

a. 1
b. 2
c. Same
d. Data is insufficient to decide

12 A long solenoid of N turns has a self-inductance L and area of cross-section A. When a current i flows through the solenoid, the magnetic field inside it has magnitude B. The current i is equal to

a. $\frac{BAN}{L}$
b. $BANL$
c. $\frac{BN}{AL}$
d. $\frac{B}{ANL}$

13 A thin copper wire of length 100 m is wound as a solenoid of length l and radius r. Its self-inductance is found to be L. Now, if the same length of wire is wound as a solenoid of length l but of radius $r/2$, then its self-inductance will be

a. $4L$
b. $2L$
c. L
d. $L/2$

Working Space

14 Two different coils have self-inductance $L_1 = 8$ mH and $L_2 = 2$ mH. At a certain instant the current in the two coils is increasing at the same constant rate and the power supplied to the two coils is the same. Find the ratio of energies stored in the two coils at that instant.

a. $\frac{1}{2}$ b. $\frac{1}{4}$

c. $\frac{1}{8}$ d. $\frac{1}{16}$

15 If a soft iron rod is inserted into inductive coil, then intensity of bulb will be

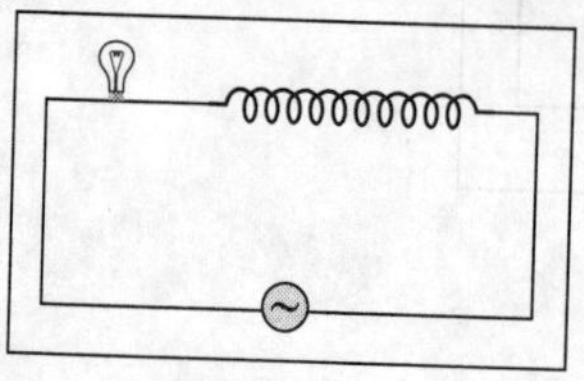

a. increase

b. decrease

c. unchanged

d. cannot say anything

16 The current in the given circuit is increasing with a rate of 4 A/s. The charge on the capacitor at an instant when the current in the circuit is 2 A will be

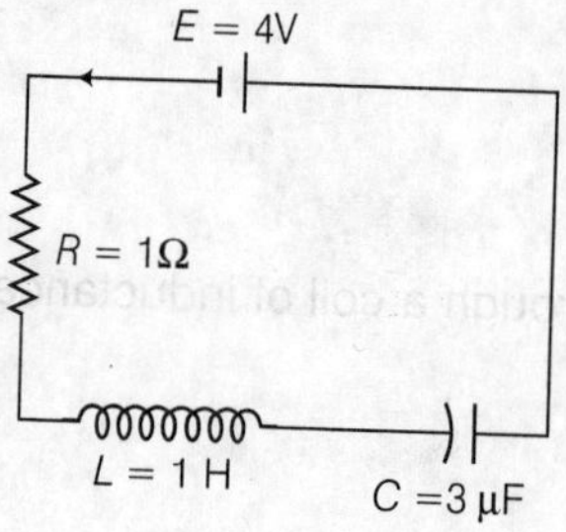

a. 4μC b. 5μC

c. 6μC d. None of these

17 The current I in an inductance coil varies with time t according to the graph shown in figure.

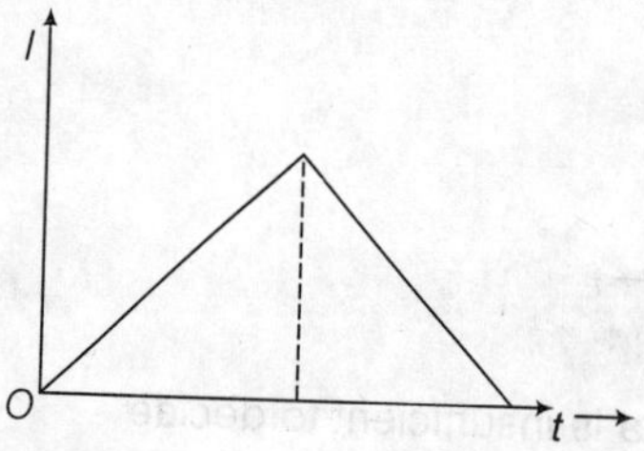

Which one of the following plots shows the variation of voltage in the coil with time ?

a. 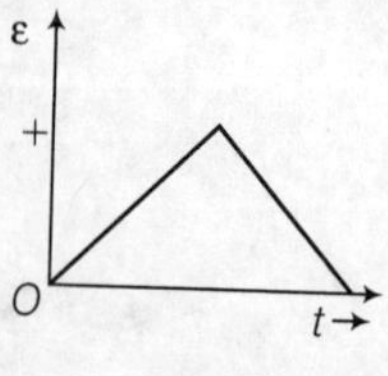

b. 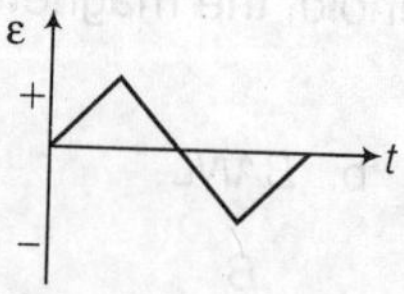

c. 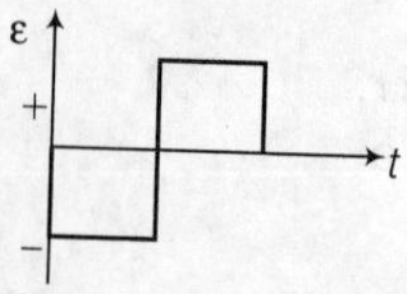

d. 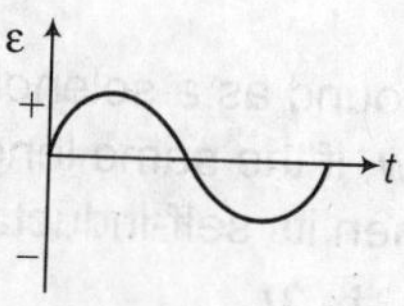

Working Space

18 A long straight wire of circular cross-section is made of a non-magnetic material. The wire is of radius *a*. The wire carries a current *I* which is uniformly distributed over its cross-section. The energy stored per unit length in the magnetic field contained within the wire is

a. $U = \frac{\mu_0 I^2}{8\pi}$
b. $U = \frac{\mu_0 I^2}{16\pi}$
c. $U = \frac{\mu_0 I^2}{4\pi}$
d. $U = \frac{\mu_0 I^2}{2\pi}$

One or More than One Options Correct Type

***Direction** (Q. Nos. 19 and 20) This section contains 2 multiple choice questions. Each question has four choices (a), (b), (c) and (d), out of which ONE or MORE THAN ONE are correct.*

19 Two different coils have self-inductance $L_1 = 8$ mH, $L_2 = 2$ mH. The current in one coil is increased at a constant rate. The current in the second coil is also increased at the same rate. At a certain instant of time, the power given to the two coils is the same. At that time, the current, the induced voltage and the energy stored in the first coil are i_1, V_1 and W_1, respectively. Corresponding values for the second coil at the same instant are i_2, V_2 and W_2, respectively. Then, **[IIT JEE 1994]**

a. $\frac{i_1}{i_2} = \frac{1}{4}$
b. $\frac{i_1}{i_2} = 4$
c. $\frac{W_2}{W_1} = 4$
d. $\frac{V_2}{V_1} = \frac{1}{4}$

20 A constant current *i* is maintained in a solenoid. Which of the following quantities will increase if an iron rod is inserted in the solenoid along axis?

a. Magnetic field at the centre
b. Magnetic flux linked with the solenoid
c. Self-inductance of the solenoid
d. Rate of Joule heating

Comprehension Type

***Direction** (Q. Nos. 21-26) This section contains 2 paragraphs, each describing theory, experiments, data, etc. Six questions related to the paragraphs have been given. Each question has only one correct answer among the four given options (a), (b), (c) and (d).*

Passage I

A coil of inductance *L*, connects the upper ends of two horizontal rails having zero resistance. A horizontal conductor of mass *m* is projected with initial velocity v_0 at $t = 0$, all the time maintaining contact with the two horizontal rails as shown in the figure. A uniform magnetic field of magnitude *B* exists in the region normal to the plain of the rails. The distance between the horizontal rails is *l*.

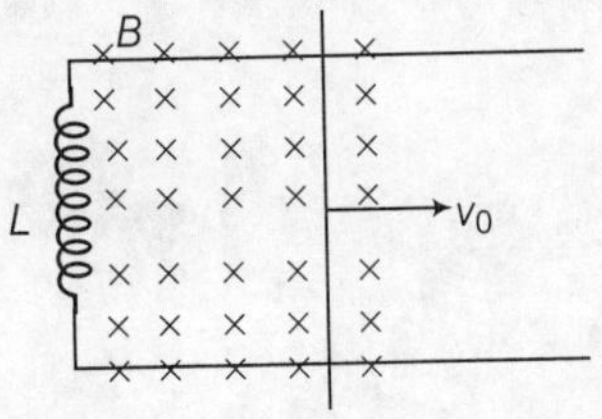

21 During the motion of the conductor

a. its velocity continuously decreases
b. it will come to rest after sometime
c. it will come to rest after sometime and then returns in opposite direction
d. its velocity remains constant

Working Space

22 The motion of conductor is

a. non-uniform with exponentially decreasing velocity
b. non-uniform with sinusoidally varying velocity
c. uniform
d. None of the above

23 Magnetic energy in the inductor

a. is maximum at $T = 0$
b. is maximum at $T = \frac{2\pi\sqrt{m}}{Bl}$
c. is maximum at $T = \frac{\pi\sqrt{m}}{Bl}$
d. None of these

Passage II

A research associate requires an air-core solenoid with self-inductance of 1 H on the outside of a hollow plastic tube 12 cm in diameter. Such a solenoid is not available in the market. So, the associate decided to make the solenoid using a copper wire of diameter 0.81 mm.

Now, help the associate in shopping different items.

24 Plastic tubes are available in following sizes pick size most suitable for the requirement so that wastage is minimum.

a. 100 m
b. 75 m
c. 50 m
d. 25 m

25 Choose the copper roll for the associate length of wire in copper roll available are

a. 12 km
b. 24 km
c. 36 km
d. 48 km

26 Please calculate the resistance of copper roll you have choosen. (Given, $\rho_{Cu} = 1.68 \times 10^{-8}\ \Omega\text{-m}$)

a. 0.5 kΩ
b. 0.6 kΩ
c. 0.7 kΩ
d. 1 kΩ

One Integer Value Correct Type

Direction *(Q. Nos. 27 and 28) This section contains 2 questions. When worked out will result in an integer from 0 to 9 (both inclusive).*

27 In the circuit as shown in the figure, when i the current is 2 A and increasing at the rate of 1 A/s, the measured potential difference $V_{AB} = 10\,V$. But when the current is 2 A and decreasing at the rate of 1 A/s, the measured potential difference is 6 V. Find the value of R (in Ω).

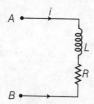

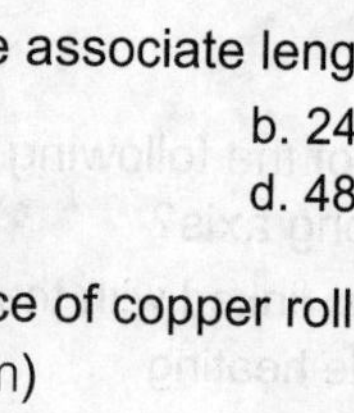

28 The network shown in figure is part of a complete circuit. If at a certain instant, the current i is 5 A and is decreasing at a rate of 10^3 A/s, then $V_B - V_A = 5k$ volts, find the value of k.

[JEE Advanced 1997]

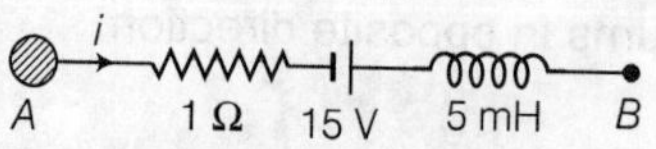

DPP-8 Growth and Decay of Current in *L*-*R* Circuit

Subjective Questions

Working Space

Direction (Q. Nos. 1-4) *These questions are subjective in nature, need to be solved completely on notebook.*

1 A metal rod *OA* of mass m and length r is kept rotating with a constant angular speed ω, with a uniform magnetic field B applied in perpendicular downward direction. An inductor L and resistance R are connected through a switch S across points O and C on the ring over which rods free end is rotating.

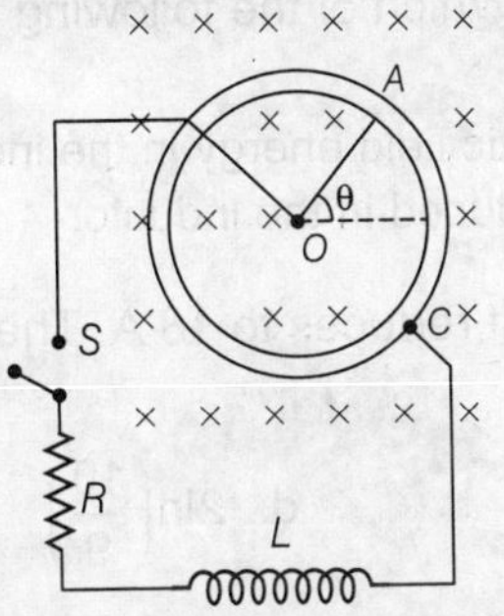

Neglect the resistance of the rod, ring and connecting wires. Obtain the value of torque required to maintain the constant speed of the rotating rod. At $t = 0$, switch is closed.

2 Consider the given circuit with a two position switch S.

Initially switch is in position 1. At $t = 0$, switch is put in position 2.

i. Find steady current in R_4.

ii. Time when current in R_4 is half of its steady value.

iii. Energy stored in inductor at above time instant.

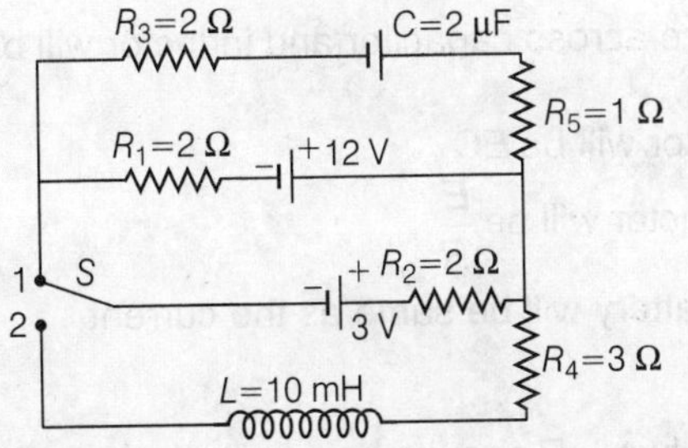

3 In the following circuit, the switch is closed at $t = 0$. Initially there is no current in inductor. Find out current in the inductor as a function of time.

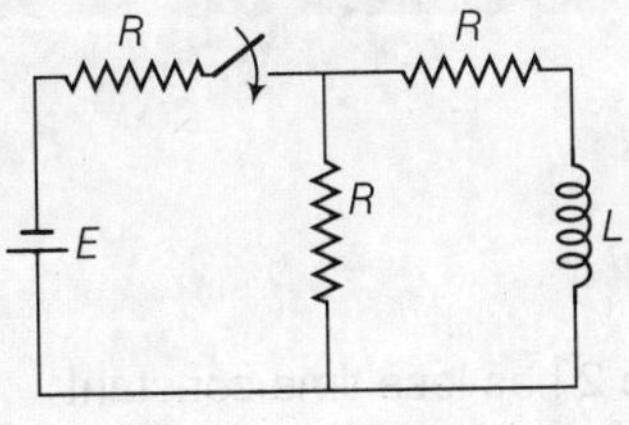

4 An inductor of inductance $L = 400\text{mH}$ and resistors of resistance $R_1 = 2\Omega$ and $R_2 = 2\Omega$ are connected to a battery of emf $E = 12\text{V}$ as shown in the figure. The internal resistance of the battery is negligible. The switch is closed at time $t = 0$.

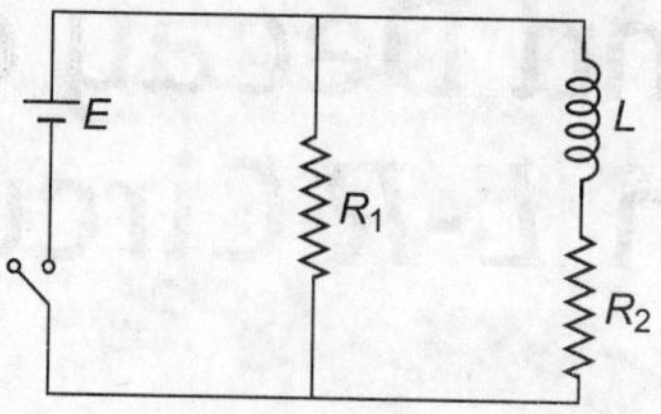

i. What is the potential drop across L as a function of time?

ii. After the steady state is reached, the switch is opened. What is the direction and the magnitude of current through R_1 as a function of time? **[IIT JEE 2001]**

Only One Option Correct Type

Direction *(Q. Nos. 5-17) This section contains 13 multiple choice questions. Each question has four choices (a), (b), (c) and (d) out of which ONLY ONE is correct.*

5 An L-R circuit with a battery is connected at $t = 0$. Which of the following quantities is not zero just after closing the switch?

a. Current in the circuit
b. Magnetic field energy in the inductor
c. Power delivered by battery
d. Emf induced in the inductor

6 In an L-R circuit, current at $t = 0$ is 20 A. After 2 s, it reduces to 18 A. The time constant of the circuit is (in s)

a. $\ln\left(\frac{10}{9}\right)$
b. 2
c. $\frac{2}{\ln\left(\frac{10}{9}\right)}$
d. $2\ln\left(\frac{10}{9}\right)$

7 In the circuit shown in figure, switch S is closed at $t = 0$. Then,

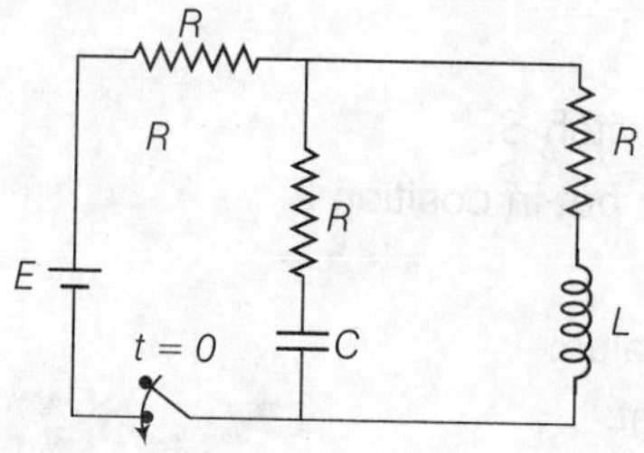

a. after a long time internal potential difference across capacitor and inductor will be zero
b. after a long time internal charge on capacitor will be EC
c. after a long time internal current in the inductor will be $\frac{E}{R}$
d. after a long time internal current through battery will be same as the current through it initially

8 Curves shown in figure shows i-t graph for two L-R circuits having batteries of equal emfs. Then,

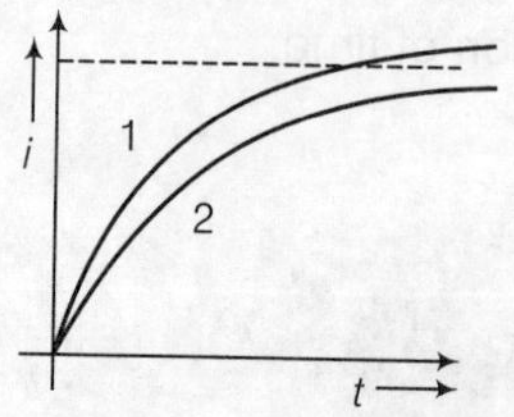

a. curve 1 has less time constant
b. curve 2 has less time constant
c. both curves has same time constant
d. Data insufficient

9 A coil of inductance 300 mH and resistance $2\,\Omega$ is connected to a source of voltage 2 V. The current reaches half of its steady value in **[AIEEE 2005]**

a. 0.05 s b. 0.1 s c. 0.15 s d. 0.3 s

10 An ideal coil of 10 H is connected in series with a resistance of $5\,\Omega$ and a battery of 5 V. After 2s the connection is made, the current flowing (in amperes) in the circuit is **[AIEEE 2007]**

a. $(1-e)$ b. e
c. e^{-1} d. $(1-e^{-1})$

11 An inductor (L = 100 mH) and resistor ($R = 100\,\Omega$) and a battery (E = 100 V) are initially connected in series as shown in the figure. After a long time, the battery is disconnected after short circuiting the points A and B. The current in the circuit 1 ms after the short circuit is **[AIEEE 2006]**

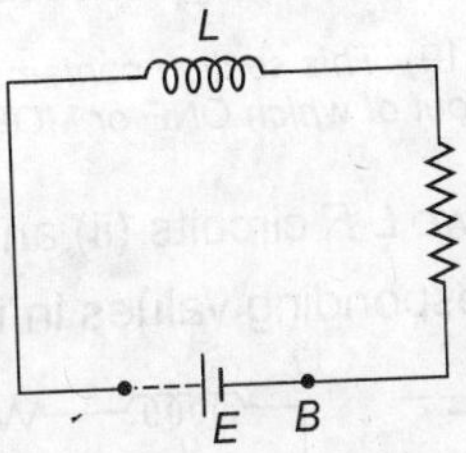

a. 0.1 A b. 1 A
c. $1/e$ A d. e A

12 In the given circuit find the ratio of i_1 to i_2, where i_1 is the initial current (t = 0) and i_2 is steady state current through the battery. Find the ratio of i_1 / i_2.

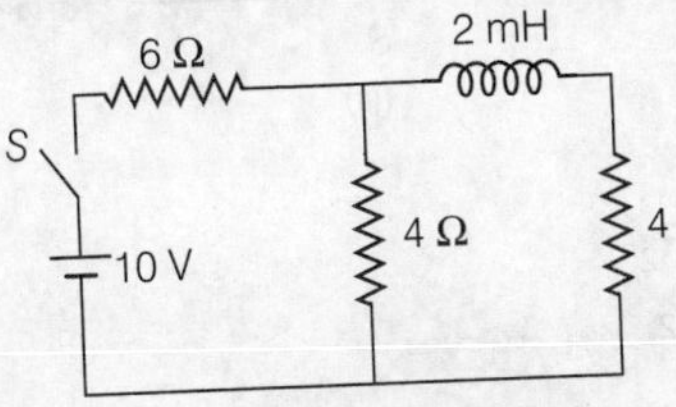

a. 1.0 b. 0.8
c. 1.2 d. 1.5

13 An inductor of inductance L = 400 mH and resistor of resistances $R_1 = 2\,\Omega$ and $R_2 = 2\,\Omega$ are connected to a battery of emf 12 V as shown in the figure. The internal resistance of the battery is negligible. The switch S is closed at t = 0. The potential drop across L as function of time is

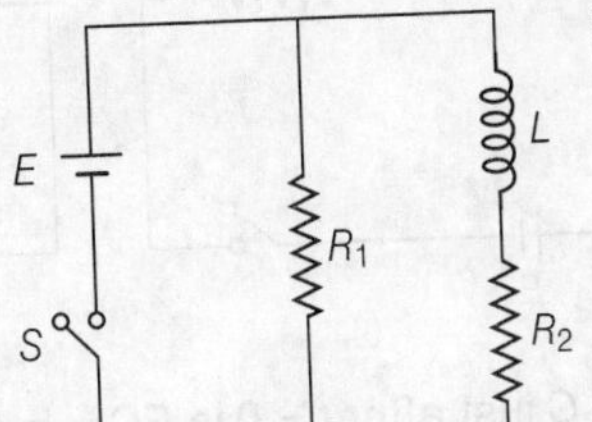

a. $6e^{-5t}$V b. $\frac{12}{t}e^{-3t}$V
c. $6(1-e^{-t/0.2})$V d. $12e^{-5t}$V

14 A coil of inductance 8.4 mH and resistance $6\,\Omega$ is connected to a 12 V battery. The current in the coil is 1.0 A at approximately the time **[IIT JEE 1999]**

a. 500 s b. 20 s
c. 35 ms d. 1 ms

15 An inductor of inductance L = 400 mH and resistors of resistances $R_1 = 2\,\Omega$ and $R_2 = 2\,\Omega$ are connected to a battery of emf 12 V as shown in the figure. The internal resistance of the battery is negligible. The switch S is closed at t = 0. The potential drop across L as a function of time is **[AIEEE 2009]**

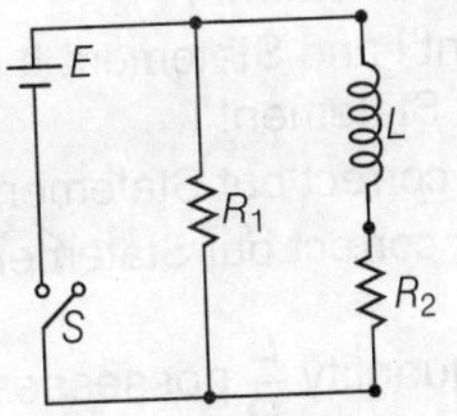

a. $6e^{-5t}$V b. $\frac{12}{t}e^{-3t}$V
c. $6\left(1-e^{\frac{-t}{0.2}}\right)$V d. $12e^{-5t}$V

16 A coil of inductance 8.4 mH and resistance $6\,\Omega$ is connected to a 12 V battery. The current in the coil is 1.0 A in the time (approx.)

a. 500 s b. 20 s
c. 35 ms d. 1 ms

Working Space

17 An inductor ($L = 100$ mH), a resistor ($R = 100\,\Omega$) and a battery ($E = 100$ V) are initially connected in series as shown in the figure. After a long time, the battery is disconnected after short circuiting the points A and B. The current in the circuit 1 ms after the short circuit is **[AIEEE 2006]**

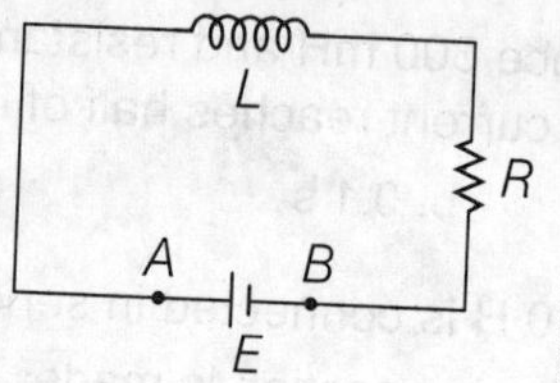

a. e A
b. 0.1 A
c. 1 A
d. $1/e$ A

One or More than One Options Correct Type

Direction *(Q. Nos. 18 and 19) This section contains 2 multiple choice questions. Each question has four choices (a), (b), (c) and (d), out of which ONE or MORE THAN ONE are correct.*

18 Current growth in two L-R circuits (ii) and (iii) as shown in Fig. (i). Let L_1 and L_2, R_1 and R_2 be the corresponding values in two circuits. Then,

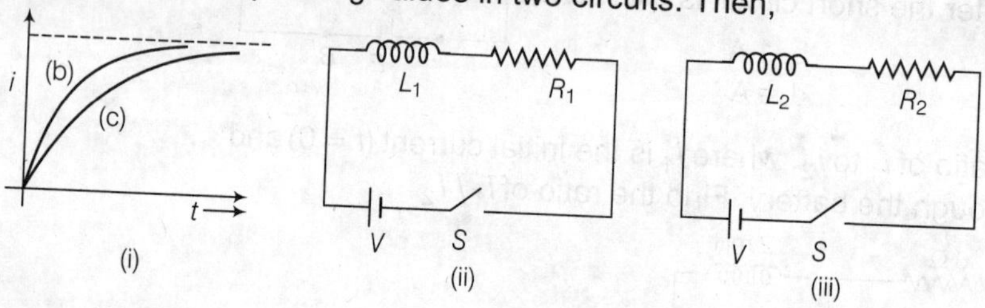

a. $R_1 > R_2$
b. $R_1 = R_2$
c. $L_1 > L_2$
d. $L_1 < L_2$

19 The switches in Fig. (i) and Fig. (ii) are closed at $t = 0$. Choose the correct options.

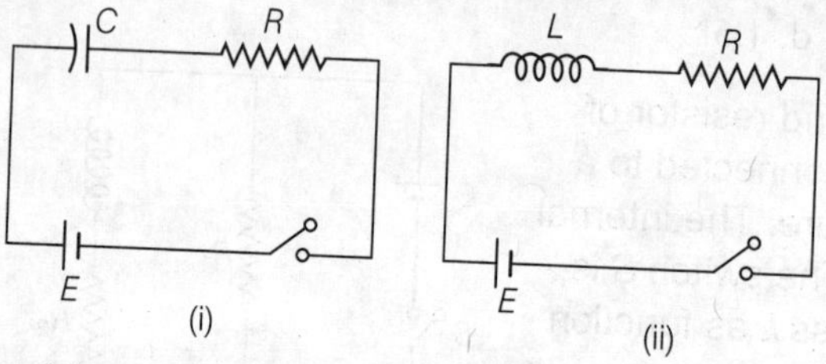

a. The charge on C just after $t = 0$ is EC
b. The charge on C long after $t = 0$ is EC
c. The current in L just after $t = 0$ is E/R
d. The current in L long after $t = 0$ is E/R

Statement Type

Direction *(Q. Nos. 20 and 21) This section is based on Statement I and Statement II. Select the correct answer from the codes given below.*

a. Both Statement I and Statement II are correct and Statement II is the correct explanation of Statement I
b. Both Statement I and Statement II but correct but Statement II is not the correct explanation of Statement I
c. Statement I is correct but Statement II is incorrect
d. Statement II is correct but Statement I is incorrect

20 **Statement I** The quantity $\frac{L}{R}$ possesses the dimension of time.

Statement II In order to reduce the rate of increase of current through a solenoid, we should increase the time constant.

21 **Statement I** In five time constants, current in an L-R circuit is within one per cent of its steady state value.

Statement II For an L-R circuit, $I = I_{max}(1 - e^{-t/\tau})$, where $\tau = \frac{L}{R}$.

Working Space

Comprehension Type

***Direction** (Q. Nos. 22-27) This section contains 2 paragraphs, each describing theory, experiments, data, etc. Six questions related to the paragraphs have been given. Each question has only one correct answer among the four given options (a), (b), (c) and (d).*

Passage I

A solenoid of resistance R and inductance L has a piece of soft iron inside it. A battery of emf E and of negligible internal resistance is connected across the solenoid as shown in the figure. At an instant, the piece of soft iron is pulled out suddenly so that inductance of the solenoid decreased to ηL with battery remain connected.

22 Find the work done to pull out the soft iron piece.

a. $\dfrac{\eta LE^2}{2R^2}$ b. $\dfrac{(1-\eta)LE^2}{2R^2}$

c. $\dfrac{(1-\eta)LE^2}{\eta R^2}$ d. $\dfrac{(1-\eta)LE^2}{2\eta R^2}$

23 Assuming $t = 0$ is the instant when iron piece has been pulled out, the current as a function of time is

a. $i = \dfrac{E}{R}\left(1-\left(1-\dfrac{1}{\eta}\right)e^{-t/\tau}\right)$

b. $i = \dfrac{E}{R}\left(1+\left(1+\dfrac{1}{\eta}\right)e^{-t/\tau}\right)$

c. $i = \dfrac{E}{R}\left(1-\left(1+\dfrac{1}{\eta}\right)e^{-t/\tau}\right)$

d. $i = \dfrac{E}{R}\left(1+\left(1-\dfrac{1}{\eta}\right)e^{-t/\tau}\right)$ $\left(\text{where, } \tau = \dfrac{\eta L}{R}\right)$

24 Find power supplied by the battery as a function of time.

a. $P = \dfrac{E^2}{R}\left(1-\left(1-\dfrac{1}{\eta}\right)e^{-t/\tau}\right)$ b. $P = \dfrac{E^2}{R}\left(1+\left(1+\dfrac{1}{\eta}\right)e^{-t/\tau}\right)$

c. $P = \dfrac{E^2}{R}\left(1-\left(1+\dfrac{1}{\eta}\right)e^{-t/\tau}\right)$ d. $P = \dfrac{E^2}{R}\left(1+\left(1-\dfrac{1}{\eta}\right)e^{-t/\tau}\right)$

Passage II

In the given circuit,

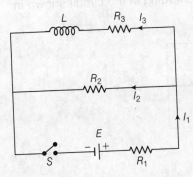

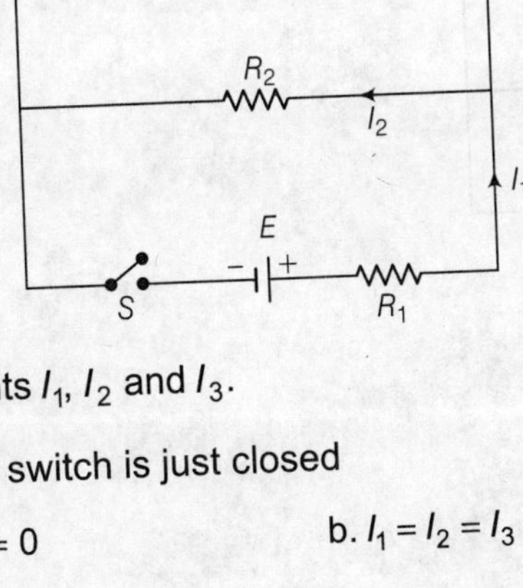

Determine the currents I_1, I_2 and I_3.

25 At the moment when switch is just closed

a. $I_1 = I_2 = \dfrac{\varepsilon}{R_1 + R_2}, I_3 = 0$ b. $I_1 = I_2 = I_3 = \dfrac{\varepsilon}{R_1 + R_2}$

c. $I_1 = I_2 = 0, I_3 = \dfrac{\varepsilon}{R_1 + R_2}$ d. $I_1 = \dfrac{\varepsilon}{R_1 + R_3}, I_2 = I_3 = 0$

Working Space

26 A long time after the switch is closed

a. $I_1 = \dfrac{\varepsilon R_3}{R_2R_3 + R_1R_2 + R_1R_3}$, $I_2 = \dfrac{\varepsilon R_2}{R_1R_2 + R_2R_3 + R_1R_3}$ and $I_3 = \dfrac{\varepsilon R_1}{R_1R_2 + R_2R_3 + R_1R_3}$

b. $I_1 = \dfrac{\varepsilon(R_3 + R_2)}{R_2R_3 + R_1R_3 + R_1R_2}$, $I_2 = \dfrac{\varepsilon R_3}{R_1R_3 + R_2R_3 + R_1R_2}$ and $I_3 = \dfrac{\varepsilon R_2}{R_1R_3 + R_2R_3 + R_1R_2}$

c. $I_1 = I_2 = I_3 = \dfrac{\varepsilon(R_1 + R_2 + R_3)}{R_1R_3 + R_2R_3 + R_1R_2}$

d. $I_1 = I_2 = \dfrac{\varepsilon(R_3 + R_2)}{R_2R_3 + R_1R_3 + R_1R_2}$, $I_3 = 0$

27 Just after the switch is opened

a. $I_1 = 0$, $I_2 = I_3 = \dfrac{\varepsilon(R_2 + R_3)}{R_1R_2 + R_2R_3 + R_3R_1}$

b. $I_1 = 0$, $I_2 = \dfrac{-\varepsilon R_2}{R_1R_2 + R_2R_3 + R_1R_3}$ and $I_3 = \dfrac{\varepsilon R_2}{R_1R_2 + R_2R_3 + R_3R_1}$

c. $I_1 = I_2 = I_3 = \dfrac{\varepsilon R_2}{R_1R_2 + R_2R_3 + R_3R_1}$

d. $I_1 = 0$, $I_2 = \dfrac{\varepsilon R_2}{R_2R_3 + R_1R_3 + R_1R_2}$ and $I_3 = \dfrac{\varepsilon R_2}{R_2R_3 + R_1R_3 + R_1R_3}$

One Integer Value Correct Type

***Direction** (Q. Nos. 28 and 29) This section contains 2 questions. When worked out will result in an integer from 0 to 9 (both inclusive).*

28 In the adjoining circuit, initially the switch is open. The switch is closed at $t = 0$. The difference between the maximum and minimum current that can flow in the circuit is K Amp. What is the value of K?

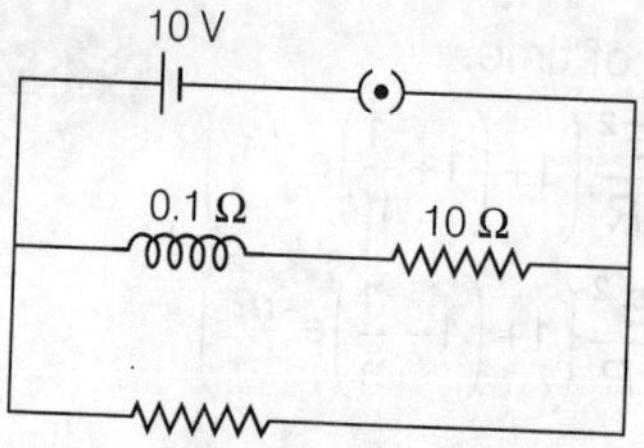

29 The ratio of time constant in charging and discharging in the circuit shown in figure is $3 : K$. Find the value of K?

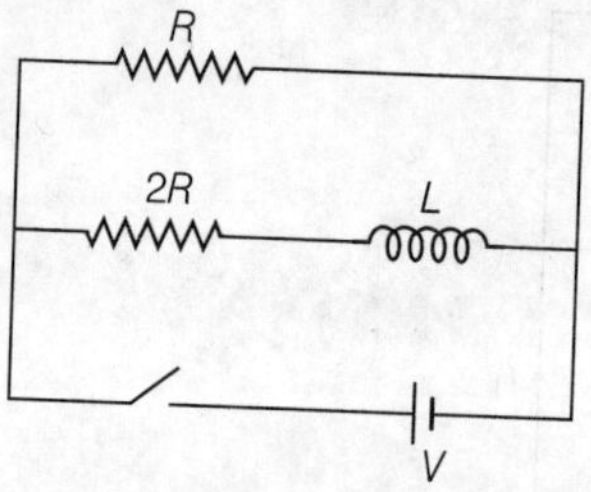

DPP-9 Mutual Induction and Inductance

Subjective Questions

Working Space

Direction (*Q. Nos.* 1-4) *These questions are subjective in nature, need to be solved completely on notebook.*

1 Two solenoids A and B are placed as shown in the figure.

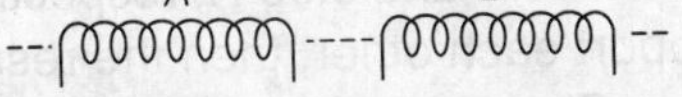

Number of turns of $A = 400$ and number of turns of $B = 700$. A current of 3.5 A in coil A produces an average flux of 300 μTm^2 through each turn of A and a flux of 90 μTm^2 through each turn of B. Calculate

i. mutual inductance of solenoids.
ii. self-inductance of solenoid A.
iii. what emf is induced in B when current in A is increasing at 0.5 A/s?

2 A short solenoid of length 4 cm, radius 2.0 cm and number of turns 100 lying inside on the axis of a long solenoid, 80 cm length and number of turns 1500. What is the flux through the long solenoid if a current of 3.0 A flows through the short solenoid? Also, obtain the mutual inductance of the two solenoids.

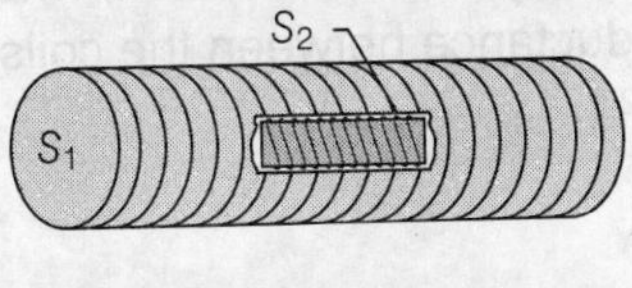

3 A circular loop of radius 0.3 cm lies parallel to a much bigger circular loop of radius 20 cm. The centre of the small loop is on the axis of the bigger loop. The distance between their centre is 15 cm.

i. What is the flux linking the bigger loop if a current of 2.0 A flows through the smaller loop?
ii. Obtain the mutual inductance of two loops.

4 A small square loop of wire of side l is placed inside a large square loop of wire of side $L(>> l)$. The loops are coplanar and their centre coincide. What is the mutual inductance of the system?

Only One Option Correct Type

Direction (*Q. Nos.* 5-17) *This section contains* 13 *multiple choice questions. Each question has four choices (a), (b), (c) and (d), out of which ONLY ONE is correct.*

5 Two coils are placed close to each other. The mutual inductance of the pair of coils depends upon **[AIEEE 2003]**

a. the relative position and orientation of the two coils
b. the materials of the wires of the coils
c. the currents in the two coils
d. the rates at which currents are changing in the two coils

Working Space

6 Two coaxial solenoids are made by winding thin insulated wire once a pipe of cross-sectional area $A = 10\ \text{cm}^2$ and length $= 20$ cm. If one of the solenoids has 300 turns and the other has 400 turns, their mutual inductance is **[AIEEE 2008]**

a. $4.8\pi \times 10^{-4}$ H
b. $4.8\pi \times 10^{-5}$ H
c. $2.4\pi \times 10^{-4}$ H
d. $2.4\pi \times 10^{-5}$ H

7 Two circular coils can be arranged in any of the three situations shown in the figure. The mutual inductance will be **[IIT JEE 2001]**

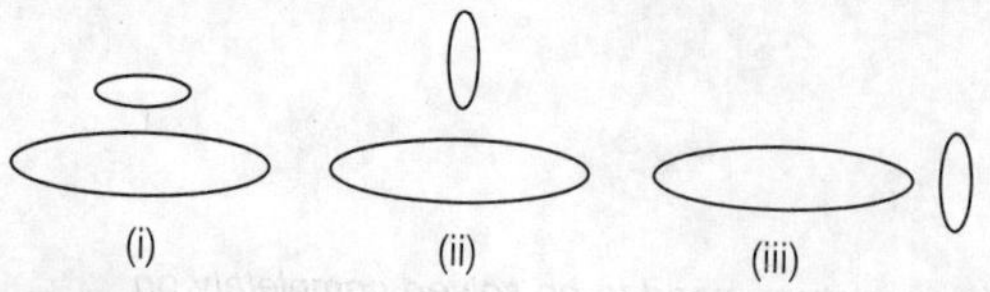

a. maximum in situation (i)
b. maximum in situation (ii)
c. maximum in situation (iii)
d. the same in all situations

8 The coefficients of self-inductance of two coils are 0.01 H and 0.03 H, respectively. When they are connected in series so as to support each other, then the resultant self-inductance becomes 0.06 H. The value of coefficient of mutual inductance will be

a. 0.02 H
b. 0.05 H
c. 0.01 H
d. zero

9 Two identical solenoid coils, each of self- inductance are connected in series. Their turns are in the same sense and the distance between them are such that the coefficient of coupling is half. Then, the equivalence inductance of the combination is

a. L
b. $2L$
c. $3L$
d. $L/2$

10 A small coil of radius r is placed at the centre of a large coil of radius R, where $R >> r$. The coils are coplaner. The coefficient inductance between the coils is

a. $\dfrac{\mu_0 \pi r}{2R}$
b. $\dfrac{\mu_0 \pi r^2}{2R}$
c. $\dfrac{\mu_0 \pi r^2}{2R^2}$
d. $\dfrac{\mu_0 \pi r}{2R^2}$

11 Two coils are at fixed locations. When coil 1 has no current and the current in coil 2 increases at the rate 15.0 A /s. The emf in coil 1 is 25.0 mV, when coil 2 has no current and coil 1 has a current of 3.6 A, flux linkage in coil 2 is

a. 16 mWb
b. 10 mWb
c. 4.00 mWb
d. 6.00 mWb

12 Two coils of self-inductance 100 mH and 400 mH are placed very close to each other. The maximum value of mutual inductance of the system is

a. 200 mH
b. 300 mH
c. $100\sqrt{2}$ mH
d. None of these

13 Two coils, X and Y are linked such that emf E is induced in Y when the current in X is changing at the rate dI/dt. If a current I_0 is now made to flow through Y, then flux linked with X will be

a. $EI_0\, dI/dt$
b. $\left(\dfrac{E}{dI/dt}\right) I_0$
c. $(E\, dI/dt)\, I_0$
d. $\dfrac{I_0\, dI/dt}{E}$

14 A small thin coil with N_2 number of turns, each of area A_2, is placed inside a long solenoid of N_1 turns and length l near its centre. Then, choose correct option.

a. Mutual inductance of coil is $\frac{\mu_0 N_1 N_2 A_2}{l}$

b. Mutual inductance of coil is $\frac{\mu_0 N_1^2 A_2}{l}$

c. Mutual inductance of coil is a constant quantity

d. Mutual inductance of coil is $\frac{\mu N_1 N_2 A_2 \sin\theta}{l}$, where θ = angle between axis of solenoid and plane of coil

15 The mutual inductance between the rectangular loop and the long straight wire as shown in figure is

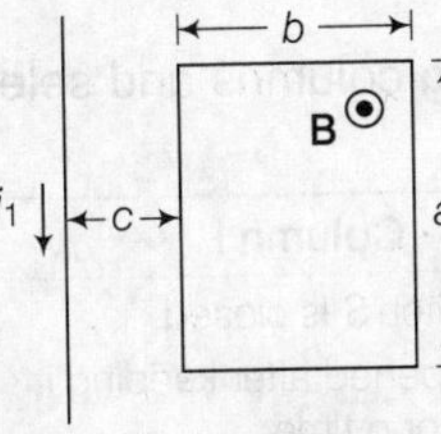

a. zero

b. $\frac{\mu_0 a}{2\pi} \ln\left(1+\frac{c}{b}\right)$

c. $\frac{\mu_0 b}{2\pi} \ln\left(\frac{a+c}{b}\right)$

d. $\frac{\mu_0 a}{2\pi} \ln\left(1+\frac{b}{c}\right)$

16 A square loop of sides a is placed in XY-plane. A very long wire is also placed in XY-plane such that side of length a of the loop is parallel to the wire. The distance between the wire and the nearest edge of the loop is d. The mutual inductance of this system is proportional to

a. a

b. a^2

c. $1/d$

d. current in wire

17 What is the mutual inductance of a two loop system as shown with centre separation l?

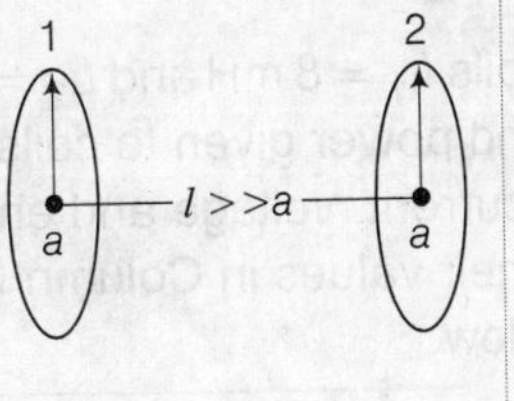

a. $\frac{\mu_0 \pi a^4}{8 l^3}$

b. $\frac{\mu_0 \pi a^4}{4 l^3}$

c. $\frac{\mu_0 \pi a^4}{6 l^3}$

d. $\frac{\mu_0 a^4}{2 l^3}$

Statement Type

Direction *(Q. Nos. 18 and 19) This section is based on Statement I and Statement II. Select the correct answer from the codes given below.*

a. Both Statement I and Statement II are correct and Statement II is the correct explanation of Statement I

b. Both Statement I and Statement II are correct but Statement II is not the correct explanation of Statement I

c. Statement I is correct but Statement II is incorrect

d. Statement II is correct but Statement I is incorrect

18 **Statement I** The mutual inductance of two concentric conducting rings of different radii is maximum if the rings are coplanar.

Statement II For two coaxial conducting rings of different radii, the magnitude of magnetic flux in one ring due to current in other ring is maximum when both rings are coplaner.

19 **Statement I** An electric lamp is connected in series with a long solenoid of copper with air core and then connected to an AC source. If an iron rod is inserted in the solenoid, then lamp becomes dim.

Statement II If an iron rod is inserted in the solenoid, then inductance of the solenoid increases.

Matching List Type

Working Space

Direction *(Q. Nos. 20 and 21) Choices for the correct combination of elements from Column I and Column II are given as options (a), (b), (c) and (d), out of which one is correct.*

20 Figure shows two coaxial coils M and N. Column I is regarding some operations done with coil M and Column II about induced current in coil N.

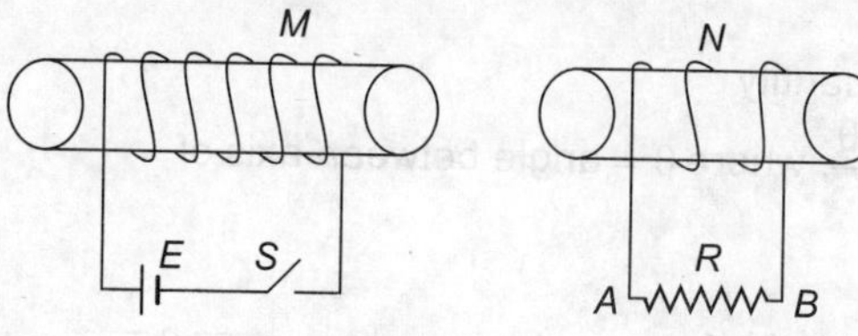

Match the following columns and select the correct option from the codes given below.

	Column I		Column II
i.	Just after switch S is closed.	p.	Current is induced from A to B.
ii.	Switch S is opened after keeping it closed for a long time.	q.	Current is induced from B to A.
iii.	After switch S is closed for a long time.	r.	No current is induced.
iv.	Just after switch S is closed while moving M away from N.		

Codes

	i	ii	iii	iv		i	ii	iii	iv
a.	q	r	p	r	b.	r	q	p	p
c.	q	q	r	p	d.	p	q	r	r

21 Two different coils $L_1 = 8$ mH and $L_2 = 2$ mH. Rate of increase of current in both coils is same and power given to coils is same. If i, V and U indicates instantaneous current, voltage and energy stored, then match the ratios in Column I with their values in Column II and select the correct option from the codes given below.

	Column I		Column II
i.	i_1 / i_2	p.	4 : 1
ii.	V_1 / V_2	q.	1 : 4
iii.	L_1 / L_2	r.	1 : 2

Codes

	i	ii	iii		i	ii	iii
a.	p	q	r	b.	r	p	q
c.	q	p	q	d.	r	q	p

Comprehension Type

Direction *(Q. Nos. 22-24) This section contains a paragraph, which describing theory, experiments, data, etc. Three questions related to the paragraph have been given. Each question has only one correct answer among the four given options (a), (b), (c) and (d).*

Passage

There is a conducting loop $ABCDEF$ of resistance λ per unit length placed near a long straight current carrying wire. The dimensions are shown in the figure. The long wire lies in the plane of the loop. The current in the long wire varies as $I = I_0 . t$, where t is instantaneous value of time

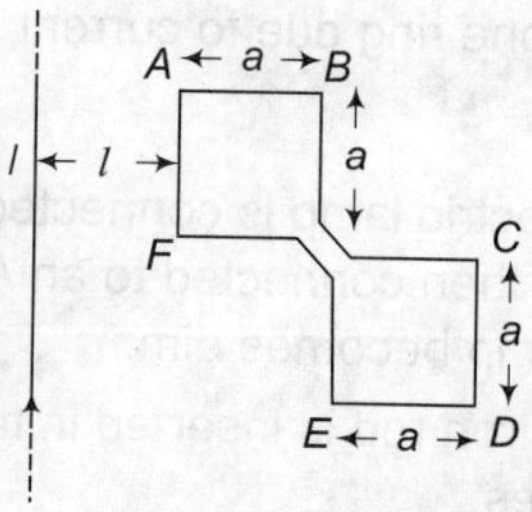

22 The mutual inductance of the pair is

a. $\frac{\mu_0 a}{2\pi}\ln\left(\frac{2a+l}{l}\right)$ b. $\frac{\mu_0 a}{2\pi}\ln\left(\frac{2a-l}{l}\right)$

c. $\frac{2\mu_0 a}{\pi}\ln\left(\frac{a+l}{l}\right)$ d. $\frac{\mu_0 a}{\pi}\ln\left(\frac{a+l}{l}\right)$

23 The emf induced in the closed loop is

a. $\frac{\mu_0 I_0 a}{2\pi}\ln\left(\frac{2a+l}{l}\right)$ b. $\frac{\mu_0 I_0 a}{2\pi}\ln\left(\frac{2a-l}{l}\right)$

c. $\frac{2\mu_0 I_0 a}{\pi}\ln\left(\frac{a+l}{l}\right)$ d. $\frac{\mu_0 I_0 a}{\pi}\ln\left(\frac{a+l}{l}\right)$

24 The heat produced in the loop in time t is

a. $\frac{\left[\frac{\mu_0 I_0}{2\pi}\ln\left(\frac{a+l}{l}\right)\right]^2 at}{4\lambda}$ b. $\frac{\left[\frac{\mu_0 I_0}{2\pi}\ln\left(\frac{2a+l}{l}\right)\right]^2 at}{8\lambda}$

c. $\frac{\left[\frac{2\mu_0 I_0}{\pi}\ln\left(\frac{a+l}{l}\right)\right]^2 at}{3\lambda}$ d. $\frac{\left[\frac{\mu_0 I_0}{2\pi}\ln\left(\frac{3a+l}{l}\right)\right]^2 at}{6\lambda}$

One Integer Value Correct Type

Direction *(Q. Nos. 25 and 26) This section contains 2 questions. When worked out will result in an integer from 0 to 9 (both inclusive).*

25 A circular wire loop of radius R is placed in the XY-plane centred at the origin O. A square loop of side a ($a << R$) having two turns is placed with its centre at $z = \sqrt{3}R$ along the axis of the circular loop as shown in figure. The plane of the loop makes an angle of 45° with respect to the Z-axis. If the mutual inductance between the loops is given by $\frac{\mu_0 a^2}{2^{P/2}\cdot R}$, then the value of P. **[JEE Advanced 2012]**

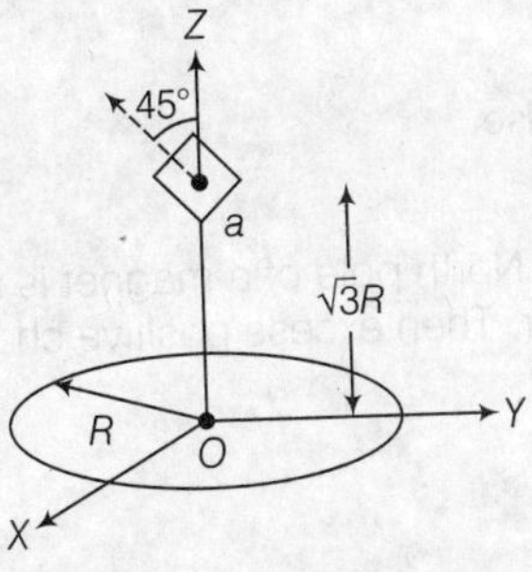

26 Two solenoids A and B are placed near to each other coaxially. Number of turns of A is 400 and number of turns of B is 700. A current of 3.5 A in coil A produces an average flux of 300 μWb through each turn of A and a flux of 90 μWb through each turn of B. Find induced emf (in millivolt) B when the current is A increases at the rate of 0.5 A/s.

SET - 1

Revisal Problems
for JEE Main

Only One Option Correct Type

***Direction** (Q. Nos. 1-24) This section contains 24 multiple choice questions. Each question has four choices (a), (b), (c) and (d), out of which ONLY ONE is correct.*

Working Space

1 A rectangular coil of size 10 cm $\times$ 20 cm has 60 turns. It is rotating in a magnetic field of 0.5 Wb / m^2 at a rate of 1800 rev/min. The maximum induced emf across the ends of the coil is

a. 111 V
b. 112 V
c. 113 V
d. 114 V

2 A closed coil of copper of 1 m $\times$ 1 m and of resistance of 2 Ω is placed perpendicular to a magnetic field of 0.10 Wb/m^2. It is rotated through 180° in 0.01 s. The induced emf and induced current in the coil will respectively be

a. 20 V, 10 A
b. 10 V, 20 A
c. 10 V, 10 A
d. 20 V, 20 A

3 A coil having 100 turns and area of 0.001 m^2 is free to rotate about an axis. The coil is placed perpendicular to a magnetic field of 1 Wb / m^2. If the coil rotates rapidly through an angle of 180°, then charge flown through coil will be (the resistance of the coil is 10 Ω).

a. 0.01 C
b. 0.02 C
c. 0.03 C
d. 0.04 C

4 A current carrying solenoid is approaching a conducting loop as shown in the figure. The direction of induced current as observed by an observer on the other side of the loop will be

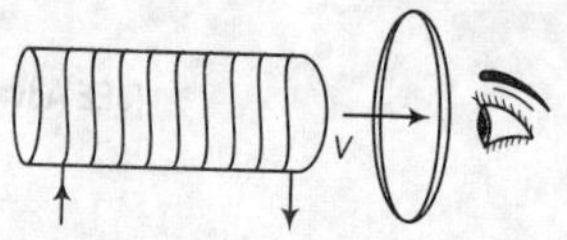

a. anti-clockwise
b. clockwise
c. East
d. West

5 Consider the arrangement shown in figure in which the North pole of a magnet is moved away from a thick conducting loop containing capacitor. Then excess positive charge will arrive on

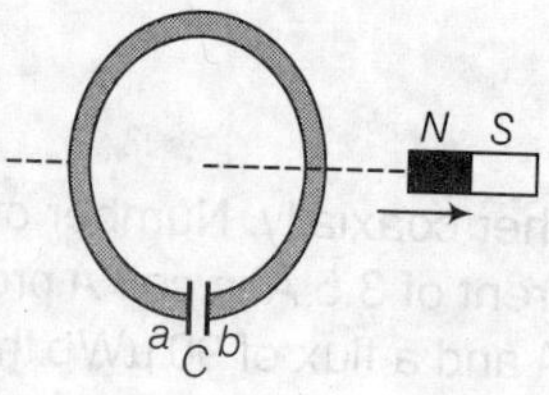

a. plate *a*
b. plate *b*
c. both plates *a* and *b*
d. neither *a* nor *b* plates

6 The current changes in an inductance coil of 100 mH from 100 mA to zero in 2 ms. The emf induced in the coil will be

a. -5V
b. 5 V
c. -50V
d. 50 V

7 When a small piece of wire passes between the magnetic poles of a horse-shoe magnet in 0.1s, emf of 4×10^{-3} V is induced in it. The magnetic flux between the poles is

a. 4×10^{-2} Wb
b. 4×10^{-3} Wb
c. 4×10^{-4} Wb
d. 4×10^{-6} Wb

Working Space

8 The normal magnetic flux passing through a coil changes with time according to following equation $\phi = 10t^2 + 5t + 1$, where ϕ is in milli webers and t is in seconds. The value of induced emf produced in the coil at $t = 5$ s will be

a. zero
b. 1 V
c. 2 V
d. 0.105 V

9 A bicycle wheel of radius 0.5 m has 32 spokes. It is rotating at the rate of 120 rev/min perpendicular to the horizontal component of earth's magnetic field $B_H = 4 \times 10^{-5}$ T. The emf induced between the rim and the centre of the wheel will be

a. 6.28×10^{-5} V
b. 4.8×10^{-5} V
c. 6.0×10^{-5} V
d. 1.6×10^{-5} V

10 A thick wire in the form of a semicircle of radius r is rotated with a frequency f in a magnetic field . What will be the peak value of emf induced?

a. $B\pi r^2 f$
b. $B\pi^2 r^2 f$
c. $2Br^2 f$
d. $2B\pi^2 r^2 f$

11 An aeroplane having a distance of 50 m between the edges of its wings is flying horizontally with a speed of 360 km/h. If the vertical component of earth's magnetic field is 4×10^{-4} Wb/m², then the induced emf between the edges of its wings will be

a. 2 mV
b. 2 V
c. 0.2 V
d. 20 V

12 An angular conductor is moving with velocity v along its angular bisector in a perpendicular magnetic field (B) as shown in the figure. The induced potential difference between its free ends will be

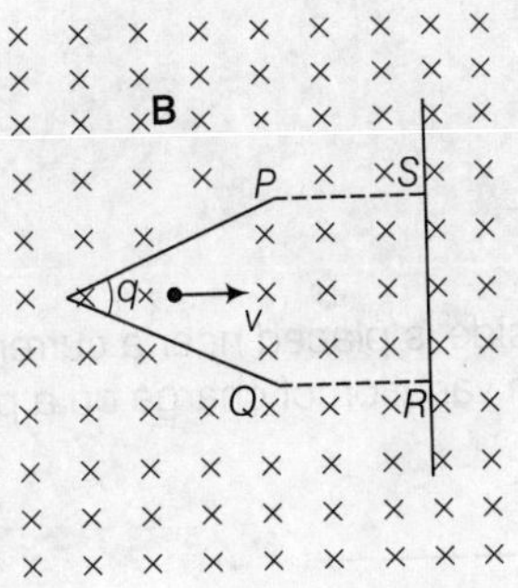

a. $2Bvl\sin\frac{\theta}{2}$
b. $2Bvl$
c. $2Bvl\sin\theta$
d. zero

13 A triangular loop lie near a straight long current carrying wire as shown in the figure.

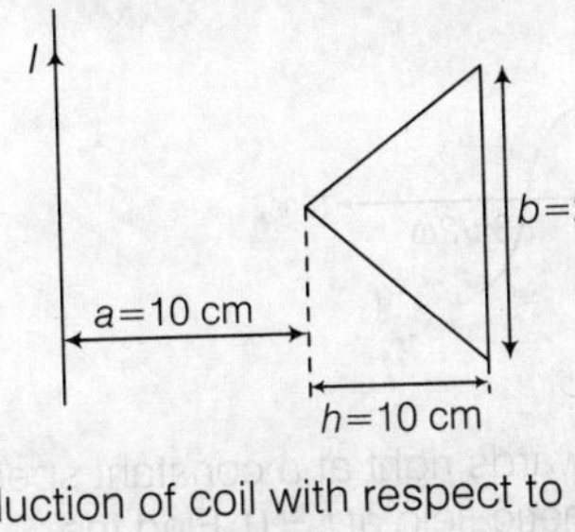

Coefficent of mutual induction of coil with respect to the wire is

a. 1.22×10^{-8} H
b. 2.55×10^{-8} H
c. 3.77×10^{-8} H
d. 5.66×10^{-8} H

14 A conducting rod is in form of parabola $y = kx^2$. It is placed in a region of uniform and perpendicular magnetic field of strength B. At $t = 0$, a conducting straight rod starts sliding up on parabola from ($x = 0, y = 0$) with a constant acceleration a and the parabolic frame starts rotating with angular speed with the frame after t second is

a. $\frac{B}{4}\sqrt{\frac{5}{k}} \cdot a^{3/2} \cdot t^2 \cos\theta$
b. $\frac{B}{3}\sqrt{\frac{2}{k}} \cdot a \cdot t^3$
c. $\frac{B}{3} \cdot \sqrt{\frac{2}{k}} \cdot a^{3/2} \cdot t^3 \sin\theta$
d. $\frac{B}{3}\sqrt{\frac{2}{k}} \cdot a^{3/2} \cdot t^3 \cdot \cos\theta$

15 In the circuit given below,

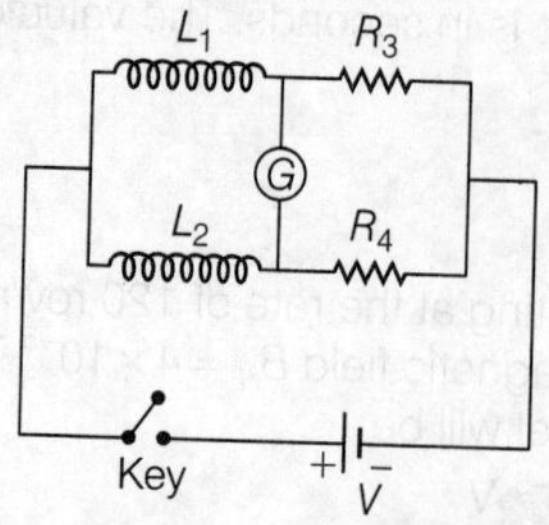

Inductors L_1 and L_2 have resistances of R_1 and R_2 are connected with R_3 and R_4. At $t = 0$ key is closed, the galvanometer always shows a zero deflection when

a. $L_1 = L_2$ and $\frac{R_1}{R_2} = \frac{R_3}{R_4}$

b. $R_1 = R_2$ and $\frac{R_3}{R_4} = \frac{L_1}{L_2}$

c. $\frac{L_1}{L_2} = \frac{R_1}{R_2} = \frac{R_3}{R_4}$

d. $\frac{L_1}{L_2} = \frac{R_2}{R_1} = \frac{R_3}{R_4}$

16 In the circuit shown in figure, $E = 10$ V, $R_1 = 1\Omega$, $R_2 = 2\,\Omega$, $R_3 = 3\Omega$ and $L = 2$H. Calculate the value of current i_1, i_2 and i_3 immediately after key S is

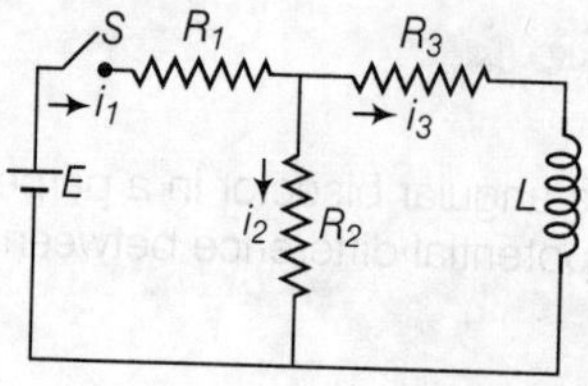

a. 3.3 A, 3.3 A, 3.3 A
b. 3.3 A, 3.3 A, 0
c. 3.3 A, 0, 0
d. 3.3 A, 3.3 A, 1.1 A

17 A square loop with a capacitor connected in one of its side is placed near a current carrying wire. If current in wire varies ar $I = I_0 \sin \omega t$, then variation of charge on a plate of capacitor is shown by

a. 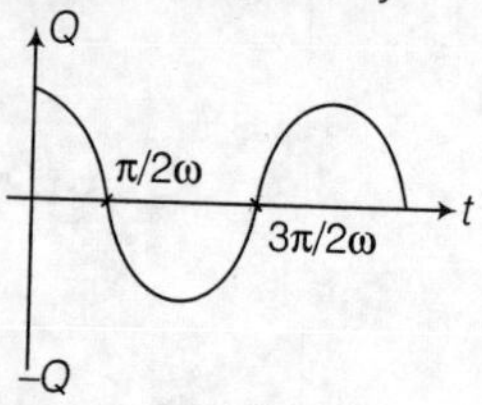

b. 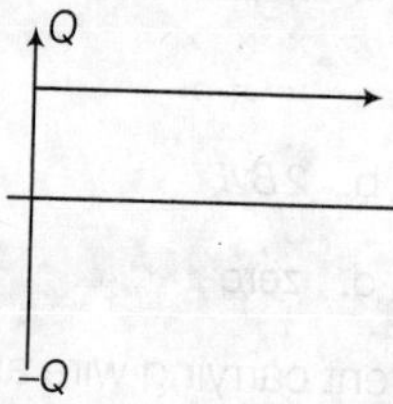

c. 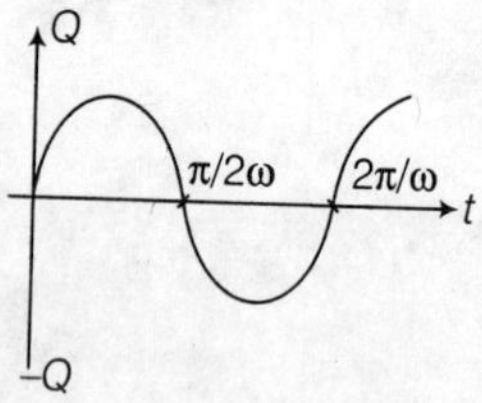

d. 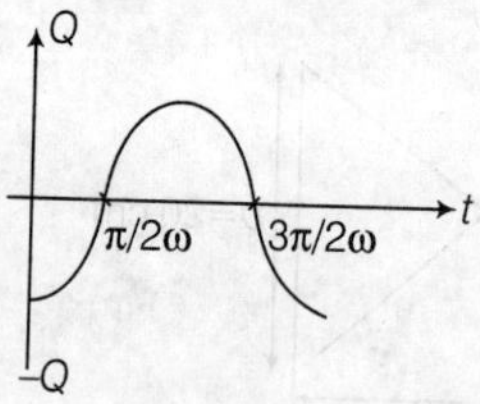

18 Figure shows a square loop of side 5 cm being moved towards right at a constant speed of 1 cm/s. The front edge just enters the 20 cm wide magnetic field at $t = 0$. Find the induced emf in the loop at $t = 2$ s and $t = 10$ s.

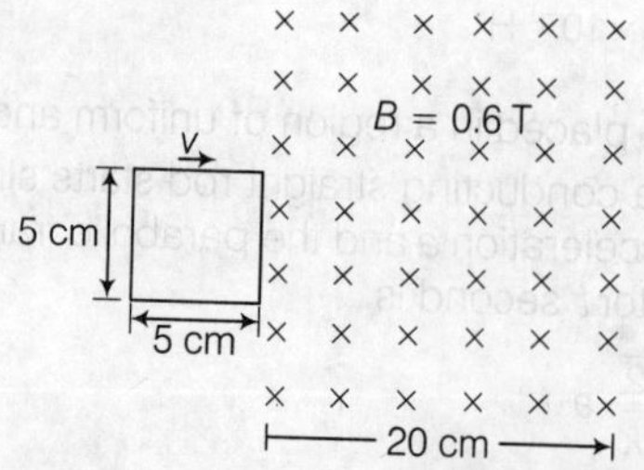

a. 3×10^{-2}, zero

b. 3×10^{-2}, 3×10^{-4}

c. 3×10^{-4}, 3×10^{-4}

d. 3×10^{-4}, zero

19 A rectangular loop sides 10 cm and 3 cm moving out of a region of uniform magnetic field of 0.5 T directed normal to the loop. If we want to move loop with a constant velocity $1\,cm\,s^{-1}$, then required mechanical force is (resistance of loop = $1\,\Omega$)

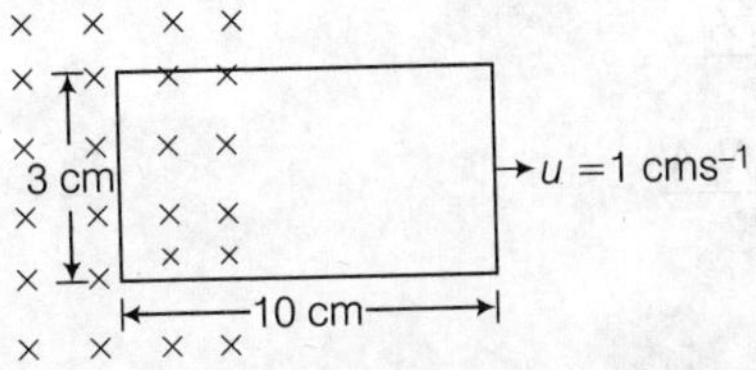

a. 2.25×10^{-8} N
b. 4.5×10^{-3} N
c. 9×10^{-3} N
d. 1.25×10^{-3} N

20 A loop of wire is placed in a magnetic field $\mathbf{B} = 0.02\,\hat{\mathbf{i}}$ T. Then, the flux through the loop is, if its area vector is $\mathbf{A} = 30\,\hat{\mathbf{i}} + 16\,\hat{\mathbf{j}} + 23\,\hat{\mathbf{k}}\,cm^2$

a. 60 μWb
b. 32 μWb
c. 46 μWb
d. 138 μWb

21 The magnetic flux passing perpendicular to the plane of the coil and directed into the paper is varying according to the relation.

$$\phi = 3t^2 + 2t + 3$$

where, ϕ is in milliwebers and t is in second. Then, the magnitude of emf induced in the loop, when $t = 2$ s is

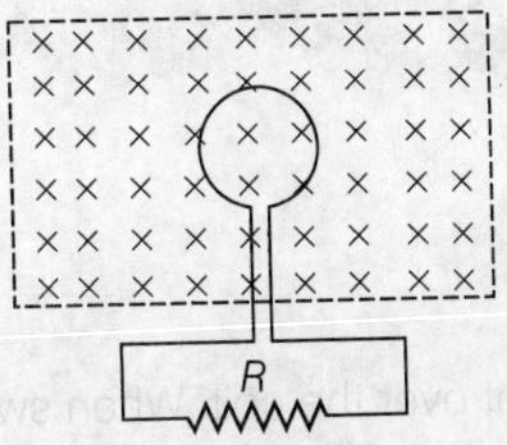

a. 31 mV
b. 19 mV
c. 14 mV
d. 6 mV

22 A conducting wire in the shape of Y with each side of length l is moving in a uniform magnetic field B, with a uniform speed v as shown in the figure. The induced emf at the two ends X and Y of the wire will be

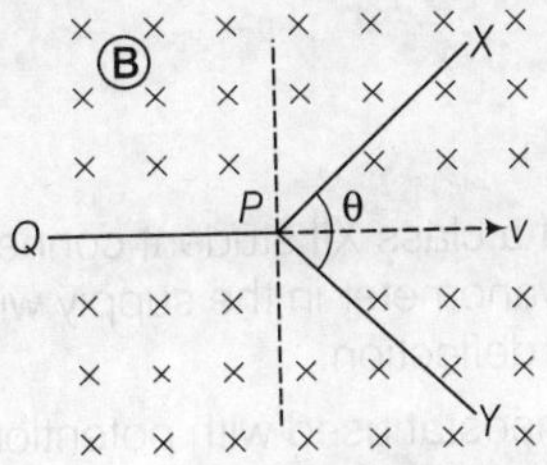

a. zero
b. $2Blv$
c. $2Blv\sin(\theta/2)$
d. $2Blv\cos(\theta/2)$

23 In figure $CODF$ is a semicircular loop of a conducting wire of resistance R and radius r. It is placed in a uniform magnetic field B, which is directed into the page (perpendicular to the plane of the loop). The loop is rotated with a constant angular speed ω about an axis passing through the centre O and perpendicular to the page. Then, the induced current in the wire loop is

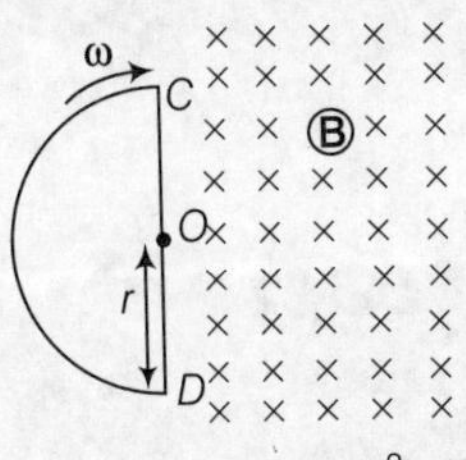

a. zero
b. $Br^2\omega/R$
c. $Br^2\omega/2R$
d. $B\pi r^2\omega/R$

Working Space

24 A small coil of N_1 turns, l_1 length, is tightly wound over the centre of a long solenoid of length l_2, area of cross-section A and number of turns N_2. If a current I flows in the small coil, then what is the flux through the long solenoid?

a. Zero
b. $\frac{\mu_0 N_1^2 A I}{l_1}$
c. Infinite
d. $\frac{\mu_0 N_1 N_2 A I}{l_2}$

Statement Type

Direction *(Q. Nos. 25-28) This section is based on Statement I and Statement II. Select the correct answer from the codes given below.*

a. Both Statement I and Statement II are correct and Statement II is the correct explanation of Statement I
b. Both Statement I and Statement II are correct but Statement II is not the correct explanation of Statement I
c. Statement I is correct but Statement II is incorrect
d. Statement II is correct but Statement I is incorrect

25 **Statement I** A vertical iron rod has a coil wound over it at its lower end.

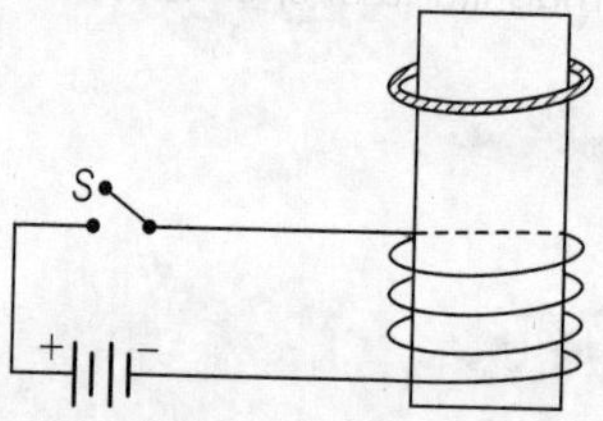

A conducting ring that can easily slide over the rod is kept over the coil. When switch S is closed, the ring jumps and float at a certain height above the coil and stays there.

Statement II An induced current in the ring interacts with horizontal component of earth's magnetic field to produce an average force in the upwards direction.

26 **Statement I** A series R-C circuit is connected to AC voltage source. Current through resistor is more when capacitor is filled with a dielectric.

Statement II Impedence of circuit is given by

$$Z = \sqrt{R^2 + (X_C)^2}$$

27 **Statement I** While doing potentiometer experiment, Amit a class XII student connected supply using multistrand wire. He also connected a galvanometer in the supply wire. When he switch off the supply, galvanometer shows a reverse deflection.

Statement II Multistrand connecting wires and coil of rheostat used with potentiometer has a non few inductance value.

28 **Statement I** A coil of metal wire is kept stationary in a non-uniform magnetic field, an emf is always induced in the coil.

Statement II Induced emf is proportional to the rate of change of flux linked with the coil.

Revisal Problems
for JEE Advanced

Only One Option Correct Type

Working Space

Direction *(Q. Nos. 1-19) This section contains 19 multiple choice questions. Each question has four choices (a), (b), (c) and (d), out of which ONLY ONE is correct.*

1 A conducting bar is pulled with a constant speed v on a smooth conducting rail. The region has a steady magnetic field of induction B as shown in the figure. If the speed of the bar is doubled, then the rate of heat dissipation will

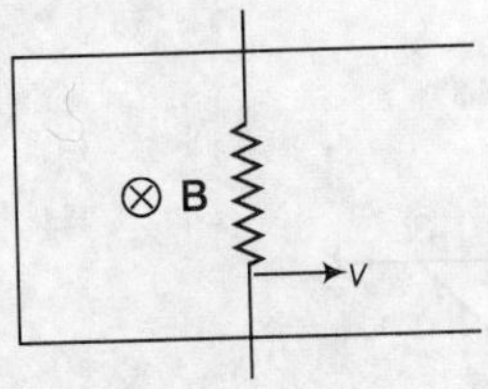

a. remain constant
b. become quarter of initial value
c. become four fold
d. get doubled

2 The coefficient of mutual inductance of the two coils is 0.5 H. If the current is increased from 2 A to 3 A in 0.01 s in one of them, then induced emf in second coil is

a. 25 V
b. 50 V
c. 75 V
d. 100 V

3 A conducting wire ABC (as shown in the figure) is moving with a constant velocity along horizontal direction. The magnetic field is perpendicular to the wire and directed into the page $AB = AC$. Find the value of angle θ made by the rod AB with horizontal at A so that emf induced at A is greater than C.

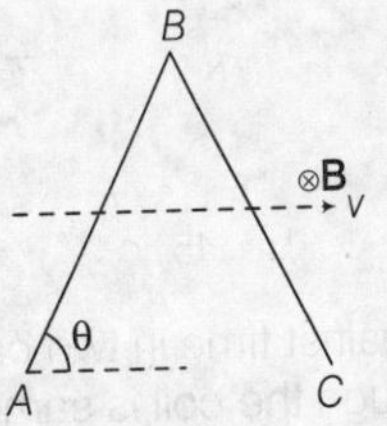

a. $\theta > 45°$
b. $\theta < 45°$
c. $\theta > 60°$
d. $\theta > 75°$

4 A uniform magnetic field exists in a square of side $2a$ (as shown in the figure).

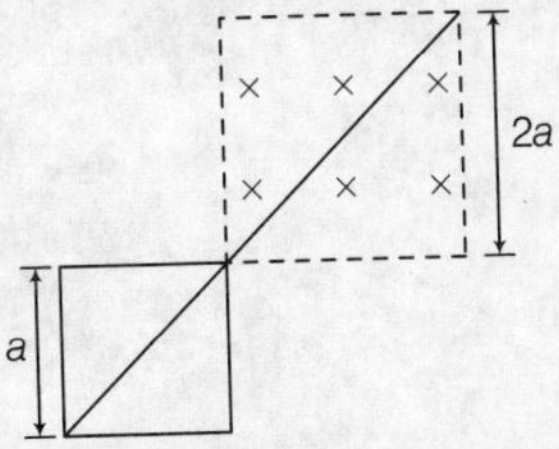

A square loop of side a enters in the field along a diagonal and leaves it at a constant speed. Draw the curve between induced emf e and distance along the diagonal, say x_0

a.
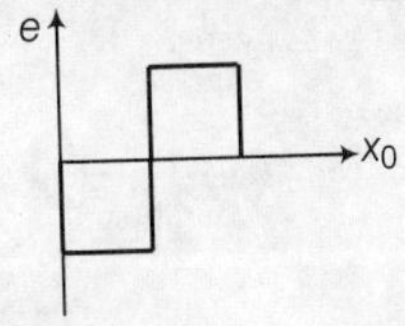

b.
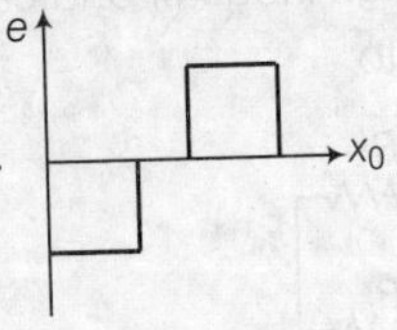

c.
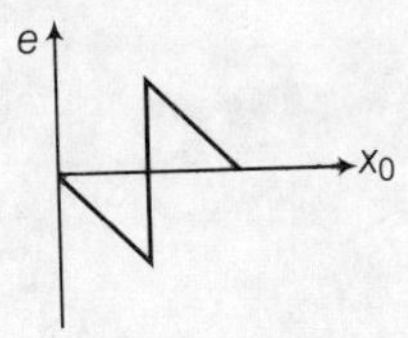

d.
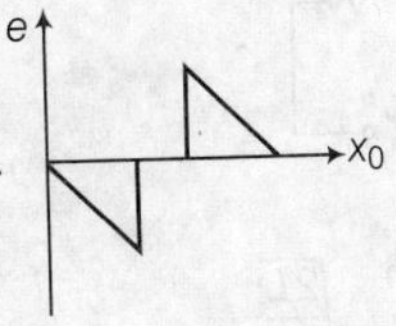

Working Space

5 A right angled triangular loop (as shown in the figure) enters uniform magnetic field (at right angle to the boundary of the field) directed into the paper. Draw the graph between induced emf e and the distance along the perpendicular to the boundary of the field (say x) along which loop moves.

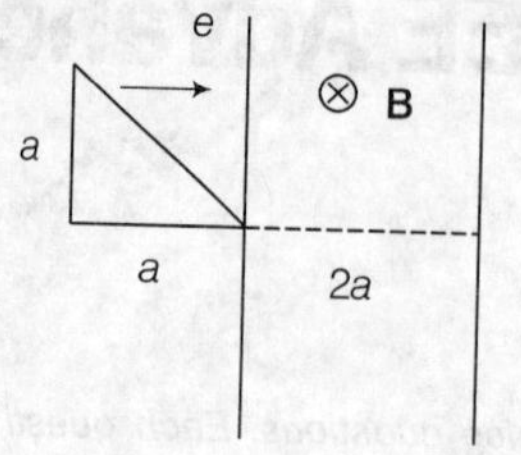

a.

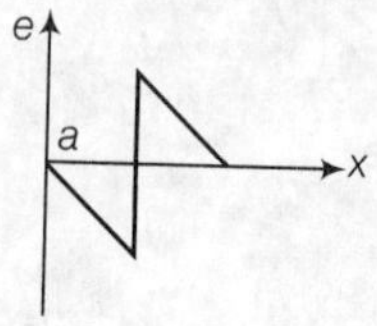

b.

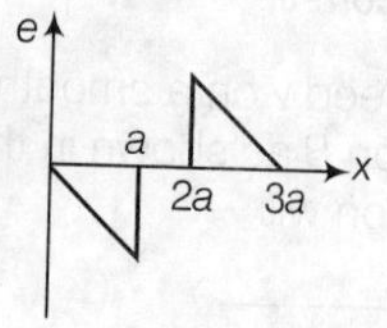

c.

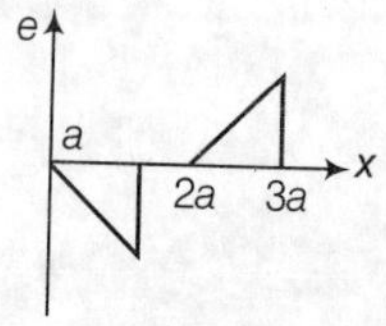

d.

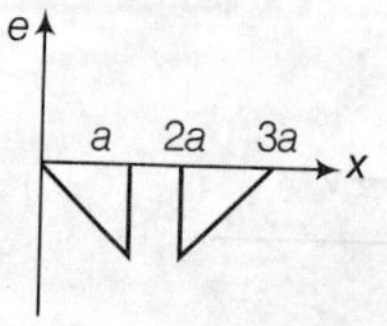

6 A conducting rod AB moves with constant velocity v on two conducting rails joined at 90° at C, when $\angle BAC = \alpha$. The magnetic field is directed into the page. The rod starts moving at vertex C at time $t = 0$. If the induced emf in the circuit at any time t is $2Bv^2t$, find the value of α.

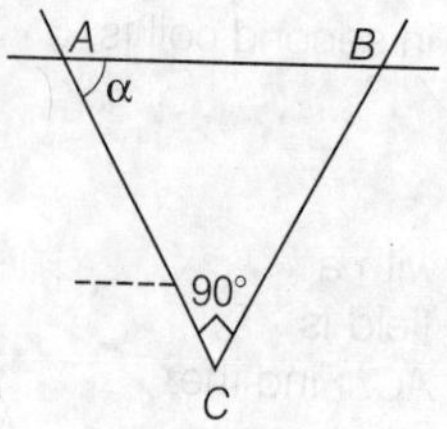

a. 30° b. 45° c. 60° d. 75°

7 Induced current against time in two cases is shown in the figure. If the magnitude of change in flux through the coil is same in both the cases, what should be the percentage decrease in resistance in case II

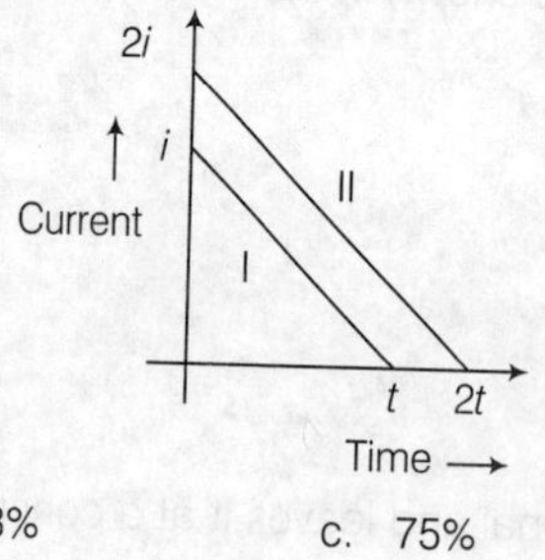

a. 25% b. 33% c. 75% d. 50%

8 The switch is closed at $t = 0$. The current through inductor and capacitor is equal after time interval $t = CR \log 2$. The resistance R is equal to

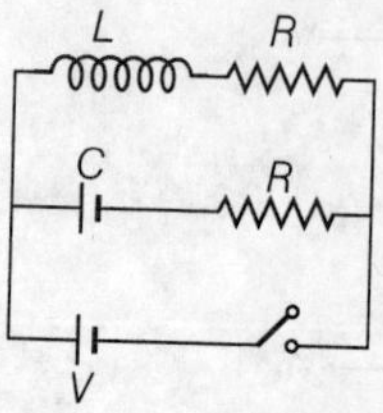

a. $\sqrt{L/2C}$ b. $\sqrt{L/C}$ c. $\sqrt{\frac{2L}{C}}$ d. $\left(\frac{L}{C}\right)^{3/2}$

9 A current of 2 A flowing through a coil of 100 turns gives rise to a magnetic flux of 5×10^{-5} Wb per turn. The magnetic energy associated with the coil is

a. 5 J
b. 0.5 J
c. 0.05 J
d. 0.005 J

10 In the circuit shown switch S is connected to position 2 for a long time and then joined to position 1. The total heat produced in resistance R_1 is

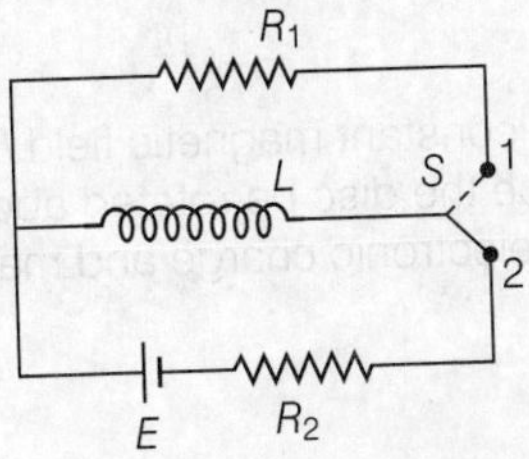

a. $\dfrac{LE^2}{2R_2^2}$
b. $\dfrac{LE^2}{2R_1^2}$
c. $\dfrac{LE^2}{2R_1R_2}$
d. $\dfrac{LE^2(R_1+R_2)^2}{2R_1^2+R_2^2}$

11 In the circuit shown S_1 and S_2 are switches. S_2 remains closed for a long time and S_1 open. Now S_1 is also closed. Just after S_1 is closed, the potential difference (V) across R is

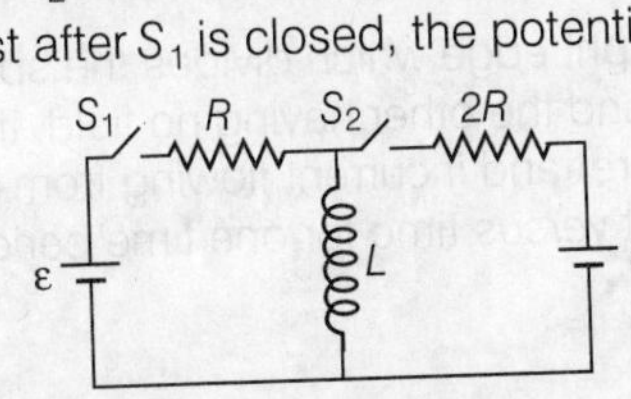

a. $\dfrac{\varepsilon}{3}$
b. 2ε
c. $\dfrac{2\varepsilon}{3}$
d. ε

12 The current through the solenoid is changing in such a way that flux through it is given by $\phi=\varepsilon t$. Then, the reading of the two AC voltmeters V_1 and V_2 differ by

a. zero
b. ε
c. $\left|\dfrac{\varepsilon(R_1-R_2)}{R_1+R_2}\right|$
d. $\dfrac{\varepsilon R_1R_2}{R_1+R_2}$

13 In the circuit shown in figure, L is ideal inductor and E is ideal cell. Switch S is closed at $t=0$. Then,

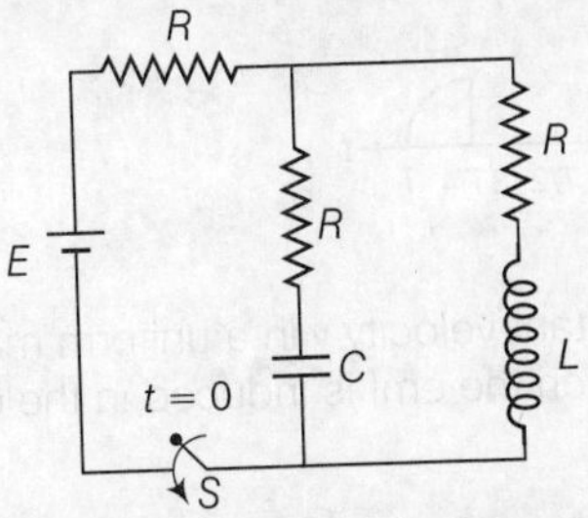

a. after a long time interval potential difference across capacitor and inductor will be equal
b. after a long time interval charge on capacitor will be EC
c. after a long time interval current in the inductor will be E/R
d. after a long time interval current through battery will be less than current through it initially

14 A triangular loop as shown in the figure is started to being pulled out at $t=0$ from a uniform magnetic with a constant velocity v. Total resistance of the loop is constant and equal to R. Then, the variation of power produced in the loop with time will be

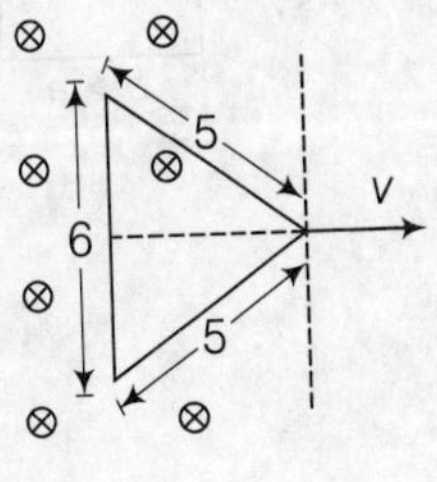

a. linearly increasing with time till whole loop comes out
b. increases parabolically till whole loop comes out
c. $P\propto t^3$ till whole loop come out
d. will be constant with time

15 A thin circular ring of area A is held perpendicular to a uniform field of induction B. A small cut is made in the ring and a galvanometer is connected across the ends such that the total resistance of the circuit is R. When the ring is suddenly squeezed to zero area, the charge flowing through the galvanometer is **[IIT JEE 1995]**

a. $\frac{BR}{A}$

b. $\frac{AB}{R}$

c. ABR

d. $\frac{B^2A}{R^2}$

16 A conducting disc of radius R is placed in a uniform and constant magnetic field B parallel to the axis of the disc. With what angular speed should be the disc be rotated about its axis such that no electric field develops in the disc. (the electronic charge and mass are e and m)

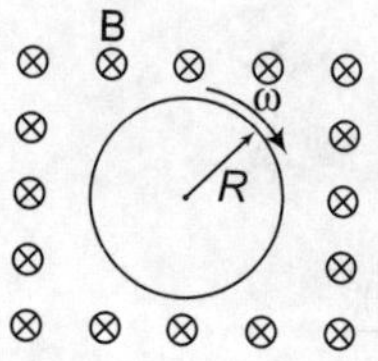

a. eB/m

b. $2eB/m$

c. $2eB/3m$

d. $eB/3m$

17 A semicircular loop of radius R is rotated about its straight edge which divides the space into two region one having a uniform magnetic field B and the other having no field. If initially the plane of loop is perpendicular to **B** (see figure) and if current flowing from O to A be taken as positive, the correct plot of induced current *versus* time for one time period is

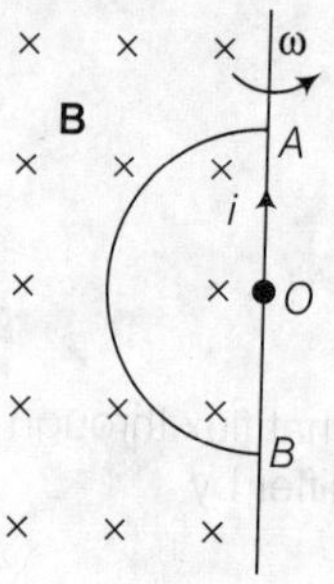

a.
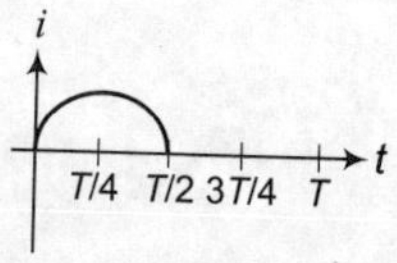

b.
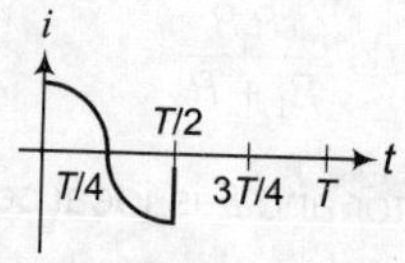

c.
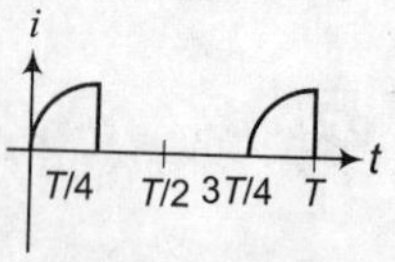

d.
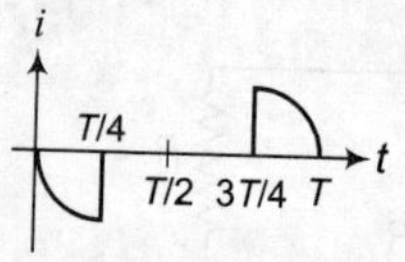

18 In the given arrangement, the loop is moved with constant velocity v in a uniform magnetic field B in a restricted region of width a. The time for which the emf is induced in the circuit is

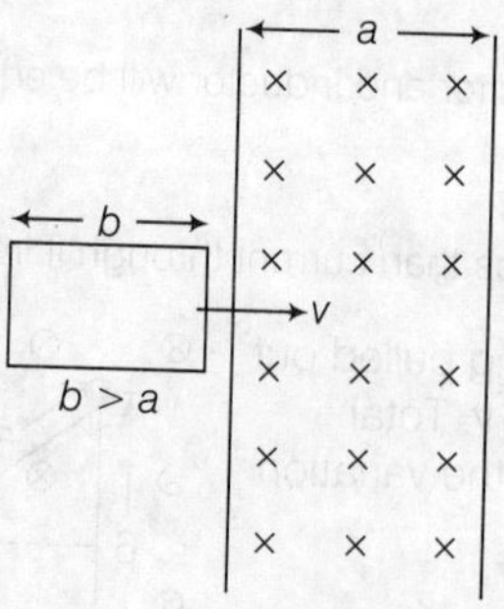

a. $\frac{2b}{v}$

b. $\frac{2a}{v}$

c. $\frac{(a+b)}{v}$

d. $\frac{2(a-b)}{v}$

Working Space

Working Space

19 Two coils have a mutual inductance 0.005 H. The current changes in the first coil according to equation $I = I_0 \sin \omega t$, where $I_0 = 10$ A and $\omega = 100\pi$ rad s^{-1}. The maximum value of emf in the second coil is

a. 2π b. 5π
c. 6π d. 12π

One or More than One Options Correct Type

***Direction** (Q. Nos. 20-26) This section contains 7 multiple choice questions. Each question has four choices (a), (b), (c) and (d), out of which ONE or MORE THAN ONE are correct.*

20 A conducting loop is pulled with a constant velocity towards a region of uniform magnetic field of induction B as shown in the figure. Then, the current involved in the loop is ($d > r$)

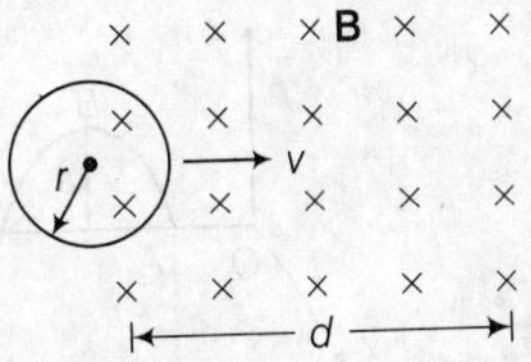

a. clockwise while entering
b. anti-clockwise while entering
c. zero when completely inside
d. clockwise while leaving

21 A loop is so that its centre lies at the origin of coordinate system. A magnetic field has the induction B pointing along Z-axis as shown in the figure. Which of the following statements is/are correct?

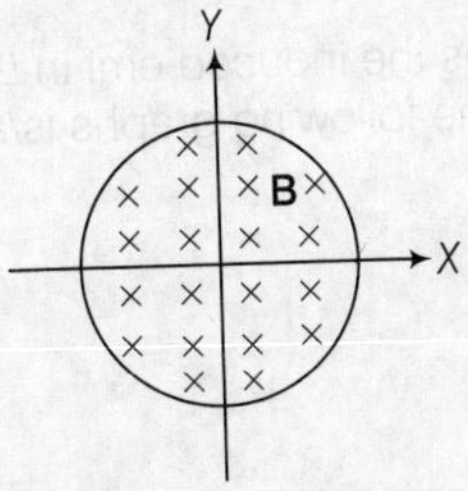

a. No emf and current will as induced in the loop if it rotates about Z-axis
b. Emf is induced but no current flows if the loop is a fibre when it rotates about Y-axis
c. Emf is induced and induced current flows in the loop is made of copper and is rotated about Y-axis
d. If the loop moves along Z-axis with constant velocity no current flows in it

22 Shown in the figure is an R-L circuit. Just after the key (K) is closed. Choose the correct option(s).

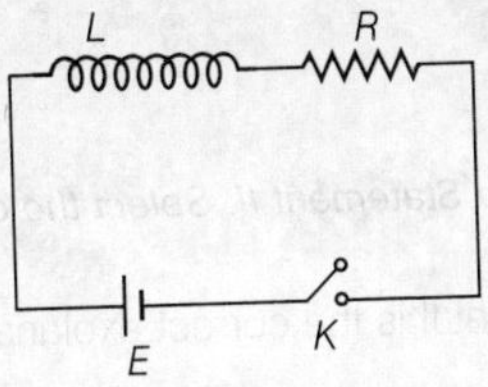

a. The current in the circuit is zero
b. Potential drop across the resistor is zero
c. Emf induced across the inductor equals the emf of the battery
d. No heat is dissipated in the circuit

23 A conducting loop of resistance R and radius r has its centre at the origin of the coordinate system in a magnetic field B. When it is rotated about Y-axis through 90°. Net charge flown in the coil is directly proportional to

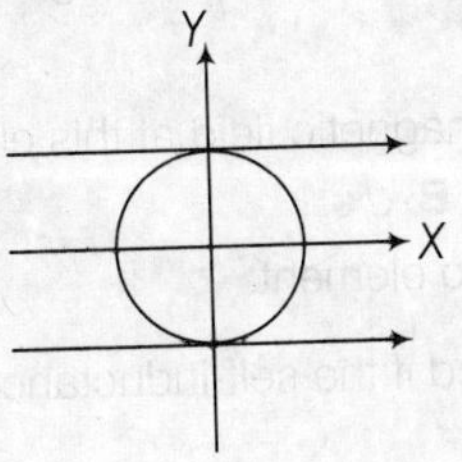

a. B b. R
c. e d. r^2

24 For the circuit shown in figure. Which of the following statement is/are correct?

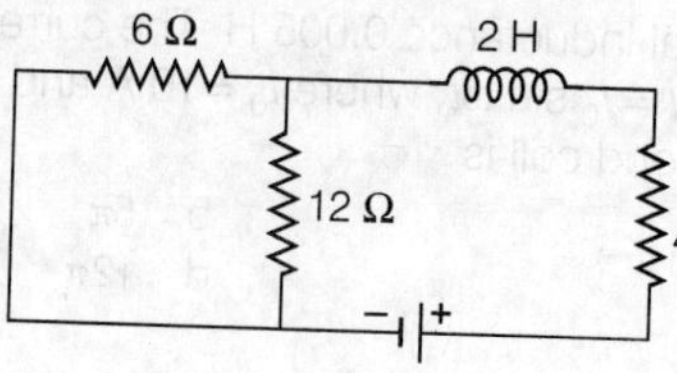

a. Its time constant is 0.25 s

b. In steady state, current through the inductance will be equal to zero

c. In steady state, current through the battery will be equal to 0.75 A

d. None of the above

25 The conductor *ABCDE* has the shape shown in figure. It lies in the *YZ*- plane, with *A* and *E* on the *Y*-axis. When it moves with a velocity *v* in a magnetic field *B*, an emf *e* is induced between *A* and *E*

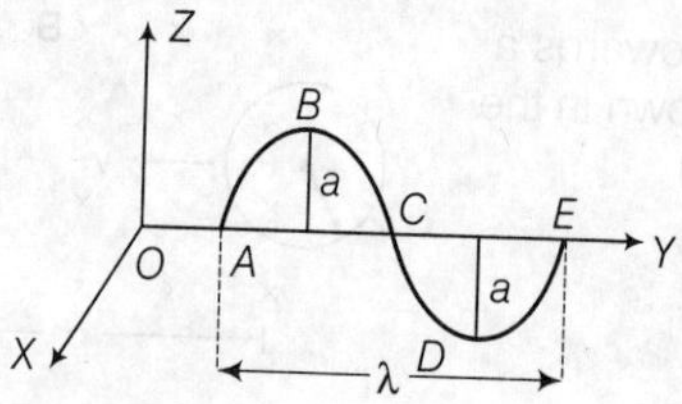

a. $e = 0$, if *v* is in *Y*-direction and *B* is in the *X*-direction

b. $e = 2Bav$, if *v* is in the *Y*-direction and *B* is in *X*-direction

c. $e = B\lambda v$, if *v* is in *Z*-direction and *B* is in *X*-direction

d. $e = B\lambda v$, if *v* is in *X*-direction

26 Switch *S* of the circuit shown is closed at $t = 0$. If *e* denotes the induced emf in *L* and *I* is the current flowing through the circuit at time *t*, which of the following graphs is/are correct?

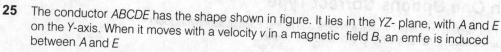

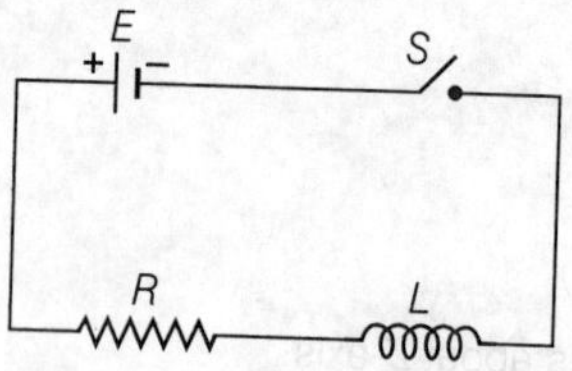

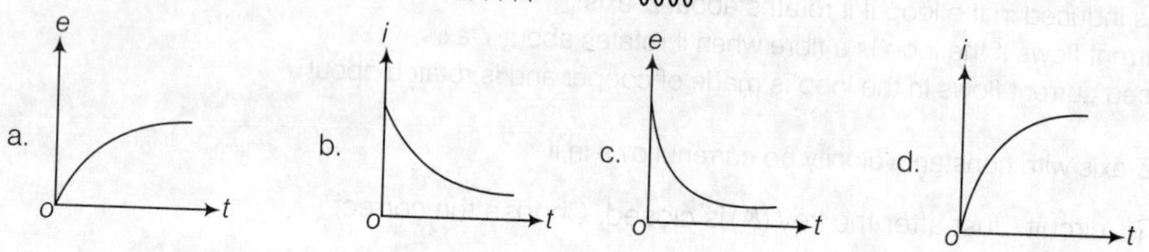

Statement Type

Direction *(Q. Nos. 27-29)* ***This section is based on Statement I and Statement II. Select the correct answer from the codes given below.***

a. Both Statement I and Statement II are correct and Statement II is the correct explanation of Statement I

b. Both Statement I and Statement II are correct but Statement II is not the correct explanation of Statement I

c. Statement I is correct but Statement II is incorrect

d. Statement II is correct but Statement I is incorrect

27 **Statement I** The phenomenon of electromagnetic induction is in accordance with Lenz's law represents the conservation of energy.

Statement II Lenz's law gives the same result as we have defined by the sign rule in Faraday's laws.

28 **Statement I** If $d\mathbf{s}$ is the area of an element and $\mathbf{B}$ is the magnetic field at this element, then magnetic field through the element is given by $d\phi_B = \mathbf{B} \cdot d\mathbf{s}$.

Statement II In general $d\phi_B$ does not vary from element to element.

29 **Statement I** The mutual inductance of two coils is doubled if the self-inductance of the primary and secondary coil is doubled.

Statement II Mutual inductance is directly proportional to the self-inductance of primary and secondary coils.

Working Space

Matching List Type

Direction *(Q. Nos. 30 and 31) Choices for the correct combination of elements from Column I and Column II are given as options (a), (b), (c) and (d), out of which one is correct.*

30 A conducting ring of radius R is rotating as well as translating on a smooth horizontal surface, where an uniform magnetic field B is present. Velocity of centre of mass is given as $v = 2R\omega$, where ω is the angular velocity of the ring. Match the following cases.

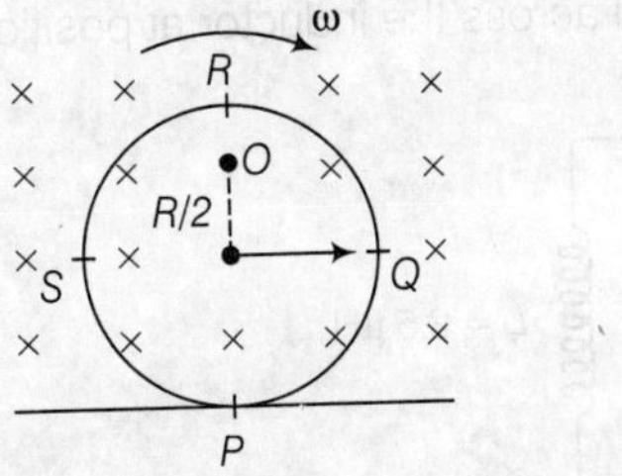

Motional Emf across

	Column I		Column II
i.	PR	p.	$5BRv/4$
ii.	PS	q.	BRv
iii.	PO	r.	$2BRv$
iv.	RO	s.	$3BRv/4$

Codes

	i	ii	iii	iv
a.	r	q	s	p
b.	p	q	r	s
c.	r	p	q	s
d.	p	r	s	q

31 In Column I some circuits are given. In all the circuits except in (i), switch S remains closed for long time and then it is opened at $t = 0$; while for (i), the situation is reversed.

	Column I		Column II
i.	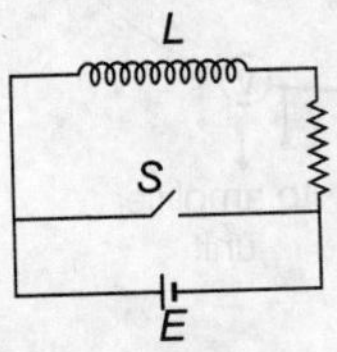	p.	Induced emf can be greater than E.
ii.	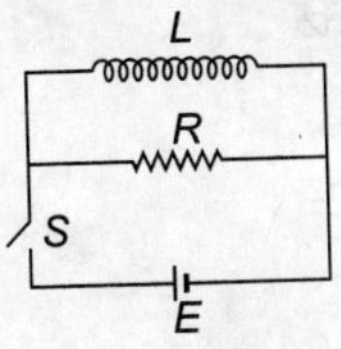	q.	Induced emf would be less than E.
iii.	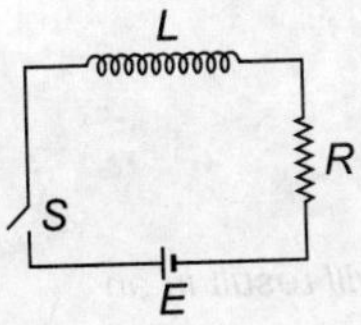	r.	Finally, energy stored in inductor is zero.
iv.	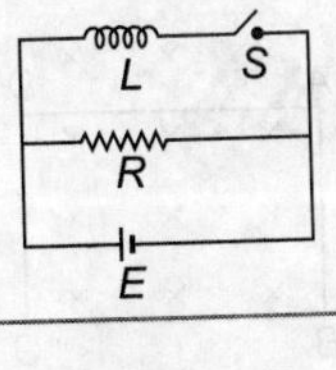	s.	Finally, energy stored in inductor is non-zero.

Codes

	i	ii	iii	iv
a.	r	r	p	p
b.	q	r	s	s
c.	p	q	r	s
d.	s	p	q	r

Comprehension Type

Working Space

***Direction** (Q. Nos. 32-37) This section contains 2 paragraphs, each describing theory, experiments, data, etc. Six questions related to the paragraphs have been given. Each question has only one correct answer among the four given options (a), (b), (c) and (d).*

Passage I

Initially the capacitor is charged to a potential of 5 V and then connected to position 1 with the shown polarity for 1 s. After 1s, it is connected across the inductor at position 2.

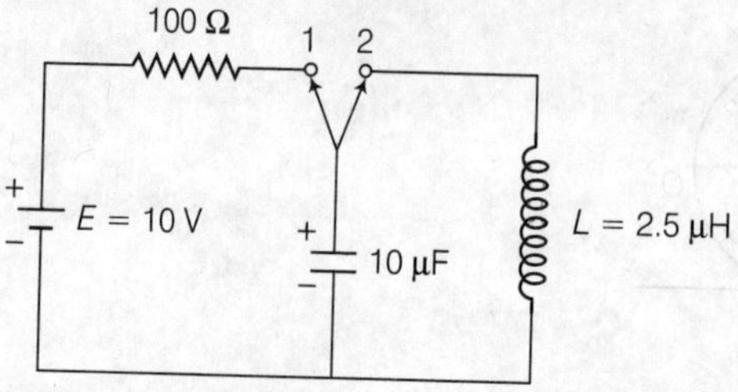

32 The potential across capacitor after 1s of its connection to position lies

a. $5\times10^3\left[2+\frac{1}{e}\right]$ V b. $5\times10^3\left[2-\frac{1}{e}\right]$ V c. $5\times10^3\left[1+\frac{2}{e}\right]$ V d. None of these

33 The maximum current flowing in the L-C circuit when the capacitor is connected across the inductor is

a. $\left[2-\frac{1}{e}\right]\times10^4$ A b. $\left[1+\frac{2}{e}\right]\times10^4$ A c. $\left[1-\frac{2}{e}\right]\times10^4$ A d. None of these

34 The frequency of LC oscillation is

a. $\left[\frac{20}{\pi}\right]$ Hz b. $\left[\frac{2}{\pi}\right]$ Hz c. $\left[\frac{40}{\pi}\right]$ Hz d. $\left[\frac{100}{\pi}\right]$ Hz

Passage II

In an electric guitar, a flexible metallic wire with linear mass density 3×10^{-3} kg/m is stretched between two fixed clamps 64 cm apart and is held under tension 267 N. A magnet is placed near the wire. Let the magnet produces a magnetic field of 4.5 mT over a 2 cm wire length at the centre of the wire.

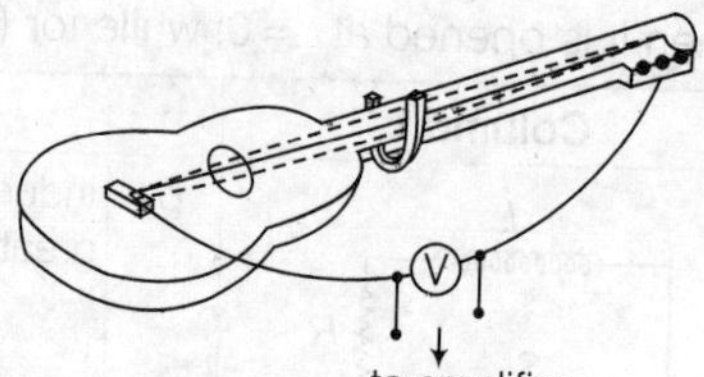

The wire is set vibrating at its lowest possible frequency.

35 Frequency of vibration of wire is

a. 334 Hz b. 290 Hz c. 233 Hz d. 370 Hz

36 Frequency of emf produced in the vibrating wire is

a. 334 Hz b. 290 Hz c. 233 Hz d. 370 Hz

37 Amplitude of emf induced between ends of wire is

a. 1.98×10^{-3} V b. 1.98×10^{-5} V c. 1.98×10^{-2} V d. 1.98 V

One Integer Value Correct Type

***Direction** (Q. Nos. 38 and 39) This section contains 2 questions. When worked out will result in an integer from 0 to 9 (both inclusive).*

38 A square coil $ABCD$ of side 2m is placed in a magnetic field $B=2t^2$, where B is in tesla and time t is in second. Find induced electric field in DC at time $t=2$ s (in V/m).

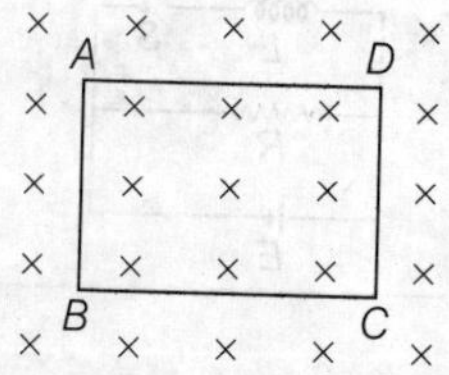

39 In the circuit as shown in the figure, charge q varies with time t as $q=(t^2-4)$, where q is in coloumbs and time t is in second. Find V_{AB} at time $t=3$ s (in V).

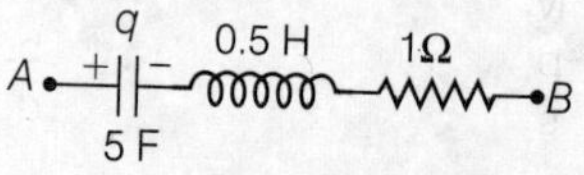

JEE Main & AIEEE Archive

(Compilation of Last 13 Years' Questions)

1 A metallic rod of length l is tied to a string of length $2l$ and made to rotate with angular speed ω on a horizontal table with one end of the string fixed at centre. If there is a vertical magnetic field B in the region, the emf induced across the ends of the rod is **(JEE Main 2013)**

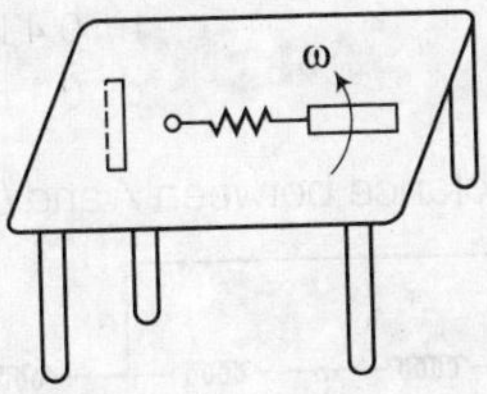

a. $\frac{2B\omega l^3}{2}$ b. $\frac{3B\omega l^3}{2}$

c. $\frac{4B\omega l^2}{2}$ d. $\frac{5B\omega l^2}{2}$

2 A circular loop of radius 0.3 cm lies parallel to a much bigger circular loop of radius 20 cm. The centre of the small loop is on the axis of the bigger loop. The distance between their centres is 15 cm. If a current of 2.0 A flows through the smaller loop, then the flux linked with bigger loop is **(JEE Main 2013)**

a. 9.2×10^{-11} Wb

b. 6×10^{-11} Wb

c. 3.3×10^{-11} Wb

d. 6.6×10^{-9} Wb

3 A coil is suspended in a uniform magnetic field with the plane of the coil parallel to the magnetic lines of force. When a current is passed through the coil, it starts oscillating, it is very difficult to stop. But if an aluminium plate is placed near to the coil, it stops. This is due to **(AIEEE 2012)**

a. development of air current when the plate is placed

b. induction of electrical charge on the plate

c. shielding of magnetic lines of force as aluminium is a paramagnetic material

d. electromagnetic induction in the aluminium plate giving rise to electromagnetic damping

4 A horizontal straight wire 20 m long extending from East to West is falling with a speed of $5.0\ \text{ms}^{-1}$, at right angles to the horizontal component of the earth's magnetic field $0.30 \times 10^{-4}\ \text{Wb/m}^2$. The instantaneous value of the emf induced in the wire will be **(AIEEE 2011)**

a. 6.0 mV

b. 3 mV

c. 4.5 mV

d. 1.5 mV

5 A boat is moving due East in a region where the earth's magnetic field is $5.0 \times 10^{-5}\ \text{NA}^{-1}\ \text{m}^{-1}$ due North and horizontal. The boat carries a vertical aerial 2m long. If the speed of the boat is $1.50\ \text{ms}^{-1}$, then magnitude of the induced emf in the wire of aerial is **(AIEEE 2011)**

a. 0.75 mV

b. 0.50 mV

c. 0.15 mV

d. 1 mV

6 Two coaxial solenoids are made by winding thin insulated wire over a pipe of cross-sectional area $A = 10\ \text{cm}^2$ and length $= 20$ cm. If one of the solenoids has 300 turns and the other 400 turns, then their mutual inductance is (given, $\mu_0 = 4\pi \times 10^{-7}\ \text{TmA}^{-1}$) **(AIEEE 2008)**

a. $2.4\pi \times 10^{-5}$ H

b. $4.8\pi \times 10^{-4}$ H

c. $4.8\pi \times 10^{-5}$ H

d. $2.4\pi \times 10^{-4}$ H

7 An ideal coil of 10 H is connected in series with a resistance of 5 Ω and a battery of 5 V. After 2 s, the connection is made, the current flowing (in A) in the circuit is **(AIEEE 2007)**

a. $(1 - e)$ b. e

c. e^{-1} d. $(1 - e^{-1})$

8 The flux linked with a coil at any instant t is given by $\phi = 10t^2 - 50t + 250$. The induced emf at $t = 3$ s is **(AIEEE 2006)**

a. −190 V

b. −10 V

c. 10 V

d. 190 V

9 A coil having n turns and resistance $R\ \Omega$ is connected with a galvanometer of resistance $4R\ \Omega$. This combination is moved for time t second from a magnetic field W_1 weber to W_2 weber. The induced current in the circuit is **(AIEEE 2004)**

a. $\frac{W_2 - W_1}{5Rnt}$

b. $-\frac{n(W_2 - W_1)}{5Rt}$

c. $-\frac{(W_2 - W_1)}{Rnt}$

d. $-\frac{n(W_2 - W_1)}{Rt}$

10 In a uniform magnetic field of induction B, a wire in the form of semi circle of radius r rotates about the diameter of the circle with angular frequency ω. If the total resistance of the circuit is R, then the mean power generated per period of rotation is **(AIEEE 2004)**

a. $\frac{B\pi r^2\omega}{2R}$

b. $\frac{(B\pi r^2\omega)^2}{8R}$

c. $\frac{(B\pi r\omega)^2}{2R}$

d. $\frac{(B\pi r^2\omega)^2}{8R}$

11 A metal conductor of length 1 m rotates vertically about one of its ends at angular velocity 5 rad s^{-1}. If the horizontal component of earth's magnetic field is 0.2×10^{-4} T, then the emf developed between the two ends of the conductor is **(AIEEE 2004)**

a. 5 μV
b. 50 μV
c. 5 mV
d. 50 mV

12 Two coils are placed close to each other. The mutual inductance of the pair of coils depends upon **(AIEEE 2003)**

a. the rates at which currents are changing in the two coils
b. relative position and orientation of the two coils
c. the materials of the wires of the coils
d. the currents in the two coils

13 When the current changes from +2 A to −2 A in 0.05 s, an emf of 8 V is induced in a coil. The coefficient of self-induction of the coil is **(AIEEE 2003)**

a. 0.2 H
b. 0.4 H
c. 0.8 H
d. 0.1 H

14 The inductance between A and D is

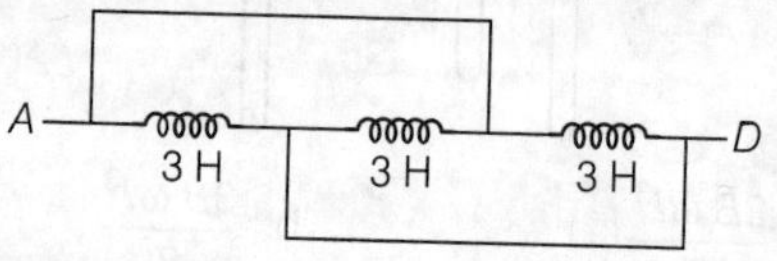

a. 3.66 H
b. 9 H **(AIEEE 2002)**
c. 0.66 H
d. 1 H

JEE Advanced & IIT JEE Archive

(Compilation of Last 10 Years' Questions)

Passage for Q. Nos. (1-2)

A point charge Q is moving in a circular orbit of radius R in the XY-plane with an angular velocity ω. This can be considered as equivalent to a loop carrying a steady current $\frac{Q\omega}{2\pi}$. A uniform magnetic field along the positive Z-axis is now switched on, which increases at a constant rate from O to B in one second. Assume that the radius of the orbit remains constant. The applications of the magnetic field induces an emf in the orbit. The induced emf is defined as the work done by an induced electric field in moving a unit positive charge around a closed loop. It is known that, for an orbiting charge, the magnetic dipole moment is proportional to the angular momentum with a proportionality constant γ.

(2013 Adv., Comprehension Type)

1. The magnitude of the induced electric field in the orbit at any instant of time during the time interval of the magnetic field change is

a. $\frac{BR}{4}$ b. $\frac{-BR}{2}$

c. BR d. $2BR$

2. The change in the magnetic dipole moment associated with the orbit, at the end of the time interval of the magnetic field change, is

a. γBQR^2 b. $-\gamma \frac{BQR^2}{2}$

c. $\gamma \frac{BQR^2}{2}$ d. γBQR^2

3. A circular wire loop of radius R is placed in the XY-plane centred at the origin O. A square loop of side a ($a << R$) having two turns is placed with its centre at $Z = \sqrt{3}\,R$ along the axis of the circular wire loop, as shown in figure. The plane of the square loop makes an angle of 45° with respect to the Z-axis. If the mutual inductance between the loops is given by $\frac{\mu_0 a^2}{2^{p/2} R}$, then the value of p is **(2012, Integer Type)**

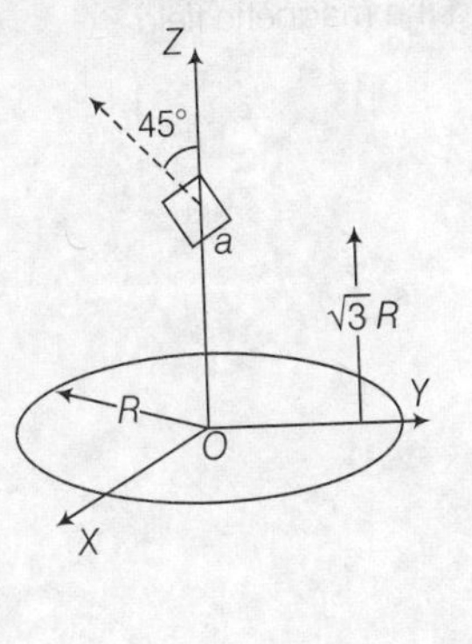

4. A current carrying infinitely long wire is kept along the diameter of a circular wire loop, without touching it. Which of the following statement(s) is/are correct?

(2012, Only One Option Correct Type)

a. The emf induced in the loop is zero, if the current is constant

b. The emf induced in the loop is finite, if the current is constant

c. The emf induced in the loop is zero, if the current decreases at a steady rate

d. The emf induced in the loop is finite, if the current decreases at a steady rate

5. A thin flexible wire of length L is connected to two adjacent fixed points and carries a current I in the clockwise direction as shown in the figure. When the system is put in a uniform magnetic field of strength B going into the plane of the paper, the wire takes the shape of a circle. The tension in the wire is

(2010, Only One Option Correct Type)

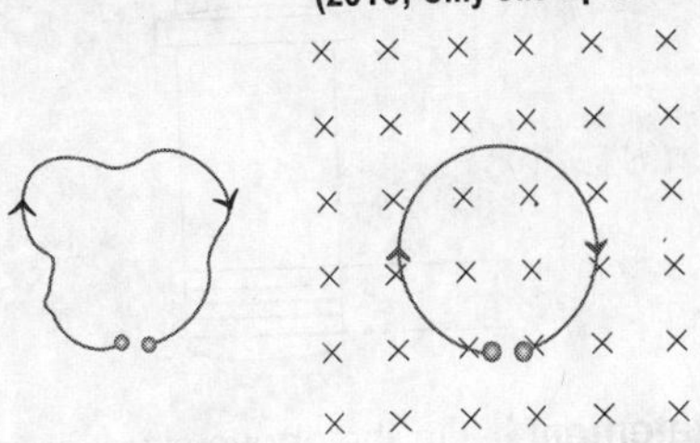

a. IBL b. $\frac{IBL}{\pi}$

c. $\frac{IBL}{2\pi}$ d. $\frac{IBL}{4\pi}$

6. The figure shows certain wire segments joined together to form a coplanar loop. The loop is placed in a perpendicular magnetic field in the direction going into the plane of the figure. The magnitude of the field increases with time. I_1 and I_2 are the currents in the segments ab and cd. Then, **(2009, Only One Option Correct Type)**

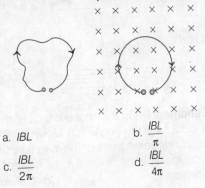

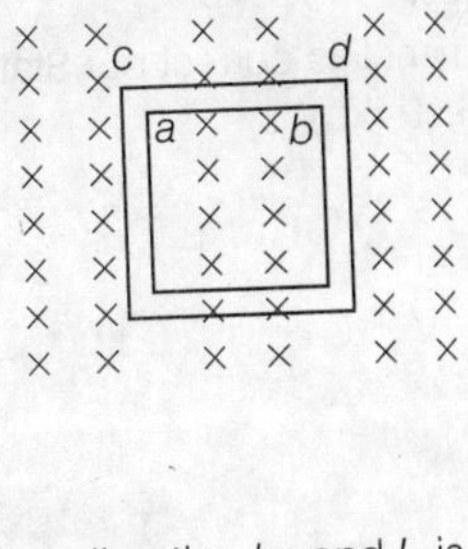

a. $I_1 > I_2$

b. $I_1 < I_2$

c. I_1 is in the direction ba and I_2 is in the direction cd

d. I_1 is in the direction ab and I_2 is in the direction dc

7. Two metallic rings A and B, identical in shape and size but having different resistivities ρ_A and ρ_B, are kept on top of two identical solenoids as shown in the figure. When current I is switched on in both the solenoids in identical manner, the rings A and B jump to heights h_A and h_B, respectively, with $h_A > h_B$. The possible relation(s) between their resistivities and their masses m_A and m_B is/are

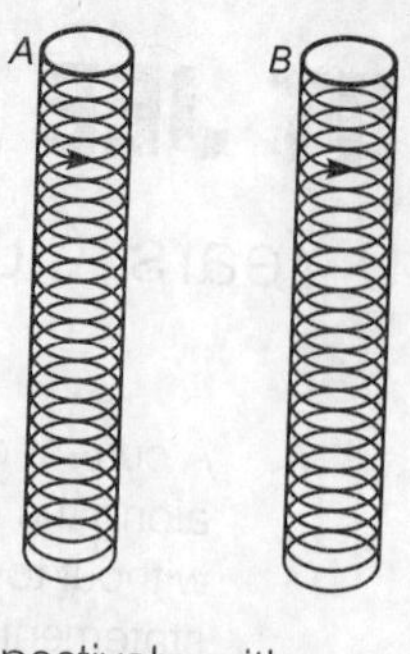

(2009, One or More than One Options Correct Type)

a. $\rho_A > \rho_B$ and $m_A = m_B$
b. $\rho_A < \rho_B$ and $m_A = m_B$
c. $\rho_A > \rho_B$ and $m_A > m_B$
d. $\rho_A < \rho_B$ and $m_A < m_B$

8. **Statement I** A vertical iron rod has a coil of wire wound over it at the bottom end. An alternating current flows in the coil. There is a conducting ring round the rod as shown in the figure. The ring can float at a certain height above the coil.

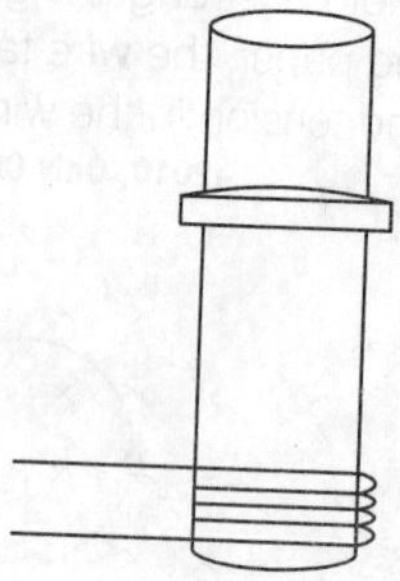

Statement II In the above situation, a current is induced in the ring which interacts with the horizontal component of the magnetic field to produce an average force in the upward direction. **(2007, Statement Type)**

a. Both Statement I and Statement II are correct and Statement II is the correct explanation of Statement I
b. Both Statement I and Statement II are correct but Statement II is not the correct explanation of Statement I
c. Statement I is correct but Statement II is incorrect
d. Statement II is correct but Statement I is incorrect

Passage for Q. Nos. (9-11)

Modern trains are based on Maglev technology in which trains are magnetically levitated, which runs its EDS Maglev system. There are coils on both sides of wheels. Due to motion of train, current induces in the coil of track which levitate it. This is in accordance with Lenz's law. If trains lower down then due to Lenz's law a repulsive force increases due to which train gets uplifted and if it goes much high then there is a net downward force due to gravity. The advantage of Maglev train is that there is no friction between the train and the track, thereby reducing power consumption and enabling the train to attain very high speeds.

Disadvantage of Maglev train is that as it slows down the electromagnetic forces decreases and it becomes difficult to keep it levitated and as it moves forward according to Lenz's law there is an electromagnetic drag force.

(2006, Comprehension Type)

9. What is the advantage of this system?
a. No friction hence no power consumption
b. No electric power is used
c. Gravitation force is zero
d. Electrostatic force draws the train

10. What is the disadvantage of this system?
a. Train experiences upward force according to Lenz's law
b. Friction froce create a drag on the train
c. Retardation
d. By Lenz's law train experience a drag

11. Which force causes the train to elevate up?
a. Electrostatic force
b. Time varying electric field
c. Magnetic force
d. Induced electric field

12. An infinitely long cylinder is kept parallel to an uniform magnetic field B directed along positive Z-axis. The direction of induced current as seen from the Z-axis will be

(2005, Only One Option Correct Type)

a. clockwise of the positive Z-axis
b. anti-clockwise of the positive Z-axis
c. zero
d. along the magnetic field

4. Alternating Current

DPP-1	Average, Peak and rms Value of AC
DPP-2	Power Consumption in an AC Circuit
DPP-3	AC Source with R-L-C Connected in Series
DPP-4	Resonance
DPP-5	Transformer
•	**Revisal Problems for JEE Main**
•	**Revisal Problems for JEE Advanced**
•	**JEE Main & AIEEE Archive**
•	**JEE Advanced & IIT JEE Archive**

DPP-1 Average, Peak and rms Value of AC

Subjective Questions

Working Space

Direction *(Q. Nos. 1-5) These questions are subjective in nature, need to be solved completely on notebook.*

1 Find the average and rms values of the waveforms given by

i.
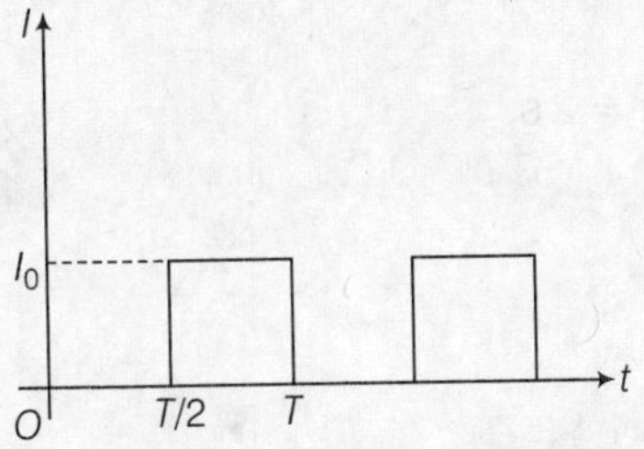

ii.
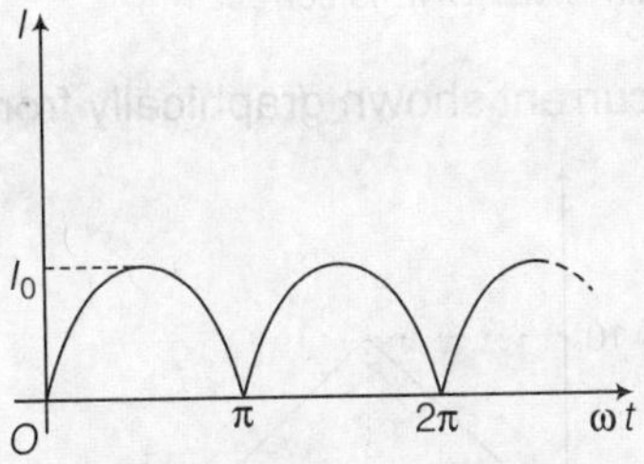

iii.
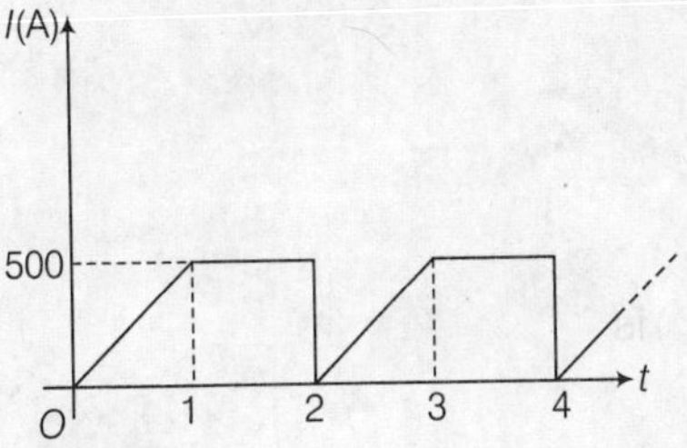

2 Find the rms and the average values of the saw-tooth waveform as shown below.

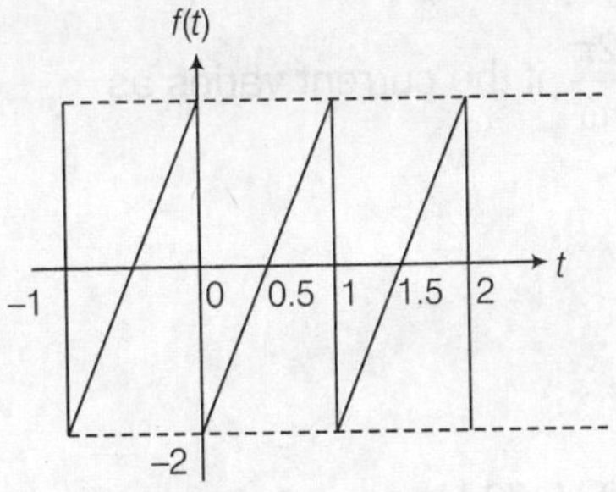

3 A resistance of 40 Ω is connected to an AC source of 220 V, 50 Hz. Find

i. the rms current.

ii. the maximum instantaneous current in the resistor.

iii. the time taken by the current to change from its maximum value to the rms value.

4 i. Ordinary moving coil galvanometer used for DC cannot be used to measure an alternating current even if its frequency is low. Explain, why?

ii. The electric current in a circuit is given by $i = i_0(t/\tau)$ for sometime. Calculate the rms current for the period $t = 0$ to $t = \tau$.

5 Consider two different currents as shown in the figure.

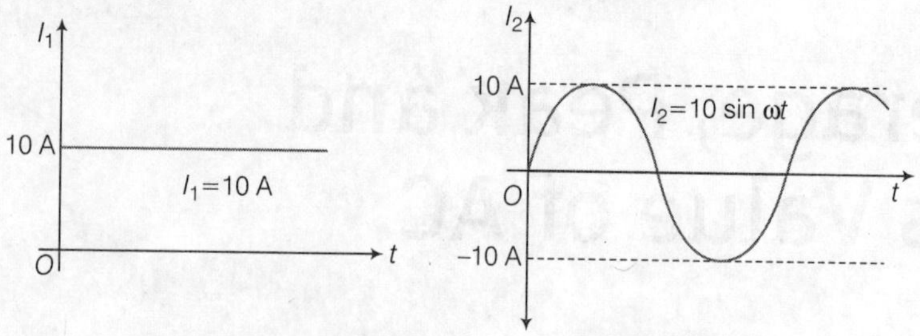

These are mixed using a suitable instrument and fed to a heating coil. Find I_{rms} and I_{av} received by the heating element.

Only One Option Correct Type

Direction *(Q. Nos. 6-15) This section contains 10 multiple choice questions. Each question has four choices (a), (b), (c) and (d), out of which ONLY ONE is correct.*

6 Find the average value of current shown graphically from $t = 0$ to $t = 2$ s.

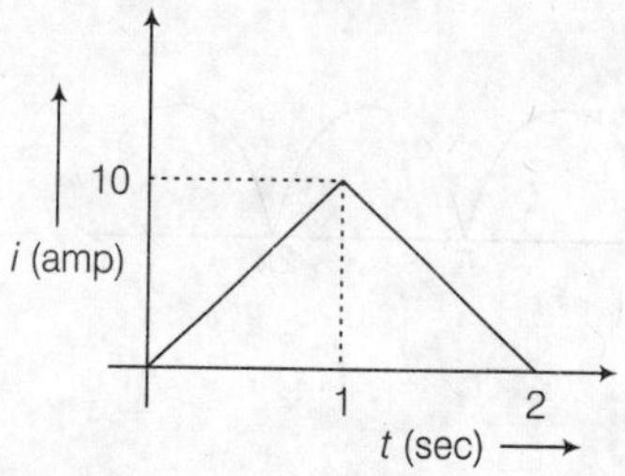

a. 3 A
b. 5 A
c. 10 A
d. 4 A

7 The effective value of current $i = 2\sin 100\pi t + 2\cos(100\pi t + 30°)$ is

a. 2 A
b. $2\sqrt{2}$ A
c. $\frac{1}{\sqrt{2}}$ A
d. $\sqrt{2}$ A

8 The average value of current from $t = 0$ to $t = \frac{2\pi}{\omega}$ if the current varies as $I = I_0 \sin \omega t$.

a. 0
b. $\frac{I_0}{\pi}$
c. $\frac{2I_0}{\pi}$
d. $\frac{I_0}{\sqrt{2}}$

9 If a domestic appliance draws 2.5 A from a 220 V, 60 Hz power supply. Find

I. the average current.
II. the average of the square of the current.
III. the current amplitude.
IV. the supply voltage amplitude.

a. 2 A, 2.5 A, 3.5 A, 311 V
b. zero, 2.5 A, 3.5 A, 311 V
c. zero, 2.5 A, 3.5 A, 210 V
d. None of these

10 An alternating voltage is given by $e = e_1 \sin \omega t + e_2 \cos \omega t$. Then, the root mean square value of voltage is given by

a. $\sqrt{e_1^2 + e_2^2}$
b. $\sqrt{e_1 e_2}$
c. $\sqrt{\frac{e_1 e_2}{2}}$
d. $\sqrt{\frac{e_1^2 + e_2^2}{2}}$

11 Two sinusoidal voltages of same frequency are shown in the figure.

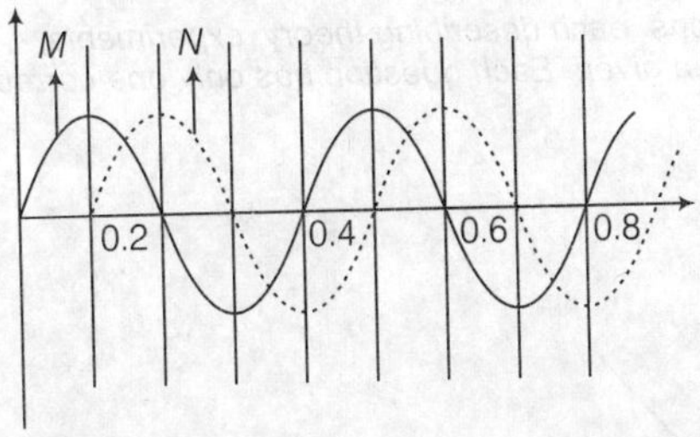

Which is the frequency and phase relationship between the voltage?

Frequency (Hz) Phase lead of N over M in rad/s

a. $0.4 - \frac{\pi}{4}$

b. $2.5 - \frac{\pi}{2}$

c. $2.5 + \frac{\pi}{2}$

d. $2.5 - \frac{\pi}{4}$

12 The rms value of the saw-tooth voltage of peak value V_0 from $t = 0$ to $t = 2T$ as shown in figure is given by

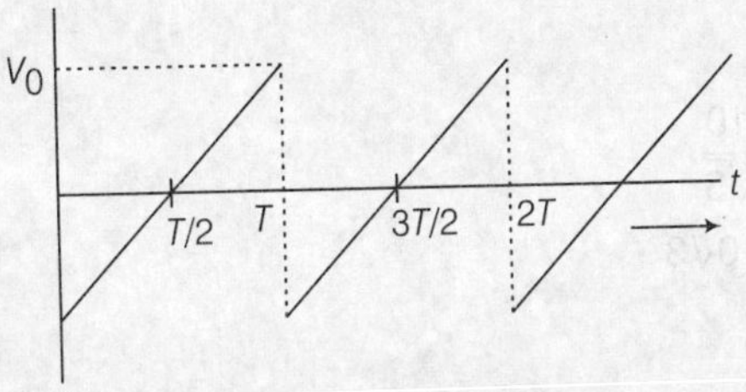

a. $\frac{V_0}{\sqrt{2}}$

b. $\frac{V_0}{\sqrt{3}}$

c. $\sqrt{2}\, V_0$

d. $\sqrt{3}\, V_0$

13 Two alternating voltage generators produce emf of the same amplitude E_0 but with a phase difference of $\pi/3$. The resultant emf is

a. $E_0 \sin[\omega t + (\pi/3)]$

b. $E_0 \sin[\omega t + \pi/6)]$

c. $\sqrt{3}\, E_0 \sin[\omega t + (\pi/6)]$

d. $\sqrt{3}\, E_0 \sin[\omega t + (\pi/2)]$

14 An AC current is given by $I = I_0 + I_1 \sin \omega t$, then its rms value will be

a. $\sqrt{I_0^2 + 0.5 I_1^2}$

b. $\sqrt{I_1^2 + 0.5 I_0^2}$

c. 0

d. $\frac{I_0}{\sqrt{2}}$

15 If a DC of value a ampere is superimposed on an AC $I = b \sin \omega t$ flowing through a wire, what is the effective value of the resulting current in the circuit?

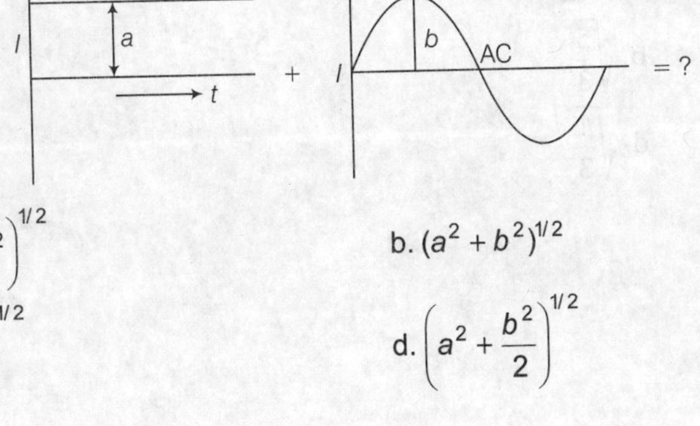

a. $\left(a^2 - \frac{1}{2} b^2\right)^{1/2}$

b. $(a^2 + b^2)^{1/2}$

c. $\left(\frac{a^2}{2} + b^2\right)^{1/2}$

d. $\left(a^2 + \frac{b^2}{2}\right)^{1/2}$

Comprehension Type

Direction *(Q. Nos. 16-20) This section contains 2 paragraphs, each describing theory, experiments, data, etc. Five questions related to the paragraphs have been given. Each question has only one correct answer among the four given options (a), (b), (c) and (d).*

Passage I

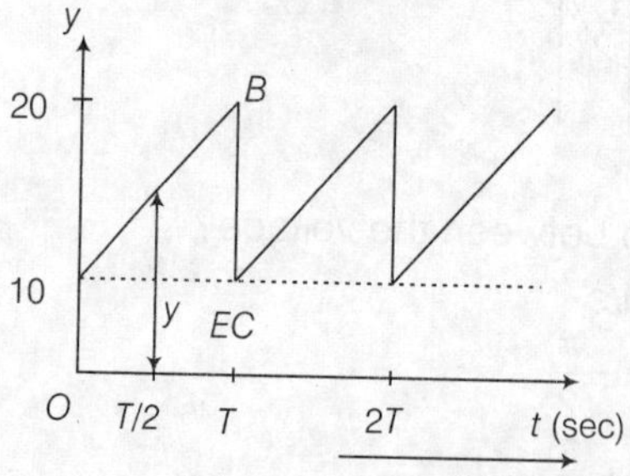

16 The average value of the waveform as shown in figure is

a. $15\sqrt{2}$
b. $10\sqrt{2}$
c. 10
d. 15

17 The rms value of the signal is

a. $10\sqrt{\frac{7}{3}}$
b. $\frac{10}{\sqrt{3}}$
c. $10\sqrt{7}$
d. $10\sqrt{3}$

Passage II

Output of a faulty inverter is expressed for one cycle as

$$\text{emf} = E(\theta) = \begin{cases} \frac{E_m}{\alpha}\theta, & 0 < \theta \le \alpha \\ E_m, & \alpha \le \theta \le \pi - \alpha \\ \frac{-E_m}{\alpha}\theta, & \pi - \alpha \le \theta \le \pi \end{cases}$$

18. The ratio of average value of emf when $\alpha = \frac{\pi}{6}$ and when $\alpha = \frac{\pi}{2}$ is

a. 1
b. $\frac{2}{3}$
c. $\sqrt{2}$
d. $\frac{5}{3}$

19. For $\alpha = 0$, rms value of emf is

a. $\frac{E_m}{\sqrt{2}}$
b. $\frac{E_m}{\sqrt{3}}$
c. E_m
d. $\frac{2}{\pi}E_m$

20. The ratio of rms values for $\alpha = \frac{\pi}{6}$ and for $\alpha = \frac{\pi}{2}$ is

a. $\sqrt{\frac{2}{3}}$
b. $\sqrt{\frac{5}{3}}$
c. $\sqrt{\frac{7}{3}}$
d. $\sqrt{\frac{1}{3}}$

One or More than One Options Correct Type

Direction *(Q. Nos. 21-24) This section contains 4 multiple choice questions. Each question has four choices (a), (b), (c) and (d), out of which ONE or MORE THAN ONE are correct.*

21 The average value of AC current in a half-time period may be

a. positive
b. negative
c. zero
d. None of the above

22 Which of the following statements are correct? Heat produced in a current carrying conductor depends upon

a. the time for which the current flows in the conductor
b. the resistance of the conductor
c. strength of the current
d. the nature of current (AC or DC)

23 Output of an AC source is $V = 100 \sin 100\pi t \cos 100\pi t$. Which of these are correct?

a. Peak voltage of source is 100 V
b. Peak voltage of source is $50\sqrt{2}$ V
c. Peak voltage of source is 50 V
d. Frequency of the source is 100 Hz

24 For the saw-tooth, voltage variate (shown) for an AC circuit.

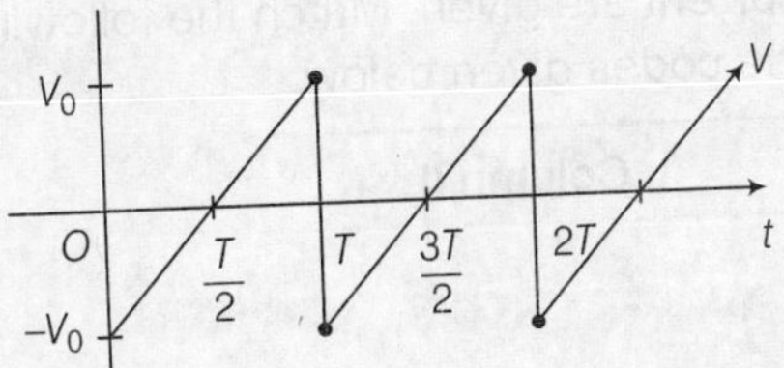

Choose correct option(s).

a. $V = V_0\left(\frac{2t}{T} - 1\right)$

b. $V_{av}\left(\frac{1}{2}\text{cycle}\right) = \frac{V_0}{2}$

c. $V_{rms} = \frac{V_0}{\sqrt{3}}$

d. $V_0 > V_{rms} > V_{av}$

One Integer Value Correct Type

Direction *(Q. Nos. 25-27) This section contains 3 questions. When worked out will result in an integer from 0 to 9 (both inclusive).*

25 The ratio of rms value of current $i = 4 \sin \omega t$ from

I. $t = 0$ to $t = \frac{\pi}{\omega}$

II. $t = \frac{\pi}{2\omega}$ to $t = \frac{3\pi}{2\omega}$ is given by $k : 1$.

Find the value of k.

Working Space

26. The current $i(t)$ shown passes through $1\mu F$ capacitor.

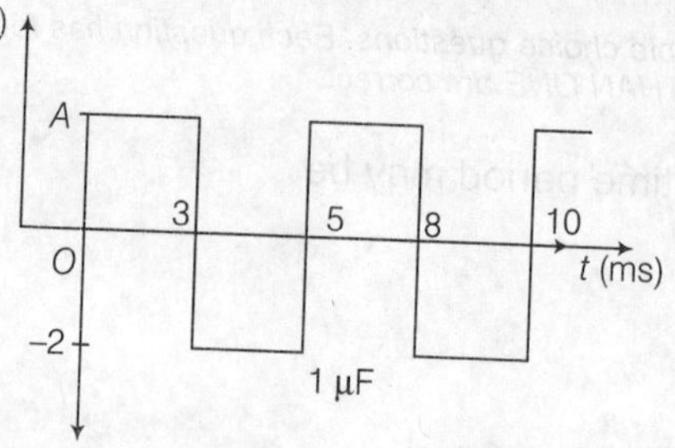

Let I_{DC} = value of constant current which can produce same voltage across the capacitor at $t = 5k$ (ms) when applied at $t > 0$, where $k = 0, 1, 2, 3...$ Find the value of $10^3 (5I_{DC})$.

27. Household electricity is supplied at 220 V rms and 50 Hz frequency. Time in which voltage can change from the rms value to zero is $N \times 2.5$ ms. Find the value of N.

Matching List Type

Direction *(Q. No. 28) Choices for the correct combination of elements from Column I and Column II are given as options (a), (b), (c) and (d), out of which one is correct.*

28 In Column I, the variation of current i with time t is given in the figure. In Column II, root mean square current i_{rms} and average current are given. Match the following columns and select the correct option from the codes given below.

	Column I		Column II
i.	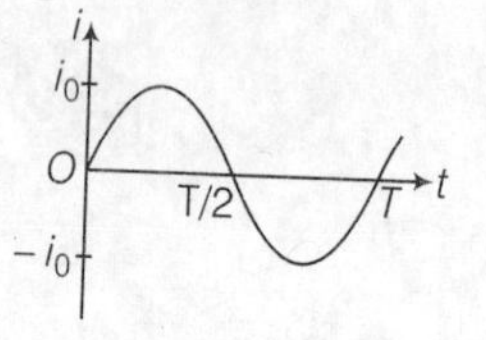	p.	$i_{rms} = \frac{i_0}{\sqrt{3}}$
ii.	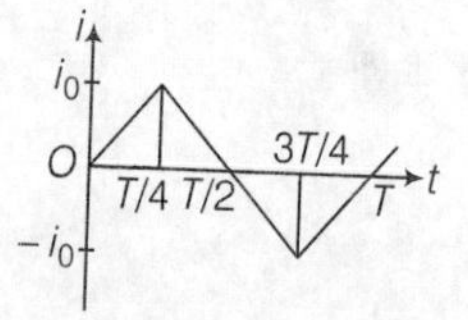	q.	Average current for positive half cycle is i_0.
iii.	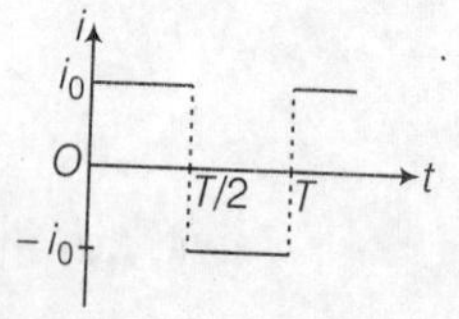	r.	Average current for positive half cycle is $\frac{i_0}{2}$.
iv.	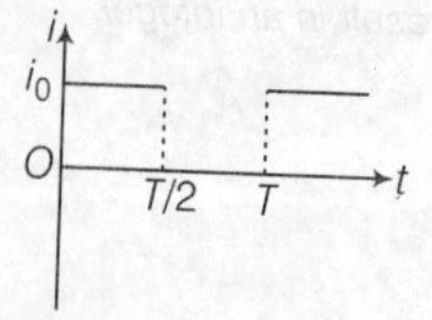	s.	Full cycle average current is zero.

Codes

	i	ii	iii	iv
a.	s	p,r,s	s	q
b.	s	p,s	q,s	q
c.	q,s	p,s	s,r	q
d.	p,q	r,s	q,s	r,p

DPP-2 Power Consumption in an AC Circuit

Subjective Questions

Working Space

Direction (Q. Nos. 1-4) *These questions are subjective in nature, need to be solved completely on notebook.*

1 For the circuit given, source voltage is $\Delta V_{rms} = 100$ V at an angular frequency of 1000 rad s^{-1}. Determine

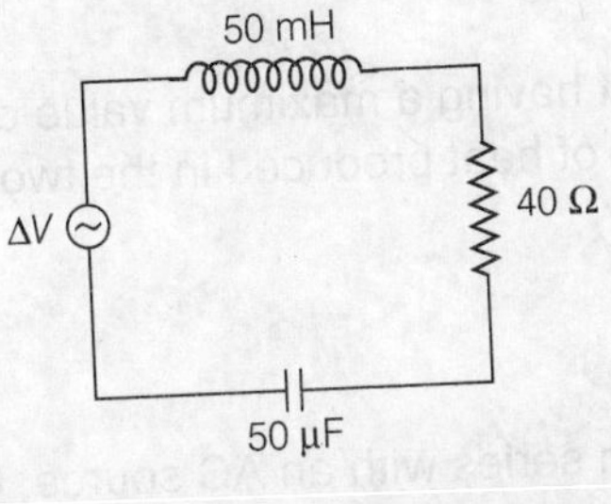

i. current in circuit.
ii. power supplied by source.
iii. power used up by resistor.

2 For an AC circuit, current and voltage are respectively $i = 2\sin(250\pi t)$ amperes and $V = 10\sin\left(250\pi t + \frac{\pi}{3}\right)$ volts. Find the

i. instantaneous power drawn from the source at $t = \frac{2}{3}$ ms.

ii. average power consumed by circuit.

3 For an *R*-*C* series circuit, the rms voltage of source is 200 V and its frequency is 50 Hz. If $R = 100\ \Omega$ and $C = \frac{100}{\pi}\ \mu F$. Find the

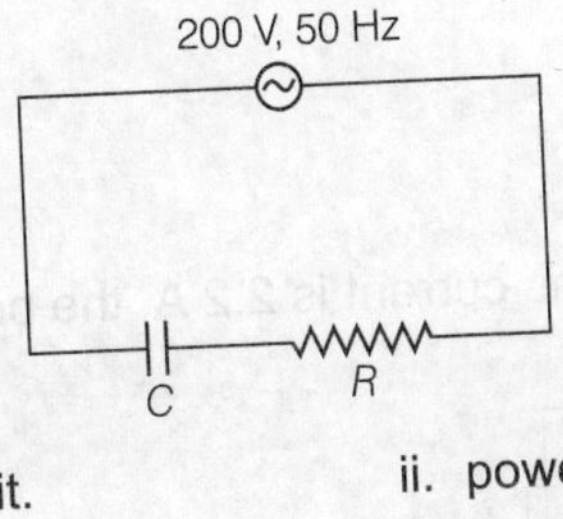

i. impedance of the circuit.
ii. power factor angle.
iii. power factor.
iv. current.
v. maximum current.
vi. voltage across *R* and *C*.
vii. maximum voltage across *C* and *R*.

4 An emf $E = 100 \sin 314t$ is applied across a pure capacitor of 637 μF. Find the

i. instantaneous current *I*.
ii. instantaneous power *P*.
iii. frequency of power.
iv. maximum energy stored in the capacitor.

Only One Option Correct Type

Working Space

Direction *(Q. Nos. 5-14) This section contains 10 multiple choice questions. Each question has four choices (a), (b), (c) and (d), out of which ONLY ONE is correct.*

5 The potential difference V across the current I flowing through an instrument in an AC circuit are given by

$$V = 5 \cos \omega t \text{ volts}$$
$$I = 2 \sin \omega t \text{ amperes}$$

The power dissipated in the instrument is

a. zero
b. 5 W
c. 10 W
d. 2.5 W

6 A 100 μF capacitor is charged with a 50 V source supply. Then, source supply is removed and the capacitor is connected across the inductance as a result of which current of amplitude 5 A flows through the inductance. Calculate the value of the inductance.

a. 0.04 H
b. 0.01 H
c. 0.05 H
d. 0.09 H

7 A direct current of 2 A and an alternating current having a maximum value of 2 A flow through two identical resistances. The ratio of heat produced in the two resistances in the same time interval will be

a. 1 : 1
b. 1 : 2
c. 2 : 1
d. 4 : 1

8 An electric bulb and a capacitor are connected in series with an AC source. On increasing the frequency of the source, the brightness of the bulb

a. increases
b. decreases
c. remains unchanged
d. None of the above

9 The power factor of the circuit is $1/\sqrt{2}$. The capacitance of the circuit is equal to

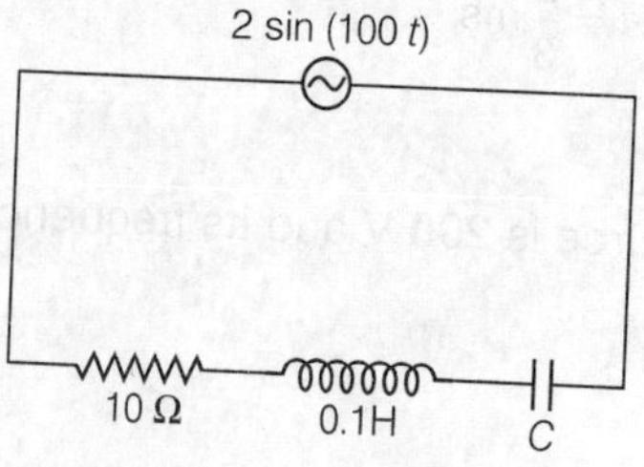

a. 400 μF
b. 300 μF
c. 500 μF
d. 200 μF

10 For the circuit as shown in figure, if the value of rms current is 2.2 A, the power factor of the box is

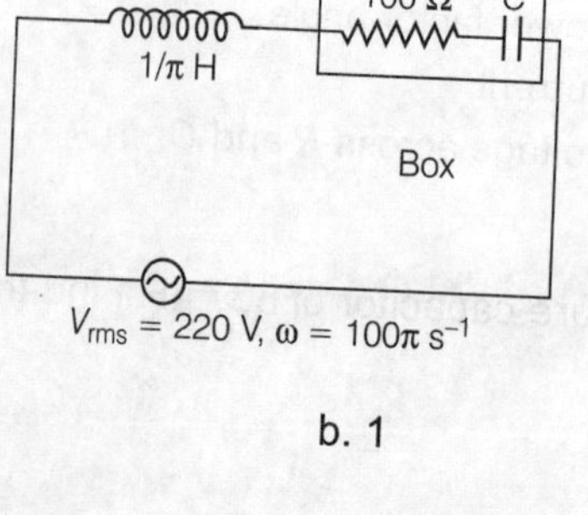

a. $\frac{1}{\sqrt{2}}$
b. 1
c. $\frac{\sqrt{3}}{2}$
d. $\frac{1}{2}$

11 For an AC circuit, the voltage applied is $E = E_0 \sin \omega t$. The resulting current in the circuit is $I = I_0 \sin\left(\omega t - \frac{\pi}{2}\right)$. The power consumption in the circuit is given by

a. $P = \frac{E_0 I_0}{\sqrt{2}}$
b. $P =$ zero
c. $P = \frac{E_0 I_0}{2}$
d. $P = \sqrt{2} E_0 I_0$

12 A coil has an inductance of 0.7 H and is joined in series with a resistance of 220 Ω. When an alternating emf of 220 V at 50 cps is applied to it, then the wattless component of the current in the circuit is (Given, $0.7\pi = 2.2$)

a. 5 A
b. 0.5 A
c. 0.7 A
d. 7 A

13 A sinusoidal AC of peak value I_0 passes through a heater of resistance R. What is the mean power output of the heater?

a. $I_0^2 R$
b. $\frac{I_0^2 R}{2}$
c. $2I_0^2 R$
d. $\sqrt{2}\, I_0^2 R$

14 When an AC source of emf $e = E_0 \sin(100t)$ is connected across a circuit, the phase difference between emf e and current i in the circuit is observed to be $\pi/4$, as shown in figure. If the circuit consists possibly only of R-C or R-L or L-C in series, find the relationship between the two elements.

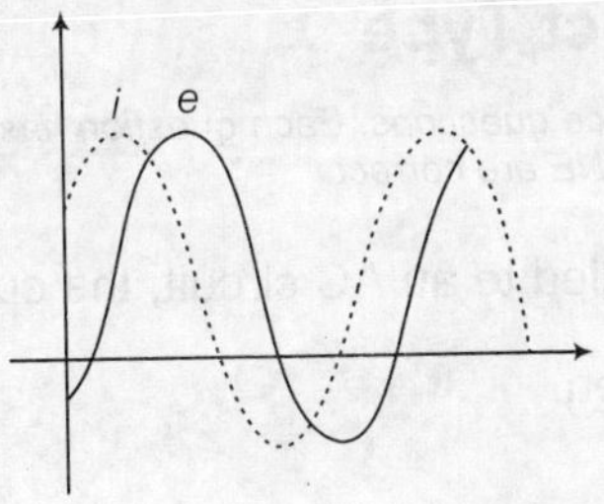

a. $R = 1\text{k}\Omega$, $C = 100\,\mu\text{F}$
b. $R = 1\text{k}\Omega$, $C = 1\,\mu\text{F}$
c. $R = 1\text{k}\Omega$, $L = 10\,\text{H}$
d. $R = 1\text{k}\Omega$, $L = 1\text{H}$

Comprehension Type

Direction *(Q. Nos. 15-19) This section contains 2 paragraphs, each describing theory, experiments, data, etc. Five questions related to the paragraphs have been given. Each question has only one correct answer among the four given options (a), (b), (c) and (d).*

Passage I

Consider a series L-C-R circuit with inductance of 20 mH, capacitance of 100 μF and resistance of 50 Ω. Circuit gets a supply through a generator producing emf of $10 \sin 314t$ volts.

15 The energy dissipated in the circuit in 20 min is

a. 960 J
b. 900 J
c. 250 J
d. 500 J

16 If resistance is removed from the circuit and the value of inductance is doubled, the variation of current with time in the new circuit is

a. $0.52 \cos 314t$
b. $0.52 \sin 314t$
c. $0.52 \sin\left(314t + \frac{\pi}{3}\right)$
d. None of these

Working Space

Passage II

If we look closely at the *L-C* oscillating system and an oscillating spring block system, we can see an analogy between the forms of two pairs of energies dash potential and kinetic energies of spring block system and electrical and magnetic energies of the *L-C* oscillator.

Now, consider an *L-C* circuit with $L = 1.25\text{H}$ containing an energy of $5.70\ \mu\text{J}$. Maximum value of charge on the capacitor is $175\ \mu\text{C}$.

17 What is the value of mass of block on a corresponding analogous spring block system.

a. 2.69×10^{-3} kg
b. 37.2kg
c. 1.25kg
c. 5.70×10^{-6} kg

18 Spring constant of spring block system will be

a. $372 \dfrac{\text{N}}{\text{m}}$
b. $37.2 \dfrac{\text{N}}{\text{m}}$
c. $3.72 \dfrac{\text{N}}{\text{m}}$
d. $0.372 \dfrac{\text{N}}{\text{m}}$

19 Maximum speed attained by the block is

a. $3\ \text{cms}^{-1}$
b. $5\ \text{cms}^{-1}$
c. $0.302\ \text{cms}^{-1}$
d. $0.14\ \text{cms}^{-1}$

One or More than One Options Correct Type

Direction *(Q. Nos. 20-22) This section contains 3 multiple choice questions. Each question has four choices (a), (b), (c) and (d), out of which ONE or MORE THAN ONE are correct.*

20 When a voltage $V_s = 200\sqrt{2} \sin(\omega t + 15^\circ)$ is applied to an AC circuit, the current in the circuit is found to be $i = 2 \sin\left(\omega t + \dfrac{\pi}{4}\right)$. Then,

a. average power consumed is $200\sqrt{6}$ W
b. average power consumed is $100\sqrt{6}$ W
c. power factor is $\dfrac{\sqrt{3}}{2}$
d. power factor is $\dfrac{1}{2}$

21 In an *R-L-C* series circuit shown in the figure, the readings of voltmeters V_1 and V_2 are 100 V and 120 V, respectively. The source voltage is 130 V. For this situation, mark out the correct statement (s).

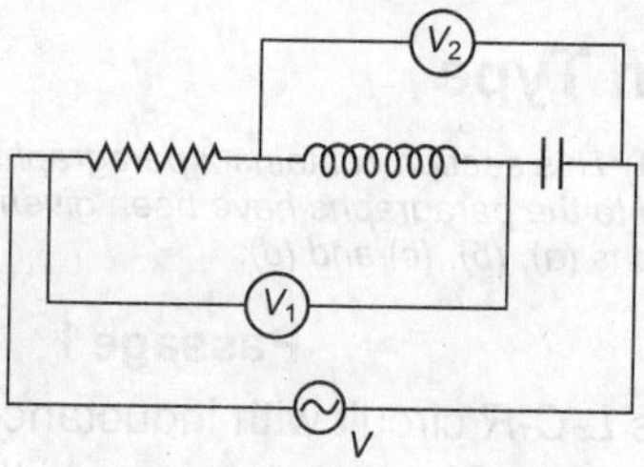

a. Voltage across resistor, inductor and capacitor are 50 V, $50\sqrt{3}$ V and $120 + 50\sqrt{3}$ V, respectively
b. Voltage across resistor, inductor and capacitor are 50 V, $50\sqrt{3}$ V and $120 - 50\sqrt{3}$ V, respectively
c. Power factor of the circuit is $\dfrac{5}{13}$
d. The circuit is capacitive in nature

22 For an *L-C-R* circuit, the power transferred from the driving source to the driven oscillator is $P = I^2 Z \cos\phi$.

a. Here, the power factor $\cos\phi \geq 0$, $P \geq 0$
b. The driving force can give no energy to the oscillator ($P = 0$) in some cases
c. The driving force cannot syphon out $P \not< 0$ the energy of oscillator
d. The driving force can take away energy out of the oscillator

Matching List Type

Direction *(Q. Nos. 23 and 24) Choices for the correct combination of elements from Column I and Column II are given as options (a), (b), (c) and (d), out of which one is correct.*

23 For series *R-L-C* circuit, $R = 100\ \Omega$, $C = \dfrac{100}{\pi}\ \mu F$ and $L = \dfrac{100}{\pi}$ mH are connected to an AC source as shown in the figure. The rms value of AC voltage is 220 V and its frequency is 50 Hz. In Column I, some physical quantities are mentioned while in Column II information about quantities are provided.

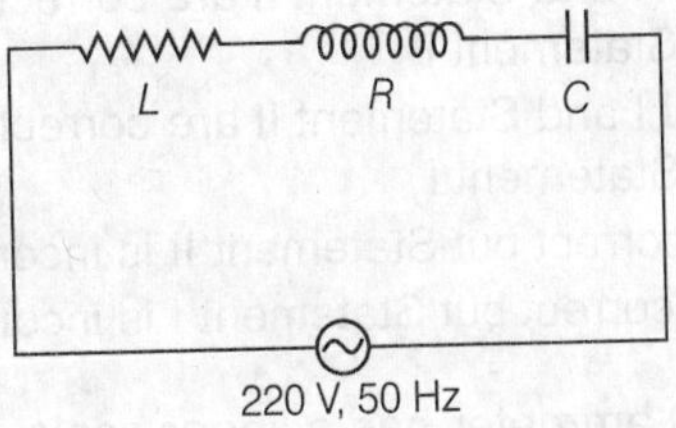

Match the following columns and select the correct option from the codes given below.

	Column I		Column II
i.	Average power dissipated in the resistor is	p.	zero
ii.	Average power dissipated in the inductor is	q.	non-zero
iii.	Average power dissipated in the capacitor is	r.	163 SI units
iv.	rms voltage across the capacitor is	s.	265.7 SI Units

Codes

	i	ii	iii	iv
a.	q,s	p	p	q,r
b.	q	r	r,s	p
c.	s,q	r	p,q	s
d.	p,q	r	p,r	s

24 Match the following graphs in Column I with the correct entries in Column II. In Column II, first quantity listed is shown on *Y*-axis and second quantity listed is on *X*-axis of the graph.

	Column I		Column II
i.	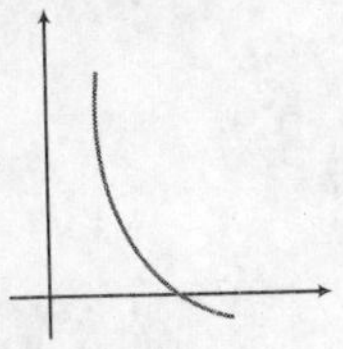	p.	$\dfrac{I}{X_L}$ *versus f* for an inductor
ii.	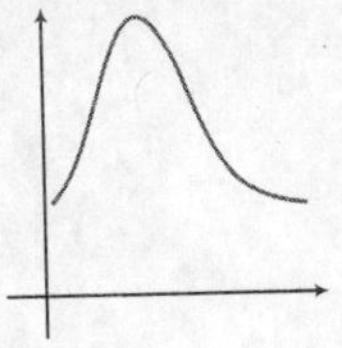	q.	*I versus f* for a capacitor
iii.	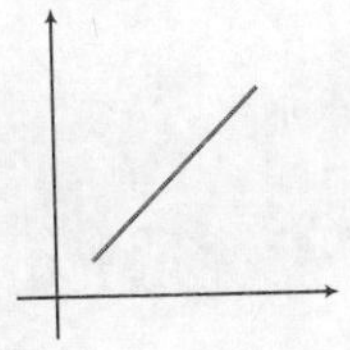	r.	*I versus f* for *L-C-R*
iv.	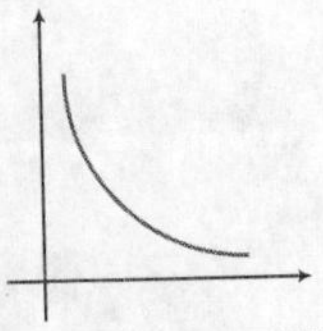	s.	*z versus f* for an *L-C-R*

Codes

	i	ii	iii	iv
a.	p	q	r	s
b.	s	r	q	p
c.	s	r	p	q
d.	r	s	q	p

Statement Type

Working Space

Direction *(Q. Nos. 25 and 26) This section is based on Statement I and Statement II. Select the correct answer from the codes given below.*

a. Both Statement I and Statement II are correct and Statement II is the correct explanation of Statement I
b. Both Statement I and Statement II are correct but Statement II is not the correct explanation of Statement I
c. Statement I is correct but Statement II is incorrect
d. Statement II is correct but Statement I is incorrect

25 **Statement I** An AC ammeter has a linear scale.

Statement II AC measuring instruments are hot wire instruments.

26 **Statement I** Average power consumed by a circuit depends only on the total impedance of circuit.

Statement II Power = I^2R

One Integer Value Correct Type

Direction *(Q. Nos. 27 and 28) This section contains 2 questions. When worked out will result in an integer from 0 to 9 (both inclusive).*

27 A coil of 0.01 H and 1 Ω is connected to 220 V, 50 Hz AC supply. If time lag between maximum alternating voltage and current is k (s), then find the value of 1000 k.

28 An alternating voltage $\Delta V = 100\sin\omega t$ (volts) is applied across an L-C-R circuit with $L = 2$ H, $C = 10\mu$F, $R = 10\,\Omega$. If ω_1 and ω_2 are two frequencies at which power is half of its maximum value, then find the value of $|\omega_2 - \omega_1|$.

DPP-3 AC Source with R-L-C Connected in Series

Subjective Questions

Working Space

Direction *(Q. Nos. 1-4) These questions are subjective in nature, need to be solved completely on notebook.*

1 A resistor of 200 Ω and a capacitor of 15 μF are connected in series to a 220 V, 50 Hz, AC source. Calculate the current in the circuit and rms voltage across the resistor and the capacitor. Is the algebraic sum of these voltages more than the source voltage? If yes, resolve the paradox.

2 A resistor *R*, inductor *L* and capacitor *C* are connected in parallel with an AC source with emf $V = V_0 \sin \omega t$.

i. Describe phasor diagram.

ii. Determine current as a function of time in

a. the inductor b. the capacitor c. the resistor

iii. Determine circuit impedance.

iv. Find power factor.

3 i. Find the fraction of energy lost (as thermal energy) in a lightly damped ($4L/C >> R^2$) *L-C-R* circuit.

ii. If an *L-C-R* circuit has two resistors, 2 capacitors and 2 inductors, find Z_{eq} for this circuit.

4 For the series circuit of figure, suppose $R = 300\ \Omega$, $L = 60$ mH, $C = 0.5\ \mu F$, source amplitude is $E_0 = 50$ V and $\omega = 10000$ rad/s. Find the reactances X_L and X_C, the impedance *Z*, the current amplitude I_0, the phase angle ϕ and the voltage amplitude across each circuit element.

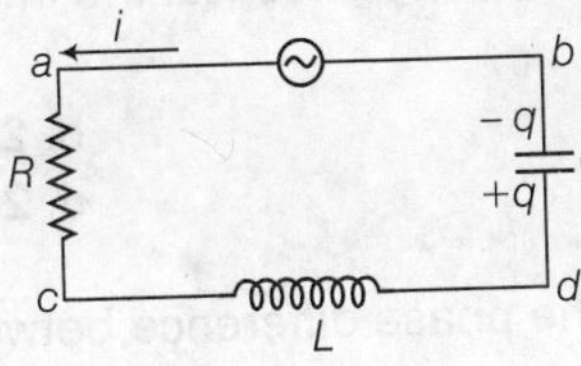

Only One Option Correct Type

Direction *(Q. Nos. 5-10) This section contains 6 multiple choice questions. Each question has four choices (a), (b), (c) and (d), out of which ONLY ONE is correct.*

5 For the series *L-C-R* circuit, the voltmeter and ammeter readings are

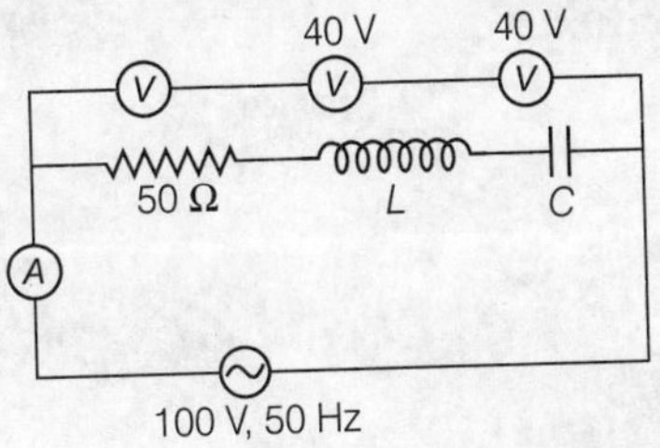

a. $V = 100$ V, $I = 2$ A
b. $V = 100$ V, $I = 5$ A
c. $V = 1000$ V, $I = 2$ A
d. $V = 300$ V, $I = 1$ A

Working Space

6 An AC circuit having supply voltage E consists of a resistor of resistance 3 Ω and an inductor of reactance 4 Ω as shown in the figure. The voltage across the inductor at $t = T/2$ is

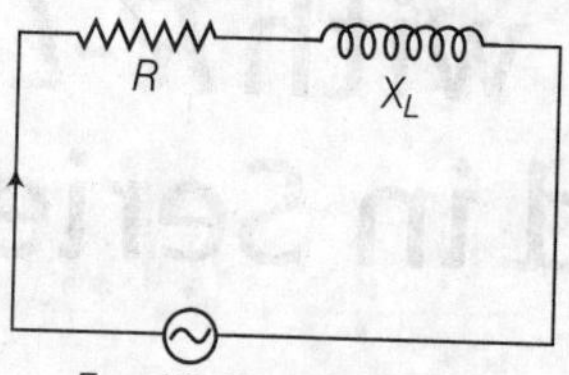

$E = 10 \sin \omega t$

a. 2 V
b. 10 V
c. zero
d. 4.8 V

7 Let $f = 50$ Hz and $C = 100\ \mu F$ in an AC circuit containing a capacitor only. If the peak value of the current in the circuit is 1.57 A at $t = 0$. The expression for the instantaneous voltage across the capacitor will be

a. $E = 50 \sin (100\pi t - \pi/2)$
b. $E = 100 \sin (50\pi t)$
c. $E = 50 \sin (100\pi t)$
d. $E = 50 \sin \left(100\pi t + \frac{\pi}{2}\right)$

8 For a series L-C-R circuit, the voltage across the resistance, capacitance and inductance is 10 V each. If the capacitance is short circuited, the voltage across the inductance will be

a. 10 V
b. $\frac{10}{\sqrt{2}}$ V
c. $\left(\frac{10}{3}\right)$ V
d. 20 V

9 When 100 V DC is applied across a solenoid, a current of 1 A flows in it. When 100 V AC is applied across the same coil, the current drops to 0.5 A. If the frequency of the AC source is 50 Hz, the impedance and inductance of solenoid are

a. 100 Ω, 0.93 H
b. 200 Ω, 1.0 H
c. 10 Ω, 0.86 H
d. 200 Ω, 0.55 H

10 For an AC circuit, the phase difference between current and potential is $\frac{\pi}{4}$. The variation of I and E *versus* t is represented in the graph. If $E = E_0 \cos (100t)$, then components of the circuit are

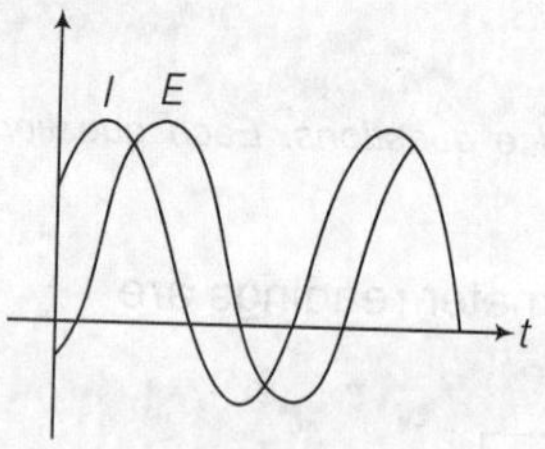

a. $R = 100\ \Omega$, $C = 1\ \mu F$
b. $R = 1 k\Omega$, $C = 10\ \mu F$
c. $R = 100\ \Omega$, $L = 1$ H
d. $R = 1 k\Omega$, $L = 10$ H

Working Space

One or More than One Options Correct Type

Direction *(Q. Nos. 11-15) This section contains 5 multiple choice questions. Each question has four choices (a), (b), (c) and (d), out of which ONE or MORE THAN ONE are correct.*

11 For the circuit shown in figure, the AC source has $\omega = 100\ \text{rads}^{-1}$. Considering the inductor and capacitor to be ideal.

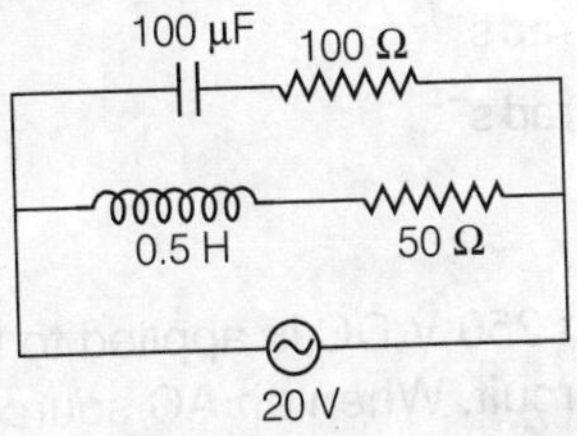

Correct alternatives are

a. the current through the circuit I is 0.3 A

b. the current through the circuit I is $0.3\sqrt{2}$ A

c. the voltage across 100 Ω resistor is $10\sqrt{2}$ V

d. the voltage across 50 Ω resistor is 10 V

12 When an AC voltage of 220 V is applied to the capacitor C,

a. the maximum voltage between plates is 220 V

b. the current is in phase with applied voltage

c. the charge on plates is in phase with applied voltage

d. power delivered to the capacitor is zero

13 As the frequency of an AC circuit increases the current first increases and then decreases. What combination of circuit elements is most likely forms the circuit?

a. Inductor and capacitor

b. Resistor and inductor

c. Resistor and capacitor

d. Resistor, inductor and capacitor

14 The phase difference between the alternating current and emf is $\frac{\pi}{2}$. Which of the following cannot be the constituent of circuit?

a. C alone

b. R, L

c. L, C

d. L alone

15 A series R-C circuit is connected to AC voltage source. Consider two cases (i) when C is without a dielectric medium and (ii) when C is filled with dielectric of constant 4. The current I_R through the resistor and voltage V_C across the capacitor are compared in the two cases. Which of the following is/are true? **[IIT JEE 2011]**

a. $I_R^A > I_R^B$

b. $I_R^A < I_R^B$

c. $V_C^A > V_C^B$

d. $V_C^A < V_C^B$

Comprehension Type

Direction *(Q. Nos. 16-21) This section contains 2 paragraphs, each describing theory, experiments, data, etc. Six questions related to the paragraphs have been given. Each question has only one correct answer among the four given options (a), (b), (c) and (d).*

Passage I

A series L-C-R circuit containing a resistance of 120 Ω has angular resonance frequency of $4 \times 10^5\ \text{rad s}^{-1}$. At resonance, the voltages across resistance and inductance are 60 V and 40 V, respectively.

16 The value of inductance L is

a. 0.1 mH

b. 0.2 mH

c. 0.35 mH

d. 0.4 mH

Working Space

17 The value of capacitance C is

a. $\frac{1}{32}\,\mu F$
b. $\frac{1}{16}\,\mu F$
c. $32\,\mu F$
d. $16\,\mu F$

18 At what frequency, the current in the circuit lags the voltage by 45°?

a. 4×10^5 rad s^{-1}
b. 3×10^5 rad s^{-1}
c. 8×10^5 rad s^{-1}
d. 2×10^5 rad s^{-1}

Passage II

A box contains a combination of L, C and R. When 250 V DC is applied to the terminals of the box, a current of 1A flows in the circuit. When an AC source of 250 V at 2250 rad s^{-1} is connected, a current of 1.25 A flows in the circuit. When frequency of AC is increased it is observed that current increases with frequency and becomes maximum at 4500 rad s^{-1}.

19 Circuit must be

a.
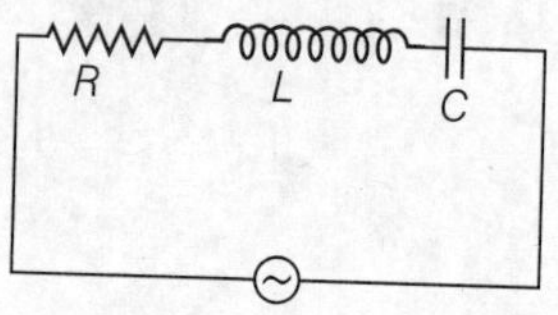

b.
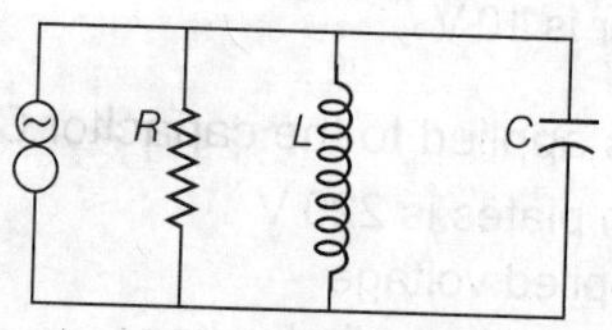

c.
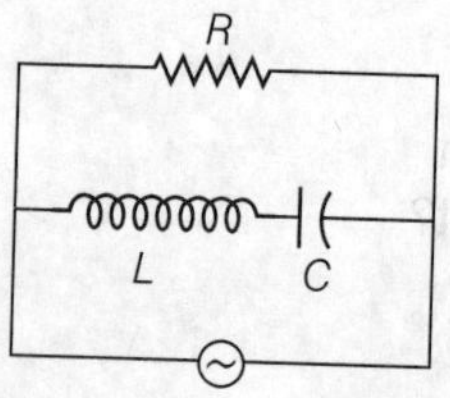

d.
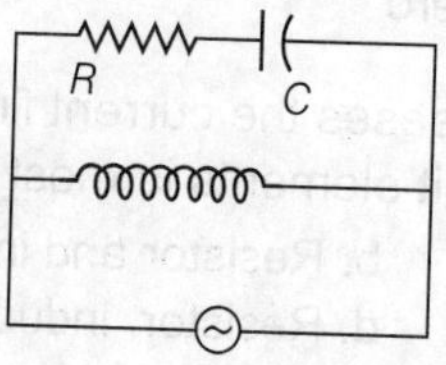

20 Total circuit reactance is

a. 333.33 Ω
b. 222.22 Ω
c. 111.11 Ω
d. 444.44 Ω

21 Capacitance of capacitor used in the circuit is

a. $2\,\mu F$
b. $2.5\,\mu F$
c. $1.5\,\mu F$
d. $1\mu F$

One Integer Value Correct Type

Direction *(Q. Nos. 22 and 23) This section contains 2 questions. When worked out will result in an integer from 0 to 9 (both inclusive).*

22 A series R-C combination is connected to an AC voltage of angular frequency $\omega = 500$ rad s^{-1}. If the impedance of the R-C circuit is $R\sqrt{1.25}$, find the time constant (in millisecond) of the circuit.

23 At a certain frequency ω_1, the reactance of a certain capacitor equals to that of a certain inductor. If the frequency is changed to $\omega_2 = 2\omega_1$, find the ratio of the reactance of the inductor to that of the capacitor.

Matching List Type

Direction *(Q. No. 24) Choices for the correct combination of elements from Column I and Column II are given as options (a), (b), (c) and (d), out of which one is correct.*

24 You are given many resistances, capacitors and inductors. These are connected to variable DC voltage source (the first two circuits) or an AC voltage source of 50 Hz frequency (next three circuits) in different ways as shown in Column II. When a current I (steady state for DC and rms for AC) flows through the circuit, the corresponding voltage V_1 and V_2 (indicated in circuits) are related as shown in Column I. Match the following columns and select the correct option from the codes given below.

	Column I		Column II
i.	$I \neq 0$, V_1 is proportional to I	p.	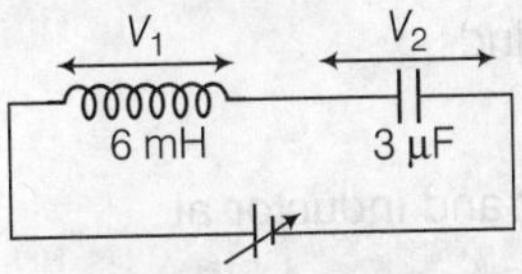
ii.	$I \neq 0$. $V_2 > V_1$	q.	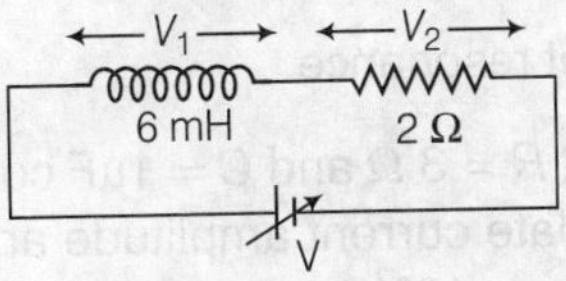
iii.	$V_1 = 0$, $V_2 = V$	r.	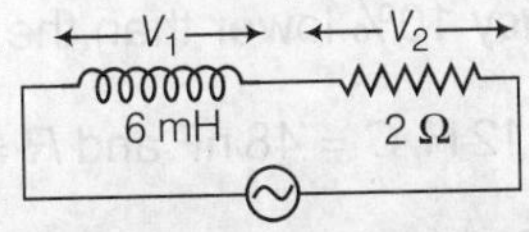
iv.	$I \neq 0$, V_2 is proportional to I	s.	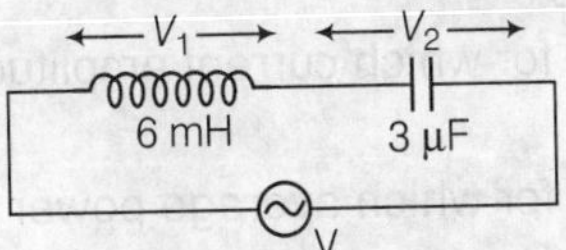
		t.	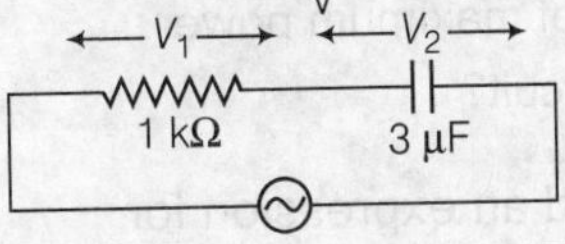

Codes

	i	ii	iii	iv
a.	r, s, t	q, r, s, t	p, q	q, r, s, t
b.	p, q	r, s, q	t, s	r, t, s
c.	t, s	q, r	q, t	t, s
d.	r, s	q, s, r, t	q	r, s, t

Statement Type

Direction *(Q. Nos. 25 and 26) This section is based on Statement I and Statement II. Select the correct answer from the codes given below.*

a. Both Statement I and Statement II are correct and Statement II is the correct explanation of Statement I

b. Both Statement I and Statement II are correct but Statement II is not the correct explanation of Statement I

c. Statement I is correct but Statement II is incorrect

d. Statement II is correct but Statement I is incorrect

25 **Statement I** An inductor is a low pass filter and a capacitor is a high pass filter.

Statement II Reactance of inductor is less at low frequency signal and reactance of capacitor is high for high frequency signals.

26 **Statement I** Main characteristics of series resonant circuit is voltage magnification.

Statement II At resonance, potential drop across inductor or capacitor is Q times of applied voltage.

DPP-4 Resonance

Subjective Questions

Direction *(Q. Nos. 1-4) These questions are subjective in nature, need to be solved completely on notebook.*

Working Space

1 For given *L-C-R* series circuit. Find

i. resonant frequency.

ii. voltage drops across capacitor and inductor at resonance.

iii. voltage drop across resistor at resonance.

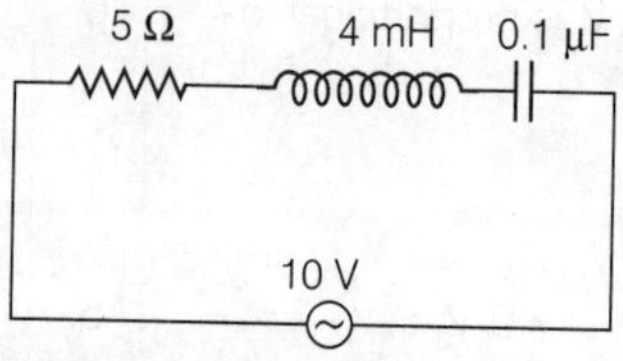

2 An *L-C-R* circuit has $L = 10$ mH, $R = 3\ \Omega$ and $C = 1\ \mu$F connected in series to a source of $15 \cos \omega t$ volts. Calculate current amplitude and average power dissipated per cycle at a frequency 10% lower than the resonant frequency.

3 A series *L-C-R* circuit with $L = 0.12$ H, $C = 48$ nF and $R = 23\ \Omega$ is connected to a 230 V variable frequency supply.

i. What is the source frequency for which current amplitude is maximum? Find this maximum value.

ii. What is the source frequency for which average power absorbed by the circuit is maximum? Obtain the value of maximum power.

iii. What is the *Q*-factor of the circuit?

4 Three students *X,Y,Z* performed an expression for studying the variation of AC with angular frequency in a series *L-C-R* circuit and obtained the graph below. They are used AC sources of the same rms value and inductances of the same value.

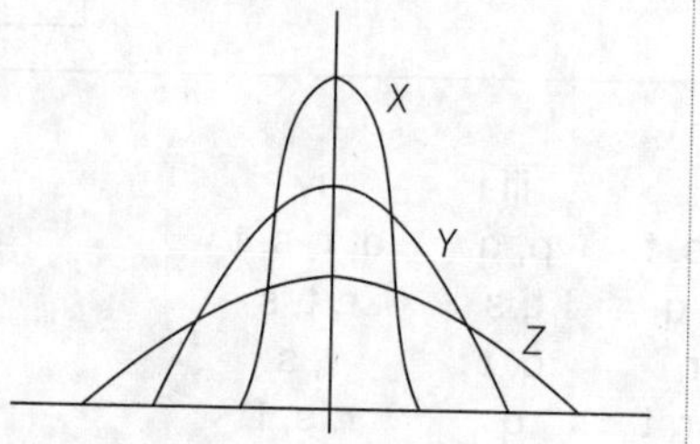

What can we conclude that

i. capacitance values

ii. resistances

used by them? In which case, will the quality factor be maximum? What can we conclude about nature of impedance of the set up at the frequency ω_0?

Only One Option Correct Type

Direction *(Q. Nos. 5-15) This section contains 11 multiple choice questions. Each question has four choices (a), (b), (c) and (d), out of which ONLY ONE is correct.*

5 A series *L-C-R* circuit containing a resistance of $120\ \Omega$ has angular resonance frequency 4×10^3 rad/s. At resonance, the voltage across resistance and inductance are 60 V and 40 V, respectively. The values of *L* and *C* are respectively

a. 20 mH, $\frac{25}{8}\ \mu$F

b. 2 mH, $\frac{1}{35}\ \mu$F

c. 20 mH, $\frac{1}{40}\ \mu$F

d. 2 mH, $\frac{25}{8}\ \mu$F

Working Space

6 For an *L-C-R* circuit, the capacitance is made one fourth, when in resonance. Then, what should be the change in inductance so that the circuit remains in resonance?

a. 4 times
b. $\frac{1}{4}$ times
c. 8 times
d. 2 times

7 For series *L-C-R* circuit, the plot of I_{max} *versus* ω is shown in figure. The bandwidth will be

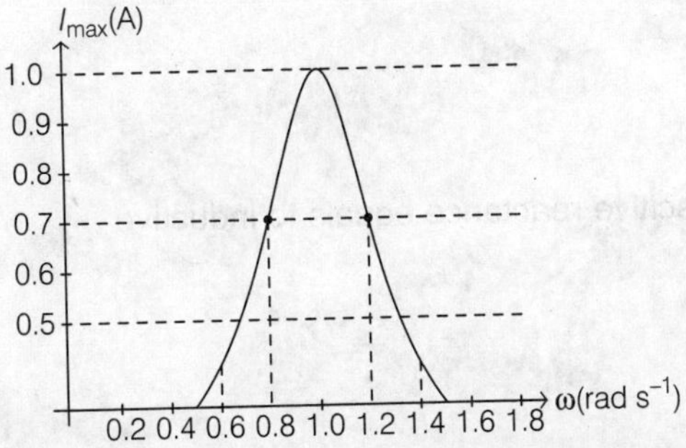

a. 0.2 rad s^{-1}
b. 0.3 rad s^{-1}
c. 0.4 rad s^{-1}
d. 0.5 rad s^{-1}

8 A coil, a capacitor and AC source of rms voltage 24 V and connected in series. By varying the frequency of the source, a maximum rms current of 6 A is observed. If coil is connected to a battery of emf 12 V and internal resistance 4 Ω, then current through it in steady state is

a. 2.4 A
b. 1.8 A
c. 1.5 A
d. 1.2 A

9 For an *L-C-R* series AC circuit, the voltage across each of the components *L*, *C* and *R* is 50 V. The voltage across the *L-C* combination will be

a. 50 V
b. $50\sqrt{2}$ V
c. 100 V
d. 0

10 For an *L-C-R* circuit, the capacitance is changed from *C* to 2*C*. For the resonant frequency to remain unchange, the inductance should be changed from *L* to

a. 4*L*
b. 2*L*
c. $\frac{L}{2}$
d. $\frac{L}{4}$

11 For a series resonant *L-C-R* circuit, the voltage across *R* is 100 V and $R = 1\text{k}\Omega$ with $C = 2\,\mu\text{F}$. The resonant frequency ω is 200 rad/s. At resonance, the voltage across *L* is

a. 250 V
b. 4×10^{-3} V
c. 2.5×10^{-2} V
d. 40 V

12 A radio can be tuned over a frequency range from 500 kHz to 1.5 MHz. If its *L-C* circuit has an effective inductance 400 μH, what is the range of its variable capacitor?

a. 20 pF-250 pF
b. 28 pF-253 pF
c. 30 pF-260 pF
d. 40 pF-270 pF

13 A capacitor of capacitance 240 pF is connected in parallel with a coil having inductance of 1.6×10^{-2} H and resistance 20 Ω. Calculate

I. the resonance frequency and
II. the circuit impedance at resonance

a. 3.2×10^6 Hz, 7.96×10^4 Ω
b. 7.96×10^4 Hz, 3.2×10^6 Ω
c. 4.5×10^7 Hz, 3.8×10^9 Ω
d. 3.8×10^9 Hz, 4.5×10^7 Ω

14 A capacitor of capacitance 250 pF is connected in parallel with a choke coil having inductance of 1.6×10^{-2} H and resistance 20 Ω. Calculate

I. the resonance frequency and

II. the circuit impedance at resonance.

a. 12×10^3 Hz, 3×10^4 Ω
b. 7.96×10^4 Hz, 3.2×10^6 Ω
c. 9×10^4 Hz, 9×10^2 Ω
d. 12×10^3 Hz, 6.2×10^6 Ω

15 For an *L-C-R* series circuit with an AC source of angular frequency ω, then

a. circuit will be capacitive if $\omega > \frac{1}{\sqrt{LC}}$

b. circuit will be inductive if $\omega = \frac{1}{\sqrt{LC}}$

c. power factor of circuit will be unity if capacitive reactance equals to inductive reactance

d. current will be leading voltage if $\omega > \frac{1}{\sqrt{LC}}$

One or More than One Options Correct Type

Direction *(Q. Nos. 16-20) This section contains 5 multiple choice questions. Each question has four choices (a), (b), (c) and (d), out of which ONE or MORE THAN ONE are correct.*

16 A resistor *R* and inductor *L*, a capacitor *C* and voltmeter are connected to an oscillator in the circuit as shown in the figure. When the frequency of the oscillator is increased up to resonance frequency, then

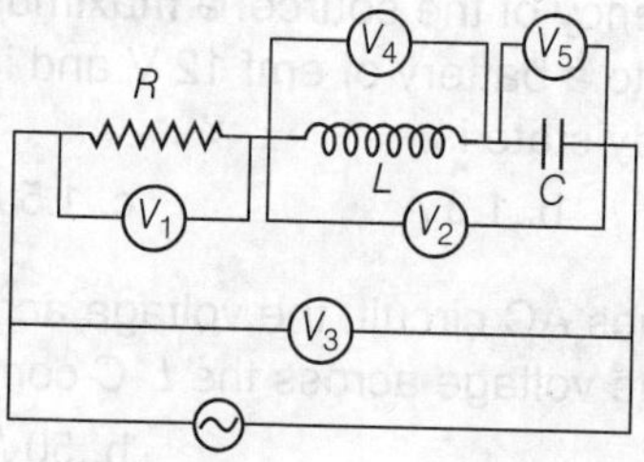

a. voltmeter reading, $V_1 = 0$
b. voltmeter reading, $V_2 = 0$
c. voltmeter reading, $V_3 = 0$
d. voltmeter, V_4 = voltmeter, V_5

17 For an AC circuit shown in figure, the supply voltage has constant rms value *V* but variable frequency *f*. At resonance, the circuit

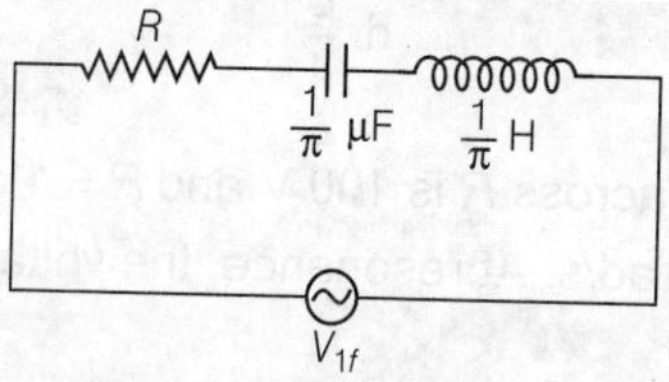

a. has a current *I* given by $I = \frac{V}{R}$

b. has a resonance frequency 500 Hz

c. has a voltage across the capacitor which is 180° out of phase with that across the inductor

d. has a current given by $I = \frac{V}{\sqrt{R^2 + \left(\frac{1}{\pi} + \frac{1}{\pi}\right)^2}}$

18 For the high pass filter shown, $R = 0.5\ \Omega$, choose the correct option(s).

a. For $X_C = (\sqrt{3}/2)\ \Omega$, $V_{out} = \frac{1}{2} V_{in}$
b. At low input frequency, $\frac{\Delta V_{out}}{\Delta V_{in}} < 1$
c. At low input frequency, $\frac{\Delta V_{out}}{\Delta V_{in}} > 1$
d. At high frequencies, $\frac{\Delta V_{out}}{\Delta V_{in}} \approx 1$

19 An AC source is connected to a network consisting of two diodes and four resistors each of resistance R.

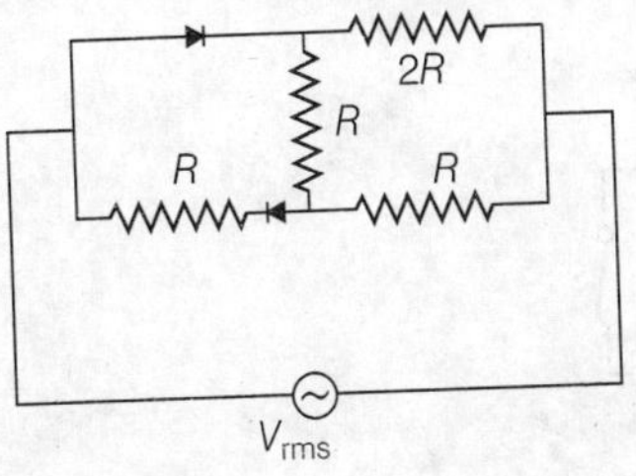

Now, choose the correct option(s).

a. For one half-time period, power consumed by circuit is $\frac{V^2_{rms}}{R}$

b. For one half-time period, power consumed by circuit is $\frac{4}{7} \cdot \frac{V^2_{rms}}{R}$

c. Average power consumed during one complete cycle is $\frac{11}{14} \cdot \frac{V^2_{rms}}{R}$

d. Ratio of equivalent resistance of circuit in each half of input cycle is 7/4

20 For an AC voltage, $\Delta V = 100 \sin 1000\, t$ is applied to a series R-L-C circuit. Given, $R = 400\ \Omega$, capacitance is $5\ \mu$F and the inductance is 0.5 H. Then,

a. current amplitude in the circuit is 2 A
b. current amplitude in the circuit is 0.2 A
c. average power dissipated in the circuit is 40 W
d. average power dissipated in the circuit is 8 W

Comprehension Type

Direction *(Q. Nos. 21-25) This section contains 2 paragraphs, each describing theory, experiments, data, etc. Five questions related to the paragraphs have been given. Each question has only one correct answer among the four given options (a), (b), (c) and (d).*

Passage I

An inductance of 2 H, a capacitance of $18\ \mu$F and a resistance of $10\ k\Omega$ are connected to an source of 20 V with adjustable frequency.

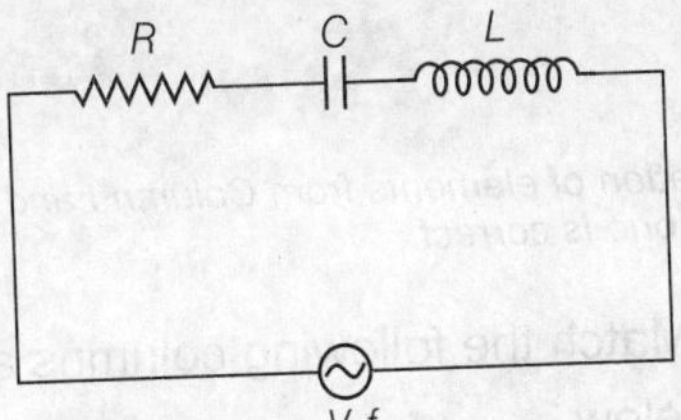

21 What frequency should be chosen to maximum current (rms) in the circuit?

a. $\frac{250}{7\pi}$ Hz
b. $\frac{250}{3\pi}$ Hz
c. $\frac{220}{3\pi}$ Hz
d. $\frac{400}{3\pi}$ Hz

22 What is the value of this maximum current?

a. 1 mA
b. 2 mA
c. 3 mA
d. 4 mA

Passage II

Consider a series L-C-R circuit with switches S_1 and S_2 which can short circuit inductor and capacitor.

Working Space

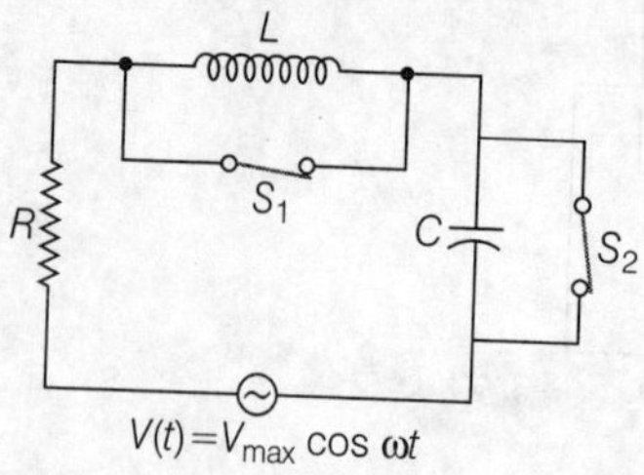

23 When only switch 1 is opened, then current in the circuit is

a. $i = \frac{V_{max}}{R} \cos \omega t$

b. $i = \frac{V_{max}}{\sqrt{R^2 + \omega^2 L^2}} \cos\left(\omega t + \tan^{-1}\left(\frac{\omega L}{R}\right)\right)$

c. $i = \frac{V_{max}}{\sqrt{R^2 + \left(\omega L - \frac{L}{C\omega}\right)^2}} \cos\left(\omega t + \tan^{-1}\left(\frac{R}{C\omega}\right)\right)$

d. $i = \frac{V_{max}}{\sqrt{\left(R^2 + \frac{1}{C^2\omega^2}\right)}} \cos\left(\omega t + \tan^{-1}\left(\frac{1}{C\omega R}\right)\right)$

24 Maximum energy stored in the capacitor when both switches S_1 and S_2 opened, will be

a. $\frac{V_{max}^2 L}{2R^2}$ b. $\frac{V_{max}^2 C}{2R^2}$ c. $\frac{V_{max}^2 LC}{2R^3}$ d. $\frac{V_{max}^2 L}{R^2}$

25 Maximum energy stored in the inductor when both switches S_1 and S_2 opened will be

a. $\frac{V_{max}^2 L}{2R^2}$ b. $\frac{V_{max}^2 L}{R^2}$ c. $\frac{V_{max}^2 LC}{2R^3}$ d. $\frac{V_{max}^2 C}{2R^2}$

Matching List Type

Direction *(Q. Nos. 26 and 27) Choices for the correct combination of elements from Column I and Column II are given as options (a), (b), (c) and (d), out of which one is correct.*

26 Consider all possibilities (L, C, R are non-zero). Match the following columns and select the correct option from the codes given below.

	Column I		Column II
i.	In L-R series, AC circuit	p.	Current lags in inductor voltage by $\pi/2$.
ii.	In R-C series, AC circuit	q.	Current lags voltage by an angle less than $\frac{\pi}{2}$.
iii.	In L-C-R series, AC circuit (consider all possibilities)	r.	Current leads voltage by an angle less than $\frac{\pi}{2}$.
iv.	In purely resistive AC circuit	s.	Current and voltage are in phase.

Codes

	i	ii	iii	iv
a.	p,q	r	p,q,r,s	s
b.	p	r	p,q	r,s
c.	p,q	q	r,s	q,r
d.	q	r,s	q,s	p,r

27 There are various applications of a series resonant circuit, where supply frequency is fixed and either L or C is varied to obtain the condition of resonance,

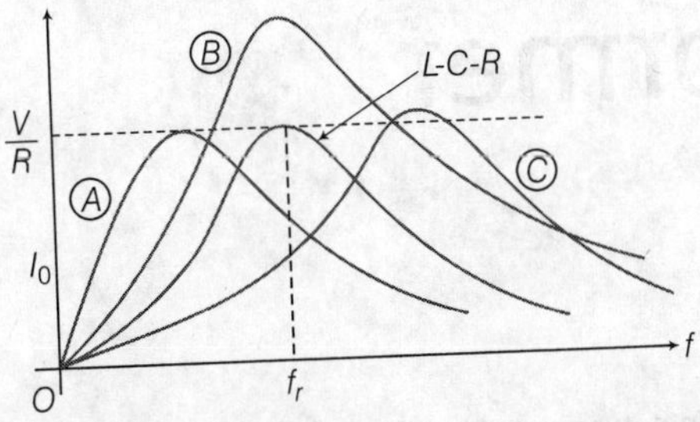

Curve for such an L-C-R circuit is marked, f_r is resonating frequency of source. Match the following columns and select the correct option from the codes given below.

	Column I		Column II
i.	$R_1 < R$	p.	Curve A
ii.	$C_1 > C$	q.	Curve B
iii.	$L_1 < L$	r.	Curve C

Codes

	i	ii	iii		i	ii	iii
a.	p	q	r	b.	q	r	p
c.	q	p	r	d.	r	p	q

One Integer Value Correct Type

Direction *(Q. Nos. 28 and 29) This section contains 2 questions. When worked out will result in an integer from 0 to 9 (both inclusive).*

28 A series L-C-R circuit with $R = 120\ \Omega$ has an angular frequency 4×10^5 rad s^{-1}. At resonance, voltages across resistance and inductor are 60 V and 40 V respectively. If frequency at which the circuit current lags behind the voltage by a phase of $\frac{\pi}{4}$ radian is $k \times 10^5$ rad s^{-1}, then find the value of k.

29 A coil has inductance of 1.3 mH and resonates at 600 kHz and its Q value is 30. If for bandwidth of 50 kHz resistor required is $49 \times N$ kΩ, then find the value of N.

Statement Type

Direction *(Q. Nos. 30 and 31) This section is based on Statement I and Statement II. Select the correct answer from the codes given below.*

a. Both Statement I and Statement II are correct and Statement II is the correct explanation of Statement I
b. Both Statement I and Statement II are correct but Statement II is not the correct explanation of Statement I
c. Statement I is correct but Statement II is incorrect
d. Statement II is correct but Statement I is incorrect

30 **Statement I** When Q-value of an L-C-R circuit is high, then power dissipation is high.

Statement II Q-value is defined as ratio of maximum energy stored and energy dissipated per cycle.

31 **Statement I** If half power point frequencies for an resonant L-C-R circuit are 2 s^{-1} and 8 s^{-1} respectively, then resonance occurs at 4 s^{-1}.

Statement II Bandwidth of circuit is $|f_2 - f_2|$, where f_1 and f_2 are half power point frequencies.

DPP-5 Transformer

Subjective Questions

Direction (Q. Nos. 1-4) These questions are subjective in nature, need to be solved completely on notebook.

Working Space

1 A transformer has 400 primary and 1000 secondary turns. The net cross-sectional area of core is 60 cm^2. Primary is connected to 50 Hz supply at 500 V. Calculate the

i. peak value of flux density in the core.

ii. voltage induced in the secondary winding.

2 At a hydroelectric power plant, the water pressure head is at a height of 300 m and the water flow available is 100 m^3s^{-1}. If the turbine generator efficiency is 60%. Estimate the electric power available from the plant. (Given, $g = 9.8\ ms^{-2}$)

3 A hydroelectric power plant generating electrical power at 440 V supply to a small scale industrial area. Industrial area gets power from a 4000 – 220 V step-down transformer. Power station is 15 km away from industrial area and resistance of two wire line used for transmission is 0.5 Ωkm^{-1}. Industrial area requires 800 kW of electric power at 220 V. Then,

i. estimate line power loss.

ii. power supplied by generating plant.

iii. characterise the step-up transformer at plant.

4 The number of turns in the primary and secondary coils of an ideal transformer are 2000 and 50, respectively. The primary coil is connected to a main supply of 120 V and secondary to a night bulb of 0.6 Ω. Calculate the

i. voltage across the secondary.

ii. current in the bulb.

iii. current in primary coil.

iv. power in primary and secondary coils.

Only One Option Correct Type

Direction (Q. Nos. 5-12) This section contains 8 multiple choice questions. Each question has four choices (a), (b), (c) and (d), out of which ONLY ONE is correct.

5 A power (step-up) transformer with 1 : 8 turn ratio has 60 Hz, 120 V across the primary, the load in the secondary is 10^4 Ω. The current in the secondary is

a. 96 A
b. 0.96 A
c. 9.6 A
d. 96 mA

6 A transformer is used to light a 140 W, 24 V lamp from 240 V AC mains. The current in the main cable is 0.7 A. The efficiency of the transformer is

a. 48%
b. 63.8%
c. 83.3%
d. 90%

7 A transformer has an efficiency of 80%. It works at 4 kW and 100 V. If the secondary voltage is 240 V. Find the primary and secondary currents.

a. 20 A, 13.3 A
b. 40 A, 13.3 A
c. 50 A, 40 A
d. 30 A, 20 A

8 A small DC motor operating at 200 V draws a current of 5.0 A at its full speed of 3000 rpm. The resistance of the armature of the motor is 8.5 Ω. Determine the efficiency of motor.

a. 85%
b. 78.75%
c. 76%
d. 72%

9 In a step-up transformer, the voltage in the primary is 220 V and the current is 5 A. The secondary voltage is found to be 22000 V. The current in secondary is

a. 5 A
b. 50 A
c. 500 A
d. 0.05 A

10 To transmit electrical energy from a generator to distant consumers,

a. high voltage and low current are transmitted
b. high voltage and high current are transmitted
c. low voltage and low current are transmitted
d. low voltage and high current are transmitted

11 In a step-up transformer in comparison with the secondary, the primary coil has

a. less voltage and high current
b. high voltage and less current
c. less voltage and less current
d. high voltage and high current

12 You have two copper cables of equal length for carrying current. One of them has a single wire of area of cross-section A, the other has ten wires of cross-section $\frac{A}{10}$. Judge their suitability for transporting AC and DC,

a. only single strand for DC, only multiple strands for AC
b. only multiple strands either for DC or for AC
c. only single strand either for AC or for DC
d. only single strand either for DC or for AC

One or More than One Options Correct Type

Direction (Q. Nos. 13-17) *This section contains 5 multiple choice questions. Each question has four choices (a), (b), (c) and (d), out of which ONE or MORE THAN ONE are correct.*

13 Which of the following is true for an ideal transformer?

a. Total magnetic flux linked with primary coil equals to the flux linked with secondary coil
b. Flux per turn in primary is equal to flux per turn in secondary
c. Induced emf in secondary coil equals induced emf in primary
d. Power associated with primary coil at any moment equals to power associated with secondary coil

14 Electrical energy is transmitted over large distances at high alternating voltages. Which of the following statement(s) is(are) correct?

a. For a given power level, there is a lower current
b. Lower current implies less power loss
c. Transmission line can be made thinner
d. It is easy to reduce the voltage at the receiving end using step-down transformer

15 In a transformer, number of turns in primary are 140 and that in the secondary are 280. If current in primary is 4 A, then that in the secondary is **[AIEEE 2002]**

a. 4 A
b. 2 A
c. 6 A
d. 10 A

Working Space

16 A transmission line with resistance 4.5×10^{-4} Ω/m is used to transmit 5 MW over 400 miles (6.44×10^5m). The output voltage of generator is 4.5 kV.

Choose the correct option(s).

a. If power is transmitted at 500 kV, then line loss is 29 kW
b. Fraction of power loss is 20%
c. 5 MW power cannot be transmitted at generated voltage of 4.5 kV
d. Copper loss occurred in transformer is less than 20%

17 A mobile charging unit has an input of 120 V AC, 8 W and an output of 9 V DC , 300 mA. Choose correct options about the device.

a. Efficiency of device is more than 50%
b. Efficiency of device is less than 50%
c. Power wasted by device is 5.3 W
d. Power wasted by device is 2.7 W

Comprehension Type

***Direction** (Q. Nos. 18-22) This section contains 2 paragraphs, each describing theory, experiments, data, etc. Five questions related to the paragraphs have been given. Each question has only one correct answer among the four given options (a), (b), (c) and (d).*

Passage I

A thermal power plant produces electrical power of 600 kW at 4000 V, which is to be transported to a place 20 km away from the power plant for consumer's usage. It can be transported either directly with a cable large current carrying capacity or by using a combination of step-up and step-down transformer at the two ends. The drawback of the direct transmission is the large energy dissipation. In the method using transformers, the dissipation is much smaller.

In this method, a step-up transformer is used at the plant side so that current is reduced to a smaller value. At consumer end, a step-down transformer is used to supply power to the consumer at the specified lower voltage. It is reasonable to assume that power cable is purely resistive and transformer are ideal with power factor unity. All currents and voltage mentioned are rms values.

18 For the method using transformers, assume that the ratio of the number of turns in the primary to that in secondary in the step-up transformer is 1: 10. If the power to the consumers has to be supplied at 200 V, the ratio of number of turns in the primary to that in the secondary in the step-down transformer is

a. 200 : 1
b. 150 : 1
c. 100 : 1
d. 50 : 1

19 For the direct transmission method with a cable of resistance 0.4 kΩ/km is used, the power dissipation (in %) during transmission is

a. 20
b. 30
c. 40
d. 50

Passage II

A transformer is rated for an output of 9000 W. Primary voltage is 1000 V. The ratio of turns of primary and secondary is 5 : 1. The iron losses are 700 W at fall load. Primary coil has a resistance of 1 Ω. Given, the efficiency of transformer as 90%.

20 The secondary voltage is

a. 1000 V
b. 5000 V
c. 200 V
d. 0 V

Working Space

21 Copper losses in secondary coil are

a. 100 W
b. 700 W
c. 200 W
d. 1000 W

22 Resistance of secondary coil is nearly

a. $0.01\,\Omega$
b. $0.1\,\Omega$
c. $0.2\,\Omega$
d. $0.4\,\Omega$.

One Integer Value Correct Type

Direction *(Q. Nos. 23-26) This section contains 4 questions. When worked out will result in an integer from 0 to 9 (both inclusive).*

23 A transformer of efficiency 90% draws an input power of 4 kW. If electrical appliance connected across the secondary draws a current of 6 A, the impedance of the circuit is $(600/n)\,\Omega$. The value of n is

24 In a transformer ratio of secondary turns (N_2) and primary turns (N_1), i.e. $\frac{N_2}{N_1} = 4$. If the voltage applied in primary is 200 V, 50 Hz. The voltage induced in secondary is $K \times 10^2$ V. Find the value of K.

25 Primary coil of a transformer has a current of 5 A. The resulting fluxes linked with primary and secondary are 0.2 mWb and 0.4 mWb, respectively. If primary has 500 turns and secondary has 1500 turns and self-inductance of secondary coil is 90 k, then find the value of k.

26 A distribution transformer is used to supply a household as

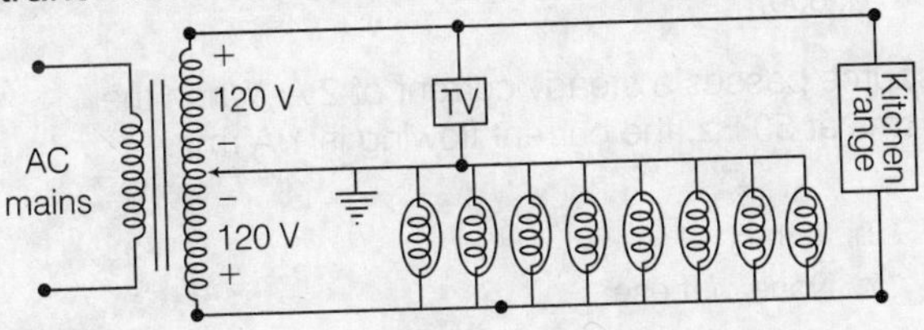

Load consists of eight 100 W bulbs. One 200 W TV. A 15 kW kitchen range. If secondary side of transformer has 72 turns and primary current is 6.67 A, then find the value of $\frac{N_P}{120}$, where N_P = number of turns of primary coil.

Matching List Type

Direction *(Q. No. 27) Choices for the correct combination of elements from Column I and Column II are given as options (a), (b), (c) and (d), out of which one is correct.*

27 Match the following columns and select the correct option from the codes given below.

	Column I		Column II
i.	AC generator	p.	Best way of reducing AC
ii.	DC generator	q.	Work on mutual induction
iii.	Choke coil	r.	Slip ring arrangement
iv.	Transformer	s.	Split ring arrangement

Codes

	i	ii	iii	iv
a.	s	r	q	p
b.	p	q	r	s
c.	s	r	p	q
d.	r	s	p	q

SET - 1

Revisal Problems for JEE Main

Only One Option Correct Type

Direction (Q. Nos. 1-19) *This section contains 19 multiple choice questions. Each question has four choices (a), (b), (c) and (d), out of which ONLY ONE is correct.*

Working Space

1 The instantaneous emf in an AC circuit is given by $E = 50 \sin (314t)$ V, where t is in seconds. In how much time the emf will become 25 V starting from zero?

a. $\frac{1}{50}$ s
b. $\frac{1}{200}$ s
c. $\frac{1}{314}$ s
d. $\frac{1}{600}$ s

2 The phase difference between current and voltage in an AC circuit is $\pi/4$ rad. If the frequency of AC is 50 Hz, then the phase difference is equivalent to the time difference

a. 0.78 s
b. 15.7 ms
c. 0.25 s
d. 2.5 ms

3 The number of poles in an AC generator is 10 and the coil is rotating at the rate of 600 revolutions per minute. Then, the frequency of AC current produced by the generator is (in Hz)

a. 10
b. 50
c. 1000
d. 600

4 A long solenoid connected to a 12 V DC source passes a steady current of 2 A. When the solenoid is connected to a source of 12 V rms at 50 Hz, the current flowing is 1 A rms. Then, the inductance of the solenoid is

a. 11 mH
b. 22 mH
c. 33 mH
d. None of these

5 A 110 V, 60 W lamp is run from a 220 V AC mains using a capacitor in series with the lamp instead of a resistor, then the voltage across the capacitor is about

a. 110 V
b. 190 V
c. 220 V
d. 311 V

6 A 100 V AC source of frequency 500 Hz is connected to a L-C-R circuit with $L = 8.1$ mH, $C = 12.5$ μF and $R = 10\,\Omega$ all connected in series. The potential difference across the resistance will be

a. 10 V
b. 100 V
c. 50 V
d. 500 V

7 An AC circuit draws 5 A at 160 V and the power consumption is 600 W. Then, the power factor is

a. 1
b. 0.75
c. 0.50
d. zero

8 A step-down transformer operates on a 2.5 kV line and supplies 80 A to a load. The ratio of the primary winding to the secondary winding is 20 : 1. Assuming 100 % efficiency, the output power is

a. 200 kW
b. 100 kW
c. 10 kW
d. None of these

9 An AC source of emf $E = 200 \sin (100t)$ is connected to a choke coil of inductance 1 H and resistance 100 Ω. The average power consumed is

a. zero
b. 200 W
c. 141 W
d. None of these

Working Space

10 A $2.5/\pi$ μF capacitor and a 3000 Ω resistance are joined in series to an AC source of 200 V and 50 s^{-1} frequency. The power factor of the circuit and the power dissipated in it will respectively be

a. 0.6, 0.06 W
b. 0.06, 0.6 W
c. 0.6, 4.8 W
d. 4.8, 0.6 W

11 A circuit drawn a power of 550 W from a source of 220 V, 50 Hz. The power factor of the circuit is 0.8 and the current lags in phase behind the potential difference. To make the power factor of circuit is 1.0, the capacitance required to be connected with it, will be

a. 70.4 μF
b. 75 μF
c. 7.5 μF
d. 750 μF

12 An L-C-R circuit has $L = 10$ mH, $R = 3\,\Omega$ and $C = 1\,\mu$F connected in series to a source of $15\cos \omega t$ V. The current-amplitude and the average power dissipated per cycle at a frequency 10% lower than the resonant frequency will be respectively

a. 0.704 A, 0.744 W
b. 0.704 A, 0.704 W
c. 7.04 A, 7.44 W
d. 70.4 A, 74.4 W

13 A 750 Hz, 20 V source is connected to a resistance of 100 Ω, an inductance of 0.1803 H and a capacitance of 10 μF all in series. The time in which the resistance (thermal capacity = 2 J/°C) will get heated by 10°C will be

a. 20 s
b. 200 s
c. 348 s
d. 448 s

14 Kanha wants to calculate the current and power dissipated in an L-C-R series circuit. He connected 100 Ω resistance to an AC source of peak value 200 V and angular frequency 300 rad/s. When he removed only the capacitance, the current was found to be lagging behind the voltage by 60°. While on removing the inductance he found, the current leading the voltage by 60°. The value of peak current and the power dissipated obtained by him will be

a. 2 A, 200 W
b. 4 A, 100 W
c. 3 A, 120 W
d. None of these

15 For an L-C-R circuit, the capacitance is changed from C to $4C$. For the same resonant frequency, the inductance should be changed from L to

a. $2L$
b. $4L$
c. $\frac{L}{2}$
d. $\frac{L}{4}$

16 A coil of self-inductance 0.16 H is connected to a condenser of capacity 0.81 μF. The frequency of AC (in cps) that should be applied so that there is a resonance in the circuit (the resistance of the circuit is negligible) should be

a. 50
b. 60
c. 442
d. 342

17 In an oscillatory circuit the value of self-inductance of the connected coil is 10 mH. If the oscillatory frequency of the circuit is 1.0 megacycle/s, then the capacity of the condenser connected in the circuit will be

a. 2.5 pF
b. 2.5 μF
c. 0.25 pF
d. 0.25 μF

18 In a torsion type AC ammeter, a current of 25 A gives a deflection of 90°.

For a deflection of 180°, the current will be

a. 50 A
b. 20 A
c. $25\sqrt{2}$ A
d. 30 A

19 An induction coil with $L = 1.3$ mH, resonates at 600 kHz. Q-value of coil is 30. If a bandwidth of 50 kHz is required, then resistor used must be of

a. 98 kΩ
b. 49 Ω
c. 0
d. capacitor used

Statement Type

Working Space

Direction *(Q. Nos. 20-25) This section is based on Statement I and Statement II. Select the correct answer from the codes given below.*

a. Both Statement I and Statement II are correct and Statement II is the correct explanation of Statement I

b. Both Statement I and Statement II are correct but Statement II is not the correct explanation of Statement I

c. Statement I is correct but Statement II is incorrect

d. Statement II is correct but Statement I is incorrect

20 **Statement I** At resonance ratio of potential drop across inductor and of potential drop across resistor is more than 1.

Statement II At resonance, phase difference between current and voltage is zero.

21 **Statement I** When two alternating currents

$$i_1 = 20\sin(314t + 30^\circ)\text{ A}$$
$$i_2 = 40\sin(314t + 45^\circ)\text{ A}$$

are mixed, then resultant current has a magnitude of 60 A.

Statement II In AC, currents are added by using vector addition laws.

22 **Statement I** Impedance of circuit is not treated as a phasor.

Statement II $Z = \sqrt{R^2 + X^2}$

23 **Statement I** Charge amplitude of a capacitor of an *L-C-R* circuit is minimum at resonance.

Statement II At resonance circuit impedance is minimum and hence, current in circuit is maximum.

24 **Statement I** Output of an amplifier has an impedance of 45kΩ. When it is connected to an 8 Ω loudspeaker coil. The ratio of number of turns of two coils must be 75.

Statement II For maximum power transmitted from one device to other, their impedances must be matched.

25 **Statement I** When current in an *R-L* circuit is decaying, 99% of energy decays in an interval of 2.3 time constant.

Statement II Energy stored in inductor at any time, $U = \frac{1}{2}LI^2 e^{-2\left(\frac{R}{L}\right)t}$.

Revisal Problems
for JEE Advanced

Only One Option Correct Type

Working Space

***Direction** (Q. Nos. 1-6) This section contains 6 multiple choice questions. Each question has four choices (a), (b), (c) and (d), out of which ONLY ONE is correct.*

1 For the given AC circuit, which of the following is incorrect?

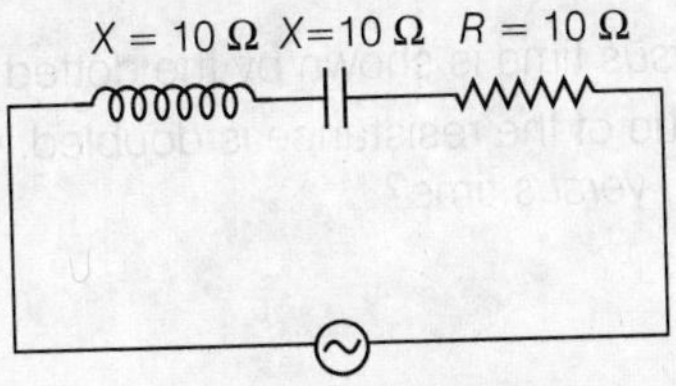

$E = 100 \sin (100\pi t + \pi/2)$

a. Voltage across resistance is lagging by 90° than voltage across capacitor
b. Voltage across capacitor is lagging by 180° than voltage across inductor
c. Voltage across inductor is leading by 90° than voltage across resistance
d. Resistance of the circuit is equal to resistance of circuit

2 For the series *L-C-R* circuit as shown in figure, the heat developed in 80 s and amplitude of wattless current *i*.

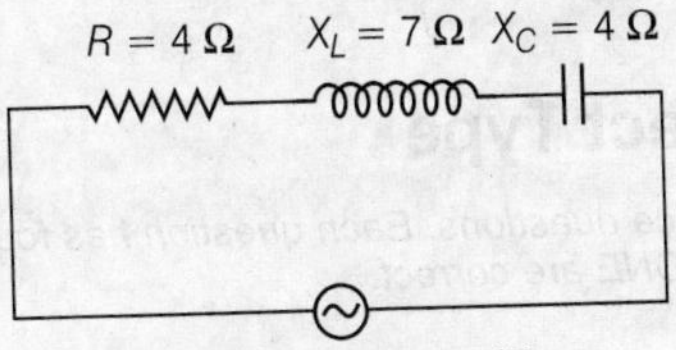

$V = 25 \sin (100\pi t + \pi/2)$

a. 4000 J, 3 A
b. 8000 J, 3 A
c. 4000 J, 4 A
d. 8000 J, 5 A

3 For the given circuit as summing inductor and source to be ideal, the phase difference between currents I_1 and I_2 is

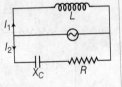

a. $\tan^{-1}\left(\frac{X_C}{R}\right) - \frac{\pi}{2}$
b. $\tan^{-1}\left(\frac{X_C}{R}\right)$
c. $\tan^{-1}\left(\frac{X_C}{R}\right) + \frac{\pi}{2}$
d. $\frac{\pi}{2}$

4 The current *I* in the inductance is varying with time according to the plot shown in figure. Which one of the following is the correct variation of voltage with time in the coil?

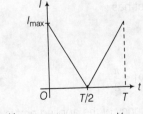

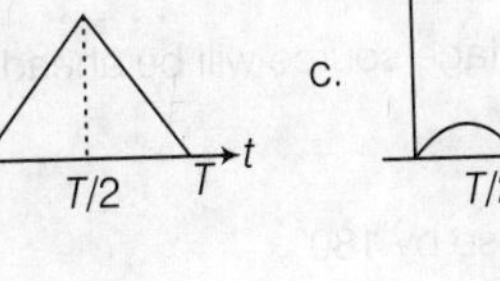

a.

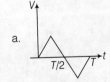

b. 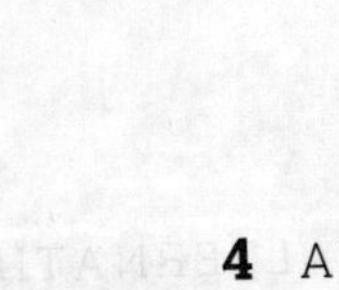

c. 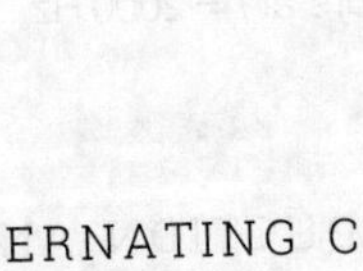

d. 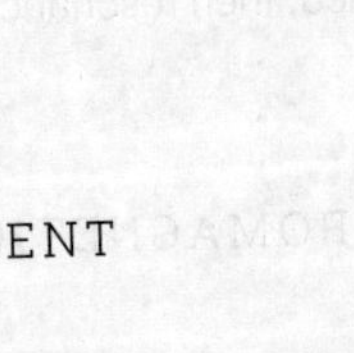

5 Figure shows an experimental plot for discharging of a capacitor in an *R-C* circuit. The time constant τ of this circuit lies between

Working Space

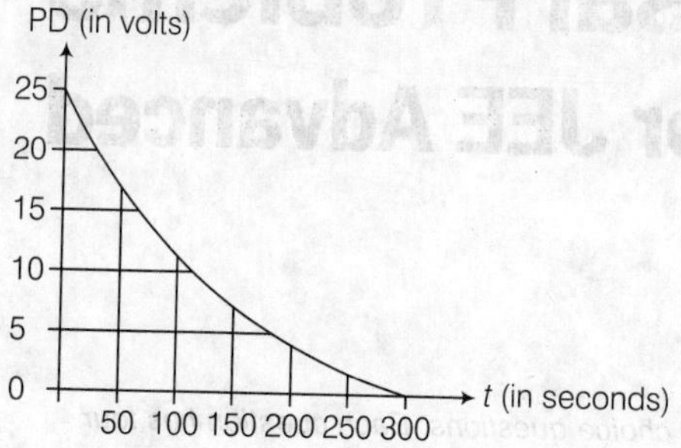

a. 150 s and 200 s
b. 0 s and 50 s
c. 50 s and 100 s
d. 100 s and 150 s

6 For an *R-C* circuit while charging, the graph of ln *I* *versus* time is shown by the dotted line in the below figure, where *I* is the current. When the value of the resistance is doubled, which of the solid curves best represents the variation of ln *I* *versus* time?

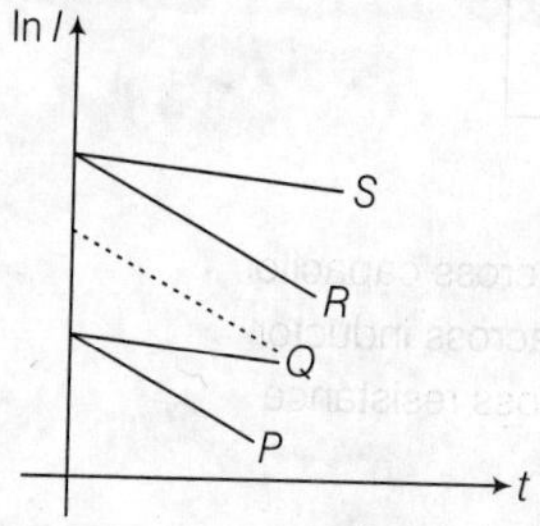

a. *P*
b. *Q*
c. *R*
d. *S*

One or More than One Options Correct Type

***Direction** (Q. Nos. 7-10) This section contains 4 multiple choice questions. Each question has four choices (a), (b), (c) and (d), out of which ONE or MORE THAN ONE are correct.*

7 For the figure shown, $R = 100\ \Omega$, $L = \frac{2}{\pi}$ H and $C = \frac{8}{\pi}\ \mu$F are connected in series with an AC source of 200 V and frequency *f*. V_1 and V_2 are two hot wire voltmeters. If the readings of V_1 and V_2 are same, then

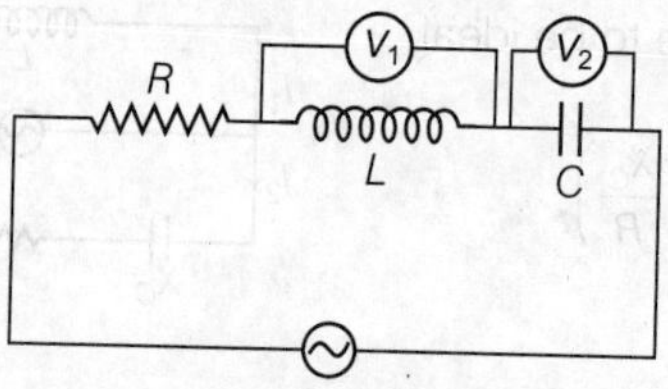

a. $f = 125$ Hz
b. $f = 250\pi$ Hz
c. current through *R* is 2 A
d. $V_1 = V_2 = 1000$ V

8 The line that draws power supply to your house from street has

a. zero average current
b. 220 V average voltage
c. voltage and current out of phase by 90°
d. voltage and current possibly differing in phase ϕ such that $|\phi| < \frac{\pi}{2}$

9 Resonance occurs in a series *L-C-R* circuit when the frequency of the applied emf is 1000 Hz. Which of the following statement(s) is/are correct?

a. When frequency = 900 Hz, then the current through the voltage source will be ahead of emf of the source
b. The impedance of the circuit is minimum at $f = 1000$ Hz
c. At only resonance, the voltage across *L* and *C* differ in phase by 180°
d. If the value of *C* is doubled, then resonance occurs at $f = 2000$ Hz

Working Space

10 For an AC circuit shown alongside in figure, the supply voltage has a constant rms value V but variable frequency f. At resonance, the circuit

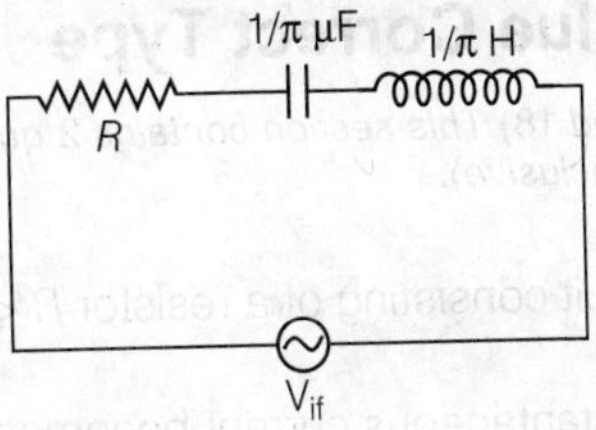

a. has current given by $I = \dfrac{V}{R}$

b. has a resonance frequency of 500 Hz

c. has a voltage across the capacitor which is 180° out of phase with that across the inductor

d. has a current given by $I = \dfrac{V}{\sqrt{R^2 + \left(\dfrac{1}{\pi} + \dfrac{1}{\pi}\right)^2}}$

Comprehension Type

***Direction** (Q. Nos. 11-16) This section contains 2 paragraphs, each describing theory, experiments, data, etc. Six questions related to the paragraphs have been given. Each question has only one correct answer among the four given options (a), (b), (c) and (d).*

Passage I

There is a series L-C-R circuit (as shown). An alternating source of emf having voltage $V = V_0 \sin \omega t$ is applied between M and N.

Here, $V_M - V_N = V_0 \sin \omega t$ and $\dfrac{1}{\omega C} - \omega L = R$

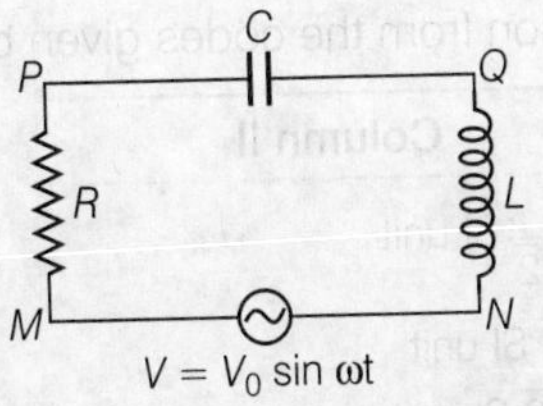

11 The potential difference across R has got a phase difference θ from that of alternating source of emf, where

a. $\theta = \dfrac{\pi}{2}$ b. $\theta = 0$ c. $\theta = \dfrac{\pi}{4}$ d. $\theta = \pi$

12 The rms value of potential difference across capacitor will be

a. $\dfrac{V_0}{\sqrt{2}}$ b. $V_0 R\omega C$

c. zero d. $\dfrac{V_0}{2R\omega C}$

13 Potential difference across inductor, i.e. $V_Q - V_R$ is

a. $\dfrac{V_0}{\sqrt{2}R}\omega L \sin\left(\omega t + \dfrac{3\pi}{4}\right)$ b. $\dfrac{V_0 \omega L}{\sqrt{2}R} \sin \omega t$

c. $\dfrac{V_0 \omega L}{\sqrt{2}R} \sin\left(\omega t + \dfrac{\pi}{2}\right)$ d. $\dfrac{V_0 \omega L}{\sqrt{2}R} \cos \omega t$

Passage II

An L-C oscillatory circuit consists of an inductor of 2 mH and a capacitor of 5 μF. Maximum value of instantaneous charge Q on the capacitor is 200 μC.

14 If I = instantaneous current in the inductor, the value of $\dfrac{dI}{dt}$ when $Q = 100\ \mu C$ is

a. $10^4\ As^{-1}$ b. $10^{-4}\ As^{-1}$ c. $10^2\ As^{-1}$ d. $10^{-2}\ As^{-1}$

15 Maximum value of current in the inductor is

a. 2.5 A b. 3.5 A c. 2 A d. 3 A

16 When $I = \dfrac{1}{2} I_{max}$, then Q at that instant is

a. $10^4 \times \sqrt{3}$ C b. $10^{-4}\sqrt{3}$ C c. $10^{-2} \times \sqrt{3}$ C d. $10^2 \times \sqrt{3}$ C

One Integer Value Correct Type

Working Space

Direction *(Q. Nos. 17 and 18) This section contains 2 questions. When worked out will result in an integer from 0 to 9 (both inclusive).*

17 A series *R-L-C* circuit consisting of a resistor $R << \sqrt{\frac{44}{C}}$, is set up in the oscillations. In time $t = \log 2 \times \left(\frac{2L}{R}\right)$, instantaneous current becomes $\frac{1}{K}$ times of maximum current. Find the value of *K*.

18 Initially, the capacitor in an *L-C* circuit is charged. A switch is charged at $t = 0$, allowing the capacitor to discharge and at time *t* the energy stored in the capacitor is one-fourth of its initial value. Inductance of inductor is $\frac{kt^2}{\pi^2 C}$. Find the value of *k*.

Matching List Type

Direction *(Q. Nos. 19 and 20) Choices for the correct combination of elements from Column I and Column II are given as options (a), (b), (c) and (d), out of which one is correct.*

19 The instantaneous voltage and current in an *L-R* circuit are given by $V = 100 \sin 100t$ $I = 10 \sin\left(100t - \frac{\pi}{4}\right)$.

Match the following columns and select the correct option from the codes given below.

	Column I		Column II
i.	R	p.	$\frac{1}{10\sqrt{2}}$ SI unit
ii.	X_L	q.	$5\sqrt{2}$ SI unit
iii.	L	r.	$10\sqrt{2}$ SI unit
iv	Average power in 1 cycle	s.	None of these

Codes

	i	ii	iii	iv
a.	q	p	q	r
b.	r	p	q	s
c.	p	q	r	s
d.	q	q	p	s

20 A fully charged capacitor is connected at $t = 0$ with an inductor (inductance *L*, capacitance *C*).

Match the following columns and select the correct option from the codes given below.

	Column I		Column II
i.	Charge in capacitor is maximum	p.	$\frac{\pi}{2}\sqrt{LC}$
ii.	Current in inductor is maximum	q.	$2\pi\sqrt{LC}$
iii.	Electrical energy in system is maximum	r.	$\frac{3}{2}\pi\sqrt{LC}$
iv.	Magnetic energy in system is maximum	s.	$4\pi\sqrt{LC}$

Codes

	i	ii	iii	iv
a.	p	q	r	s
b.	s	p	q	r
c.	q,s	p,r	q,s	p,r
d.	p	p,q	r	p,s

JEE Main & AIEEE Archive

(Compilation of Last 13 Years' Questions)

1 For the circuit shown here, the point C is kept connected to point A till the current flowing through the circuit becomes constant. Afterward, suddenly point C is disconnected from point A and connected to point B at time $t = 0$. Ratio of the voltage across resistance and the inductor at $t = L/R$ will be equal to **(JEE Main 2014)**

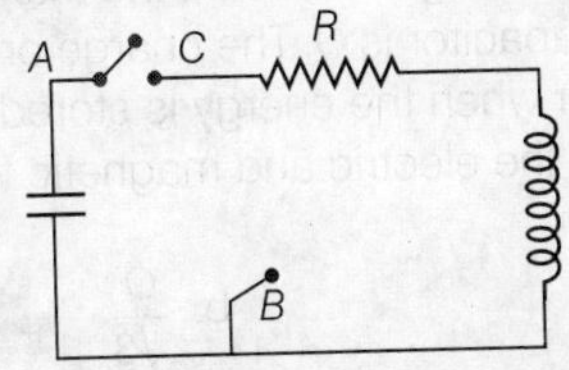

a. $\frac{e}{1-e}$ b. 1

c. -1 d. $\frac{1-e}{e}$

2 For an L-C-R circuit as shown below, both switches are open initially. Now, switch S_1 is closed and S_2 kept open (q is charge on the capacitor and $\tau = RC$ is capacitance time constant). Which of the following statement is correct? **(JEE Main 2013)**

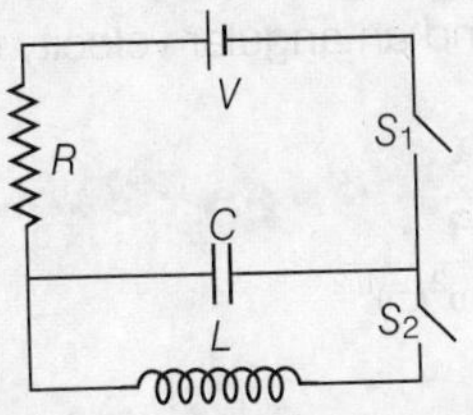

a. Work done by the battery is half of the energy dissipated in the resistor

b. At $t = \tau, q = CV/2$

c. At $t = 2\tau, q = CV(1-e^{-2})$

d. At $t = \frac{\tau}{2}, q = CV(1-e^{-1})$

3 The figure shows an experimental plot discharging of a capacitor in an R-C circuit. The time constant τ of this circuit lies between **(JEE Main 2012)**

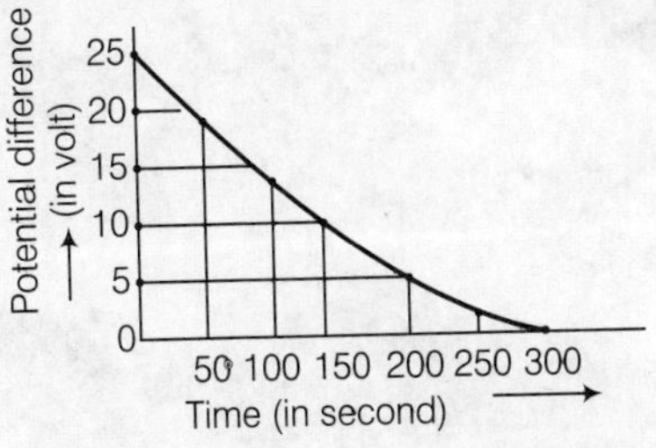

a. 150 s and 200 s b. 0 and 50 s

c. 50 s and 100 s d. 100 s and 150 s

4 For a series L-C-R circuit, $R = 200\ \Omega$ and the voltage and the frequency of the main supply is 220 V and 50 Hz, respectively. On taking out only the capacitance from the circuit, the current lags behind the voltage by 30°. On taking out only the inductor from the circuit, the current leads the voltage by 30°. The power dissipated in the L-C-R circuit is **(AIEEE 2010)**

a. 305 W b. 210 W

c. zero d. 242 W

5 An inductor of inductance $L = 400$ mH and resistors of resistances $R_1 = 4\ \Omega$ and $R_2 = 2\ \Omega$ are connected to battery of emf 12 V as shown in the figure. The internal resistance of the battery is negligible. The switch S is closed at $t = 0$. The potential drop across L as a function of time is **(AIEEE 2009)**

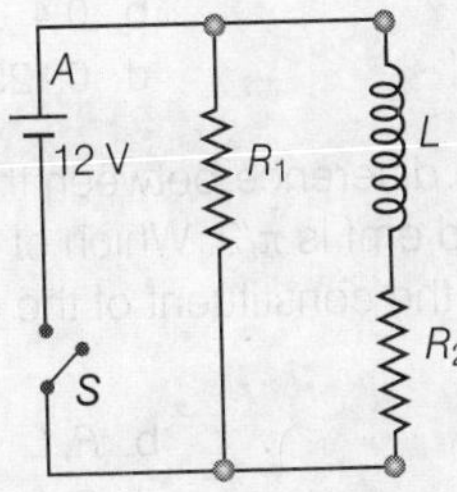

a. $6e^{-5t}$ V b. $\frac{12}{t}e^{-3t}$ V

c. $6(1-e^{-t/0.2})$ V d. $12e^{-5t}$ V

6 For an AC circuit, the voltage applied is $E = E_0 \sin \omega t$. The resulting current in the circuit is $I = I_0 \sin\left(\omega t - \frac{\pi}{2}\right)$. The power consumption in the circuit is given by **(AIEEE 2007)**

a. $P = \frac{E_0 I_0}{\sqrt{2}}$ b. $P = 0$

c. $P = \frac{E_0 I_0}{2}$ d. $P = \sqrt{2} E_0 I_0$

7 An inductor ($L = 100$ mH), a resistor ($R = 100\ \Omega$) and a battery ($E = 100$ V) are initially connected in series as shown in the figure. After a long time, the battery is disconnected after short circuiting the points A and B. The current in the circuit 1 millisecond after the short circuit is **(AIEEE 2006)**

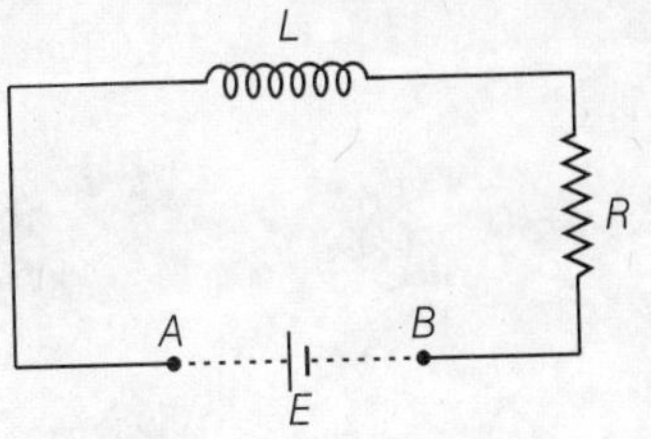

a. $1/e$ A b. e A c. 0.1 A d. 1 A

8 In an AC generator, a coil with N turns, all of the same area A and total resistance R, rotates with frequency ω in a magnetic field B. The maximum value of emf generated in the coil is **(AIEEE 2006)**

a. $NABR\omega$ b. NAB
c. $NABR$ d. $NAB\omega$

9 In a series resonant L-C-R circuit, the voltage across R is 100 V and $R = 1\text{k}\Omega$ with $C = 2\,\mu\text{F}$. The resonant frequency ω is 200 rad/s. At resonance, the voltage across L is **(AIEEE 2006)**

a. 2.5×10^{-2} V b. 40 V
c. 250 V d. 4×10^{-3} V

10 A coil of inductance 300 mH and resistance 2 Ω is connected to a source of voltage 2 V. The current reaches half of its steady state value in **(AIEEE 2005)**

a. 0.05 s b. 0.1 s
c. 0.15 s d. 0.3 s

11 A circuit has a resistance of 12 Ω and an impedance of 15 Ω. The power factor of the circuit will be **(AIEEE 2005)**

a. 0.8 b. 0.4
c. 1.25 d. 0.125

12 The phase difference between the alternating current and emf is $\pi/2$. Which of the following cannot be the constituent of the circuit? **(AIEEE 2005)**

a. C alone b. R, L
c. L, C d. L alone

13 The self-inductance of the motor of an electric fan is 10 H. In order to impart maximum power at 50 Hz, it should be connected to a capacitance of **(AIEEE 2005)**

a. 4 μF b. 8 μF
c. 1 μF d. 2 μF

14 Alternating current cannot be measured by DC ammeter because **(AIEEE 2004)**

a. AC cannot pass through DC ammeter
b. AC changes direction
c. average value of current for complete cycle is zero
d. DC ammeter will get damaged

15 For an L-C-R series AC circuit, the voltage across each of the components L, C and R is 50 V. The voltage across the L-C combination will be **(AIEEE 2004)**

a. 50 V b. $50\sqrt{2}$ V
c. 100 V d. zero

16 In an L-C-R circuit, the capacitance is changed from C to $2C$. For the resonant frequency to remain unchanged, the inductance should be changed from L to **(AIEEE 2004)**

a. $4L$ b. $2L$
c. $L/2$ d. $L/4$

17 In an oscillating L-C circuit, the maximum charge on the capacitor is Q. The charge on the capacitor when the energy is stored equally between the electric and magnetic fields is **(AIEEE 2003)**

a. $\frac{Q}{2}$ b. $\frac{Q}{\sqrt{3}}$
c. $\frac{Q}{\sqrt{2}}$ d. Q

18 The core of any transformer is laminated so as to **(AIEEE 2003)**

a. reduce the energy loss due to eddy currents
b. make it light weight
c. make it robust and strong
d. increase the secondary voltage

19 The power factor of an AC circuit having resistance R and inductance L (connected in series) and an angular velocity ω is **(AIEEE 2002)**

a. $\frac{R}{\omega L}$
b. $\frac{R}{(R^2 + \omega^2 L^2)^{1/2}}$
c. $\frac{\omega L}{R}$
d. $\frac{R}{(R^2 - \omega^2 L^2)^{1/2}}$

20 For a transformer, number of turns in the primary coil are 140 and that in the secondary coil are 280. If current in primary coil is 4 A, then that in the secondary coil is **(AIEEE 2002)**

a. 4 A b. 2 A
c. 6 A d. 10 A

JEE Advanced & IIT JEE Archive

(Compilation of Last 10 Years' Questions)

Passage for Q. Nos. (1-2)

A thermal power plant produces electric power of 600 kW at 4000 V, which is to be transported to a place 20 km away from the power plant for consumers' usage. It can be transported either directly with a cable of large current carrying capacity or by using a combination of step-up and step-down transformers at the two ends.

The drawback of the direct transmission is the large energy dissipation. In the method using transformer, the dissipation is much smaller. In this method, a step-up transformer is used at the plant side so that the current is reduced to a smaller value.

At the consumers' end, a step-down transformer is used to supply power to the consumers at the specified lower voltage. It is reasonable to assume that the power cable is purely resistive and the transformers are ideal with a power factor unity. All the current and voltages mentioned are rms values.

(2013 Adv., Comprehension Type)

1 If the direct transmission method with a cable of resistance $0.4\ \Omega\ \text{km}^{-1}$ is used, the power dissipation (in %) during transmission is

a. 20
b. 30
c. 40
d. 50

2 In the method using the transformers, assume that the ratio of the number of turns in the primary coil to that in the secondary coil in the step-up transformer is 1 : 10. If the power to the consumers has to be supplied at 200 V, the ratio of the number of turns in the primary coil to that in the secondary coil in the step-down transformer is

a. 200 : 1
b. 150 : 1
c. 100 : 1
d. 50 : 1

3 At time $t = 0$, terminal A in the circuit shown in the figure is connected to B by a key and an alternating current $I(t) = I_0 \cos(\omega t)$, with $I_0 = 1$ A and $\omega = 500$ rad s^{-1} starts flowing in it with the initial direction shown in the figure. At $t = 7\pi/6\omega$, the key is switched from B to D.

Now, onwards only A and D are connected. A total charge Q flows from the battery to charge the capacitor fully. If $C = 20\ \mu$F, $R = 10\ \Omega$ and the battery is ideal with emf of 50 V, identify the correct statement(s).

(2013 Adv., One or More than One Options Correct Type)

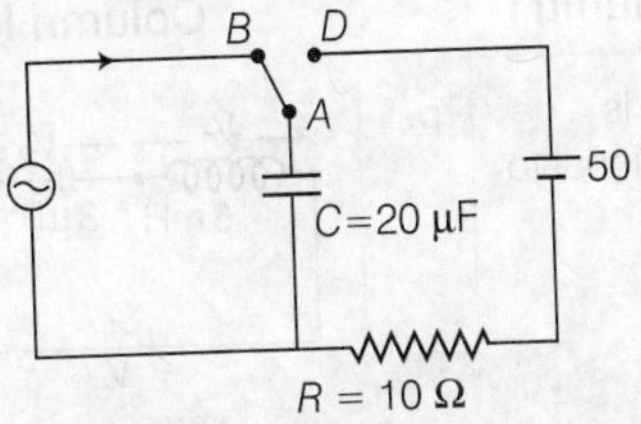

a. Magnitude of the maximum charge on the capacitor before $t = \dfrac{7\pi}{6\omega}$ is 1×10^{-3} C

b. The current in the left part of the circuit just before $t = \dfrac{7\pi}{6\omega}$ is clockwise

c. Immediately after A is connected to D, the current in R is 10 A

d. $Q = 2 \times 10^{-3}$ C

4 A series R-C circuit is connected to AC voltage source. Consider two cases; (A) when C is without a dielectric medium and (B) when C is filled with dielectric of constant 4. The current I_R through the resistor and voltage V_C across the capacitor are compared in the two cases. Which of the following is/are true?

(2011, One or More than One Options Correct Type)

a. $I_R^A > I_R^B$
b. $I_R^A < I_R^B$
c. $V_C^A > V_C^B$
d. $V_C^A < V_C^B$

5 A series R-C combination is connected to an AC voltage of angular frequency $\omega = 500$ rad/s. If the impedance of the R-C circuit is $R\sqrt{1.25}$, the time constant (in millisecond) of the circuit is

(2011, Integer Type)

6 An AC voltage source of variable angular frequency ω and fixed amplitude V_0 is connected in series with a capacitance C and an electric bulb of resistance R (inductance zero). When ω is increased

(2010, Only One Option Correct Type)

a. the bulb glows dimmer
b. the bulb glows brighter
c. total impedance of the circuit is unchanged
d. total impedance of the circuit increases

7 You are given many resistances, capacitors and inductors. These are connected to a variable DC voltage source (the first two circuits) or an AC voltage source of 50 Hz frequency (the next three circuits) in different ways as shown in Column II. When a current I (steady state for DC or rms for AC) flows through the circuit, then corresponding voltage V_1 and V_2 (indicated in circuits) are related as shown in Column I.

(2010, Matching Type)

	Column I		Column II
i.	$I \neq 0$, V_1 is proportional to I	p.	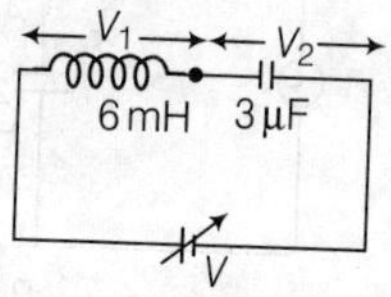
ii.	$I \neq 0$, $V_2 > V_1$	q.	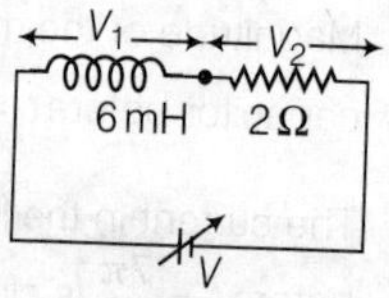
iii.	$V_1 = 0$, $V_2 = V$	r.	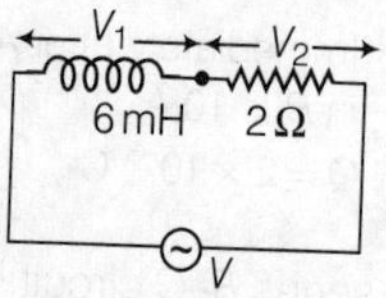
iv.	$I \neq 0$, V_2 is proportional to I	s.	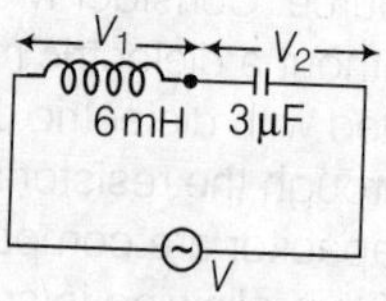
		t.	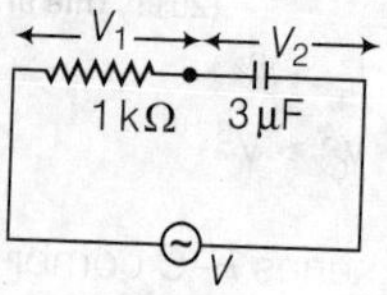

Codes

	i	ii	iii	iv
a.	r,s,t	q,r,s,t	q, or p, q	q, r, s, t
b.	p, q	r	s, t	p
c.	p	q	r	s
d.	t	p or q, r	p,q,r	s,t

Passage for Q. Nos. (8-10)

The capacitor of capacitance C can be charged (with the help of a resistance R) by a voltage source V by closing switch S_1 while keeping switch S_2 open. The capacitor can be connected in series with an inductor L by closing switch S_2 and opening S_1.

(2006, Comprehension Type)

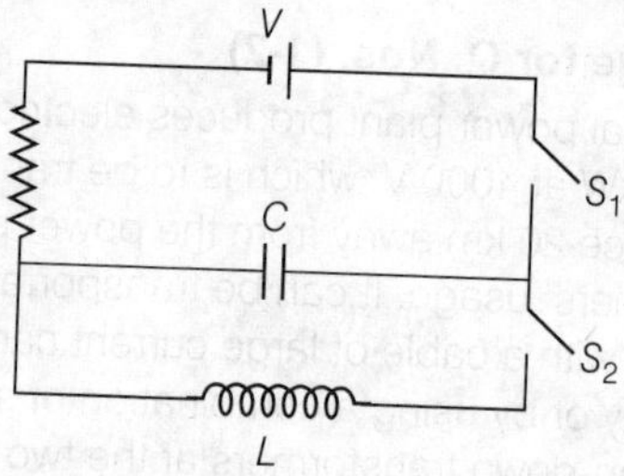

8 Initially, the capacitor was uncharged. Now, switch S_1 is closed and S_2 is kept open. If time constant of this circuit is τ, then

a. after time interval τ, charge on the capacitor is $CV/2$

b. after time interval 2τ, charge on the capacitor is $CV(1 - e^{-2})$

c. the work done by the voltage source will be half of the heat dissipated when the capacitor is fully charged

d. after time interval 2τ, charge on the capacitor is $CV(1 - e^{-1})$

9 After the capacitor gets fully charged, S_1 is opened and S_2 is closed so that the inductor is connected in series with the capacitor. Then,

a. at $t = 0$, energy stored in the circuit is purely in the form of magnetic energy

b. at any time $t > 0$, current in the circuit is in the same direction

c. at $t > 0$, there is no exchange of energy between the inductor and capacitor

d. at any time $t > 0$, maximum instantaneous current in the circuit may be $V\sqrt{\frac{C}{L}}$

10 If the total charge stored in the LC circuit is Q_0, then for $t \geq 0$, which of these is correct?

a. The charge on the capacitor is $Q = Q_0 \cos\left(\frac{\pi}{2} + \frac{t}{\sqrt{LC}}\right)$

b. The charge on the capacitor is $Q = Q_0 \cos\left(\frac{\pi}{2} - \frac{t}{\sqrt{LC}}\right)$

c. The charge on the capacitor is $Q = -LC\frac{d^2Q}{dt^2}$

d. the charge on the capacitor is $Q = -\frac{1}{\sqrt{LC}}\frac{d^2Q}{dt^2}$

Answers *with* Explanations

Answers

1. Magnetic Effect of Current

DPP-1 Magnetic Force and Motion of Charged Particle in Magnetic Field

1. (i) $\frac{-8}{3}$ ms^{-1} (ii) $-0.4\,\hat{\mathbf{j}}$ T (iii) $\mathbf{v}(t) = (v_0\hat{\mathbf{i}} - v_0\hat{\mathbf{k}}) + \frac{B_0 q v_0}{m}\cdot\hat{\mathbf{j}}$, $\mathbf{r}(t) = (v_0\hat{\mathbf{i}} - v_0\hat{\mathbf{k}})t + \frac{t^2}{2}\left(\frac{B_0 q}{m}\right)v_0\hat{\mathbf{j}}$

2. (i) $\frac{2mv_0}{Bq}\alpha$ (ii) $\frac{2m\alpha}{Bq}$ (iii) $\frac{2mv_0}{Bq}\sin\alpha$ (iv) $\alpha = \beta$ (v) 1 (vi) $2\,mv_0\sin\alpha$

3. (i) $\theta = 45°$ (ii) 0.1 m, 0.1414 m **4.** $\frac{r_p}{r_\alpha} = 1$ **5.** *a* **6.** *b* **7.** *a* **8.** *a* **9.** *a*

10. *a* **11.** *b* **12.** *b* **13.** *d* **14.** *c* **15.** *c* **16.** *c* **17.** *a* **18.** *d* **19.** *c*

20. *d* **21.** *a,b,c* **22.** *a,c,d* **23.** *b,d* **24.** *c* **25.** *d* **26.** *a* **27.** *d* **28.** *a*

DPP-2 Biot-Savart's Law and Its Applications

1. (i) $\frac{\mu_0 I}{4\pi x}$ (ii) $\left(1+\frac{1}{\pi}\right)\frac{\mu_0 I}{2R}$ (iii) $\frac{\mu_0 I}{8R}$ (iv) $\frac{\mu_0 I}{12}\left(\frac{1}{a}-\frac{1}{b}\right)$

2. (i) $i = 3$ A (ii) 1.3×10^{-6} T (iii) 2.9×10^{-6} N

3. 0 **4.** 10^{-4} T, outwards **5.** *d* **6.** *a* **7.** *b* **8.** *c* **9.** *d* **10.** *d*

11. *b* **12.** *d* **13.** *a* **14.** *d* **15.** *d* **16.** *b* **17.** *b* **18.** *c* **19.** *b*

20. *a,b,c,d* **21.** *a,c* **22.** 7 **23.** *a*

DPP-3 Ampere's Circuital Law and Its Applications

1. $B_A = 200\,\mu$T, $B_B = 133\,\mu$T **2.** (i) $B = \frac{\mu_0 Ir}{2\pi a^2}$ (ii) $B = \frac{\mu_0 I}{2\pi r}$ **3.** (i) $B = 0.04$ T (ii) 0 (iii) 0

4. (i) $B = \frac{\mu_0\mu_r kr^3}{4}$ (ii) $B = \frac{\mu_0 ka^4}{4r}$ **5.** *b* **6.** *d* **7.** *c* **8.** *b* **9.** *b* **10.** *d* **11.** *d*

12. *d* **13.** *d* **14.** *a* **15.** *c* **16.** *a* **17.** *a* **18.** *a,d* **19.** *c,d* **20.** *b,c* **21.** *a*

22. 1 **23.** 9 **24.** 3 **25.** *a* **26.** *c*

DPP-4 Magnetic Force on a Current Carrying Wire

1. $ab = 0$, $bc = -40\times10^3\,\hat{\mathbf{i}}$, $cd = -40\times10^{-3}\,\hat{\mathbf{k}}$, $da = 40\times10^{-3}(\hat{\mathbf{k}}+\hat{\mathbf{i}})$

2. 0 **3.** $F = BIR$ **4.** (i) $a = 0, b = -B_0 il$, $c = B_0 il$ (ii) $\sqrt{2}B_0 il$ **5.** *b* **6.** *d* **7.** *d*

8. *b* **9.** *a* **10.** *b* **11.** *b* **12.** *a* **13.** *d* **14.** *a* **15.** *b* **16.** *b*

17. *a,b,c* **18.** *a,b* **19.** *a,c* **20.** *b* **21.** *d* **22.** 2 **23.** 4 **24.** 5 **25.** *a* **26.** *d*

DPP-5 Magnetic Dipole, Magnetic Moment and Torque

1. (i) $M = \frac{\sqrt{3}Va^2}{R}$ (ii) $M = \frac{3Va^2}{R}$ (iii) $M = 3\sqrt{3}a^2\frac{V}{R}$

2. (i) MB (ii) $-MB$ (iii) 0

3. (i) $i = \frac{ev}{2\pi r}$ (ii) $M = \frac{evr}{2}$ **4.** (i) $\tau = 25\,\hat{\mathbf{k}}$ (ii) PE = 0 **5.** *b* **6.** *a* **7.** *a* **8.** *b*

9. *b* **10.** *a* **11.** *c* **12.** *a* **13.** *b* **14.** *a,b,d* **15.** *b,d* **16.** *a,b* **17.** *b* **18.** *a*

19. *b* **20.** 4 **21.** 6 **22.** 2

DPP-6 Moving Coil Galvanometer and Its Conversion

1. (i) $\tau = ni(\mathbf{A}\times\mathbf{B})$, $\tau = niAB\sin\theta$ (ii) $\tau = niAB$ (iii) $k\theta = niAB$

2. (i) $I_g = nI_s$ (ii) $I_g = 0.6\,\text{mA}$ (iii) Voltage sensitivity $=\dfrac{NAB}{kR}$ (remains same)

3. (i) $I_g = 0.2\,\text{mA}$ (ii) $R = 75\,\text{k}\Omega$ **4.** (i) $k = NBA$ (ii) $k = \dfrac{2NiBA}{\pi}$ (iii) $\theta_{max} = Q\sqrt{\dfrac{NAB\pi}{2Ii}}$ **5.** *c* **6.** *d*

7. *b* **8.** *d* **9.** *b* **10.** *a* **11.** *b* **12.** *b* **13.** *b* **14.** *d* **15.** *c* **16.** *b*
17. *b* **18.** *a* **19.** *a,b* **20.** *b,c* **21.** *d* **22.** *a* **23.** *a* **24.** *d* **25.** 5 **26.** 4

Revisal Problems (JEE Main)

1. *b* **2.** *d* **3.** *b* **4.** *a* **5.** *a* **6.** *c* **7.** *a* **8.** *a* **9.** *c* **10.** *a*
11. *b* **12.** *b* **13.** *c* **14.** *b* **15.** *a* **16.** *d* **17.** *b* **18.** *b* **19.** *c* **20.** *a*
21. *d* **22.** *a* **23.** *a* **24.** *a* **25.** *d* **26.** *b* **27.** *b* **28.** *a* **29.** *d* **30.** *b*

Revisal Problems (JEE Advanced)

1. *a* **2.** *b* **3.** *a* **4.** *b* **5.** *a* **6.** *c* **7.** *d* **8.** *d* **9.** *b* **10.** *b*
11. *b* **12.** *d* **13.** *a* **14.** *a,b,c* **15.** *a,c,d* **16.** *a,c* **17.** *a,b,c* **18.** *b,c,d* **19.** *a,b* **20.** *a,c*
21. *b,d* **22.** *c* **23.** *d* **24.** *c* **25.** *b* **26.** *a* **27.** *c* **28.** *b* **29.** *d* **30.** *a*
31. *b* **32.** *a* **33.** *a* **34.** *c* **35.** *a* **36.** 4 **37.** 3 **38.** 5 **39.** 3

JEE Main & AIEEE Archive

1. *b* **2.** *b* **3.** *a* **4.** *d* **5.** *d* **6.** *a* **7.** *b* **8.** *b* **9.** *c* **10.** *b*
11. *c* **12.** *c* **13.** *a* **14.** *b* **15.** *a* **16.** *a* **17.** *c* **18.** *c* **19.** *d* **20.** *d*
21. *b* **22.** *b* **23.** *a* **24.** *c* **25.** *b* **26.** *a* **27.** *d* **28.** *b* **29.** *a* **30.** *d*
31. *a* **32.** *a* **33.** *a* **34.** *a* **35.** *c* **36.** *d* **37.** *c* **38.** *b*

JEE Advanced & IIT JEE Archive

1. *c* **2.** *b* **3.** 3 **4.** *a,c* **5.** *b,c* **6.** *c,d* **7.** *b* **8.** *d* **9.** 5 **10.** *c*
11. *a* **12.** *b,d* **13.** 7 **14.** *a,c,d* **15.** *c* **16.** *b* **17.** *a* **18.** *a,c* **19.** *c* **20.** *b,d*
21. (i) $k = NBA$ (ii) $k = \dfrac{2BiNA}{\pi}$ (iii) $Q_{max} = Q\sqrt{\dfrac{BN\pi A}{2li}}$

2. Magnetism

DPP-1 Bar Magnet and Magnetic Field Lines

1. (i) $B_N = 2.5\times10^{-4}\,\text{T}$, $B_S = 0.3\times10^{-4}\,\text{T}$ (ii) $B_{net} = 2.2\times10^{-4}\,\text{T}$ (directed away)

2. (ii) (a) $\dfrac{M'}{M} = \dfrac{1}{2}, \dfrac{T'}{T} = \dfrac{1}{2}$ (b) $\dfrac{M'}{M} = \dfrac{1}{2}, \dfrac{T'}{T} = 1$ **3.** $B = 2.24\times10^{-2}\,\text{T}$ **4.** $a = 3R$ **5.** *a*

6. *d* **7.** *d* **8.** *b* **9.** *c* **10.** *c* **11.** *c* **12.** *d* **13.** *b* **14.** *c* **15.** *c*
16. *c* **17.** *a* **18.** *c* **19.** *c* **20.** *a*

DPP-2 Earth's Magnetism

1. (i) $B = 3\times10^{-5}\,\text{T}, \phi = 0$ (ii) $B = 5.5\times10^{-5}\,\text{T}, \phi = \tan^{-1}(2\sqrt{3})$ (iii) $B = 6\times10^{-5}\,\text{T}, \phi = 90°$

2. (i) $R = 5\,\text{cm}$ (ii) $r = 6.3\,\text{cm}$

3. (iii) For P, dip $= 0$, declination $= 0$; For Q, dip $= 0$, declination $= 11.3°$

4. $\dfrac{B_2}{B_1} = 1.83$ **5.** *a* **6.** *d* **7.** *b* **8.** *b* **9.** *b* **10.** *c* **11.** *b* **12.** *a* **13.** *b*
14. *b* **15.** *d* **16.** *c* **17.** *c* **18.** *c* **19.** *b,c,d* **20.** *c,d* **21.** *b,d* **22.** *a* **23.** *c*
24. *d* **25.** 2 **26.** 2

DPP-3 Magnetic Materials and Their Properties

1. (i) The value of χ is small but positive quantity for paramagnetic substances and negative quantity for diamagnetic substances. (ii) $\mathbf{B} = \mu_0 \mathbf{H}$

2. $I = \dfrac{B}{\mu_0} - ni$

3. $21 \times 10^{-3}\%$

4. (i) $I = 5 \times 10^4\ \text{Am}^{-1}$ from South to North (ii) $H = 199\ \text{Am}^{-1}$ towards South pole
(iii) $B = 6.26 \times 10^{-2}$ T towards North pole

5. *a*	6. *d*	7. *b*	8. *b*	9. *d*	10. *b*	11. *b*	12. *b*	13. *a*	14. *a*
15. *b*	16. *c*	17. *c*	18. *b*	19. *b*	20. *a*	21. *b*	22. *a*	23. 2	24. 8
25. 2	26. *a,d*	27. *a,b,c*	28. *a,d*	29. *a*					

Revisal Problems (JEE Main)

1. *b*	2. *c*	3. *b*	4. *d*	5. *d*	6. *a*	7. *a*	8. *a*	9. *b*	10. *c*
11. *c*	12. *d*	13. *d*	14. *b*	15. *d*	16. *a*	17. *d*	18. *a*	19. *d*	20. *b*

Revisal Problems (JEE Advanced)

1. *a*	2. *d*	3. *c*	4. *c*	5. *a*	6. *c*	7. *d*	8. *b*	9. *d*	10. *c*
11. *c,d*	12. *a,b*	13. *b,c,d*	14. *a,b,d*	15. *a,b,c,d*		16. *b,c,d*	17. *a*	18. *b*	19. *a*
20. *b*	21. *a*	22. *b*	23. *c*	24. *d*	25. *c*	26. 8	27. 7	28. 4	29. 5

JEE Main & AIEEE Archive

1. *c*	2. *b*	3. *b*	4. *b*	5. *b*	6. *b*

JEE Advanced & IIT JEE Archive

1. *a*	2. *b*

3. Electromagnetic Induction

DPP-1 Magnetic Flux and Faraday's Law

1. $\phi = \dfrac{\mu_0 a I}{2\pi} \log_e\left(1 + \dfrac{a}{b}\right)$

2. (i) 12 m (ii) Unethical

3. (i) 3 nV (*ii*) 20 mV

4. (i) $E_{\text{net}} = 4Bv\,(l - 2vt)$ (ii) $I = \dfrac{Bv}{\lambda}$ (iii) $F = \dfrac{B^2 v^2}{\lambda}(l - 2vt)$ (iv) $\dfrac{4B^2v^2}{\lambda}(l - 2vt)$ (v) $\dfrac{4B^2v^2}{\lambda}(l - 2vt)$

5. *b*	6. *c*	7. *c*	8. *d*	9. *b*	10. *a*	11. *d*	12. *d*	13. *d*	14. *d*
15. *a*	16. *c*	17. *a*	18. *a*	19. *a,c*	20. *b,d*	21. *b,c,d*	22. *a*	23. *d*	24. 9
25. 1	26. *c*								

DPP-2 Lenz's Law and Its Applications

5. *d*	6. *b*	7. *a*	8. (i) *a* (ii) *b*	9. *b*	10. *a*	11. *c*	12. *d*	13. *d*
14. *b,c*	15. *a,b*	16. *a,c*	17. *d*	18. 2	19. 2	20. *a*	21. *d*	

DPP-3 Induced EMF in a Moving Rod in Uniform Magnetic Field and Circuit Problems

1. (iii) $\phi = Blv$ (iv) $P_{\text{electrical}} = \dfrac{B^2 l^2 v^2}{R}$

2. $e = 4.5$ mV

3. $W = 10^{-6}$ J

4. $E = 2B\omega r^2$

5. *d*	6. *a*	7. *a*	8. *c*	9. *d*	10. *b*	11. *c*	12. *c*	13. *a*	14. *b,d*
15. *a,d*	16. *a,b,d*	17. *a*	18. *c*	19. *b*	20. 2	21. 4	22. *c*	23. *d*	

DPP-4 EMF Induced in a Rod or Loop in Non-Uniform Magnetic Field

1. $v = ue^{-kt}$

2. (i) $e = \frac{\mu_0 b}{2\pi} \ln\left[\frac{(a+d)(b+a)}{a(b+a+d)}\right]\frac{di}{dt}$ (ii) Clockwise

3. (i) $\phi = 2.772 \times 10^{-7}$ Wb (ii) $e = 1.386 \times 10^{-5}$ V (iii) Clockwise

4. $V = \frac{\mu_0 Iv}{2\pi} \ln\left[\frac{d + 2r\cos\theta}{d - 2r\cos\theta}\right]$

5. *b* **6.** *d* **7.** *b* **8.** *b* **9.** *a* **10.** *b* **11.** *d* **12.** *a* **13.** *a* **14.** *c*
15. *b* **16.** *a* **17.** *a* **18.** *a* **19.** *b* **20.** *a,c,d* **21.** *a,c* **22.** *a,c* **23.** *b,c*
24. *a,b,d*

DPP-5 Induced EMF in Rod, Ring, Disc Rotated in a Uniform Magnetic Field

1. 9.096×10^{-11} Wb **2.** $x = \frac{R}{2}$ **3.** $E = 0.46$ mV

4. (i) $\varepsilon = 125$ mV (clockwise) (ii) 0.02 A (clockwise)

5. *d* **6.** *c* **7.** *c* **8.** *c* **9.** *b* **10.** *d* **11.** *c* **12.** *b* **13.** *b* **14.** *c*
15. *d* **16.** 5 **17.** 9 **18.** *a,d* **19.** *b,d* **20.** *b* **21.** *c* **22.** *d*

DPP-6 Loop in a Time Varying Magnetic Field and Induced EMF

1. (i) (a) $E = \frac{r}{2}\left|\frac{dB}{dt}\right|$ (b) $E = \frac{R^2}{2r}\left|\frac{dB}{dt}\right|$ (ii) Anti-clockwise

2. $\varepsilon = \sqrt{R^2 - l^2}\left(\frac{dB}{dt}\right)l$

3. Current in segment *AE* is $\left(\frac{7}{22}\right)$ A from *E* to *A*, current in segment *BE* is $\left(\frac{6}{22}\right)$ A from *B* to *E* and current in segment *EF* is $\left(\frac{1}{22}\right)$ A from *F* to *E*.

4. $\omega = \frac{qB}{2m}$ **5.** *d* **6.** *a* **7.** *a* **8.** *b* **9.** *d* **10.** *d* **11.** *b* **12.** *b* **13.** *c*
14. *a,b,c* **15.** *b,d* **16.** *d* **17.** *d* **18.** *a* **19.** *a* **20.** *a* **21.** 2 **22.** 5 **23.** *c*
24. *b*

DPP-7 Self-Induction, Self-Inductance and Magnetic Energy Density

1. 2.5 **2.** 45 log (2) H **3.** $V_R = -\frac{\varepsilon}{3}$ **4.** $\frac{\mu_0}{\pi} \ln\left(\frac{d-a}{a}\right)$

5. *d* **6.** *a* **7.** *d* **8.** *a* **9.** *d* **10.** *c* **11.** *a* **12.** *a* **13.** *c* **14.** *b*
15. *b* **16.** *c* **17.** *c* **18.** *b* **19.** *a,c,d* **20.** *a,b,c* **21.** *c* **22.** *b* **23.** *d* **24.** *c*
25. *b* **26.** *c* **27.** 4 **28.** 3

DPP-8 Growth and Decay of Current in *L-R* Circuit

1. $\frac{1}{2} mgr\cos\theta + \frac{1}{4} \cdot \frac{B^2 r^4 \omega}{R}$

2. (i) $i = i_4 (1 - e^{-RL})$ (ii) 1.386×10^{-3} s (iii) 4.5×10^{-4} J

3. $i = \frac{E}{3R}(1 - e^{-3Rt/2L})$

4. (i) $V = 12e^{-5t}$ V (ii) Clockwise, $6e^{-10t}$

5. *d* **6.** *c* **7.** *d* **8.** *a* **9.** *b* **10.** *d* **11.** *c* **12.** *b* **13.** *d* **14.** *d*
15. *d* **16.** *d* **17.** *d* **18.** *b,d* **19.** *b,d* **20.** *b* **21.** *a* **22.** *d* **23.** *a* **24.** *a*
25. *a* **26.** *b* **27.** *b* **28.** 1 **29.** 2

DPP-9 Mutual Induction and Inductance

1. (i) $M = 1.8 \times 10^{-2}$ H (ii) $L = 3.43 \times 10^{-2}$ H (iii) $\varepsilon_B = 9 \times 10^{-3}$ V
2. 8.9×10^{-4} Wb, $M_{12} = M_{21} = 2.96 \times 10^{-4}$ H
3. (i) 9.1×10^{-11} Wb (ii) 4.55×10^{-11} H
4. $M = \dfrac{2\sqrt{2}\,\mu_0 l^2}{\pi L}$ 5. *a* 6. *c* 7. *a* 8. *c* 9. *c* 10. *b* 11. *d* 12. *a*
13. *b* 14. *d* 15. *b* 16. *a* 17. *d* 18. *a* 19. *b* 20. *c* 21. *c* 22. *a*
23. *a* 24. *b* 25. 7 26. 9

Revisal Problems (JEE Main)

1. *c* 2. *a* 3. *b* 4. *a* 5. *b* 6. *b* 7. *c* 8. *d* 9. *a* 10. *b*
11. *b* 12. *a* 13. *a* 14. *d* 15. *c* 16. *b* 17. *d* 18. *d* 19. *a* 20. *a*
21. *c* 22. *c* 23. *c* 24. *d* 25. *d* 26. *a* 27. *d* 28. *b*

Revisal Problems (JEE Advanced)

1. *c* 2. *b* 3. *b* 4. *d* 5. *c* 6. *b* 7. *c* 8. *b* 9. *d* 10. *a*
11. *a* 12. *c* 13. *d* 14. *b* 15. *b* 16. *a* 17. *d* 18. *b* 19. *b*
20. *b,c,d* 21. *a,c,d* 22. *a,b,c,d* 23. *a,d* 24. *a,c* 25. *a,c,d* 26. *c,d* 27. *c* 28. *c* 29. *c*
30. *a* 31. *a* 32. *b* 33. *a* 34. *d* 35. *c* 36. *c* 37. *a* 38. 4 39. 8

JEE Main & AIEEE Archive

1. *d* 2. *a* 3. *d* 4. *b* 5. *c* 6. *d* 7. *d* 8. *b* 9. *b* 10. *b*
11. *b* 12. *b* 13. *d* 14. *d*

JEE Advanced & IIT JEE Archive

1. *b* 2. *b* 3. 7 4. *c* 5. *c* 6. *c* 7. *b,d* 8. *a* 9. *a* 10. *d*
11. *b* 12. *c*

4. Alternating Current

DPP-1 Average, Peak and rms Value of AC

1. (i) $I_{rms} = \dfrac{I_0}{\sqrt{2}}, I_{av} = \dfrac{I_0}{2}$ (ii) $I_{av} = 0.637\, I_0, I_{rms} = 0.707\, I_0$ (iii) $I_{av} = 375$ A, $I_{rms} = \dfrac{500\sqrt{2}}{\sqrt{3}}$ A
2. $\sqrt{\dfrac{4}{3}}$
3. (i) $I_{rms} = 5.5$ A (ii) $I_0 = 7.8$ A (iii) 2.5ms
4. (ii) $\dfrac{i_0}{\sqrt{3}}$
5. $I_{rms} = 12.247$ A, $I_{av} = 11.06$ A
6. *b* 7. *d* 8. *a* 9. *b* 10. *d* 11. *b* 12. *b* 13. *c* 14. *a* 15. *d*
16. *d* 17. *a* 18. *d* 19. *c* 20. *c* 21. *a,b,c* 22. *a,b,c* 23. *c,d* 24. *a,b,c,d*
25. (i) $i_{rms} = \dfrac{i_m}{\sqrt{2}}$, (ii) $K = 1$ 26. 8 27. $N = 3$ 28. *a*

DPP-2 Power Consumption in an AC Circuit

1. (i) $I_{rms} = 2$ A (ii) $P_s = 160$ W (iii) $P_R = 160$ W
2. (i) 10 W (ii) 5 W
3. (i) $100\sqrt{2}\,\Omega$ (ii) 45° (iii) $\dfrac{1}{\sqrt{2}}$ (iv) $\sqrt{2}$ A (v) 2 A (vi) $R = 100\sqrt{2}$ V, $C = 100\sqrt{2}$ V, (vii) $C = 200$ V, $R = 200$ V
4. (i) $I = 20\cos 314t$ A (ii) $P = 100\sin 628t$ W (iii) $f_P = 100$ Hz (iv) $U_0 = 3.185$ J
5. *a* 6. *b* 7. *c* 8. *a* 9. *c* 10. *a* 11. *b* 12. *b* 13. *b* 14. *a*
15. *a* 16. *a* 17. *c* 18. *a* 19. *c* 20. *a,c* 21. *a,c,d* 22. *a,b,c* 23. *a* 24. *b*
25. *d* 26. *d* 27. 4 28. 5

DPP-3 AC Source with *R-L-C* Connected in Series

1. 200 V **2.** (ii) $\tan\phi = \frac{R}{X_C} - \frac{R}{X_L}$ (iii) $T = \frac{R}{\sqrt{1+\left(R\omega C - \frac{R}{\omega L}\right)^2}}$ (iv) $\cos\phi = \frac{1}{\sqrt{1+\left(R\omega C - \frac{R}{\omega L}\right)^2}}$

3. (i) $\frac{\Delta U}{U} = \frac{2\pi}{Q}$ (ii) $Z = \sqrt{(R_1+R_2)^2 + \left(\omega L_1 + \omega L_2 - \frac{1}{\omega C_1} - \frac{1}{\omega C_2}\right)^2}$

4. $\phi = 53°, V_{RO} = 30\text{ V}, V_{LO} = 60\text{ V}, V_{CO} = 20\text{ V}$ **5.** *a* **6.** *d* **7.** *c* **8.** *b* **9.** *d* **10.** *b*

11. *a,c* **12.** *c,d* **13.** *a,d* **14.** *b,c* **15.** *b,c* **16.** *b* **17.** *a* **18.** *c* **19.** *c* **20.** *a*

21. *d* **22.** 4 **23.** 4 **24.** *a* **25.** *a* **26.** *a*

DPP-4 Resonance

1. (i) $\omega_r = \frac{5}{2\pi} \times 10^4$ Hz (ii) $V_C = 400$ V (iii) $V_R = 10$ V **2.** $I_m = 0.704$ A, $P_{av} = 5.16 \times 10^{-4}$ J/cycle

3. (i) $I_{max} = 14.14$ A (ii) $P_{max} = 2300$ W, frequency $= 663.14$ Hz (iii) $Q = 21.74$

4. (i) $C_X = C_Y = C_Z$ (ii) $R_X < R_Y < R_Z$ **5.** *a* **6.** *a* **7.** *c* **8.** *c* **9.** *d* **10.** *c*

11. *a* **12.** *b* **13.** *b* **14.** *b* **15.** *c* **16.** *b,d* **17.** *a,b,c* **18.** *a,b,d* **19.** *a,c* **20.** *b,d*

21. *b* **22.** *b* **23.** *b* **24.** *a* **25.** *a* **26.** *a* **27.** *c* **28.** 8 **29.** 2 **30.** *d*

31. *b*

DPP-5 Transformer

1. $B_m = 0.938$ Wb/m^2 (ii) $V_S = 1250$ V **2.** $P = 176.4$ MW

3. (i) 600 kW (ii) 1400 kW (iii) 440-7000 V **4.** (i) $E_S = 3$ V (ii) $I_S = 5$ A (iii) $I_P = 0.125$ A (iv) 15 W

5. *d* **6.** *c* **7.** *b* **8.** *c* **9.** *d* **10.** *a* **11.** *a* **12.** *b* **13.** *b,d*

14. *a,b,c,d* **15.** *b* **16.** *a,c* **17.** *b,c* **18.** *a* **19.** *b* **20.** *c* **21.** *c* **22.** *b*

23. 6 **24.** 8 **25.** 6 **26.** 6 **27.** *d*

Revisal Problems (JEE Main)

1. *d* **2.** *d* **3.** *b* **4.** *c* **5.** *b* **6.** *b* **7.** *b* **8.** *c* **9.** *d* **10.** *c*

11. *b* **12.** *a* **13.** *c* **14.** *a* **15.** *d* **16.** *c* **17.** *a* **18.** *c* **19.** *a* **20.** *b*

21. *d* **22.** *b* **23.** *d* **24.** *a* **25.** *a*

Revisal Problems (JEE Advanced)

1. *a* **2.** *a* **3.** *c* **4.** *d* **5.** *d* **6.** *b* **7.** *a,c,d* **8.** *a,d* **9.** *a,b* **10.** *a,c*

11. *c* **12.** *d* **13.** *a* **14.** *a* **15.** *c* **16.** *b* **17.** 2 **18.** 9 **19.** *d* **20.** *c*

JEE Main & AIEEE Archive

1. *b* **2.** *a* **3.** *a* **4.** *d* **5.** *d* **6.** *b* **7.** *a* **8.** *d* **9.** *c* **10.** *b*

11. *a* **12.** *c* **13.** *c* **14.** *c* **15.** *d* **16.** *c* **17.** *c* **18.** *a* **19.** *b* **20.** *b*

JEE Advanced & IIT JEE Archive

1. *b* **2.** *a* **3.** *a,c,d* **4.** *b,c* **5.** 4 **6.** *b* **7.** *a* **8.** *b* **9.** *d* **10.** *c*

1. Magnetic Effect of Current

DPP-1 Magnetic Force and Motion of Charged Particle in Magnetic Field

1. (i) As, $\mathbf{F} \perp \mathbf{B} \Rightarrow \mathbf{a} \perp \mathbf{B}$

or $\mathbf{a} \cdot \mathbf{B} = 0$

$\Rightarrow (x\hat{i} + 2\hat{j}) \cdot (3\hat{i} + 4\hat{j}) \times 10^{-2} = 0$

$\Rightarrow 3x + 8 = 0$

$\Rightarrow x = -\dfrac{8}{3} \text{ ms}^{-1}$

(ii) $\mathbf{F} = q(\mathbf{v} \times \mathbf{B})$

$-1.28 \times 10^{-13}\hat{k} = 1.6 \times 10^{-19} \times [(2\hat{i} + 3\hat{j}) \times 10^6 \times (-B_0\hat{j})]$

$\Rightarrow B_0 = \dfrac{1.28}{3.2} = 0.4$

So, $\mathbf{B} = -0.4\hat{j}$ T

(iii) Force on the particle is given by

$$\mathbf{F} = q\,(\mathbf{v} \times \mathbf{B}) = q\,[(v_0\hat{i} - v_0\hat{k}) \times (-B_0\hat{k})] = q(B_0 v_0 \hat{j})$$

Acceleration of the particle is given by

$$\mathbf{a} = \frac{qB_0v_0}{m} \cdot \hat{j}$$

Velocity of particle after t second is

$$\mathbf{v}(t) = \mathbf{v}(t = 0) + \mathbf{a}(t)$$

$$\mathbf{v}(t) = v_0\hat{i} - v_0\hat{k} + \frac{B_0q}{m} \cdot v_0\hat{j}$$

Position of the particle after t second is

$$\mathbf{r}(t) = ut + \frac{1}{2}at^2$$

$$= (v_0\hat{i} - v_0\hat{k})t + \frac{t^2}{2}\left(\frac{B_0q}{m}\right)v_0\hat{j}$$

2. (i) Radius of circular path is

$$r = \frac{mv_0}{Bq}$$

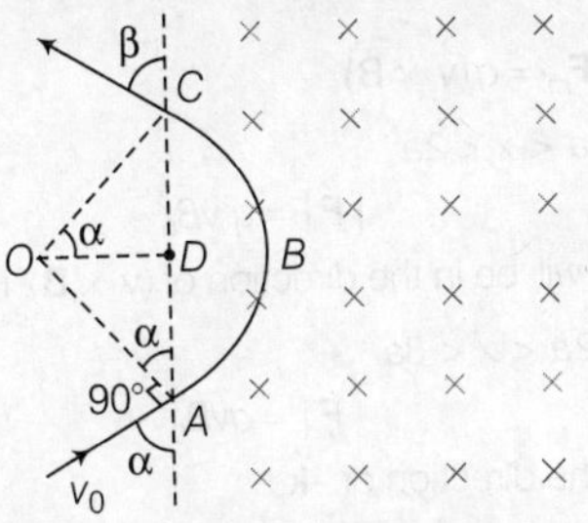

Let O be the centre of circular path, then

$\angle OCD = \angle OAD$

$\Rightarrow 90° - \beta = 90° - \alpha$

$\Rightarrow \beta = \alpha$

$\Rightarrow \angle AOC = 2\alpha$

So, length of arc $ABC = r(2\alpha) = \dfrac{2mv_0}{Bq}\alpha$

(ii) Time $(ABC) = \dfrac{\text{Arc length } ABC}{\text{Speed}} = \dfrac{2m\alpha}{Bq}$

(iii) Distance, $AC = 2 \times CD = 2OC \sin\alpha$

$= \dfrac{2mv_0}{Bq} \sin \alpha$

(iv) $\alpha = \beta$

(v) KE of particle remains same.

$\therefore$ Ratio is 1.

(vi) At A and C, the change in vertical component of momentum

$= mv_0\cos\alpha - mv_0\cos\alpha = 0$

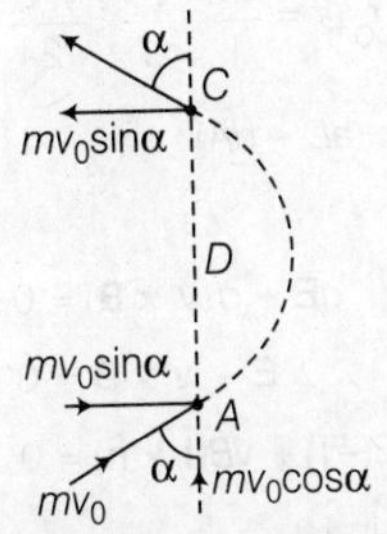

And change in horizantal component of momentum

$= mv_0\sin\alpha - (-mv_0\sin\alpha)$

$= 2mv_0\sin\alpha$

3. (i) Due to Lorentz magnetic force, the particle will follow a circular arc EDF of radius r inside the magnetic field. AE is tangent to circle $\angle AEO = 90°$.

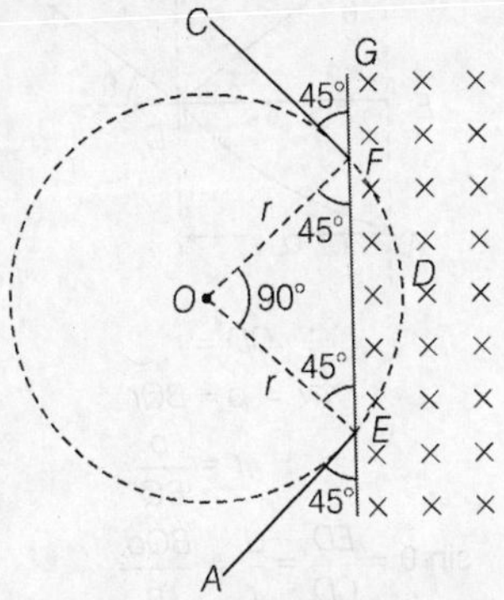

Also, $\angle OEF = 45°$

Since, equal sides have their opposite angles equal.

$\Rightarrow \angle OFE = 45°$

Also, $\angle OFC = 90°$

From geometry, we get

$\theta (= \angle CFG) = 45°$

(ii) $\dfrac{mv^2}{r} = qvB$

$\Rightarrow r = \dfrac{mv}{Bq}$

$\Rightarrow r = \dfrac{1.6 \times 10^{-27} \times 10^7}{1 \times 1.6 \times 10^{-19}}$ m $= 0.1$ m

$EF = \sqrt{r^2 + r^2} = \sqrt{2}r$

$\Rightarrow EF = \sqrt{2} \times 0.1$ m $= 0.1414$ m

4. Since, $r = \dfrac{mv}{Bq} = \dfrac{\sqrt{2mK}}{Bq}$

where, K = kinetic energy of the particle

$\Rightarrow \dfrac{r_p}{r_\alpha} = \sqrt{\dfrac{m_p}{m_\alpha}}\left(\dfrac{q_\alpha}{q_p}\right) = \sqrt{\dfrac{1}{4}} \cdot \left(\dfrac{2}{1}\right) = 1$

5. (a) The particle will move in a circular path with radius d if it is to just miss the wall.

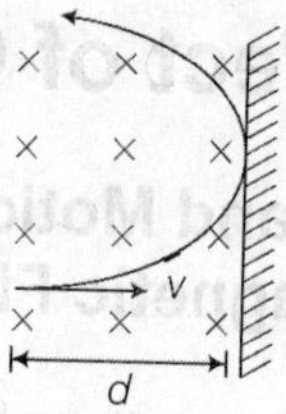

$$mv = Bqr$$
$$r = d$$
or
$$B = \frac{v}{(q/m)d} = \frac{v}{sd}$$

6. (b)
$$r = \frac{\sqrt{2mE}}{qB}$$
$$\Rightarrow \quad r \propto \frac{\sqrt{m}}{q}$$
$$\Rightarrow \quad r_{H^+} : r_{He^+} : r_{O^{+2}} = \frac{\sqrt{1}}{1} : \frac{\sqrt{4}}{1} : \frac{\sqrt{16}}{2} = 1:2:2$$

7. (a) $\mathbf{F} \propto (\mathbf{v} \times \mathbf{B}) = \hat{\mathbf{k}}\,(aD - dA)$

8. (a) $\mathbf{F}_e + \mathbf{F}_m = 0$
$$\Rightarrow \quad q\mathbf{E} + q(\mathbf{v} \times \mathbf{B}) = 0$$
$$\Rightarrow \quad \mathbf{E} + \mathbf{v} \times \mathbf{B} = 0$$
$$E(-\hat{\mathbf{j}}) + vB(\hat{\mathbf{i}} \times \hat{\mathbf{n}}) = 0$$
$\Rightarrow \hat{\mathbf{i}} \times \hat{\mathbf{n}}$ should be in $\hat{\mathbf{j}}$,

$\therefore$ $\hat{\mathbf{n}} = \hat{\mathbf{k}}$

So, $\hat{\mathbf{n}} = \hat{\mathbf{k}}$ and $E = vB$

9. (a) A to D is a part of a circle with centre C.

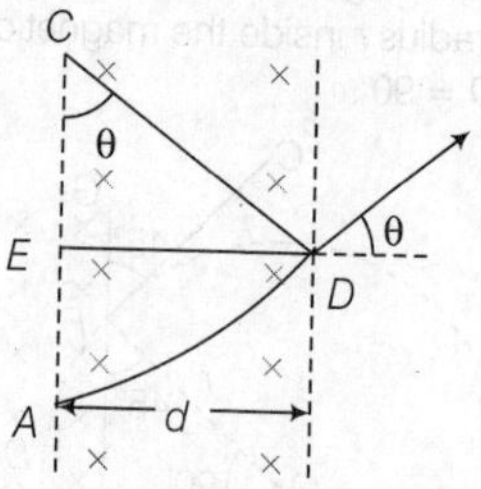

$$CD = r$$
$$mv = p = BQr$$
or
$$r = \frac{p}{BQ}$$
$$\sin\theta = \frac{ED}{CD} = \frac{d}{r} = \frac{BQd}{p}$$

10. (a) As radius is decreasing, then particle is slowing down.
$$r = \frac{mv}{Bq}$$

11. (b) Magnetic force, $F_m = qvB$ and directed along $\mathbf{v} \times \mathbf{B}$, i.e. radially inwards.

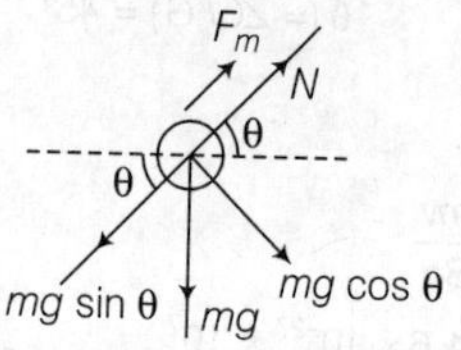

$$N - mg\sin\theta + F_m = \frac{mv^2}{R}$$
$$\Rightarrow \quad N = \frac{mv^2}{R} + mg\sin\theta - qvB \qquad \text{...(i)}$$

From conservation of energy,

Loss in gravitational potential energy = gain in kinetic energy
$$\Rightarrow \quad mgR\sin\theta = \frac{1}{2}mv^2 \qquad \text{...(ii)}$$

From Eqs. (i) and (ii), we get
$$N = 3mg\sin\theta - qB\sqrt{2gR\sin\theta}$$

At $\theta = \frac{\pi}{2}$, $\quad N_{max} = 3mg - Bq\sqrt{2gR}$

12. (b) Magnetic force is always perpendicular to the velocity or is zero.
$$F_m = q(\mathbf{v} \times \mathbf{B})$$
So, it can only produce a change in direction of velocity. As there is an electric field in Y-direction, particle experiences a net acceleration $\frac{qE}{m}$ in Y-direction. So, speed of particle depends on y-coordinate, whereas v_x and v_z remain same.

13. (d) $R_1 < R_2$

and $\quad R = \frac{mv}{qB}$

or $\quad \left(\frac{m}{q}\right)_1 < \left(\frac{m}{q}\right)_2$

14. (c) A charged particle enters and leaves a uniform magnetic field symmetrically.

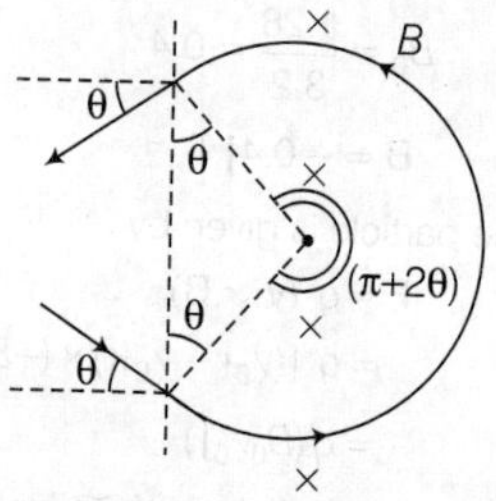

$$\text{Time spent} = \frac{\text{Angle}}{\text{Angular speed}} = \frac{\pi + 2\theta}{\omega} = \frac{T(\pi + 2\theta)}{2\pi}$$

15. (c) B and v are neither parallel nor perpendicular, so path is a helical.

16. (c) Work done by electric field = ΔK
$$qE_0x_0 = \frac{1}{2}mv^2$$
$$x_0 = \frac{1}{2}\frac{mv^2}{qE_0}$$
$$\mathbf{v} = 4\hat{\mathbf{i}} + 3\hat{\mathbf{j}} \quad \Rightarrow \quad v = 5$$
So,
$$x_0 = \frac{25}{2\alpha E_0}$$
where, $\alpha = \frac{q}{m}$ (given)

17. (a) $\because \quad \mathbf{F}_m = q(\mathbf{v} \times \mathbf{B})$

For region, $a < x < 2a$
$$|\mathbf{F}| = q\,vB_0$$
The force $\mathbf{F}$ will be in the direction of $(\mathbf{v} \times \mathbf{B})$, i.e. along $(\hat{\mathbf{i}} \times \hat{\mathbf{j}}) = \hat{\mathbf{k}}$

For region, $2a < x < 3a$
$$|\mathbf{F}| = qvB_0$$
$\mathbf{F}$ will be in the direction of $-\hat{\mathbf{k}}$.

Hence, the direction of circular motion will be as shown in graph (a).

18. (d) $\alpha = \frac{q}{m}$, path of particle will be helix of time period,

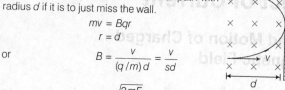

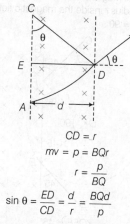

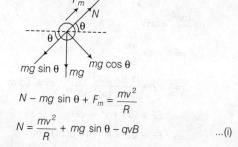

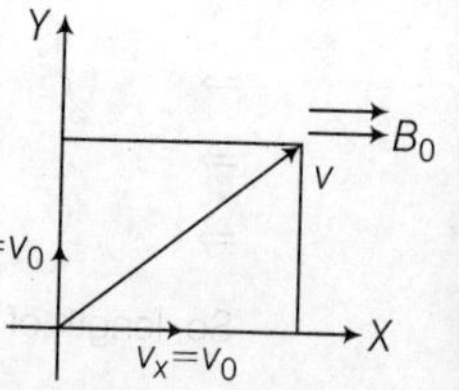

$$T = \frac{2\pi m}{B_0 q} = \frac{2\pi}{B_0\alpha}$$
The given time, $t = \frac{\pi}{B_0\alpha} = \frac{T}{2}$

Coordinates of particle at time, $t = \frac{T}{2}$ would be $\left(v_x\frac{T}{2}, 0, -2r\right)$.

Here, $\quad r = \frac{mvr}{B_0q} = \frac{v_0}{B_0\alpha}$

Coordinates are $\left(\frac{v_0\pi}{B_0\alpha}, 0, \frac{-2v_0}{B_0\alpha}\right)$.

19. (c) $y = 2r = \dfrac{2mv_0}{B_0 q} = \dfrac{2v_0}{B_0 \alpha}$

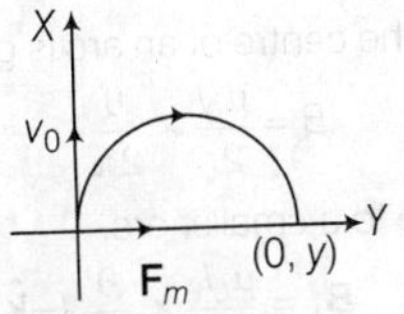

20. (d) Magnetic force on charge

$$q = \mathbf{F} = q(\mathbf{v} \times \mathbf{B})$$

Since, electron has negative charge, **F** will be opposite to $(\mathbf{v} \times \mathbf{B})$, i.e. y will be less than zero.

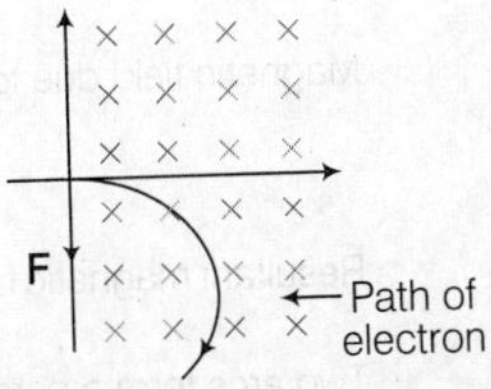

The trajectory will be shown as in the figure. As, $\mathbf{F} \perp \mathbf{v}$, $v = u$ and y = negative.

21. (a,b,c) If $x = y$, then $\mathbf{B} \| \mathbf{v}$, i.e. $f_{mag} = 0$, $f_{mag} = q$ $(\mathbf{v} \times \mathbf{B}) = q(x^2 - y^2)\,\hat{\mathbf{k}}$, $x > y$, force is along Z-axis.

22. (a, c, d) $\mathbf{v} \perp \mathbf{B}$ in Region II. Therefore, path of particle is circle in Region II.

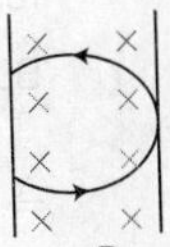

Particle enters in Region III, if radius of circular path, $r > l$

or $\dfrac{mv}{Bq} > l$ $\qquad v = \dfrac{Bqe}{m}$

or $v > \dfrac{Bql}{m}$

If $v = \dfrac{Bql}{m}$, $r = \dfrac{mv}{Bq} = l$, particle will turn back and path length will be maximum. If particle returns to Region I, time spent in the Region II will be

$$t = \frac{T}{2} = \frac{\pi m}{Bq}$$

which is independent of v.

23. (b,d) Electric field can deviate the path of the particle in the shown direction only when it is along negative Y-direction. In the given option, **E** is either zero or along X-direction. Hence, it is the magnetic field which is really responsible for its curved path. Option (a) cannot be accepted as the path will be helix in that case (when the velocity vector makes an angle other than 0°, 180° or 90° with the magnetic field, path is a helix). Only in option (b), the particle can move in XY- plane.

$$\mathbf{F}_{net} = q\mathbf{E} + q(\mathbf{v} \times \mathbf{B})$$

Initial velocity is along X-direction. So, let $\mathbf{v} = v\hat{\mathbf{i}}$

In option (b),

$$\mathbf{F}_{net} = q(a\hat{\mathbf{i}}) + q\,[(v\hat{\mathbf{i}}) \times (c\hat{\mathbf{k}} + a\hat{\mathbf{j}})] = qa\hat{\mathbf{i}} - qvc\hat{\mathbf{j}}$$

24. (c) $\mathbf{F} = q(\mathbf{v} \times \mathbf{B})$

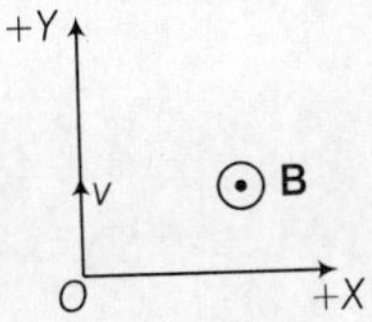

Let **B** be directed out of the plane of the paper.

For positive charges, i.e. A and D, force experienced will be in the direction of $\mathbf{v} \times \mathbf{B}$, i.e. towards $+X$-direction.

For negative charge, i.e. B and C, force will be along $-X$-direction.

Now, $\qquad r = \dfrac{mv}{Bq} = \dfrac{v}{B\left(\dfrac{q}{m}\right)}$

$\Rightarrow \qquad r \propto \dfrac{1}{s}$ where, $s = \dfrac{q}{m}$

Since, magnitude of charge is same

$\Rightarrow \qquad r \propto m$

Hence, (i) → (r), (ii) → (p), (iii) → (q), (iv) → (s)

25. (d) $v_A = \sqrt{2as} = v$ (say)

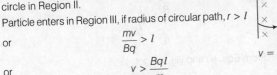

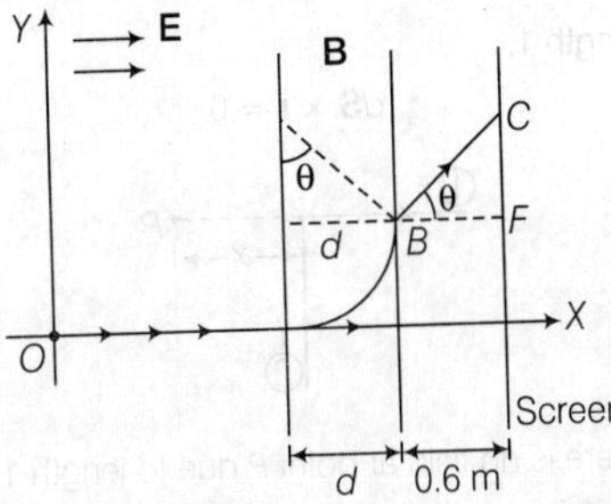

$x = 1.8\text{ m} = 2.4\text{ m} = 3.0\text{ m}$

or $\qquad v = \sqrt{2\left(\dfrac{qE}{m}\right)s}$

$$= \sqrt{\frac{2 \times 1.0 \times 10 \times 1.8}{1}} = 6\text{ms}^{-1}$$

In magnetic field, speed does not change. Hence, particle will collide with 6 ms^{-1}.

In magnetic field, path of the particle is a circle. Radius of circular particle is

$$r = \frac{mv}{Bq} = \frac{(1)(6)}{(5)(1)} = 1.2\text{m}$$

$$d = (2.4 - 1.8)\text{ m} = 0.6\text{ m}$$

Since, $d < r$ $\qquad \theta = \sin^{-1}\left(\dfrac{d}{r}\right) = \sin^{-1}\left(\dfrac{0.6}{1.2}\right) = 30°$

$$AE = r(1 - \cos\theta) = 1.2\left(1 - \frac{\sqrt{3}}{2}\right) = 0.6\,(2 - \sqrt{3})$$

$$FC = BF\tan\theta = \frac{0.6}{\sqrt{3}}$$

$\therefore$ y-coordinate $= AE = FC$

$$= 0.6\left(2 - \sqrt{3} + \frac{1}{\sqrt{3}}\right) = \frac{1.2\,(\sqrt{3} - 1)}{\sqrt{3}}\text{ m}$$

26. (a) Total time, $t = t_{OA} + t_{AB} + t_{BC}$

$$t_{OA} = \sqrt{\frac{2s}{A}} = \sqrt{\frac{2s}{\frac{qE}{m}}} = \frac{\sqrt{2ms}}{qE} = \sqrt{\frac{2 \times 1 \times 1.8}{(1)(10)}} = 0.6\text{ s} = \frac{3}{5}\text{ s}$$

$$t_{AB} = \left(\frac{30°}{360°}\right)T = \left(\frac{1}{12}\right)\left(\frac{2\pi m}{Bq}\right) = \frac{(2\pi)(1)}{(12)(5)(1)} = \frac{\pi}{30}\text{ s}$$

$$t_{BC} = \frac{Bc}{v} = \frac{0.6\sec\theta}{v} = \frac{(0.6)\left(\frac{2}{\sqrt{3}}\right)}{6} = \frac{1}{5\sqrt{3}}\text{ s}$$

$$\therefore \quad t = \frac{1}{5}\left(3 + \frac{\pi}{6} + \frac{1}{\sqrt{3}}\right)\text{s}$$

27. (d) Statement I is incorrect but Statement II is correct.

Force on particle inside magnetic field which causes centripetal acceleration is

$$F_m = Bqv\sin\theta$$

Hence, its acceleration is

$$a_m = B\left(\frac{q}{m}\right)\sin\theta v$$

or $\qquad a_m \propto$ proportional to v

So, when velocity is made twice, then acceleration is twice.

28. (a) A magnetic field interacts with a moving charge as a moving charge produces a magnetic field.

$\therefore$ Both Statements I and II are correct and Statement II explains Statement I.

DPP-2 **Biot-Savart's Law and Its Applications**

1. (i) For length 1,

$$d\mathbf{S} \times \hat{\mathbf{r}} = 0$$

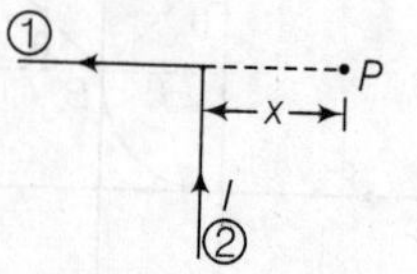

So, there is no field at point P due to length 1.

Length 2 is only semi-infinite thus,

$$B = \frac{1}{2}\left(\frac{\mu_0 I}{2\pi x}\right) = \frac{\mu_0 I}{4\pi x}, \text{ directed into the paper.}$$

(ii) Total field is superposition of field due to an infinite wire and due to a circular loop.

$$\therefore \quad B_{net} \text{ at } P = \frac{\mu_0 I}{2\pi R} + \frac{\mu_0 I}{2R}$$

$$\Rightarrow \quad B = \left(1 + \frac{1}{\pi}\right)\frac{\mu_0 I}{2R}$$

into plane of paper.

(iii) For straight sections, $d\mathbf{S} \times \hat{\mathbf{r}} = 0$

The quarter circle is one-fourth of loop

$$\therefore \quad B = \frac{1}{4}\left(\frac{\mu_0 I}{2R}\right) = \frac{\mu_0 I}{8R}$$

(iv) $B_{net} = B(a) - B(b)$

$$B = \frac{\mu_0 I}{4\pi}\left\{\frac{\frac{1}{6}\times 2\pi a}{a^2} - \frac{\frac{1}{6}\times 2\pi b}{b^2}\right\} = \frac{\mu_0 I}{12}\left(\frac{1}{a} - \frac{1}{b}\right)$$

directed out of plane of paper.

2. (i) For the magnetic field at P to be zero, the current in the wire b should be coming out of the plane of the figure so that the fields due to a and b may be opposite at P. The magnitude of these fields should be equal so that

$$\frac{\mu_0 (9.6 \text{ A})}{2\pi\left[2 + \frac{10}{11}\right]} = \frac{\mu_0 i}{2\pi\left[\frac{10}{11}\right]}$$

$$i = 3.0 \text{ A}$$

(ii) As, we know $(ab)^2 = 4 \text{ m}^2, (as)^2 = 2.56 \text{ m}^2$ and $(bs)^2 = 1.44 \text{ m}^2$, so that $(ab)^2 = (as)^2 + (bs)^2$ and $\angle asb = 90°$.

The magnetic field at s due to wire

$$a = \frac{\mu_0 (9.6)}{2\pi \times 1.6} = \frac{\mu_0}{2\pi} \times 6 \text{ Am}^{-1}$$

and due to the wire,

$$b = \frac{\mu_0}{2\pi} \times \frac{3}{1.2} = \frac{\mu_0}{2\pi} \times 2.5 \text{ Am}^{-1}$$

These fields are at 90° to each other so that their resultant will have a magnitude

$$= \sqrt{\left(\frac{\mu_0}{2\pi} \times 6 \text{ Am}^{-1}\right)^2 + \left(\frac{\mu_0}{2\pi} \times 2.5 \text{ Am}^{-1}\right)^2}$$

$$= \frac{\mu_0}{2\pi}\sqrt{36 + 6.25} \text{ Am}^{-1}$$

$$= 2 \times 10^{-7} \text{ T mA}^{-1} \times 6.5 \text{ Am}^{-1}$$

$$= 1.3 \times 10^{-6} \text{ T}$$

(iii) The force per unit length on the wire,

$$b = \frac{\mu_0 i_1 i_2}{2\pi d}$$

$$= \frac{2 \times 10^{-7} \times 9.6 \times 3}{2} = 2.9 \times 10^{-6} \text{ N}$$

3. Magnetic field at the centre of an arc is given by

$$B = \frac{\mu_0 I}{2r} \times \frac{\theta}{2\pi}$$

Magnetic field due to a smaller arc,

$$\mathbf{B}_1 = \frac{\mu_0 I_1}{2r} \times \frac{\theta}{2\pi}(-\hat{\mathbf{k}})$$

Magnetic field due to a larger arc,

$$\mathbf{B}_2 = \frac{\mu_0 I_2}{2r} \times \frac{(2\pi - \theta)}{2\pi}(+\hat{\mathbf{k}})$$

$$\text{Resultant magnetic field} = \left[\frac{-\mu_0 I_1 \theta}{4\pi r} + \frac{\mu_0 I_2 (2\pi - \theta)}{4\pi r}\right]\hat{\mathbf{k}} \quad \text{...(i)}$$

Two arcs form a parallel combination of resistors.

$$\text{Here,} \quad I_1 R_1 = I_2 R_2$$

$$\Rightarrow \quad \frac{I_1}{I_2} = \frac{R_2}{R_1} \quad \text{...(ii)}$$

Here, R_1 and R_2 are the resistances of smaller and bigger arcs, respectively.

$$\text{But} \quad \frac{R_2}{R_1} = \frac{l_2}{l_1}$$

$$\Rightarrow \quad \frac{I_1}{R_1} = \frac{I_2}{R_2} \quad \text{...(iii)}$$

From Eqs. (ii) and (iii), we get

$$I_1 \theta = I_2 (2\pi - \theta) \quad \text{...(iv)}$$

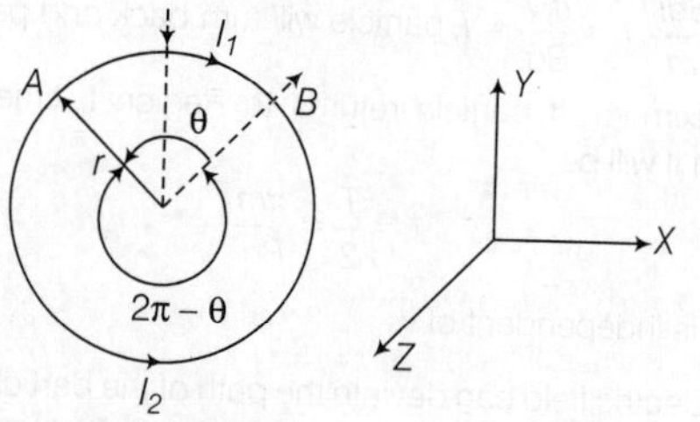

Hence, from Eqs. (i) and (iv), we get net magnetic field at the centre of coil will be zero.

4. Magnetic field at O due to LS and MO part = 0 and net field is due to QS and PR part.

$$B_{PR} = \frac{1}{2}\left[\frac{\mu_0}{2\pi} \times \frac{i}{OR}\right]$$

$$= \frac{(10^{-7})(10)}{0.02} = 5 \times 10^{-5} \text{ T}$$

(perpendicular to paper outwards)

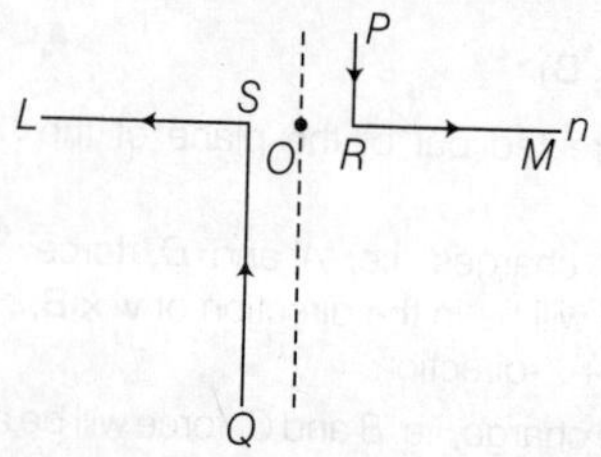

$$\text{Also,} \quad B_{QS} = \frac{1}{2}\left[\frac{\mu_0}{2\pi} \times \frac{i}{OS}\right]$$

$$= \frac{(10^{-7}) \times 10}{0.02} = 5 \times 10^{-5} \text{ T}$$

(perpendicular to paper outwards)

Since, both the fields are in same direction, so net field will be sum of these two.

$$\therefore \quad B_{net} = B_1 + B_2$$

$$= 5 \times 10^{-5} + 5 \times 10^{-5} = 10^{-4} \text{ T}$$

5. (d) Magnetic field at current carrying wire of radius R, i.e.

$$B_1 = \frac{\mu_0 i}{4\pi R} \times \frac{\pi}{2}$$

Similarly, magnetic field at current carrying of radius R', i.e.

$$B_2 = \frac{\mu_0}{4\pi} \times \frac{i}{R'} \times \frac{3\pi}{2}$$

Net magnetic field at point O, i.e. $B = B_1 + B_2$

$$B = \frac{\mu_0 i}{4\pi R} \times \frac{\pi}{2} + \frac{\mu_0}{4\pi} \cdot \frac{i}{R'} \times \frac{3\pi}{2}$$

$$= \frac{\mu_0 i}{8}\left(\frac{1}{R} + \frac{3}{R'}\right)$$

6. (a) As, we know

$$\frac{x}{d} = \sin 45^\circ \Rightarrow x = \frac{d}{\sqrt{2}}$$

We can extend the wire and then subtract the field of extended part.

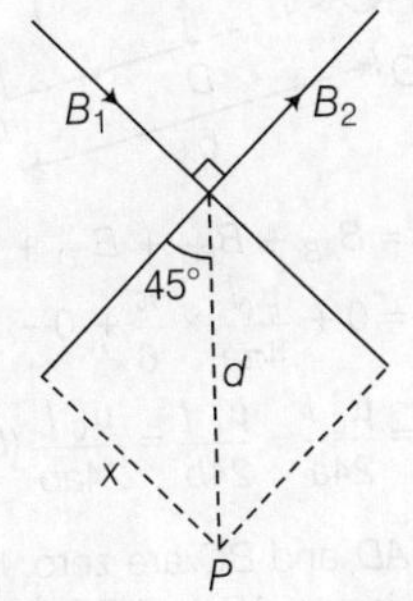

Magnetic field, i.e.

$$B_1 = \frac{\mu_0 I}{4\pi d/\sqrt{2}}(\sin 0^\circ + \sin 90^\circ) - \frac{\mu_0 I}{4\pi d/\sqrt{2}}(\sin 0^\circ + \sin 45^\circ)$$

$$= \frac{\mu_0 I}{2\sqrt{2}\pi d}\left(1 - \frac{1}{\sqrt{2}}\right)$$

and similarly, $\quad B_2 = \frac{\mu_0 I}{2\sqrt{2}\pi d}\left[1 - \frac{1}{\sqrt{2}}\right]$

Then, net magnetic field at point P

$$B = (B_1 + B_2) = 2B = \frac{\mu_0 I}{\sqrt{2}\pi d}\left(1 - \frac{1}{\sqrt{2}}\right)$$

7. (b) Given, $I_1 = 8$ A, $I_2 = 5$ A

and $r = 4$ cm $= 0.04$ m

Force per unit length on two parallel wire carrying current,

$$F = \frac{\mu_0}{4\pi} \cdot \frac{2I_1 I_2}{r} = \frac{10^{-7} \times 2 \times 8 \times 5}{0.04}$$

$$= 2 \times 10^{-4} \text{ N}$$

The force on A of length 10 cm is

$$F' = F \times 0.1 \qquad (\because 1 \text{ m} = 100 \text{ cm})$$

$$\Rightarrow \quad F' = 2 \times 10^{-4} \times 0.1$$

$$= 2 \times 10^{-5} \text{ N}$$

8. (c) Magnetic field induction at O due to current through ACB is

$$B_1 = \frac{\mu_0 i\theta}{4\pi r}$$

It is acting perpendicular to the paper downwards. Magnetic field induction at O due to current through ABD is

$$B_2 = \frac{\mu_0}{4\pi} \cdot \frac{i(2\pi - \theta)}{r}$$

It is acting perpendicularly to paper upwards.

$\therefore$ Total magnetic field at O due to current loop is

$$B = B_2 - B_1 = \frac{\mu_0}{4\pi} \cdot \frac{i}{r}(2\pi - \theta) - \frac{\mu_0}{4\pi} \cdot \frac{i}{r}\theta$$

$$= \frac{\mu_0}{2\pi} \cdot \frac{i}{r}(\pi - \theta)$$

9. (d) Perpendicular distance to O from PQ.

$$PQ = r \sin \frac{\theta}{2}$$

Magnetic field induction at O due to current through PQ and QR is

$$B = \frac{\mu_0}{4\pi} \cdot \frac{i}{a}[\sin(90^\circ - \theta/2) + \sin 90^\circ] \times 2$$

$$= \frac{\mu_0}{2\pi r} \cdot \frac{i}{\sin \frac{\theta}{2}}\left(\cos \frac{\theta}{2} + 1\right)$$

$$= \frac{\mu_0}{2\pi} \cdot \frac{i}{r} \cdot \frac{\left(1 + \cos \frac{\theta}{2}\right)}{\sin \frac{\theta}{2}}$$

10. (d) Resistance of arm $PQRS$ is 3 times the resistance of arm PS. If resistance of arm $PS = r$, then resistance of arm $PQRS = 3r$.

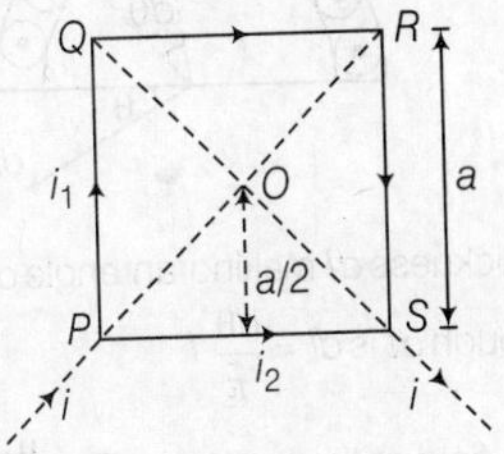

Potential difference across P and S is

$$V_{PS} = i_2 r = i_1 3r$$

$$\Rightarrow \quad i_2 = 3i_1$$

Magnetic field induction at O due to wire lengths PQ, QR and RS is

$$B_{PQRS} = B_{PQ} + B_{QR} + B_{RS}$$

$$= 3 \times \frac{\mu_0}{4\pi} \cdot \frac{i_1}{a}(\sin 45^\circ + \sin 45^\circ)$$

$$= 3 \times \frac{\mu_0}{4\pi} \times \frac{i_1}{a} \times \frac{2}{\sqrt{2}} = \frac{3\sqrt{2}\mu_0 i_1}{4\pi a} \otimes$$

downwards into plane of paper.

Magnetic field due to wire length $PS = B_{PS}$

$$= \frac{\mu_0}{4\pi} \cdot \frac{i_2}{a}(\sin 45^\circ + \sin 45^\circ)$$

$$= \frac{\sqrt{2}\mu_0}{4\pi} \cdot \frac{i_2}{a} \text{ upwards out of plane of paper}$$

$$= \frac{\sqrt{2}\mu_0}{4\pi a} \times 3i_1 \text{ upwards}$$

So, $\quad B_{\text{net}} = B_{PQRS} - B_{PS} = 0$

11. (b) As shown in the figure take an element dl at C of wire, where $OC = l$.

Let $\quad PC = r$ and $\angle OPC = \phi$

According to Biot-Savart's law, the magnitude of magnetic field at P due to current element at C is

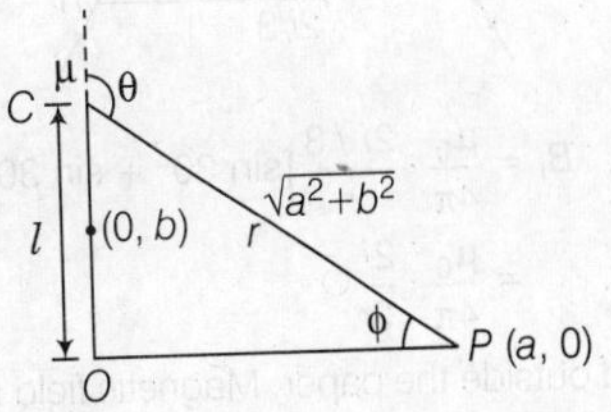

$$dB = \frac{\mu_0}{4\pi} \cdot \frac{idl \sin \theta}{r^2}$$

Here, $\quad \theta = 90^\circ + \phi, r = a \sec \phi$

and $\quad l = a \tan \phi$

$\therefore \quad dl = a \sec^2 \phi \, d\phi \quad$ (after differentiating)

$$\therefore dB = \frac{\mu_0}{4\pi} \cdot \frac{ia(\sec^2 \phi \, d\phi) \sin(90^\circ + \phi)}{a^2 \sec^2 \phi} = \frac{\mu_0}{4\pi} \cdot \frac{i}{a} \cos \phi \, d\phi$$

The magnetic field induction at point P is

$$B = \int dB = \int_{90^\circ}^{\phi} \frac{\mu_0}{4\pi} \cdot \frac{i}{a} \cos \phi \, d\phi$$

$$= \frac{\mu_0}{4\pi} \cdot \frac{i}{a} [-\sin \phi]_{90^\circ}^{\phi}$$

$$= \frac{\mu_0}{4\pi} \cdot \frac{i}{a} (1 - \sin \phi)$$

$$= \frac{\mu_0 i}{4\pi a} \left(1 - \frac{b}{\sqrt{a^2 + b^2}}\right)$$

12. (d) Consider wire to be made up of large number of thin wires of infinite length.

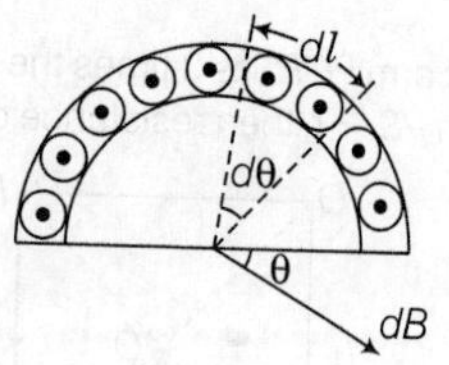

Consider thickness dl making an angle $d\theta$ at centre,

Current through dl is $dI = \frac{d\theta}{\pi} I$

$\therefore$ Magnetic field at the centre $= dB = \frac{\mu_0}{4\pi} \cdot \frac{2dI}{R} = \frac{\mu_0 I}{2\pi^2 R} d\theta$

$$\therefore \quad B = \int_{-\pi/2}^{\pi/2} dB \cos \theta \, d\theta$$

$$= \int_{-\pi/2}^{\pi/2} \frac{\mu_0 I}{2\pi^2 R} \cos \theta \, d\theta = \frac{\mu_0 I}{\pi^2 R}$$

13. (a) Magnetic field along a current carrying wire of radius r, i.e.

$$\mathbf{B}_1 = \frac{\mu_0}{4\pi} \times \frac{I}{r} \times \pi = \frac{\mu_0 I}{4r}$$

Similarly, magnetic field along a current carrying wire of radius $2r$, i.e.

$$\mathbf{B}_2 = \frac{\mu_0}{4\pi} \times \frac{I}{2r} \times \pi = \frac{\mu_0 I}{8r}$$

So, net magnetic field at point P,

$$\mathbf{B} = \mathbf{B}_1 + \mathbf{B}_2 = \frac{\mu_0 I}{4r} + \frac{\mu_0 I}{8r} = \frac{3}{8} \cdot \frac{\mu_0 I}{r}$$

i.e. perpendicular to the plane of the paper and directed inwards.

14. (d) Magnetic field at O due to PR,

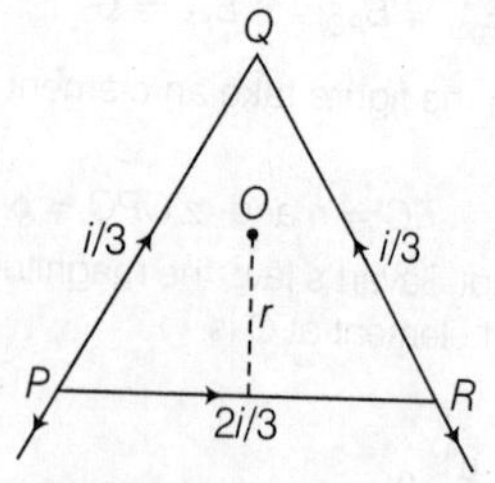

$$B_1 = \frac{\mu_0}{4\pi} \cdot \frac{2i/3}{r} [\sin 30^\circ + \sin 30^\circ]$$

$$= \frac{\mu_0}{4\pi} \cdot \frac{2i}{3r} \odot$$

It is directed outside the paper. Magnetic field at O due to PQR (i.e. for the wire PQ and QR)

$$B_2 = 2 \times \frac{\mu_0}{4\pi} \cdot \frac{\left(\frac{i}{3}\right)}{r} [\sin 30^\circ + \sin 30^\circ]$$

$$= \frac{\mu_0}{4\pi} \cdot \frac{2i}{3r} \otimes$$

It is directed inside the paper.

$\therefore$ Resultant magnetic field at O is given by

$$B = B_1 - B_2 = 0$$

15. (d) At the centre C, magnetic field due to wires RQ and PS will be zero.

Due to wire QP,

$$B_1 = \frac{1}{2}\left[\frac{\mu_0 I}{2R_1}\right] = \frac{\mu_0 I}{4R_1} \odot \quad \text{(perpendicular to paper outwards)}$$

and due to wire SR,

$$B_2 = \frac{1}{2}\left[\frac{\mu_0 I}{2R_2}\right] = \frac{\mu_0 I}{4R_2} \otimes \quad \text{(perpendicular to paper inwards)}$$

$\therefore$ Net magnetic field would be, $B = B_1 - B_2$

$$B = \frac{\mu_0 I}{4}\left[\frac{1}{R_1} - \frac{1}{R_2}\right] \odot \quad \text{(perpendicular to paper outwards)}$$

16. (b) Net magnetic field due to loop $ABCD$ at O is

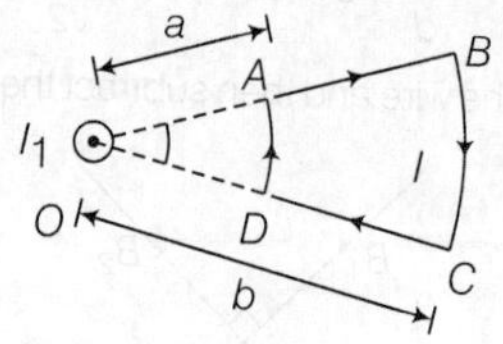

$$B = B_{AB} + B_{BC} + B_{CD} + B_{DA}$$

$$= 0 + \frac{\mu_0 I}{4\pi a} \times \frac{\pi}{6} + 0 - \frac{\mu_0 I}{4\pi b} \times \frac{\pi}{6}$$

$$= \frac{\mu_0 I}{24a} - \frac{\mu_0 I}{24b} = \frac{\mu_0 I}{24ab}(b - a)$$

17. (b) The forces on AD and BC are zero, because magnetic field due to a straight wire on AD and BC is parallel to length of the loop.

18. (c) $B_R = B$ due to ring

$\Rightarrow$ $B_1 = B$ due to wire 1

$B_2 = B$ due to wire 2

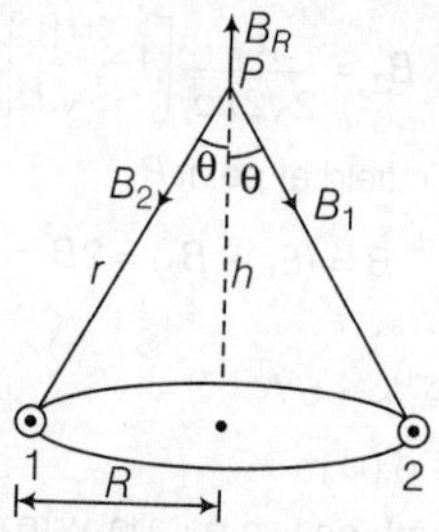

In magnitudes, $B_1 = B_2 = \frac{\mu_0 I}{2\pi r}$

Resultant of B_1 and B_2

$$= 2B_1 \cos \theta = 2\left(\frac{\mu_0 I}{2\pi r}\right)\left(\frac{h}{r}\right) = \frac{\mu_0 I h}{\pi r^2}$$

$$B_R = \frac{\mu_0 I R^2}{2(R^2 + x^2)^{3/2}} = \frac{2\mu_0 I \pi a^2}{4\pi r^3}$$

As, $R = a$, $x = h$ and $a^2 + h^2 = r^2$

For zero magnetic field at point P,

$$\frac{\mu_0 I h}{\pi r^2} = \frac{2\mu_0 I \pi a^2}{4\pi r^3}$$

$$\Rightarrow \quad \pi a^2 = 2rh \quad \Rightarrow \quad h \approx 1.2a$$

19. (b) Magnetic field at mid-point of two wires

$= 2$ (magnetic field due to one wire)

$$= 2\left[\frac{\mu_0}{2\pi} \cdot \frac{I}{d}\right] = \frac{\mu_0 I}{\pi d}$$

Magnetic moment of loop,

$$M = IA = I\pi a^2$$

$$\text{Torque on loop} = MB \sin 150^\circ = \frac{\mu_0 I^2 a^2}{2d}$$

20. (a,b,c,d)

(a) $B = 0$ for all points on X-axis.

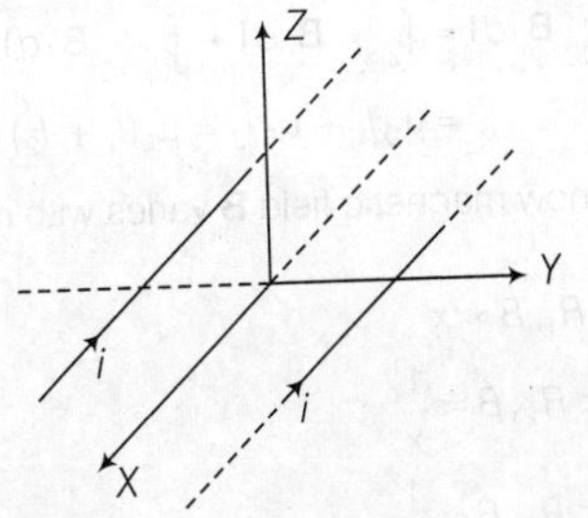

(b) At all points on the Y-axis, excluding the origin, B has only a z-component.

(c) At all points on Z-axis, excluding the origin, B has only a y-component.

(d) B can't have an x-component.

21. (a,c) $\mathbf{F}_{BA} = 0$, because magnetic lines are parallel to this wire.

$\mathbf{F}_{CD} = 0$, because magnetic lines are anti-parallel to this wire.

$\mathbf{F}_{CB}$ is perpendicular to paper outwards and $\mathbf{F}_{AD}$ is perpendicular to paper inwards. These two forces (although calculated by integration) cancel each other but produce a torque which tend to rotate the loop in clockwise direction about an axis OO'.

22. (7) Here, $B_{PR} = 0$ and $B_{PQ} = 0$

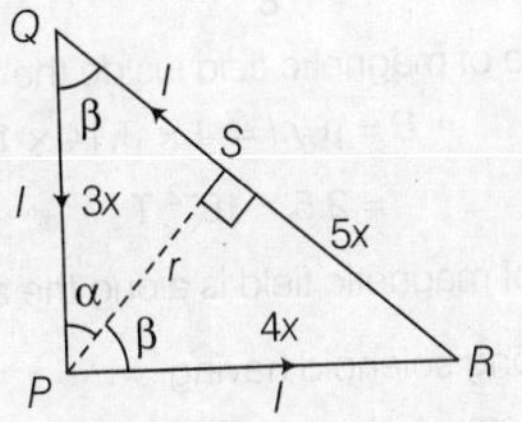

$$\sin\alpha = \frac{3}{5}, \quad \sin\beta = \frac{4}{5}$$

Also, $$\sin\alpha = \frac{r}{4x} = \frac{3}{5}$$

$$r = \frac{12x}{5}$$

Magnetic field along QR,

$$B = B_{QR} = \frac{\mu_0 I}{4\pi r}(\sin\alpha + \sin\beta)$$

$$= \frac{\mu_0}{4\pi} \cdot \frac{I}{12x/5}\left(\frac{3}{5} + \frac{4}{5}\right) = 7\left(\frac{\mu_0 I}{48\pi x}\right)$$

So, $k = 7$

23. (a) At point P, the magnetic field due to AB and CD will be zero.

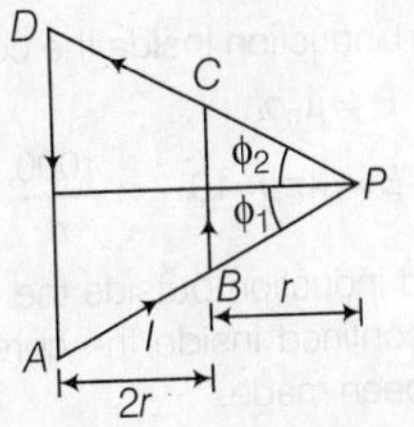

Magnetic field at BC,

$$\mathbf{B}_{BC} = \frac{\mu_0}{4\pi} \cdot \frac{I}{r}[\sin\phi_1 + \sin\phi_2](-\mathbf{k})$$

and $$\mathbf{B}_{DA} = \frac{\mu_0}{4\pi} \cdot \frac{I}{3r}[\sin\phi_1 + \sin\phi_2](\hat{\mathbf{k}})$$

Net field at point P,

$$\mathbf{B} = \mathbf{B}_{BC} + \mathbf{B}_{DA}$$

$$= \frac{2\mu_0}{4\pi} \times \frac{I}{3r}[\sin\phi_1 + \sin\phi_2]\hat{\mathbf{k}}$$

Hence, (i) → (q,r), (ii) → (p,s), (iii) → (p), (iv) → (r)

DPP-3 Ampere's Circuital Law and Its Applications

1. By Ampere's circuital law, magnetic field at point A is given by

$$B_A = \frac{\mu_0 I_A}{2\pi r_A}$$

Given, $I_A = 1$ A (out of paper) and $r_A = 1$ mm

$$\therefore \quad B_A = \frac{4\pi \times 10^{-7} \times 1}{2\pi\,(1 \times 10^{-3})} = 200\,\mu\text{T}$$

(towards top of page)

At point B, similarly we have

$$B_B = \frac{\mu_0 I_B}{2\pi r_B}$$

$$I_B = 1 - 3 = -2\text{ A}$$

or $$I_B = 2\text{ A}$$

$$r_B = 3 \times 10^{-3}\text{ m}$$

$$\therefore \quad B_B = \frac{4\pi \times 10^{-7} \times 2}{2\pi \times 3 \times 10^{-3}}$$

$= 133\,\mu\text{T}$ (towards bottom of page)

2. (i) Taking the case, when $r < a$. Let the point P be inside the wire at a distance r from the axis of wire. Consider an amperian loop labelled 1 is a circle of radius r concentric with cross-section such that point P_1 lies on this loop.

Current enclosed by the loop,

$$I' = \frac{I}{\pi a^2} \times \pi r^2 = \frac{Ir^2}{a^2}$$

Let B be magnetic field induction at P_1. Line integral of $\mathbf{B}$ over the amperian loop is

$$\int \mathbf{B} \cdot d\mathbf{l} = B 2\pi r$$

According to Ampere's circuital law,

$$\oint \mathbf{B} \cdot d\mathbf{l} = \mu_0 \times I_{\text{enclosed}} = \frac{\mu_0 I r^2}{a^2}$$

$$\therefore \quad B\,2\pi l = \frac{\mu_0 I r^2}{a^2} \quad \Rightarrow \quad B = \frac{\mu_0 I r}{2\pi a^2}$$

It means that, $B \propto r$ (when $r < a$)

(ii) Taking the case when $r > a$. Take point P outside the wire at a distance r from the axis of wire.

Consider an amperian loop labelled 2 is a circle concentric with cross-section of wire such that point P lies on this loop. Current enclosed by this loop is I.

Let B be the magnetic field induction at point P, line integral of $\mathbf{B}$ over the amperian loop

$$= \int \mathbf{B} \cdot d\mathbf{l} = B 2\pi r$$

According to Ampere's circuital law,

$$\oint \mathbf{B} \cdot d\mathbf{l} = \mu_0 \times I_{\text{enclosed}} = \mu_0 I$$

$$\therefore \quad B 2\pi r = \mu_0 I \quad \text{or} \quad B = \frac{\mu_0 I}{2\pi r}$$

It means that, $B \propto \frac{1}{r}$ (when $r > a$)

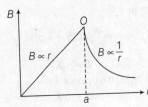

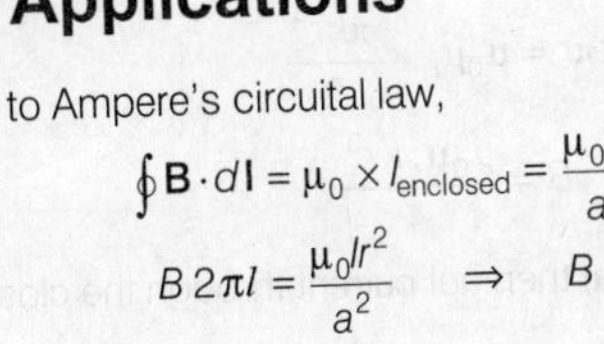

3. Here, inner radius, $r_1 = 20$ cm

Outer radius, $r_2 = 22$ cm

and $I = 10$ A

$\therefore$ Mean radius of toroid,

$$r = \frac{r_1 + r_2}{2} = \frac{20 + 22}{2}$$

$$= 21 \text{ cm or } 0.21 \text{ m}$$

Total length of toroid = Circumference of toroid = $2\pi r$

$$= 2 \times \pi \times 0.21 = 0.42\pi \text{ m}$$

Total number of turns, $N = 4200$

$\therefore$ Number of turns per unit length will be

$$n = \frac{4200}{0.42\pi} = \frac{1000}{\pi} \text{ m}^{-1}$$

(i) Magnetic field induction inside the core of toroid,

$$B = \mu_0 nI$$

$$B = 4\pi \times 10^{-7} \times \frac{1000}{\pi} \times 10 = 0.04 \text{ T}$$

(ii) Magnetic field induction outside the toroid is zero. Since, the field is only confined inside the core of the toroid on which winding has been made.

(iii) Magnetic field induction in the empty space surrounded by toroid is also zero.

4. (i) Consider an elementary ring of radius r, thickness dr whose centre lies on the axis of the conductor. Area of the elementary ring, $dA = 2\pi r\, dr$. Current passing through the elementary ring is

$$dI = I\, dA = (kr^2)(2\pi r\, dr) = k\, 2\pi r^3\, dr$$

Total current passing through the closed path of radius r is

$$I = \int_0^r k\, 2\pi r^3\, dr = \frac{k\,\pi r^4}{2}$$

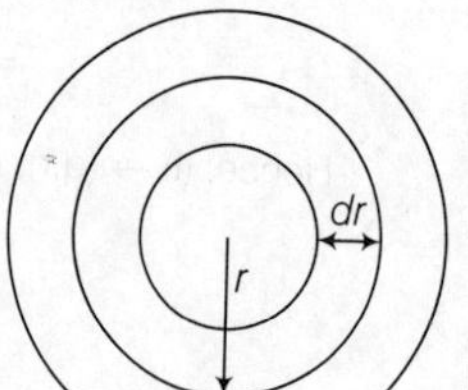

Using Ampere's circuital law, we have

$$\oint \mathbf{B} \cdot d\mathbf{l} = \mu_0 \mu_r I$$

$$B 2\pi r = \mu_0 \mu_r \times \frac{\pi k r^4}{2}$$

or $$B = \frac{\mu_0 \mu_r k r^3}{4}$$

(ii) If $r > a$, then net current through the closed path of radius r is

$$I = \int_0^a k 2\pi r^3\, dr = \frac{\pi k a^4}{2}$$

Using Ampere's circuital law, we have

$$\int \mathbf{B} \cdot d\mathbf{l} = \mu_0 I$$

or $$B \cdot 2\pi r = \frac{\mu_0 \pi k a^4}{2}$$

$\Rightarrow$ $$B = \frac{\mu_0 k a^4}{4r}$$

Thus, net magnetic field when distance, i.e. $r > a$ is

$$B = \frac{\mu_0 k a^4}{4r}$$

5. (b) In a hollow cylinder having infinite length and carrying a uniform current per unit length λ along the circumference. According to Ampere's circuital law, we get

$$\int \mathbf{B} \cdot d\mathbf{l} = \mu_0 \times dI$$

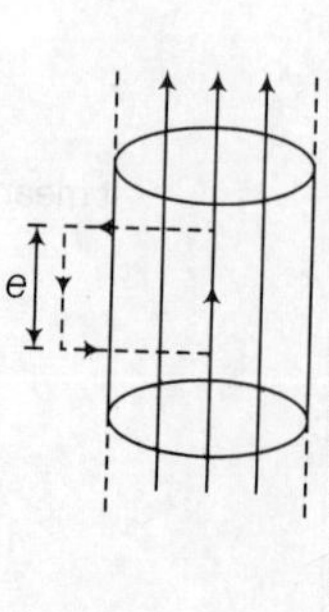

(Linear density i.e. $\lambda = \frac{dI}{dl}$, so $dI = \lambda\, dl$)

i.e. $$B = \mu_0 \lambda$$

Magnetic field inside the cylinder, i.e.

$$B = \mu_0 \lambda$$

6. (d) According to Ampere's circuital law, we have

$\oint \mathbf{B} \cdot d\mathbf{l} = \mu_0 nI$

$$\oint_{ABCD} \mathbf{B} \cdot d\mathbf{l} = \oint_{ABCA} \mathbf{B} \cdot d\mathbf{l} + \oint_{CDAC} \mathbf{B} \cdot d\mathbf{l}$$

$$= \mu_0 I_1 + \mu_0 I_2 = \mu_0 (I_1 + I_2)$$

7. (c) As, we know magnetic field **B** varies with respect to radius R, such that

(i) $0 < x < R_1, B \propto x$

(ii) $R_1 < x < R_2, B \propto \frac{1}{x}$

(iii) $R_2 < x < R_3, B \propto \frac{1}{x}$

So, (ii) and (iii) decrease hyperbolically but with different slopes as the media are different.

8. (b) The length of solenoid, $l = 80$ cm $= 0.8$ m

Number of layers = 5

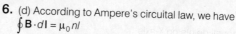

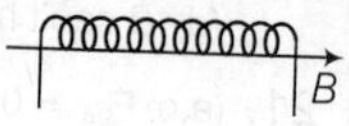

Number of turns per layer = 400

Diameter of solenoid = 1.8 cm

Current in solenoid, $I = 8$ A

$\therefore$ The total number of turns,

$$N = 400 \times 5 = 2000$$

and number of turns/length,

$$n = \frac{2000}{0.8} = 2500$$

The magnitude of magnetic field inside the solenoid

$$B = \mu_0 nI = 4 \times 3.14 \times 10^{-7} \times 2500 \times 8$$

$$= 2.5 \times 10^{-2} \text{ T}$$

The direction of magnetic field is along the axis of solenoid.

9. (b) ***Case I*** A long solenoid having

$n_1 = 200$ turns/cm,

$B_1 = 6.28 \times 10^{-2}$ Wb $= 200 \times 10^{-2}$ turns/m

According to Ampere's circuital law, we get

$$B_1 = \mu_0 n_1 I_1$$

$$6.28 \times 10^{-2} = 4\pi \times 10^{-7} \times 200 \times 10^{-2} \times I_1$$

$\Rightarrow$ $$I_1 = \frac{6.28 \times 10^{-2}}{4\pi \times 10^{-7} \times 200 \times 10^{-2}}$$

Case II When another long solenoid having $n_2 = 100$ turns/cm,

$$n_2 = 100 \times 10^{-2} \text{ turns/m}, I_2 = \frac{i}{3}$$

According to Ampere's circuital law, we have

$B_2 = \mu_0 n_2 I_1$

$$= \frac{4\pi \times 10^{-7} \times 100 \times 10^{-2} \times 6.28 \times 10^{-2}}{4\pi \times 10^{-7} \times 200 \times 10^{-2}}$$

$$= 1.046 \times 10^{-2} \text{ T} \approx 1.05 \times 10^{-2} \text{ Wb/m}^2$$

10. (d) Consider a long straight wire of radius a carries a steady current I.

i.e. $$\mathbf{B} = \frac{\mu_0 I}{2\pi a^2} \times r$$

Case I Magnetic field at $r = \frac{a}{2}$ is

$$B_1 = \frac{\mu_0 I}{2\pi a^2} \times \frac{a}{2} = \frac{\mu_0 I}{4\pi a} \quad \text{...(i)}$$

Case II Magnetic field at $r = 2a$ is

$$B_2 = \frac{\mu_0 I}{2\pi r} = \frac{\mu_0 I}{2\pi (2a)} = \frac{\mu_0 I}{4\pi a} \quad \text{...(ii)}$$

On dividing Eq. (i) by Eq. (ii), we get

$$\frac{B_1}{B_2} = 1$$

11. (d) Take any point *P* inside the thin walled pipe. Consider a circular loop through this point and apply Ampere's circuital law. As, net current inside the loop is zero, the magnetic field at any point inside the loop will be zero.

$$\oint \mathbf{B}\cdot d\mathbf{l} = \mu_0 nI = \mu_0 \times 0 = 0$$

12. (d) For $x < \frac{R}{2}$, $|\mathbf{B}| = 0$

For $\frac{R}{2} \le x < R$

According to Ampere's circuital law, we get

$$\int \mathbf{B}\cdot d\mathbf{l} = \mu_0 I$$

$$|\mathbf{B}|\, 2\pi x = \mu_0 \left[\pi r^2 - \pi\left(\frac{R}{2}\right)^2\right]\hat{\mathbf{i}}$$

$$|\mathbf{B}| = \frac{\mu_0 J}{2x}\left[x - \frac{R^2}{4}\right]$$

For $x \ge R$, $\int \mathbf{B}\cdot d\mathbf{l} = \mu_0 I$

$$|\mathbf{B}|\, 2\pi r = \mu_0 I$$

$$|\mathbf{B}| = \frac{\mu_0 I}{2xr}$$

Thus, graph is

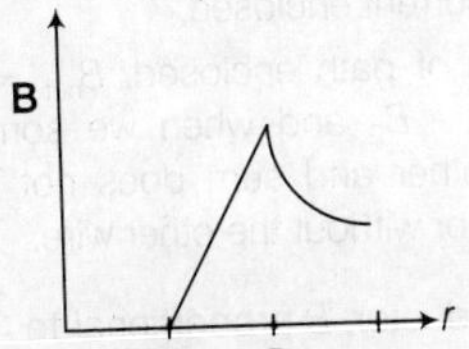

13. (d) Consider an element of thickness *dr* at a distance *r* from the centre. The number of turns in this element,

$$dN = \left(\frac{N}{b-a}\right)dr$$

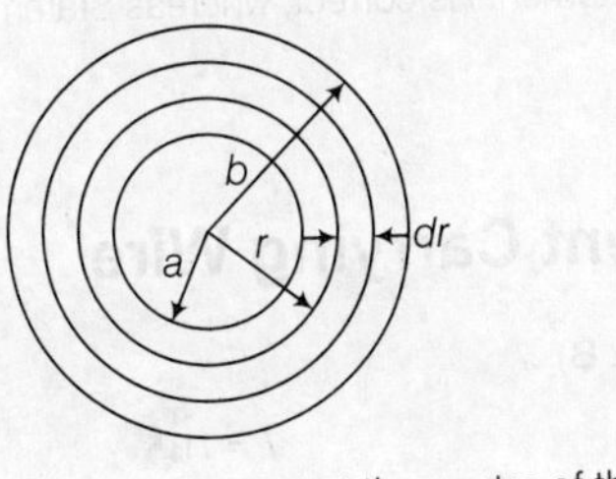

Magnetic field due to this element at the centre of the coil will be

$$dB = \frac{\mu_0 (dN) I}{2r}$$

$$= \frac{\mu_0 I}{2}\cdot\frac{N}{b-a}\cdot\frac{dr}{r}$$

$\therefore$ Net magnetic field on coil,

$$B = \int_{r=a}^{r=b} dB = \frac{\mu_0 NI}{2(b-a)} \ln\left(\frac{b}{a}\right)$$

14. (a) If we take a small strip of *dr* at distance *r* from centre, then number of turns in this strip would be

$$dN = \left(\frac{N}{b-a}\right)dr$$

Magnetic field due to this element at the centre of the coil will be

$$dB = \frac{\mu_0 (dN)\, I}{2r} = \frac{\mu_0 NI}{(b-a)}\cdot\frac{dr}{r}$$

$$\therefore \quad B = \int_{r=a}^{r=b} dB = \frac{\mu_0 NI}{2(b-a)} \ln\left(\frac{b}{a}\right)$$

15. (c) As, magnetic field inside the wire,

$$B_{\text{inside}} = \frac{\mu_0 Ir}{2\pi R^2}$$

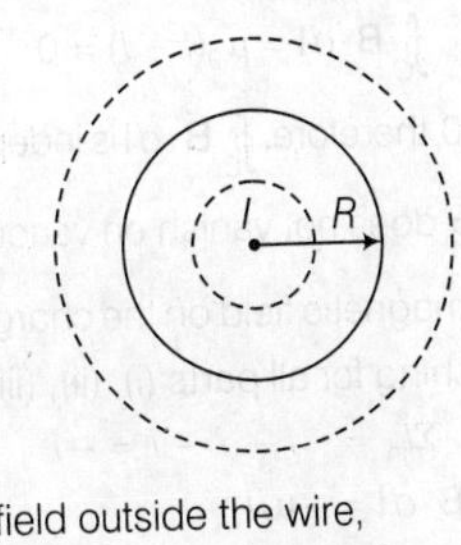

and magnetic field outside the wire,

$$B_{\text{outside}} = \frac{\mu_0 I}{2\pi r}$$

So, from the surface nature of magnetic field charges. Hence, it is clear from the graph that wire *o* has greatest radius.

16. (a) Inside the wire, $B(r) = \frac{\mu_0}{2\pi}\cdot\frac{I}{R^2}\, r$

$$B \propto E$$

It can be seen that magnitude of magnetic field is maximum at the surface of wire *m*.

17. (a) Inside the wire,

$$B(r) = \frac{\mu_0 Ir}{2\pi R^2}$$

$$\frac{dB}{dr} = \frac{\mu_0 I}{2\pi R^2} = \frac{\mu_0 J}{2\pi R^2} = \frac{\mu_0}{2\pi}\hat{\mathbf{j}}$$

i.e. Slope $\propto \hat{\mathbf{j}}$

current density (slope of curve)$_n$ > (slope of curve)$_m$

It can be seen that slope of curve for wire *n* is greater than wire *m*.

18. (a,d) A long thick conducting cylinder of radius *R* carries a uniformly distributed current over its cross-section.

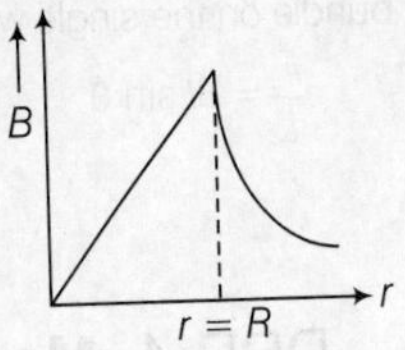

As, strength of magnetic field inside conducting cylinder, i.e. $B_{in} \propto r$

So, magnetic field strength is maximum on the surface, i.e. $r = R$

As, magnetic field **B** is lying outside the cylinder.

So, $B \propto \frac{1}{r}$. Hence, the energy density of magnetic field outside the cylinder varies as inversely proportional to $\frac{1}{r^2}$.

19. (c,d) Cylinder magnetic field *B* is

$$B = \frac{\mu_0 ir}{2\pi R^2}, \text{ for } r < a$$

$$B = \frac{\mu_0 i}{2\pi r}, \text{ for } r \ge a$$

We can consider the given cylinder as a combination of two cylinders. One of cylinder of radius *R* carrying current *I* in one direction and other of radius $\frac{R}{2}$ carrying $\frac{I}{3}$ in other direction.

At point *A*, $B = \frac{\mu_0 (I/3)}{2\pi\,(R/2)} + 0 = \frac{\mu_0 I}{3\pi R}$

At point *B*, $B = \frac{\mu_0 (4I/3)}{2(\pi R^2)}\,\frac{R}{2} + 0 = \frac{\mu_0 I}{3\pi R}$

20. (b,c) Consider a simple amperian loop passing once through both the identical current carrying co-axial loop.

(i) According to Ampere's circuital law,

$$\oint_C \mathbf{B} \cdot d\mathbf{l} = \mu_0(I - I) = 0$$

(ii) As $\oint_C \mathbf{B} \cdot d\mathbf{l} = 0$, therefore, $\oint_C \mathbf{B} \cdot d\mathbf{l}$ is independent of sense of C.

(iii) The value of B does not vanish on various point of C.

21. (a) Work done by magnetic field on the charge = 0 in any part of its motion. 's' is matching for all parts (i), (ii), (iii), (iv).

For loop 1, $\Sigma I_{in} = -i + i - i = -i$

$\therefore$ $\oint \mathbf{B} \cdot d\mathbf{l} = -\mu_0 i$

For loop 2, $\Sigma I_{in} = i - i + i = i$

$\therefore$ $\oint \mathbf{B} \cdot d\mathbf{l} = \mu_0 i$

For loop 3, $\Sigma I_{in} = -i + i = 0$

$\therefore$ $\oint \mathbf{B} \cdot d\mathbf{l} = 0$

For loop 4, $\Sigma I_{in} = +i - i = 0$

$\therefore$ $\oint \mathbf{B} \cdot d\mathbf{l} = 0$

Hence, (i) → (q,s), (ii) → (p,s), (iii) → (r,s), (iv) → (r,s)

22. (1) As the solenoids are identical, the currents in Q and R will be the same and will be half the current in P. The magnetic field within a solenoid is given by $B = \mu_0 nI$. Hence, the field in Q will be equal to the field in R and will be half the field in P, i.e. 1.0 T.

23. (9) One wire feels force due to the field of other ninty nine.

$$B = \frac{\mu_0 I r}{2\pi R^2} = \frac{4\pi \times 10^{-7} \times 99 \times 2 \times 0.2 \times 10^{-2}}{2\pi \times (0.5 \times 10^{-2})^2}$$

$$= 352 \times 9 \times 10^{-6}\ \text{T} = 3.168 \times 10^{-3}\ \text{T}$$

This field points tangent to a circle of radius 0.2 cm and exerts force

$$F = BIL \sin\theta$$

towards centre of bundle on the single wire.

And $\frac{F}{L} = BI \sin\theta$

$$= 2 \times 352 \times 9 \times 10^{-6}\ \text{N}$$
$$= 704 \times 9 \times 10^{-6}\ \text{N}$$

$\therefore$ $N = 9$

24. (3) Using Ampere's law,

$$\int \mathbf{B} \cdot d\mathbf{l} = \mu_0 I = \mu_0 \int J dA \quad \text{...(i)}$$

For $r > R$, Eq. (i) becomes

$$B \int dS = \mu_0 \int_0^R (br)(2\pi r\, dr)$$

$$B \times 2\pi r = \frac{2\pi\mu_0 bR^3}{3} \Rightarrow B = \frac{\mu_0 bR^3}{3r}$$

Hence, $N = 3$

25. (a) Both Statement I and II are correct and Statement II explains the Statement I.

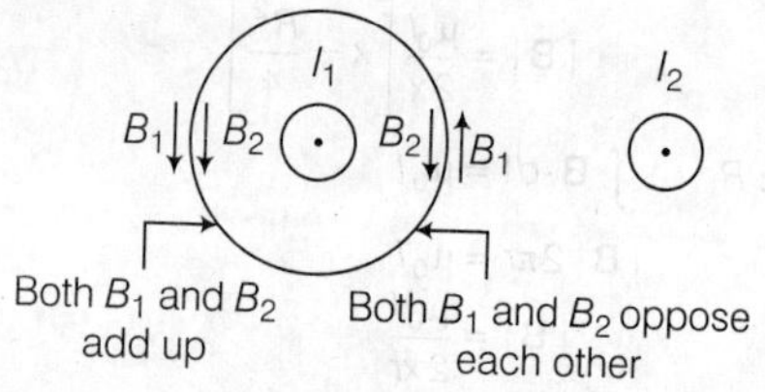

In $\mathbf{B} \cdot d\mathbf{l} = \mu_0 I_{net}$, $\mathbf{B}$ the magnetic field exists due to both wires whereas I_{net} is current enclosed.

At some points of path enclosed, $B_{net} = B_1 + B_2$ and at some points $B_{net} = B_1 - B_2$ and when we some $\mathbf{B} \cdot d\mathbf{l}$, these effects balance each other and sum does not have effect of B_2. So, $\mathbf{B} \cdot d\mathbf{l} = \mu_0 I_1$ with or without the other wire.

26. (c) Inside a conductor, B proportional to $\frac{1}{r}$ is correct but current enclosed (by an Amperian loop) is given by

$$I_{enclosed} = JA_{enclosed}$$

$\Rightarrow$ $I_{enclosed} = J\pi r^2$

or $I \propto r^2$

So, Statement I is correct, whereas Statement II is incorrect.

DPP-4 Magnetic Force on a Current Carrying Wire

1. For each segment,

$$I = 5\ \text{A}, \mathbf{B} = 0.02\hat{\mathbf{j}}\ \text{T}$$

Force on a segment is given by $\mathbf{F} = I(\mathbf{l} \times \mathbf{B})$

Segment	l (m)	$\mathbf{F} = I(\mathbf{l} \times \mathbf{B})$ (N)
ab	$-0.4\hat{\mathbf{j}}$	0
bc	$0.4\hat{\mathbf{k}}$	$-40 \times 10^{-3}\hat{\mathbf{i}}$
cd	$0.4\hat{\mathbf{i}} + 0.4\hat{\mathbf{j}}$	$-40 \times 10^{-3}\hat{\mathbf{k}}$
da	$0.4\hat{\mathbf{i}} - 0.4\hat{\mathbf{k}}$	$40 \times 10^{-3}(\hat{\mathbf{k}} + \hat{\mathbf{i}})$

2. Since, each free electron is in thermal motion it has some velocity ($\mathbf{v}$) directed randomly and hence direction of force when placed in a magnetic field ($\mathbf{B}$) given by

$$\mathbf{F} = q(\mathbf{v} \times \mathbf{B}) = -e(\mathbf{v} \times \mathbf{B})$$

The net force due to all such random motion electron cancels out and hence the force on the wire is zero in the absence of current.

3. The force acting on a current carrying wire joining two fixed points a and b in a uniform magnetic field is independent of the shape of the wire.

Hence, for all the three cases (i), (ii) and (iii), the magnetic force (F) is given by

$$F = BIR$$

4. $\mathbf{F} = I(\mathbf{l} \times \mathbf{B})$

Here, $\mathbf{l} = l\hat{\mathbf{i}}$

$$\mathbf{B} = B_0(\hat{\mathbf{i}} + \hat{\mathbf{j}} + \hat{\mathbf{k}})$$
$$I = i$$

$\Rightarrow$ $\mathbf{F} = i\,[l\hat{\mathbf{i}} \times B_0(\hat{\mathbf{i}} + \hat{\mathbf{j}} + \hat{\mathbf{k}})]$

or $\mathbf{F} = i\,[B_0 l\hat{\mathbf{k}} + B_0 l(-\hat{\mathbf{j}})] = B_0 il\,[\hat{\mathbf{k}} - \hat{\mathbf{j}}]$ ...(i)

(i) Comparing Eq. (i) with $\mathbf{F} = a\hat{\mathbf{i}} + b\hat{\mathbf{j}} + c\hat{\mathbf{k}}$

We get

$$a = 0, b = -B_0 il, c = B_0 il$$

(ii) $|\mathbf{F}| = \sqrt{a^2 + b^2 + c^2}$

$$= \sqrt{(B_0 il)^2 + (B_0 il)^2} = \sqrt{2}\, B_0 il$$

5. (b) $d\mathbf{F} = I(d\mathbf{l} \times \mathbf{B})$ or $|\mathbf{F}| = BIl \sin\theta$

The force on BC and DE are equal and opposite and cancel out. The force on CD is BIL in negative Z-direction.

$\because$ Force on CD = $\mathbf{F} = I\,[L(-\hat{\mathbf{i}}) \times B\hat{\mathbf{j}}]$

or $\mathbf{F} = BIL(-\hat{\mathbf{k}}) = -BIL\hat{\mathbf{k}}$

6. (d) A closed current carrying loop of any shape placed in any uniform magnetic field experiences no force.

7. (d) Net force on a current carrying loop in a uniform magnetic field is zero. So, magnetic force cannot balance its weight.

8. (b) Force per unit length between two parallel conductors

$$= \frac{F}{l} = \frac{\mu_0 I_1 I_2}{2\pi d}$$

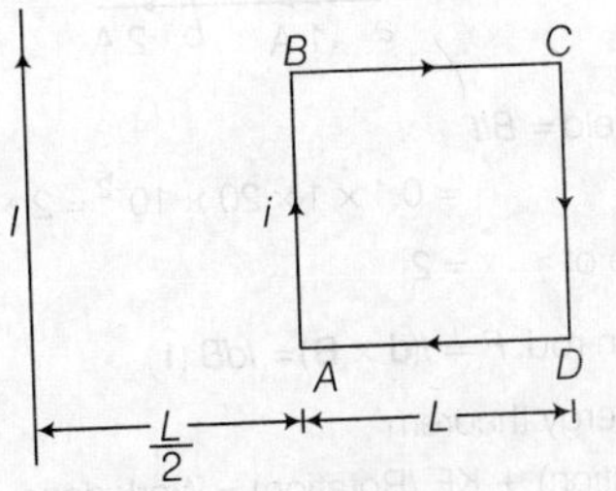

$$\therefore \quad F_{AB} = \frac{\mu_0 Ii}{2\pi\left(\frac{L}{2}\right)}$$

and $$F_{CD} = \frac{\mu_0 Ii}{2\pi\left(\frac{3}{2}L\right)}$$

So, force on AB > force on CD.

Also, force is attractive if current flows in same direction and is repulsive when current is in opposite direction.

Let current in loop is in clockwise sense, then AB is attracted and CD will be repelled.

$$\therefore \quad F_{net} = F_{AB} - F_{CD}$$

and so, loop $ABCD$ is attracted towards the wire.

9. (a) $$mg = kx_0 \quad \text{...(i)}$$

$$mg + ILB = k(2x) \quad \text{...(ii)}$$

$$ILB = mg$$

$$\Rightarrow \quad B = \frac{mg}{IL}$$

$$\Rightarrow \quad B = \frac{mgR}{\varepsilon L} \quad (\because I = \varepsilon/R)$$

10. (b) $y^2 = 2x$

$\Rightarrow$ So, $y = \pm 2$ at $x = 2$ and $\mathbf{L} = -4\hat{\mathbf{j}}$

($\because$ force on parabola is same as force on **AB**)

$$\mathbf{F} = i(\mathbf{L} \times \mathbf{B})$$
$$= 2[-4\hat{\mathbf{j}} \times -4\hat{\mathbf{k}}]$$
$$= 32(\hat{\mathbf{j}} \times \hat{\mathbf{k}}) = 32\hat{\mathbf{i}}$$

11. (b) $\mathbf{F} = I(\mathbf{l} \times \mathbf{B})$, force will be towards Q.

12. (a) $$\mathbf{B} = \begin{cases} \left|\dfrac{B_0 x}{2R}\right| \hat{\mathbf{k}} \text{ when } & x > 0 \\ \left|\dfrac{B_0 x}{2R}\right| (-\hat{\mathbf{k}}) \text{ when } & x < 0 \end{cases}$$

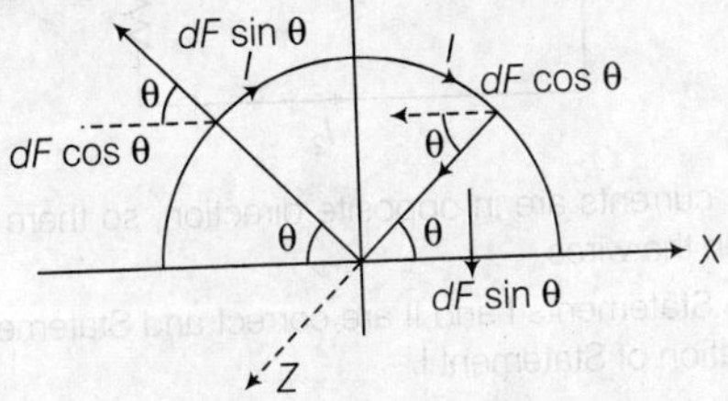

$$\mathbf{F}_{net} = F_x\hat{\mathbf{i}} + F_y\hat{\mathbf{j}}$$

$$F_y = \Sigma dF_y = 0 \quad \text{(from figure)}$$

$$F_x = \Sigma dF_x = -2\int_0^{\pi/2} dF\cos\theta \quad \text{(from figure)}$$

Hence, $\mathbf{F}_{net}$ is along negative X-axis.

13. (d) $dF = 2T\sin\frac{\theta}{2} = 2T\cdot\frac{\theta}{2} = T\theta \quad \left(\because \sin\frac{\theta}{2} \approx \frac{\theta}{2}\right)$

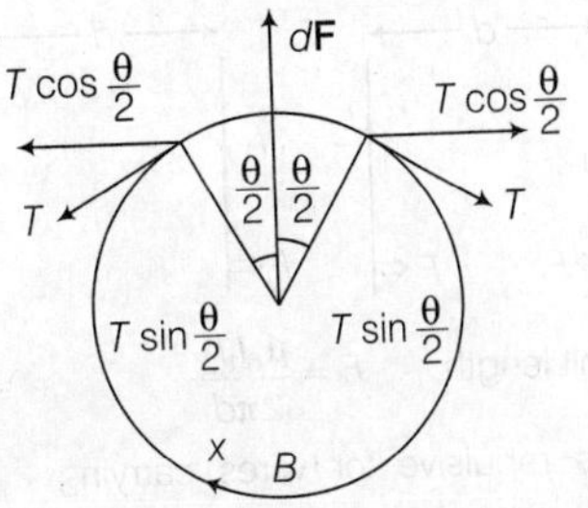

Consider force **F** on small element $dl = r\theta$

$$IdlB = T\theta$$
$$IR\theta B = T\theta$$
$$T = IRB$$
$$T = 100 \times 0.5 \times 0.2 = 10 \text{ N}$$

14. (a) $$F = \frac{\mu_0}{4\pi}\cdot\frac{2I_1I_2}{d}$$

$$F = \frac{\mu_0}{4\pi}\cdot\frac{2 \times 2 \times 2}{d} \times l$$

$$F' = \frac{\mu_0}{4\pi}\cdot\frac{2 \times 1 \times 1}{d} \times l = \frac{F}{4}$$

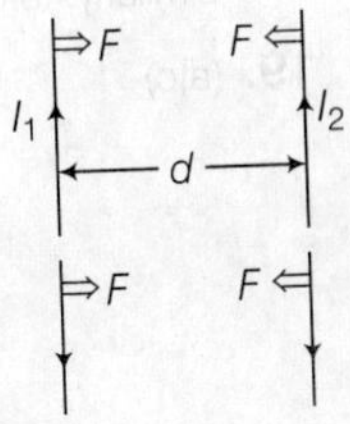

15. (b) Rod will move downward with constant velocity if net force on it is zero.

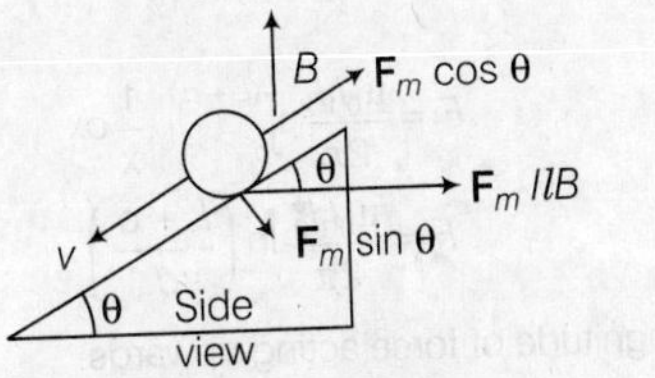

or $$F_m\cos\theta = mg\sin\theta$$

$$IlB\cos\theta = mg\sin\theta$$

$$\therefore \quad B = \frac{mg}{Il}\tan\theta$$

16. (b) $$a = \frac{\text{Force on wire } ABC}{\text{Mass}} = \frac{\text{Force on } AC}{\text{Mass}}$$

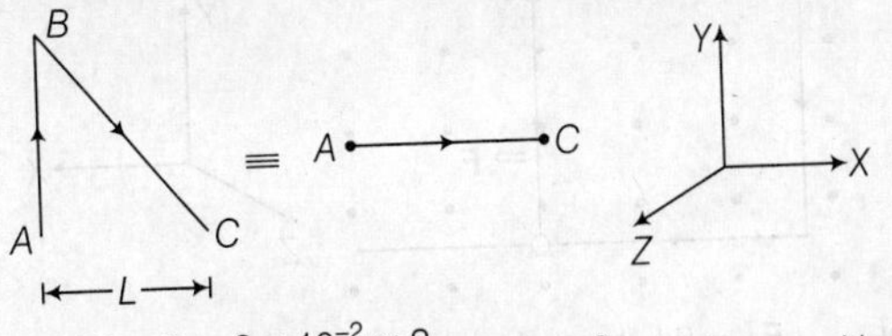

$$= \frac{iLB}{m} = \frac{2 \times 3 \times 10^{-2} \times 2}{10 \times 10^{-3}} = 12 \text{ ms}^{-2} \quad \text{(along positive } Y\text{-axis)}$$

17. (a,b,c) Given, $\mathbf{B} = B_0\hat{\mathbf{j}}$

$$\because \quad \mathbf{F} = I(l \times \mathbf{B})$$

$\therefore$ Force on wire 1,

$$\mathbf{F}_1 = i(a\hat{\mathbf{k}} \times B_0\hat{\mathbf{j}}) = -iaB_0(\hat{\mathbf{i}})$$

Force on wire 2,

$$\mathbf{F}_2 = i[a(-\hat{\mathbf{i}}) \times B_0\hat{\mathbf{j}}] = iaB_0(-\hat{\mathbf{k}})$$

Force on wire 3,

$$\mathbf{F}_3 = i(a\hat{\mathbf{j}} \times B_0\hat{\mathbf{j}}) = 0$$

Force on wire 4,

$$\mathbf{F}_4 = i[a(-\hat{\mathbf{k}}) \times B_0\hat{\mathbf{j}}] = iaB_0(\hat{\mathbf{i}})$$

i.e. force on wire 4 is iaB_0 in positive X-direction.

18. (a,b) Force is attractive for wires carrying current in same direction.

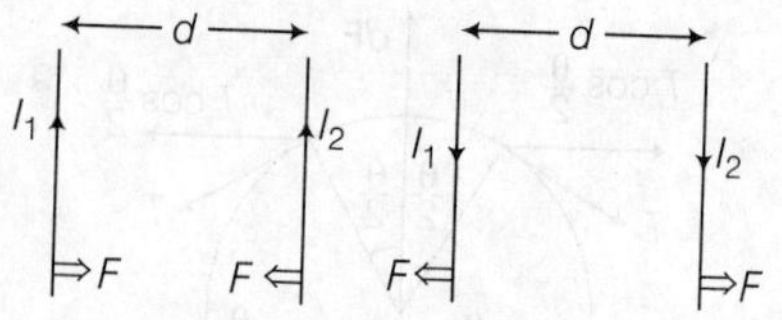

Force per unit length $F = \dfrac{\mu_0 I_1 I_2}{2\pi d}$

The force is repulsive for wires carrying current in opposite direction.

$$F = \frac{\mu_0 I_1 I_2}{2\pi d}$$

Hence, wire 1 is attracted by wire 3 at a distance $2d$ (say) and repelled by wires 2 and 4 at distance d and $3d$, respectively.

Wire 4 is symmetrical in situation as wire 1.

Similarly, forces on 2 and 3 are also equal.

19. (a,c) $d\mathbf{F} = I_2(d\mathbf{l} \times \mathbf{B})$

$$|d\mathbf{F}| = I_2\, dx \frac{\mu_0 I_1}{2\pi x}$$

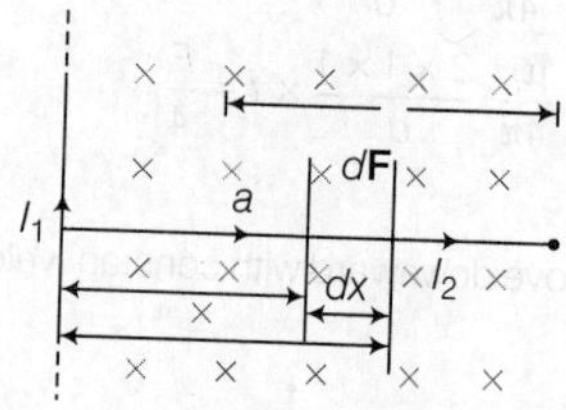

$\Rightarrow$
$$F = \frac{\mu_0 I_1 I_2}{2\pi} \int_d^{(L+d)} \frac{1}{x}\, dx$$

$$F = \frac{\mu_0 I_1 I_2}{2\pi} \ln\left(\frac{L+d}{d}\right)$$

is the magnitude of force acting upwards.

Since, force near end a is more than that away from a towards b, there is a tendency of wire to rotate clockwise. Hence, τ is clockwise.

20. (b) $\mathbf{F} = I(\mathbf{L} \times \mathbf{B})$

$$= I(L\hat{\mathbf{j}} + B\hat{\mathbf{k}}) = BIL(\hat{\mathbf{i}})$$

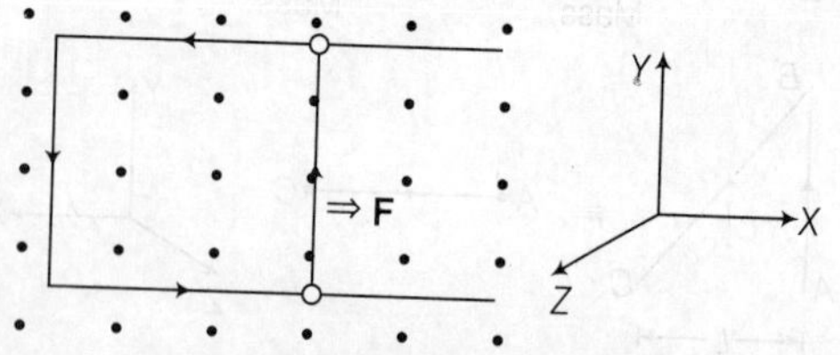

or $\quad |\mathbf{F}| = BIL$ (towards right)

21. (d) Let v be the final velocity attained at $x = d$ under the influence of constant force $F = BIL$

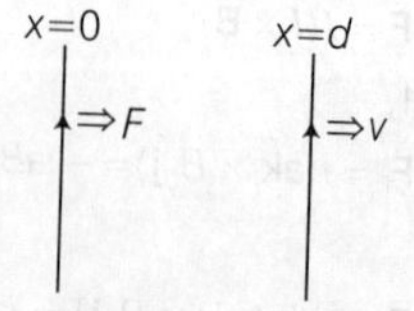

$\therefore$ Acceleration, $a = \dfrac{F}{m} = \dfrac{BIL}{m}$

Using $\quad v^2 = u^2 + 2as,$

$\Rightarrow \quad v^2 = 0 + 2 \times \dfrac{BIL}{m} d$

or $\quad d = \dfrac{v^2 m}{2BIL}$

22. (2) Current divides across branches as shown in the figure,

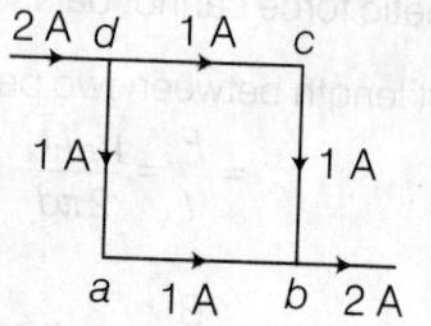

Magnetic field $= BIl$ (for all four sides)

$$= 0.1 \times 1 \times 20 \times 10^{-2} = 2 \times 10^{-2}\ \text{N}$$

Thus, value of $\quad x = 2$

23. (4) Force on rod, $F = I(\mathbf{d} \times \mathbf{B}) = IdB\,(\hat{\mathbf{i}})$

By work-energy theorem

KE (Translation) + KE (Rotation) = Work done

$\Rightarrow \quad F \times L = \dfrac{1}{2} mv^2 + \dfrac{1}{2}(I\omega^2)$

$\Rightarrow \quad IdBL = \dfrac{1}{2} mv^2 + \dfrac{1}{2}\left(\dfrac{1}{2} mR^2\right)\left(\dfrac{v}{R}\right)^2$

$\Rightarrow \quad IdBL = \dfrac{3}{4} mv^2$

or $\quad v = \sqrt{\left(\dfrac{4IdBL}{3m}\right)}$

$$= \sqrt{\frac{4 \times 16 \times (10 \times 10^{-2}) \times 0.5 \times 1}{\left(3 \times \frac{1}{3}\right)}}$$

$$= \sqrt{64 \times \frac{1}{2} \times 10^{-1}} = 4 \times \frac{1}{\sqrt{5}}\ \text{ms}^{-1}$$

24. (5) Force on a small element of wire is

$$d\mathbf{F} = i\,d\mathbf{x} \times \mathbf{B}$$

$$= \left[iB_0\left(1 + \frac{x}{a}\right)dx\right][\hat{\mathbf{i}} \times \hat{\mathbf{k}}]$$

$$= iB_0\left(1 + \frac{x}{a}\right)dx \cdot (-\hat{\mathbf{j}})$$

So, magnitude of total force on wire will be

$$F_B = \int_{x=a}^{x=2a} dF = \int_{x=a}^{x=2a} iB_0\left(1 + \frac{x}{a}\right)dx$$

$$= iB_0\left(x^2 + \frac{x^2}{2a}\right)\Bigg|_{x=a}^{x=2a}$$

$$= iB_0\left(\frac{5a}{2}\right) \text{ in negative } Y\text{-axis.}$$

25. (a) $\dfrac{F}{l} = \dfrac{\mu_0 I_1 I_2}{2\pi d} = \dfrac{4\pi \times 10^{-7} \times 3 \times 3}{2\pi \times 1.8 \times 10^{-2}} = 10^{-4}\ \text{Nm}^{-1}$

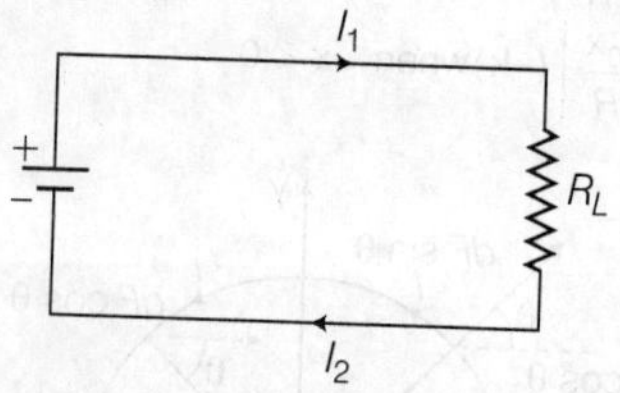

and as currents are in opposite direction, so there is a repulsion between the wires.

$\therefore$ Both Statements I and II are correct and Statement II is correct explanation of Statement I.

26. (d) As magnetic field of current I_1 is parallel to current and length of conductor II and so

$$\mathbf{F} = I(\mathbf{l} \times \mathbf{B}) = 0, \text{ as } \mathbf{I} \parallel \mathbf{B}$$

Hence, Statement I is incorrect and Statement II is correct.

DPP-5 Magnetic Dipole, Magnetic Moment and Torque

1. For a current carrying coil, magnetic dipole moment is

$$\mathbf{M} = NI\mathbf{A}$$

Now, number of turns in coil,

Case I $N = \dfrac{\text{Total length of wire}}{\text{Perimeter of coil}} = \dfrac{12a}{3a} = 4$

Current in coil is $I = \dfrac{V}{R}$

Area of coil, $A = \dfrac{\sqrt{3}}{4} \cdot a^2$

So, **M** magnetic dipole moment associated with the coil,

$$= 4 \times \frac{V}{R} \times \frac{\sqrt{3}}{4} a^2 = \sqrt{3}\,\frac{Va^2}{R}$$

Case II $A = a^2, I = \dfrac{V}{R}$

and $N = \dfrac{12a}{4a} = 3$

$\therefore$ $M = 3 \times \dfrac{V}{R} \times a^2 = \dfrac{3Va^2}{R}$

Case III $A = \dfrac{6\sqrt{3}}{4} a^2, I = \dfrac{V}{R}$

$N = \dfrac{12a}{6a} = 2$

$\therefore$ $M = \dfrac{6\sqrt{3}}{4} a^2 \times 2 \times \dfrac{V}{R} = 3\sqrt{3}\, a^2 \dfrac{V}{R}$

2. PE $= -\mathbf{M} \cdot \mathbf{B} = -MB\cos\theta$

(i) For maximum potential energy $\theta = 180°$, i.e. angle between **M** and **B** must be 180°. So, Case II has maximum PE.

$\Rightarrow$ $(PE)_{max} = -MB\cos 180° = MB$

(ii) For minimum PE,

$\theta = 0°$

So, Case III has minimum PE.

$\Rightarrow$ $(PE)_{min} = -MB\cos 0° = -MB$

(iii) For zero PE, $\theta = \dfrac{\pi}{2}$

So, Case I has zero PE.

$\Rightarrow$ PE $= -MB\cos 90° = 0$

3. (i) The orbiting electron behaves as a current loop of current i.

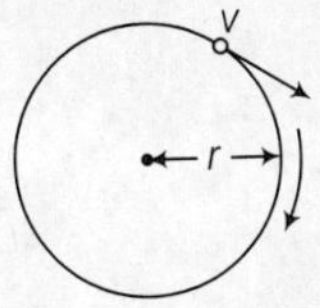

Here, $i = \dfrac{e}{T}$

where, T = time period

$\Rightarrow$ $i = \dfrac{e}{\dfrac{2\pi r}{v}} = \dfrac{ev}{2\pi r}$

(ii) Magnetic moment, $M = iAN$

where, N = number of turns of the current loop

i = current

A = area of the loop.

$\Rightarrow$ $M = \left(\dfrac{ev}{2\pi r}\right) \times (\pi r^2) \times (1) = \dfrac{evr}{2}$

4. Given, $\mathbf{M} = 4\hat{\mathbf{i}} - 3\hat{\mathbf{j}}$ and $\mathbf{B} = 3\hat{\mathbf{i}} + 4\hat{\mathbf{j}}$

(i) $\tau = \mathbf{M} \times \mathbf{B}$

$$\tau = \begin{vmatrix} \hat{\mathbf{i}} & \hat{\mathbf{j}} & \hat{\mathbf{k}} \\ 4 & -3 & 0 \\ 3 & 4 & 0 \end{vmatrix} = \hat{\mathbf{k}}\,[16 - (-9)] = 25\,\hat{\mathbf{k}}$$

(ii) Potential energy $= -\mathbf{M} \cdot \mathbf{B} = -(4\hat{\mathbf{i}} - 3\hat{\mathbf{j}}) \cdot (3\hat{\mathbf{i}} + 4\hat{\mathbf{j}})$

or PE $= -[12 - 12] = 0$

5. (b) $M = IA = I$ (Area)

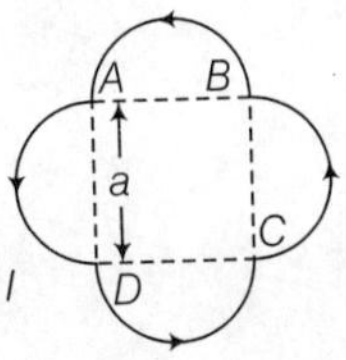

where, Area (A) = Area of square $ABCD$

+ 4 (Area of semi-circle)

$$= \left(a^2 + \frac{2\pi a^2}{4}\right) I = a^2 \cdot \left(1 + \frac{\pi}{2}\right) I$$

6. (a) PE $= -MB\cos\theta$

where, θ is the angle between area vector and B.

Case I $\theta = 180°$ $\Rightarrow$ PE $= MB$ (max)

Case II $\theta = 90°$ $\Rightarrow$ PE $= 0$

Case III $\theta > 90°$, i.e. obtuse

$\Rightarrow \cos\theta$ = negative or PE = positive

Case IV $\theta < 90°$, i.e. acute

$\Rightarrow$ $\cos\theta$ = positive or PE = negative

7. (a) $\tau = I\alpha$

$iAB = I\alpha$ $\quad (\because \tau = \mathbf{M} \times \mathbf{B})$

$i\pi R^2 B = \dfrac{MR^2}{2}\alpha$

or $\alpha = \dfrac{2i\pi B}{M} = 40\,\pi$ rad s^{-2} $\quad \left(\because I = \dfrac{MR^2}{2}\right)$

8. (b) $M = IA$

$I = \dfrac{qv}{2\pi r}$ $\quad \left(\because I = \dfrac{\Delta Q}{\Delta t} = \dfrac{q}{2\pi r / v}\right)$

So, $M = IA = \dfrac{qv}{2\pi r} \times \pi r^2 = \dfrac{qvr}{2}$

9. (b) $\tau = NIAB\sin\theta$

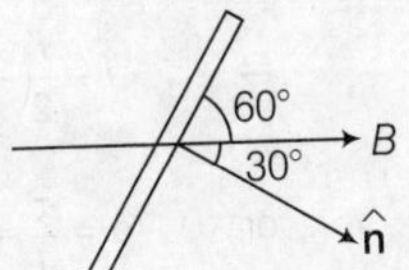

$[\tau = \mathbf{M} \times \mathbf{B}$ or $|\,\tau\,| = MB\sin\theta]$

$\Rightarrow$ $\tau = (2n)\,(2l) \times (2a) \times B\sin 30°$

$= 4nlB$ or $8Bnal\cos 60°$

10. (a) Area vector will be perpendicular to plane. Area vector is at 60° angle with X-axis and 30° angle with $-Z$-axis.

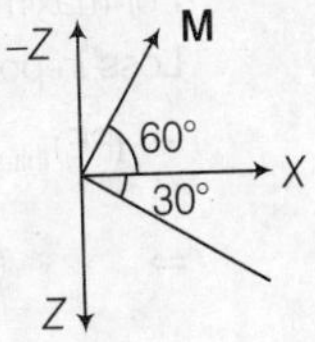

$\mathbf{M} = M\cos 60\hat{\mathbf{i}} - M\sin 60°\,\hat{\mathbf{k}}$

$\Rightarrow$ $\mathbf{M} = \left(\dfrac{M}{2}\right)\hat{\mathbf{i}} - \left(\dfrac{M\sqrt{3}}{2}\right)\hat{\mathbf{k}} = \dfrac{M}{2}\,[\hat{\mathbf{i}} - \sqrt{3}\,\hat{\mathbf{k}}]$

$= 0.05\,[\hat{\mathbf{i}} - \sqrt{3}\,\hat{\mathbf{k}}]$ A-m^2

where, $M = iAN = 10 \times (0.1)^2 = 0.1$ A-m^2

11. (c) Taking moments about B to be zero.

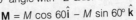

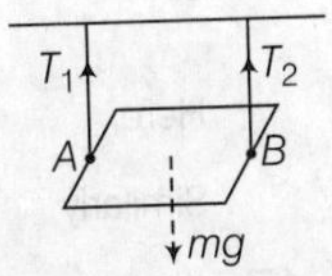

$T_1 l + IlB = mg\,\dfrac{l}{2}$

$T_1 = \dfrac{mg - 2IbB}{2}$

12. (a) The gravitational torque must be counter balanced by the magnetic torque about O for the equilibrium of the sphere.

$\tau_m = \pi I r^2 B\sin\theta$

$\tau_{gravity} = mgr\sin\theta$

For equilibrium, $\tau_m = \tau_{gravity}$

or $\pi I r^2 B\sin\theta = mgr\sin\theta$

or $B = \dfrac{mg}{\pi I r}$

13. (b) $W = -\mathbf{M} \cdot \mathbf{B} = -MB\cos\theta$

$\Delta W = W_f - W_i = -(-MB) - (-MB) = 2MB$

$= 2 \times (iAN) \times B = 2 \times 0.5 \times \pi \times (0.1)^2 \times 100 \times 2$

$= 2\pi$ J

14. (a,b,d) $\tau = \mathbf{M} \times \mathbf{B}$

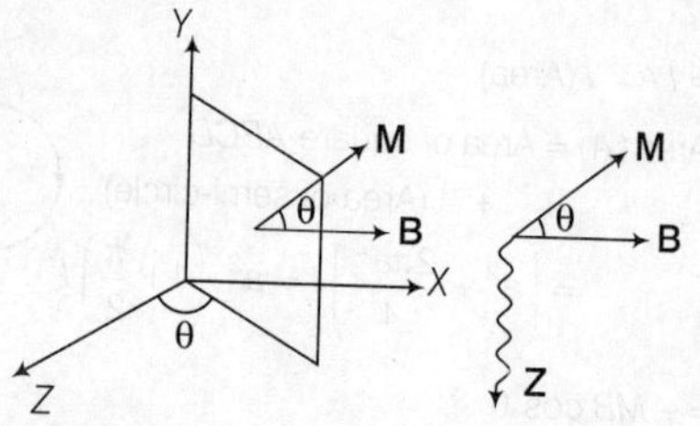

$$|\tau| = |\mathbf{M}||\mathbf{B}|\sin\theta = iAB_0 \sin\theta = i(ab)B_0\sin\theta$$

τ is in the direction of $\mathbf{M} \times \mathbf{B}$, i.e. along the negative Y-direction.

Since, τ is in negative Y-direction, the direction of rotation of loop is such that θ decreases.

15. (b,d) The magnetic dipole moment of the current carrying coil is given by

$$\mathbf{M} = NiA\hat{\mathbf{n}} = 100 \times 0.5 \times (0.08 \times 0.04)\,\hat{\mathbf{i}}$$

or $\mathbf{M} = 16 \times 10^{-2}(\hat{\mathbf{i}})$ A-m^2

The torque acting on the coil is

$$\tau = \mathbf{M} \times \mathbf{B} = MB\,(\hat{\mathbf{i}} \times \hat{\mathbf{j}}) = 1.6 \times 10^{-2} \times \frac{0.05}{\sqrt{2}}\,\hat{\mathbf{k}}$$

or $\quad \tau = 5.66 \times 10^{-5}$ (N-m) $\hat{\mathbf{k}}$

16. (a,b) $\tau = \mathbf{M} \times \mathbf{B}$

At $t = 0$, $\quad \tau = (3\hat{\mathbf{i}} - 4\hat{\mathbf{j}}) \times (4\hat{\mathbf{i}} + 3\hat{\mathbf{j}}) = 25\,\hat{\mathbf{k}}$

$$|\tau| = \tau = 25$$

Since, angular acceleration $\alpha = \dfrac{\tau}{I}$

where, I = moment of inertia about axis of rotation, i.e. about direction parallel to τ.

$$\Rightarrow \quad I = \frac{I'}{2} = \frac{2 \times 10^{-2}}{2} = 1 \times 10^{-2} \text{ kg-m}^2$$

(from perpendicular axis theorem)

or $\quad \alpha = \dfrac{\tau}{I} = \dfrac{25}{10^{-2}}$ rad/s^2 = 2500 rad s^{-2}

For maximum angular velocity,

Loss in potential energy = Gain in kinetic energy (rotational)

$$\Rightarrow (PE)_{initial} - (PE)_{final} = \frac{1}{2} I\omega_{max}^2$$

$$\Rightarrow \quad (0) - (-MB) = \frac{1}{2} I\omega_{max}^2$$

$$\Rightarrow \quad MB = \frac{1}{2} I\omega_{max}^2 \quad \text{or} \quad \omega_{max}^2 = \frac{2MB}{I}$$

or $$\omega_{max} = \sqrt{\frac{2MB}{I}} = \sqrt{\frac{2 \times 5 \times 5}{1 \times 10^{-2}}} = 50\sqrt{2} \text{ rad s}^{-1}$$

Here, $\quad |\mathbf{M}| = M = |3\hat{\mathbf{i}} - 4\hat{\mathbf{j}}| = \sqrt{3^2 + 4^2} = 5$

Similarly, $\quad |\mathbf{B}| = B = |4\hat{\mathbf{i}} + 3\hat{\mathbf{j}}| = \sqrt{4^2 + 3^2} = 5$

17. (b) $\tau = \mathbf{M} \times \mathbf{B}$

$$\mathbf{M} = IA\hat{\mathbf{k}}$$

$$\mathbf{B} = B\cos 45^\circ\,\hat{\mathbf{i}} + B\sin 45^\circ\,\hat{\mathbf{j}}$$

$$\Rightarrow \quad \tau = \frac{IAB}{\sqrt{2}}[\hat{\mathbf{k}} \times (\hat{\mathbf{i}} + \hat{\mathbf{j}})] = \frac{IAB}{\sqrt{2}}[\hat{\mathbf{j}} - \hat{\mathbf{i}}]$$

$$= \frac{I_0 L^2 B}{\sqrt{2}}[-\hat{\mathbf{i}} + \hat{\mathbf{j}}]$$

or $\quad |\tau| = \tau = I_0 L^2 B$

Thus, τ is acting along $(-\hat{\mathbf{i}} + \hat{\mathbf{j}})$ or SQ.

$$\tau \perp \mathbf{B}$$
$$\tau \perp \mathbf{M}$$

direction of torque is given by right hand cross-product rule.

18. (a) $\theta = \omega_0 t + \dfrac{1}{2}\alpha t^2$ ...(i)

Here, $\omega_0 = 0$ and $\alpha = \dfrac{\tau}{I}$

$$\tau = |\mathbf{M} \times \mathbf{B}| = \left|\frac{I_0 L^2 B}{\sqrt{2}}(-\hat{\mathbf{i}} + \hat{\mathbf{j}})\right| = \frac{I_0 L^2 B}{\sqrt{2}} \times \sqrt{2} = I_0 L^2 B \quad \text{...(ii)}$$

I = moment of inertia of coil $PQRS$ about SQ

$$= \frac{M' L^2}{6} = \frac{4ML^2}{6} \text{ or } \frac{2ML^2}{3} \quad \text{...(iii)}$$

Substituting Eqs. (ii) and (iii) in Eq. (i), we get

$$\theta = \frac{1}{2} \times \frac{I_0 L^2 B}{(2ML^2/3)} \times (\Delta t)^2 = \frac{3I_0 L^2 B}{4ML^2}(\Delta t)^2 = \frac{3I_0 B}{4M}(\Delta t)^2$$

19. (b)

(i) As, the current in XY-plane is anti-clockwise, so moment will be along Z-axis by right hand thumb rule.

(ii) $\tau = \mathbf{M} \times \mathbf{B}$

$$\Rightarrow |\tau| = |\mathbf{M}||\mathbf{B}|\sin\theta = \frac{5 \times \pi \times (0.1)^2 \times 100\sqrt{2}}{\sqrt{2}}$$

(here, $\theta = 45^\circ$)

(iii) Net force on a closed loop carrying current in a uniform magnetic field is zero.

Hence, (i) → (q,r), (ii) → (q,s), (iii) → (p)

20. (4) Magnetic field at a distance (r = 2 cm)

$$= B_0 = \frac{\mu_0 I_1}{2\pi r} = \frac{\mu_0 \times 2}{2\pi \times 2 \times 10^{-2}} = \frac{50\mu_0}{\pi}\text{ T}$$

Magnetic moment of the current carrying loop = $M = iAN$

$$= 2 \times \pi \times (2 \times 10^{-2})^2 \times 100$$
$$= 2 \times \pi \times 4 \times 10^{-4} \times 100 = 8 \times 10^{-2}\pi$$

⇒ Torque acting on the loop = $|\tau| = |\mathbf{M} \times \mathbf{B}| = |\mathbf{M}||\mathbf{B}|\sin\theta$

$$= MB_0 \sin 90^\circ = MB_0$$

or $$\tau = 8 \times 10^{-2}\pi \times \frac{50\mu_0}{\pi} = 400 \times 10^{-2}\mu_0 = 4\mu_0$$

Thus, value of $x = 4$

21. (6)

$$U_{min} = -\mu B\cos 0^\circ$$
$$= (-10 \times 10^{-3}\text{ A-m}^2)(55 \times 10^{-6}\text{ T})$$
$$= -55 \times 10^{-8} = -5.5 \times 10^{-7}\text{ J}$$
$$U_{max} = -\mu B\cos 180^\circ$$
$$= (-10 \times 10^{-3})(55 \times 10^{-6}) \times -1$$
$$= 5.5 \times 10^{-7}\text{J}$$
$$U_{max} - U_{min} = 2 \times 5.5 \times 10^{-7}$$
$$= 10.0 \times 10^{-7} = 10^{-6}\text{ J}$$

22. (2) Sphere is in translational equilibrium.

$\therefore \quad f_s - mg\sin\theta = 0$ ...(i)

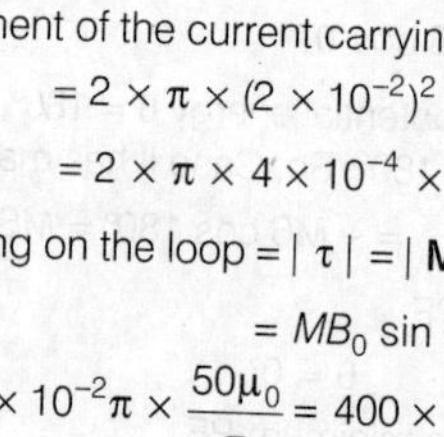

The sphere is also in rotational equilibrium.

$\therefore \quad f_s R - \mu B\sin\theta = 0$ ...(ii)

From Eqs. (i) and (ii), we get

$$\mu B = MgR$$

and $\quad \mu = NI\pi R^2$

So, $$I = \frac{Mg}{\pi NBR} = \frac{0.1 \times 10}{\pi \times 5 \times 0.5 \times 0.2} = \frac{2}{\pi}\text{ A}$$

DPP-6 Moving Coil Galvanometer and Its Conversion

1. (i) The torque acting on a closed loop having n turns is given by

$$\tau = \mathbf{M} \times \mathbf{B}$$

where, $\mathbf{M}$ = magnetic moment vector

Also, $\mathbf{M} = ni\mathbf{A}$

$\Rightarrow$ $\tau = ni(\mathbf{A} \times \mathbf{B})$

τ is directed in the direction of $\mathbf{A} \times \mathbf{B}$ obtained using cross-product rule. Also, magnitude of torque is

$$|\tau| = ni|\mathbf{A} \times \mathbf{B}|$$
$$= niAB\sin\theta$$

where, θ = angle between **A** and **B**.

(ii) The pole pieces are made cylindrical. As a result, the magnetic field at the arms of the coil remains parallel to the plane of the coil everywhere even as coil rotates.

Hence, $\theta = 90^\circ$ (i.e. $\mathbf{A} \perp \mathbf{B}$)

So, $\tau = niAB$

(iii) The restoring torque = $k\theta$

The coil will stay at deflection θ,

Deflecting torque = Restoring torque

$\Rightarrow$ $niAB = k\theta$

2. (i) Maximum current that can be measured, i.e.

I_g = Current sensitivity × number of divisions

$= I_s \times n = nI_s$

(ii) Full scale deflection,

$$I_g = 20 \times 30$$
$$= 600\mu A = 0.6 \text{ mA}$$

(iii) When only number of turns are increased, ratio $\frac{N}{R}$ remains nearly constant as resistance of coil of galvanometer also increases.

Hence, the voltage sensitivity $S_v = \frac{BNA}{kR}$ remains same.

But current sensitivity increases as $S_i = \frac{BNA}{k}$

Voltage sensitivity = $\frac{\text{Current sensitivity}}{R} = \frac{NAB}{kR}$

3. (i) Resistance per volt is another way of specifying the current at full scale deflection. The grading of 5000 ΩV^{-1} at full scale deflection means that current required at full scale deflection is

$$I_g = \frac{V}{R} = \frac{1}{5000} \text{ A} = 0.2 \text{ mA}$$

(ii) When $V_{AB} = 20$ V,

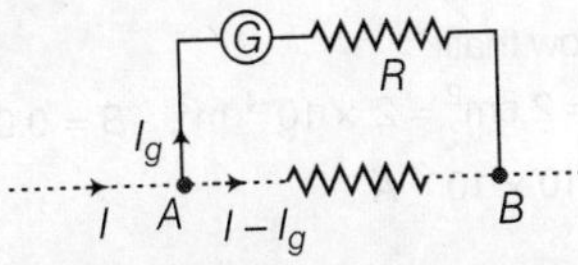

$$I_g(G + R) = 20$$

$\Rightarrow$ $0.2 \times 10^{-3}(G + R) = 20$

$\Rightarrow$ $R = 1 \times 10^5 - G$

$$= 1 \times 10^5 - 5 \times 5000$$
$$= 75000 \ \Omega$$

So for the purpose, a 75000 Ω resistor is to be put in the series with galvanometer and its grading is 75000/20 = 3750 ΩV^{-1}

(iii) The higher resistance per volt of the meter, the lesser is the current it draws from the circuit and better it is. So, 5000 ΩV^{-1} is more accurate than 2000 ΩV^{-1}.

4. (i) As we know that in the moving coil galvanometer when a coil rotates, then the suspension wire gets twisted. Now, a restoring torque is developed in the suspension wire.

In equilibrium,

Deflection torque = Restoring torque

$$\tau = MB = ki$$

$\therefore$ $k = \frac{MB}{i} = \frac{(NiA)B}{i} = NBA$

(ii) Given, for current i deflection produced is $\pi/2$, then torque on moving coil galvanometer is

$$\tau = k\theta = BiNA$$

$\therefore$ $k = \frac{2NiBA}{\pi}$ $(\because \theta = \pi/2)$

(iii) Angular impulse = $\int \tau dt = \int NABidt$

$\Rightarrow$ $\int \tau dt = NAB\int idt = NABQ$

Also, $\int \tau dt = I\omega_0 = NABQ$

$\Rightarrow$ $\omega_0 = \frac{NABQ}{I}$

Using energy conservation, at maximum deflection whole kinetic energy (rotational) will be converted into potential energy of a spring.

$\Rightarrow$ $\frac{1}{2}I\omega_0^2 = \frac{1}{2}k\theta_{max}^2$

or $\theta_{max} = \omega_0\sqrt{\frac{I}{k}}$

On substituting the values, we get

$$\theta_{max} = Q\sqrt{\frac{NAB\pi}{2Ii}} \quad \left(\because k = \frac{2NiBA}{\pi}\right)$$

where, I is the moment of inertia.

5. (c) If potential difference across C and D is 20 V and r is the resistance of ammeter, then

$$R + r = \frac{20}{4} = 5$$
$$R = 5 - r, R < 5 \ \Omega$$

6. (d) Given, if a shunt of $\frac{1}{10}$th of the coil resistance is applied to a moving coil galvanometer, then

$$\frac{I_g}{I} = \frac{S}{S + G}$$

$$\frac{I_g}{I} = \frac{(G/10)}{(G/10 + G)} = \frac{1}{11}$$

initially, sensitivity of a galvanometer,

$$\alpha_i = \theta/I_g \quad \text{...(i)}$$

finally after shunt, it becomes

$$\alpha_f = \theta/I \quad \text{...(ii)}$$

On dividing Eq. (ii) by Eq. (i), we get

$\therefore$ $\frac{\alpha_f}{\alpha_i} = \frac{\theta/I}{\theta/I_g} = \frac{1}{11}$

7. (b) We know that resistance across voltmeter

$$R = \frac{V'}{I_g} - G$$

The voltmeter gives full scale deflection for PD of V volts. Its resistance is G.

Hence, $I_g = V/G$, given that $V' = nV$

$$R = \frac{nV}{(V/G)} - G = (n - 1)G$$

8. (d) Given, $I_g = 1$ A, $G = 0.81\,\Omega$, $I = 10$ A

So, value of the required shunt,

$$S = \frac{I_g G}{I - I_g} = \frac{1 \times 0.81}{10 - 1} = 0.09\,\Omega$$

9. (b) Resistance of the galvanometer,

$$G = \frac{\text{Current sensitivity}}{\text{Voltage sensitivity}} = \frac{10}{2} = 5\,\Omega$$

Number of divisions on the galvanometer scale, $n = 150$.

Current required for full scale deflection,

$$I_g = \frac{n}{\text{Current sensitivity}} = \frac{150}{10} = 15 \text{ mA}$$

$$= 15 \times 10^{-3} \text{ A}$$

Required range of voltmeter $= 150 \times 1 = 150$ V

Required series resistance

$$R = \frac{V}{I_g} - G = \frac{150}{15 \times 10^{-3}} - 5 = 9995\,\Omega$$

10. (a) Here, current across ammeter,

$$I_g = 0.5\% \text{ of } I = 0.005\, I$$

And current across shunt resistance,

$$I_s = I - I_g = I - 0.005\, I = 0.995 I$$

$\therefore$ Resistance across shunt,

$$S = \frac{I_g G}{I_s} = \frac{0.005\, I \times G}{0.995\, I} = \frac{G}{199}$$

Thus, resistance of ammeter will be

$$= \frac{GS}{G + S} = \frac{\frac{G}{199} \times G}{G + \frac{G}{199}} = \frac{G}{200}$$

11. (b) Given, $I_g = 25 \times 4 \times 10^{-4} = 10^{-2}$ A

$G = 100\,\Omega$

Resistance across voltmeter,

$$R = \frac{V}{I_g} - G = \frac{2.5}{10^{-2}} - 100 = 150\,\Omega \text{ in series}$$

12. (b) Here, $I_1 = 200\,\mu$A, $\theta_1 = 30°$, $\theta_2 = \frac{\pi}{10}$ rad $= 18°$, $I_2 = ?$

So, $$I_1 = \frac{k}{NBA}.\theta_1 \text{ and } I_2 = \frac{k}{NBA}.\theta_2$$

$$\Rightarrow \quad \frac{I_2}{I_1} = \frac{\theta_2}{\theta_1}$$

$$\Rightarrow \quad I_2 = \frac{\theta_2}{\theta_1} \times I_1 = \frac{18}{30} \times 200 = 120\,\mu\text{A}$$

Current sensitivity of the galvanometer,

$$= \frac{\theta_2}{I_2} = \frac{18°}{120\,\mu\text{A}} = 0.15°\,\mu\text{A}^{-1}$$

13. (b) Without shunt, $I_g = I = 55k$

where, k is the figure of merit of the galvanometer.

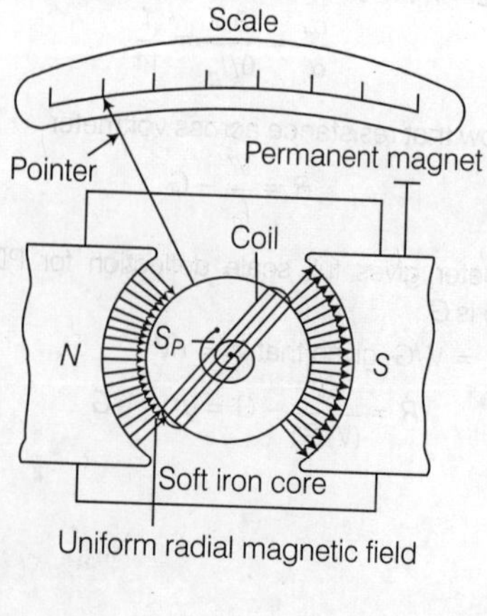

With shunt,

$$I'_g = (55 - 45)\, k = 10k$$

$$\therefore \quad \frac{I'_g}{I_g} = \frac{10\,k}{55\,k} = \frac{2}{11}$$

$$I'_g = \frac{2}{11} I_g = \frac{2}{11} I$$

When shunt of $8\,\Omega$ is used,

$$I'_g = \frac{R_s}{R_g + R_s} \times I$$

or $$\frac{8}{R_g + 8} \times I = \frac{2}{11} I$$

or $$\frac{8}{R_g + 8} \times I = \frac{2}{11} I$$

$$\Rightarrow \quad 88 = 2\, R_g + 16$$

$$\Rightarrow \quad R_g = \frac{72}{2} = 36\,\Omega$$

14. (d) Given, $R' = 2R$ and $I'_s = \frac{3}{2} I_s$

$$\because \quad \text{Voltage sensitivity} = \frac{\text{Current sensitivity}}{R}$$

$$\Rightarrow \quad V'_s = \frac{I'_s}{R'} = \frac{3}{4}\left(\frac{I_s}{R}\right) = \frac{3}{4}(V_s) = 0.75 V_s$$

Hence, voltage sensitivity decreases by a factor of $\frac{1}{4}$.

15. (c) $G = 12\,\Omega$, $I_g = 2.5 \times 10^{-3}$ A

For ammeter, $$S = \frac{G I_g}{I - I_g} = \frac{12 \times 2.5 \times 10^{-3}}{7.5 - 2.5 \times 10^{-3}}$$

$$\Rightarrow \quad S = 4 \times 10^{-3}\,\Omega$$

Net resistance of ammeter R_a is

$$\frac{1}{R_a} = \frac{1}{G} + \frac{1}{S}$$

$$\Rightarrow \quad R_a = \frac{SG}{S + G} = 4.0 \times 10^{-3}\,\Omega$$

For galvanometer as voltmeter,

$$R = \frac{V}{I_g} - G = \frac{10.0}{2.5 \times 10^{-3}} - 12 = 3988\,\Omega$$

Let R_v be the voltmeter resistance.

or $$R_v = R + G = 4000\,\Omega$$

Ratio of ammeter and voltmeter resistances,

$$\frac{R_a}{R_v} = \frac{4 \times 10^{-3}}{4 \times 10^3} = 10^{-6}$$

16. (b) As, we know that

$N = 100$, $A = 2 \text{ cm}^2 = 2 \times 10^{-4} \text{ m}^2$, $B = 0.01$ T

$I = 10 \text{ mA} = 10 \times 10^{-3}$ A

$\theta = 0.05$ rad

So, torsional constant of the spiral spring,

$$k = \frac{NABi}{\theta}$$

$$= \frac{100 \times 2 \times 10^{-4} \times 0.01 \times 10 \times 10^{-3}}{0.05}$$

$$= 4 \times 10^{-5} \text{ N-m rad}^{-1}$$

17. (b) Given, $G = 100\,\Omega$, $I_g = 10^{-5}$ A, $I = 1$ A, $S = ?$

As, $$I_g \times G = (I - I_g) \times S$$

$$\therefore \quad S = \frac{I_g}{I - I_g} G = \frac{10^{-5}}{1 - 10^{-5}} \times 100$$

$$S = \frac{10^{-3}}{1 - 0.00001} = 10^{-3}\,\Omega$$

18. (a) In moving coil galvanometer,

$$G = 100\,\Omega, S = 0.1\,\Omega$$
$$I_g = 100\,\mu\text{A} = 100 \times 10^{-6}\text{ A}$$

So, minimum current in the circuit so that ammeter shows maximum deflection.

$$I = \frac{G + S}{S} \times I_g = \left[\frac{100 + 0.1}{0.1}\right] \times 100\,\mu\text{A}$$
$$= 100100\,\mu\text{A} = 100.1\text{ mA}$$

19. (a,b) The galvanometer cannot as such be used as an ammeter to measure the value of the current in a given circuit. This supports two reasons

(i) galvanometer is a very sensitive device, it gives a full scale deflection for a current of the order of μA.

(ii) For measuring currents, the galvanometer has to be connected in series and as it has a large resistance, this will change the value of the current in the circuit.

20. (b,c) For $S = 200\text{ k}\Omega$

Potential drop across voltmeter,

$$V = I_g\,(G + S)$$
$$= 50 \times 10^{-6}\,(100 + 200000) = 10\text{ V}$$

For $S = 1\,\Omega$,

Current flows across ammeter,

$$I = \left[\frac{G + S}{S}\right] I_g$$
$$= \left[\frac{100 + 1}{1}\right] \times 50 \times 10^{-6}$$
$$= 5\text{ mA}$$

21. (d) For a DC ammeter or voltmeter, a moving coil galvanometer is used. So, deflection of coil $\theta = \frac{BNA}{k} I$

or $\theta \propto I$

Also, $\theta = \frac{BNA}{kR} IR$

$$= \frac{BNA}{kR} V \quad \text{or} \quad \theta \propto V$$

For an AC ammeter or voltmeter, a hot wire deflection instrument is used.

$\therefore$ $\theta = kI^2R$

$\Rightarrow$ $\theta \propto I^2$

and $\theta = k\frac{V^2}{R}$

$\Rightarrow$ $\theta \propto V^2$

Hence, (i) $\rightarrow$ (s), (ii) $\rightarrow$ (r), (iii) $\rightarrow$ (q), (iv) $\rightarrow$ (p).

22. (a) When a uniform field is used,

$$\tau = BINA\cos\theta = k\theta$$

$\Rightarrow$ $\frac{\theta}{I} = BNA\cos\theta$

$\therefore$ In a uniform field, torque decreases as θ increases. So, scale will be non-linear in case (a).

23. (a) $\tau_1 = NBIA\cos\theta$

$$\tau_2 = NBIA\cos(90° + \theta) = NBIA\sin\theta$$

$\Rightarrow$ $\tau_1^2 + \tau_2^2 = (NBIA)^2(\cos^2\theta + \sin^2\theta)$

$\Rightarrow$ $B = \frac{\sqrt{\tau_1^2 + \tau_2^2}}{NIA}$

24. (d) A non-linear instrument's scale is difficult to read and calibrate e.g. when the pointer is between 1 and 2, it can be read as 1.5 but the true value may be 1.4 mA.

25. (5) If a voltmeter has resistance G and reads V volt at full scale, then it will be graded as $\frac{G}{V} = 5000$

$\Rightarrow$ $G = 5 \times 5000$

$\Rightarrow$ $G = 25000\,\Omega$

$\therefore$ $I_g = \frac{5}{25000} = 2 \times 10^{-4}\text{ A}$

If resistance R is connected in series to make range up to 20 V, then

$$I_g = \frac{V}{R + G}$$
$$2 \times 10^{-4} = \frac{20}{R + 25000}$$
$$R = 75000\,\Omega$$

Resistance of new voltmeter will be more

$$= 75000 + 25000$$
$$= 100000\,\Omega$$

So, resistance per volt of the new voltmeter,

$$= \frac{100000}{20}$$
$$= 5000\,\Omega/\text{V} = 5 \times 10^3\,\Omega/\text{V}$$

26. (4)

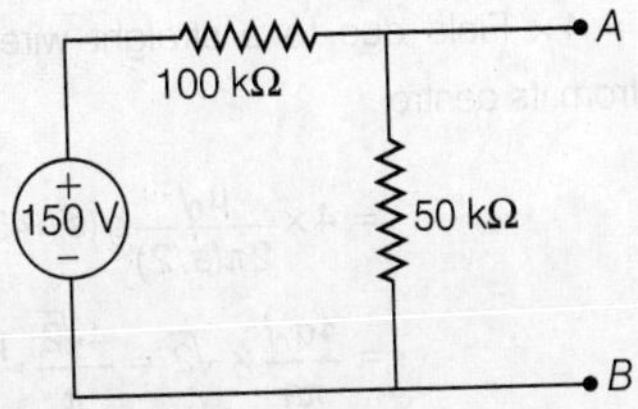

Potential difference between points A and B = potential drop across 50 kΩ resistance

$\Rightarrow$ $V_{AB} = 50\text{ k}\Omega \times \frac{150\text{ V}}{(100 + 50)\text{ k}\Omega}$

$$= 50\text{ V}$$

Now, voltmeter has a sensitivity of 1 kΩ/V, hence resistance offered by voltmeter,

$$R = \frac{1\text{ k}\Omega}{\text{V}} \times 50 = 50\text{ k}\Omega$$

So, the circuit is

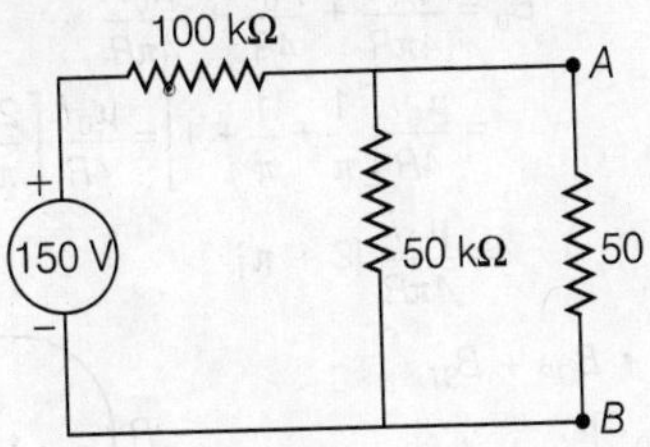

The equivalent resistance of 2 – 50 kΩ resistances in parallel is

$$\frac{1}{R_{eq}} = \frac{1}{50} + \frac{1}{50}$$

$\Rightarrow$ $R_{eq} = 25\text{ k}\Omega$

So, $V_{AB} = \frac{25 \times 150}{125} = 30\text{ V}$

$\therefore$ $\%\text{ Error} = \frac{50 - 30}{50} \times 100$

$$= 40\%$$
$$= 0.4 = 0.1 \times 4$$

Revisal Problems (JEE Main)

1. (b) $dB = \frac{\mu_0 i dl \sin\theta}{4\pi r^2}$

When $\theta = 90°$, then dB will be maximum.

2. (d) $B = \frac{\mu_0}{2\pi r}$. At null-point, the value of B must be equal to the horizontal component of earth's magnetic field (B_H) but its direction must opposite to that of B_H.

$$\therefore \quad H = \frac{\mu_0 i}{2\pi r}$$

$$\Rightarrow \quad 2 \times 10^{-5} = \frac{4\pi \times 10^{-7} \times 30}{2 \times \pi \times r}$$

$$\Rightarrow \quad r = 0.3 \text{ m}$$

3. (b) $B_2 > B_1$

For a circular coil of radius r,

$$B_{\text{centre}} = B_1 = \frac{\mu_0 l}{2r} = \frac{\pi\mu_0 l}{2\pi r}$$

As, $2\pi r = L$

$$B_1 = \frac{\pi\mu_0 l}{L} \text{ T}$$

For a square coil of side a,

$B_2 = B_{\text{centre}} = 4 \times$ Field due to a straight wire of length a at a distance $\frac{a}{2}$ from its centre.

$$= 4 \times \frac{\mu_0 l}{2\pi(a/2)} g(\sin 45° + \sin 45°)$$

$$= \frac{4\mu_0 l}{\pi a} \times \sqrt{2} = \frac{4\sqrt{2}}{\pi} \cdot \frac{\mu_0 l}{a}$$

As, $L = 4a,\ a = \frac{L}{4}$

So, $B_2 = \frac{16\sqrt{2}}{\pi} \cdot \frac{\mu_0 l}{4} \text{T}$

4. (a) $B_O = B_{CD} + B_{DEF} + B_{FG}$,

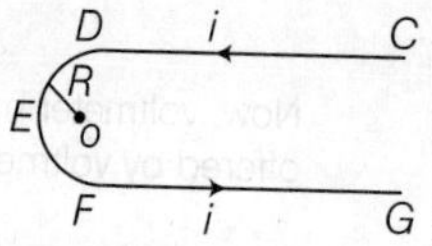

$$B_{CD} = B_{FG} = \frac{1}{2}\left[\frac{\mu_0 l}{2\pi R}\right],$$

$$B_{DEF} = \frac{1}{2}\left[\frac{\mu_0 l}{2R}\right]$$

$$\therefore \quad B_O = \frac{\mu_0}{4\pi R} + \frac{\mu_0 i}{4R} + \frac{\mu_0 i}{4\pi R}$$

$$= \frac{\mu_0 i}{4R}\left[\frac{1}{\pi} + \frac{1}{\pi} + 1\right] = \frac{\mu_0 i}{4R}\left[\frac{2}{\pi} + 1\right]$$

$$= \frac{\mu_0 i}{4\pi R}[2 + \pi]$$

5. (a) $B_O = B_{PQ} + B_{QR} + B_{ST}$,

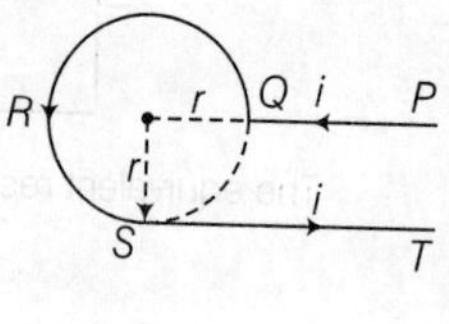

$$B_{ST} = \frac{1}{2}\left[\frac{\mu_0 i}{2\pi r}\right]$$

$$B_{PQ} = 0,\ B_{QRS} = \frac{3}{4} \times \frac{\mu_0 i}{2\pi} \times 2\pi r$$

$$B_O = \frac{\mu_0 i}{4\pi r} + \frac{3}{4} \cdot \frac{\mu_0 i}{2r}$$

$$= \frac{\mu_0 i}{4r}\left[\frac{3}{2} + \frac{1}{\pi}\right] = \frac{\mu_0 i}{4\pi r}\left[\frac{3\pi}{2} + 1\right] \text{T}$$

6. (c) Field at centre of a complete coil $= B = \frac{\mu_0 l}{2r}$ T

An arc of central angle $\frac{\pi}{4}$ is $\frac{1}{8}$th of complete coil.

$\therefore$ Magnetic field due to an arc of angle $\frac{\pi}{4}$ is given by

$$= \frac{1}{8} \times \frac{\mu_0 l}{2r} = \frac{\mu_0 l}{16r} \text{T}$$

7. (a) $l = (2\pi r)\, n$ or $n = \frac{l}{2\pi r}$

$$B = \frac{\mu_0 ni}{2r} = \frac{\mu_0 il}{4\pi r^2}$$

or $$B = \frac{4\pi \times 10^{-7} \times 6.28 \times 1}{2 \times 2 \times \pi \times (0.1)^2} = 6.28 \times 10^{-5} \text{T}$$

8. (a) The field at C due to the straight part of the conductor is

$$B_1 = \frac{\mu_0}{4\pi} \cdot \frac{2i}{r}$$

or $$B_1 = \frac{10^{-7} \times 2 \times 8}{0.1} = 16 \times 10^{-6} \text{ Wb/m}^2$$

acting vertically downwards.

The field at C due to the circular part of the conductor is

$$B_2 = \frac{\mu_0 ni}{2r} = \frac{4\pi \times 10^{-7} \times 1 \times 8}{2 \times 0.1}$$

$$= 16\pi \times 10^{-6} \text{ Wb/m}^2$$

acting vertically upwards.

Thus, the net field at C is

$$B = (16\pi \times 10^{-6} - 16 \times 10^{-6}) \text{ Wb/m}^2$$

acting vertically upwards.

or $$B = 3.424 \times 10^{-5} \text{ Wb/m}^2$$

acting vertically upwards.

9. (c) The potential energy of a magnetic dipole m placed in an external magnetic field is $U = -MB$. Therefore, work done in rotating the dipole is

$$W = \Delta U = 2MB = 2 \times 5.4 \times 10^{-6} \times 0.8$$

$$= 8.6 \times 10^{-6} \text{J} = 8.6\ \mu\text{J}$$

10. (a) $M = NiA = 100 \times 4 \times \pi r^2$

$$= 400 \times 3.14 \times 25 \times 10^{-4} = 3.14 \text{ A-m}^2$$

11. (b) M = Current × Area

$$= i\left(\frac{1}{2}\pi a^2 + \frac{1}{2}\pi b^2\right) = \frac{1}{2} i\pi (a^2 + b^2)$$

12. (b) KE $= qV$

$$\therefore \quad E_K = qV$$

$$\therefore \quad E_K \propto q$$

$$\therefore \quad V = \text{constant}$$

$$E_{K_p} : E_{K_d} : E_{K_\alpha} :: 1:1:2$$

13. (c) $E_K = \frac{q^2 r^2 B^2}{2m}$

$$\therefore \quad E_K \propto \frac{q^2}{m}, \frac{E_{K\alpha}}{E_{K_p}} = \frac{q_\alpha^2}{m_\alpha} \times \frac{m_p}{q_p^2}$$

$$E_{K_\alpha} = \frac{4}{4} \times \frac{1}{1} \times E_{K_p} = 8 \text{ eV}$$

14. (b) $\therefore \quad \mathbf{F} = q(\mathbf{v} \times \mathbf{B}) = 2\, evB \sin 90° = 2\, evB$

15. (a) The electron will pass undeviated if the electric force and magnetic force are equal and opposite. Thus,

$$Ee = Bev \text{ or } B = \frac{E}{v}$$

But $$E = \frac{V}{d}$$

Therefore, $$B = \frac{V}{vd} = \frac{600}{3 \times 10^{-3} \times 2 \times 10^6}$$

$$\therefore \quad B = 0.1 \text{ Wb/m}^2$$

The direction of field is perpendicular to the plane of paper vertically downward.

16. (d) $\mathbf{F} = q(\mathbf{v} \times \mathbf{B})$

$$\mathbf{v} \times \mathbf{B} = \begin{vmatrix} \hat{\mathbf{i}} & \hat{\mathbf{j}} & \hat{\mathbf{k}} \\ 3 & 2 & 0 \\ 5 \times 10^5 & 0 & 0 \end{vmatrix} = \hat{\mathbf{k}}\,(-10 \times 10^5) = \hat{\mathbf{k}}\,(-10^6)$$

$$q = 2e = 2 \times 1.6 \times 10^{-19} = 3.2 \times 10^{-19}\ \text{C}$$

$$\mathbf{F} = 3.2 \times 10^{-19}\,(-\hat{\mathbf{k}} \times 10^6)$$

$$\Rightarrow \quad \mathbf{F} = -3.2 \times 10^{-13}\,\hat{\mathbf{k}}$$

$$\therefore \quad |\mathbf{F}| = 3.2 \times 10^{-13}\ \text{N}$$

17. (b) The component of velocity of the beam of protons, parallel to the field direction

$$= v\cos\theta = 4 \times 10^5 \times \cos 60^\circ = 2 \times 10^5\ \text{ms}^{-1}$$

and the component of velocity of the proton beam at right angle to the direction of field

$$= v\sin\theta = 4 \times 10^5 \times \sin 60^\circ = 2\sqrt{3} \times 10^5\ \text{ms}^{-1}$$

Therefore, the radius of circular path = $(mv\sin\theta / Be)$

or $$r = \frac{1.7 \times 10^{-27} \times 2\sqrt{3} \times 10^5}{0.3 \times 1.6 \times 10^{-19}} = 12.26 \times 10^{-3}\ \text{m}$$

or $$r = 1.226 \times 10^{-2}\ \text{m}$$

Pitch of the helix = $v\cos\theta \times (2\pi m / Be)$

$$\therefore \quad \text{Pitch} = \frac{2 \times 10^5 \times 2 \times 3.14 \times 1.7 \times 10^{-27}}{0.3 \times 1.6 \times 10^{-19}}$$

$$= 44.5 \times 10^{-3}\ \text{m} = 4.45 \times 10^{-2}\ \text{m}$$

18. (b) In order to make a proton circulate the earth along the equator, the minimum magnetic field induction **B** should be horizontal and perpendicular to equator. The magnetic force provides the necessary centripetal force.

i.e. $$qvB = \frac{mv^2}{r} \quad \text{or} \quad B = \frac{mv}{qr}$$

Here, $$m = 1.7 \times 10^{-27}\ \text{kg}$$

$$v = 1.0 \times 10^7\ \text{ms}^{-1}$$

$$q = e = 1.6 \times 10^{-19}\ \text{C}$$

$$r = 6.37 \times 10^6\ \text{m}$$

$$B = \frac{1.7 \times 10^{-27} \times 1.0 \times 10^7}{1.6 \times 10^{-19} \times 6.37 \times 10^6}$$

$$= 1.67 \times 10^{-8}\ \text{Wb/m}^2$$

19. (c) We have $F = qvB = \dfrac{mv^2}{r}$

or $$v = \frac{qBr}{m} = \frac{3.2 \times 10^{-19} \times 1.2 \times 0.45}{6.8 \times 10^{-27}} = 2.6 \times 10^7\ \text{ms}^{-1}$$

The frequency of rotation,

$$n = \frac{v}{2\pi r} = \frac{2.6 \times 10^7}{2 \times 3.14 \times 0.45} = 92 \times 10^6\ \text{s}^{-1}$$

Kinetic energy of α-particle,

$$E_K = \frac{1}{2} \times 6.8 \times 10^{-27} \times (2.6 \times 10^7)^2 = 2.3 \times 10^{-12}\ \text{J}$$

$$= \frac{2.3 \times 10^{-12}}{1.6 \times 10^{-19}}\ \text{eV} = 14 \times 10^6\ \text{eV} = 14\ \text{MeV}$$

If V is accelerating potential of α-particle, then kinetic energy = qV

14×10^6 eV = 2 eV (since, charge on α-particle = $2e$)

$$\therefore \quad V = \frac{14 \times 10^6}{2} = 7 \times 10^6\ \text{V}$$

20. (a) For L length of wire to balance,

$$F_{\text{magnetic}} = mg \Rightarrow ILB = mg$$

Therefore, $$B = \frac{mg}{IL} = (m/L)\,g/I = \frac{45 \times 10^{-3} \times 9.8}{30}$$

$$= 1.47 \times 10^{-2}\ \text{T} = 147\ \text{G}$$

21. (d) The equivalent magnetic moment is

$$M = iA = ef(\pi r^2)$$

But $$f = \frac{v}{2\pi r}$$

$$\therefore \quad M = \frac{ev}{2\pi r}\,\pi r^2 = \frac{evr}{2}$$

22. (a) Magnetic moment on account of orbital motion of an electron

$$M = \frac{evr}{2}$$

From Bohr's quantum condition, $mvr = \dfrac{nh}{2\pi}$, but $n = 1$

$$\therefore \quad vr = \frac{h}{2\pi m}$$

$$\therefore \quad M = \frac{evr}{2} = \frac{eh}{4\pi m}$$

23. (a) $\tau_{max} = MB = niAB = ni(l \times b)\,B$

$$\tau_{max} = 600 \times 10^{-5} \times 5 \times 10^{-2} \times 12 \times 10^{-2} \times 0.10$$

$$= 3.6 \times 10^{-6}\ \text{N-m}$$

24. (a) $W = 2MB = 2\pi i N a^2 B$

or $W = 3.14 \times 2 \times 0.1 \times 100 \times (0.05)^2 \times 1.5 = 0.236$ J

25. (d) Torque acting on coil,

$$\tau = NBIA\sin\theta$$

Here, $$\theta = 0^\circ$$

$$\therefore \quad \tau = 0$$

26. (b) The direction of **F** is along (**v** × **B**) which is towards the right. Thus, the beam deflects to your right side.

27. (b) The particle is moving clockwise which shows that force on the particle is opposite to given by right hand palm rule of Fleming's left hand rule. These two laws are used for positive charge. Here, since laws are disobeyed, we can say that charge is negative.

28. (a) Both Statements I and II are correct and Statement II explains Statement I.

For the Amperian loop *abcda* as current enclosed is zero.

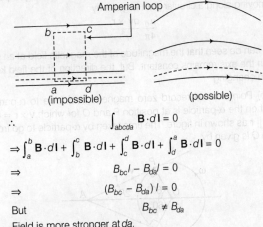

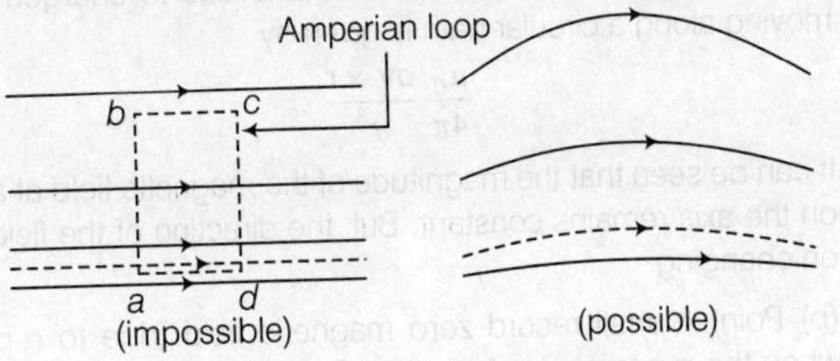

$$\therefore \quad \int_{abcda} \mathbf{B} \cdot d\mathbf{l} = 0$$

$$\Rightarrow \int_a^b \mathbf{B} \cdot d\mathbf{l} + \int_b^c \mathbf{B} \cdot d\mathbf{l} + \int_c^d \mathbf{B} \cdot d\mathbf{l} + \int_d^a \mathbf{B} \cdot d\mathbf{l} = 0$$

$$\Rightarrow \quad B_{bc} l - B_{da} l = 0$$

$$\Rightarrow \quad (B_{bc} - B_{da})\,l = 0$$

But $$B_{bc} \neq B_{da}$$

Field is more stronger at *da*.

So, $$\int \mathbf{B} \cdot d\mathbf{l} \neq 0$$

whereas, $I_{net} = 0$

∴ It is a contradiction.

29. (d) Statement I is incorrect but Statement II is correct.

Alternating currents have negligible effect on the compass needle, due to rapid change of direction and magnitude. If there is DC current in wires, then it can deflect a compass needle.

30. (b) Statement I is correct and Statement II is also correct but Statement II does not explain Statement I.

Magnetic forces exerted by a wire on other tend to align the wires. The net force on each wire is zero but net torque is not zero.

Revisal Problems (JEE Advanced)

1. (a) $|\mathbf{B}| = \dfrac{\mu_0}{4\pi}\cdot\dfrac{I\Delta l r\sin\theta}{r^3} = \dfrac{\mu_0}{4\pi}\cdot\dfrac{\Delta q}{\Delta t}\cdot\dfrac{\Delta l\sin\theta}{r^2}$

$$= \frac{\mu_0}{4\pi}\Delta q\left(\frac{\Delta l}{\Delta t}\right)\frac{\sin\theta}{r^2}$$

$$= \frac{\mu_0}{4\pi}\Delta q\, v\frac{\sin\theta}{r^2}$$

$$= \frac{10^{-7}\times 2\times 100\times \sin 30^\circ}{4} = 25\times 10^{-7}$$

$$= 2.5\times 10^{-6}\,\text{T} = 2.5\,\mu\text{T}$$

2. (b) $B = \dfrac{\mu_0 i}{2R}$

where, $\quad i = \dfrac{q}{T} = \dfrac{qv}{2\pi r}$

Also, $\quad v = \dfrac{e}{\sqrt{4\pi\varepsilon_0 mr}}$

$\therefore \quad B = \dfrac{\mu_0 e^2}{4\pi r^2\sqrt{4\pi\varepsilon_0 mr}} = \dfrac{\mu_0 e^2}{8\pi r^2\sqrt{\pi\varepsilon_0 mr}}$

3. (a) The point charge moves in circle as shown in the figure. The magnetic field vectors at a point P on axis of circle are $\mathbf{B}_A$ and $\mathbf{B}_C$ at the instants the point charge is at A and C respectively as shown in the figure.

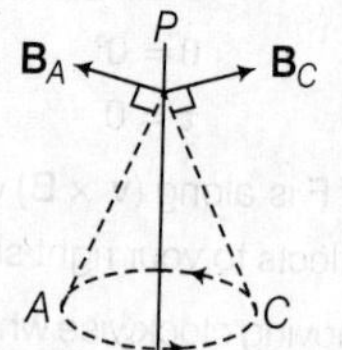

Hence, as the particles rotate in circle, only magnitude of magnetic field remains constant at the point on axis P but its direction changes.

Alternate Method

The magnetic field at point on the axis due to charged particle moving along a circular path is given by

$$\frac{\mu_0}{4\pi}\frac{q\mathbf{v}\times\mathbf{r}}{r^3}$$

It can be seen that the magnitude of the magnetic field at an point on the axis remains constant. But, the direction of the field keeps on changing.

4. (b) Point A shall record zero magnetic field (due to α-particle) when the α-particle is at position P and Q for which $\mathbf{v}\times\mathbf{r} = 0 \Rightarrow \mathbf{v}\parallel\mathbf{r}$ as shown in figure. The time taken by α-particle to go from P to Q is given by

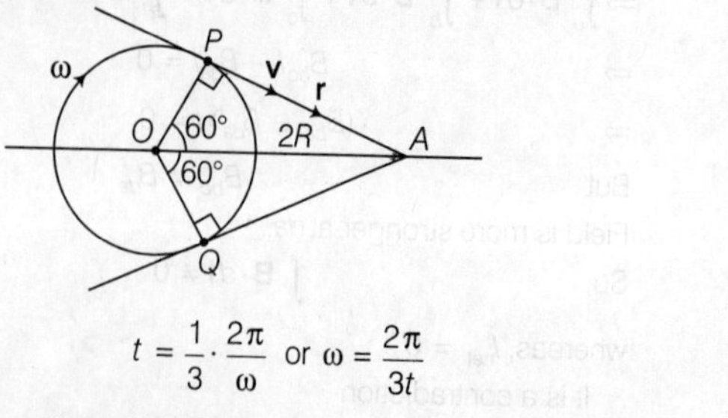

$$t = \frac{1}{3}\cdot\frac{2\pi}{\omega} \quad \text{or} \quad \omega = \frac{2\pi}{3t}$$

5. (a) $\quad \mathbf{B} = \dfrac{\mu_0}{4\pi}q\dfrac{\mathbf{v}\times\mathbf{r}}{r^3}$

and $\quad \mathbf{E} = \dfrac{1}{4\pi\varepsilon_0}\cdot\dfrac{q\mathbf{r}}{r^3}$

$\therefore \quad \mathbf{B} = \mu_0\varepsilon_0(\mathbf{v}\times\mathbf{E}) = \dfrac{\mathbf{v}\times\mathbf{E}}{c^2}$

$$= \frac{(\hat{i}-3\hat{j})\times 2\hat{k}}{c^2} = \frac{6\hat{i}-2\hat{j}}{c^2}$$

6. (c) $\mathbf{B}_{\text{due to first loop}} = 4\dfrac{\mu_0 i}{4\pi\dfrac{a}{2}}(\cos 45^\circ + \cos 45^\circ)$

$$= \frac{2\sqrt{2}\,\mu_0 i}{\pi a}$$

$\mathbf{B}_{\text{due to second loop}} = -\dfrac{4\mu_0 i}{4\pi\dfrac{2a}{2}}(\cos 45^\circ + \cos 45^\circ)$

$$= \frac{-\sqrt{2}\,\mu_0 i}{\pi a}$$

$$\mathbf{B} = \frac{2\sqrt{2}\,\mu_0 i}{\pi a}\left(1 - \frac{1}{2} + \ldots\infty\right) = \frac{2\sqrt{2}\,\mu_0 i}{\pi a}\ln 2$$

7. (d) The particle will move in a non-uniform helical path with increasing pitch as shown in the figure.

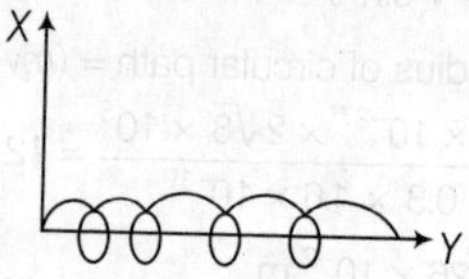

Its time period will be

$$T = \frac{2\pi m}{qB} = 2\pi \text{ second}$$

Changing the view, the particle is seemed to move in a circular path in XZ-plane as shown below:

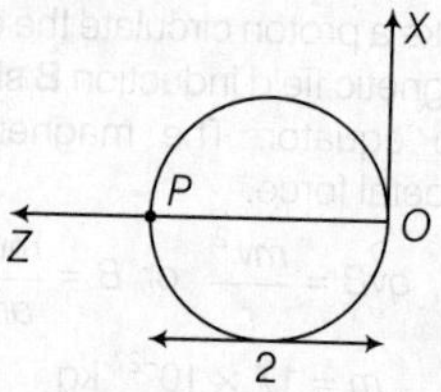

After π second, the particle will be at point P, hence x-coordinate will be zero.

For linear motion along Y-direction,

$$Y(\pi) = 0(\pi) + \frac{1}{2}\cdot\frac{Eq}{m}(\pi)^2$$

$$Y(\pi) = \frac{\pi^2}{2} \text{ and } OP = 2$$

Hence, the coordinates $\left(0, \dfrac{\pi^2}{2}, 2\right)$

8. (d) For $0 \le x \le d$, magnetic induction is negative.

For $d \le x \le 2d$, magnetic field due to the two planes cancel each other, hence becomes zero.

For $2d \le x \le 3d$ magnetic field B is positive.

9. (b)

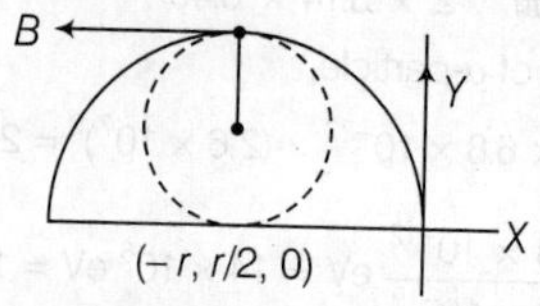

10. (b) The charged particle moves in a circle of radius $\dfrac{a}{2}$.

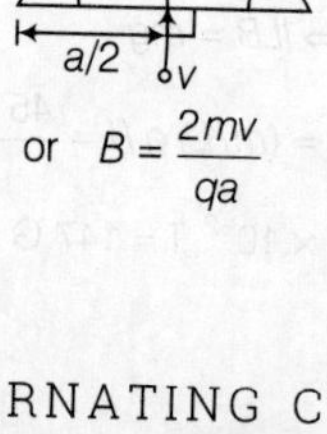

$\therefore \quad qvB = \dfrac{mv^2}{a/2} \quad \text{or} \quad B = \dfrac{2mv}{qa}$

11. (b) Given, square loops are $A \equiv (0, 0, 0)$

$$B \equiv (0, 0, a)$$

$$C \equiv \left(\frac{a}{\sqrt{2}}, \frac{a}{\sqrt{2}}, a\right)$$

and $$D \equiv \left(\frac{a}{\sqrt{2}}, \frac{a}{\sqrt{2}}, 0\right)$$

As, we know that magnetic moment of the loop *ABCDA*

$$\mu = I\mathbf{A} = I(\mathbf{AB} \times \mathbf{BC})$$

$$= I\left[a\hat{\mathbf{k}} \times \left(\frac{a}{\sqrt{2}}\hat{\mathbf{i}} + \frac{a}{\sqrt{2}}\hat{\mathbf{j}}\right)\right]$$

$$= \left[\frac{a^2}{\sqrt{2}}\hat{\mathbf{j}} - \frac{a^2}{\sqrt{2}}\hat{\mathbf{i}}\right] I$$

12. (d) Since, the current in *AB* and *AC* will be same in magnitude and produce equal and opposite magnetic field at the centre. There will be no current in *BC*. Therefore, magnetic field at *O* will be zero.

13. (a) Magnetic field due to the circular loop at centre *O*,

$$B_1 = \frac{\mu_0 I_c}{2R}$$

Magnetic field due to the straight wire at point *O*,

$$B_2 = \frac{\mu_0 I_e}{2\pi H}$$

As these two fields act in opposite directions, so

$$B = B_1 - B_2 = 0 \quad \text{or} \quad B_1 = B_2$$

or $$\frac{\mu_0 I_c}{2R} = \frac{\mu_0 I_e}{2\pi H} \quad \text{or} \quad H = \frac{I_e R}{I_c \pi}$$

14. (a,b,c) Using $-e(\mathbf{v} \times \mathbf{B})$ for the region outside the plates, the direction of magnetic field can be found. Inside the plates, net force on the electron is zero hence electric force is opposite to that of magnetic force. The direction of electric field between the plates is opposite to that of direction of force on the negative (electron) charge.

15. (a,c,d) All are possible, because neither mass nor charge is specified, $r = mv/Bq$. Only option (b) is not possible.

16. (a,c) Field due to each plate $= \frac{1}{2}\mu_0 K = 2\ \mu T$

At *A*, fields add up being in the same directions whereas at *B*, cancel out due to opposite direction.

17. (a,b,c) Consider a charged particle of charge *q* and mass *m* is moving with a velocity *v* as shown in a uniform magnetic field along negative *Z*-direction.

$$\text{Length of arc } AB = \frac{\pi r}{3} = \frac{\pi mv}{qB}$$

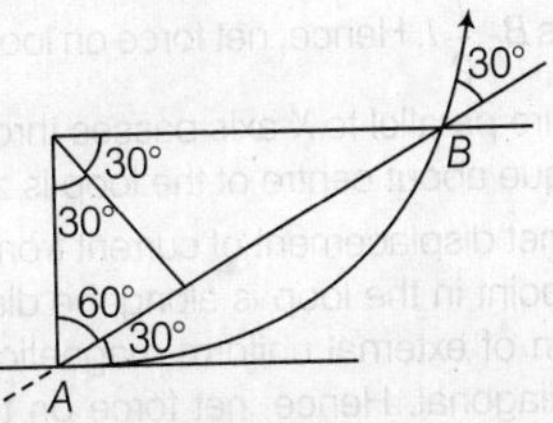

Time taken by a charged particle *q*,

$$t = \frac{\theta}{\omega} = \frac{\pi/3}{2\pi/T} = \frac{T}{6} = \frac{\pi m}{3qB} \quad \left(\because T = \frac{2\pi m}{qB}\right)$$

Distance travelled in a uniform magnetic field = *vt*

$$= v \times \frac{\pi m}{3qB} = \frac{\pi mv}{3qB}$$

18. (b,c,d) If $\mathbf{v} \cdot \mathbf{B} = 0$

$\Rightarrow$ Force is perpendicular to velocity as $\mathbf{v} \perp \mathbf{B}$.

If $\mathbf{v} \times \mathbf{B} = 0 \Rightarrow F = 0$, so it is a straight line.

The speed is constant.

19. (a,b) In going from point *P* to *Q*, increase in kinetic energy from work-energy theorem,

$$WF_e = \Delta K$$

or $$(qE)(2a) = \frac{1}{2}m(4v^2 - v^2)$$

$$E = \frac{3}{4}\left(\frac{mv^2}{qa}\right)$$

At *P*, rate of work done by electric field

$$= \mathbf{F}_e \cdot \mathbf{v} = (qE)v$$

$$= qv\left[\frac{3}{4}\left(\frac{mv^2}{qa}\right)\right] = \frac{3}{4}\left(\frac{mv^3}{a}\right)$$

20. (a,c) As, we know that net force acting on *BA*,

$$\mathbf{F}_{BA} = 0$$

because magnetic lines are parallel to this wire.

and force acting in *CD*,

$$\mathbf{F}_{CD} = 0$$

because magnetic lines are anti-parallel to this wire.

Torque acting on it tends to rotate it in clockwise direction about an axis *OO′*.

21. (b,d) Time taken by a proton to travel in a uniform magnetic field,

$$t_p = \frac{2\theta \times R_p}{v} = \frac{2\theta \times m_p v}{eBv} = \frac{2\theta m_p}{eB}$$

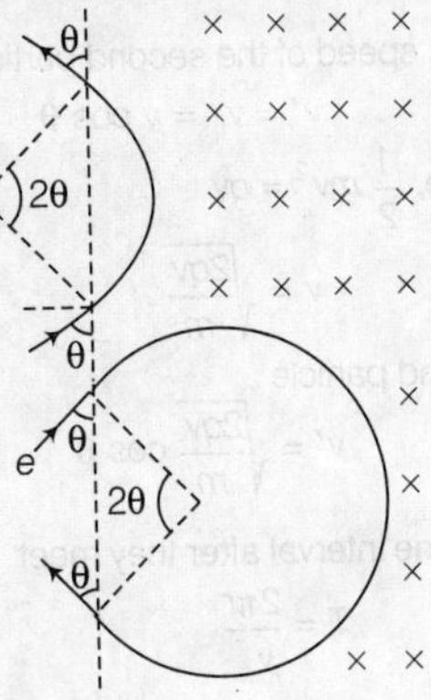

Time taken by a electron to travel in a uniform magnetic field,

$$t_e = \frac{(2\pi - 2\theta) \times R_e}{v}$$

$$= \frac{(2\pi - 2\theta)\, m_e v}{eBv} = \frac{(2\pi - 2\theta)\, m_e}{eB}$$

i.e. $$t_e \neq t_p$$

22. (c) The component of velocity in the direction of magnetic field doesn't produce any magnetic force, so it cannot produce circular motion, it can only produce translational motion.

$$\mathbf{v} = (8\hat{\mathbf{i}} - 6\hat{\mathbf{j}} + 4\hat{\mathbf{k}}) \times 10^6\ \text{m/s}$$

and $$\mathbf{B} = -0.4\hat{\mathbf{k}}\ \text{T}$$

So, the $v_z = 4\hat{\mathbf{k}}$ will produce translation motion and $v_x = 8\hat{\mathbf{i}}$ and $v_y = -6\hat{\mathbf{j}}$ will produce circular motion as they are perpendicular to the magnetic field.

23. (d) $|q(\mathbf{v}\times\mathbf{B})| = 1.6$

$\Rightarrow |q(8\hat{\mathbf{i}} - 6\hat{\mathbf{j}} + 4\hat{\mathbf{k}})\times(-0.4\hat{\mathbf{k}})]\times 10^6| = 1.6$

$\Rightarrow \quad 4\times 10^6 \times q = 1.6$

$$q = 0.4\times 10^{-6}\ \text{C}$$

$$\omega = \frac{qB}{m} = \frac{0.4\times 10^{-6}\times 0.4}{4\times 10^{-15}}$$

$$= 4\times 10^7\ \text{rad s}^{-1}$$

24. (c) As, the coordinates are asked after $3T$ time. During $3T$ time, it completes 3 circles completely and attain its origin position back with respect to its circular motion but due to translational motion it will move forward. There is no change in x, y-coordinates. z-coordinate after time

$$3T = 3\times\frac{2\pi m}{qB}\times 4\times 10^6$$

$$= 1.884\ \text{m}$$

25. (b) The first particle will have a helical path and the second particle will move rectilinearly along the field. For the two particles to meet again and again

$$v_{\|}T = v'T$$

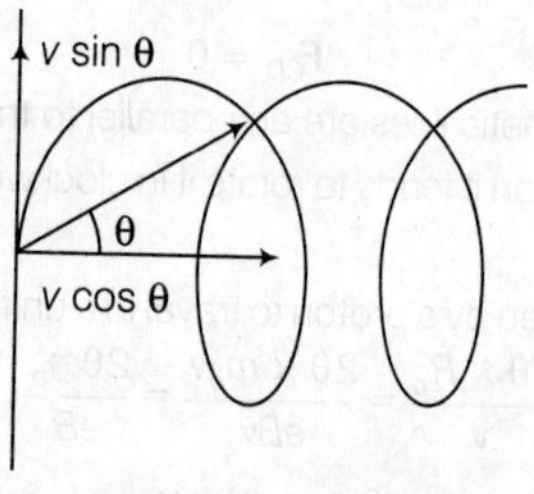

where, v' is the speed of the second particle.

So, $\quad v' = v_{\|} = v\cos\theta$

KE of a particle, $\frac{1}{2}mv^2 = qv$

$\Rightarrow \quad v = \sqrt{\frac{2qv}{m}}$

Speed of second particle

$$v' = \sqrt{\frac{2qv}{m}}\cos\theta$$

26. (a) Minimum time interval after they meet

$$T = \frac{2\pi r}{v_{\perp}}$$

$$= \frac{2\pi\times mv\sin\theta}{v\sin\theta\times qB}$$

$$= \frac{2\pi m}{qB}$$

27. (c) Distance travelled by second particle

$$= v'\times T = \sqrt{\frac{2qv}{m}}\cos\theta\times\frac{2\pi m}{qB}$$

$$\left(\because v' = \sqrt{\frac{2qv}{m}}\cos\theta \text{ and } T = \frac{2\pi m}{qB}\right)$$

$$= \frac{2\pi}{B}\cos\theta\times\frac{m}{q}\sqrt{\frac{2qv}{m}}$$

$$= \frac{2\pi}{B}\cos\theta\sqrt{\frac{2qv}{m}\times\frac{m^2}{q^2}}$$

$$= \frac{2\pi}{B}\cos\theta\times\sqrt{\frac{2vm}{q}}$$

28. (b) Magnetic field at any point on Ampere's loop can be due to all currents passing through inside or outside the loop. But, net contribution in the left hand side will come from inside current only.

29. (d) For $l < a$, current passing through within the cylinder of radius r is given by

$$\int_0^r J\,dA = \int_0^r kr^2\, 2\pi r\, dr = 2\pi k\int_0^r r^3 dr = \frac{k\pi r^4}{2}$$

Now, using Ampere's law, we get

$$B2\pi r = \mu_0 I = \frac{\mu_0 k\pi r^4}{2}$$

$$\Rightarrow \quad B = \frac{\mu_0 kr^3}{4}$$

30. (a) For $r > a$, $\quad I = \frac{k\pi a^4}{2}$

$\therefore$ Magnetic field induction at a point distance r from the axis,

$$B = \frac{\mu_0 k\pi a^4}{4r} \quad \text{(by putting } r = a\text{)}$$

31. (b)

(i) A uniform electric field exerts a constant force on the charged particle, hence the particle may move in a straight line or a parabolic path.

(ii) Under action of a uniform magnetic field, the charged particle may move in a straight line when projected along or opposite to direction of magnetic field. The charged particle moves in circle when it is projected perpendicular to the magnetic field. If the initial velocity of the charged particle makes an angle between 0° and 180° (except 90°) with magnetic field, the particle moves along a helical path of a uniform pitch.

(iii) If charged particle is shot parallel to both fields, it moves along a straight line. If the charged particle is shot at any angle with both the field (except 0° and 180°), then particle moves along a helix with non-uniform pitch.

(iv) From results of A and B, all the given paths are possible.

Hence, (i) → (p,q), (ii) → (p,r,s), (iii) → (p,s), (iv) → (p,q,r,s)

32. (a)

(i) Because the magnetic field is parallel to X-axis, the force on wire parallel to X-axis is zero. The force on each wire parallel to Y-axis is $B_0\frac{i}{2}l$. Hence, net force on loop is $B_0 il$. Since, force on each wire parallel to Y-axis passes through centre of the loop net torque about centre of the loop is zero.

(ii) Because the magnetic field is parallel to Y-axis, the force on wire parallel to Y-axis is zero. The force on each wire parallel to X-axis is $B_0\frac{i}{2}l$. Hence, net force on loop is $B_0 il$. Since, force on each wire parallel to X-axis passes through centre of the loop, net torque about centre of the loop is zero.

(iii) Since, net displacement of current from entry point in the loop to exit point in the loop is along the diagonal of the loop. The direction of external uniform magnetic field is also along the same diagonal. Hence, net force on the loop is zero. Since, force on each wire on the loop passes through centre of the loop net torque about centre of the loop is zero.

(iv) The net displacement of current from entry point in the loop to exit point in the loop is along the diagonal (of length $\sqrt{2}l$) of the loop. The direction of external uniform magnetic field is also perpendicular to the same diagonal. Hence, magnitude of net force on the loop is $B_0 i(\sqrt{2}l)$. Since, force on each wire on the loop passes through centre of the loop, net torque about centre of the loop is zero.

Hence, (i) → (r,s), (ii) → (r,s), (iii) → (q,r), (iv) → (p,r)

33. (a) The magnetic field is along negative Y-direction in case (p), (q) and (r). In case (s), there will be component of magnetic field along negative X-axis.

Thus, **(i) → (p,q,r)**

z-component of magnetic field is zero in all cases.

Thus, **(ii) → (p,q,r,s)**

The magnetic field at P is $\frac{\mu_0 I}{4\pi d}$. In case (r) only.

Thus, **(iii) → (r)**

The magnetic field at P is less than $\frac{\mu_0 I}{2\pi d}$ for all cases.

Thus, **(iv) → (p,q,r,s)**

34. (c) (i) Magnetic moment of a circular current carrying coil,

$$M = NIA = 100 \times 5 \times \pi\,(0.1)^2 = 5\pi \text{ A-m}$$

As the current in XY-plane is along anti-clockwise direction, so moment will be along Z-axis by right hand thumb rule.

i.e. $\mathbf{M} = 5\pi\hat{\mathbf{k}}$

(ii) Torque, $\tau = \mathbf{M} \times \mathbf{B} = 5\pi\hat{\mathbf{k}} \times (-\hat{\mathbf{i}} + \hat{\mathbf{k}}) = -5\pi\hat{\mathbf{j}}$

(iii) Net force on a closed current carrying loop in a uniform magnetic field is zero.

(iv) Magnetic field at centre of loop due to current in loop will be along positive Z-axis from right hand thumb rule.

Hence, (i) → (q,p), (ii) → (r,s), (iii) → (p,r), (iv) →(q,s)

35. (a) (i) → (p), (ii) → (p,q,s), (iii) → (q,s), (iv) → (q,r)

36. (4) Torque of magnetic force about PQ,

$$\tau_m = (ILB)\,L\cos\theta = IL2B\cos\theta \quad \text{...(i)}$$

Torque of gravitational force about PQ, ...(ii)

$$\tau_g = [(\lambda L)\,gL\sin\theta + 2(\lambda L)\,g(1/2)\,L\sin\theta]$$

$$= 2\lambda L^2 g\sin\theta$$

$$\tau_m = \tau_g$$

$$\Rightarrow \quad \tan\theta = \frac{IB}{2\lambda g}$$

$$= \frac{10\sqrt{3} \times 2}{2 \times \sqrt{3} \times 10} = 1$$

$$\Rightarrow \quad \theta = 45^\circ$$

$$\therefore \quad k = 4$$

37. (3) Let at time t, particle be at point P(x, y) and its velocity be

$$\mathbf{v} = (v_x\hat{\mathbf{i}} + v_y\hat{\mathbf{j}})$$

$$|\mathbf{v}| = |\mathbf{v}_0| \Rightarrow v_0^2 = v_x^2 + v_y^2$$

(work done by magnetic field is always zero, so change in magnitude of velocity)

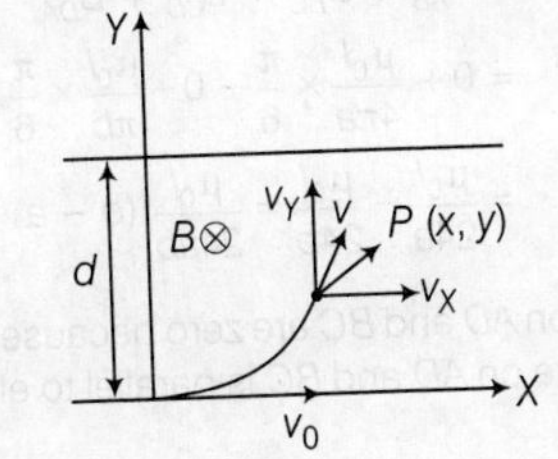

Then, magnetic force on the particle at point P is

$$\mathbf{F} = q(v_x\hat{\mathbf{i}} + v_y\hat{\mathbf{j}}) \cdot B_0\left(1 + \frac{y}{d}\right)(-\hat{\mathbf{k}})$$

$$\Rightarrow \quad -qB_0\left[1 + \frac{y}{d}\right]dy = m\,dv_x$$

Now, when the particle will be coming out of the at that point y = d. Let the velocity in X-direction be v_x, then integrating we get

$$\int_{v_0}^{v_x} dv_x = -\frac{qB_0}{m}\int_0^d\left[1 + \frac{y}{d}\right]dy$$

$$= -\frac{qB_0}{m}\left[d + \frac{d^2}{2d}\right] = -\frac{3qB_0 d}{2m}$$

So, $$v_x = v_0 - \frac{3qB_0 d}{2m}$$

$$\therefore \quad k = 3$$

38. (5) As, we know net magnetic field in the shaded part of a cylindrical cavity.

$$B_R = B_T - B_C$$

where, R = remaining portion

T = total portion and

C = cavity

$$B_R = \frac{\mu_0 I_T}{2a\pi} - \frac{\mu_0 I_C}{2(3a/2)\,\pi} \quad \text{...(i)}$$

$$I_T = J(\pi a^2)$$

$$I_C = J\left(\frac{\pi a^2}{4}\right)$$

Substituting the values in Eq. (i), we have

$$B_R = \frac{\mu_0}{a\pi}\left[\frac{I_T}{2} - \frac{I_C}{3}\right]$$

$$= \frac{\mu_0}{a\pi}\left[\frac{\pi a^2 J}{2} - \frac{\pi a^2 J}{12}\right] = \frac{5\mu_0 aJ}{12}$$

$$\therefore \quad N = 5$$

39. (3) Magnetic field across the wire 2 is

$$B_2 = \frac{\mu_0 I}{2\pi x_1} + \frac{\mu_0 I}{2\pi\,(x_0 - x_1)}$$

(when currents are in opposite directions)

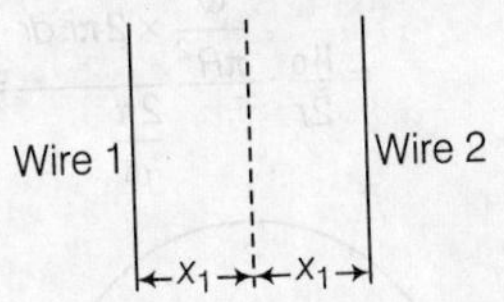

$$B_1 = \frac{\mu_0 I}{2\pi x_1} - \frac{\mu_0 I}{2\pi\,(x_0 - x_1)}$$

(when currents are in same direction)

Substituting $x_1 = \frac{x_0}{3}\left(\text{as } \frac{x_0}{x_1} = 3\right)$ in above equation, we get

magnetic field across wire 1,

$$B_1 = \frac{3\mu_0 I}{2\pi x_0} - \frac{3\mu_0 I}{4\pi x_0} = \frac{3\mu_0 I}{4\pi x_0}$$

Radius of current carrying wire 1,

$$R_1 = \frac{mv}{qB_1} \text{ and } B_2 = \frac{9\mu_0 I}{4\pi x_0}$$

and radius along wire 2,

$$R_2 = \frac{mv}{qB_2}$$

⇒ Ratio of radius of both wires,

$$\frac{R_1}{R_2} = \frac{B_2}{B_1} = \frac{9}{3} = 3$$

JEE Main & AIEEE Archive

1. (b) Force exerted on a current carrying conductor,

$$F_{ext} = BIL$$

Work done $= BIldx$

Average power $= \dfrac{\text{Work done}}{\text{Time taken}}$

$$P = \frac{1}{t}\int_0^2 F_{ext}dx = \frac{1}{t}\int_0^2 BIldx$$

$$= \frac{1}{5 \times 10^{-3}}\int_0^2 3 \times 10^{-4} e^{-0.2x} \times 10 \times 3dx$$

$$= 9\,[1 - e^{-0.4}]$$

$$= 9\left[1 - \frac{1}{e^{0.4}}\right] = 2.967 \approx 2.97 \text{ W}$$

2. (b) For charged particle in magnetic field,

$$\text{Radius, } r = \frac{mv}{qB} = \frac{\sqrt{2Km}}{qB}$$

$$\left(\text{as } mv = p \Rightarrow K = \frac{1}{2}mv^2 = \frac{m^2v^2}{2m} = p^2/2m,\ p = \sqrt{2\,Km}\right)$$

$$\Rightarrow \quad r \propto \frac{\sqrt{m}}{q}$$

or

$$m_d = 2m_p \text{ and } q_d = q_p$$

$$m_\alpha = 4m_p \text{ and } q_\alpha = 2q_p$$

$$\Rightarrow r_p : r_d : r_\alpha = \frac{\sqrt{m_p}}{q_p} : \frac{\sqrt{2m_p}}{q_p} : \frac{\sqrt{4m_p}}{2q_p} = 1 : \sqrt{2} : 1$$

$$\Rightarrow \quad r_\alpha = r_p < r_d$$

3. (a) Taking an elemental ring of radius r and thickness dr, we obtain magnetic field at the centre of the ring,

$$dB = \frac{\mu_0}{2r}\,dl = \frac{\mu_0}{2r}\cdot\frac{dq}{T}$$

$$= \frac{\mu_0}{2r}\cdot\frac{\dfrac{Q}{\pi R^2} \times 2\pi r\,dr}{\dfrac{2\pi}{\omega}} = \frac{\mu_0 Q\omega}{2\pi R^2}\,dr$$

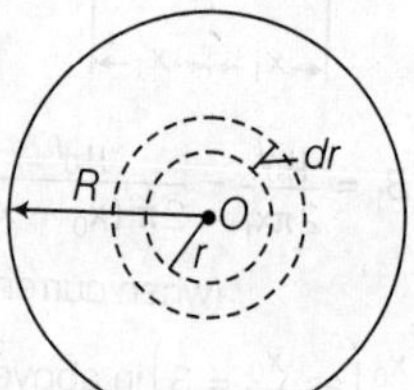

So, net magnetic field at the centre of disc,

$$B = \int_0^R dB = \frac{\mu_0\omega Q}{2\pi R^2}\int_0^R dr = \frac{\mu_0\omega Q}{2\pi R}$$

i.e. $\qquad B \propto \dfrac{1}{R}$

4. (d) Let us consider the disc to be made up of large number of concentric elementary rings.

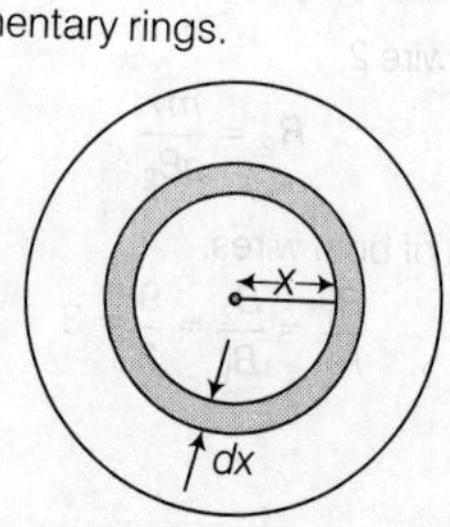

Consider one such ring of radius x and thickness dx.

Charge on this elementary ring,

$$dq = \sigma \times 2\pi x dx = 2\pi\sigma x dx$$

Current associated with this elementary ring,

$$dI = \frac{dq}{dt} = dq \times f = \sigma\omega x dx \quad (\because f \text{ is frequency and } \omega = 2\pi f)$$

Magnetic moment of this elementary ring,

$$dM = dI\pi x^2 = \pi\sigma\omega\, x^3 dx$$

$\therefore$ Magnetic moment of the entire disc,

$$M = \int_0^R dM = \pi\sigma\omega\int_0^R x^3\,dx = \frac{1}{4}\pi R^4\,\sigma\omega$$

5. (d) Consider the wire to be made up of large number of thin wires of infinite length. Consider such wire of thickness dl subtending an angle $d\theta$ at centre.

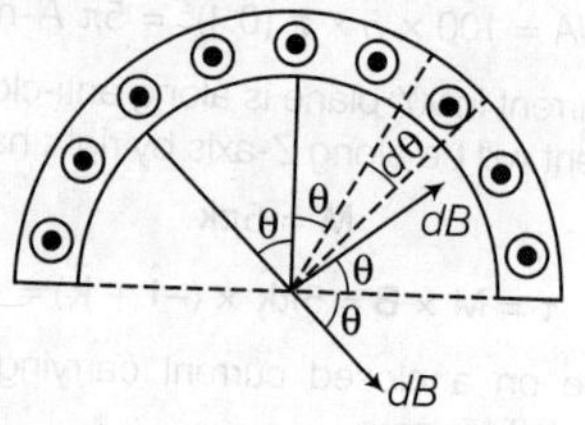

Current through this wire,

$$dI = \frac{d\theta}{\pi}I$$

$\therefore$ Magnetic field at centre due to this portion,

$$dB = \frac{\mu_0}{4\pi}\cdot\frac{2dI}{R} = \frac{\mu_0 I}{2\pi^2 R}\,d\theta$$

Net magnetic field at the centre,

$$B = \int_0^\pi d\,B\cos\theta = \frac{\mu_0 I}{\pi^2 R}$$

6. (a) The magnetic field in between wires will be in opposite direction.

$$\therefore \qquad B_{\text{in between}} = \frac{\mu_0 i}{2\pi x}\hat{\mathbf{j}} - \frac{\mu_0 i}{2\pi\,(2d - x)}(-\hat{\mathbf{j}})$$

$$= \frac{\mu_0 i}{2\pi}\left[\frac{1}{x} - \frac{1}{2d - x}\right](\hat{\mathbf{j}})$$

At $x = d$, $\quad B_{\text{in between}} = 0$

For $x < d$, $\quad B_{\text{in between}} = (\hat{\mathbf{j}})$

For $x > d$, $\quad B_{\text{in between}} = (-\hat{\mathbf{j}})$

Towards x, net magnetic field will add up and direction will be $(-\hat{\mathbf{j}})$.

Towards x', net magnetic field will add up and direction will be $(\hat{\mathbf{j}})$.

7. (b) Net magnetic field due to loop $ABCD$ at O is

$$B = B_{AB} + B_{BC} + B_{CD} + B_{DA}$$

$$= 0 + \frac{\mu_0 I}{4\pi a} \times \frac{\pi}{6} + 0 - \frac{\mu_0 I}{\pi b} \times \frac{\pi}{6}$$

$$= \frac{\mu_0 I}{24a} - \frac{\mu_0 I}{24b} = \frac{\mu_0 I}{24ab}(b - a)$$

8. (b) The forces on AD and BC are zero because magnetic field due to a straight wire on AD and BC is parallel to elementary length of the loop.

9. (c) $B = \dfrac{\mu_0 I}{2\pi R}$

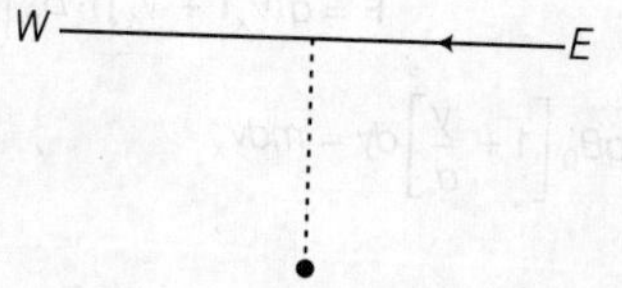

$$B = \frac{2 \times 10^{-7} \times 100}{4}\text{ T} = 5 \times 10^{-6}\text{ T, Southward}$$

10. (b) The magnetic field inductions at a point P at a distance d from O in a direction perpendicular to the plane of the wires due to currents through AOB and COD are perpendicular to each other is given by

$$B = \sqrt{B_1^2 + B_2^2} = \left[\left(\frac{\mu_0}{4\pi}\frac{2I_1}{d}\right)^2 + \left(\frac{\mu_0}{4\pi}\frac{2I_2}{d}\right)^2\right]^{1/2} = \frac{\mu_0}{2\pi d}\sqrt{(I_1^2 + I_2^2)}$$

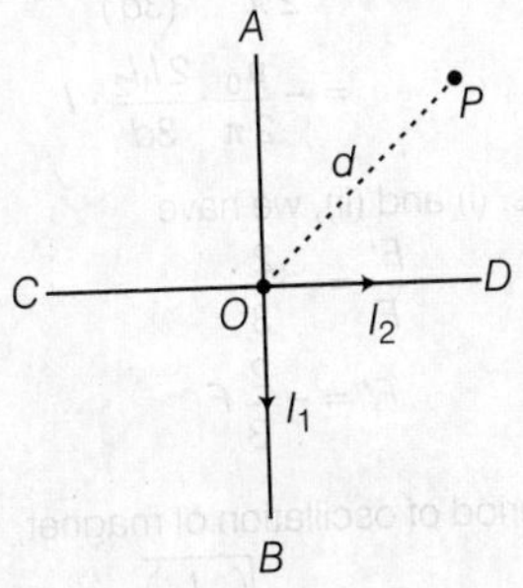

11. (c) Current density, $J = \dfrac{I}{\pi a^2}$

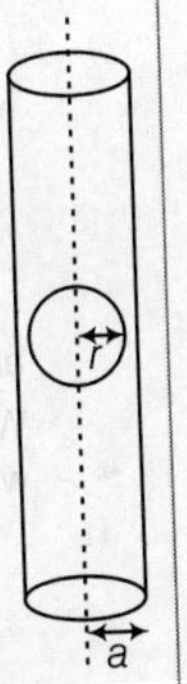

From Ampere's circuital law,

$$\oint B \cdot d\mathbf{l} = \mu_0 \cdot I_{\text{enclosed}}$$

For $r < a$,

$$B \times 2\pi r = \mu_0 \times J \times \pi r^2$$

$\Rightarrow$ $B = \dfrac{\mu_0 I}{\pi a^2} \times \dfrac{r}{2}$

At $r = \dfrac{a}{2}$, $B_1 = \dfrac{\mu_0 I}{4\pi a}$

For $r > a$, $B \times 2\pi r = \mu_0 I$

$\Rightarrow$ $B = \dfrac{\mu_0 I}{2\pi r}$

At $r = 2a$, $B_2 = \dfrac{\mu_0 I}{4\pi a}$

$\therefore$ $\dfrac{B_1}{B_2} = 1$

12. (c) By using Ampere's circuital law, the magnetic field at any point inside the pipe is zero.

13. (a) In case of motion of a charged particle perpendicular to the motion, i.e. displacement, then work done

$$W = \int \mathbf{F} \cdot d\mathbf{s} = \int F ds \cos\theta = 0 \qquad (\text{as } \theta = 90^\circ)$$

and by work-energy theorem, $W = \Delta KE$, the kinetic energy and hence speed v remains constant. But, **v** changes, so momentum changes.

14. (b) Let **E** and **B** be along X-axis. When a charged particle is released from rest, it will experience an electric force along the direction of electric field or opposite to the direction of electric field depending on the nature of charge.
Due to this force, it acquires some velocity along X-axis. Due to this motion of charge, magnetic force is zero because angle between **v** and **B** would be either 0° or 180°.
So, only electric force is acting on particle and hence it will move along a straight line.

15. (a) Magnetic field due to a long solenoid is given by

$$B = \mu_0 nI$$

From given data,

$$6.28 \times 10^{-2} = \mu_0 \times 200 \times 10^2 \times I \qquad \ldots\text{(i)}$$

and $$B = \mu_0 \times 100 \times 10^2 \times \left(\frac{I}{3}\right) \qquad \ldots\text{(ii)}$$

On solving Eqs. (i) and (ii), we get

$$B \approx 1.05 \times 10^{-2}\ \text{Wb/m}^2$$

16. (a) The force per unit length between the two wires is

$$\frac{F}{l} = \frac{\mu_0}{4\pi} \cdot \frac{2I^2}{d} = \frac{\mu_0 I^2}{2\pi d}$$

The force will be attractive as current directions in both are same.

17. (c) Magnetic field or magnetic induction at P,

$$B_P = \frac{\mu_0 I_2}{2R} = \frac{4\pi \times 10^{-7} \times 4}{2 \times 0.02\pi}$$
$$= 4 \times 10^{-5}\ \text{Wb/m}^2$$

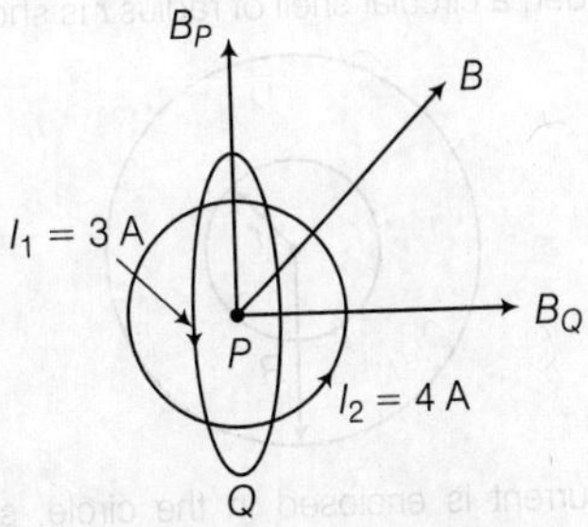

and $$B_Q = \frac{\mu_0 I_1}{2R} = \frac{4\pi \times 10^{-7} \times 3}{2 \times 0.02\pi}$$
$$= 3 \times 10^{-5}\ \text{Wb/m}^2$$

$\therefore$ $$B = \sqrt{B_P^2 + B_Q^2}$$
$$= \sqrt{(4 \times 10^{-5})^2 + (3 \times 10^{-5})^2}$$
$$= 5 \times 10^{-5}\ \text{Wb/m}^2$$

18. (c) Magnetic needle is placed in a non-uniform magnetic field. It experiences force and torque both due to unequal forces acting on poles.

19. (d) Magnetic force,

$$F = qvB \qquad (\because \theta = 90^\circ) \ldots\text{(i)}$$

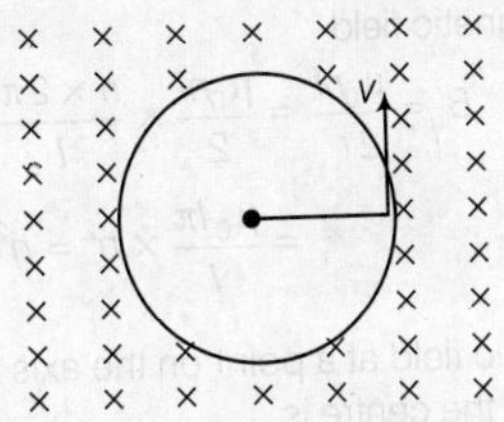

Centripetal force,

$$F = \frac{mv^2}{r} \qquad \ldots\text{(ii)}$$

From Eqs. (i) and (ii), we get

$$\frac{mv^2}{r} = qvB$$

or $$r = \frac{mv}{qB}$$

The time taken by the particle to complete one revolution,

$$T = \frac{2\pi r}{v} = \frac{2\pi mv}{vqB} = \frac{2\pi m}{qB}$$

20. (d) Full scale deflection current $= \dfrac{150}{10}$ mA

$$= 15\ \text{mA}$$

Full scale deflection voltage $= \dfrac{150}{2}$ mV $= 75$ mV

Galvanometer resistance, $G = \dfrac{75\ \text{mV}}{15\ \text{mA}} = 5\ \Omega$

Required full scale deflection voltage,

$$V = 1 \times 150 = 150\ \text{V}$$

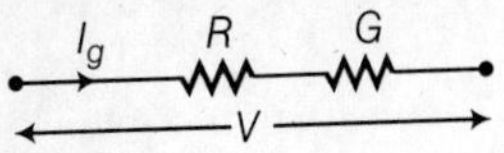

Let resistance to be connected in series be R. Then,

$$V = I_g(R + G)$$

$\therefore$ $$150 = 15 \times 10^{-3}(R + 5)$$

or $$10^4 = R + 5$$

or $$R = 10000 - 5 = 9995\,\Omega$$

21. (b) Let R be the radius of a long thin cylindrical shell.

To calculate the magnetic induction at a distance r, ($r < R$) from the axis of cylinder, a circular shell of radius r is shown in the figure.

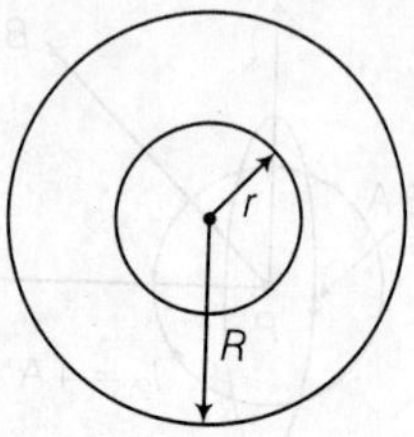

Since, no current is enclosed in the circle, so from Ampere's circuital law, magnetic induction is zero at every point of circle. Hence, the magnetic induction at every point inside the infinitely long straight thin walled tube (cylindrical) is zero.

22. (b) The magnetic field at the centre of circular coil is

$$B = \frac{\mu_0 I}{2r}$$

where, r = radius of circle $= \dfrac{l}{2\pi}$ $\qquad (\because l = 2\pi r)$

$\therefore$ $$B = \frac{\mu_0 I}{2} \times \frac{2\pi}{l} = \frac{\mu_0 I\pi}{l} \qquad \ldots(i)$$

When wire of length l bents into a circular loop of n turns, then

$$l = n \times 2\pi r' \Rightarrow r' = \frac{l}{n \times 2\pi}$$

Thus, new magnetic field

$$B' = \frac{\mu_0 nI}{2r'} = \frac{\mu_0 nI}{2} \times \frac{n \times 2\pi}{l}$$

$$= \frac{\mu_0 I\pi}{l} \times n^2 = n^2 B \qquad \text{[from Eq. (i)]}$$

23. (a) The magnetic field at a point on the axis of a circular loop at a distance x from the centre is

$$B = \frac{\mu_0 IR^2}{2(R^2 + x^2)^{3/2}} \qquad \ldots(i)$$

Given, $B = 54\ \mu\text{T}$, $x = 4$ cm, $R = 3$ cm

Putting the given values in Eq. (i), we get

$$54 = \frac{\mu_0 I \times (3)^2}{2(3^2 + 4^2)^{3/2}}$$

$\Rightarrow$ $$54 = \frac{9\mu_0 I}{2(25)^{3/2}} = \frac{9\mu_0 I}{2 \times (5)^3}$$

$\therefore$ $$\mu_0 I = \frac{54 \times 2 \times 125}{9}$$

$\Rightarrow$ $$\mu_0 I = 1500\ \mu\text{T-cm} \qquad \ldots(ii)$$

Now, putting $x = 0$ in Eq. (i), magnetic field at the centre of loop is

$$B = \frac{\mu_0 IR^2}{2R^3} = \frac{\mu_0 I}{2R} = \frac{1500}{2 \times 3} \qquad \text{[from Eq. (ii)]}$$

$$= 250\ \mu\text{T}$$

24. (c) Force acting between two current carrying conductors

$$F = \frac{\mu_0}{2\pi} \cdot \frac{I_1 I_2}{d} l \qquad \ldots(i)$$

where, d = distance between the conductors

l = length of small conductor

Again, $$F' = \frac{\mu_0}{2\pi} \cdot \frac{(-2I_1)(I_2)}{(3d)} \cdot l$$

$$= -\frac{\mu_0}{2\pi} \cdot \frac{2I_1 I_2}{3d} \cdot l \qquad \ldots(ii)$$

Thus, from Eqs. (i) and (ii), we have

$$\frac{F'}{F} = -\frac{2}{3}$$

$\Rightarrow$ $$F' = -\frac{2}{3}F$$

25. (b) The time period of oscillation of magnet,

$$T = 2\pi\sqrt{\left(\frac{I}{MH}\right)} \qquad \ldots(i)$$

where, I = moment of inertia of magnet

$= \dfrac{mL^2}{12}$ (m being the mass of magnet)

M = pole strength $\times L$

and H = horizontal component of the earth's magnetic field.

When the three equal parts of magnet are placed on one another with their like poles together, then

$$I' = \frac{1}{12}\left(\frac{m}{3}\right) \times \left(\frac{L}{3}\right)^2 \times 3$$

$$= \frac{1}{12} \cdot \frac{mL^2}{9} = \frac{I}{9}$$

and $$M' = \text{pole strength} \times \frac{L}{3} \times 3 = M$$

Hence, $$T' = 2\pi\sqrt{\left(\frac{I/9}{MH}\right)}$$

or $$T' = \frac{1}{3} \times T$$

or $$T' = \frac{2}{3}\text{ s}$$

26. (a) $W = MB(1 - \cos\theta)$

$\Rightarrow$ $W = MB(1 - \cos 60°)$ $\qquad (\because \theta = 60°)$

or $W = \dfrac{MB}{2}$ or $MB = 2W$

Torque, $\tau = MB\sin 60°$

$$= \frac{MB\sqrt{3}}{2} = \frac{2W\sqrt{3}}{2} = W\sqrt{3}$$

27. (d) Inside bar magnet, lines of force are from South to North.

28. (b) When particle describes circular path in a magnetic field, its velocity is always perpendicular to the magnetic force.

Power, $P = \mathbf{F} \cdot \mathbf{v} = Fv\cos\theta$

Here, $\theta = 90°$

$\therefore$ $P = 0$

But, $P = \dfrac{W}{t}$

$\Rightarrow$ $W = Pt$

Hence, work done $W = 0$ (everywhere)

29. (a) The force on a particle is

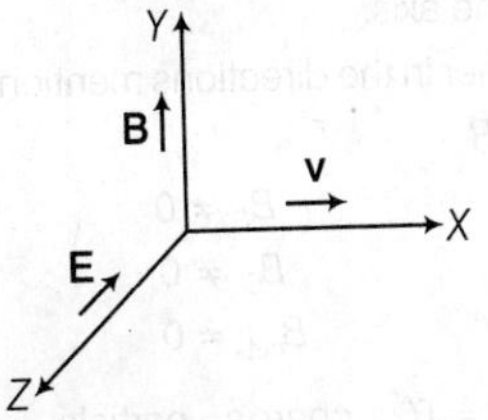

So, $\mathbf{F} = q(\mathbf{E} + \mathbf{v} \times \mathbf{B})$

or $\mathbf{F} = \mathbf{F}_e + \mathbf{F}_m$

$\therefore$ $\mathbf{F}_e = q\mathbf{E} = -16 \times 10^{-18} \times 10^4(-\hat{\mathbf{j}}) = 16 \times 10^{-14}\ \hat{\mathbf{k}}$

and $\mathbf{F}_m = -16 \times 10^{-18}(10\hat{\mathbf{i}} \times B\hat{\mathbf{j}}) = -16 \times 10^{-17} \times B(+\hat{\mathbf{k}})$

$= -16 \times 10^{-17} B\hat{\mathbf{k}}$

Since, particle will continue to move along positive X-axis, so resultant force is equal to zero.

$$\mathbf{F}_e + \mathbf{F}_m = 0$$

$\Rightarrow$ $16 \times 10^{-14} = 16 \times 10^{-17} B$

$\Rightarrow$ $B = \dfrac{16 \times 10^{-14}}{16 \times 10^{-17}} = 10^3$

or $B = 10^3\ \text{Wb/m}^2$

30. (d) To increase the range of ammeter, we have to connect a small resistance in parallel (shunt), let its value be R.

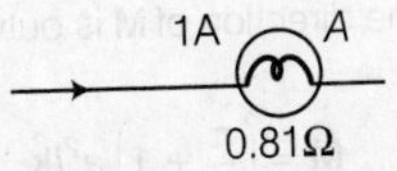

Apply KCL at junction to divide the current.

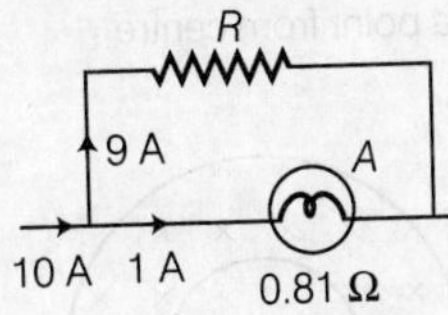

Voltage across R = Voltage across ammeter

$9R = 0.81 \times 1$ or $R = \dfrac{0.81}{9} = 0.09\ \Omega$

31. (a) Magnetic field in circular coil A is $B_A = \dfrac{\mu_0 NI}{2R}$

where, R is radius and I is current flowing in coil.

Similarly, $B_B = \dfrac{\mu_0 N(2I)}{2(2R)} = \dfrac{\mu_0 NI}{2R}$

$\therefore$ $\dfrac{B_A}{B_B} = 1$

32. (a) Since, momenta are same and masses of electron and proton are different, so they will attain different velocities and hence experience different forces. But, radius of circular path is depending on momentum, so both will be moving on same trajectory (curved path).

33. (a) In a circular motion of a uniform magnetic field, the necessary centripetal force to the charged particle is provided by the magnetic force.

i.e. $\dfrac{mv^2}{r} = qvB$ or $r = \dfrac{mv}{qB}$

Thus, the time period T is

$$T = \frac{2\pi r}{v} = \frac{2\pi}{v}\left(\frac{mv}{qB}\right) = \frac{2\pi m}{qB}$$

So, T is independent of its speed.

34. (a) Electrons, protons and helium atoms are deflected in magnetic field, so the compound can emit electrons, protons and He^{2+}.

35. (c) The component $dl \cos\theta$ of element dl is parallel to the length of the wire 1. Hence, force on this elemental component

$$F = \frac{\mu_0}{4\pi} \cdot \frac{2I_1I_2}{r}(dl \cos\theta) = \frac{\mu_0 I_1 I_2\, dl \cos\theta}{2\pi r}$$

36. (d) As, the side BC is outside the field, no emf is induced across BC. Since, AB and CD are not cutting any flux, the emf induced across these two sides will also be zero.

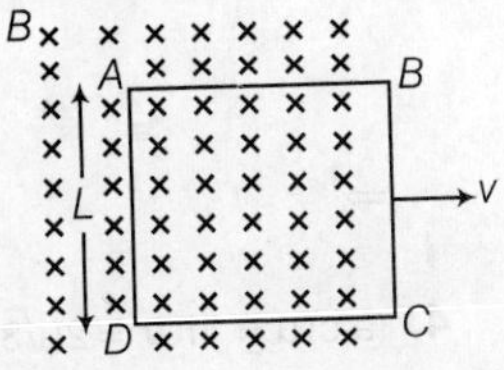

The side AD is cutting the flux and emf induced across this side is BvL with corner A at higher potential.

(according to Lorentz force, force on the charges will be towards A and on negative charges will be towards D)

37. (c) A voltmeter is a high resistance device and is always connected in parallel with the circuit. While an ammeter is a low resistance device and is always connected in series with the circuit.

So, to use ammeter in place of voltmeter, a high resistance must be connected in series with the ammeter to make its resistance high.

38. (b) Due to flow of current in same direction in two adjacent sides, an attractive magnetic force will be produced due to which spring will get compressed.

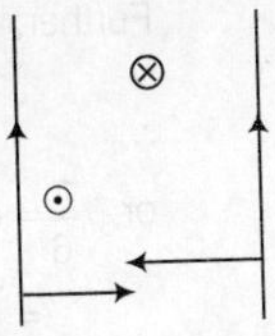

JEE Advanced & IIT JEE Archive

1. (c) B_R = B due to ring, B_1 = B due to wire 1

$\Rightarrow$ B_2 = B due to wire 2

In magnitudes, $B_1 = B_2 = \dfrac{\mu_0 I}{2\pi r}$

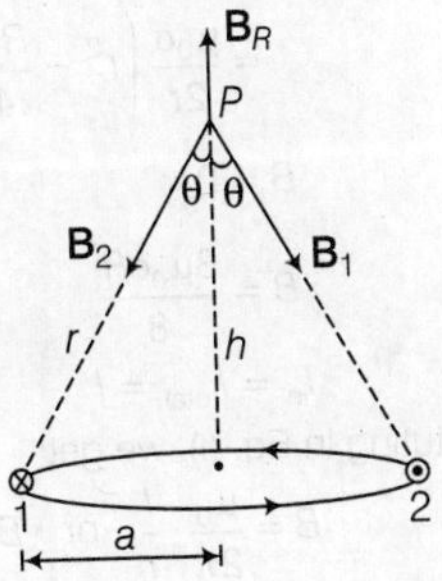

Resultant of B_1 and $B_2 = 2B_1 \cos\theta$

$$= 2\left(\frac{\mu_0 I}{2\pi r}\right)\left(\frac{h}{r}\right) = \frac{\mu_0 Ih}{\pi r^2}$$

$$B_R = \frac{\mu_0 IR^2}{2(R^2 + x^2)^{3/2}}$$

$$= \frac{2\mu_0 Ia^2}{2r^3}$$

As, $R = a,\ x = h$

and $a^2 + h^2 = r^2$

For zero magnetic field at P,

$\Rightarrow$ $h \approx 1.2\ a$

2. (b) Magnetic field at mid-point of two wires

$= 2$ (magnetic field due to one wire)

$= 2\left[\frac{\mu_0}{2\pi}\frac{I}{d}\right] = \frac{\mu_0 I}{\pi d} \otimes$

Magnetic moment of loop,

$M = IA = I\pi a^2$

Torque on loop $= MB \sin 30^\circ = \frac{\mu_0 I^2 a^2}{2d}$

3. (3) $B_2 = \frac{\mu_0 I}{2\pi x_1} + \frac{\mu_0 I}{2\pi (x_0 - x_1)}$

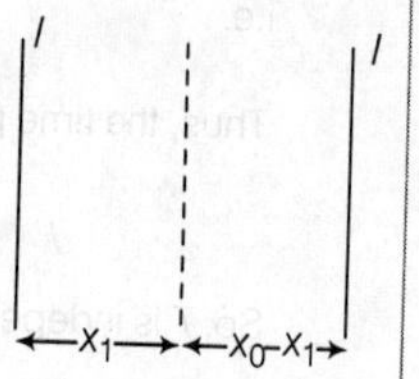

(when currents are in opposite directions)

$B_1 = \frac{\mu_0 I}{2\pi x_1} - \frac{\mu_0 I}{2\pi (x_0 - x_1)}$

(when currents are in same direction)

Substituting $x_1 = \frac{x_0}{3}$, (as $\frac{x_0}{x_1} = 3$)

$$B_1 = \frac{3\mu_0 I}{2\pi x_0} - \frac{3\mu_0 I}{4\pi x_0} = \frac{3\mu_0 I}{4\pi x_0}$$

$$R_1 = \frac{mv}{qB_1}$$

and $$B_2 = \frac{9\mu_0 I}{4\pi x_0}$$

$$R_2 = \frac{mv}{qB_2}$$

$\Rightarrow$ $$\frac{R_1}{R_2} = \frac{B_2}{B_1} = \frac{9}{3} = 3$$

4. (a,c) $\mathbf{u} = 4\hat{\mathbf{i}}, \mathbf{v} = 2(\sqrt{3}\hat{\mathbf{i}} + \hat{\mathbf{j}})$

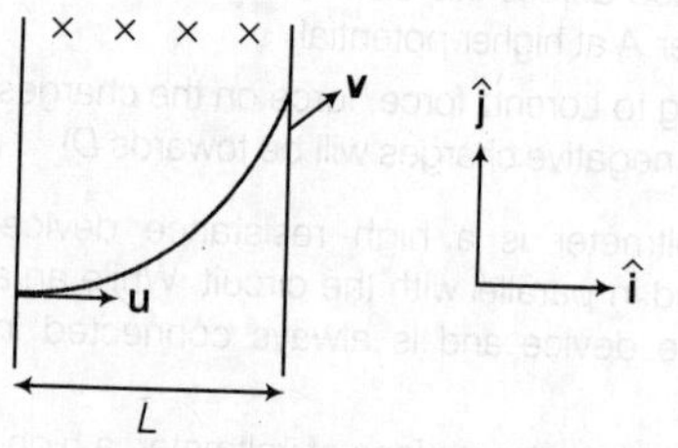

According to the figure, magnetic field should be in $\otimes$ direction or along $-Z$-direction.

Further, $\tan\theta = \frac{v_y}{v_x} = \frac{2}{2\sqrt{3}} = \frac{1}{\sqrt{3}}$

$\therefore$ $\theta = 30^\circ$

or $\frac{\pi}{6}$ = angle of **v** with X-axis

= angle rotated by the particle

$= Wt = \left(\frac{BQ}{M}\right)t$

$\therefore$ $B = \frac{\pi M}{6Qt} = \frac{50\pi M}{3Q}$ units (as $t = 10^{-3}$ s)

5. (b,c) In the region, $0 < r < R$

$B_P = 0$, $B_Q \neq 0$, along the axis

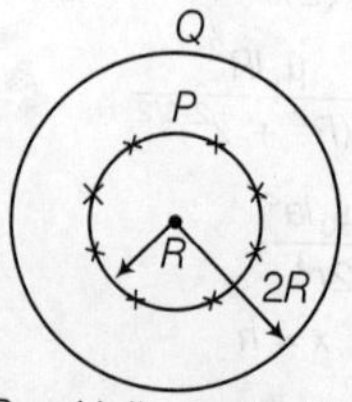

$P \rightarrow$ Hollow cylindrical conductor

$Q \rightarrow$ Solenoid

$\therefore$ $B_{net} \neq 0$

In the region, $R < r < 2R$

$B_P \neq 0$, tangential to the circle of radius r centred on the axis.

$B_Q \neq 0$, along the axis.

$\therefore B_{net} \neq 0$, neither in the directions mentioned in options (b) or (c).

In region, $r > 2R$

$B_P \neq 0$

$B_Q \neq 0$

$\therefore$ $B_{net} \neq 0$

6. (c,d) When $\theta = 0^\circ$, charge particle rotates under influence of magnetic field and its motion along Y-axis is accelerating due to electric field. So, its radius of path is constant but its pitch keeps on increasing due to increase in velocity along Y-axis. So, option (c) is correct. Also, option (d) is correct.

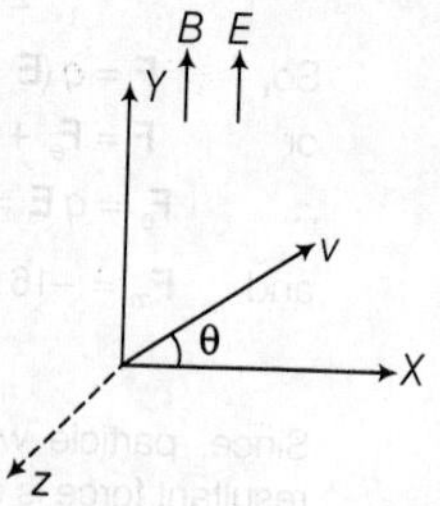

7. (b) Area of the given loop is

A = (area of two circles of radius $\frac{a}{2}$

+ area of a square of side a)

$$= 2\pi\left(\frac{a}{2}\right)^2 + a^2 = \left(\frac{\pi}{2} + 1\right)a^2$$

$$|\mathbf{M}| = IA = \left(\frac{\pi}{2} + 1\right)a^2 I$$

From screw law, the direction of **M** is outwards or in positive Z-direction.

$\therefore$ $$\mathbf{M} = \left(\frac{\pi}{2} + 1\right)a^2 I\hat{\mathbf{k}}$$

8. (d) r = distance of a point from centre

For $r \leq R/2$

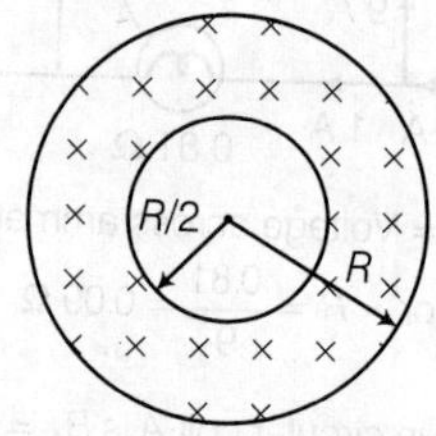

Using Ampere's circuital law,

$\oint \mathbf{B} \cdot d\mathbf{l}$ or $Bl = \mu_0 (I_{in})$

or $B(2\pi r) = \mu_0 (I_{in})$

or $B = \frac{\mu_0}{2\pi} \cdot \frac{I_{in}}{r}$ ...(i)

Since, $I_{in} = 0$

$\therefore$ $B = 0$

For $\frac{R}{2} \leq r \leq R$, $I_{in} = \left[\pi r^2 - \pi\left(\frac{R}{2}\right)^2\right]\sigma$

Here, σ = current per unit area. Substituting in Eq. (i), we get

$$B = \frac{\mu_0}{2\pi}\frac{\left[\pi r^2 - \pi\frac{R^2}{4}\right]\sigma}{r}$$

$$= \frac{\mu_0 \sigma}{2r}\left(r^2 - \frac{R^2}{4}\right)$$

At $r = \frac{R}{2}$, $B = 0$

At $r = R$, $B = \frac{3\mu_0 \sigma R}{8}$

For $r \geq R$, $I_{in} = I_{total} = I$ (say)

Therefore, substituting in Eq. (i), we get

$$B = \frac{\mu_0}{2\pi} \cdot \frac{I}{r} \text{ or } B \propto \frac{1}{r}$$

9. (5) $B_R = B_T - B_C$

where, R = remaining portion, T = total portion

and C = cavity

$$B_R = \frac{\mu_0 I_T}{2a\pi} - \frac{\mu_0 I_C}{2\,(3a/2)\pi} \qquad \ldots(\text{i})$$

$$I_T = J\,(\pi a^2)$$

$$I_C = J\left(\frac{\pi a^2}{4}\right)$$

Substituting the values in Eq. (i), we have

$$B_R = \frac{\mu_0}{a\pi}\left[\frac{I_T}{2} - \frac{I_C}{3}\right]$$

$$= \frac{\mu_0}{a\pi}\left[\frac{\pi a^2 J}{2} - \frac{\pi a^2 J}{12}\right] = \frac{5\,\mu_0 a J}{12}$$

$\therefore$ $N = 5$

10. (c) Correct answer is (c), because induced electric field lines (produced by change in magnetic field) and magnetic field lines form closed loops.

11. (a) If we take a small strip of dr at distance r from centre, then number of turns in this strip would be

$$dN = \left(\frac{N}{b-a}\right)dr$$

Magnetic field due to this element at the centre of the coil will be

$$dB = \frac{\mu_0\,(dN)I}{2\,r} = \frac{\mu_0 NI}{(b-a)}\cdot\frac{dr}{r}$$

$$\therefore \quad B = \int_{r=a}^{r=b} dB = \frac{\mu_0 NI}{2(b-a)}\ln\left(\frac{b}{a}\right)$$

12. (b,d) $r = \dfrac{mv}{Bq}$ or $r \propto m$

$\therefore$ $r_e < r_p$ as $m_e < m_p$

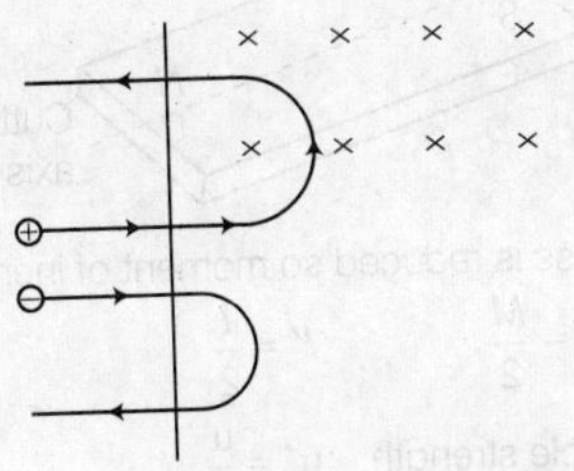

Further $T = \dfrac{2\pi m}{Bq}$ or $T \propto m$

$\therefore$ $T_e < T_p$, $t_e = \dfrac{T_e}{2}$ and $t_p = \dfrac{T_p}{2}$

or $t_e < t_p$

13. (7) Magnetic field at point P due to wires RP and RQ is zero. Only wire QR will produce magnetic field at P.

$$r = 3x\cos 37° = (3x)\left(\frac{4}{5}\right) = \frac{12x}{5}$$

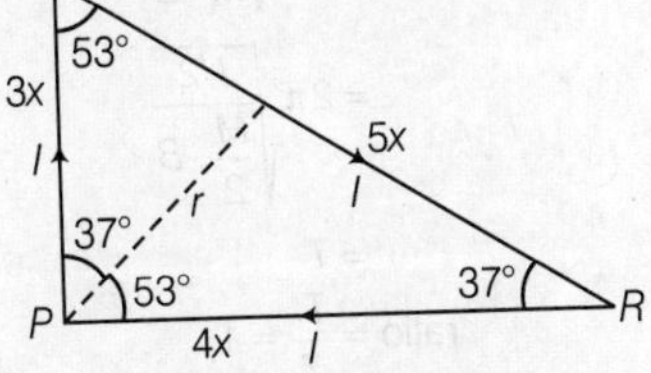

Now, $B = \dfrac{\mu_0}{4\pi}\cdot\dfrac{I}{12\,x/5}[\sin 37° + \sin 53°]$

$$= 7\left(\frac{\mu_0 I}{48\pi x}\right)$$

$\therefore$ $k = 7$

14. (a,c,d) $\mathbf{v} \perp \mathbf{B}$ in Region II. Therefore, path of particle is circle in Region II.

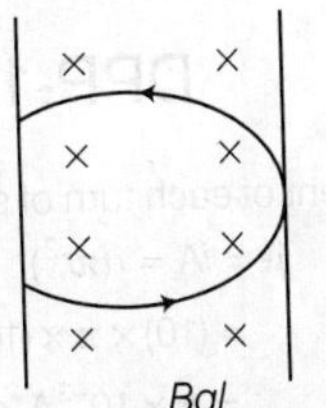

$$v = \frac{Bql}{m}$$

Particle enters in Region III if radius of circular path, $r > l$

or $\dfrac{mv}{Bq} > l$

or $v > \dfrac{Bql}{m}$

If $v = \dfrac{Bql}{m}$, $r = \dfrac{mv}{Bq} = l$, particle will turn back and path length will be maximum. If particle returns to Region I, time spent in Region II will be

$$t = \frac{T}{2} = \frac{\pi m}{Bq}$$

which is independent of v.

15. (c) $\phi = BINA$

$$\therefore \quad \phi = \left(\frac{BNA}{C}\right)I$$

Using iron core, value of magnetic field increases. So, deflection increases for same current. Hence, sensitivity increases. Soft iron can be easily magnetised or demagnetised.

16. (b) (i) $\rightarrow$ (q,r), (ii) $\rightarrow$ (p), (iii) $\rightarrow$ (q,r), (iv) $\rightarrow$ (q,s or q)

17. (a) $\mathbf{F}_m = q\,(\mathbf{v} \times \mathbf{B})$

18. (a,c) $\mathbf{F}_{BA} = 0$, because magnetic lines are parallel to this wire.

$\mathbf{F}_{CD} = 0$, because magnetic lines are anti-parallel to this wire. $\mathbf{F}_{CB}$ is perpendicular to paper outwards and $\mathbf{F}_{AD}$ is perpendicular to paper inwards. These two forces (although calculated by integration) cancel each other but produce a torque which tend to rotate the loop in clockwise direction about an axis OO'.

19. (c) (i) $\rightarrow$ (p), (ii) $\rightarrow$ (p,q,s), (iii) $\rightarrow$ (q,s), (iv) $\rightarrow$ (q,r)

20. (b,d) Electrostatic and gravitational field do not make closed loops.

21. (i) $\tau = MB = ki$

$$\therefore \quad k = \frac{MB}{i} = \frac{(NiA)B}{i} = NBA$$

(ii) $\tau = k\cdot\theta = BiNA$

$$\therefore \quad k = \frac{2BiNA}{\pi} \qquad (\text{as } \theta = \pi/2)$$

(iii) $\tau = BiNA$

or $\int_0^t \tau\,dt = BNA\int_0^t i\,dt$

$$I\omega = BNAQ$$

or $$\omega = \frac{BNAQ}{I} \qquad \ldots(\text{i})$$

At maximum deflection, whole kinetic energy (rotational) will be converted into potential energy of spring.

Hence, $\dfrac{1}{2}I\omega^2 = \dfrac{1}{2}k\theta_{max}^2$

Substituting the values, we get

$$\theta_{max} = Q\sqrt{\frac{BN\pi A}{2Ii}}$$

2. Magnetism

DPP-1 Bar Magnet and Magnetic Field Lines

1. (i) The dipole moment of each turn of solenoid,

$$\mu = iA = i(\pi r^2)$$
$$= (10) \times \pi \times (10^{-4})$$
$$= \pi \times 10^{-3} \text{A-m}^2$$

If each current loop is replaced by a dipole having pole strength M and separation between the poles d, we have

$$\mu = nd$$

As, $n = 200$

$\Rightarrow$ $200\,d = 10$

or $d = 5 \times 10^{-4}$m

$$\Rightarrow \quad M = \frac{\mu}{d} = \frac{\pi \times 10^{-3} \text{A-m}^2}{5 \times 10^{-4}\text{ m}}$$
$$= 2\pi \text{ A-m}$$

The solenoid is replaced by a bar magnet having same pole strength.

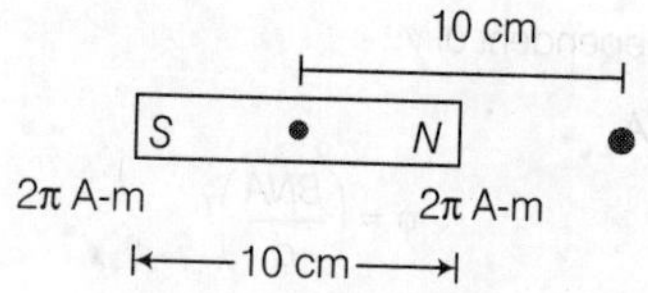

$$\Rightarrow \quad B_N = B_P \text{ (due to North pole)}$$
$$= \frac{\mu_0}{4\pi} \frac{2\pi \text{ A-m}}{(5\text{ cm})^2}$$
$$= 2.5 \times 10^{-4}\text{T}$$

Also, $B_S = B_P$ (due to South pole)

$$= \frac{\mu_0}{4\pi} \frac{2\pi \text{ A-m}}{(15\text{cm})^2}$$
$$= 0.3 \times 10^{-4}\text{T}$$

(ii) The field B_N is away from the poles and B_S is towards the poles. Hence, $B_{\text{net}} = B_N - B_S = 2.2 \times 10^{-4}$ T (directed away).

2. (i)

(a) As, system of magents does not show any motion.

$\therefore$ $m_{\text{net}} = 0$

This is possible when dipole moments of II and III form a resultant which is opposite to the dipole moment of III.

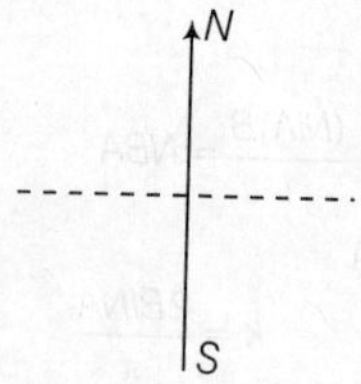

This can be achieved if one position 2 other magnets as shown in the figure.

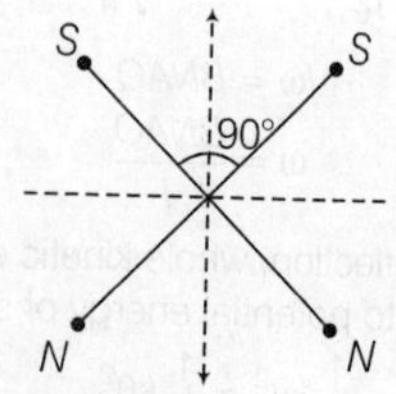

Resultant dipole moment

(b) Also, the external field must be uniform with respect to position (not time).

(ii)

(a) When magnet is cut perpendicular to its length,

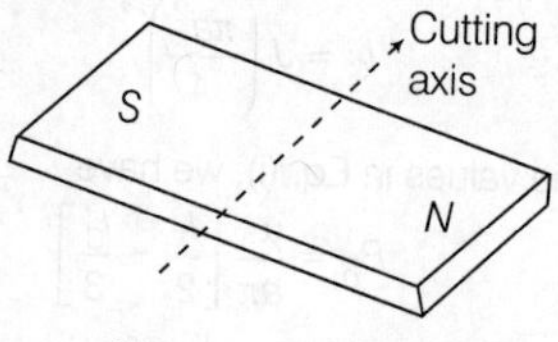

$$l = \frac{l}{2}, \quad M' = \frac{M}{2}, \quad I' = \frac{I}{8}$$

Pole strength μ remains same.

So, magnetic dipole moment of new $\frac{1}{2}$ part of magnet will

$$\mathbf{M} = \mu' l' = \mu . \frac{l}{2} = \frac{\mu l}{2} = \frac{M}{2}$$

So, $\text{ratio} = \frac{M'}{M} = \frac{1}{2}$

and $T' = 2\pi \sqrt{\frac{I'}{M \cdot B}}$

$$= 2\pi \sqrt{\frac{\frac{I}{8}}{\frac{M}{2} \cdot B}} = \frac{1}{2}(T)$$

$$\Rightarrow \quad \frac{T'}{T} = \frac{1}{2}$$

(b) Cutting along the length reduces pole strength but length remains same.

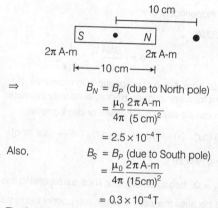

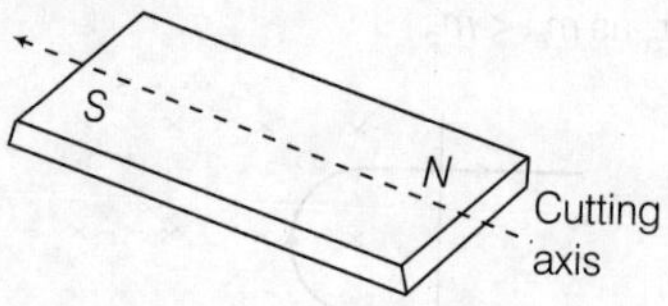

Also, mass is reduced so moment of inertia is also reduced.

So, $M' = \frac{M}{2}$, $I' = \frac{I}{2}$

$l' = l$, pole strength, $\mu' = \frac{\mu}{2}$

As, dipole moment, $\mathbf{M} = \mu\, l$

and $\mathbf{M}' = \mu' l' = \frac{\mu}{2} \cdot l$

$$= \frac{\mu l}{2} = \frac{\mathbf{M}}{2}$$

As, time period of oscillation is

$$T = 2\pi \sqrt{\frac{I}{MB}}$$
$$T' = 2\pi \sqrt{\frac{I'}{M' B}}$$
$$= 2\pi \sqrt{\frac{I/2}{\frac{M}{2} \cdot B}}$$
$$= T$$

So, $\text{ratio} = \frac{T'}{T} = 1$

$$\frac{\mathbf{M}'}{\mathbf{M}} = \frac{1}{2}$$

3. Here, magnetic moment, $M_1 = M_2 = 12.5\,\text{A-m}^2$

As, $O_1O_2 = 10\,\text{cm}$

So, P is mid-way between O_1 and O_2.

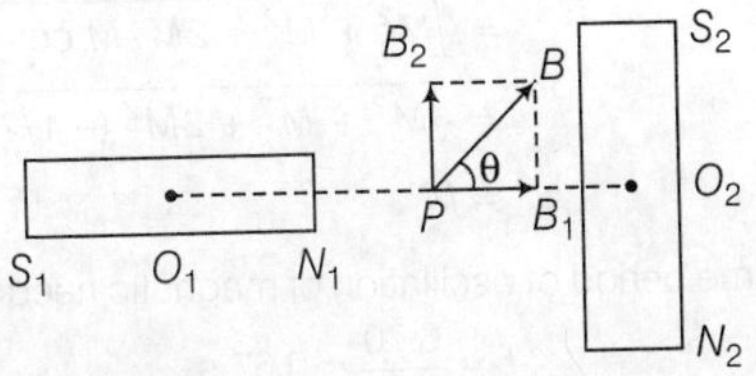

$\therefore \quad d = O_1P = O_2P = 5\,\text{cm} = 5 \times 10^{-2}\,\text{m}$

As, P lies on axial line of N_1S_1,

$$\therefore \quad B_1 = \frac{\mu_0}{4\pi} \cdot \frac{2M_1}{d^3} = \frac{10^{-7} \times 2 \times 12.5}{(5 \times 10^{-2})^3} = 0.02\ \text{T (along } O_1PO_2).$$

Again, P lies on equatorial line of N_2S_2.

$$\therefore \quad B_2 = \frac{\mu_0}{4\pi} \cdot \frac{M}{d^3} = \frac{10^{-7} \times 12.5}{(5 \times 10^{-2})^3} = 0.01\,\text{T (perpendicular to } PO_2).$$

Resultant magnetic field strength at P,

$$B = \sqrt{B_1^2 + B_2^2} = \sqrt{(0.02)^2 + (0.01)^2} = 2.24 \times 10^{-2}\ \text{T}$$

If θ is the angle which B_2 makes with B_1, then

$$\tan\theta = \frac{B_2}{B_1} = \frac{0.01}{0.02} = 0.5,\ \theta = \tan^{-1}(0.5)$$

$$\Rightarrow \quad \theta = 26.6^\circ$$

4. For circular coil C_1, $n_1 = \dfrac{L}{2\pi R}$

Magnetic moment, $M_1 = n_1 i A_1 = \dfrac{L}{2\pi R} \times i \times \pi R^2 = \dfrac{LiR}{2}$

For square coil C_2, $n_2 = \dfrac{L}{4a}$

Magnetic moment, $M_2 = n_2 i A_2 = \dfrac{L}{4a} \times i \times a^2 = \dfrac{Lia}{4}$

Moment of inertia of circular coil about the diameter as axis,

$$I_1 = \frac{\text{Mass} \times (\text{radius})^2}{2} = \frac{MR^2}{2}$$

Moment of inertia of square coil about an axis passing through its centre parallel to breadth,

$$I_2 = \frac{Ma^2}{12}$$

Time period of oscillation of the magnet in magnetic field is given by

$$T = 2\pi\sqrt{\frac{I}{MB}} \quad \text{and} \quad \omega = \frac{2\pi}{T} = \sqrt{\frac{MB}{I}}$$

$$\omega_1^2 = \frac{M_1 B}{I_1}$$

and $\quad \omega_2^2 = \dfrac{M_2 B}{I_2}$

Given, $\quad \omega_1^2 = \omega_2^2$

$$\Rightarrow \quad \frac{M_1}{I_1} = \frac{M_2}{I_2}$$

or $\quad \dfrac{LiR/2}{MR^2/2} = \dfrac{Lia/4}{Ma^2/12}$

On solving, we get

$$a = 3R$$

5. (a) Work done in turning the dipole through an angle of 60°.

$$W = mB(\cos 0^\circ - \cos 60^\circ) = mB(1 - 0.5)$$

or $\quad mB = 2W$

Torque needed to maintain the needle in the position,

$$\tau = mB \sin 60^\circ = 2W \times \frac{\sqrt{3}}{2} = \sqrt{3}\,W$$

6. (d) As, magnetic moment of a bar magnet $M = mL$ and magnetic force,

$$F = mB \Rightarrow F = \frac{M}{L} \times B$$

Length of the magnet,

$$L = \frac{M}{F} \times B = \frac{3\ \text{A-m}}{6 \times 10^{-4}\ \text{N}} \times 2 \times 10^{-5}\ \text{T} = 0.1\,\text{m}$$

7. (d) Force between two bar magnets, $F \propto \dfrac{1}{d^4}$

Case I Force between bar magnets whose centres are r metre apart is

$$\because \quad F_1 \propto \frac{1}{r^4} \quad \text{...(i)}$$

Case II When distance between two bar magnets is increased by $2r$, we get

$$F_2 \propto \frac{1}{(2r)^4} \quad \text{...(ii)}$$

On dividing Eq. (i) and Eq. (ii), we get

$$\frac{F_1}{F_2} \propto \frac{d_2^4}{d_1^4}$$

$$\Rightarrow \quad \frac{F_1}{F_2} \propto \frac{(2r)^4}{(r)^4}$$

$$\Rightarrow \quad F_2 = \frac{4.8}{(2)^4} = 0.3\,\text{N}$$

8. (b) Magnetic field at equatorial line having distance of 2 m with their axis.

i.e. $\quad B_1 = \dfrac{\mu_0}{4\pi} \times \dfrac{m}{(d/2)^3}$

and magnetic field at axial to the plane,

i.e. $\quad B_2 = \dfrac{\mu_0}{4\pi} \times \dfrac{2m}{(d/2)^3}$

Net resultant magnetic field at a point,

$$B = \sqrt{B_1^2 + B_2^2} = \frac{\mu_0 m}{4\pi\,(d/2)^3}\sqrt{(1)^2 + (2)^2}$$

$$= \frac{10^{-7} \times 1}{\left(\frac{2}{2}\right)^3} \times \sqrt{5} = \sqrt{5} \times 10^{-7}\ \text{T}$$

9. (c) Given, pole strength, $m = 120$ CGS units $= 12$ A-m

Magnetic length is $2l = 10$ cm or $l = 0.05$ m

$$d = 20\,\text{cm} = 0.2\,\text{m}$$

$$\Rightarrow \quad B = \frac{\mu_0}{4\pi} \cdot \frac{2Md}{(d^2 - l^2)^2} = \frac{\mu_0}{4\pi} \cdot \frac{4mld}{(d^2 - l^2)^2} \quad (\because M = 2ml)$$

$$= \frac{10^{-7} \times 4 \times 12 \times 0.05 \times 0.2}{[(0.2)^2 - (0.05)^2]^2} = 3.4 \times 10^{-5}\ \text{T}$$

10. (c) The resultant field on axial line at P is given by

$$B = B_N - B_S = \frac{\mu_0}{4\pi} \cdot \frac{2Md}{(d^2 - l^2)^2}$$

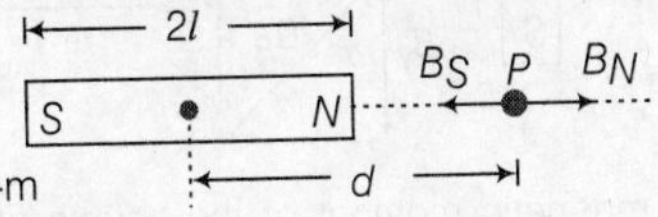

Here, $d = r$

$$\Rightarrow \quad B = \frac{\mu_0}{4\pi} \cdot \frac{4Mr}{(r^2 - l^2)^2}$$

Only for a particular case when $d >> r$, then

$$B = \frac{\mu_0 2M}{4\pi r^3}$$

or $\quad B \propto \dfrac{1}{r^3}$

11. (c) Given, magnetic moment, $m = 20$ CGS unit, magnetic field, $B = 0.3$ CGS unit.

So, amount of work done in deflection of a magnet by an angle of 30°, i.e.

$$W = +mB(\cos\theta_1 - \cos\theta_2)$$
$$= +20 \times 0.3 \times (\cos 0° - \cos 30°)$$
$$= +6 \times \left[1 - \frac{\sqrt{3}}{2}\right] = +6 \times \left[\frac{2-\sqrt{3}}{2}\right]$$
$$= 3(2 - \sqrt{3})$$

12. (d) Magnetic field at a point x on a axial plane,

$$B_{\text{axial}} = \frac{\mu_0}{4\pi} \times \frac{2m}{x^3} \quad \dots(\text{i})$$

Magnetic field at an equatorial plane,

$$B_{\text{equatorial}} = \frac{\mu_0}{4\pi} \times \frac{m}{y^3} \quad \dots(\text{ii})$$

Given, $B_{\text{axial}} = B_{\text{equatorial}}$ ...(iii)

On comparing Eqs. (i), (ii) and (iii), we get

$$\frac{\mu_0}{4\pi} \times \frac{m}{y^3} = \frac{\mu_0}{4\pi} \times \frac{2m}{x^3}$$

or $\frac{x}{y} = (2)^{1/3}$

13. (b) When a bar magnet is suspended freely in the earth's magnetism, it always aligns itself in the direction of field.

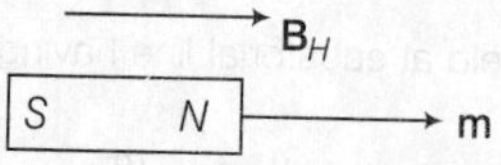

(i.e. along N-S direction)

So, by potential energy stored in a bar magnet,

$$U = -mB_H \cos\theta$$
$$U = -mB_H \cos 0°$$
$$= -2 \times 25 = -50\,\mu J$$

Loss in PE = Gain in KE

KE of magnet at N-S position $= U_f - U_i$

$$= 0 - (-50)$$
$$= +50\,\mu J$$

14. (c) As, magnetic field on end-on position,

$$B_{\text{axial}} = 9 = \frac{2M}{x^3} \quad \dots(\text{i})$$

and magnetic field on broadside $B_{\text{equatorial}} = \frac{M}{\left(\frac{x}{2}\right)^3} = \frac{8M}{x^3}$ ...(ii)

On comparing Eqs. (i) and (ii), we get

$$B_{\text{equatorial}} = 36 \text{ gauss}$$

15. (c) Consider two equal bar magnets are kept as shown in the figure. The direction of resultant magnetic field as given by

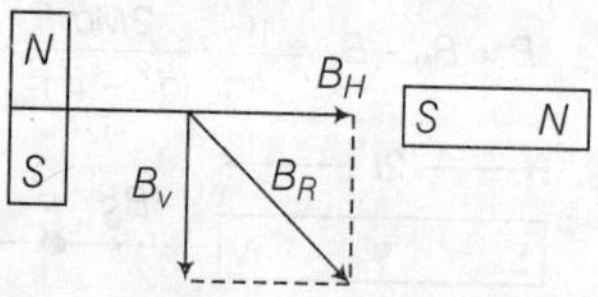

16. (c) Net magnetic moment of the couple of magnets is $\sqrt{M^2 + M^2} = M\sqrt{2}$ at 45° (mid-way). So, point P lies on axial line of net magnetic moment. Therefore,

$$B = \frac{\mu_0}{4\pi} \frac{2\sqrt{2}\,M}{d^3}$$

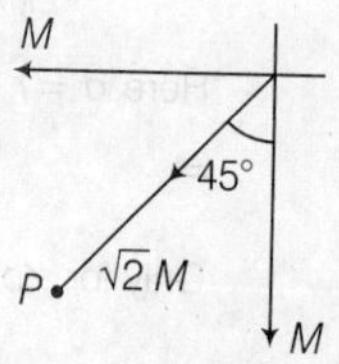

17. (a) As, magnetic moment are directed along SN, angle between magnetic dipole moments is 120°.

Resultant magnetic moment

$$= \sqrt{M^2 + M^2 + 2M \cdot M \cos 120°}$$
$$= \sqrt{M^2 + M^2 + 2M^2(-1/2)}$$
$$= M$$

18. (c) As, time period of oscillation of magnetic needle

i.e. $T = \frac{6.70}{10} = 0.67$ s

Magnetic field on a needle,

$$B = \frac{4\pi^2 I}{MT^2}$$
$$= \frac{4 \times (3.14)^2 \times (7.5 \times 10^{-6})}{6.7 \times 10^{-2} \times (0.67)^2}$$
$$= 0.01 \text{ T}$$

19. (c) Magnetic field lines form closed loop. Outside the magnet, these are directed from North pole towards South pole and inside the magnet. These are directed from South pole towards North pole.

20. (a) Couple between two small magnets.

Case I The magnets being in the end-on position.

$$\text{Couple} = \frac{\mu_0}{4\pi} \cdot \frac{2MM'}{x^3}$$

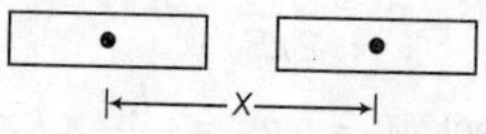

Case II The magnets being in the broadside-on position (two magnets are at right angle).

$$\text{Couple} = \frac{\mu_0}{4\pi} \cdot \frac{MM'}{y^3}$$

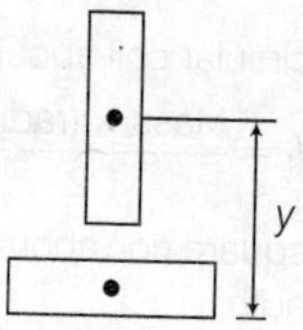

Force between two small magnets.

Case III The magnet being in the end-on position with respect to deflecting magnet.

$$F = \frac{\mu_0}{4\pi}\left[\frac{6MM'}{x^4}\right]$$

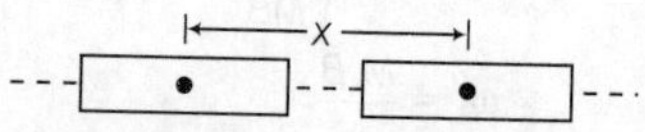

Case IV The magnet being in the broadside-on position with respect to the deflecting magnet.

$$F = \frac{\mu_0}{4\pi}\left[\frac{3MM'}{y^4}\right]$$

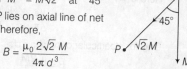

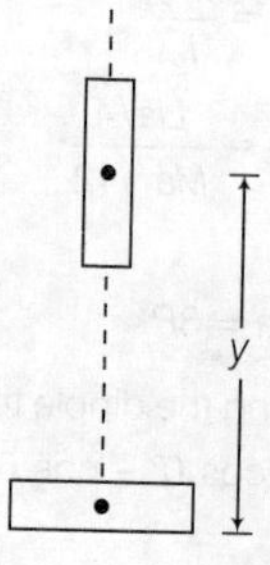

DPP-2 Earth's Magnetism

1. Magnetic field of earth can be approximated or magnetic field of a dipole (dipole moment M).

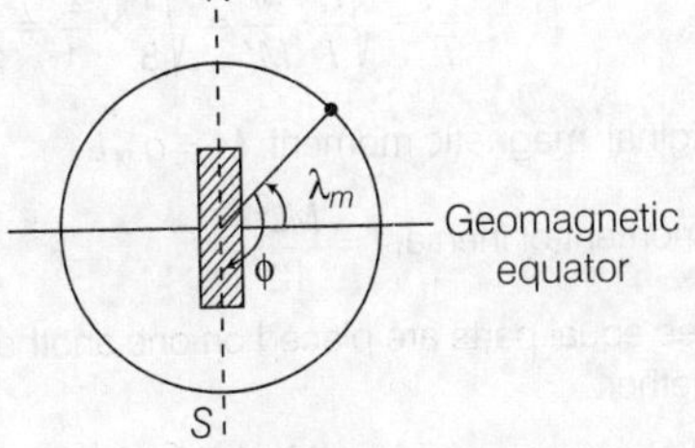

So, magnitude of earth's magnetic field at latitude λ_m is given by

$$B = \frac{\mu_0}{4\pi r^3} \cdot \sqrt{1 + 3\sin^2 \lambda_m}$$

Also, inclination angle ϕ is related to λ_m as $\tan\phi = 2\tan\lambda_m$.

(i) At geomagnetic equator, $\lambda_m = 0$

$$\therefore \quad B = \frac{\mu_0 M}{4\pi r^3} = \frac{4\pi \times 10^{-7} \times 8 \times 10^{22}}{4\pi \times 6.4 \times 10^6}$$

$$= 3 \times 10^{-5} \text{ T}$$

and $\quad \tan\phi = 2\tan\lambda_m = 2\tan 0 = 0$

$\Rightarrow \quad \phi = 0$

(ii) At $\lambda_m = 60°$,

$$B = \frac{\mu_0 M}{4\pi r^3} \sqrt{1 + 3\sin^2 \lambda_m}$$

$$= \frac{4\pi \times 10^{-7} \times 8 \times 10^{22}}{4\pi \times 6.4 \times 10^6} \times \sqrt{1 + 3\sin^2 60} = 5.5 \times 10^{-5} \text{ T}$$

Also, ϕ = inclination

$$= \tan^{-1}(2\tan 60°)$$

$$= \tan^{-1}(2\sqrt{3})$$

(iii) At North pole, $\quad \lambda_m = 90°$

$$\Rightarrow \quad B = \frac{\mu_0 M}{4\pi r^2} \sqrt{1 + 3} = 6 \times 10^{-5} \text{ T}$$

and $\quad \phi = \tan^{-1}(2\tan\lambda_m)$

$$= \tan^{-1}(\infty) = 90°$$

2. Magnetic moment of bar magnet, $M = 5.25 \times 10^{-2} \text{ JT}^{-1}$

Magnitude of earth's magnetic field, $H = 0.42 \times 10^{-4}$ T

(i) At a distane r on normal bisector of bar magnet $B = \dfrac{\mu_0 M}{4\pi r^3}$

When resultant magnetic field is inclined at 45°, $B = H$

$$\Rightarrow \quad \frac{\mu_0 M}{4\pi R^3} = H = 0.42 \times 10^{-4}$$

$$\Rightarrow \quad R^3 = \frac{\mu_0 M}{0.42 \times 10^{-4} \times 4\pi}$$

$$= \frac{4\pi \times 10^{-7} \times 5.25 \times 10^{-2}}{0.42 \times 10^{-4} \times 4\pi}$$

$$= 12.5 \times 10^{-5}$$

$$\Rightarrow \quad R = 0.05 \text{m} = 5 \text{ cm}$$

(ii) Magnetic field at distance r from centre on the axis is

$$B = \frac{\mu_0 2M}{4\pi r^3}$$

As resultant field is at 45° with the earth's field.

$$\therefore \quad B = 4$$

$$\Rightarrow \quad \frac{\mu_0 2M}{4\pi r^3} = 4$$

$$\Rightarrow \quad r^3 = \frac{4\pi \times 10^{-7} \times 2 \times 5.25 \times 10^{-2}}{4\pi \times 0.42 \times 10^{-4}}$$

$$= 25 \times 10^{-5}$$

$$\Rightarrow \quad r = 0.063 \text{ m} = 6.3 \text{ cm}$$

3. (i) $|\mathbf{B}| = \dfrac{\mu_0}{4\pi} \cdot \dfrac{M}{R^3}(4\cos^2\theta + \sin^2\theta)^{1/2} = \dfrac{\mu_0 M}{4\pi R^3}(3\cos^2\theta + 1)^{1/2}$

$\therefore$ $|\mathbf{B}|$ is minimum at $\theta = \dfrac{\pi}{2}$.

$\Rightarrow |\mathbf{B}|$ is minimum at magnetic equator.

(ii) $\tan\delta = \dfrac{B_V}{B_H} = 2\cot\theta$

$\therefore$ At $\theta = \dfrac{\pi}{2}$, $\delta = 0$

which gives magnetic equator as locus.

(iii) From given figure,

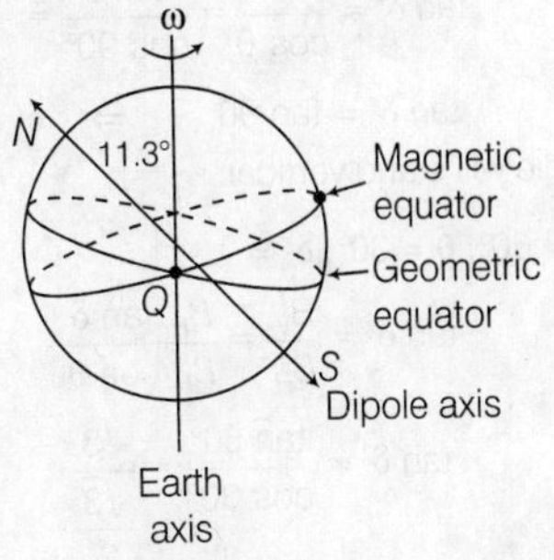

P is on magnetic equator.

$\therefore \quad$ dip = 0

Q is on magnetic equator.

$\therefore \quad$ dip = 0

For P is in plane S.

$\therefore \quad$ Declination is zero and for Q, declination is 11.3°.

4. The time period of oscillation is given by $T = 2\pi\sqrt{\dfrac{I}{MB_H}}$

where, B_H is magnetic field along horizontal component, I is moment of inertia and M is magnetic moment.

So, the time period at the first place is $T_1 = \dfrac{1}{20}$ min = 3.0 s and at the second place, it is $T_2 = \dfrac{1}{30}$ min = 2.0 s.

If the total magnetic field at the first place is B_1, the horizontal component of the field is $B_{H_1} = B_1 \cos 45° = \dfrac{B_1}{\sqrt{2}}$

Similarly, if the total magnetic field at the second place is B_2, then the horizontal component is

$$B_{H_2} = B_2 \cos 30° = B_2 \times \frac{\sqrt{3}}{2} = \frac{\sqrt{3}B_2}{2}$$

We have

$$T_1 = 2\pi\sqrt{\frac{1}{MB_{H_1}}} \text{ and } T_2 = 2\pi\sqrt{\frac{1}{MB_{H_2}}}$$

Thus, $\quad \dfrac{T_1}{T_2} = \sqrt{\dfrac{B_{H_2}}{B_{H_1}}}$ or $\dfrac{B_{H_2}}{B_{H_1}} = \dfrac{T_1^2}{T_2^2}$

or $\quad \dfrac{\frac{B_2\sqrt{3}}{2}}{\frac{B_1}{\sqrt{2}}} = \dfrac{T_1^2}{T_2^2}$

$$\Rightarrow \quad \frac{B_2}{B_1} = \sqrt{\frac{2}{3}} \times \frac{T_1^2}{T_2^2} = \sqrt{\frac{2}{3}} \times \frac{9}{4} = 1.83$$

5. (a) Here, $\phi_1 = 30°$ and $\phi_2 = 45°$. Ratio of horizontal magnetic field B_H.

$$B_H = B\cos\phi = \frac{(B_H)_1}{(B_H)_2} = \frac{B\cos 30°}{B\cos 45°} = \frac{\sqrt{3}/2}{1/\sqrt{2}} = \sqrt{\frac{3}{2}}$$

6. (d) Suppose B_H and B_V be the horizontal and vertical components of the earth's magnetic field **B**. Since, δ is the true angle of dip.

Therefore, $$\tan\delta = \frac{B_V}{B_H}$$

$$\tan 60^\circ = \frac{B_V}{0.38\times10^{-4}}$$

$$\sqrt{3}\times 0.38\times10^{-4} = B_V$$

$$\Rightarrow \quad B_V = 0.658\times10^{-4}\text{ T}$$

7. (b) When the dip circle is at right angles to magnetic meridian, vertical component of the earth's magnetic field is effective. The dip needle stands vertical. Therefore, apparent dip, $\delta = 90^\circ$

8. (b) As angle of dip,

$$\tan\delta' = \frac{\tan\delta}{\cos\theta} = \frac{\tan\delta}{\cos 90^\circ} = \frac{\tan\delta}{0} = \infty$$

$$\tan\delta' = \tan 90^\circ \quad \Rightarrow \quad \delta' = 90^\circ$$

So, dip needle will stand vertical.

9. (b) Here, $\delta = 60^\circ$, $\theta = 30^\circ$, $\delta' = ?$

As, $$\tan\delta' = \frac{B_V}{B_H} = \frac{B_H\tan\delta}{B_H\cos\theta}$$

$$\tan\delta' = \frac{\tan 60^\circ}{\cos 30^\circ} = \frac{\sqrt{3}}{\frac{\sqrt{3}}{2}}$$

$$\tan\delta' = 2 \quad \Rightarrow \quad \delta' = \tan^{-1}(2)$$

10. (c) At neutral point magnetic field due to magnet = B_H

$$\Rightarrow \quad \frac{\mu_0}{4\pi}\frac{2m}{r^3} = B_H \Rightarrow \frac{10^{-7}\times 2m}{r^3} = B_H$$

$$m = \frac{B_H\times r^3}{2\times10^{-7}} = \frac{0.3\times10^{-4}\times(0.20)^3}{2\times10^{-7}}$$

$$= 1.2\text{ A-m}^2$$

11. (b) For equilibrium of the system torques on M_1 and M_2 due to B_H must counter balance each other, i.e. $\mathbf{M}_1\times\mathbf{B}_H = \mathbf{M}_2\times\mathbf{B}_H$

$$M_1B_H\sin\theta = M_2\,B_H\sin(90^\circ-\theta)$$

$$\Rightarrow \quad \tan\theta = \frac{M_2}{M_1} = \frac{1}{3}$$

$$\Rightarrow \quad \theta = \tan^{-1}\left(\frac{1}{3}\right)$$

12. (a) A compass needle will experience a torque due to horizontal component of the earth's magnetic field, i.e.

$$\tau = MB_H\sin\theta$$

where, θ is the angle between geographic and magnetic meridian called angle of declination.

$$\Rightarrow \quad 1.2\times10^{-3} = 40\times10^{-6}\times 60\times\sin\theta$$

$$\Rightarrow \quad \frac{1.2\times10^{-3}}{60\times40\times10^{-6}} = \sin\theta$$

$$\Rightarrow \quad \frac{1.2\times10^{-3}\times10^6}{24\times10^2} = \sin\theta = \frac{1}{2}$$

$$\Rightarrow \quad \theta = \sin^{-1}(1/2) = 30^\circ$$

13. (b) As, time period of oscillation of thin rectangular magnet, i.e.

$$T = 2\pi\sqrt{\frac{I}{MB_H}}$$

***Case* I** $$T = 2\pi\sqrt{\frac{I}{MB_H}} \quad \text{...(i)}$$

where, I is moment of inertia of a bar magnet.

$$I = \frac{ml^2}{12}, \text{ where } m = \text{mass}$$

***Case* II** Magnet is cut into two identical pieces

$$T' = 2\pi\sqrt{\frac{I'}{M'B_H}} \quad \text{...(ii)}$$

Comparing ratio of Eq. (ii) by Eq. (i), we get

$$\therefore \quad \frac{T'}{T} = \sqrt{\frac{I'}{I}\cdot\frac{M}{M'}} = \sqrt{\frac{1}{8}\times\frac{2}{1}} = \frac{1}{2} \qquad \left(\because \frac{I'}{I} = \frac{1}{8}\right)$$

14. (b) As original magnetic moment, $M = q_M L$

Original moment of inertia, $I = \dfrac{ML^2}{12}$

When three equal parts are placed on one another with their like poles together.

$$I' = \frac{1}{12}\times\frac{M}{3}\left(\frac{L}{3}\right)^2\times 3 = \frac{1}{12}\cdot\frac{ML^2}{9} = \frac{1}{9}I$$

As, $$M' = 9M\times\frac{L}{3}\times 3 = M$$

where, M' = new pole strength

$$\therefore \quad T = 2\pi\sqrt{\frac{I}{MB_M}}$$

$$T' = 2\pi\sqrt{\frac{I'}{M'B_M}}$$

$$\therefore \quad \frac{T'}{T} = \sqrt{\frac{I'}{I}\cdot\frac{M}{M'}} = \sqrt{\frac{1}{9}\cdot\frac{1}{1}} = \frac{1}{3}$$

$$\Rightarrow \quad T' = \frac{1}{3}T$$

$$\Rightarrow \quad T' = \frac{1}{3}\times 2 = \frac{2}{3}\text{ s}$$

15. (d) As, time period of an oscillating magnetic needle,

$$T = 2\pi\sqrt{\frac{I}{MB_H}}$$

When needle oscillates in vertical plane, it oscillates in total earth's magnetic field (B).

Hence, $$T' = 2\pi\sqrt{\frac{I}{MB}}$$

$$\Rightarrow \quad \frac{T'}{T} = \sqrt{\frac{B_H}{B}} = \sqrt{\frac{B\cos\phi}{B}} = \sqrt{\cos 60^\circ}$$

$$\Rightarrow \quad T' = \frac{T}{\sqrt{2}}$$

16. (c) As, time period of vibration of magnetic needle

$$T_1 = 2\pi\sqrt{\frac{I}{MV}} \quad \text{and} \quad T_2 = 2\pi\sqrt{\frac{I}{MH}}$$

$$\therefore \quad \frac{T_2}{T_1} = \sqrt{\frac{V}{H}} = \sqrt{\tan\theta}$$

or $$\tan\theta = \left(\frac{T_2}{T_1}\right)^2 = \left(\frac{2}{2}\right)^2 = 1$$

$$\Rightarrow \quad \theta = 45^\circ$$

17. (c) Let θ be the declination at the place. As it is clear from figure,

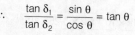

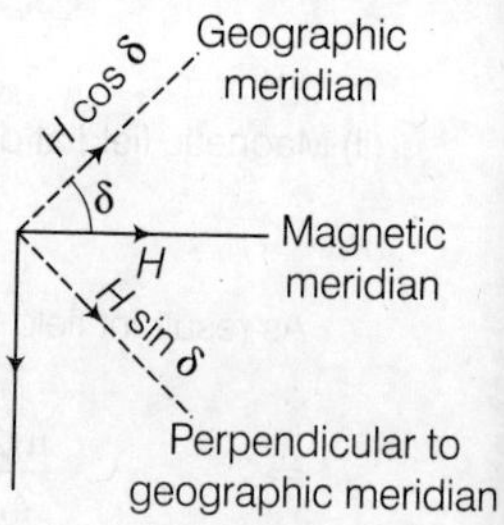

$$\tan\delta_1 = \frac{V}{H\cos\theta}$$

and $$\tan\delta_2 = \frac{V}{H\cos(90^\circ-\theta)}$$

$$= \frac{V}{H\sin\theta}$$

$$\therefore \quad \frac{\tan\delta_1}{\tan\delta_2} = \frac{\sin\theta}{\cos\theta} = \tan\theta$$

$$\Rightarrow \quad \theta = \tan^{-1}\left(\frac{\tan\delta_1}{\tan\delta_2}\right)$$

18. (c) When mass is quadrupled, i.e. made 4 times.
As, I becomes 4 times. As, $T \propto \sqrt{I}$

$\therefore$ T becomes twices, i.e. motion remains SHM with time period $= 2T$

19. (b,c,d) As, the angle of dip at a point on the geographical equator satisfies following conditions.

(i) It can be zero at specific point.

(ii) It can be positive or negative.

(iii) It is bounded by the magnetic field of the earth.

20. (c,d) Time period of oscillation when magnets are tied with like poles together

$$T_1 = 2\pi\sqrt{\frac{I}{(M_1 + M_2)B_H}}$$

When unlike poles are tied.
Time period is

$$T_2 = 2\pi\sqrt{\frac{I}{(M_1 + M_2)B_H}}$$

$$\Rightarrow \quad \frac{T_1}{T_2} = \sqrt{\frac{M_1 - M_2}{M_1 + M_2}}$$

$$\Rightarrow \quad \frac{M_1}{M_2} = \frac{T_2^2 + T_1^2}{T_2^2 - T_1^2}$$

$$= \frac{\left(\frac{60}{4}\right)^2 + \left(\frac{60}{12}\right)^2}{\left(\frac{60}{4}\right)^2 - \left(\frac{60}{12}\right)^2}$$

$$= \frac{15^2 + 5^2}{15^2 - 5^2} = \frac{5}{4}$$

21. (b,d) At neutral point,

$$B_{\text{magnet}} + B_{\text{earth}} = 0$$

$$\Rightarrow \quad B_{\text{magnet}} = -B_{\text{earth}}$$

$$\Rightarrow \quad \frac{2\mu_0 M}{4\pi r^3} = B_H$$

$$\Rightarrow \quad r = 20 \text{ cm}$$

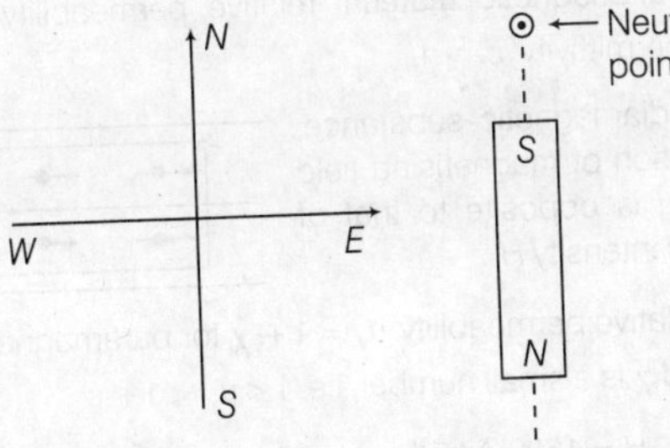

22. (a) As vertical component of the earth's magnetic field,

$$V = H \tan\delta$$

i.e. $$\tan\delta = \frac{V}{H} = 1$$

$$\tan\delta = \tan 45°$$

$$S = 45°$$

So, angle of dip, $\delta = 45°$

23. (c) Magnetic declination is of the order of 20° West at a place on the earth.

24. (d) Isogonic lines joint places of equal declination. Agonic lines pass through zero declination points. Isoclinic lines join points of equal dip. Aclinic line is a magnetic equator (zero dip). Isodynamic lines join places with same value of horizontal component.

25. (2) If ϕ is the correct angle of dip at some place and let it is measured in a plane making angle α with the magnetic meridian.

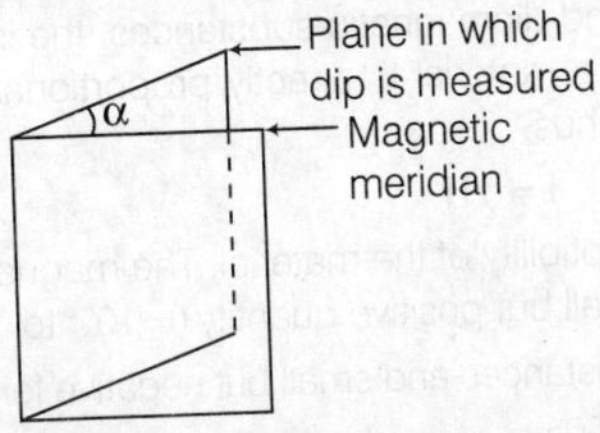

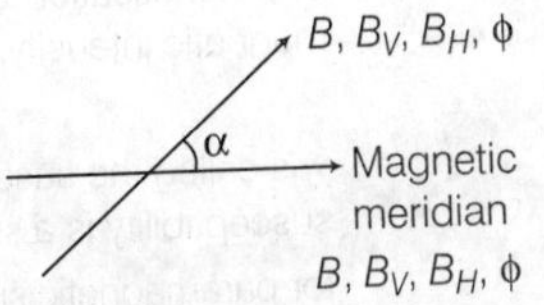

From above figures, it is clear that values of B, B_H and ϕ depend on orientation of plane but value of B_V remains same in all planes.

So, $$B_{H_1} = B_H \cos\alpha$$

$$\frac{B_V}{B_H} = \tan\theta$$

and $$\frac{B_V}{B_{H_1}} = \tan\theta_1$$

Solving above set of equation, we have

$$\frac{B_V}{B_H \cos\alpha} = \tan\theta_1$$

and $$\frac{\tan\theta}{\cos\alpha} = \tan\theta_1$$

$$\Rightarrow \quad \tan\theta_1 = \tan\theta . \sec\alpha$$

If ϕ_1 and ϕ_2 are dip values at two different planes, then

$$\cos\alpha = \tan\phi \cot\phi_1 \quad \text{...(i)}$$

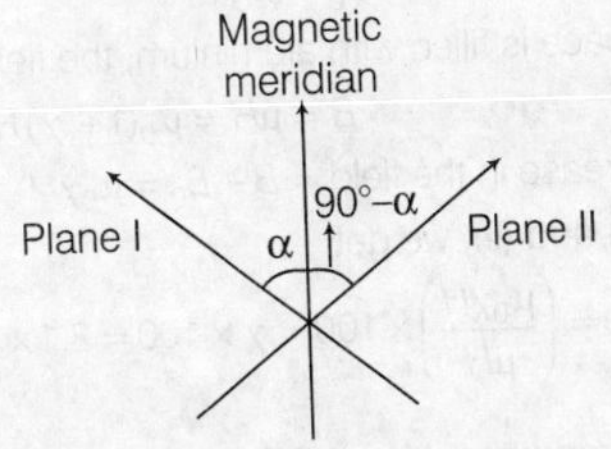

As both given planes are perpendicular so far IInd plane angle is $(90° - \alpha)$.

For IInd plane

$$\cos(90° - \alpha) = \tan\phi \cos\phi_2$$

$$\Rightarrow \quad \sin\alpha = \tan\phi \cot\phi_2 \quad \text{...(ii)}$$

From Eqs. (i) and (ii), we have

$$\cot^2\phi = \cot^2\phi_1 + \cot^2\phi_2$$

$$= \cot^2 45° + \cot^2(90° - 45°) = 2$$

26. (2) Let (r, θ) be the coordinate of point w.r.t. dipole.

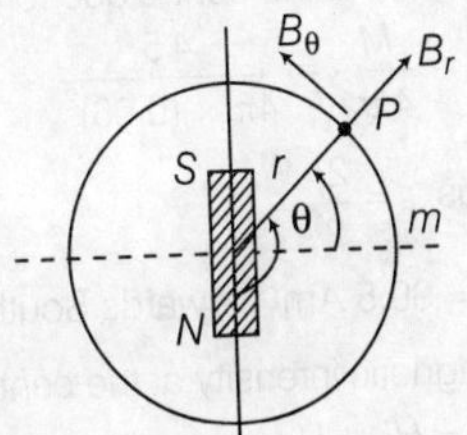

Then, components of field at P are

$$B_r = \frac{\mu_0}{4\pi} \cdot \frac{2m\cos\theta}{r^3}$$

$$B_\theta = \frac{\mu_0}{4\pi} \cdot \frac{m\sin\theta}{r^3}$$

$$\Rightarrow \quad \tan\phi = \frac{B_V}{B_H} = \frac{-B_r}{B_\theta}$$

$$\Rightarrow \quad \tan\phi = -2\cot\theta$$

From figure, $\theta = 90° + \lambda$

$$\Rightarrow \quad \tan\phi = 2\tan\lambda$$

DPP-3 Magnetic Materials and Their Properties

1. (i) For paramagnetic and diamagnetic substances, the intensity of magnetisation of a material is directly proportional to the magnetic intensity. Thus,

$$\mathbf{I} = \chi \mathbf{H}$$

χ is called the susceptibility of the material. The magnetic susceptibility is a small but positive quantity ($\approx 10^{-3}$ to 10^{-5}) for paramagnetic substances and small but negative for diamagnetic substances.

(ii) When a magnetic field is applied to a material, the actual magnetic field inside the material is the sum of the applied magnetic field and the magnetic field due to magnetisation.

Hence, $\mathbf{B}_{net} = \mu_0 \mathbf{H} + \mathbf{I}$, where μ_0 is the permeability of vacuum.

2. The magnetic field produced by the current is

$$B_0 = \mu_0 n i$$

$$\Rightarrow \quad H = \frac{B_0}{\mu_0} = ni$$

The ring gets magnetised and produces an extra field due to magnetisation.

$$\text{Net field } = B = \mu_0(H + I) \quad \text{(given)}$$

$$\text{or} \quad I = \frac{B_0}{\mu_0} - H = \frac{B}{\mu_0} - ni$$

3. Without aluminium, the magnetic field is given by

$$B_0 = \mu_0 H \quad \text{...(i)}$$

When the space is filled with aluminium, the field becomes

$$B = \mu H = \mu_0(1 + \chi) H$$

$$\therefore \quad \text{Increase in the field} = B - B_0 = \mu_0 \chi H \quad \text{...(ii)}$$

From Eqs. (i) and (ii), we get

$$\%\text{ increase} = \left(\frac{\mu_0 \chi H}{\mu H}\right) \times 100 = \chi \times 100 = 2.1 \times 10^{-3}\%$$

4. Here, pole strength $M = 4.5$ A-m

$$\Rightarrow \quad 2l = 12 \text{ cm} = 0.12 \text{ m}$$

Cross-sectional area, $a = 0.9 \text{ cm}^2 = 0.9 \times 10^{-4} \text{ m}^2$

(i) As, we know that intensity of magnetisation

$$I = \frac{\text{Pole strength } (M)}{\text{Cross-sectional area } (A)} = \frac{4.5}{0.9 \times 10^{-4}}$$

$$= 5 \times 10^4 \text{ Am}^{-1} \text{ from } S \text{ to } N$$

(ii) Magnetic intensity at the centre due to its North pole.

$$H_1 = \frac{M}{4\pi r^2} = \frac{4.5}{4\pi \times (0.06)^2}$$

where radius, $r = \dfrac{2l}{2}$

$$= 99.5 \text{ Am}^{-1} \text{ towards South pole}$$

Resultant magnetic intensity at the centre,

$$H = H_1 + H_2 = 99.5 + 99.5 = 199 \text{ Am}^{-1} \text{ towards South pole}$$

(iii) Now, magnetic field at the centre of magnet,

$$B = \mu_0 (H + I) = 4\pi \times 10^{-7} (-199 + 5 \times 10^4)$$

$$= 6.26 \times 10^{-2} \text{ T towards North pole}$$

5. (a) For a diamagnetic substances, susceptibility is very small and negative, i.e. $-1 \le \chi \le 0$.

6. (d) The Meissner effect is the expulsion of a magnetic field from a superconductor during its transition to the superconducting state when it is cooled below their superconducting critical temperature (T_C) i.e. $T < T_C$. So, magnet is levitated over disc became of repulsion of superconducting disc.

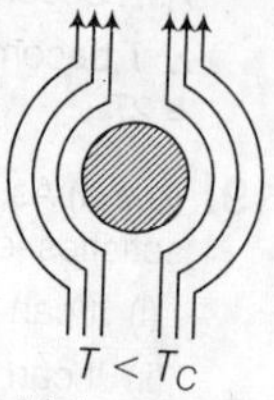

Meissner effect

7. (b) At high enough temperature, a ferromagnetic substance becomes a paramagnetic substance. This temperature is known as Curie temperature. After this, susceptibility varies with temperature as $\chi = \dfrac{C'}{T - T_c}$.

where, T_C is the Curie point and C' is a constant.

8. (b) As soft iron is preferred for making electromagnet. It has low retentivity and low coercivity and low hysteresis loop loss.

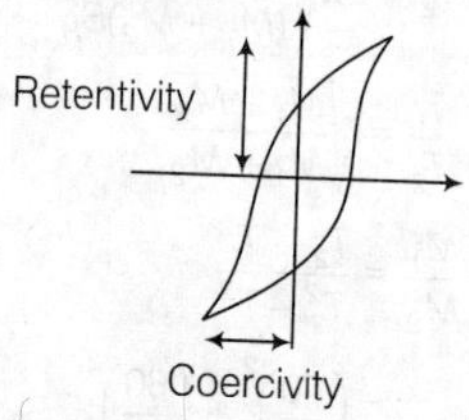

9. (d) As, permanent magnets are made of hard ferromagnetic materials which retains magnetisation even after the removal of the external magnetising field.

Consequently, they have high retentivity, high coercivity and large hysteresis loss.

10. (b) A magnet strongly attracts a ferromagnetic substance, weakly attracts a paramagnetic substances and weakly repels a diamagnetic substance.

11. (b) In permanent magnet, it has high retentivity, high coercivity and large hysteresis loss. So, it retains magnetisation even after the removal of the external magnetising field.

12. (b) For diamagnetic material relative permeability $\mu_r < 1$ and relative permittivity $\varepsilon_r > 1$.

13. (a) In a diamagnetic substance, the direction of magnetising field intensity I is opposite to that of magnetic intensity H.

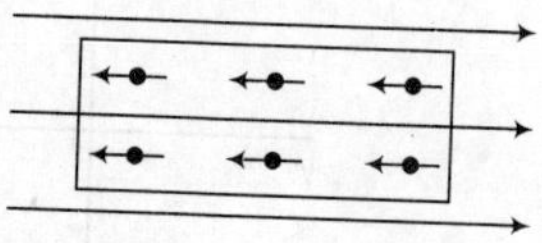

14. (a) As, relative permeability, $\mu_r = 1 + \chi$ for paramagnetic material $\mu_r > 1$ and χ is a small number, i.e. $1 < \mu_r < 1 + \varepsilon$

15. (b) Given, $H = 1600 \text{ Am}^{-1}$, $\phi = 2.4 \times 10^{-5}$ Wb

$$A = 0.2 \text{ cm}^2 = 0.2 \times 10^{-4} \text{ m}^2$$

As magnetic induction,

$$B = \frac{\phi}{A} = \frac{2.4 \times 10^{-5}}{0.2 \times 10^{-4}} = 12 \times 10^{-1} \text{ Wb/m}^2 = 1.2 \text{ Wb/m}^2$$

$$\text{Permeability,} \quad \mu = \frac{B}{H} = \frac{1.2}{1600} = 7.5 \times 10^{-4} \text{ mA}^{-1}$$

$$\text{As,} \quad \mu = \mu_0 (1 + \chi_m)$$

So, susceptibility of an iron bar,

$$\chi_m = \frac{\mu}{\mu_0} - 1 = \frac{7.5 \times 10^{-4}}{4\pi \times 10^{-7}} - 1$$

$$= 0.597 \times 10^3 - 1 = 597 - 1 = 596$$

16. (c) Here, $d = 7500 \text{ kg/m}^3$, $m = 0.075$ kg, $M = 8 \times 10^{-7}$ A-m^2

So, intensity of magnetisation is

$$I = \frac{M}{V} = \frac{Md}{m} = \frac{8 \times 10^{-7} \times 7500}{0.075} = 0.08 \text{ Am}^{-1}$$

17. (c) Here, coercivity of a small bar magnet is 4×10^3 Am^{-1}.

$n = 500, l = 1$ m

So, $n = \frac{500}{1} = 500$ turns/m

As, $H = nI$, then current $I = \frac{H}{n} = \frac{4 \times 10^3}{500} = 8$ A

18. (b) For diamagnetic substance, the intensity of magnetisation of a material is directly proportional to the magnetic intensity.

Thus, $$\mathbf{I} = \chi \mathbf{H}$$

Since, magnetic susceptibility (χ) is small but negative, hence *OC* curve lest represents the variation.

19. (b) When a bar of soft iron is placed in the uniform magnetic field which is parallel to it because of large permeability of soft iron, magnetic lines of force prefer to pass through it. Concentration of lines in soft iron bar increases as shown in figure.

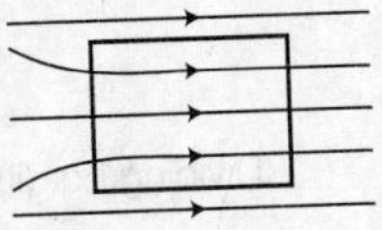

20. (a) As temperature of ferromagnetic material is raised, its susceptibility χ remains constant first and then decreases as shown in figure.

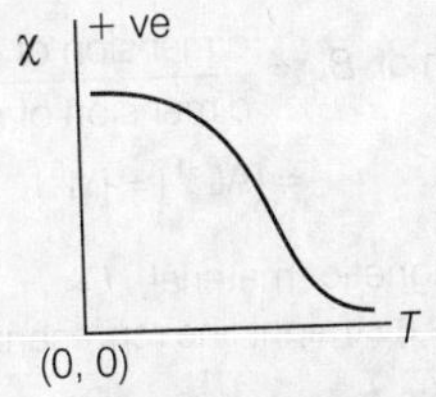

21. (b) Here, $I_1 = 8$ Am^{-1}, $B_1 = 0.6$ T, $t_1 = 4$ K

$I_2 = ?, B_2 = 0.2$ T, $t_2 = 16$ K

According to Curie law, $I \propto \frac{B \text{ (magnetic field induction)}}{t \text{ (in kelvin)}}$

Ratio of magnetisation, $\frac{I_2}{I_1} = \frac{B_2}{B_1} \times \frac{t_1}{t_2} = \frac{0.2}{0.6} \times \frac{4}{16} = \frac{1}{12}$

or $I_2 = I_1 \times \frac{1}{12} = \frac{8}{12} = \frac{2}{3}$ Am^{-1}

22. (a) Nickel exhibits ferromagnetism because of a quantum physics effect called exchange coupling in which the electron spins of one atoms interact with those of neighbouring atoms. The result is alignment of the magnetic dipole moments of the atoms, in spite of the randomising tendency of atomic collisions. This persistent alignment is what gives ferromagnetic materials their permanent magnetism.

If the temperature of a ferromagnetic material is raised above a certain critical value called the Curie temperature, the exchange coupling ceases to be effective. Most such materials then become simply paramagnetic, i.e. the dipoles still tend to align with an external field but much more weakly and thermal agitation can now easily disrupt the alignment.

23. (2) As coercivity $OA = H = ni = 120$ Am^{-1}

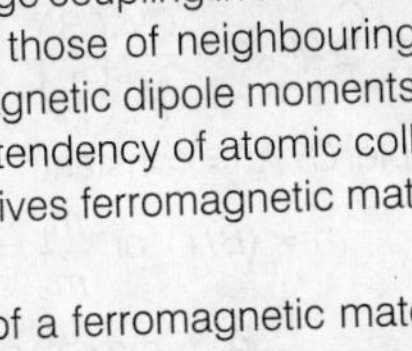

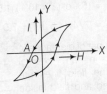

So, $n = \frac{\text{Number of turns}}{\text{Length of bar magnet}} = \frac{72}{1.2 \text{ m}} = 60$

$\therefore$ Current flowing through the solenoid, i.e.

$$i = \frac{H}{n} = \frac{120}{60} = 2 \text{ A}$$

24. (8) Volume of cube = (side)^3 = $(10^{-6})^3 = 10^{-18}$ m^3 = 10^{-12} cm^3

Mass of cube = Volume × density = 7.9×10^{-12} g

Number of atoms present in sample $N = N_A \times \frac{\text{Sample mass}}{\text{Molar mass}}$

$$= \frac{7.9 \times 10^{-12} \times 6.023 \times 10^{23}}{55}$$

$$= 8.56 \times 10^{10} \text{ atoms}$$

Total dipole moment of sample = $10\%(N \times \mu)$

$$= \frac{1}{10} \times 8.65 \times 10^{10} \times 9.27 \times 10^{-24}$$

$$= 8 \times 10^{-14} \text{ A-m}^2$$

$\therefore$ Magnetisation of sample = $\frac{\text{Magnetic dipole moment}}{\text{Volume}}$

$\Rightarrow$ $$= \frac{8 \times 10^{-14}}{10^{-18}} = 8 \times 10^4 \text{ A-m}^{-1}$$

25. (2) For a charged particle associated with a magnetic dipole moment is

$$\mu = \frac{evr}{2}$$

Also there is a magnetic force on the particle,

$$F_m = Bqv = \frac{mv^2}{r} \quad \Rightarrow \quad r = \frac{mv}{eB}$$

So, $$\mu = \frac{1}{2} ev\left(\frac{mv}{eB}\right) = \frac{1}{B}\left(\frac{1}{2}mv^2\right) = \frac{K}{B}$$

As charge is cancelled out.

$\therefore$ This relation is true for both electrons and positive ions both rotates in opposite sense inside magnetic field so, the dipole moments of ions and electrons are in same direction.

Now, magnetisation of sample is

M = magnetisation of electrons + magnetisation of ions

$= \mu_e n_e + \mu_i n_i = n(\mu_e + \mu_i)$ (as, $n_i = n_e$ in sample)

$$= n\left(\frac{K_e}{B} + \frac{K_i}{B}\right) = \frac{n}{B}(K_e + K_i)$$

$$= \frac{5 \times 10^{21}}{0.2}(0.4 \times 10^{-21} + 7.6 \times 10^{-21})$$

$$= \frac{40}{0.2} = 200 \frac{\text{A}}{\text{m}}$$

26. (a,d) In the solenoid, $H = nI$ = a constant and $B = \mu_0 \mu_r nI$ changes due to variation in $\mu_r (\mu_r = 1 + \chi)$ and beyond Curie temperature $\chi \times \frac{1}{T - T}$. When temperature of the iron core of solenoid (which is ferromagnetic material) is raised beyond Curie temperature, then soft iron core behaves as paramagnetic material. We know that $(\chi_m)_{\text{para}} \approx 10^{-5} \cdot (\chi_m)_{\text{Ferro}} \approx 10^3$

$\therefore$ $$\frac{(\chi_m)_{\text{Ferro}}}{(\chi_m)_{\text{Para}}} = \frac{10^3}{10^{-5}} = 10^8$$

Thus, magnetisation of the core diminishes by a factor 10^8.

27. (a,b,c) For a ferromagnetic sample in the presence of an external field, few domains that are oriented favourably grows in size and there oriented less favourably reduces in size. Also, orientation of domain changes as they tend to align with the field.

28. (a,d) For a paramagnetic sample, $M = C\frac{B_0}{T}$

29. (a) $[\mathbf{B}] \equiv [\text{MA}^{-1} \text{T}^2]$

$$[\mathbf{M}] \equiv [\text{AL}^2]$$

$$\left[\frac{\mathbf{M}}{\mathbf{L}}\right] \equiv [\text{ATM}^{-1}]$$

$$\sqrt{\varepsilon_0 \mu_0} = \frac{1}{C} \equiv \frac{1}{[\text{LT}^{-1}]} \equiv [\text{L}^{-1}\text{T}]$$

Revisal Problems (JEE Main)

1. (b) Suppose Q be the energy dissipated per unit volume per hysteresis cycle in the given sample. Then, total energy lost by the volume V of the sample in time will be

$$W = Q \times V \times n \times t$$

where, n is the number of hysteresis cycles per second.

Here, volume, $V = \dfrac{\text{Mass}}{\text{Density}} = \dfrac{12}{7500}\ \text{m}^3$

Hysteresis loss, $W = 300 \times \dfrac{12}{7500} \times 50 \times 3600\ \text{J} = 86400\ \text{J}$

2. (c) In a permanent magnet at room temperature, domains of a magnet are partially aligned due to thermal agitation.

3. (b) Given, $I_1 = 8\ \text{Am}^{-1}$, $B_1 = 0.6\ \text{T}$, $t_1 = 4\ \text{K}$, $I_2 = ?$

$B_2 = 0.2\ \text{T}$, $t_2 = 16\ \text{K}$

According to Curie's law, we get

$$I \propto \frac{B\ (\text{magnetic field induction})}{t\ (\text{temperature in kelvin})}$$

$$\Rightarrow \quad \frac{I_2}{I_1} = \frac{B_2}{B_1} \times \frac{t_1}{t_2} = \frac{0.2}{0.6} \times \frac{4}{16} = \frac{1}{12}$$

Similarly, $I_2 = I_1 \times \dfrac{1}{12} = 8 \times \dfrac{1}{12} = \dfrac{2}{3}\ \text{Am}^{-1}$

4. (d) For diamagnetic substances, the magnetic susceptibility is negative and it is independent of temperature.

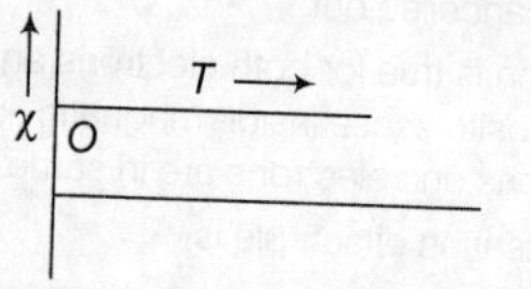

5. (d) At neutral point P,

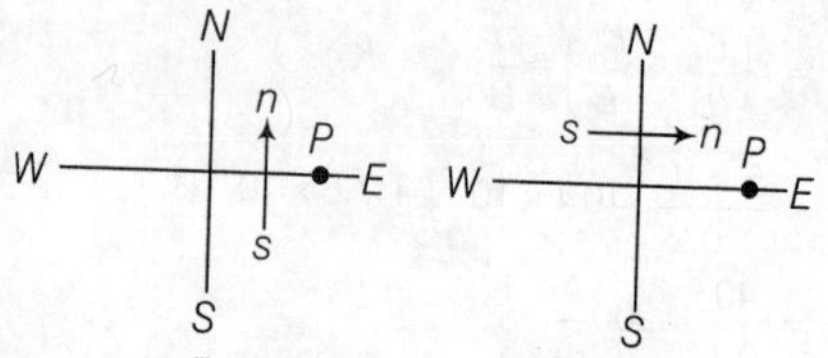

(i) (ii)

In Fig. (ii), net magnetic induction at P = resultant of

$$\frac{\mu_0\, 2m}{4\pi d^3} = 2B_H \text{ along horizontal}$$

and B_H along vertical, i.e. $B_R = \sqrt{(2B_H)^2 + (B_H)^2} = \sqrt{5}\, B_H$

6. (a) As we know magnetic induction, $B = \mu_0 (I + H)$

Intensity of magnetisation,

$$I = \frac{M}{V} = \frac{6}{30 \times 10^{-6}} = 2 \times 10^5\ \text{Am}^{-1}$$

$$H = 5\ \text{oersted} = \frac{5}{4\pi \times 10^{-3}}\ \text{Am}^{-1}$$

$$\mu_0 = 4\pi \times 10^{-7}\ \text{Wb/A-m}$$

$$B = 4\pi \times 10^{-7}\left[2 \times 10^5 + \frac{5}{4\pi \times 10^{-3}}\right] = 0.2517\ \text{T}$$

7. (a) As magnetic flux, $\phi = BA$ ...(i)

$\because$ $B = \mu_0 (I + H)$ ...(ii)

From Eqs. (i) and (ii), $\phi = \mu_0 (I + H) A$

$\because$ $I = \dfrac{m_p}{A}$ ...(iii)

From Eqs. (ii) and (iii), $\phi = \mu_0 A \left[\dfrac{m_p}{A} + H\right] = \mu_0 [m_p + AH]$

8. (a) Two bar magnets having same geometry with magnetic moments are m and $2m$, then

***Case* I** In sum of their positions,

$$T_1 = 2\pi \sqrt{\frac{I_1 + I_2}{(m_1 + m_2) B_H}} = 2\pi \sqrt{\frac{I + I}{(m + 2m) B_H}}$$

$$= 2\pi \sqrt{\frac{2I}{3m \times B_H}} \quad \text{...(i)}$$

***Case* II** In difference of their positions,

$$T_2 = 2\pi \sqrt{\frac{I_1 + I_2}{(m_2 - m_1) B_H}} = 2\pi \sqrt{\frac{I + I}{(2m - m) B_H}}$$

$$= 2\pi \sqrt{\frac{2I}{mB_H}} \quad \text{...(ii)}$$

Dividing Eq. (ii) by Eq. (i), we get

$$\frac{T_2}{T_1} = \frac{\sqrt{3}}{1} > 1$$

$$\Rightarrow \quad T_1 < T_2$$

9. (b) $\because B_H = \dfrac{\mu_0}{4\pi} \cdot \dfrac{m}{r^3}$

Hence, dimension of $B_H = \dfrac{\text{dimension of } m}{\text{dimension of } r^3} = \dfrac{[\text{A-L}^2]}{[\text{L}^3]}$

$$= [\text{AL}^{-1}] = [\text{M}^0\text{L}^{-1}\text{AT}^0]$$

10. (c) In a paramagnetic material, $I \propto H$. Therefore, the graph between H and I is a straight line represented by in figure.

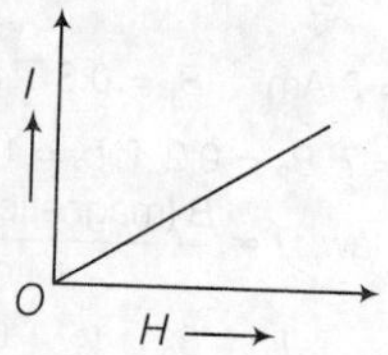

11. (c) According to Curie's law, $\chi_m = \dfrac{C}{T}$

As, magnetic susceptibility

$$\chi_m = \frac{M}{H},\ M = \frac{m}{V} \text{ and } H = \frac{B}{\mu},$$

$$\frac{m/V}{B/\mu} = \frac{C}{T} \text{ or } m = \frac{CV}{\mu}\left(\frac{B}{T}\right)$$

For a given sample, CV/μ = constant

Thus, $m \propto (B/T)$ or $\dfrac{m_1}{m_2} = \dfrac{B_1/T_1}{B_2/T_2}$

Given, $B_1 = 0.84\ \text{T}$, $B_2 = 0.98\ \text{T}$

$T_1 = 4.2\ \text{K}$, $T_2 = 2.8\ \text{K}$

Thus, $\dfrac{m_1}{m_2} = \dfrac{0.84/4.2}{0.98/2.8} = 4/7$

or $m_2 = (7/4)\, m_1$

Initial total magnetic moment of the sample, i.e.

$$m_1 = 15\% \text{ of } (2 \times 10^{24})(1.5 \times 10^{-23}\ \text{J/T}) = 4.5\ \text{J/T}$$

Thus, $m_2 = (7/4)(4.5\ \text{J/T}) = 7.9\ \text{J/T}$

12. (d) Figure shows a bar of diamagnetic material placed in an external magnetic field. The field lines are repelled or expelled and field inside the material is reduced. This reduction is slight being one part in 10^5. When placed in non-uniform magnetic field, bar will move from high to low field.

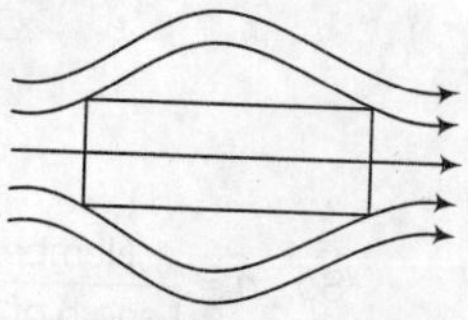

13. (d) Given, $V = (10 \times 0.5 \times 0.2)\ cm^3$

$$= (10 \times 0.5 \times 0.2) \times (10^{-2})^3$$
$$= 10^{-6}\ m^3$$
$$m = 5\ A\text{-}m^2,\ H = 5000\ Am^{-1}$$

Magnetic field B of a ferromagnetic substance is given by

$$B = \mu_0 (I + H) = \mu_0 \left(\frac{m}{V} + H\right)$$
$$= 4 \times 3.14 \times 10^{-7} \times \left[\frac{5}{10^{-6}} + 5000\right]$$
$$= 6.28\ T$$

14. (b) $R = 15\ cm = 15 \times 10^{-2}\ m$,

$N = 3500$, turns $\mu_r = 800$, $I = 1.2\ A$

Clearly, n (number of turns per unit length) $= N/2\pi R$

Since, magnetic field

$$B = \mu H = \mu_r \mu_0 nI \qquad (\text{as } \mu_r = \mu/\mu_0 \text{ and } H = nI)$$
$$B = 800\,(4\pi \times 10^{-7}\ Tm/A) \times \frac{3500}{2\pi\,(15 \times 10^{-2}\ m)} \times (1.2\ A)$$
$$= 4.48\ T$$

15. (d) The volume of cubic domain is

$$V = (10^{-6})^3\ m^3 = 10^{-18}\ m^3 = 10^{-12}\ cm^3$$

Sample mass = Volume × Density

$$= 7.9\ g\ cm^{-3} \times 10^{-12}\ cm^3$$
$$= 7.9 \times 10^{-12}\ g$$

It is given that Avogadro number, i.e. 6.023×10^{23} of iron atoms have a mass of 55 g. Hence, the number of atoms in the domain is given by

$$N = \frac{7.9 \times 10^{-12} \times 6.023 \times 10^{23}}{55}$$
$$= 8.65 \times 10^{10}\ \text{atoms}$$

16. (a) As we know that magnetic field H is independent of the material of core and

$$H = nI = 1000 \times 2.0 = 2 \times 10^3\ Am^{-1}$$

Magnetisation, $m = (B - \mu_0 H)/\mu_0$

$$= (\mu_r \mu_0 H - \mu_0 H)/\mu_0 \qquad (\because B = \mu_0 \mu_r H)$$
$$= (\mu_r - 1)\, H = 399 \times H$$
$$= 399 \times 2 \times 10^3 = 8 \times 10^5\ Am^{-1}$$

17. (d) Statement I is incorrect and Statement II is correct.

$$\mathbf{L}_n = \mathbf{m}V_n r_n = n\left(\frac{h}{2\pi}\right)$$
$$\therefore \quad L_n \propto n$$

So, angular moment is reduced in given transition.

Also, $\mathbf{M}_n = i_n A = i_n \pi r_n^2$

Now, $i_n \propto \frac{z^2}{n^3}, r_n \propto \frac{n^2}{z} \Rightarrow \mathbf{M}_n \propto \frac{n^4}{n^3}$

$$\Rightarrow \quad \mathbf{M}_n \propto n$$

So, magnetic dipole moment decreases in the transition.

18. (a) Area of hysteresis curve ∝ work done in one cycle of magnetisation and demagnetisation. So, a soft ferromagnetic core is preferred as it accounts for low hysteresis loss. So, Statement II explains Statement I.

19. (d) Curie's law is valid for a paramagnetic sample only till the saturation is reached. But, 100% magnetisation is not possible. So, Statement I is incorrect and Statement II is correct.

20. (b) Statement I and II both are correct but Statement II is not the applicable to ferromagnetic materials.

Above Curie temperature, random thermal motion of molecules prevents the formation of domain.

Revisal Problems (JEE Advanced)

1. (a) $F = \frac{\mu_0}{4\pi} \times \frac{m_1 m_2}{r^2} = \frac{10^{-7} \times 50 \times 100}{(10 \times 10^{-2})^2} = 50 \times 10^{-3}\ N$

2. (d) $MB \sin 30° = F \times$ perpendicular distance

$$F = 6.25\ N$$

3. (c) $B = \frac{\mu_0}{4\pi} \frac{2M}{(l^2 + x^2)^{3/2}} = \frac{\mu_0}{4\pi} \frac{2M}{x^3}$

$$\therefore \quad \frac{B_1}{B_2} = \left(\frac{x_2}{x_1}\right)^3 = 8 : 1 \text{ approximately.}$$

4. (c) According to tangent law,

$$B_A = B_B \tan \theta$$
$$\Rightarrow \quad \frac{\mu_0}{4\pi} \frac{2M}{d_1^3} = \frac{\mu_0}{4\pi} \frac{M}{d_2^3} \tan \theta$$

or
$$\left(\frac{d_2}{d_1}\right)^3 = \frac{\tan\theta}{2}$$

or
$$\frac{d_1}{d_2} = (2 \cot \theta)^{1/3}$$

5. (a) $T = 2\pi \sqrt{\frac{I}{MB}} = 2\pi \sqrt{\frac{ml^2}{12 \times m_p l B}} = 4\ s$

$$T = 2\pi \sqrt{\frac{\frac{m}{2} l^2}{12 \times \frac{m_p l}{2} IB}} = 4\ s$$

6. (c) $T = 2\pi \sqrt{\frac{I}{MB}}$

or
$$T \propto \frac{1}{\sqrt{M}}$$
$$\frac{T_1}{T_2} = \sqrt{\frac{3M - 2M}{3M + 2M}}$$

or
$$T_2 = 5\sqrt{5}\ s$$

7. (d) $T = 2\pi \sqrt{\frac{I}{MB}} = 2\pi \sqrt{\frac{ml^2}{12\ m_p l B}}$

$$T \propto \sqrt{ml}$$
$$\Rightarrow \quad \frac{T'}{T} = \left(\frac{ml}{n^2 \times ml}\right)^{1/2}$$

or
$$T' = \frac{T}{n}$$

8. (b) As the magnet is placed with its South pole pointing South, hence the neutral point lies on the equatorial line.

Hence, $B = \frac{\mu_0}{4\pi} \cdot \frac{M}{(r^2 + l^2)^{3/2}} = B_H$

$$B_H = 10^{-7} \times \frac{1.34}{[(0.15)^2 + (0.10)^2]^{3/2}}$$
$$= 0.34 \times 10^{-4}\ T$$

9. (d) In medium **B** and **H** are in opposite directions.

10. (c) Number of unpaired electrons

$$= \frac{\text{Total magnetic moment}}{\text{Magnetic moment of one electron}}$$

$$= \frac{8.00 \times 10^{22} \text{ A-m}^2}{9.27 \times 10^{-24} \text{ A-m}^2}$$

$$= 8.63 \times 10^{45}$$

Each iron atom has two unpaired electrons so that the number of iron atoms required is $N = \frac{1}{2}(8.63 \times 10^{45})$

$$= 4.31 \times 10^{45} \text{ atoms}$$

Mass of iron required

$$M = \frac{\text{Number of atoms} \times \text{Density}}{\text{Number of atoms per unit volume}}$$

$$= \frac{4.31 \times 10^{45} \times 7900}{8.63 \times 10^{28}}$$

$$= 3.95 \times 10^{20} \text{ kg}$$

11. (c,d) In diamagnetic materials, all the electrons are paired so there is no permanent net magnetic moment per atom. Paramagnetic and ferromagnetic properties are due to the presence of some unpaired electrons so their atoms have a net magnetic moment.

12. (a,b) Electrons and protons have spin magnetic moment.

13. (b,c,d) When a ferromagnetic material goes through hysteresis loop on the application of external field, the magnetic susceptibility may vary between negative value to infinite value.

14. (a,b,d) In a multi-electron atom, the electrons usually pair up with their spins opposite to each other. Atoms containing an odd number of atoms, however, must have at least one unpaired electron and have some spin magnetic moment.

15. (a,b,c,d) In ferromagnetic sample, domain size is up to 10^{-8} m^3 with nearly 10^{17} to 10^{21} atoms in each domain. Thermal energies can cause domain destruction.

16. (b,c,d) Diamagnetism is present in all substances but in para and ferromagnetic substances other effects are more dominant.

17. (a) Neutral point is a point where the magnetic field of a magnet is equal and opposite to the horizontal component of the earth's magnetic field.

18. (b) The susceptibility of a diamagnetic substance is independent of temperature.

19. (a) Magnetic moment is a vector quantity having the direction pointing from South to North pole of the magnet.
Unit of magnetic moment is

$$\frac{\text{N-m}}{\text{T}} \text{ or } \frac{\text{N-m}^3}{\text{Wb}}$$

Magnetic induction is also a vector quantity with unit as NA^{-1}m^{-1} which can be verified by relation $F = BIl$.

20. (b)

	Dia-magnetic substances	Para-magnetic substances	Ferro-magnetic substances
1. Susceptibility	$-1 \le \chi < 0$	$0 < \chi < \varepsilon$	$\chi >> 1$
2. Relative permeability (μ_r)	$0 \le \mu_r \le 1$	$1 < \mu_r < 1 + \varepsilon$	$\mu_r >> 1$
3. Permeability of a medium	$\mu < \mu_0$	$\mu > \mu_0$	$\mu >> \mu_0$

21. (a) If $B_2 > B_1$, critical temperature, (at which resistance of semiconductors abruptly become zero) in case 2 will be less than compared to case 1.

22. (b) With increase in temperature, T_C is decreasing.

$$T_C(0) = 100\text{K}$$
$$T_C = 75\text{K at } B = 7.5\text{T}$$

Hence, at $B = 5\text{T}$, T_C should lie between 75K and 100 K.

23. (c) Total flux around the magnetic circuit is constant. Since, the area of cross-section of air gap is same or that of material, it follows that $B_{gap} = B_{material}$.

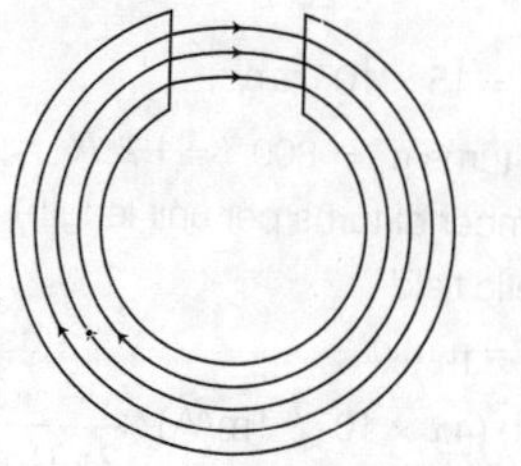

If gap is small, then fringing is insignificant and $B_m \approx B_g$.

24. (d) Using $B = \mu H$,

$$B_g = \mu_0 H_g$$
$$B_m = \mu H_m$$

Taking a closed loop and by Ampere's circuital rule,

$$H_g l + H_m L = Ni$$

$$\Rightarrow \quad \frac{B_g l}{\mu_0} + \frac{B_m L}{\mu} = Ni$$

As, gap is given small, then

$$\therefore \quad B_m = B_g$$

and $$B_g\left(\frac{l}{\mu_0} + \frac{L}{\mu}\right) = Ni$$

$$\frac{B_g}{\mu_0 \mu_r}(l\mu_r + L) = Ni$$

$$\Rightarrow \quad B_g = \frac{Ni}{\frac{1}{\mu_0 \mu_r}(l\mu_r + L)}$$

So, magnetic flux is given by $\phi = B_g A$

$$= \frac{\mu_0 \mu_r NiA}{(l\mu_r + L)}$$

25. (c) We have

$$N = 500, \quad I = 20 \text{ A},$$
$$l = 1 \text{ cm}, \quad L = 6 \text{ cm},$$
$$\mu = \mu_0 \mu_r = 3000 \times 4\pi \times 10^{-7}$$

So, from solution of previous question,

$$B_g = \frac{500 \times 20 \times 4\pi \times 10^{-7}}{(0.01 + 0.0002)}$$

$$\approx 1.3 \text{ Wb/m}^2$$

and when there is no air gap, $l = 0$, so

$$B = \frac{\mu NI}{L}$$

$$= \frac{500 \times 20 \times 3000 \times 4\pi \times 10^{-7}}{0.6}$$

$$\approx 53 \text{ Wb/m}^2$$

$$\text{Ratio} = \frac{B}{B_g} = \frac{53}{1.3} \approx 40$$

26. (8) As, the magnet is placed with its North pole pointing South, neutral points are obtained on the axial line.
As the neutral points and the magnetic field B due to magnet become equal and opposite to horizontal component of the earth's magnetic field B_H.

$$B = \frac{\mu_0}{4\pi} \frac{2Mr}{(r^2 - l^2)^2} = B_H$$

$$M = \frac{4\pi}{\mu_0} \frac{B_H (r^2 - l^2)^2}{2r} = 8.0 \text{ A-m}^2$$

27. (7) Resultant field is due to cable and due to earth's piled, $B_r = 7$ T.

28. (4) As, $B_0 = \mu_0 H$ and $M = C\dfrac{B_0}{T}$,

We have $\quad M = \dfrac{C}{T}\mu_0 H$

Also, $\quad \chi_B = \dfrac{M}{H} = \dfrac{C}{T}\mu_0$

$$\Rightarrow \quad C = \frac{\chi_B T}{\mu_0} = \frac{3 \times 10^{-4} \times 300}{4\pi \times 10^{-7}}$$

$$= 6.45 \times 10^4 \text{ KAT}^{-1} \text{ m}^{-1}$$

29. (5) For coil A,

$$H = nI = \frac{N}{l} I = \frac{200 \times 15}{0.5} = 6000 \text{ ATm}^{-1}$$

So, $\quad B_A = \mu_0 \mu_r H$

$$= 4\pi \times 10^{-7} \times 250 \times 6000$$

$$= 6 \times 10^6 \times \pi \times 10^{-7}$$

$$= \frac{6\pi}{10} \text{ Wbm}^{-2}$$

Hence, $\quad \phi_A$ = flux due to coil A

$$= BA$$

$$= \frac{6\pi}{10} \times 5 \times 10^{-4}$$

$$= 3\pi \times 10^{-4} \text{ Wb}$$

Now, induced emf in coil

$B = \varepsilon_B$ = Rate of change of flux

$$= N\frac{\Delta\phi}{\Delta t} = \frac{500 \times 3\pi \times 10^{-4}}{\pi}$$

$$= 1500 \times 10^{-4} \text{ V} = 150 \times 10^{-3} \text{ V}$$

$$= 150 \text{ mV} = 5 \times 30 \text{ mV}$$

JEE Main & AIEEE Archive

1. (c) For solenoid, the magnetic field needed to magnetise the magnet is given by $B = \mu_0 nI$.
where, n is number of turns per unit length $= N/l$

and $\quad N = 100$

$$l = 10 \text{ cm} = \frac{10}{100} \text{ m} = 0.1 \text{ m}$$

$$\Rightarrow 3 \times 10^3 = \frac{100}{0.1} \times I$$

$$\Rightarrow \quad I = 3 \text{ A}$$

2. (b) Net magnetic field,

$$B_{net} = B_1 + B_2 + B_H$$

$$B_{net} = \frac{\mu_0}{4\pi} \frac{(M_1 + M_2)}{r^3} + B_H$$

$$= \frac{10^{-7}(1.2 + 1)}{(0.1)^3} + 3.6 \times 10^{-5}$$

$$= 2.56 \times 10^{-4} \text{ Wb/m}^2$$

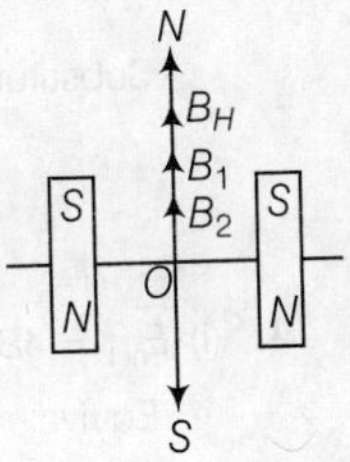

3. (b) For diamagnetic material, $0 < \mu_r < 1$ and for any material, $\varepsilon_r > 1$.

4. (b) Ferromagnetic substances have strong tendency to get magnetised (induced magnetic moment) in the same direction as that of applied magnetic field, so magnet attracts N_1 strongly.
Paramagnetic substances get weakly magnetised (magnetic moment induced is small) in the same direction as that of applied magnetic field, so magnet attracts N_2 weakly.
Diamagnetic substances also get weakly magnetised when placed in an external magnetic field but in opposite direction and hence N_3 is weakly repelled by magnet.

5. (b) Electromagnets are made of soft iron. The soft iron has low retentivity and low coercivity.

6. (b) When magnet is divided into two equal parts, the magnetic dipole moment

$$M' = \text{Pole strength} \times \frac{l}{2} = \frac{M}{2}$$

($\because$ pole strength remains same.)

Also, the mass of magnet becomes half

$$m' = \frac{m}{2}$$

Moment of inertia of magnet, $I = \dfrac{ml^2}{12}$

New moment of inertia,

$$I' = \frac{1}{12}\left(\frac{m}{2}\right)\left(\frac{l}{2}\right)^2 = \frac{ml^2}{12 \times 8}$$

$$\therefore \quad I' = \frac{I}{8}$$

Now, $\quad T = 2\pi\sqrt{\left(\dfrac{I}{MB}\right)}$

$$T' = 2\pi\sqrt{\left(\frac{I'}{M'B}\right)} = 2\pi\sqrt{\left(\frac{I/8}{MB/2}\right)}$$

$$\therefore \quad T' = \frac{T}{2}$$

$$\Rightarrow \quad \frac{T'}{T} = \frac{1}{2}$$

JEE Advanced & IIT JEE Archive

1. (a) If $B_2 > B_1$, critical temperature (at which resistance of semiconductors abruptly becomes zero) in case 2 will be less than compared to case 1.

2. (b) With increase in temperature, T_C is decreasing.

$T_C(0) = 100$ K, $T_C = 75$ K at $B = 7.5$ T

Hence, at $B = 5$ T, T_C should lie between 75K and 100K.

3. Electromagnetic Induction

DPP-1 Magnetic Flux and Faraday's Law

1. Magnetic field due to straight current carrying wire is downwards and its magnitude at a distance r from the wire is

$$B = \frac{\mu_0 I}{2\pi r}\ \text{T}$$

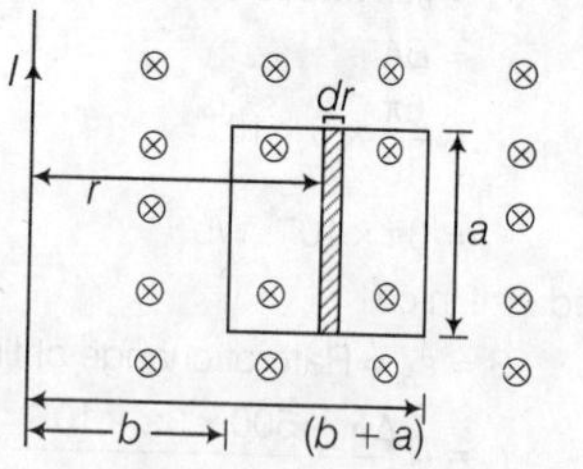

So, magnetic flux through area dA of strip of width dr and length a is

$$d\phi = BdA = \frac{\mu_0 I}{2\pi r}\cdot a\,dr$$

So, magnetic flux through complete loop is

$$\phi = \int_b^{b+a} B\,dA = \int_b^{b+a} \frac{\mu_0 Ia}{2\pi}\cdot\left(\frac{dr}{r}\right)$$

$$= \frac{\mu_0 aI}{2\pi}\log_e r\Big|_b^{b+a} = \frac{\mu_0 aI}{2\pi}[\log_e(b+a) - \log_e b]$$

$$= \frac{\mu_0 aI}{2\pi}\log_e\left(1+\frac{a}{b}\right)$$

2. (i) Magnetic field due to power line is time varying. Its magnitude at a distance r from the line is

$$B = \frac{\mu_0 I_0 \cos(2\pi ft)}{2\pi r}$$

where, current in wire is $I = I_0\cos(2\pi ft)$

I_0 = peak current, f = frequency

So, flux linked with the wire loop (of length l) is

$$\phi = \int B\,dA$$

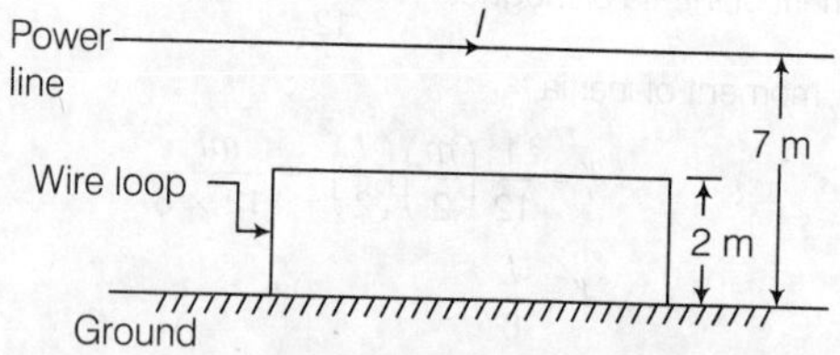

$$\therefore \quad \phi = \int_5^7 \frac{\mu_0 I_0}{2\pi r}\cos(2\pi ft)\cdot l\cdot dr = \frac{\mu_0 I_0 l}{2\pi}\cos(2\pi ft)\int_5^7 \frac{dr}{r}$$

$$= \frac{\mu_0 I_0 l}{2\pi}\log_e(1.4)\cdot\cos(2\pi ft)$$

Induced emf in coil must be

$$\therefore \quad \varepsilon = -N\frac{d\phi}{dt}$$

$$\Rightarrow \quad \varepsilon = \frac{-N\mu_0 I_0 l}{2\pi}\log(1.4)\cdot\frac{d}{dt}\cos(2\pi ft)$$

$$= N\mu_0 I_0 fl\cdot\log(1.4)\cdot\sin(2\pi ft)$$

When $2\pi ft = 90°$, $\varepsilon = 170$ V

$$\Rightarrow \quad \varepsilon_0 = N\mu_0 I_0 fl\log(1.4)$$

$$\Rightarrow \quad l = \frac{\varepsilon_0}{N\mu_0 I_0 f\log(1.4)} = 12\ \text{m}$$

(ii) Technique is unethical because the current in rectangle creates a back emf in the power line. This results in a power loss to electrical supply company, just as if the wire has been physically connected.

3. (i) Centripetal force required for circular motion of electron is generated by a radial electric field caused by the redistribution of the electrons in the disc.

$$F = eE = mr\omega^2$$

$$\Rightarrow \quad E = \frac{mr\omega^2}{e}$$

From $dV = -E\,dr$, we have

$$dV = -\frac{m\omega^2}{e}r\,dr$$

$$\Rightarrow \quad \int_{V_1}^{V_2} dV = -\frac{m\omega^2}{e}\int_0^R r\,dr$$

$$V_1 - V_2 = \frac{m\omega^2 R^2}{2e} = \frac{(9.1\times10^{-31})(130)^2(0.25)^2}{(2)(1.6\times10^{-19})}$$

$$= 3.0\times10^{-9}\ \text{V} = 3.0\ \text{nV}$$

$V_1 > V_2$, i.e. potential at centre is more than the potential at edge.

(ii) A disc may be assumed to be made up of a large number of radial, conducting, differential elements rotating with angular velocity ω about the centre of the disc O. Thus,

$$V_{centre} - V_{edge} = \frac{1}{2}BR^2\omega$$

Now, $V_{centre} > V_{edge}$ for anti-clockwise rotation and $V_{edge} > V_{centre}$ for clockwise rotation.

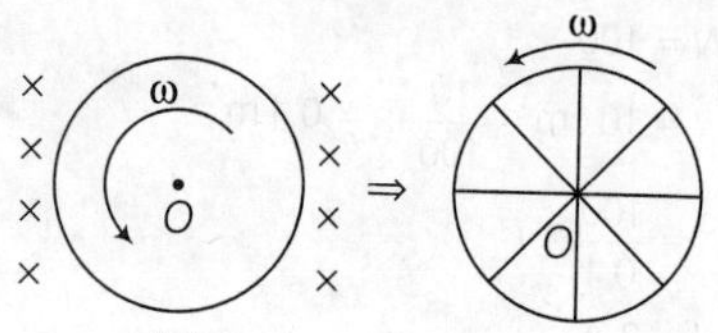

(a) (b)

Substituting the values, we have

$$V_{centre} - V_{edge} = \frac{1}{2}\times5.0\times10^{-3}\times0.25\times0.25\times130$$

$$= 0.02\ \text{V} = 20\ \text{mV}$$

4. (i) $E_{net} = 4Bl'v$

Equivalent circuit

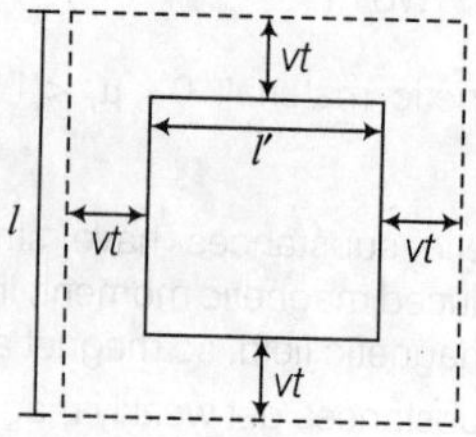

$$\therefore \quad E_{net} = 4B(l-2vt)\,v$$

$$E_{net} = 4Bv(l-2vt)$$

(ii) $$I = \frac{E_{net}}{4r} = \frac{4Bl'v}{4\lambda l'} \Rightarrow I = \frac{Bv}{\lambda}$$

(iii) Force required for moving each wire = $Il'B$

$$\text{Force, } F = \frac{B^2 v}{\lambda}(l-2vt)$$

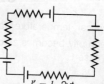

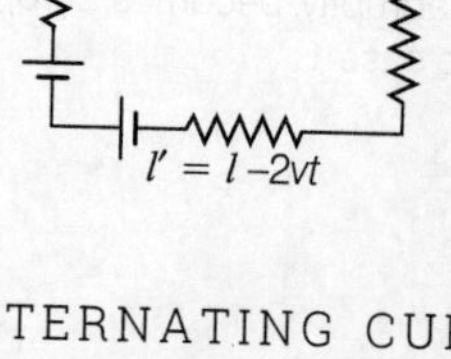

$l' = l - 2vt$

(iv) Total instantaneous power required to maintain constant velocity.

$$P = 4Fv = \frac{4B^2v^2}{\lambda}(l - 2vt)$$

(v) Instantaneous thermal power developed in the circuit

$$= 4I^2r = 4I^2\lambda(l - 2vt)$$

$$= 4\frac{B^2v^2}{\lambda^2}\lambda(l - 2vt)$$

$$\text{Thermal power} = \frac{4B^2v^2}{\lambda}(l - 2vt)$$

5. (b) Two emfs are induced in the closed circuit each of value Blv. These two emf's are additive, so

$$E_{net} = 2Blv$$

6. (c) Flux varies with time and coefficent $\frac{4}{T}$ is common.

$$\therefore \quad e = -N\frac{d\phi}{dt} = N\frac{4}{T}.\phi_m$$

$$= 100 \times \frac{4}{1/50} \times 0.02$$

$$= 200\text{ V}.$$

7. (c) At $t = 0$, inductor behave as broken wire, then

$$I = \frac{V}{R_2}$$

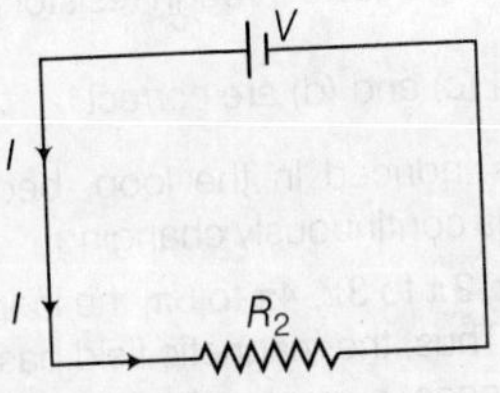

At $t = \infty$ inductor behave as conducting wire

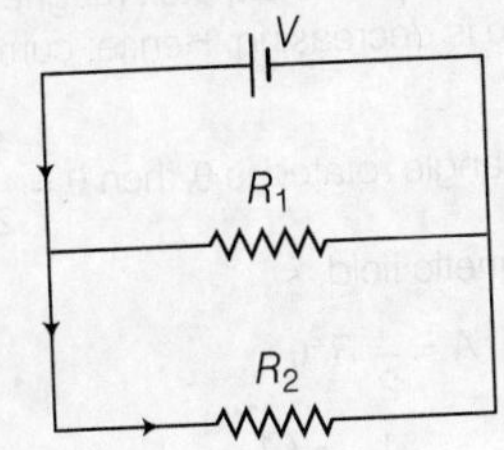

$$I = \frac{V}{R_1R_2/R_1 + R_2} = \frac{V(R_1 + R_2)}{R_1R_2}$$

8. (d) Induced emf $= -Blv$, where $l = 2R$

$$\therefore \quad \varepsilon = -B(2R)v$$

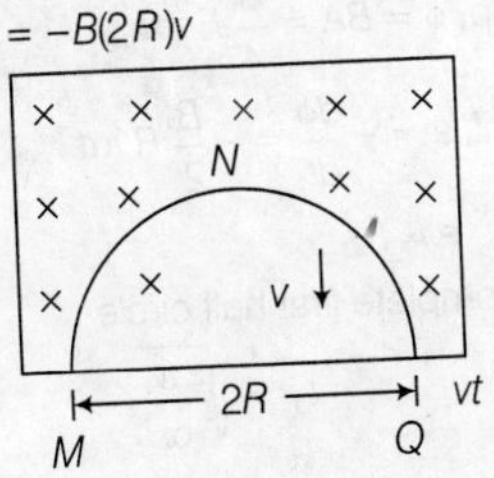

The induced current in the ring must generate magnetic field in upward direction. Thus, Q is at higher potential.

9. (b) $\phi = \phi_m \sin\omega t$,

$$\varepsilon = -\frac{d\phi}{dt} = -\phi_m\omega\cos\omega t$$

$$= \phi_m\omega\sin\left(\omega t - \frac{\pi}{2}\right)$$

So, induced voltage lags behind the flux by $\frac{\pi}{2}$ radians.

10. (a) Inward magnetic field is increasing. Therefore, induced current in both the loops should be anti-clockwise. But as the area of loop on right side is more, induced emf in this will be more compared to the left side loop $\left(e = -\frac{d\phi}{dt} = -A\frac{dB}{dt}\right)$. So, net current in the complete loop will be as given

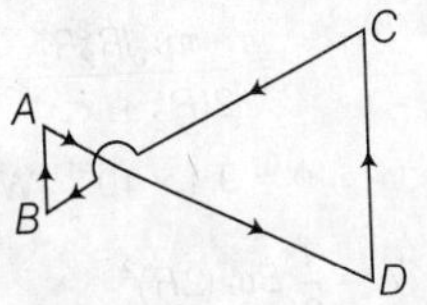

11. (d) $P = \frac{e^2}{R}$; $\quad e = -\frac{d}{dt}(BA)$

$$= -A\frac{d}{dt}(B_0e^{-t}) = AB_0e^{-t}$$

$$P = \frac{1}{R}(AB_0e^{-t})^2 = \frac{A^2B_0^2e^{-2t}}{R}$$

At the time of starting, $t = 0$

$$\text{So,} \quad P = \frac{A^2B_0^2}{R}$$

$$P = \frac{(\pi r^2)^2 B_0^2}{R} = \frac{B_0^2\pi^2r^4}{R}$$

12. (d) When conductor moves a distance dx (indirection of **v**) in time dt, the component of displacement dx moved perpendicular to field is $dx\sin\theta$. And hence, induced emf

$$E = -Bl\frac{dx}{dt}\sin\theta = -Blv\sin\theta = -l(\mathbf{v} \times \mathbf{B}) \text{ or } E = l(\mathbf{B} \times \mathbf{v})$$

13. (d) $e = \int_{2l}^{3l} (\omega x) B\,dx = B\omega\left[\frac{x^2}{2}\right]_{2l}^{3l}$

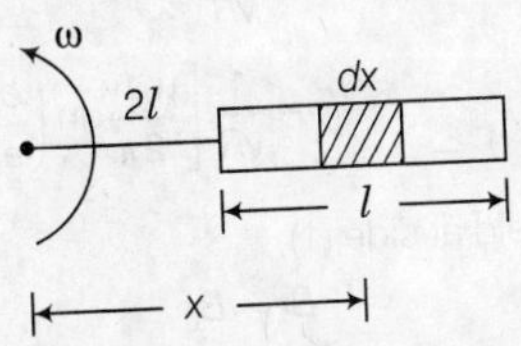

$$= \frac{B\omega}{2}[(3l)^2 - (2l)^2] = \frac{5Bl^2\omega}{2}$$

14. (d)

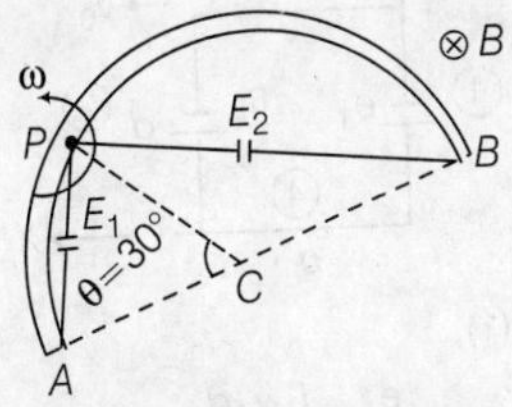

Let emf's induced in lengths PA and PB are E_1 and E_2, then potential difference between points A and C is $|E_1 - E_2|$.

Now, effective lengths of wires are

$$l_1 = AP = 2R\sin\frac{\theta}{2}$$

$$\text{and} \quad l_2 = BP = 2R\sin\left(\frac{\pi}{2} - \frac{\theta}{2}\right)$$

$$= 2R\cos\frac{\theta}{2}$$

$$\text{So,} |E_1 - E_2| = \frac{1}{2}B\omega\left(4R^2\sin^2\frac{\theta}{2} - 4R^2\cos^2\frac{\theta}{2}\right)$$

$$= 2B\omega R^2\left(\sin^2\frac{\theta}{2} - \cos^2\frac{\theta}{2}\right) = 2B\omega R^2\cos\theta$$

$$= 2B\omega R^2 \times \frac{1}{2} \qquad \left(\text{as, } \theta = 30°, \cos 30° = \frac{1}{2}\right.$$

$$= B\omega R^2$$

15. (a) Magnetic field at the centre of a small loop

$$B = \frac{\mu_0 I R_2^2}{2(R_2^2 + x^2)^{3/2}}$$

Area of smaller loop $S = \pi R_1^2$

Flux through smaller loop = BS

$$= \frac{\pi\mu_0 I R_2^2 R_1^2}{2(R_2^2 + x^2)^{3/2}}$$

$$\Rightarrow \quad \phi = 9.1 \times 10^{-11} \text{ Wb}$$

16. (c)

$$V_2 - V_1 = \frac{1}{2} B\omega (2R)^2$$

$$= 2BR(\omega R) = 2BRv$$

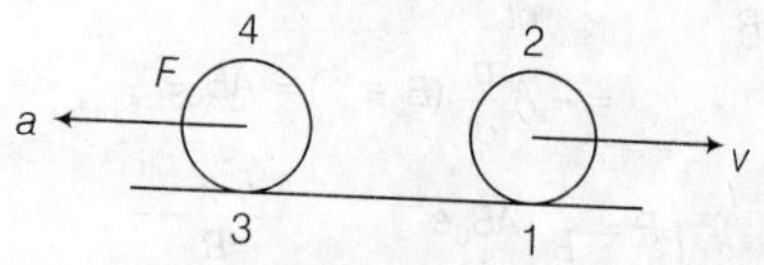

$$V_3 - V_4 = 2BRv$$

But $\quad V_1 = V_3$

So, $\quad V_2 - V_4 = 4BRv$

17. (a) Induced emf

$$\int_a^b BV\,dx = \int_a^b \frac{\mu_0 I}{2\pi x} \times v\,dx$$

$$\text{Induced emf, } E = \frac{\mu_0 I v}{2\pi} \ln\left(\frac{b}{a}\right)$$

$$\text{Power dissipated} = \frac{E^2}{R}$$

Also, $\quad$ power = Fv

$$F = \frac{E^2}{vR}$$

$$F = \frac{1}{vR}\left[\frac{\mu_0 I v}{2\pi} \ln\left(\frac{b}{a}\right)\right]^2$$

18. (a) Magnetic field at side (1),

$$B_1 = B_0$$

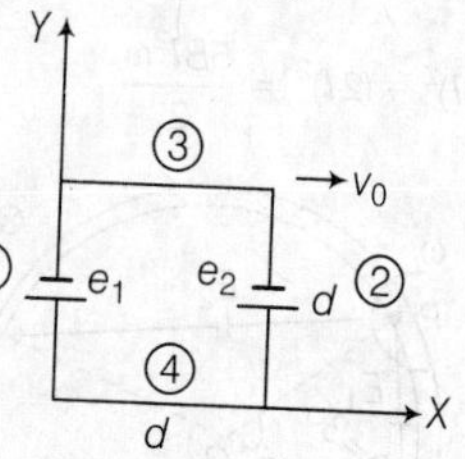

Induced emf in (1),

$$e_1 = B_0 v_0 d$$

Magnetic field at side (2),

$$B_2 = B_0\left(1 + \frac{d}{a}\right)$$

Induced emf in (2),

$$e_2 = B_0\left(1 + \frac{d}{a}\right) v_0 d$$

Induced emf in (3) and (4) will be zero.

$$\text{Net emf, } e = e_2 - e_1$$

$$= \frac{B_0 v_0 d^2}{a}$$

19. (a,c) Due to the current in the straight wire, net magnetic flux from the circular loop is zero. Because, in half of the circle magnetic field is inwards and in other half, magnetic field is outwards. Therefore, change in current will not cause any change in magnetic flux from the loop. Therefore, induced emf under all conditions through the circular loop is zero.

20. (b,d) Induced emf, $e = -\frac{d\phi}{dt}$

For identical rings induced emf will be same.

But current will be different. Given,

$$h_A > h_B$$

Hence, $\quad v_A > v_B \quad$ (as $v^2 = 2gh$)

When $\rho_A < \rho_B, I_A > I_B \Rightarrow h_A > h_B$ is fulfilled only when $m_A = m_B$.

If $\rho_A < \rho_B$, then $I_A > I_B$ this can be fulfilled if $m_A \le m_B$.

21. (b,c,d) Induced emf in loop, $\varepsilon = -\frac{d\phi}{dt} = \frac{-d}{dt}(BA)$

Induced current in loop,

$$I = \frac{\varepsilon}{R} = -\frac{1}{R} \cdot \frac{d}{dt} BA$$

$$= -\frac{1}{R} \cdot B \cdot \frac{d}{dt} a^2 \quad \text{(where, } a = \text{side of loop)}$$

$$= -\frac{1}{R} \cdot B \cdot 2a \cdot \frac{da}{dt}$$

$$= -\frac{2Ba}{R} \cdot v$$

where, v = speed of pulling of loop.

Also, $\quad \frac{dq}{dt} = \frac{-B}{R} \cdot \frac{dA}{dt}$

$$\Rightarrow \quad dq = \frac{-B}{R} dA$$

$\Rightarrow \quad Q$ = charge flows through resistor $R = -\frac{BA}{R}$

So, options (b), (c) and (d) are correct.

22. (a) The emf is induced in the loop, because area inside the magnetic field is continuously changing.

From $\theta = 0$ to π, 2π to 3π, 4π to 5π, the loop begins to enter the magnetic field. Thus, the magnetic field passing through the loop is increasing. Hence, current in the loop is anti-clockwise and for $\theta = \pi$ to 2π, 3π to 4π, 5π to 6π, etc., magnetic field passing through the loop is decreasing. Hence, current in the loop is clockwise.

Let at any time, angle rotated is θ, then $\theta = \frac{1}{2}\alpha t^2$

Area inside magnetic field,

$$A = \frac{1}{2} R^2 \theta$$

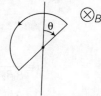

$$= \frac{1}{2} R^2 \left(\frac{1}{2}\alpha t^2\right)$$

$$= \frac{1}{4} R^2 \alpha t^2$$

Flux in the loop, $\phi = BA = \frac{B}{4} R^2 \alpha t^2$

$$\text{emf, } e = -\frac{d\phi}{dt} = -\frac{B}{2} R^2 \alpha t$$

$$\Rightarrow \quad e \propto t$$

Time taken to complete first half circle

$$t_1 = \sqrt{\frac{2\pi}{\alpha}}$$

When the loop starts coming out,

$$\theta = \frac{1}{2}\alpha t^2$$

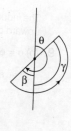

$$\beta = \theta - \pi, \gamma = \pi - \beta$$

$$= \pi - \theta + \pi = 2\pi - \theta$$

Area within magnetic field,

$$A = \frac{1}{2} R^2 \gamma = \frac{1}{2} R^2 (2\pi - \theta)$$

$$= \pi R^2 - \frac{R^2 \theta}{2}$$

Flux, $\phi = BA = B\left(\pi R^2 - \dfrac{R^2\theta}{2}\right)$

$$\text{emf}, e = -\frac{d\phi}{dt} = \frac{B}{2}R^2\alpha t$$

$$e = \frac{BR^2}{2}\alpha t \quad \Rightarrow e \propto t$$

23. (d) Time taken to complete second half revolution

$$t_2 = \sqrt{\frac{4\pi}{\alpha}} - \sqrt{\frac{2\pi}{\alpha}}$$

We see that $t_2 < t_1$.

We can write induced emf as

$$e = (-1)^n\left[\frac{1}{2}BR^2\alpha t\right]$$

where, $n = 1, 2, 3, \ldots$ is the number of half revolutions completed by loop. Smaller time will be taken to complete the second half revolution as compared to the previous half revolution.

24. (9) For an elemental strip of area $dA = a\,dx$ distant x from left hand side wire, magnitude of magnetic field is

$$B = \frac{\mu_0 I}{2\pi x} + \frac{\mu_0}{2\pi}\frac{I}{(d-x)}$$

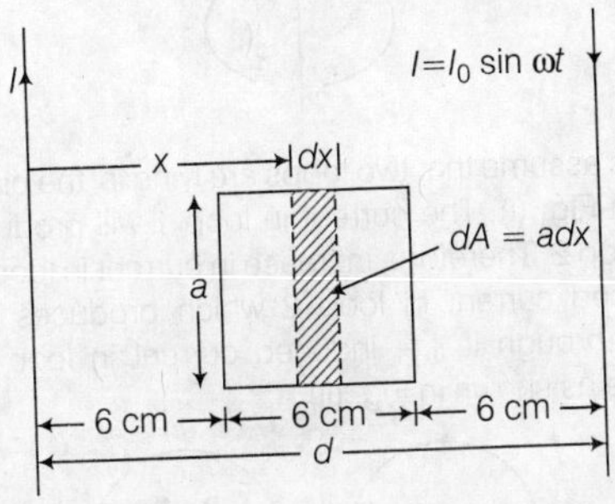

$$\Rightarrow \quad B = \frac{\mu_0 I}{2\pi}\left\{\frac{1}{x} + \frac{1}{(18-x)}\right\} \times \frac{1}{10^{-2}}$$

$$\Rightarrow \quad B = \frac{\mu_0 I \times 100}{2\pi}\left\{\frac{1}{x} + \frac{1}{18-x}\right\}$$

So, flux linked with this strip is

$$d\phi = B\,dA$$

$$= \frac{\mu_0 I \times 100}{2\pi}\left\{\frac{1}{x} + \frac{1}{18-x}\right\} \times a\,dx$$

$$= \frac{\mu_0 I \times 100}{2\pi}\left\{\frac{1}{x} + \frac{1}{18-x}\right\} \times 6 \times 10^{-2} \times dx$$

$$= \frac{3\mu_0 I}{\pi}\left\{\frac{1}{x} + \frac{1}{18-x}\right\} dx$$

Total flux associated with complete wire loop

$$\phi_B = \int_{6\text{cm}}^{12\text{cm}} d\phi = \frac{3\mu_0 I}{\pi}\int_{6\text{cm}}^{12\text{cm}}\left(\frac{1}{x} + \frac{1}{18-x}\right)dx$$

$$\Rightarrow \quad \phi_B = \frac{3\mu_0 I}{\pi}\log 2$$

$$\Rightarrow \quad \phi_B = \frac{15\mu_0 \cdot \log 2}{\pi} \cdot \sin(120\pi t)$$

So, induced emf is

$$\varepsilon = \frac{d\phi_B}{dt} = \frac{d}{dt}\left[\frac{15\mu_0 \log 2}{\pi}\sin(120\pi t)\right]$$

$$= \frac{15\mu_0(\log 2) \times 120\pi \times \cos(120\pi t)}{\pi}$$

Hence, maximum value of induced emf is

$$\varepsilon_{max} = 1800\mu_0 (\log 2)\text{ V}$$

$$= 9 \times \mu_0 (\log 4) \times 10^2 \text{ V}$$

$$\therefore \quad k = 9$$

25. (1) Given, $v = 1\text{ms}^{-1}$

Let v be the terminal velocity of the slider then, emf developed

$$= e = Blv$$

Power dissipated will be

$$P = i^2R = \left(\frac{e}{R}\right)^2 \cdot R = ei$$

When slider achieve terminal velocity, net downward force on slider is zero.

or magnetic force = gravitational force

$$\Rightarrow \quad Bil = mg$$

$$\Rightarrow \quad i = \frac{mg}{Bl} = \frac{0.2 \times 9.8}{1.0 \times 0.6} = 3.27\text{ A}$$

As power developed $P = ie$

$$\Rightarrow \quad \text{Emf induced}, e = \frac{P}{i}$$

$$\Rightarrow \quad e = \frac{1.96}{3.27} = 0.6\text{ V}$$

But $e = Blv$

$$\Rightarrow \quad v = \frac{E}{Bl} = \frac{0.6}{0.6 \times 1.0} = 1\text{ ms}^{-1}$$

26. (c) Charge on capacitor $q = CV_c$

$$= 2 \times 3 \times e^{-2t}$$

$$= 6e^{-2t}$$

$$i_c = \text{current through capacitor} = \frac{dq}{dt} = -12e^{-2t}$$

By applying junction rule at junction,

$$i_L = i_1 + i_2 + i_c$$

$$= 10e^{-c} + 4 - 12e^{-2t}$$

$$\Rightarrow \quad i_L = 2 + 2(1 - e^{-2t})$$

So, at $t = 0$, $i_L = 2\text{A}$

and $t = \infty$, $i_L = 4\text{A}$ (so, i_L is represented by graph t)

and $V_L = V_{od} = L\dfrac{di}{dt}$

$$= 4 \times \frac{d}{dt}(4 - 2e^{-2t})$$

$$= 16e^{-2t}$$

So, V_L decreases from 16v to 0. (graph s)

Also, by applying KVL in branch ac, we get

$$V_a - i_1R_1 + i_2R_2 = V_c$$

or $V_{ac} = V_a - V_c = i_1R_1 - i_2R_2$

$$= (10e^{-2t})(2) - (4)(3)$$

$$= 20e^{-2t} - 12\text{ V}$$

So, $V_{ac} = 8$ V at $t = 0$ and at $t = \infty$,

$$V_{ac} = -12\text{ V}$$ (graph r)

By KVL in branch ab,

$$V_a - i_1R_1 + V_c = V_b$$

$$\Rightarrow \quad V_{ab} = V_a - V_b = i_1R_1 - V_c$$

$$= 10e^{-2t}(2) - 3(e^{-2t})$$

$$= 17e^{-2t}\text{ V}$$

$\therefore$ V_{ab} decreases from $17V$ to 0. (graph q)

By KVL in branch cd,

$$V_c - i_2R_2 - V_L = V_a$$

$$\Rightarrow \quad V_{cd} = V_c - V_d = i_2R_2 + V_L$$

$$= 4 \times 3 + 16e^{-2t}$$

$$= 12 + 16e^{-2t}$$

So, V_{cd} decreases exponentially from 28 V to 12V. (graph p)

Hence, (i) → (t), (ii) → (s), (iii) → (r), (iv) → (q), (v) → (p)

DPP-2 Lenz's Law and Its Applications

1. (i) When N-pole approaches coil, end of coil also develops a N-polarity to oppose approach of N-pole.

Hence, current flows from point A to B.

(ii) Resistance decreases causing increase of current and also the flux.

$\therefore$ The emf induced in A is clockwise to oppose field of outer coil.

(iii) For loop A flux is increasing.

$\therefore$ Induced current is in anti-clockwise sense.

For loop B, flux is decreasing.

$\therefore$ Induced current is in clockwise sense.

(iv) There is no induced current in loop A is flux linked with A remains constant.

For loop B, flux is reducing.

$\therefore$ Induced current in B is clockwise.

2. (i) When switch is closed, current in P grows, induced current in S will oppose this growth. Hence, the current in S must be anti-clockwise.

(ii) When switch is opened, current in P decays from maximum to zero. Induced current in S will oppose the decay. Hence, current in S must be clockwise.

3. When A approaches C, flux linked with B is increasing and $\therefore$ current in B is anti-clockwise when looked from C.

4. The current induced in the loop will be clockwise. According to right hand thumb rule, magnetic field passing through the loop due to current in AB is perpendicular to plane of paper and outwards. The direction of current induced in the loop must be clockwise, so as to oppose the increase in this field.

5. (d) As the rod moves through the magnetic field, an emf will be built up across the rod but no current flows.

Without the current, there is no force to oppose the motion of rod. So, the rod travels at a constant speed over the rails.

6. (b) When N-pole of magnet is moved towards the coil, the upper face of the coil acquires North polarity. Therefore, work has to be done against the force of repulsion, in bringing the magnet closer to the coil.

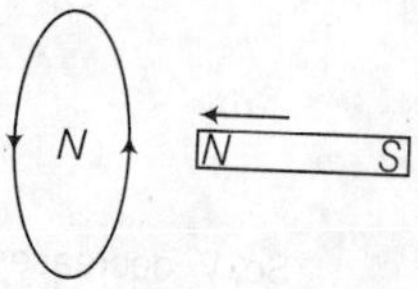

7. (a) The current induced in the loop will be clockwise in A and anti-clockwise in B. According to right hand thumb rule, magnetic field B, is perpendicular to the plane of paper, for A inwards for B outwards and as current is increasing, field of wire is increasing.

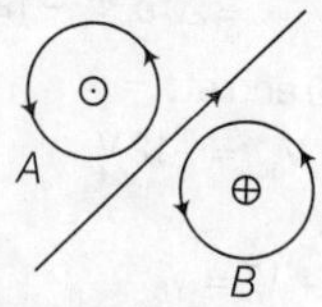

8. (i) (a) According to the Ohm's law, as temperature increases, R increases and hence current decreases. Therefore, flux linked with B is decreasing so, induced current in B is anti-clockwise.

(ii) (b) According to Fleming's right hand rule, as the flux through loop decreases (inward), so current will be clockwise.

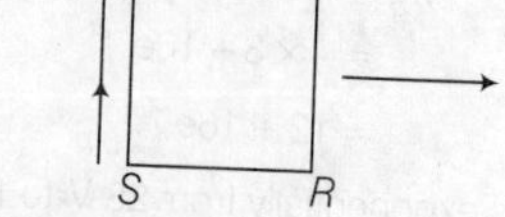

9. (b) As, we know induced current flows in clockwise direction. Therefore, magnetic field is into the plane of the paper. As it opposes the increasing inducing field, then inducing field must be outwards of the plane of the paper.

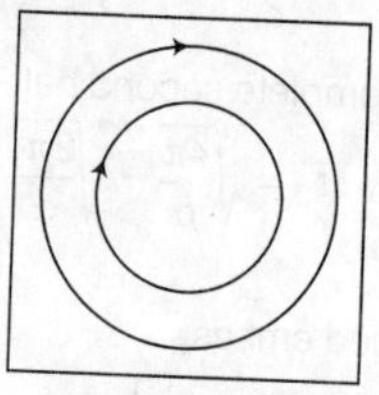

10. (a) Acceleration of falling magnet will remain equal to g. As, there is no induced current through the ring. Therefore, there will be no opposing force on the falling magnet.

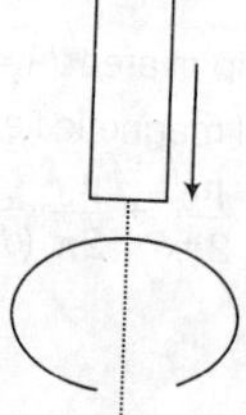

11. (c) Let us assume that two loops are lying in the plane of paper as shown in Fig. (i). The current in loop 1 will produce 0 magnetic field in loop 2. Therefore, increase in current in loop 1 will produce an induced current in loop 2 which produces magnetic field passing through it, i.e. induced current in loop 2 will also be clockwise as shown in Fig. (ii)

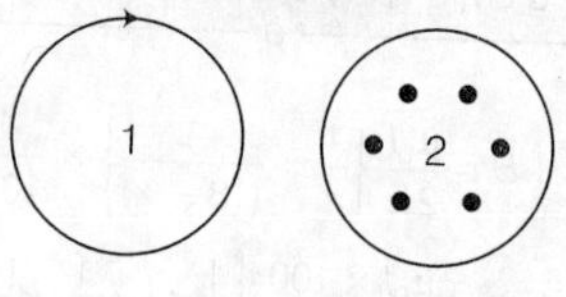

Fig. (i)

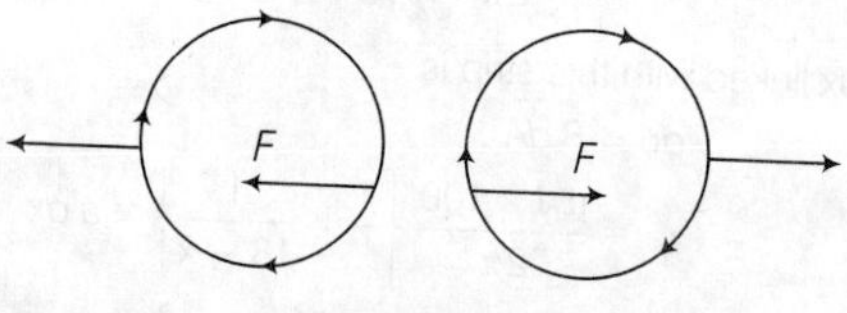

Fig. (ii)

The loops will now repel each other as the currents at the nearest and farthest point of the loops flow in opposite directions.

12. (d) When switch S is closed magnetic field lines passing through θ increases in the direction from right to left. So, according to Lenz's law induced current in θ, i.e. I_θ will flow in such a direction, so that the magnetic field lines due to I_θ pass from left to right through θ.

13. (d) Cross $\times$ magnetic field passing from the closed loop is increasing. Therefore, from Lenz's law induced current will produce dot magnetic field. Hence, induced current is anti-clockwise.

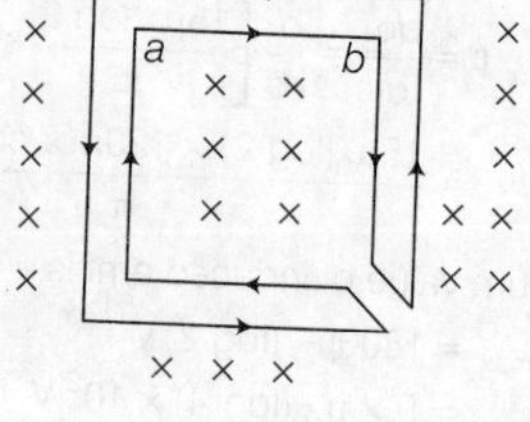

14. (b,c) Given case is related to part (ii) and part (iii).

Clockwise I

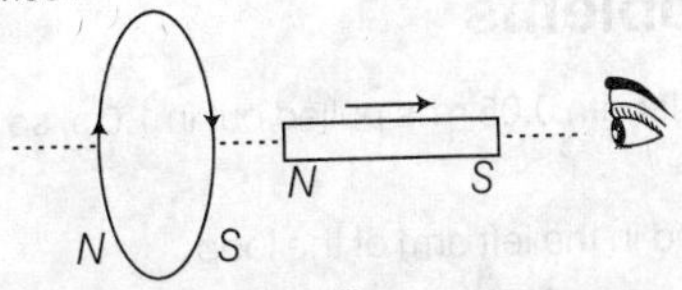

(i)

Anti-clockwise I

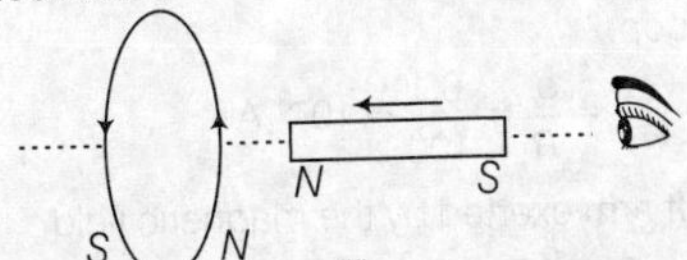

(ii)

Anti-clockwise I

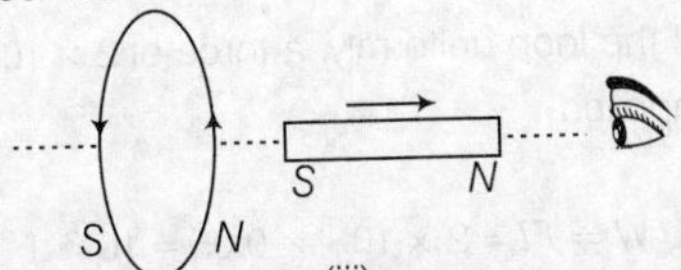

(iii)

Clockwise I

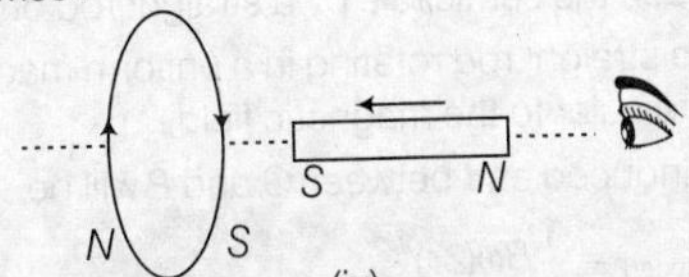

(iv)

15. (a, b) Due to repulsion between two coils A and B, then rate of change of flux increases, then induced current (i) is increased. While attraction between both coils i.e. A and B. Then, rate of change of flux decreases, then induced current (i) is decreased.

16. (a,c) According to Lenz's law, induced current at point P and Q is clockwise.

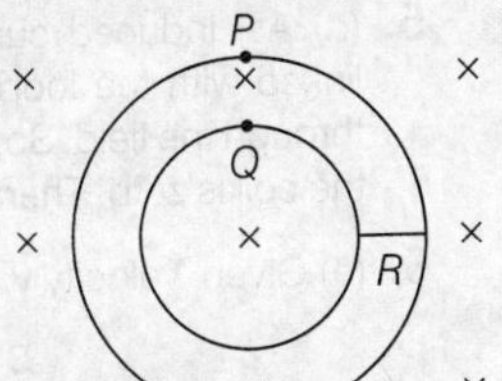

Also, $\varepsilon \propto \frac{d}{dt}\phi$

$\Rightarrow \quad \varepsilon \propto A \cdot \frac{d}{dt}B$

$\Rightarrow \quad \varepsilon \propto A$

$\Rightarrow \quad \varepsilon_P > \varepsilon_Q$

17. (d) For a constant current, force on rod is constant

$$F = m\frac{dv}{dt} \Rightarrow \int_0^v dv = \int_0^t \frac{IlB}{m} \cdot dt$$

$$\Rightarrow \quad v = \frac{IlB}{m} \cdot t = \frac{E_0 lBt}{mR}$$

For a constant emf E_0, the current produced by induced emf opposes the current produced by battery.

$$F = m\frac{dv}{dt} = IlB = \left(\frac{E_0 - Blv}{R}\right) lB$$

$$\Rightarrow \quad \frac{dv}{v - E_0/Bl} = -\left(\frac{B^2l^2}{mR}\right) dt$$

$$\text{Integrating, } \int_0^v \frac{dv}{v - E_0/Bl} = -\frac{B^2l^2}{mR}\int_0^t dt$$

$$\Rightarrow \quad \log_e\left(\frac{v - E_0/Bl}{-E_0/Bl}\right) = -\frac{B^2l^2t}{mR}$$

$$\Rightarrow \quad v = \frac{E_0}{Bl}\left(1 - e^{-\frac{B^2l^2t}{mR}}\right)$$

With a constant current, the acceleration is a constant, so velocity does not reach terminal speed.

However, with constant emf, the increasing motional emf decreases the applied force resulting in a limiting or terminal speed of $v = \frac{E_0}{Bl}$

Hence, (i) → (s), (ii) → (r), (iii) → (p), (iv) → (q)

18. (2) Given, circuit is

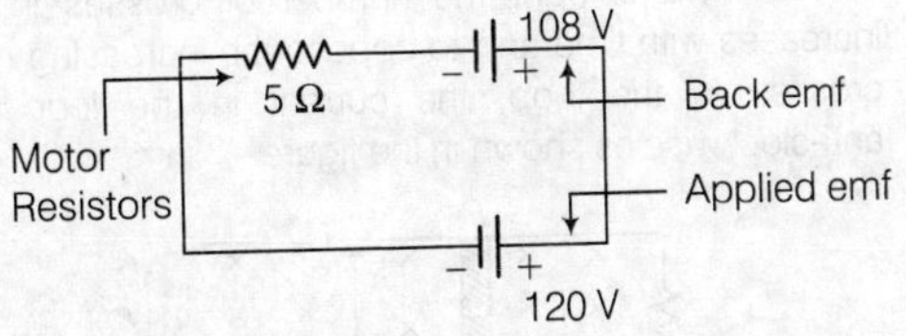

At $t = 0$, there is no back emf.

$$\therefore \quad I_1 = \frac{V}{R} = \frac{120}{5} = 24\text{ A}$$

After a long time, at full speed the back emf is effective in the circuit.

$$\therefore \quad I_2 = \frac{V_{\text{effective}}}{R} = \frac{120 - 108}{5} = 2.4\text{ A}$$

and the ratio, $\frac{I_1}{I_2} = \frac{24}{2.4} = 10 = 5 \times 2$

So, $k = 2$

19. (2) Induced emf in loops $AEFD$ and $EBCF$ will be

$$\varepsilon_1 = \frac{d\phi_1}{dt} = A_1 \cdot \frac{dB}{dt} = 1 \times 1 \times 1 = 1\text{ V}$$

$$\varepsilon_2 = A_2\frac{dB}{dt} = 0.5 \times 1 \times 1 = 0.5\text{ V}$$

ε_1 and ε_2 both are in anti-clockwise sense. Considering resistances of wire loops, we have

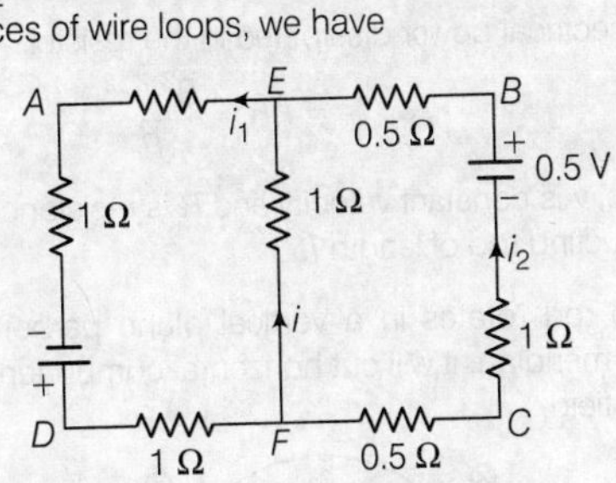

By using KVL, we have

$$i_1 = i + i_2 \quad \text{...(i)}$$

$$3i_1 + i = 1 \quad \text{...(ii)}$$

$$2i_2 - i = 0.5 \quad \text{...(iii)}$$

On solving these equations, we get

$$i = \frac{1}{22}\text{ A}$$

$\therefore \quad p = 2$

20. (a) Both Statements I and II are correct and Statement II explains Statement I. Force on free charge (electron) is as directed

So, upper end of rod will be negative.

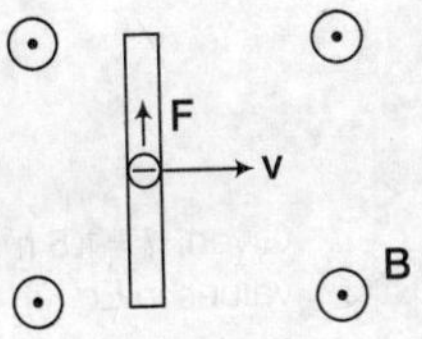

21. (d) Statement I is incorrect.

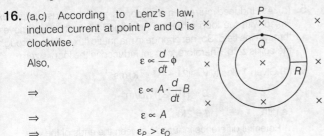

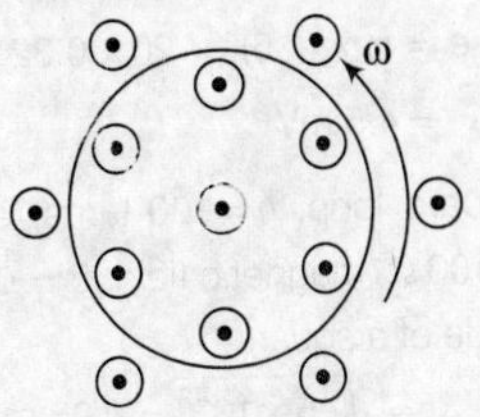

As disc is rotating counter clockwise, emf increases with increasing radius and rim is more positive with respect centre of the disc.

DPP-3 Induced EMF in a Moving Rod in Uniform Magnetic Field and Circuit Problems

1. (i) The change in area produces the change in magnetic flux.

(ii) The direction of current in the loop is anti-clockwise. As, the rod moves towards right the number of crosses in the loop increases with time and to oppose the increasing number of crosses in the loop, the current in the loop must be anti-clockwise as shown in the figure.

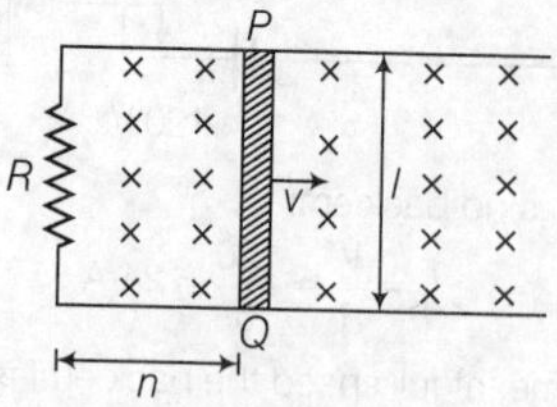

Induced current by changing the area of the rectangular loop.

(iii) According to Faraday's law,

The magnitude of induced emf, i.e.

$$E = \frac{d\phi}{dt} = \frac{d}{dt}(BA)$$

As, we know area, i.e. $A = ln$

$$\Rightarrow \quad \frac{d}{dt}(B\,ln) = Bl\left[\frac{dx}{dt}\right] = Blv$$

(iv) As, magnitude of current, i.e.

$$I = \frac{e}{R} = \frac{Blv}{R}$$

So, electrical power dissipated in the resistor.

i.e. $$P_{\text{electrical}} = I^2R = \frac{B^2l^2v^2}{R} \quad \left(\because I = \frac{Blv}{R}\right)$$

where, v is constant velocity and R is resistance across conducting rod of length l.

2. When the rod rotates in a vertical plane perpendicular to the magnetic meridian, it will cut horizontal component of the earth's magnetic field.

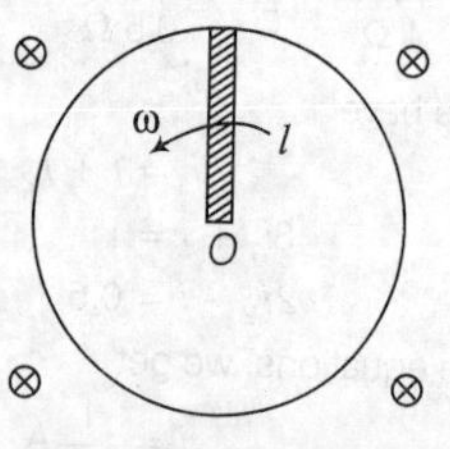

Induced emf between the ends of the rod

i.e. $$e = \frac{1}{2}B_H l^2 \omega$$

$$= \frac{1}{2} \times B_H \times l^2 \times 2\pi f \quad (\text{as, } \omega = 2\pi f)$$

$$= \pi l^2 f B_H \quad \ldots\text{(i)}$$

Given, $l = 1.5$ m, $B_H = 0.32$ G, $f = 20$ rev/s, so, substitute these values in Eq. (i), we get

The emf induced between the ends of the rod, i.e.

$$e = \pi \times (1.5)^2 \times 20 \times 0.32 \times 10^{-4}$$

$$= 4.5 \text{ mV}$$

3. Given, in a square loop, $n = 100$ turns, area $A = 2.5 \times 10^{-3} \text{m}^2$, resistance $R = 100\ \Omega$, magnetic field, $B = 0.40$ T

As, we know, side of a square

$$l = \sqrt{2.5 \times 10^{-3}} = 0.05 \text{ m}$$

As, the loop of length 0.05 m is pulled out in 1.0 s, so speed of the loop $= 0.05 \text{ ms}^{-1}$.

So, emf induced in the left arm of the loop,

$$e = NBlv$$

$$= 100 \times 0.40 \times 0.05 \times 0.05 = 0.1 \text{ V}$$

Current in the loop,

$$I = \frac{e}{R} = \frac{0.1}{100} = 10^{-3} \text{ A}$$

Force on the left arm exerted by the magnetic field,

$$F = BIl = 10^{-3} \times 0.05 \times 0.04$$

$$= 2 \times 10^{-5} \text{ N acting toward left.}$$

In order to pull the loop uniformly, a force of 2×10^{-5} N towards right must be applied.

i.e. Work done

$$W = Fl = 2 \times 10^{-5} \times 0.05 = 10^{-6} \text{ J}$$

4. It is given that the magnetic field is uniform. Join the end joints O and P and replace the semicircle by a straight rod of length $2r$.

Now, we have a straight rod rotating in a uniform magnetic field in a plane perpendicular to the magnetic field.

Therefore, the induced emf between O and P will be

$$E_{\text{induced}} = \frac{1}{2}B\omega(2r)^2$$

$$= \frac{B}{2} \times \omega \times 4 \times r^2$$

$$= 2B\omega r^2$$

From the right hand rule, we see that electrons will accumulate at end O. Therefore, end P is at a higher potential than O.

5. (d) As, induced current in the loop is zero, because total flux linked with the loop remains unchanged when the loop moves through the field. So, net change in magnetic flux passing through the coil is zero. Therefore, current induced in the loop is zero.

6. (a) Given, velocity $\mathbf{v} = (2\hat{\mathbf{i}} + 3\hat{\mathbf{j}} + \hat{\mathbf{k}}) \text{ms}^{-1}$,

$$\mathbf{B} = (\hat{\mathbf{i}} + 2\hat{\mathbf{j}}) \text{Wb/m}^2$$

Length, $$l = 2\hat{\mathbf{k}}$$

$\therefore$ Potential difference induced between the ends of the wire,

$$e = (\mathbf{v} \times \mathbf{B})\, l$$

$$= [(2\hat{\mathbf{i}} + 3\hat{\mathbf{j}} + \hat{\mathbf{k}}) \times (\hat{\mathbf{i}} + 2\hat{\mathbf{j}})] \cdot 2\hat{\mathbf{k}}$$

$$= (4\hat{\mathbf{k}} - 3\hat{\mathbf{k}} - \hat{\mathbf{j}} - \hat{\mathbf{i}}) \cdot 2\hat{\mathbf{k}}$$

$$= (\hat{\mathbf{k}} - \hat{\mathbf{j}} - \hat{\mathbf{i}}) \cdot 2\hat{\mathbf{k}}$$

$$= 2 \text{ V}$$

7. (a) Given, $l = 30$ cm, 0.3 cm

$$B = 4.0 \text{ T}$$

$$v = 10 \text{ ms}^{-1}$$

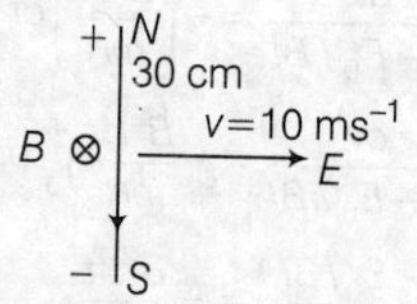

Then, induced potential at the North end of the bar is

$$e = Blv = 4 \times 0.3 \times 10$$

$$= 12 \text{ V}$$

8. (c) As, Wheatstone bridge is balanced. Then, current through arm AC is zero. Then, effective resistance R of bridge, i.e.

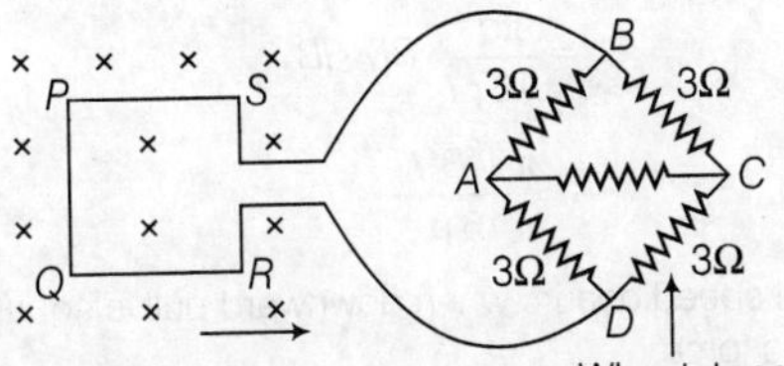

Wheatstone bridge

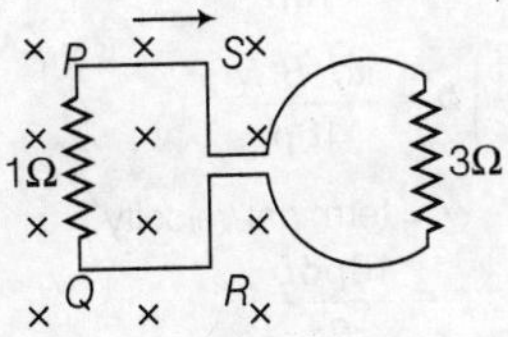

$$\frac{1}{R} = \frac{1}{6} + \frac{1}{6} = \frac{1}{3}$$

$\Rightarrow$ $R = 3\,\Omega$

Total resistance, $R = 3 + 1 = 4\Omega$

So, induced emf, $i = \frac{e_{\text{induced}}}{R_{\text{total}}} \Rightarrow \frac{Blv}{4}$

$$= \frac{2 \times 0.1 \times v}{4} = 1 \times 10^{-3}$$

$\therefore$ Speed of the loop,

$$v = 2 \times 10^{-2}\ \text{ms}^{-1}$$
$$= 2\ \text{cms}^{-1}$$

9. (d) The electric field/emf is induced in sides AD and BC both in the same direction. However, there will be no current in loop as the potential of A is same as that of B and potential of D is same as that of C.

10. (b) For an infinite single strip on length x and width dr parallel to wire, distant r,

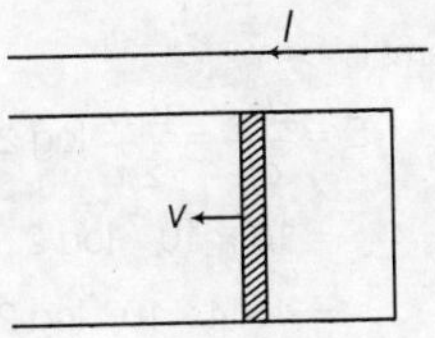

Area of strip = xdr

Flux through strip is

$$d\phi = \frac{\mu_0 I x}{2\pi r} dr$$

$\therefore$ Total flux linked with area enclosed by rod and coils

$$\phi = \frac{\mu_0 I x}{2\pi} \int_a^{a+b} \frac{dr}{r}$$
$$= \frac{\mu_0 I x}{2\pi} \log\left(\frac{a+b}{a}\right)$$

So, induced emf in the rod,

$$\varepsilon = \frac{d\phi}{dt} = \frac{\mu_0 I}{2\pi} \log\left(\frac{a+b}{a}\right)\frac{dx}{dt}$$
$$= \frac{\mu_0 I v}{2\pi} \log\left(\frac{a+b}{a}\right)$$

So, induced current $i = \frac{\varepsilon}{R}$

$= 6 \times 10^{-4}$ A (after substituting values)

So, force required to move rod with a constant speed v is

$$F = BIl$$
$$= \left[\frac{\mu_0 I}{2\pi} \int_a^{a+b} \frac{dr}{r}\right] \cdot \frac{\varepsilon}{R} \cdot l$$
$$= \frac{4\pi \times 10^{-7} \times 6 \times 10^{-4} \times 100}{2\pi} \times \log\left(\frac{11}{1}\right)$$
$$= 12 \log 11 \times 10^{-9}\ \text{N}$$
$$= 12 \log 11\ \text{nN}$$

11. (c) Induced current in the circuit,

$$I = \frac{BIv}{R + \frac{R}{2}} = \frac{2Blv}{3R}$$

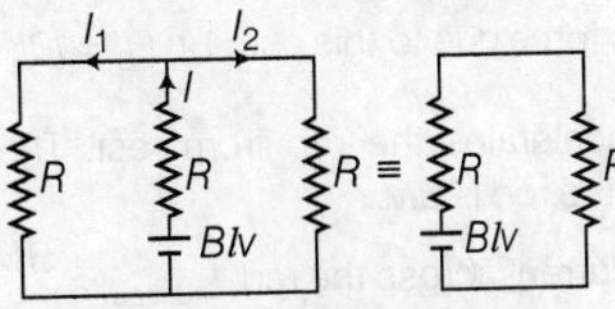

So, induced currents across conductor,

$$I_1 = I_2 = \frac{I}{2}$$
$$= \frac{2Blv}{3R \times 2} = \frac{Blv}{3R}$$

12. (c) In uniform magnetic field, as the cylinder is kept stationary. So, the flux linked with the cylinder is not changing. So, no current is induced in the cylinder.

13. (a) As, Faraday's laws are of EMI (Electromagnetic Induction) Lenz's law obeys the law of energy conservation. Fleming's right hand rule gives us the direction of induced current and Fleming's left hand rule gives us the direction of force on current carrying conductor in magnetic field.

14. (b,d) The field lines is given to be closed loop. Thus, gravitational and electrostatic fields do not form closed loops.

15. (a,d) ***Case I***

Induced emf, $e = Blv$

$$= 1 \times 0.2 \times 5 \times 10^{-2}$$

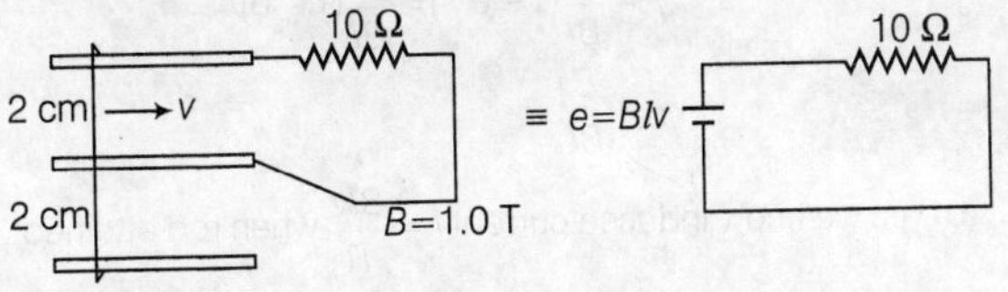

Induced current at positions,

$$i = \frac{e}{R} = 0.1\ \text{mA}$$

Case II

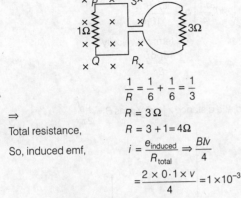

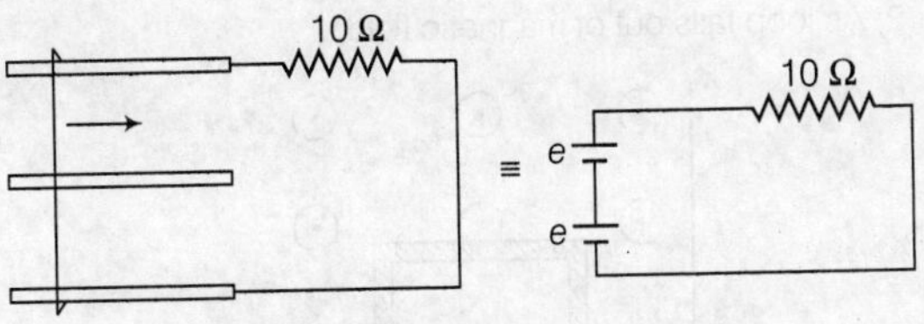

Induced current at positions

$$i = \frac{2e}{R}$$
$$= 0.2\ \text{mA}$$

16. (a,b,d) As we know, conducting rod of length l is moved at constant velocity v_0, we get

Work external force = Thermal energy dissipated in $R = \dfrac{B^2l^2v^2}{R}$

So, we conclude that

(a) Thermal power dissipated in the resistor is equal to the rate of work done by an external person pulling the rod.

(b) If applied external force is doubled, then a part of the external power increases the velocity of the rod.

(d) If resistance R is doubled, then power required to maintain the constant velocity v_0 becomes half.

17. (a) The current of the battery at any instant $I = \dfrac{E}{R}$

The magnetic force due to this current, $F_B = IBL = \dfrac{EBL}{R}$

This force accelerates the rod from rest. Then, motional emf developed in the rod is Blv.

So, induced current across the rod $I_{\text{induced}} = \dfrac{Blv}{R}$

The magnetic force due to induced current

$$F_{\text{induced}} = I_{\text{induced}} \cdot Bl = \frac{B^2l^2v}{R}$$

Then, net force acting on a rod

$$\Rightarrow \quad F_B - F_{\text{induced}} = m\frac{dv}{dt}$$

$$\frac{EBl}{R} - \frac{B^2l^2v}{R} = m\frac{dv}{dt}$$

$$\int_0^v \frac{dv}{E - Blv} = \frac{Bl}{mR}\int_0^t dt$$

$$\frac{E - Blv}{E} = e^{-\frac{B^2l^2}{mR}t}$$

Speed of a rod, $v = \dfrac{E}{Bl}(1 - e^{-t/\tau})$,

where, $\tau = \dfrac{mR}{B^2l^2}$

18. (c) The rod will attain terminal velocity at $t \to \infty$, we get

$$v_T = \frac{E}{Bl}(1 - e^{\infty}) = \frac{E}{Bl}(1 - 0)$$

$$v_T = \frac{E}{Bl}$$

19. (b) As, we know induced current $I = \dfrac{Blv}{R}$ when rod attained v_T.

So, $$I_{\text{induced}} = \frac{Blv_T}{R} \times \frac{Bl}{R} \times \left(\frac{E}{Bl}\right) = \frac{E}{R} \qquad \left(\because v_T = \frac{E}{Bl}\right)$$

Thus, $I_{\text{induced}} = I_{\text{battery}}$, hence net current through the circuit is zero.

20. (2) As loop falls out of magnetic field,

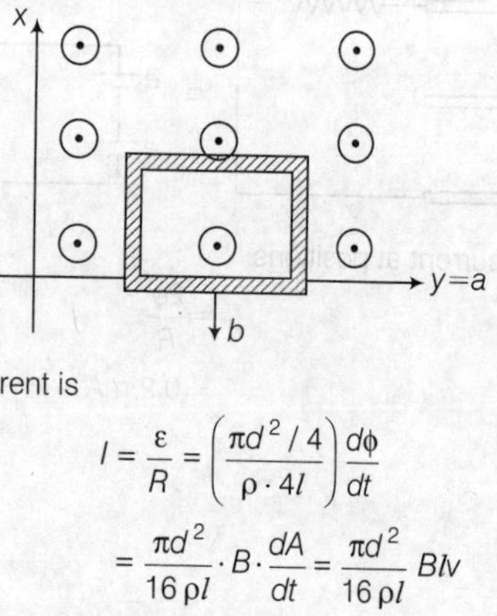

Induced current is

$$I = \frac{\varepsilon}{R} = \left(\frac{\pi d^2/4}{\rho \cdot 4l}\right)\frac{d\phi}{dt}$$

$$= \frac{\pi d^2}{16\,\rho l} \cdot B \cdot \frac{dA}{dt} = \frac{\pi d^2}{16\,\rho l} Blv$$

Force due to interaction of induced current and field is

$$F = IlB$$

$$= \frac{\pi d^2}{16\,\rho l} \cdot Blv \cdot lB$$

$$= \frac{\pi d^2 B^2 lv}{16\,\rho}$$

Terminal speed occurs when downward pull is equal to upward magnetic force.

$$\Rightarrow \quad m \times b = \frac{\pi d^2 B^2 lv}{16\rho}$$

$$\Rightarrow \quad \sigma\left(4\pi l\frac{d^2}{4}\right)b = \frac{\pi d^2 B^2 lv}{16\rho}$$

when, v = terminal velocity

$$= \frac{16\rho\sigma b}{B^2}$$

So, $k = 2$

21. (4) Field at a distance x from wire is $B = \dfrac{\mu_0 I}{2\pi x}$

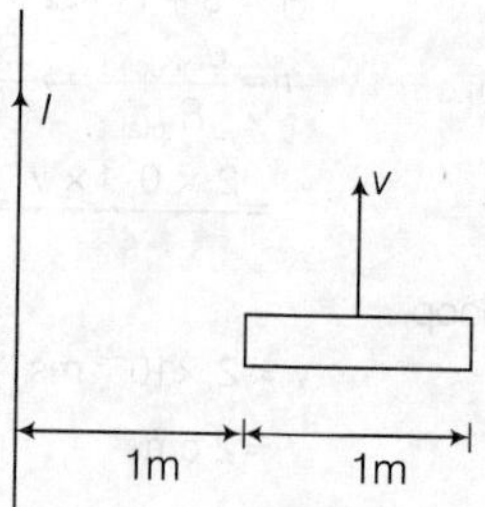

and $$s\phi_B = \int B \cdot dA$$

$$\therefore \quad \phi_B = \int_l^{2l} \frac{\mu_0 lvt}{2\pi}\left(\frac{dx}{x}\right)$$

$$= \frac{\mu_0 lvt}{2\pi}\log 2$$

Hence, induced emf is

$$E = \frac{d\phi_B}{dt} = \frac{\mu_0 lv}{2\pi}\log 2$$

$$= 16 \times 10^{-7}\log 2$$

$$= 4 \times 4 \times 10^{-7}\log 2$$

So, $$N = 4 = 4 \times 10^{-7}\log 16$$

22. (c) Statement I is correct and Statement II is also correct but Statement II does not explain Statement I.

To connect a light bulb across the wings, we have to form a close loop and flux through this closed loop is constant and zero emf is generated.

23. (d) Statement I is incorrect and Statement II is correct.

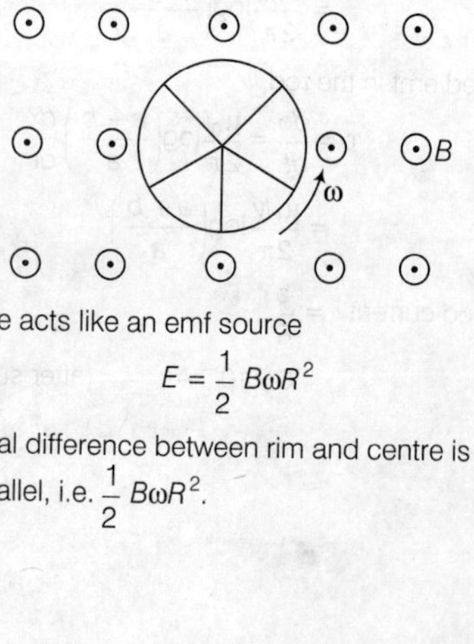

Each spoke acts like an emf source

$$E = \frac{1}{2}B\omega R^2$$

So, potential difference between rim and centre is due to N such cells in parallel, i.e. $\dfrac{1}{2}B\omega R^2$.

DPP-4 **EMF Induced in a Rod or Loop in Non-Uniform Magnetic Field**

1. Let at any instant the velocity of rod is v, induced emf in rod will be $\varepsilon = Blv$.

The electrical equivalent circuit is shown in the figure.

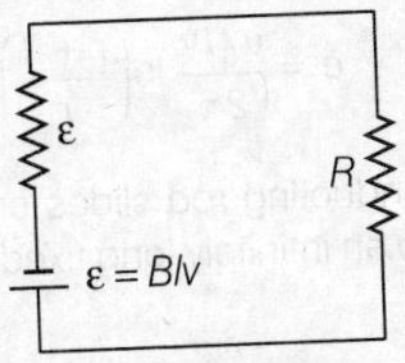

Circuit current at that instant is

$$I = \frac{\varepsilon}{R + r} = \frac{Blv}{R + r}$$

Magnetic force opposing motion of rod is

$$F_m = BIl = -m\frac{dv}{dt}$$

Substituting for I and rearranging, we get

$$\frac{dv}{v} = \frac{-B^2l^2}{(R + r)m} \cdot dt$$

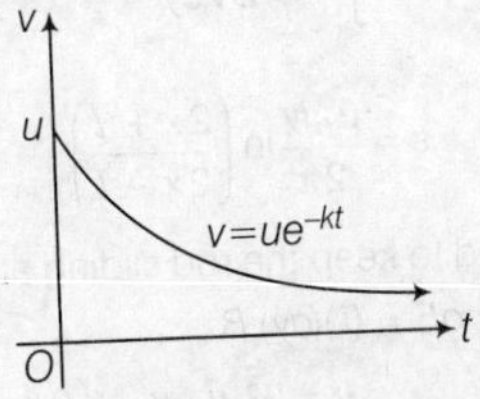

$$\Rightarrow \quad \int_u^v \frac{dv}{v} = \frac{-B^2l^2}{(R + r)m} \cdot \int_0^t dt$$

$$\Rightarrow \quad \log\left(\frac{v}{u}\right) = \frac{-B^2l^2}{(R + r)m} \cdot t$$

$$\Rightarrow \quad v = u \cdot e^{-kt}$$

where, $$k = \frac{B^2l^2}{(R + r)m}$$

So, velocity reduces exponentially.

2. (i) Flux linked with the loop,

$$\phi = \frac{\mu_0}{2\pi} ib \ln\left(\frac{b + a}{a}\right) - \frac{\mu_0 ib}{2\pi} \ln\left[\frac{a + b + d}{a + d}\right]$$

$$= \frac{\mu_0 ib}{2\pi} \ln\left[\frac{(a + d)(b + a)}{a(a + b + c)}\right]$$

So, induced emf in the loop (if the current in both the wires is changing at the rate of di/dt.) is

$$e = \frac{d\phi}{dt} = \frac{\mu_0 b}{2\pi} \ln\left[\frac{(a + d)(b + a)}{a(b + a + d)}\right] \frac{di}{dt}$$

(ii) As, we know $\frac{di}{dt} > 0$, then the direction of induced current is clockwise.

$\therefore$ Force on the loop will be away from wires.

3. (i) Here, current in long wire, $i = 10$ A.

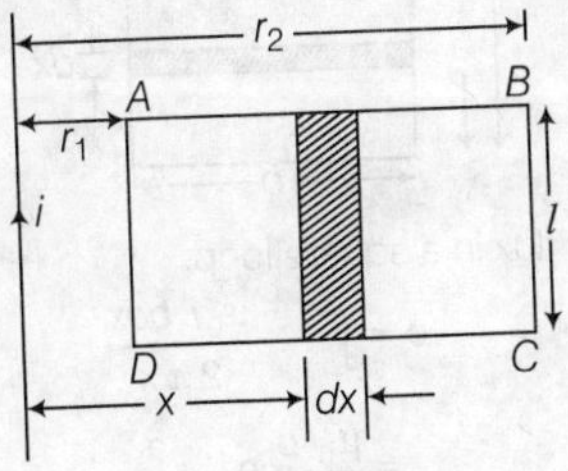

For the rectangular circuit $ABCD$, $l = 0.2$ m, $r_1 = 0.05$m, $r_2 = 0.10$m.

Consider a strip of rectangular circuit at a distance x from the wire and having width dx.

Magnetic field intensity at location of the strip, $B = \frac{\mu_0 i}{2\pi x}$

Area of the strip, $A = l dx$

$\therefore$ Magnetic flux linked with the strip

$$d\phi = BA = \frac{\mu_0 i}{2\pi x} l dx$$

Total magnetic flux linked with the rectangular circuit,

$$\phi = \int_{r_1}^{r_2} \frac{\mu_0 i}{2\pi x} l dx = \frac{\mu_0 il}{2\pi} \log_e\left(\frac{r_2}{r_1}\right)$$

$$= \frac{\mu_0 il}{2\pi} \log_e\left(\frac{0.10}{0.05}\right)$$

So, $$\phi = \frac{\mu_0 il}{2\pi} \times \log_e 2 = \frac{4\pi \times 10^{-7} \times 10 \times 0.2 \times 0.63}{2\pi}$$

$$= 2.772 \times 10^{-7} \text{ Wb}$$

(ii) The emf induced in the circuit

$$e = -\frac{d\phi}{dt} = \frac{2.772 \times 10^{-7}}{0.02} = 1.386 \times 10^{-5} \text{ V}$$

The direction of magnetic field due to current i in the wire is perpendicular to the plane of the loop and inwards. As current reduces to zero, this field reduces to zero.

(iii) The direction of current induced in rectangular circuit is such as to maintain this field. Hence, induced current is clockwise.

4. Join the end points O and B and replace the two semicircles by a straight rod length $4r$.

Induced emf in straight rod OB will be same as in the actual conductor. Now, this straight rod can be replaced by the combination of two rods OA and AB. Induced emf in OB will be same as induced emf in OA + induced emf in AB.

But induced emf in AB will be zero, because its velocity will be parallel to its length.

Hence, induced emf in OA will be net induced emf in actual conductor.

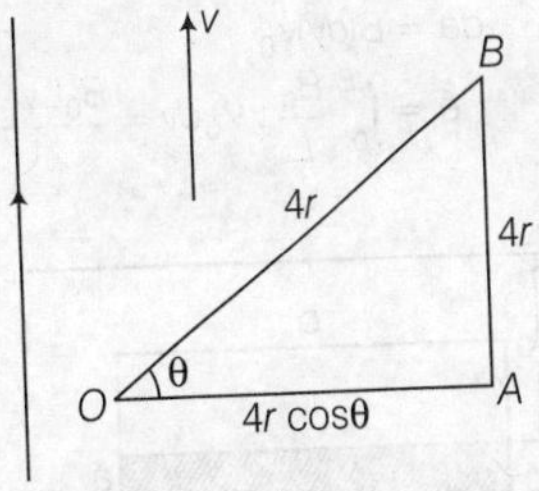

Consider a small element of the wire of length dx located at a distance x from the wire. The magnetic field at this element is B

i.e. $$B = \frac{\mu_0 I}{2\pi x}$$

The potential difference across this element is

$$dV = \frac{\mu_0 I}{2\pi x} v dx.$$

So, potential difference across the ends of the rod is

$$V = \int dV = \int_{d - 2r\cos\theta}^{d + 2r\cos\theta} \frac{\mu_0 I}{2\pi x} v dx$$

$$\Rightarrow \quad V = \frac{\mu_0 Iv}{2\pi} \ln\left[\frac{d + 2r\cos\theta}{d - 2r\cos\theta}\right]$$

5. (b) Emf is induced in the ring and it oppose the motion. Then, due to the resistance of the ring all energy dissipates. So, the swings through the field, the pendulum will be come to rest very soon.

6. (d) Magnetic field intensity at a distance r from the straight wire carrying current is

$$B = \frac{\mu_0 i}{2\pi r}$$

As, area of loop, $A = a^2$

and magnetic flux $\phi = BA$

$\therefore$
$$\phi = \frac{\mu_0 i a^2}{2\pi r}$$

The induced emf in the loop is

$$e = \left|\frac{d\phi}{dt}\right| = \left|\frac{d}{dt}\frac{\mu_0 i a^2}{2\pi r}\right|$$

$\Rightarrow$
$$e = \frac{\mu_0 i a^2}{2\pi r^2}\frac{dr}{dt} = \frac{\mu_0 i a^2 v}{2\pi r^2}$$

where, $v = \frac{dr}{dt}$ is velocity.

7. (b) Consider a conducting rod moves with constant velocity v perpendicular to the long straight wire, i.e. induced emf,

$$e = Blv$$
$$e = \frac{\mu_0 I}{2\pi r} lv$$

8. (b) As some instant t, flux linked with a strip with coordinate y is

$$d\phi = Bldy = 4lt^2 ydy$$

where, l = length of side of square loop.

So, total flux linked with the square loop is

$$\phi = \int_0^l d\phi = \int_0^l 4lt^2 ydy = 2l^3t^2$$

Hence, induced emf in the loop is

$$\varepsilon = \frac{d\phi}{dt} = \frac{d}{dt} 2l^3t^2 = 4l^3t$$

$\therefore$ ε at $t = 2.5$ s $= 8 \times 10^{-5}$ V clockwise

9. (a) Given, magnetic field in a region is given by

$$\mathbf{B} = \frac{B_0}{L} y\hat{\mathbf{k}}$$

where, L is a fixed length.

Then, induced emf between the ends of the rod is

$$de = B(dy)v_0$$
$$e = \int_0^L \frac{B_0}{L} y v_0 dy = \frac{B_0 L v_0}{2}$$

10. (b)

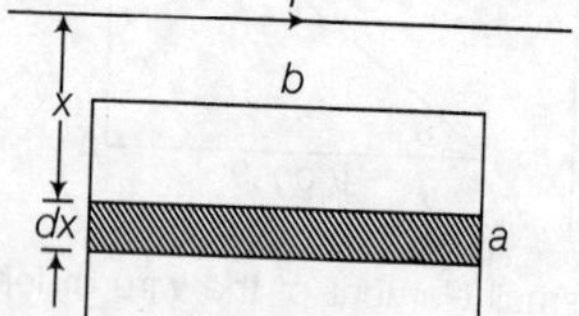

Consider a long straight wire which is parallel to edge, such that change in flux, i.e.

$$d\phi = B(bdx)$$
$$d\phi = \frac{\mu_0 I}{2\pi x} bdx$$
$$\phi = \int_d^{d+a} \frac{\mu_0 Ib\, dx}{2\pi x}$$
$$\phi = \frac{\mu_0 Ib}{2\pi} \ln\left(\frac{d+a}{d}\right)$$

As, $I = I_0 e^{-t/\tau}$ and induced emf in the loop,

$$E = -\frac{d\phi}{dt}$$
$$E = \frac{\mu_0 b I_0}{2\pi\tau} \ln\left[\frac{d+a}{d}\right] e^{-t/\tau}$$

11. (d) Induced emf in the rod,

$$de = (B)(dx)v$$
$$de = \frac{\mu_0 I}{2\pi x} v dx$$
$$e = \int_r^{r+l} \frac{\mu_0 Iv}{2\pi x} dx$$
$$e = \frac{\mu_0 Iv}{2\pi} \ln\left(\frac{r+l}{r}\right)$$

12. (a) Consider a conducting rod slides on a pair of thick metallic rails laid parallel to an infinitely long fixed wire such that

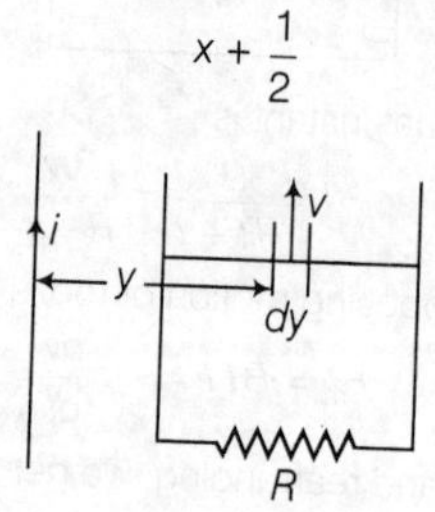

Induced emf in the rod, i.e.

$$\varepsilon = \int_{x-\frac{l}{2}}^{x+\frac{l}{2}} Bvdy$$
$$\varepsilon = \frac{\mu_0 iv}{2\pi} \ln\left(\frac{2x+l}{2x-l}\right)$$

Net force needed to keep the rod sliding at a constant speed v is

$$dF = (i)(dy)B$$
$$F = \int_{x-\frac{l}{2}}^{x+l/2} \frac{\mu_0 iv}{2\pi R} \ln\left(\frac{2x+l}{2x-l}\right)\frac{\mu_0 i}{2\pi y} dy$$

$$\left[\because \text{current } (i) = \frac{e}{R}\right]$$

$$= \frac{v}{R}\left(\frac{\mu_0 i}{2\pi} \ln\frac{2x+l}{2x-l}\right)^2$$

13. (a) As, we know emf induced in a current carrying rod

i.e.
$$\varepsilon = \frac{\mu_0 iv}{2\pi} \ln\left[\frac{2x+l}{2x-l}\right]$$

So, induced current in a rod,

$$I = \frac{\varepsilon}{R} = \frac{\mu_0 iv}{2\pi R} \ln\left[\frac{2x+l}{2x-l}\right]$$

14. (c) Consider a small length of loop, i.e. dx having magnetic field **B**, such that flux induced in a loop,

$$d\phi = B(bdx)$$
$$d\phi = \frac{\mu_0 i\, bdx}{2\pi x}$$

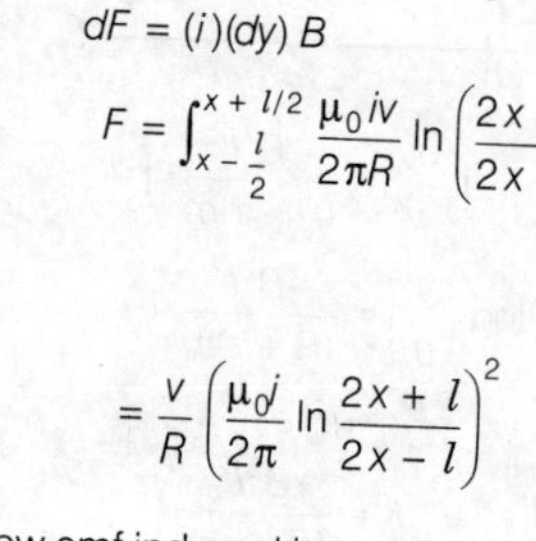

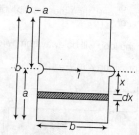

Net magnetic flux in a square loop,

$$\phi = \int_{b-a}^{a} \frac{\mu_0 i\, bdx}{2\pi x}$$
$$= \frac{\mu_0 i b}{2\pi} \ln\left(\frac{a}{b-a}\right)$$

15. (b) Induced emf across rod PQ is

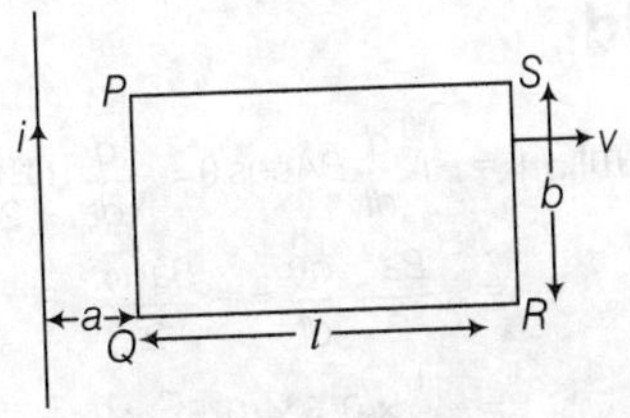

$$e_{PQ} = Bl_{PQ}v = \frac{\mu_0 i\, bv}{2\pi a}$$

The emf across SR

$$e_{SR} = Bl_{SR}v = \frac{\mu_0 i\, bv}{2\pi(a+l)}$$

Net emf induced across rear end is at a distance a from the wire is

$$e = e_{PQ} - e_{SR} = \frac{\mu_0 i\, lvb}{2\pi a(a+l)}$$

16. (a)

(i) If current is increased, flux in the loop will increase in inside direction, then due to Lenz's law induced emf in the loop will be in anti-clockwise direction. Due to this current, the current in the nearer side of loop to the wire will be opposite to that of wire. Hence, there will be repulsion.

(ii) This situation is just opposite to part (i).

(iii) If loop is moved away, then flux decreases and this becomes similar to part (ii).

(iv) Similar to part (i).

17. (a) Magnetic flux through $PQROP$ when current in wire AB is I such that

$$\phi = \int BdA$$

$$= \int_0^r \frac{\mu_0 I}{2\pi(r+x)}(2\sqrt{r^2 - x^2})dx = \frac{\mu_0 IR(\pi - 2)}{2\pi}$$

18. (a) As, induced emf

$$\varepsilon = d\phi/dt$$

$$\varepsilon = \frac{-\mu_0 r(\pi - 2)}{2\pi}\frac{dI}{dt}$$

Now, area of triangle = total charge

$$Q = I_0 T/2$$

So, induced current, $I_0 = \dfrac{2Q}{T}$

Current at time $t = I_0 - \left(\dfrac{I_0}{T}\right)t$

$$\Rightarrow \quad I = \frac{2Q}{T}\left(1 - \frac{t}{T}\right)$$

$$\Rightarrow \quad \frac{dI}{dt} = -\frac{2Q}{T^2}$$

Induced emf in $PQROP$ region,

$$\varepsilon = \frac{\mu_0 r(\pi - 2)}{2\pi}\left(\frac{2Q}{T^2}\right) = \frac{\mu_0 r(\pi - 2)Q}{\pi T^2} \quad \left[\because \text{Induced current } (I_1) = \frac{\varepsilon}{R}\right]$$

Induced current in region $PQROP$

$$I_1 = \frac{\varepsilon}{R} = \frac{\mu_0 r(\pi - 2)}{\pi R T^2}Q$$

19. (b) Heat generated in region $PQROP$ in function of time T,

$$H = \int_0^T I_1^2 R\, dt = I_1^2 RT$$

$$= \frac{\mu_0^2 r^2 (\pi - 2)^2 Q^2}{\pi^2 R T^3}$$

20. (a,c,d)

Current in loop,

$$I = d\phi = \frac{dC}{dt}V = \frac{d}{dt}CV_0(1 - e^{-t/\tau})$$

$$= \frac{CV_0}{\tau}.e^{-t/\tau} = \frac{V_0}{R}e^{-t/\tau}$$

Also, if ε = Induced emf for circuit

Then, $\varepsilon = IR + V_c$

$$= \frac{V_0}{R}e^{-t/\tau}\cdot R + V_0(1 - e^{-t/\tau})$$

$$= V_0 = \frac{d}{dt}\phi_B = A\cdot\frac{dB}{dt}$$

$$\Rightarrow \quad \frac{dB}{dt} = \frac{V_0}{\pi r^2}$$

21. (a,c) Each coil has a pulse of voltage tending to produce anti-clockwise current as magnet approaches and then a pulse of clockwise voltage or magnet moves away.

Speed of magnet is

$$v = \frac{l}{t} = \frac{1.5}{2.4 \times 10^{-3}}\,\text{ms}^{-1}$$

$$= 625\,\text{ms}^{-1}$$

22. (a,c)

(a) Induced current $= \dfrac{E}{R} = \dfrac{Blv}{R} = \dfrac{0.360}{0.4} = 0.9$ A

(b) Force on axle $= F = IlB = 0.108$ N

(c) As flux is decreasing as axle rolls, induced current flows through R from B to A.

(d) When axle rolls pasts resistor R, magnetic flux through closed loop increases.

Hence, counter clockwise current will flow to produce upward flux. So, current still flows from B to A in R.

23. (b,c)

(a) Flux linked with loop $= BA\cos\theta$

$$= (a + bt)(\pi r^2)\cos 0^\circ$$

$$= \pi(a + bt)r^2 \text{ weber}$$

(b) Induced emf in loop

$$E = -\frac{d\phi}{dt}$$

$$= -\frac{d}{dt}\pi r^2(a + bt)$$

$$= -\pi b r^2 \text{ volts}$$

(c) Power $= EI = \dfrac{E^2}{R} = \dfrac{\pi^2 b^2 r^4}{R}$ watts

(d) As, flux is increasing with time, front face of loop is a N-pole.

24. (a,b,d)

(a) Induced emf $= -\dfrac{d}{dt}NBA$

$$= -\pi a^2\left(\frac{d}{dt}B\right) = \pi a^2 k$$

(b) Charge stored

$$Q = CE = C\pi a^2 k$$

(c) B points into plane o

f paper and is decreasing so induced emf also produces a field that points into the paper.

$\therefore$ Current flows clockwise and so upper plate is positive.

(d) Changing B induces E and this pushes on charges in the wire.

DPP-5 Induced EMF in Rod, Ring Disc Rotated in a Uniform Magnetic Field

1. Here, radius of smaller loop, $r_1 = 0.3 \text{ cm} = 0.3 \times 10^{-2} \text{ m}$

Radius of bigger loop, $r_2 = 20 \text{ cm} = 20 \times 10^{-2} \text{ m}$

Distance between their centres,

$$x = 15 \text{ cm} = 15 \times 10^{-2} \text{ m}$$

$$I = 24 \text{ A}, \phi_2 = ?$$

As, symmetry shows that flux linking the bigger loop due to current in smaller loop is same as flux linked with smaller loop due to same current in bigger loop.

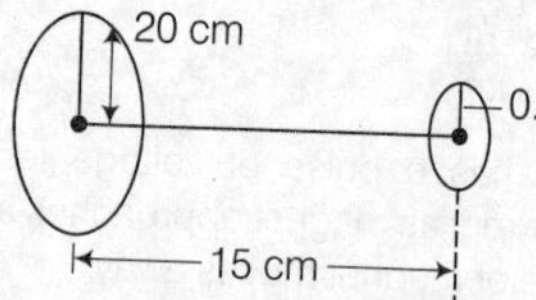

Now, magnetic field at the centre of smaller loop,

$$B_2 = \frac{\mu_0}{4\pi} \times \frac{2\pi I r_2^2}{(r_2^2 + x^2)^{3/2}}$$

$$= \frac{10^{-7} \times 2\pi \times 2 \times (20 \times 10^{-2})^2}{[(20 \times 10^{-2}) + (15 \times 10^{-2})^2]^{3/2}}$$

$$= 3.217 \times 10^{-6} \text{T}$$

Magnetic flux linked with smaller coil

i.e. $\phi_2 = B_2 \times A_1 = B_2(\pi r_1^2)$

$$= 3.217 \times 10^{-6} \times \frac{22}{7} \times (0.3 \times 10^{-2})^2$$

$$= 9.096 \times 10^{-11} \text{ Wb}$$

2. $\phi_A = B_A A$ magnetic flux across a circular loop A,

$$\phi_A = \frac{\mu_0 I \pi R^2 \times \pi r^2}{2\pi(R^2 + x^2)^{3/2}}$$

So, emf induced across a loop A, $e_A = -\frac{d\phi}{dt}$

$$= \frac{\mu_0 I \pi}{2} R^2 r^2 (-3/2)(R^2 + x^2)^{-5/2} \times 2x$$

As, e_A is maximum, when $\frac{de_A}{dx} = 0$

$$\Rightarrow \quad \frac{d}{dx}\left[\frac{x}{(R^2 + x^2)^{5/2}}\right] = 0$$

$$\Rightarrow \quad (R^2 + x^2)^{5/2} \times \left(-\frac{5x}{2}\right)(R^2 + x^2)^{3/2} \times 2x = 0$$

$$\Rightarrow \quad R^2 + x^2 - 5x^2 = 0$$

$$\Rightarrow \quad R^2 - 4x^2 = 0$$

$$\Rightarrow \quad (R)^2 - (2x)^2 = 0$$

or $\Rightarrow x = R/2$

So, distance between the two loop (x) when induced emf in loop A is maximum at $x = \frac{R}{2}$.

3. Let there be an element dx of rod at a distance x from the wire. Now, emf developed in the element, $dE = Bdxv$

$$\therefore \quad dE = \left(\frac{\mu_0}{4\pi}\frac{2I}{x}\right)dx \cdot v$$

$$\therefore \quad E = \frac{\mu_0 I v}{2\pi}\int_a^b \frac{dx}{x} = \frac{\mu_0 I v}{2\pi}\log_e \frac{b}{a}$$

$$\Rightarrow \quad E = \frac{4\pi \times 10^{-7} \times 100 \times 5}{2\pi} \times \log_e \frac{100}{1}$$

$$= 4.6 \times 10^{-4} \text{ V} = 0.46 \text{ mV}$$

So, net induced emf in the rod is 0.46 mV.

4. (i) Induced emf, $\varepsilon = -N\frac{d}{dt}BA\cos\theta = -\frac{d}{dt}B\frac{\theta a^2}{2}\cos 0^\circ \quad (\because N = 1)$

$$= -\frac{Ba^2}{2}\cdot\frac{d\theta}{dt} = -\frac{Ba^2\omega}{2}$$

$$= -\frac{1}{2} \times 0.5 \times (0.5)^2 \times 2$$

$$= -0.125\text{V} = 125 \text{ mV (clockwise)}$$

(ii) When $t = 0.25$ s

Angle turned

$$\theta = \omega t = 2 \times 0.25$$

$$\theta = 0.5 \text{ rad}$$

The arc PQ has length

$$l = r\theta = 0.5 \times 0.5 = 0.25 \text{m}$$

Length of circuit is

$$= 0.5\text{m} + 0.5\text{m} + 0.25\text{m} = 6.25\Omega$$
$$= 1.25\text{m}$$

Resistance of circuit $= 1.25 \times 5 = 6.25\ \Omega$

Current in the circuit $= \frac{0.125}{6.25} = 0.02\text{A}$ (clockwise)

5. (d) All batteries are in parallel then same emf induced across each battery which is equal to e.

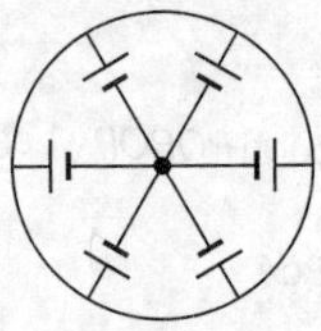

6. (c) $x_2 = 0.10$ m

The emf induced between the point A and B of rod,

$$e = \int_{x_1 = 0.07}^{x_2 = 0.10} B(\omega x)dx$$

$$= \omega B\int_{0.07}^{0.1} x dx = \omega B\left[\frac{x^2}{2}\right]_{0.07}^{0.1}$$

$$= 2 \times 10 \times \left[\frac{(0.1)^2}{2} - \frac{(0.07)^2}{2}\right]$$

$$= 0.05092 \approx 0.051 \text{ V}$$

7. (c) Given, $l = 1$ m, $\omega = 5 \text{ rads}^{-1}$

$$B = 0.2 \times 10^{-4} \text{ T}$$

The emf developed between the two ends of the conductors.

$$e = \frac{B\omega l^2}{2} = \frac{0.2 \times 10^{-4} \times 5 \times 1^2}{2} = 50\ \mu\text{V}$$

8. (c) Kinetic energy of slider is dissipated in form of heat in resistor.

$$\therefore \quad -\frac{d}{dt}(\text{KE}) = \frac{\varepsilon^2}{R}$$

$$\Rightarrow \quad -\frac{d}{dt}\left(\frac{1}{2}mv^2\right) = \frac{B^2l^2v^2}{R}$$

$$\Rightarrow \quad \frac{dv}{v} = \frac{-B^2l^2}{mR}\cdot dt$$

$$\text{or} \quad \int_{v_0}^{v}\frac{dv}{v} = \frac{-B^2l^2}{mR}\int_0^t dt$$

$$\Rightarrow \quad \log v = \frac{-B^2l^2}{mR}\cdot t + \log v_0$$

which is best shown by graph (c).

9. (b) Consider a rod rotates with a small angle but uniform angular velocity ω about its perpendicular bisector.

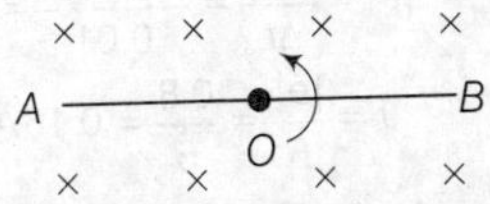

Now, potential difference across OB = potential difference across OA

$$= \frac{B\omega\left(\frac{l}{2}\right)^2}{2} = \frac{B\omega l^2}{8}$$

10. (d) Potential difference across AB, i.e.

$$V_{AB} = V_A - V_B = \frac{B\omega l^2}{2} \quad \ldots\text{(i)}$$

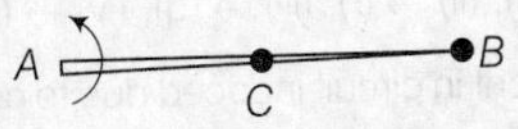

Potential difference across AC, i.e.

$$V_{AC} = V_A - V_C = \frac{B\omega\left(\frac{l}{2}\right)^2}{2} = \frac{B\omega l^2}{8} \quad \ldots\text{(ii)}$$

On subtracting Eq. (ii) from Eq. (i)

So, potential difference developed between the mid-point C of the rod and end B is

$$V_C - V_B = \frac{3B\omega l^2}{8}$$

11. (c) Potential drop across V_{OA}, $V_O - V_A = \dfrac{B\omega l^2}{2}$ ...(i)

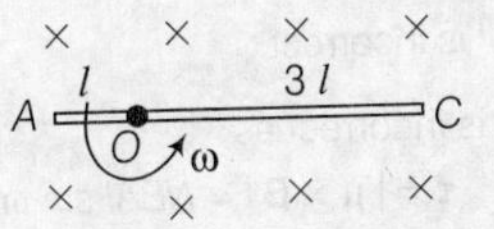

Potential drop across V_{OC}, $V_O - V_C = \dfrac{B\omega(3l)^2}{2} = \dfrac{9B\omega l^2}{2}$ ...(ii)

On subtracting Eq. (ii) from Eq. (i), we get

$$V_A - V_C = 4B\omega l^2$$

12. (b) As, we know, magnetic flux, $\phi = BA\cos\omega t$

Induced emf, $e = -\dfrac{d\phi}{dt} = BA\omega\sin\omega t$

Induced current across magnetic field B,

$$i = \frac{BA\omega\sin\omega t}{R}$$

Power generated across a circle $P_{inst} = i^2R$

Average power generated per period of rotation,

$$P_{avg} = \frac{\int_0^T P_{inst} \times dt}{\int_0^T dt}$$

$$P_{avg} = \frac{(\omega BA)^2}{2R}, \; A = \frac{\pi r^2}{2}$$

13. (b) A motional emf, $e = Blv$ is induced in the rod. Or we can say a potential difference is induced between the two ends of the rod AB, with A at higher potential and B at lower potential. Due to this potential difference, there is an electric field in the rod.

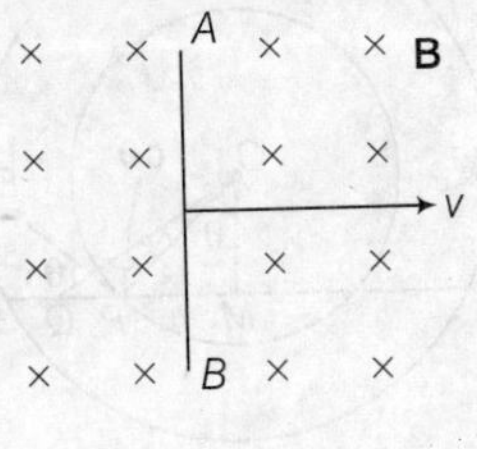

14. (c) Potential drop across V_{OP} and V_{OQ} such that

$$V_O - V_P = V_O - V_R = \frac{B\omega(\sqrt{2}R)^2}{2} = B\omega R^2$$

$$V_Q - V_O = \frac{B\omega(2R)^2}{2} = 2B\omega R^2$$

15. (d) As, potential drop across the ring is zero.

So, induced current, $i = \dfrac{\varepsilon}{R} = 0$

Thus, no current flows in a ring.

16. (5) The total flux through N turns of the coil

i.e. $\phi_{total} = NBA\cos\theta$

According to Faraday's law of electromagnetic induction, we get

$$E_{induced} = -\frac{d\phi}{dt} = -\frac{d}{dt}[NBA\cos\theta]$$

$$= -NA\cos\theta\frac{dB}{dt}$$

∴ The current induced in the coil,

$$I_{induced} = \frac{E_{induced}}{R}$$

$$= \frac{100 \times 100 \times 10^{-4} \times \cos 60° \times 2}{0.2}$$

$$= 5\text{ A}$$

17. (9) Flux through loop is

$$\phi = \int_0^{2a} \frac{\mu_0}{4\pi} 2Ia\left(\frac{1}{x} + \frac{1}{3a - x}\right)dx$$

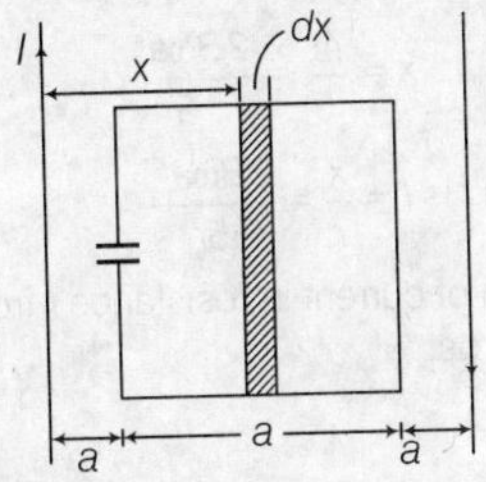

So, induced emf is

$$\varepsilon = -\frac{d\phi}{dt}$$

$$= \frac{-\mu_0}{\pi} \cdot aI_0\omega \cdot \log 2 \cdot \cos\omega t$$

Charge on the capacitor,

$$Q = C\varepsilon = \frac{C\mu_0}{\pi} \cdot aI_0\omega \cdot \log 2 \cdot \cos\omega t$$

So, current in the loop,

$$= \frac{dQ}{dt} = \frac{d}{dt}\frac{C\mu_0}{\pi} \cdot I_0\omega \cdot a\log 2 \cdot \cos\omega t$$

$$= \frac{\mu_0}{\pi}CI_0\omega^2 a\log 2 \cdot \sin\omega t$$

So, maximum current in loop is

$$I_{max} = \frac{\mu_0}{\pi} \cdot CI_0\omega^2 \cdot a\log 2$$

Substituting the values, we have

$$I_{max} = 4 \times 10^{-7} \times 1 \times 10^{-6} \times 5 \times 2500 \times \pi^2 \times 2 \times 10^{-2} \times \log 2$$

$$= 10^{-11} \times \pi^2 \times \log 2$$

So, $k - 2 = 9$

18. (a,d) As, we know, angular displacement $\theta = \omega t$. Only half circular part will be involved in inducing emf, so effective area,

$$A = \frac{\pi a^2}{2}$$

As, magnetic flux induced in a circular loop, i.e

$$\phi = BA\cos\theta$$

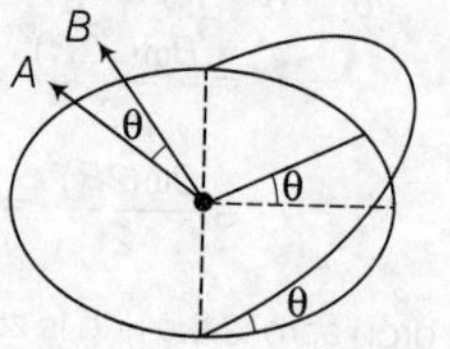

Then, emf induced in a circular loop,

$$e = \frac{-d\phi}{dt} = BA\sin\theta\left(\frac{d\theta}{dt}\right) = \frac{B\pi a^2}{2}\omega\sin\theta$$

$\therefore$ Induced current, $I = \frac{e}{R} = \frac{B\pi a^2\omega}{2R}\sin\theta$

Clearly, $I = 0$, when $\theta = 0°$ and

when $\theta = \frac{\pi}{2}$, $I = \frac{B\pi a^2\omega}{2R}$

19. (b,d)

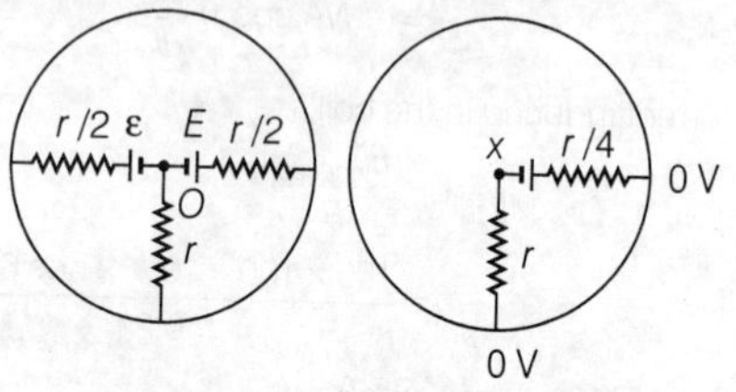

$$\text{Induced emf, } \varepsilon = \frac{B\omega x^2}{2}$$

By modal equation, $4\left(\frac{x-\varepsilon}{r}\right) + \left(\frac{x-0}{r}\right) = 0$

$$x = \frac{4\varepsilon}{5} = \frac{2B\omega a^2}{5}$$

Current through r is $I = \frac{x}{r} = \frac{2B\omega a^2}{5r}$

So, the direction of current in resistance r from circumference to centre, i.e. inwards.

20. (b) At $t = 0.01$s,

$$|e| = \frac{\Delta\phi_B}{\Delta t} = \frac{8\times10^{-3}}{0.01} = 0.8\text{V}$$

$$i = \frac{|e|}{R} = \frac{0.8}{5} = 0.16\text{A}$$

$$\Delta q = i\Delta t = 0.16\times0.01 = 1.6\times10^{-3}\text{C}$$

At $t = 0.02$s

$$|e| = \frac{8\times10^{-3}}{0.02} = 0.4\text{V}$$

$$i = \frac{|e|}{R} = 0.08\text{A}$$

$$\Delta q = i\Delta t = 0.08\times0.02 = 1.6\times10^{-3}\text{C}$$

Hence, (i) → (s), (ii) → (r), (iii) → (q), (iv) → (q)

21. (c) Let I = current in circuit induced due to collapse of loop in time Δt second.

$$I = \frac{E}{R} = \frac{BA}{R\Delta t} = \frac{B}{R}\cdot\frac{A}{\Delta t}$$

where, $I\,\Delta t$ = charge q through the circuit

So, $q = I\,\Delta t = \frac{BA}{R}$

$$= \frac{15\times10^{-6}\times(0.2)^2}{0.5} = 1.2\ \mu\text{C}$$

This charge is stored by capacitor.

So, Statement I is correct.

Also, $C \neq \varepsilon Q$ in this case, as $C = \frac{\varepsilon_0 A}{d}$, it is a constant.

So, Statement II is incorrect.

22. (d) Statement I is incorrect as

$$\tau = |\mu\times\mathbf{B}| = NBAI\sin\omega t$$

So, torque required is not uniform.

Also, Statement II is correct.

DPP-6 Loop in a Time Varying Magnetic Field and Induced EMF

1. (i) Consider a magnetic field where B (magnitude of the magnetic field is a function of r ($r < R$), the distance of the point from O. For the circular path as shown,

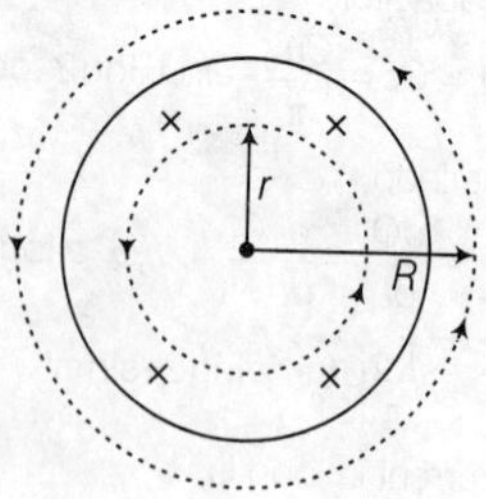

By Faraday's law, we have $\oint \mathbf{E}\cdot d\mathbf{s} = -\frac{d}{dt}\phi_B$

***Case* I** For $r < R$

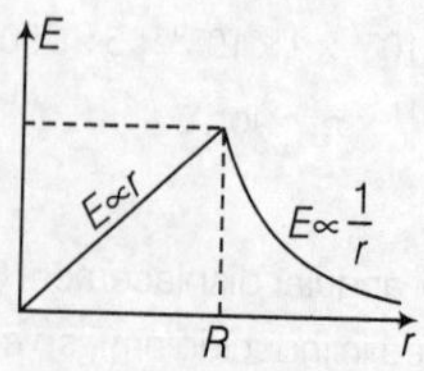

$$E(2\pi r) = \pi r^2\left|\frac{dB}{dt}\right|$$

$$\Rightarrow \quad E = \frac{r}{2}\left|\frac{dB}{dt}\right|$$

***Case* II** For $r > R$

$$E(2\pi r) = \pi R^2\left|\frac{dB}{dt}\right|$$

$$\Rightarrow \quad E = \frac{R^2}{2r}\left|\frac{dB}{dt}\right|$$

(ii) Direction of E can be easily obtained as it would be responsible for the induced current when a conducting loop is placed on the given path, e.g. in the present case, for $\frac{dB}{dt} > 0$, path is in anti-clockwise direction.

2. Consider a point on the circumference of a circle of radius r ($r < R$). Let E be the induced electric field along the tangents to the circle. Then, $E = \frac{r}{2}\cdot\frac{dB}{dt}$

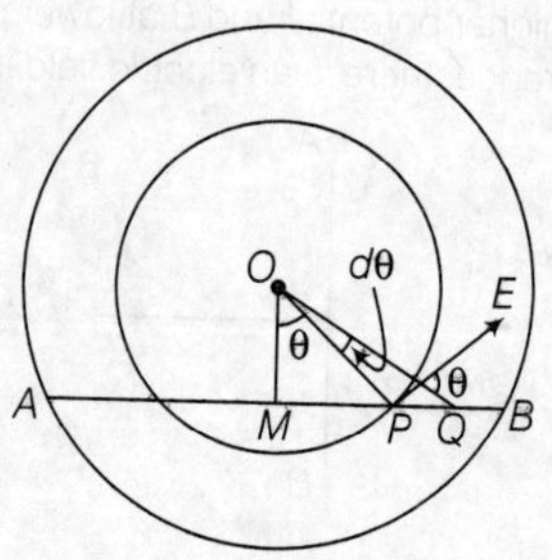

Now, consider a point P in the rod and a small distance dl along AB.

So, $\varepsilon = \int \mathbf{E} \cdot d\mathbf{l} = \int \frac{l}{2} \cdot \frac{dB}{dt} dl \cos\theta$

$$= \frac{(OM)}{2} \cdot \frac{dB}{dt} \int dl$$

$$= \frac{\sqrt{R^2 - l^2}}{2} \left(\frac{dB}{dt}\right) 2l$$

$$\Rightarrow \quad \varepsilon = \sqrt{R^2 - l^2} \left(\frac{dB}{dt}\right) l \qquad (\because \int dl = 2l)$$

3. Induced emf in loops $AEFD$ and $EBCF$ would be

$$e_1 = \left|\frac{d\phi_1}{dt}\right| = S_1 \left|\frac{dB}{dt}\right|$$

$$= (1 \times 1) = 1\text{V}$$

Similarly, $e_2 = \left|\frac{d\phi_2}{dt}\right| = S_2 \left|\frac{dB}{dt}\right|$

$$= (0.5 \times 1) \times 1\text{V}$$

$$= 0.5\text{V}$$

Now current is increasing, induced current will produce the magnetic field in direction, hence e_1 and e_2 will be applied as shown.

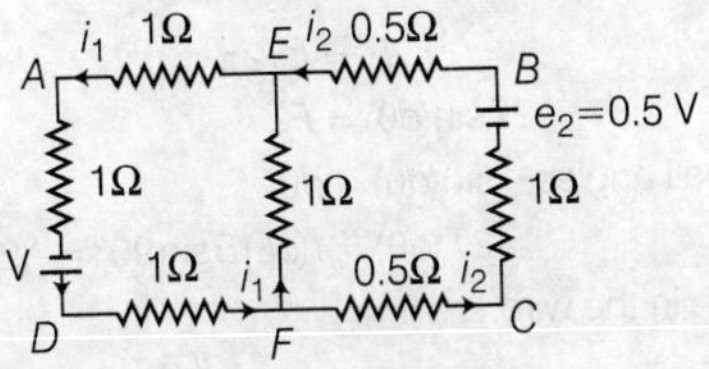

By Kirchhoff's first law at junction F, we get

$$i_1 = i + i_2 \qquad \text{...(i)}$$

Kirchhoff's second law is loop $FEADF$

$$3i_1 + i = 1 \qquad \text{...(ii)}$$

Kirchhoff's second law is loop $FEBCF$

$$2i_2 - i = 0.5 \qquad \text{...(iii)}$$

On solving Eqs. (i), (ii) and (iii), we get

$$i_1 = \left(\frac{7}{22}\right)\text{A}$$

$$i_2 = \left(\frac{6}{22}\right)\text{A}$$

$$i = \left(\frac{1}{22}\right)\text{A}$$

Therefore, current in segment AE is $\left(\frac{7}{22}\right)$A from E to A, current in segment BE is $\left(\frac{6}{22}\right)$A from B to E and current in segment EF is $\left(\frac{1}{22}\right)$A from F to E.

4. Due to sudden change in flux an electric field is set up and the ring experiences an impulsive torque and suddenly acquires an angular velocity.

i.e. $E = \frac{r}{2} \frac{dB}{dt}$

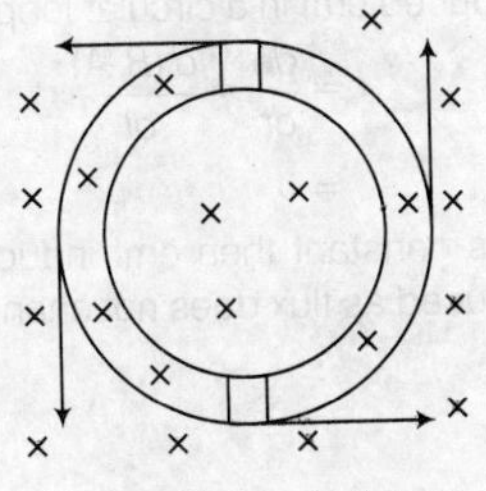

Force experienced by an element of ring

$$= dF = dqE$$

Torque experienced by this element

$$d\tau = dF \cdot r = dq\, E \cdot r$$

Total torque experienced by ring, $\tau = qEr$

$$\Rightarrow \quad \tau = q \frac{r}{2} \frac{dB}{dt} r = \frac{q}{2} r^2 \frac{dB}{dt}$$

Angular impulsive velocity,

$$L = \int \tau dt = \frac{qr^2}{2} \int \frac{dB}{dt} dt = \frac{qr^2}{2} B$$

$$\Rightarrow \quad L = I\omega = mr^2\omega = \frac{qr^2 B}{2}$$

$$\Rightarrow \quad \omega = \frac{qB}{2m}$$

5. (d) Induced electric field at point P,

$$E = \frac{R}{2} \frac{dB}{dt} \text{ towards right.}$$

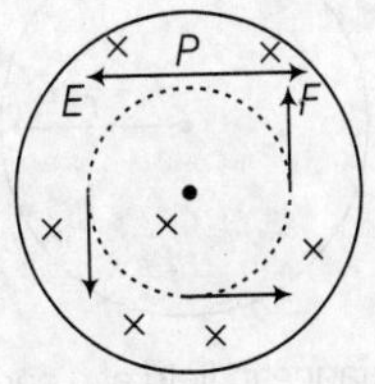

So, acceleration of an electron,

$$a = \frac{eE}{m} = \frac{eR}{2m} \cdot \frac{dB}{dt} \text{ towards left.}$$

6. (a) As, $r < R$, then net electric field at point P,

$$E = \frac{r}{2} \frac{dB}{dt}$$

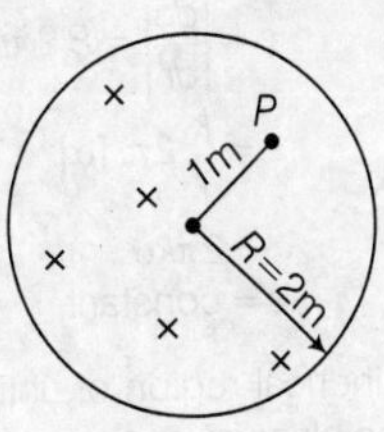

$$E = \frac{1}{2} \frac{d}{dt} (16 - 4t^2)$$

$$E = \frac{1}{2} [0 - 8t]$$

At $t = 2$ s, $E = -8\text{V/m}$

7. (a) Induced emf,

$$\varepsilon = \left|\frac{d\phi}{dt}\right| = \frac{d}{dt}(B \cdot A)$$

$$= \frac{AdB}{dt}$$

$$= \sqrt{3} \times \sqrt{3}$$

$$= 3\text{V}$$

Induced current $= \frac{3}{5} = 0.6\text{A}$

Potential difference between the point A and B,

$$\Delta V_{AB} = 1 - (0.6)(1) = 0.4\text{V}$$

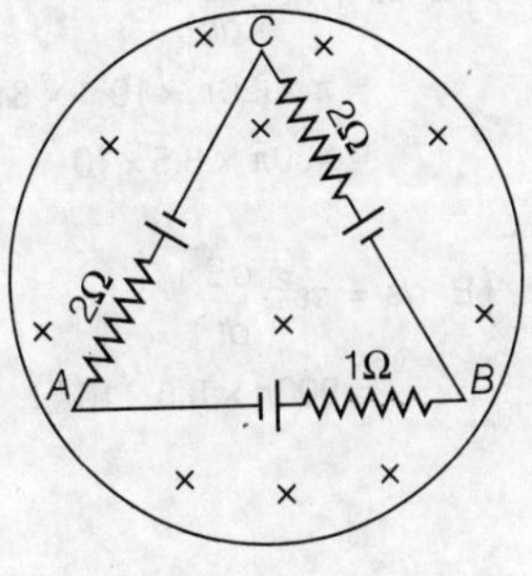

8. (b) Consider a cylindrical region of uniform magnetic field which is perpendicular to the plane.

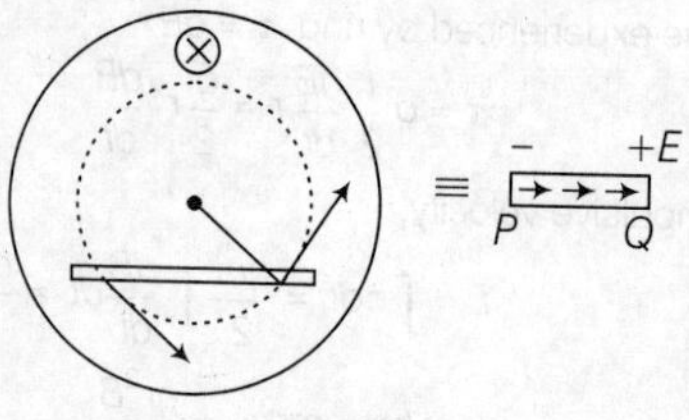

So, change in magnetic flux induced emf at high potential, i.e. positive across Q while low potential, i.e. negative across P in the conducting rod PQ.

9. (d) Consider a flexible wire loop in the shape of a circle has a radius r that grows linearly with time.

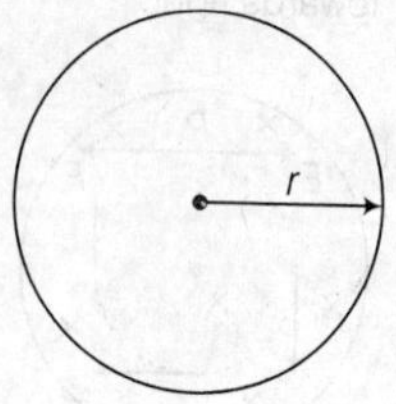

So, magnitude of magnetic field at a centre of the loop,

$$B = \frac{k}{r}$$

Radius, $r = r_0 + \alpha t$

Magnetic flux in a circular loop,

$$\phi = B\pi r^2$$

Thus, emf induced w.r.t. time,

$$\varepsilon = \left|\frac{d\phi}{dt}\right| = B\,2\pi r\,\frac{dr}{dt}$$

$$\varepsilon = \frac{k}{r}\,2\pi r[\alpha]$$

$$= 2\pi k\alpha$$

$$\varepsilon = \text{constant}$$

10. (d) Consider a cylindrical region of uniform magnetic field exist perpendicular to the plane of paper.

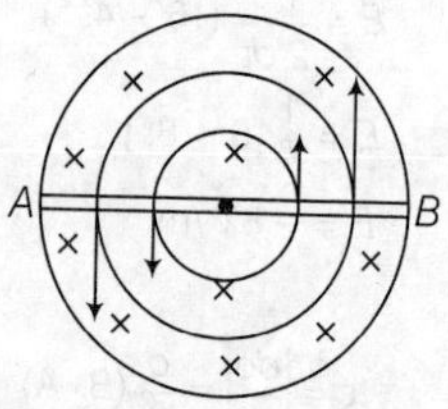

$\mathbf{E} \perp d\mathbf{l}$ at every point to the plane of paper.

So, $\quad \Delta V_{AB} = E\,dl\cos 90^\circ = \text{zero}$

Thus, induced emf between the ends of the rod is zero.

11. (b) By integral form of Faraday's law,

$$\left|\oint \mathbf{E}\cdot d\mathbf{s}\right| = \left|-\frac{d\phi_B}{dt}\right| = A.\left|\frac{dB}{dt}\right|$$

For path S_1,

$$\oint \mathbf{E}\cdot d\mathbf{s} = \pi\varepsilon^2.\frac{dB}{dt}$$

$$= \pi \times (20)^2 \times 10^{-4} \times 8.5 \times 10^{-3}$$

$$= 400\pi \times 8.5 \times 10^{-7}$$

and for path S_2,

$$\oint \mathbf{E}\cdot d\mathbf{s} = \pi\varepsilon_2^2\cdot\frac{dB}{dt}$$

$$= 900\pi \times 8.5 \times 10^{-7}$$

12. (b) As, we know,

$$\int \mathbf{E}\cdot d\mathbf{l} = \frac{d\phi}{dt} = S\left|\frac{dB}{dt}\right|$$

or $$E(2\pi r) = \pi a^2\left|\frac{dB}{dt}\right|$$

For $r \geq a$

i.e. electric field at a point P at a distance r from the centre,

$$E = \frac{a^2}{2r}\left|\frac{dB}{dt}\right|$$

Induced electric field, $\quad E \propto \frac{1}{r}.$

13. (c) As, length of a thin flexible wire,

$$L = 2\pi R$$

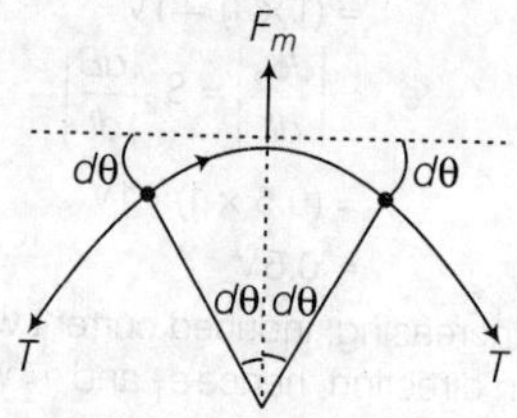

Radius of a wire,

$$R = L/2\pi$$

$$2T\sin(d\theta) = F_m$$

For small angles, $\sin(d\theta) \approx d\theta$

$\therefore \quad 2T(d\theta) = I(dL)B\sin 90^\circ = I(2R\cdot d\theta)\cdot B$

Tension in the wire,

$\therefore \quad T = IRB = \frac{ILB}{2\pi}$

14. (a,b,c) As, we know induced emf of a circular loop

$$\varepsilon = A\frac{dB}{dt} = (\pi R^2)\frac{dB}{dt} = \pi(1)^2(6) = 6\pi\text{V}$$

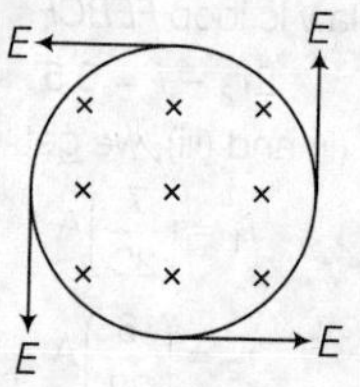

Induced current $= \frac{\varepsilon}{R} \Rightarrow \frac{6\pi}{2\pi r \times 1} = 3\text{A}$

As, shown in figure, such that electric field is in the tangential direction.

15. (b,d) A circular loop of radius r, having N turns of a wire is placed in a uniform and constant magnetic field $\mathbf{B}$.

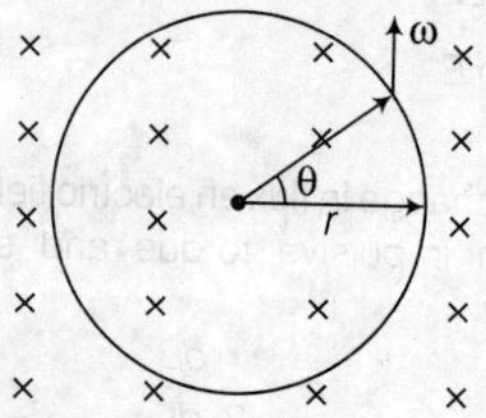

As, we know induced emf in a circular loop,

$$\varepsilon = \frac{d\phi}{dt} = \frac{d(B.A)}{dt} \qquad (\because B \text{ is constant})$$

$$= 0$$

So, flux remains constant then emf induced is zero. Thus, emf must not be induced as flux does not change with time.

16. (d) As, electric field induced on a ring

$$\int \mathbf{E} \cdot d\mathbf{l} = \frac{dB}{dt} S$$

$$E(2\pi R) = (\pi R^2)\frac{dB}{dt}$$

$$E(2\pi R) = \pi R^2 B_0 t\sqrt{3}$$

$$E = \frac{B_0 R t\sqrt{3}}{2}$$

Tangential force acting on a ring,

$$E = \frac{F}{Q} = F = QE$$

and $$F = \frac{Q B_0 R t\sqrt{3}}{2}$$

17. (d) As, torque acting due to electric field on charge Q is equal to the frictional torque.

$$FR = \mu mgR$$

$$\frac{Q B_0 R t\sqrt{3}}{2} = \frac{Q B_0 R}{2\,mg} mg$$

$$t = \frac{1}{\sqrt{3}} \text{s}$$

Thus, time taken by ring to start rotating on axis is $\frac{1}{\sqrt{3}}$ s.

18. (a) When $\tau_F > \tau_{f_{\max}}$

Net torque $\tau_F - \tau_{f_{\max}} = \frac{Q B_0 R^2 t\sqrt{3}}{2} - \mu mgR$

So, as net torque acting on a ring, $\tau = I\alpha$

$$= \frac{I d\omega}{dt} = mR^2 \frac{d\omega}{dt}$$

$$= \frac{Q B_0 R^2 (t\sqrt{3} - 1)}{2}$$

$$\int_0^{\omega} d\omega = \frac{Q B_0}{2m} \left(\int_{1/\sqrt{3}}^{\sqrt{3}} (t\sqrt{3} - 1) dt\right)$$

$$\omega = \frac{2 Q B_0}{m\sqrt{3}}$$

So, angular velocity just after the magnet is switched off at time $t = \sqrt{3}$ s is $\frac{2Q B_0}{m\sqrt{3}}$.

19. (a) Magnetic flux induced in a conducting ring,

$$\phi = B.A = \pi r^2 B$$

Area of a ring $= \pi r^2$

Induced emf in a coil, $e = \pi r^2 \frac{dB}{dt}$

and induced current in a coil, $i = \frac{e}{R} = \pi r^2 \frac{dB}{dt} \times \frac{1}{R}$

At all points, $\mathbf{E} \parallel d\mathbf{l}$

All points are at same potential.

For a small element dl on the ring, emf induced in the element

$$= de = \left(\frac{e}{2\pi r}\right) dl$$

Resistance dR of element $= \left(\frac{R}{2\pi r} dl\right)$

PD across element $= -i\,dR + de$

$$= -i \times \frac{R}{2\pi r} \times dl + \left(\frac{e}{2\pi r}\right) dl$$

$$= \left[\frac{e}{2\pi r} - \frac{iR}{2\pi r}\right] dl$$

Hence, (i) → (s), (ii) → (p), (iii) → (r), (iv) → (q)

20. (a) At 1s, flux is increasing so, flux of coil is opposing this. Hence, field points upwards and so, current is anti-clockwise.

At 5 s, flux is constant, so, current induced is zero.

At 9 s, flux is increasing (with direction reversed) so, current in loop is clockwise.

At $t = 15$ s, induced current is in anti-clockwise sense.

Hence, (i) → (q), (ii) → (r), (iii) → (p), (iv) → (q)

21. (2) As, $$\int \mathbf{E} \cdot d\mathbf{l} = -A\frac{dB}{dt}$$

Given, $$B = 17 + (0.2)\sin(\omega t + \phi)$$

$$E(2\pi r) = -\pi r^2 (0.2)\omega\cos(\omega t + \phi)$$

Induced electric field in a cylindrical magnet,

$$E = \frac{-r}{2}(0.2)\omega\cos(\omega t + \phi)$$

So, magnitude of the amplitude

$$= \frac{r}{2}(0.2)\omega = 2 \times 10^2 \text{ mN/C}$$

22. (5) For a circular path of radius 10 cm.

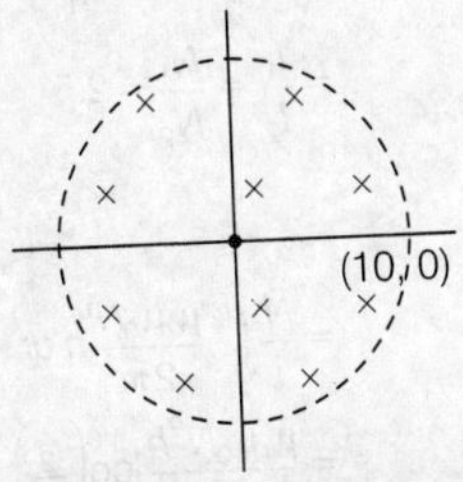

$$\oint \mathbf{E} \cdot d\mathbf{l} = E(2\pi r) = -\pi r^2 \left(\frac{dB}{dt}\right)$$

Force on charge is

$$F = QE$$

$$= -Q \frac{r}{2} \frac{dB}{dt}$$

$$= -1 \times 10^{-6} \times \frac{0.1}{2} \times (-1.0)$$

$$= 5 \times 10^{-9} \text{ N}$$

$$= 5\,n\text{N}$$

23. (c) Statement I is correct.

$$E = \text{Induced emf}$$

$$= -\frac{d\phi_B}{dt}$$

$$= -18t^2 + 36t$$

Maximum emf occurs when

$$\frac{dE}{dt} = 0$$

$\Rightarrow$ $$t = 1 \text{ s}$$

So, maximum current (at $t = 1$ s) is

$$I = \frac{E}{R}$$

$$= \frac{(-18 + 36)}{3}$$

$$= 6 \text{ A}$$

Statement II is incorrect or $\frac{dy}{dx} = 0$ also at points of minima and also at points of inflexion.

24. (b) Both Statements are correct but Statement II is just Lenz's law and it does not explain Statement I.

DPP-7 Self-Induction, Self-Inductance and Magnetic Energy Density

1. Inductance of a solenoid is given by

$$L = \frac{\mu_0 N^2 A}{l}$$

$$= \frac{\mu_0 N^2}{l} \cdot \frac{\pi d^2}{4}$$

Also, length of wire,

$$l_w = N\pi d$$

As, $d_2 = 2.5d$, $\quad N_1 = 2.5N_2$

Now, we find the ratio of two inductances

$$\frac{L_2}{L_1} = \frac{\frac{\mu_0 \pi}{4} \cdot \frac{N_2^2}{l_2} . d_2^2}{\frac{\mu_0 \pi}{4} . \frac{N_1^2}{l_1} . d_1^2}$$

$$= \frac{\frac{N_2^2 d_2^2}{l_2}}{\frac{N_1^2 d_1^2}{l_1}} = \frac{\frac{l_w/\pi}{l_2}}{\frac{l_w/\pi}{l_1}}$$

$$= \frac{l_1}{l_2} = \frac{N_1}{N_2} = 2.5$$

2. $L = \frac{N}{I} \phi_B$

$$= \frac{N}{I} \int_{r_1}^{r_2} \frac{\mu_r \mu_0 NI}{2\pi r} h\, dr$$

$$= \frac{\mu_\varepsilon \mu_0 N^2 h}{2\pi} \log\left(\frac{r_2}{r_1}\right)$$

Here, $r_2 = 10\text{cm}, r_1 = 5\text{ cm}$

$h = 1\text{ cm}$,

$N = 5000, \mu_r = 900$

$$\therefore \quad L = \frac{900 \times 4\pi \times 10^{-7} \times 25 \times 10^6 \times 1 \times 10^{-2}}{2\pi} \times \log\left(\frac{10}{5}\right)$$

$$= 45 \log\left(\frac{10}{5}\right)$$

$$= 45 \log(2)\text{ H}$$

3. Before closing S_1, current in the inductor is $i - \varepsilon/2R$. Just after closing S_1, current in inductor will remain same and other currents are as shown in the figure.

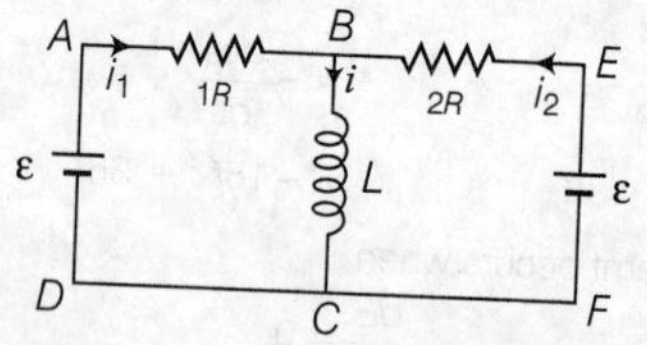

For left loop *ABCD*,

$$\varepsilon = i_1 R + L(di/dt) \quad \ldots\text{(i)}$$

For right loop *BEFC*,

$$\varepsilon = i_2(2R) + L(di/dt) \quad \ldots\text{(ii)}$$

According to current junction law at point *B*, we get

$$i_1 + i_2 = i \quad \ldots\text{(iii)}$$

On solving Eqs. (i), (ii) and (iii), we get

$$\Rightarrow \quad \frac{di}{dt} = \frac{2\varepsilon}{3L}$$

Potential difference across *R* is

$$V_R = i_1 R = \varepsilon - L(di/dt)$$

$$= \varepsilon - L\left[\frac{2\varepsilon}{3L}\right]$$

$$= \frac{\varepsilon - 2\varepsilon}{3} = \frac{-\varepsilon}{3}$$

4. Since, the wires are infinite, so the system of these two wires can be considered as a closed rectangle of infinite length and breadth equal to *d*.

So, flux through the strip,

$$\phi = \int_a^{d-a} \frac{\mu_0 I}{2\pi r}(l dr) = \frac{\mu_0 I l}{2\pi} \ln\left(\frac{d-a}{a}\right)$$

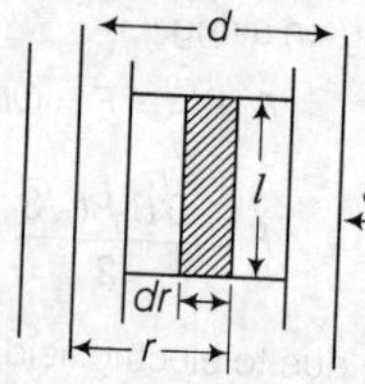

The other wire produces the same result, so the total flux through the dotted rectangle is

$$\phi_{\text{total}} = \frac{\mu_0 I l}{\pi} \ln\left(\frac{d-a}{a}\right)$$

Total inductance of length *L*

i.e. $$L = \frac{\phi_{\text{total}}}{I} = \frac{\mu_0 l}{\pi} \ln\left(\frac{d-a}{a}\right)$$

$\therefore$ Inductance per unit length

$$= \frac{L}{l} = \frac{\mu_0}{\pi} \ln\left(\frac{d-a}{a}\right)$$

5. (d) Consider two inductance L_1 and L_2 are placed far apart. Then, combined inductance in parallel combination,

$$\frac{1}{L_{eq}} = \frac{1}{L_1} + \frac{1}{L_2} \Rightarrow L_{eq} = \frac{L_1 L_2}{L_1 + L_2}$$

6. (a) Given, induced emf = 8V

$$i_2 = -2.0\text{A}, i_1 = +2\text{ A}$$

$$dt = 0.05\text{s},$$

then coefficient of self induction of the coil,

$$e = L\left|\frac{di}{dt}\right| \quad \Rightarrow \quad 8 = L\left|\frac{(2)-(-2)}{0.05}\right|$$

$$8 = L(4/0.05) \quad \Rightarrow \quad L = 0.1\text{H}$$

7. (d) In parallel combination, equivalent inductance across *AD*.

$$\frac{1}{L_{eq}} = \frac{1}{L_1} + \frac{1}{L_2} + \frac{1}{L_3}$$

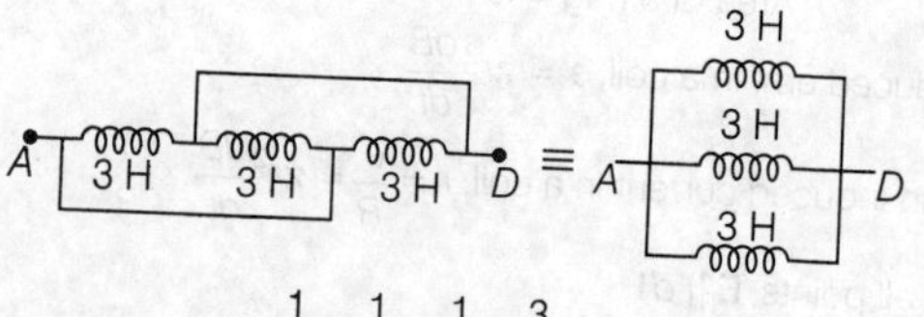

$$\frac{1}{3} + \frac{1}{3} + \frac{1}{3} = \frac{3}{3} = 1\text{H}$$

$$L_{eq} = 1\text{H}$$

8. (a) Consider a circular coil having *n* turns of length *l* and across of cross-section *A* and initial resistance be *R*.

Then, self inductance of a coil,

$$L_1 = \frac{\phi}{i} = \mu_0 n^2 l A = \frac{\mu_0 N^2 l A}{l^2} = \frac{\mu_0 N^2 A}{l} \quad \left(\because n = \frac{N}{l}\right) \ldots\text{(i)}$$

When number of turns are doubled

$$L_2 = \frac{(2N)^2 \times \mu_0 A}{l} \quad \ldots\text{(ii)}$$

On comparing Eqs. (i) and (ii), we get

$$\frac{L_1}{L_2} = \frac{1}{4} \Rightarrow L_2 = 4L_1$$

9. (d) As, self-inductance of a coil, $L = \frac{\phi}{I}$

$\therefore$ $$\text{Henry} = \frac{\text{Weber}}{\text{Ampere}}$$

Also, $$L = -\frac{\varepsilon}{\frac{dI}{dt}}$$

$\Rightarrow$ $$\text{Henry} = \frac{\text{V}}{\text{A/s}} = \text{Vs A}^{-1} = \Omega\text{s} = \text{JA}^{-2}$$

10. (c) Given, $L = 2\text{H}, i = 2\text{A}, \frac{di}{dt} = 4\text{A/s}$

As, energy stored in the inductor is

$$U = \frac{1}{2}Li^2$$

So, $$\frac{dU}{dt} = Li\frac{di}{dt} = 2 \times 2 \times 4 = 16 \text{ J/s}$$

11. (a) As, we know, induced emf across self-inductance, i.e. $e = L\frac{dI}{dt}$

$$\left(\frac{dI}{dt}\right)_1 > \left(\frac{dI}{dt}\right)_2$$

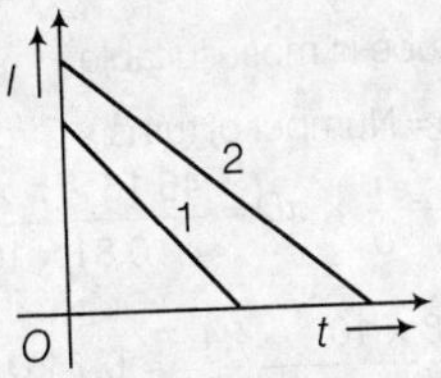

$\therefore e_1 > e_2$. It means self-induced emf across 1 is greater 2.

12. (a) Consider a long solenoid of N turns has a self-inductance L and area of cross-section A,

So, $$\phi = Li \qquad \text{...(i)}$$

But flux due to magnetic field having cross-section area A, is

$$\phi = NBA \qquad \text{...(ii)}$$

So, $$NBA = Li$$

Induced current in a long solenoid, $i = \frac{NBA}{L}$

13. (c) Given, $l = 100$ m, radius r, then self inductance of thin copper wire,

$$L = \frac{\mu_0 N^2 A}{l} \qquad \text{...(i)}$$

i.e. $$2\pi r N = 100$$

Similarly, same length of wire is wound as a solenoid of length l of radius $r/2$, then its self inductance will be

$$L' = \frac{\mu_0 N'^2 (A/4)}{l}$$

So, $$\frac{2\pi r}{2}N' = 100 \quad \Rightarrow \quad N' = 2N$$

$$L' = \frac{\mu_0 (2N)^2 (A/4)}{l} \qquad \text{...(ii)}$$

On comparing self-inductance of Eqs. (i) and (ii), we get

$$L' = L$$

14. (b) As, emf induced in a coil, $\varepsilon = L\frac{di}{dt}$

So, $$\frac{\varepsilon_1}{\varepsilon_2} = \frac{L_1}{L_2} = \frac{8\text{mH}}{2\text{mH}} = 4 \quad \left(\text{as } \frac{di}{dt} = \text{constant}\right)$$

As, power supplied $P = \varepsilon i$ = constant

So, $$\frac{i_1}{i_2} = \frac{\varepsilon_2}{\varepsilon_1} = \frac{1}{4}$$

As energy, stored in coil, $U = \frac{1}{2}LI^2$

i.e. $$\frac{U_1}{U_2} = \frac{L_1}{L_2}\left[\frac{I_1}{I_2}\right]^2 = \frac{8}{2}\left[\frac{1}{4}\right]^2 = \frac{1}{4}$$

15. (b) ***Case* I** Consider a inductive coil having flux ϕ and self-inductance L, then current, i, flows through it.

i.e. $$\phi = Li$$

***Case* II** When a soft iron rod is inserted into inductive coil, then

$$\phi = L'i'$$

then, $$\phi' = \phi$$

$$L'i' = Li$$

$$i' = \frac{Li}{L'}$$

As, we know $$L' > L$$

So, $$i' < i$$

i.e. High inductance inside a soft iron rod into inductive coil, then current will decrease. Hence, the intensity of a bulb will decrease.

16. (c) $E = 4\text{V}$

By Kirchhoff's law in loop $ABCD$, we get

$$-iR - \frac{Ldi}{dt} - \frac{q}{C} + E = 0$$

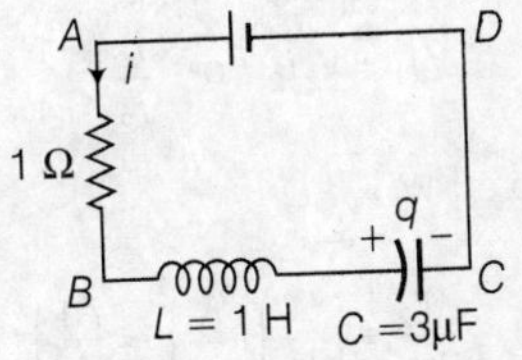

$$-2 \times 1 - 1 \times 4 - \frac{q}{3} + 4 = 0$$

$$\frac{q}{3\mu\text{F}} = 2\text{ V}$$

$$q = 6\mu\text{C}$$

So, charge on the capacitor at an instant when the current in the circuit is 2 A will 6μC.

17. (c) As, we know current across inductor, i.e. $I_1 = Kt$

$$\left\{\text{for } 0 < t < \frac{T}{2}\right\}$$

where, K is the slope.

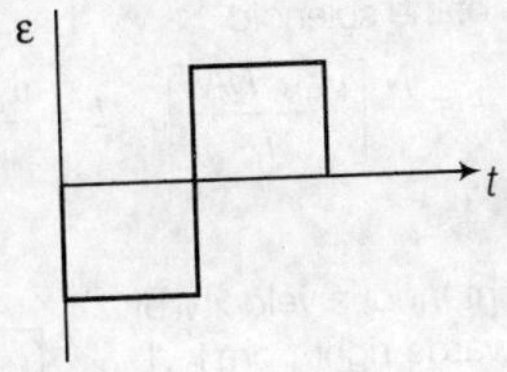

For an inductor,

Induced emf of a coil, $\varepsilon_1 = -L\frac{dI}{dt}$

$\Rightarrow$ $$\varepsilon_1 = -KL$$

So, induced current,

$$I_2 = -Kt + \text{constant}$$

$$\left\{\frac{T}{2} < t < T\right\} \Rightarrow \varepsilon_2 = KL$$

18. (b) Long straight wire

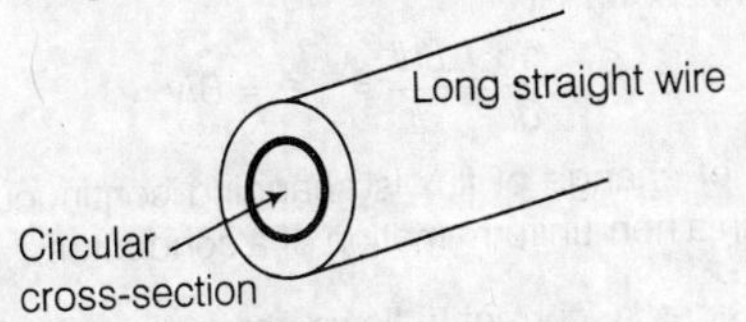

Energy density $$\left(\frac{dU}{dV}\right) = \frac{B_0^2}{2\mu_0}$$

i.e. $$U = \int_0^R \frac{B^2}{2\mu_0}dV$$

Magnetic field in a straight wire, i.e. $B = \frac{\mu_0 Ir}{2\pi R^2}$

$$dV = 2\pi r dr$$

where, r is a radius of circular cross-section of wire.

$$U = \int_0^R \left(\frac{\mu_0^2 I^2 r^2}{4\pi^2 R^4}\right)\frac{2\pi r dr}{2\mu_0} = \int_0^R \frac{\mu_0 I^2}{4\pi} \times \frac{1}{R^4} \times r^3 dr$$

$$= \frac{\mu_0 I^2}{4\pi R^4} \times \left[\frac{r^4}{4}\right]_0^R = \frac{\mu_0}{4\pi} \times \frac{I^2}{R^4} \times \left[\frac{R^4}{4} - 0\right]$$

$$U = \frac{\mu_0 I^2}{16\pi}$$

19. (a,c,d) From Faraday's law, the induced voltage $V \propto L$, if rate of change of current is constant $\left(V = -\frac{Ldi}{dt}\right)$

$$\frac{V_2}{V_1} = \frac{L_2}{L_1} = \frac{2}{8} = \frac{1}{4} \quad \text{or} \quad \frac{V_1}{V_2} = 4$$

Power given to two coils is same, i.e.

$$V_1 i_1 = V_2 i_2 \quad \text{or} \quad \frac{i_1}{i_2} = \frac{1}{4}$$

Energy stored $W = \frac{1}{2}Li^2$

$$\frac{W_2}{W_1} = \left(\frac{L_2}{L_1}\right)\left(\frac{i_1}{i_2}\right)^2 = \left(\frac{1}{4}\right)(4)^2 = 4 \quad \text{or} \quad \frac{W_1}{W_2} = \frac{1}{4}$$

20. (a,b,c) Magnetic field in case of solenoid, $B = \mu ni$

where, μ is permeability of iron rod.

$$B = \mu_0 \mu_r ni \qquad \left(\because \mu_r = \frac{\mu}{\mu_0}\right)$$

Now, magnetic flux linked with the solenoid,

$$\phi \propto B$$

Now, magnetic flux through each turn of a solenoid is

$$\phi_B = B.A = \mu_0 \mu_r \frac{NA}{l} i \qquad \left(\because n = \frac{N}{l}\right)$$

where, A is the cross-sectional area of the solenoid. Now, self-inductance of the solenoid,

$$L = \frac{N\phi_B}{l} = \frac{N}{i}\left[\frac{\mu_0 \mu_r NAi}{l}\right], \quad L = \frac{\mu_0 \mu_r N^2 A}{l}$$

For iron rod $\mu_r > 1$.

21. (c) Let the instantaneous velocity of the rod be v towards right from KVL rule, we get

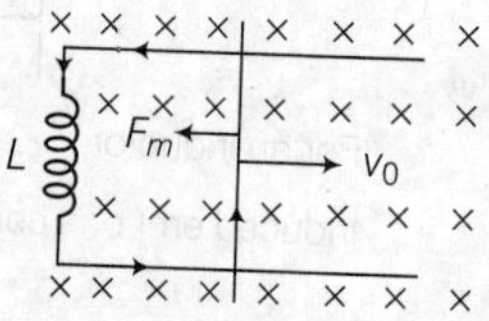

$$e = -\frac{Ldi}{dt} = Bvl = \frac{Bldx}{dt}$$

$$-Ldi = Bldx$$

Integrating, we get

$$\Rightarrow \quad Li = -Blx \Rightarrow i = -\frac{Blx}{L}$$

So, at $x = 0$, then $i = 0$, then the motion of conductor will come to rest after sometime and then return to opposite direction.

22. (b) As, we know that

$$\frac{d\phi}{dt} = \frac{Ldi}{dt} = -e = Blv$$

So, rate of change of flux is changing continuously by varying velocity in a non-uniform motion of a conductor.

23. (d) Consider a current i flows through an inductance, then magnetic flux, i.e. ϕ linked with the coils is $\phi = Li$ where L is coefficient of self-inductance of inductor.

$$Li = Blx$$

$$\Rightarrow \quad i = \frac{Bl}{L}x$$

Force exerted on a current carrying inductor in a uniform magnetic field,

$$F = iBl = \frac{B^2 l^2 x}{L} \Rightarrow m\left(\frac{d^2x}{dt^2}\right) = -\frac{B^2 l^2 x}{L}$$

$$\frac{d^2x}{dt^2} = -\omega^2 x$$

$$\frac{d^2x}{dt^2} = -\omega^2 x$$

As, angular velocity of an inductor, $\omega = \frac{Bl}{\sqrt{mL}}$

Time taken by an inductor, $T = \frac{2\pi}{Bl}\sqrt{mL}$

Magnetic energy is maximum when $v = 0$

$$\therefore \quad t = \frac{T}{4}, \frac{3T}{4}, \ldots$$

24. (c) $L = \frac{\mu_0 N^2 A}{l} = \mu_0 n^2 Al = \frac{\mu_0 Al}{d^2}$

$$\Rightarrow \quad l = \frac{Ld^2}{\mu_0 \pi r^2} = \frac{1 \times (0.81 \times 10^{-3})^2}{4\pi \times 10^{-7} \times \pi \times (0.06)^2} \approx 46 \text{ m}$$

So, a 50 m long tube is most suitable.

25. (b) Length of wire = Number of turns × Circumference of each turn

$$= \frac{l}{d} \times \pi D = \frac{46.16 \times \pi \times 0.12}{0.81 \times 10^{-3}} \approx 21 \text{ km}$$

26. (c) $R = \frac{\rho l}{A} = \frac{1.68 \times 10^{-8} \times 24}{\pi \times (0.405 \times 10^{-3})^2} \approx 0.7 \text{ k}\Omega$

27. (4) ***Case I*** Given, $i = 2$ A,

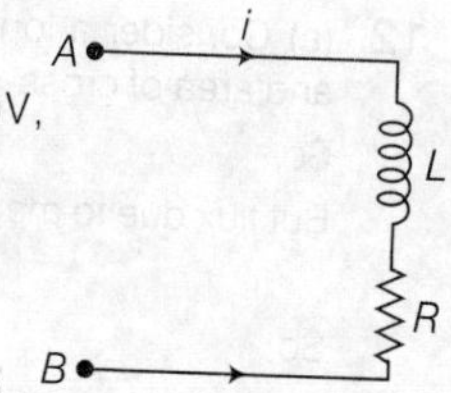

$$\frac{di}{dt} = 1 \text{ A/s}, V_{AB} = 10 \text{ V},$$

then potential drop across AB

$$V_{AB} = iR + \frac{Ldi}{dt}$$

$$10 = 2R + L \quad \ldots(i)$$

Case II Similarly, $V_{AB} = 6$ V

$$\frac{di}{dt} = -1 \text{ A/s}$$

$$i = 2 \text{ A}$$

Now, potential drop across AB i.e. $V_{AB} = iR + \frac{Ldi}{dt}$

$$\Rightarrow \quad 6 = 2 \times R + L \times (-1)$$

$$6 = 2R - L \quad \ldots(ii)$$

On solving Eqs. (i) and (ii), we get

$$R = 4\Omega$$

28. (3) Given, $\frac{di}{dt} = 10^3$ A/s

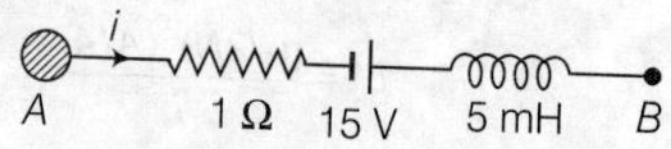

$\therefore$ Induced emf across inductance, $|e| = L\frac{di}{dt}$

So, $\quad |e| = (5 \times 10^{-3})(10^3) \text{V} = 5 \text{V}$

Since, the current is decreasing, the polarity of this emf would be so as to increase the existing current. The circuit can be redrawn as

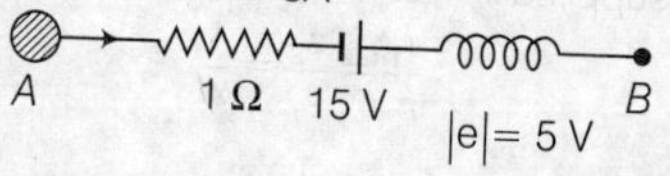

Now, $\quad V_A - 5 + 15 + 5 = V_B$

$$V_A - V_B = -15 \text{V}$$

or $\quad V_B - V_A = 15 \text{V} = 5 \times 3 \text{V}$

DPP-8 Growth and Decay of Current in *L-R* Circuit

1. Induced emf, $\varepsilon = \frac{1}{2}Br^2\omega$

Using Kirchhoff's law, $\varepsilon = L\frac{di}{dt} + iR \Rightarrow \frac{di}{\varepsilon - iR} = \frac{dt}{L}$

$$\int_0^i \frac{di}{\varepsilon - iR} = \int_0^t \frac{dt}{L} \Rightarrow i = \frac{\varepsilon}{R}(1 - e^{-\frac{R}{L}t})$$

Under steady state (when $t \to \infty$)

$$i = \frac{\varepsilon}{R}(1 - e^{-\infty}) = \frac{\varepsilon}{R} = \frac{1}{2}\left(\frac{Br^2\omega}{R}\right)$$

Force on rod due to this induced current,

$$F = BIL = B \times \frac{1}{2}\frac{Br^2\omega}{R} \times r = \frac{1}{2}\cdot\frac{B^2r^3\omega}{R}$$

Torque due to this force = $F \times$ distance

$$= F \times \frac{r}{2} = \frac{1}{4}\cdot\frac{B^2r^4\omega}{R}$$

Torque due to weight of rod at θ angular position $= mg\cos\theta \times \frac{r}{2}$

So, total torque required $= \frac{mgr\cos\theta}{2} + \frac{1}{4}\cdot\frac{B^2r^4\omega}{R}$

2. Initially in position 1 of switch, *L-R* circuit is open. At $t = 0$, switch is shifted to position 2 and *L-R* circuit is active.

(i) Steady state current in R_4 is

$$i_4 = \frac{E_2}{R_2 + R_4} = \frac{3}{2+3} = \frac{3}{5} = 0.6\text{ A}$$

Also, i = instantaneous current in *L-R* circuit

$$i = i_4(1 - e^{-\frac{R}{L}t})$$

(ii) If at time t, current is $\frac{1}{2}$ of its steady value,

$$\frac{1}{2} = \left(1 - e^{-\frac{R}{L}t}\right) \Rightarrow e^{\frac{R}{L}t} = 2$$

$$\Rightarrow \quad t = \frac{L}{R}\log 2$$

$$\Rightarrow \quad t = \frac{1}{200} \times 0.693 = 1.386 \times 10^{-3}\text{s}$$

(iii) Energy stored in inductor at this instant

$$= \frac{1}{2}Li^2 = \frac{1}{2} \times 10 \times 10^{-3} \times \left(\frac{0.6}{2}\right)^2$$

$$= 4.5 \times 10^{-4}\text{ J}$$

3.

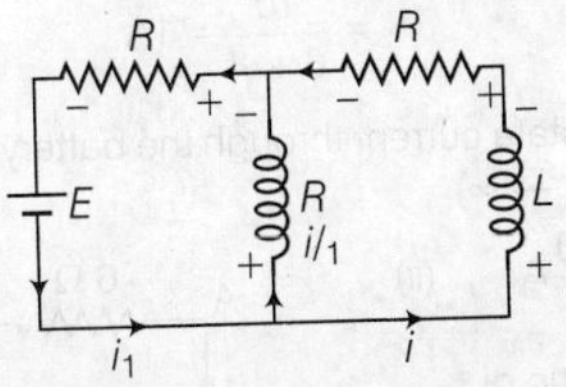

Case I Calculate i_L at $t \to \infty$

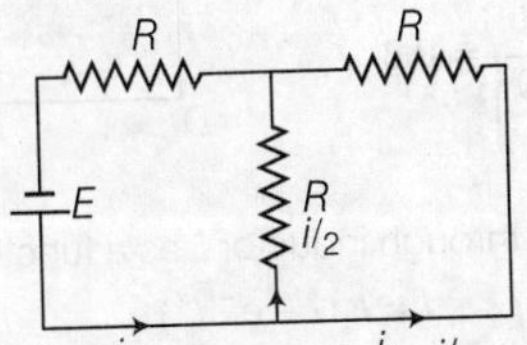

Current across inductor i_L,

$$i_L = \frac{F}{R_{eq}} = \frac{2E}{3R}$$

$$i_L = \frac{E}{3R}$$

Case II Find τ of circuit.

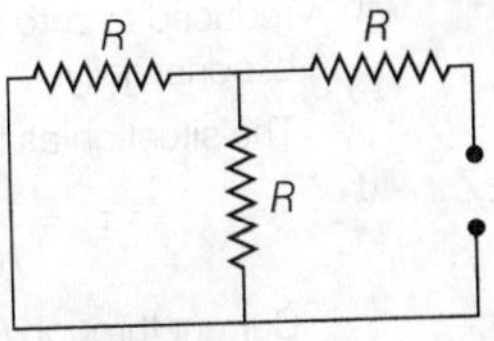

Equivalent resistance,

$$R_{eq} = \frac{3R}{2}$$

Inductive time constant, $\tau = \frac{2L}{3R}$

Growth of current in *L-R* circuit, $i = i_L(1 - e^{-t/\tau})$

$$= \frac{E}{3R}\left(1 - e^{-\frac{3Rt}{2L}}\right)$$

4. (i) Given, $R_1 = R_2 = 2\Omega$, $E = 12$ V and $L = 400$ mH = 0.4H. Two parts of the circuits are in parallel with the applied battery. So, the upper circuit can be broken as

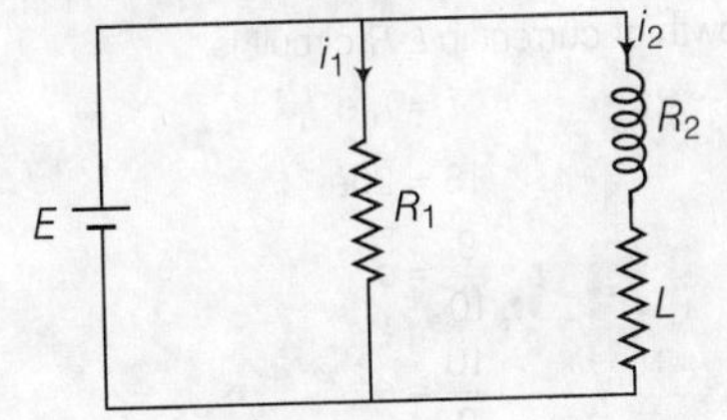

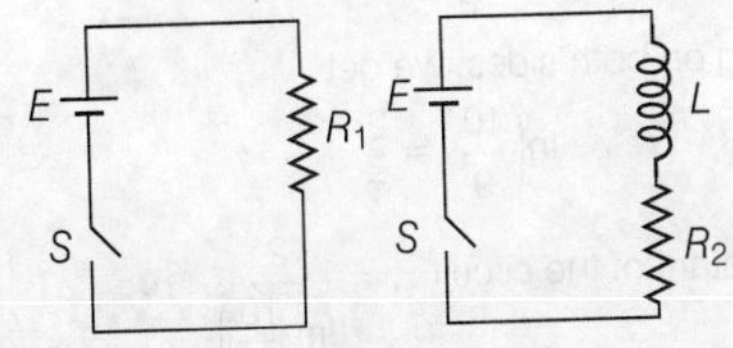

(i) (ii)

Now, refer to Fig. (ii).

This is a simple *L-R* circuit, whose time constant,

$$\tau_L = \frac{L}{R_2} = \frac{0.4}{2} = 0.2\text{ s}$$

and steady state current,

$$i_0 = \frac{\varepsilon}{R_2} = \frac{12}{2} = 6\text{A}$$

Therefore, if switch S is closed at time $t = 0$, then current in the circuit at any time t will be given by

$$i(t) = i_0(1 - e^{-t/\tau_L})$$

$$i(t) = 6(1 - e^{-t/0.2}) = 6(1 - e^{-5}) = i$$

Thus, potential drop across L at any time t is

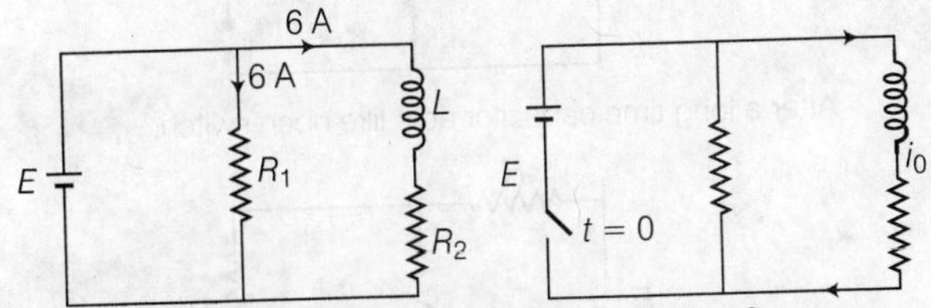

Steady state condition
(iii)

$t = 0$
S is open
(iv)

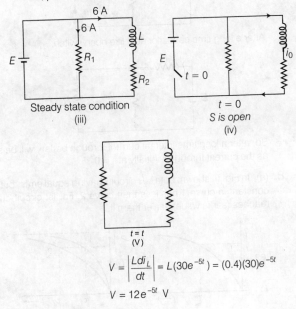

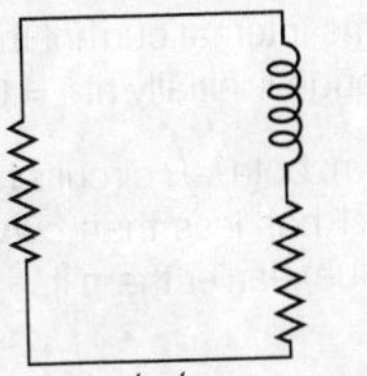

$t = t$
(V)

$$V = \left|\frac{Ldi_L}{dt}\right| = L(30e^{-5t}) = (0.4)(30)e^{-5t}$$

$$V = 12e^{-5t}\text{ V}$$

(ii) The steady state current in L or R_2 is $i_0 = 6$A

Now, as soon as the switch is opened, current in R_1 is reduced to zero immediately. But in R_L and L it decreases exponentially.

The situation as follows

$$\tau_L' = \frac{L}{R_1 + R_2} = \frac{0.4}{(2+2)} = 0.1 \text{ s}$$

Current through R_1 at any time t is

$$i = i_0 e^{-t/\tau'} \Rightarrow i = 6e^{-10t}$$

So, direction of current in R_1 (as shown in figure) is clockwise.

5. (d) Just after closing the switch, $i = 0$ hence power by battery and magnetic energy is zero. But emf induced across inductor is maximum in L-R circuit.

6. (c) Given, at $t = 0$ is 20 A. After 2 s, it reduces to 18 A.

Then, growth of current in L-R circuit is

$$i = i_0 e^{-t/\tau}$$

$$18 = 20e^{-t/\tau}$$

$$\frac{9}{10} = e^{-2/\tau}$$

$$\frac{10}{9} = e^{2/\tau}$$

Taking log on both sides, we get

$$\ln\left(\frac{10}{9}\right) = \frac{2}{\tau}$$

Time constant of the circuit, $\tau = \dfrac{2}{\ln\left(\dfrac{10}{9}\right)}$

7. (d) Initially the circuit is closed.

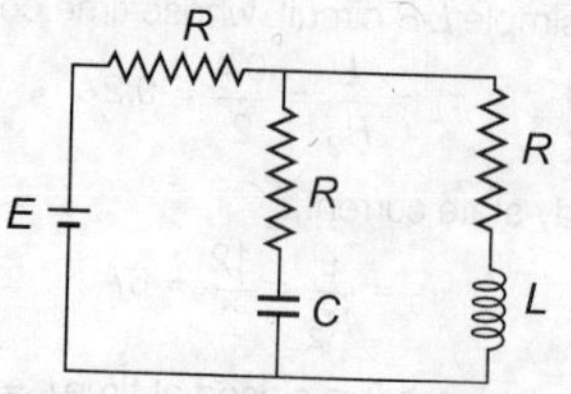

Capacitor acts like an closed path

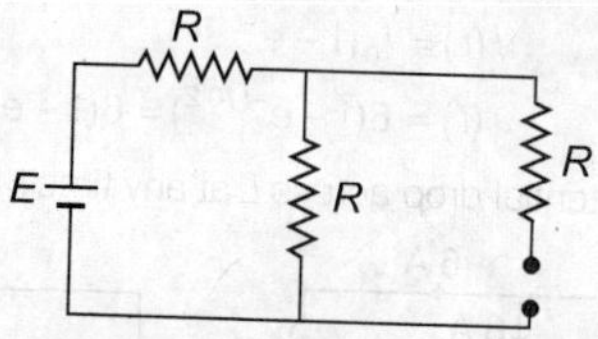

After a long time capacitor acts like open switch.

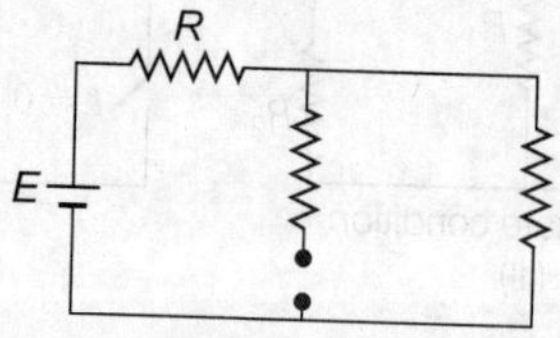

So, after a long time internal current through battery will be same as the current through it initially at $t = 0$.

8. (a) In Fig. (i) shown, both L-R circuits having equal emfs, but time constant in curve 1 has less than curve 2. For Ist circuit current reaches (same value) earlier them II.

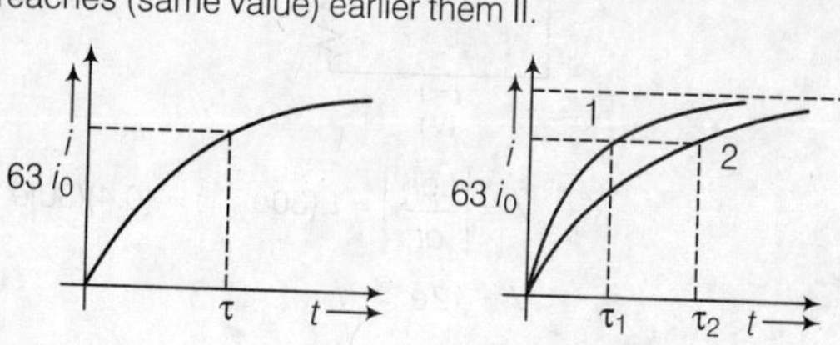

9. (b) Given, $L = 300$mH, $R = 2\Omega$, $\varepsilon = 2$V. Then, current flows across L-R circuits, is

$$i = i_0(1 - e^{-t/\tau})$$

When current reaches half of its steady value,

$$i = \frac{i_0}{2}$$

$$\frac{i_0}{2} = i_0(1 - e^{-t/\tau}) \Rightarrow e^{-t/\tau} = \frac{1}{2}$$

Taking log on both sides, we get

$$t/\tau = \ln 2$$

$$t = \tau \ln 2$$

$$t = \frac{L}{R}\ln 2 \quad \ldots\text{(i)}$$

On putting the values of L, R in Eq. (i), we get

$$t = 0.1\text{s}$$

10. (d) Here, $L = 10$H, $R = 5\Omega$, $\varepsilon = 5$V

Growth of current in an L - R circuit,

$$i = i_0(1 - e^{-t/\tau})$$

$$i = \frac{E}{R}(1 - e) \quad \left(\because \tau = \frac{L}{R} \Rightarrow \frac{10}{5} = 2\right)$$

$$i = \frac{5}{5}(1 - e^{-\frac{2}{2}})$$

$$i = (1 - e^{-1})$$

11. (c) Given, $L = 10$H, $R = 5\Omega$, $\varepsilon = 5$V, $t = 2$s

Current L-R circuit,

$$i = i_0 e^{-t/\tau}$$

$$i = \frac{E}{R}e^{-t/\tau} \quad \left(\therefore \tau = \frac{L}{R} = 1\text{ms}\right)$$

$$= \frac{100}{100}e^{-\frac{1\text{ ms}}{1\text{ms}}}$$

$$i = \frac{1}{e}$$

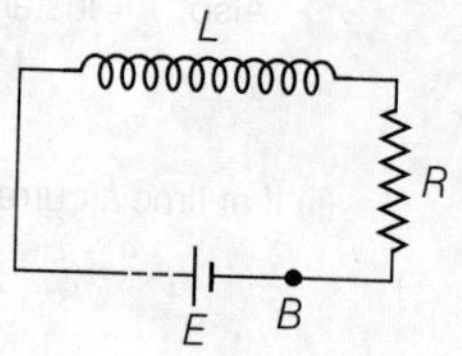

12. (b) Initially, current i_1 across loop $ABCD$ at $t = 0$

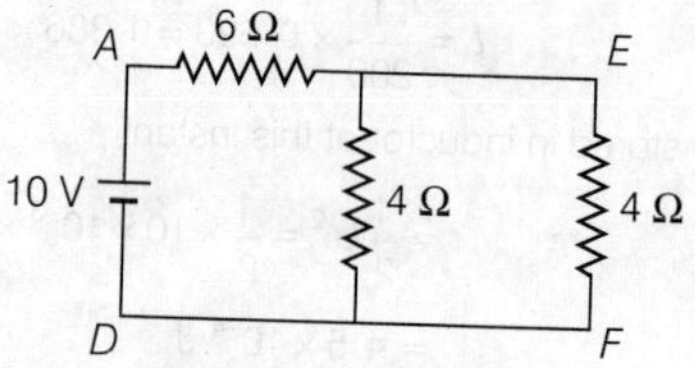

$$i_1 = \frac{10}{6+4} = 1\text{A} \quad \ldots\text{(i)}$$

Then steady state current through the battery across $ADFE$, we get, i_2 at ($t \to \infty$)

$$\Rightarrow \quad i_2 = \frac{10}{8}\text{A} \quad \ldots\text{(ii)}$$

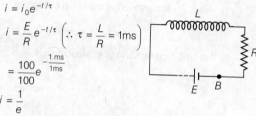

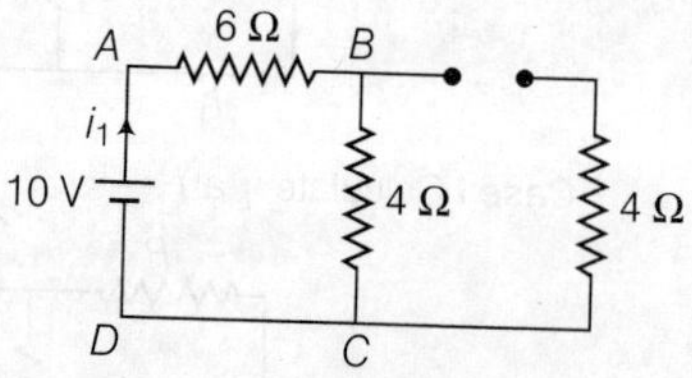

Comparing ratio of Eqs. (i) and (ii), we get

$$\frac{i_1}{i_2} = \frac{1}{\left(\frac{10}{8}\right)} = 0.8$$

13. (d) The current through inductor L as a function of time is

$$I = I_0(1 - e^{-t/\tau})$$

Now,
$$I_0 = \frac{\varepsilon}{R_2} = \frac{12\text{ V}}{2\Omega} = 6\text{A}$$

and inductive time constant,

$$\tau = \frac{L}{R_2} = \frac{400 \times 10^{-3}\text{H}}{2\Omega} = 0.2\text{s}$$

Current across inductor,

$$I = 6(1 - e^{-t/0.2}) = 6(1 - e^{-5t})$$

Potential drop across L,

$$V = \varepsilon - R_2 L = 12 - 2 \times 6\,(1 - e^{-5t}) = 12\, e^{-5t}\, \text{V}$$

14. (d) The current time $(i\text{-}t)$ equation in L-R circuit is given by

i.e. Growth of current in L-R circuit

$$\Rightarrow \quad i = i_0(1 - e^{-t/\tau_L}) \qquad \ldots\text{(i)}$$

where, $$i_0 = \frac{V}{R} = \frac{12}{6} = 2\text{ A}$$

and inductive time constant,

$$\tau_L = \frac{L}{R} = \frac{8.4 \times 10^{-3}}{6} = 1.4 \times 10^{-3}\text{s}$$

So, current in the coil, $i = 1$A

So, substituting these values in Eq. (i), we get

$$t = 0.97 \times 10^{-3}\text{s}$$

or $$t = 0.97\,\text{ms} \;\Rightarrow\; t \approx 1\text{ ms}$$

15. (d) Potential drop (across BC) $= L\dfrac{dI_2}{dt} + I_2 R_2$

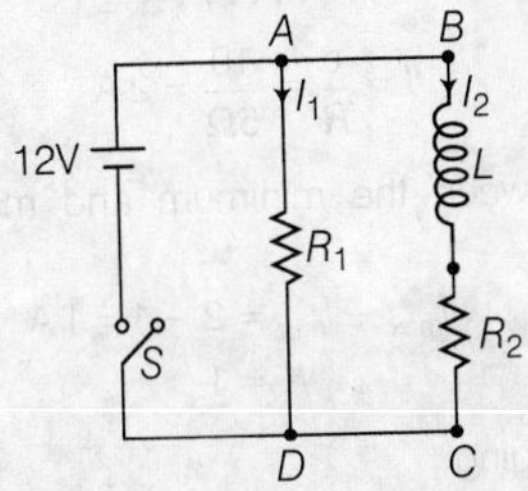

$$I_2 = I_0(1 - e^{-t/\tau})$$

$$I_0 = \frac{E}{R_2} = \frac{12}{2} = 6\text{ A}$$

$$\tau = \frac{L}{R} = \frac{400 \times 10^{-3}}{2} = 0.2\text{ s}$$

$$I_2 = 6(1 - e^{-t/0.2})$$

Potential drop across, $L = E - I_2 R_2 = 12 - 2 \times 6\,(1 - e^{-t/0.2})$

$$= 12e^{-t/0.2} = 12e^{-5t}\text{ V}$$

16. (d) $I_0 = \dfrac{12}{6} = 2$ A

Current decrease from 2 A to 1 A, i.e. becomes half in time

$$t = 0.693\,\frac{L}{R} = 0.693 \times \frac{8.4 \times 10^{-3}}{6} = 1\text{ ms}$$

17. (d) During decay of current, $I = I_0 e^{-\frac{Rt}{L}} = \dfrac{E}{R}\, e^{-\frac{Rt}{L}}$

$$= \frac{100}{100}\, e^{-\frac{100 \times 10^{-3}}{100 \times 10^{-3}}} = \frac{1}{e}\text{ A}$$

18. (b,d) Consider current growth in two L-R circuit as shown.

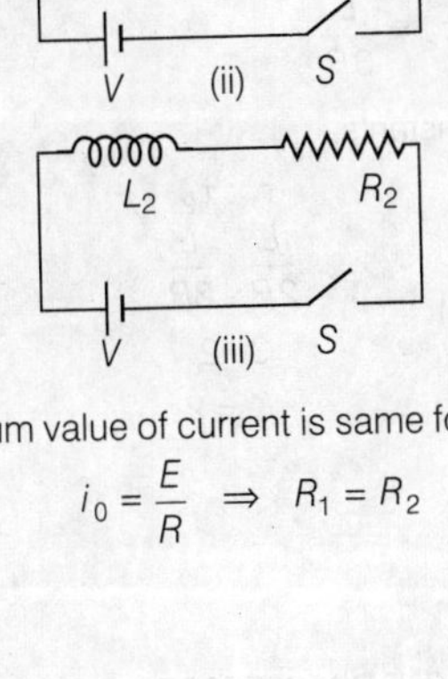

Since, maximum value of current is same for both the circuits so,

$$i_0 = \frac{E}{R} \;\Rightarrow\; R_1 = R_2$$

Also, shown in figure, i.e. $\tau_2 > \tau_1$

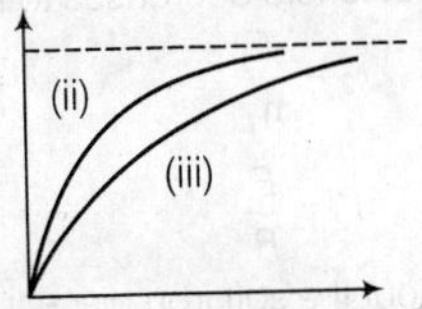

where, τ_1 and τ_2 are time constant for L - R circuits.

i.e. $$\frac{L_2}{R_2} > \frac{L_1}{R_1}$$

$$\Rightarrow \quad L_2 > L_1$$

It shows that inductance across circuit 2 is more than circuit 1.

19. (b,d)

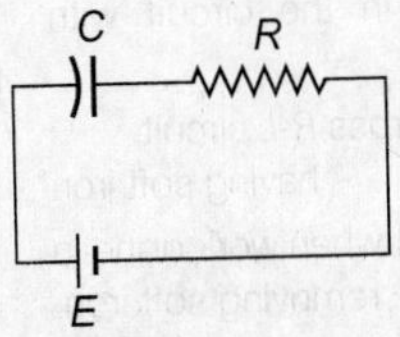

***Case* I** If $t = 0$, then C short circuit and L open circuit.

***Case* II** If $t \to \infty$, then C open circuit and L short circuit.

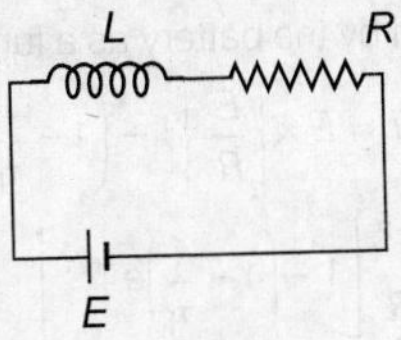

So, after a long time $V_C = E$

Then, charge on a capacitor

$$q_c = EC$$

and current in L long time after $t = 0$ is E/R.

20. (b) Induced current across inductance of a long solenoid,

$$I = \frac{\varepsilon}{R} = \frac{L}{R} \cdot \frac{dI}{dt} \qquad \left(\because \varepsilon = L\frac{dI}{dt}\right)$$

or $$\left[\frac{L}{R}\right] = I \cdot \frac{dt}{dI}$$

$$= \frac{[I] \cdot [T]}{[I]} = [T]$$

Statement I is correct.

Statement II is also correct to reduce the rate of increase of current through solenoid, we increase time constant (L/R).

21. (a) Both Statements I and II are correct and Statement II is correct explanation of Statement I.

As, $$I = I_{\text{max}}(1 - e^{-t/\tau})$$

$$\Rightarrow \quad e^{-t/\tau} = \left(1 - \frac{I}{I_{\text{max}}}\right)$$

When $I = 1\%$ of I_{max},

$$e^{-t/\tau} = (1 - 0.99)$$

$$\Rightarrow \quad t = -\tau \log(0.01)$$

$$\Rightarrow \quad t = \tau \log(100) \approx 4.61\tau$$

22. (d) $L i_1 = \eta L i_2$

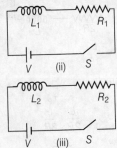

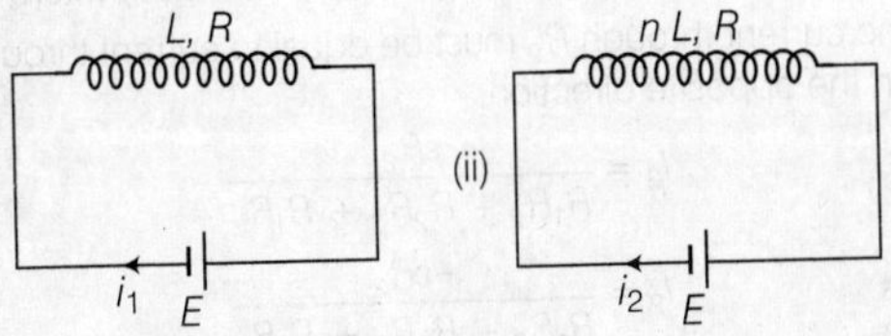

Consider a solenoid of resistance R and inductance L has a piece of soft iron inside it.

$$i_2 = \frac{i_1}{\eta} \qquad \left(\because i_1 = \frac{E}{R}\right)$$

At an instant, the piece of soft iron is pulled out suddenly, so that inductance of the solenoid decreases to η_L.

$$i_2 = \frac{E}{\eta_L} \qquad \text{[at } t = 0, \text{ Fig. (ii)]}$$

$$i_2' = \frac{E}{R} \qquad \text{[at } t \to \infty, \text{ Fig. (ii)]}$$

Work done to pull out the soft iron piece, i.e.

$$W = \Delta U = U_f - U_i = \frac{1}{2}\eta i (i_2)^2 - \frac{1}{2}L(i_1)^2$$

$$= \frac{1}{2}\left(\frac{1-\eta}{\eta}\right)\frac{E^2 L}{R^2}$$

23. (a) When iron piece has been pulled out, the current in the circuit with respect to time

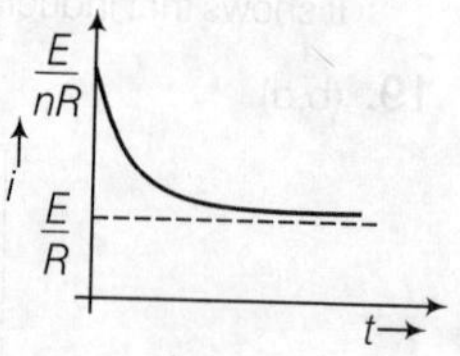

i = current across R-L circuit having soft iron + current when work done in removing soft iron.

$$= \left(\frac{E}{\eta R} - \frac{E}{R}\right)e^{-t/\tau} + \frac{E}{R} = \frac{E}{R}\left(1 - \left(1 - \frac{1}{\eta}\right)e^{-t/\tau}\right)$$

24. (a) Power supplied by the battery as a function of time, i.e.

$$P = Ei = E \times \left[\frac{E}{R}\left\{1 - \left(1 - \frac{1}{\eta}\right)e^{-t/\tau}\right\}\right]$$

$$= \frac{E^2}{R}\left[1 - \left(1 - \frac{1}{\eta}\right)e^{-t/\tau}\right]$$

25. (a) At the moment when switch is closed, no current will flow through the inductor.

$\therefore$ Resistor R_1 and R_2 can be treated as in series.

$$\varepsilon = I(R_1 + R_2)$$

$$\Rightarrow \quad I_1 = I_2 = \frac{\varepsilon}{R_1 + R_2} \text{ and } I_3 = 0$$

26. (b) A long time after the switch is closed, there is no voltage drop across the inductor, so resistors R_2 and R_3 can be treated as parallel resistors are in series with R_1.

$$I_1 = I_2 + I_3$$

$$\varepsilon = I_1 R_1 + I_2 R_2$$

and $\quad I_2 R_2 = I_3 R_3$

Hence, $\quad \dfrac{\varepsilon - I_2 R_2}{R_1} = I_2 + \dfrac{I_2 R_2}{R_3}$

$$\Rightarrow \quad I_2 = \frac{\varepsilon R_3}{R_1 R_3 + R_1 R_2 + R_2 R_3}$$

and $\quad I_3 = \dfrac{I_2 R_2}{R_3}$

and $\quad I_1 = I_2 + I_3 = \dfrac{\varepsilon R_2}{R_1 R_3 + R_1 R_2 + R_2 R_3}$

$$I_1 = \frac{\varepsilon(R_3 + R_2)}{R_1 R_3 + R_1 R_2 + R_2 R_3}$$

27. (b) Just after the switch is opened the current through the inductor continuous with the same magnitude and direction. With the open switch, no current flows through the branch with switch.

$\therefore$ The current through R_2 must be equal to current through R_3, but in the opposite direction.

$$I_3 = \frac{\varepsilon R_2}{R_1 R_2 + R_2 R_3 + R_1 R_3}$$

and $\quad I_2 = \dfrac{-\varepsilon R_2}{R_1 R_2 + R_2 R_3 + R_1 R_3}$

(opposite to direction shown)

and $\quad I_1 = 0$

28. (1) ***Case I*** When the switch is open, then minimum current flow in a circuit, i.e.

$$I_{min} \text{ at } t \to 0$$

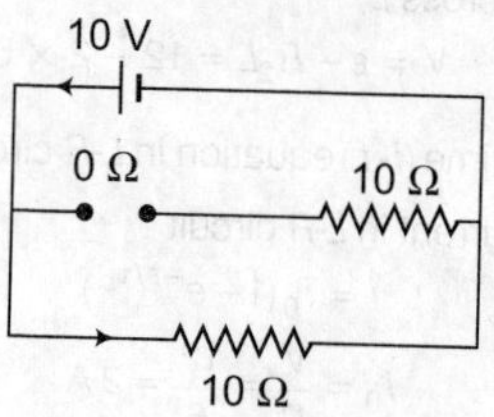

Current across circuit,

$$I = \frac{\varepsilon}{R} = \frac{10}{10} = 1\text{A}$$

Case II When the switch is closed, then maximum current.

$$I_{max} \text{ at } t \to \infty$$

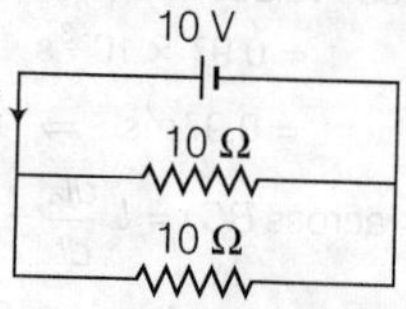

$$I = \frac{\varepsilon}{R} = \frac{10}{5\Omega} = 2\text{ A}$$

Difference between the minimum and maximum value of a current, i.e.

$$I_{max} - I_{min} = 2 - 1 = 1\text{ A}$$

$$K = 1$$

29. (2) During charging

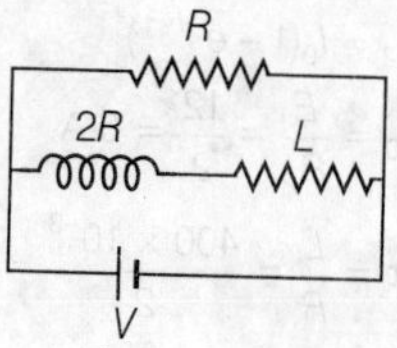

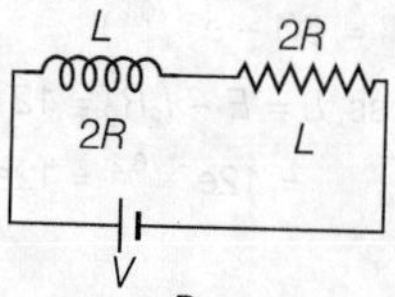

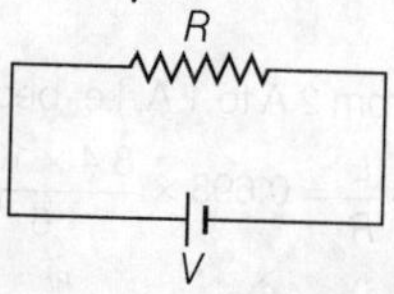

Time constant, $\tau_1 = \dfrac{L}{2R}$

During discharging

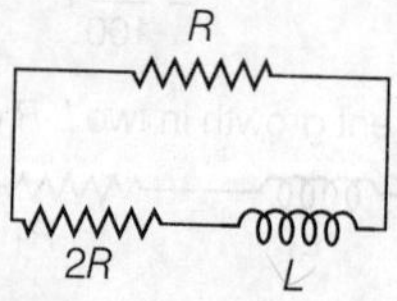

Time constant, $\tau_2 = \dfrac{L}{3R}$

Ratio of time constants,

$$\tau_1 : \tau_2$$

$$\Rightarrow \quad \frac{L}{2R} : \frac{L}{3R}$$

$$\Rightarrow \quad 3 : 2$$

So, $\quad K = 2$

DPP-9 Mutual Induction and Inductance

1. (i) Mutual inductance of B w.r.t. A

$$M = \frac{\text{Number of flux linkages}}{\text{Current}} = \frac{N_B \phi_B}{i_A}$$

$$= \frac{700 \times 90 \times 10^{-6}}{3.5} = 1.8 \times 10^{-2}\ \text{H}$$

(ii) Self-inductance of solenoid A,

$$L = \frac{N_A \phi_A}{i_A} = \frac{400 \times 300 \times 10^{-6}}{3.5} = 3.43 \times 10^{-2}\ \text{H}$$

(iii) Emf induced in B,

$$\varepsilon_B = M\left(\frac{d}{dt} i_A\right) = 1.8 \times 10^{-2} \times 0.5 = 9 \times 10^{-3}\ \text{V}$$

2. As the short solenoid produces a complicated magnetic field, so it is difficult to calculate mutual inductance and flux through the outer solenoid. For this purpose, we make use of the principle of reciprocity of mutual inductance i.e.

$$M_{12} = M_{21}$$

Suppose S_1 represents the long solenoid and S_2 represents the short solenoid. Then,

$$l_1 = 80\ \text{cm}, = 0.80\ \text{m}, N_1 = 1500$$
$$l_2 = 4\ \text{cm} = 0.04\ \text{m}, N_2 = 100$$
$$R_2 = 2.0\ \text{cm} = 0.02\ \text{m}, I_2 = 3.0\ \text{A}$$

The uniform magnetic field inside the long solenoid is given by

$$B_1 = \frac{\mu_0 N_1 I_1}{l_1}$$

Since, the short solenoid lies completely inside the long solenoid, the flux linked with it is given by

$$\phi_2 = N_2 A_2 B_1 = N_2 \cdot A_2 \cdot \frac{\mu_0 N_1 I_1}{l_1}$$

$\therefore$ Flux through each turn of short solenoid

$$= \frac{\phi_2}{N_2} = \frac{\mu_0 N_1 I_1}{l_1} \times A_2$$

$$= \frac{\mu_0 N_1 I_1}{l_1} \cdot \pi R_2^2$$

As, we know,

$$\phi_2 = M_{21} I_1$$

$$\Rightarrow \quad N_2 \cdot \frac{\mu_0 N_1 I_1 \cdot \pi R_2^2}{l_1} = M_{21} I_1$$

From the symmetry of mutual inductance, we have

$$M_{12} = M_{21} = \frac{\mu_0 \pi R_2^2 \cdot N_1 \cdot N_2}{l_1}$$

$$= \frac{4\pi \times 10^{-7} \times \pi \times (0.02)^2 \times 1500 \times 100}{0.80}\ \text{H}$$

$$= 2.96 \times 10^{-4}\ \text{H}$$

Total flux linked with the long solenoid is

$$N_1 \phi_1 = M_{12} I_2 = 2.96 \times 10^{-3} \times 3$$
$$= 8.88 \times 10^{-4}\ \text{Wb} \approx 8.9 \times 10^{-4}\ \text{Wb}$$

3. As, we know the idea of symmetry of mutual inductance between two coils, i.e.

$$M_{12} = M_{21}$$

Consider to calculate the flux through the smaller loop due to a current I_1 in the bigger loop. The area of the smaller loop is so small that we can use formula for field **B** at any point on the axis of the bigger loop.

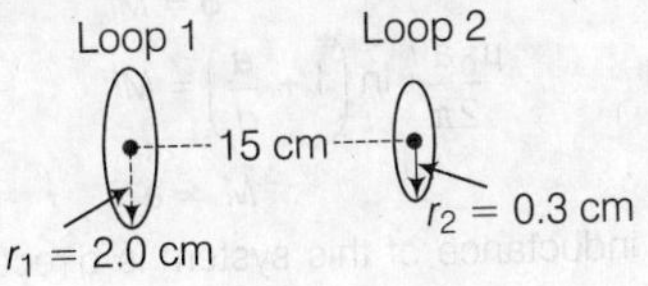

Field B_2 at 2 due to current I_1 in 1 is

$$B_2 = \frac{\mu_0 I_1 r_1^2}{2(n^2 + r_1^2)^{3/2}}$$

where, x is the distance between the centres of the two loops. Flux linked with loop 2,

$$\phi_2 = B_2 \times \text{area} = B_2 \pi r_2^2 = \frac{\mu_0 \pi r_1^2 r_2^2}{2(x^2 + r_1^2)^{3/2}} \cdot I_1$$

$$\therefore \quad M_{21} = \frac{\phi_2}{I_1} = \frac{\pi \mu_0\, r_1^2 \cdot r_2^2}{2(x^2 + r_1^2)^{3/2}} = M_{21}$$

Thus, flux linked with loop 1,

$$\phi_1 = M_{12} I_2 = \frac{\mu_0 \pi r_1^2 r_2^2}{2(x^2 + r_1^2)^{3/2}} \cdot I_2$$

Given, $\quad r_1 = 20\ \text{cm} = 0.20\ \text{m}, r_2 = 0.3\ \text{cm} = 3 \times 10^{-3}\ \text{m}$

$$x = 15\ \text{cm} = 0.15\ \text{m},$$
$$\mu_0 = 4\pi \times 10^{-7}\ \text{TmA}^{-1}$$

$$\therefore \quad M_{21} = \frac{\pi \times 4\pi \times 10^{-7} \times (0.2)^2 \times (3 \times 10^{-3})^2}{2[(0.15)^2 + (0.2)^2]^{3/2}}$$

$$= \frac{144\pi^2 \times 10^{-9}}{2 \times (625)^{3/2}} = 4.55 \times 10^{-11}\ \text{H} \qquad \text{[part (ii)]}$$

Flux linking the bigger loop is

$$\phi_1 = M_{12} \cdot I_2 = M_{21} \cdot I_1 = 4.55 \times 10^{-11} \times 2\ \text{Wb}$$
$$= 9.1 \times 10^{-11}\ \text{Wb} \qquad \text{[part (i)]}$$

4. Suppose a current I passes through the square of a loop of side L. Magnetic field at the centre O,

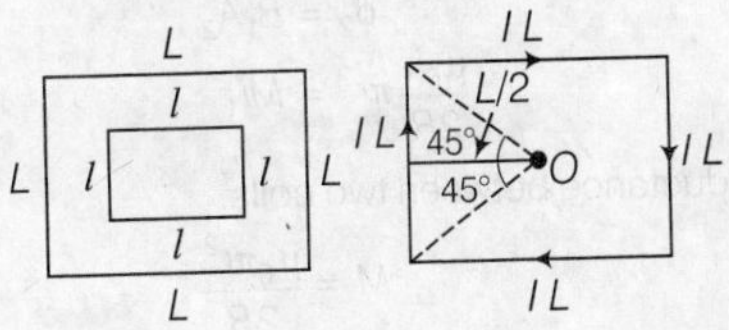

$$B = 4 \times \text{magnetic field due to each side}$$

$$= 4 \times \frac{\mu_0}{4\pi} \times \frac{I}{L/2} [\sin 45° + \sin 45°]$$

$$= \frac{2\mu_0 I}{\pi L}\left[\frac{1}{\sqrt{2}} + \frac{1}{\sqrt{2}}\right] = \frac{2\sqrt{2}\,\mu_0 I}{\pi L}$$

Magnetic flux linked with the small square loop,

$$\phi = BA = Bl^2 = \frac{2\sqrt{2}\,\mu_0 I l^2}{\pi L}$$

Mutual inductance of the system, $M = \frac{\phi}{I}$

$$= \frac{2\sqrt{2}\,\mu_0 I l^2}{\pi L I} = \frac{2\sqrt{2}\,\mu_0 l^2}{\pi L}$$

5. (a) Mutual inductance depends upon the relative position and orientation of the two coils.

6. (c) Given, $A = 10\ \text{cm}^2 = 10 \times 10^{-4}\ \text{m}^2$,

$$l = 20\ \text{cm} = 20 \times 10^{-2}\ \text{m}$$
$$N_1 = 300$$
$$N_2 = 400$$

Mutual inductance between two coaxial solenoids,

$$M = \frac{\mu_0 N_1 N_2 A}{l}$$

$$= \frac{4\pi \times 10^{-7} \times 300 \times 400 \times 10 \times 10^{-4}}{20 \times 10^{-2}}$$

$$M = 2.4\pi \times 10^{-4}\ \text{H}$$

7. (a) In situation I when the current flows in one coil, the other coil receives maximum flux due to the maximum area intercepting the flux. Hence, mutual inductance is maximum in this situation.

8. (c) Given, $L_1 = 0.01\ H_1, L_2 = 0.03$ H, when two coils are connected in series, then equivalent coefficient of inductance of both coils, i.e. $L = L_1 + L_2 + 2M$

$$0.06 = \cdot 01 + 0.03 + 2M$$

Value of coefficient of mutual inductance is $M = 0.01$ H

9. (c) Given, $L_1 = L_2 = L$ and coefficient of coupling, i.e. $K = \frac{1}{2}$.

When two coils are connected in series, then $L_{eq} = L_1 + L_2 + 2M$

As, we know,

Mutual inductance, $M = K\sqrt{L_1L_2}$

$$M = \frac{1}{2}$$

$$L_{eq} = L + L + 2\left(\frac{L}{2}\right)$$

So, equivalent inductance of the coil is

$$L_{eq} = 3L$$

10. (b) Consider a small coil of radius r is placed at the centre of a large coil of radius R, where $R >> r$.

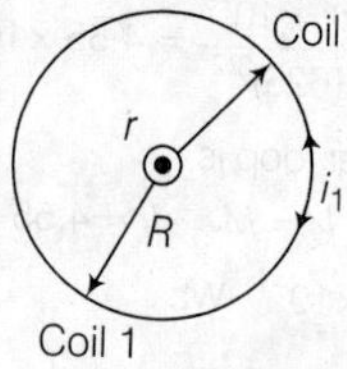

So, flux across coil 2, we get $\phi_2 = Mi_1$

$$\phi_2 = B_1A_2$$

$$\frac{\mu_0 i_1}{2R}\pi r^2 = Mi_1$$

Mutual inductance between two coils,

$$M = \frac{\mu_0 \pi r^2}{2R}$$

11. (d) Given, $e_1 = 25$ mV, $\frac{di_2}{dt} = 15$ A/s

The emf induced across coil 1, $e_1 = \frac{Mdi_2}{dt}$

$$25 \text{ mV} = M(15)$$

Mutual inductance across coil 1, $M = \frac{25}{15}$ mH

Given, $i_1 = 3.6$ A, flux linkage in coil 2, we get

$$\phi_2 = Mi_1$$

$$\phi_2 = \left(\frac{25}{15}\text{mH}\right)(3.6)$$

$$\phi_2 = 6.00 \text{ mWb}$$

12. (a) Given, $L_1 = 100$ mH, $L_2 = 400$ mH,

$$M = K\sqrt{L_1L_2}$$

For M to be maximum, $K = 1$

Maximum mutual inductance between the two coils,

$$M = \sqrt{L_1L_2} = 200 \text{ mH}$$

13. (b) Let M be the mutual inductance between X and Y

$$E = M\frac{dI}{dt}$$

$$\Rightarrow \quad M = \frac{E}{dI/dt}$$

Now, flux linked with X is

$$\phi_X = MI_0$$

$$\Rightarrow \quad \phi_X = MI_0 = \frac{E}{dI/dt} I_0$$

14. (a) If outer solenoid carries current I, then field due to solenoid (near its centre) is

$$B_1 = \mu_0 \frac{N_1}{l} \cdot I_1$$

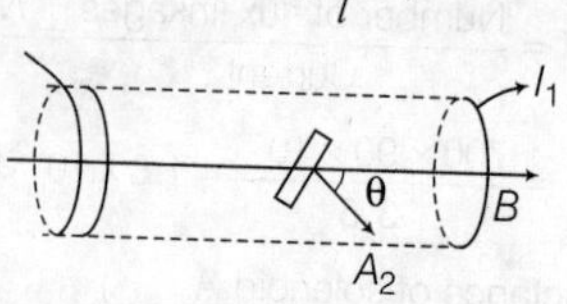

Magnetic flux linked with coil depends on orientation of coil,

$$\phi = N_2BA_2\sin\theta = \mu_0 \frac{N_1}{l} \cdot I_1 \cdot N_2A_2\sin\theta$$

$$= \frac{\mu_0 N_1 N_2 A_2 \sin\theta}{l} I$$

So, mutual Inductance of coil is $\frac{\phi}{I} = \frac{\mu_0 N_1 N_2 A_2 \sin\theta}{l}$

15. (b) Consider a rectangular loop of distance c apart from long straight wire, then net flux across current carrying straight wire, i.e.

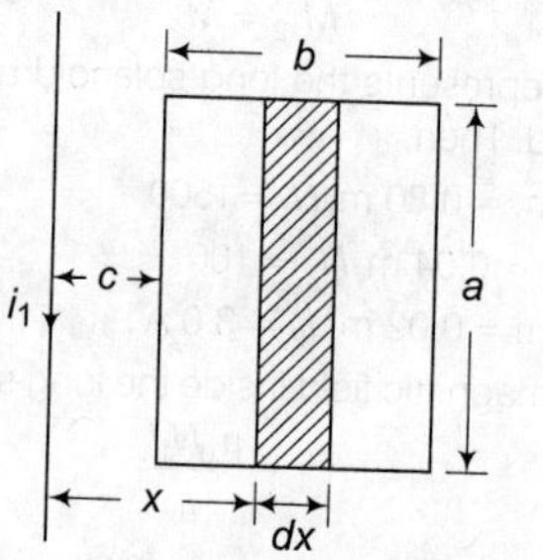

$$\phi = \int_c^{b+c} B\, dA = \int_c^{b+c} \frac{\mu_0 i_1}{2\pi x} a dx$$

$$= \frac{\mu_0 i_1 a}{2\pi} \ln\left(1 + \frac{c}{b}\right)$$

So, mutual inductance between rectangular loop and the loop straight wire is

$$\phi_2 = Mi$$

$$\frac{\mu_0 i_1 a}{2\pi} \ln\left(1 + \frac{c}{b}\right) = Mi_1$$

$$M = \frac{\mu_0 a}{2\pi} \ln\left(1 + \frac{c}{b}\right)$$

16. (a) Consider a square of loop of side a is placed in XY-plane, such that side of length a of the loop is parallel to the wire.

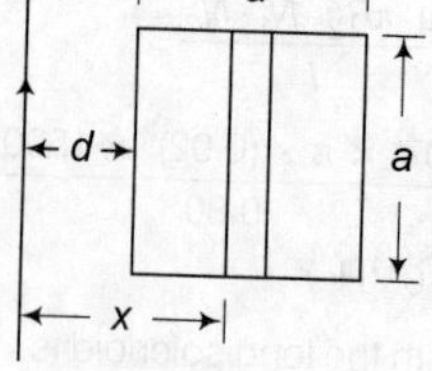

Flux linked across the square loop, i.e.

$$\phi = \int_d^{d+a} B\, dA = \int_d^{d+a} \frac{\mu_0 i}{2\pi x} a\, dx = \frac{\mu_0 ia}{2\pi} \ln\left(1 + \frac{a}{d}\right)$$

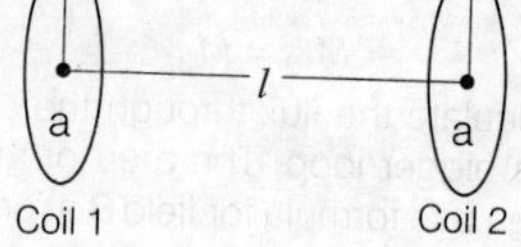

$$\phi = Mi$$

$$\frac{\mu_0 a\, i}{2\pi} \ln\left(1 + \frac{a}{d}\right) = Mi$$

$$M \propto a$$

So, mutual inductance of this system is directly proportional to side of a square, i.e. a.

17. (d) Magnetic field across coil 1,

$$B_1 = \frac{\mu_0 i a^2}{2(l^2 + a^2)^{3/2}} \quad (\text{as } l >> a)$$

$$B_1 = \frac{\mu_0 i a^2}{2l^3}$$

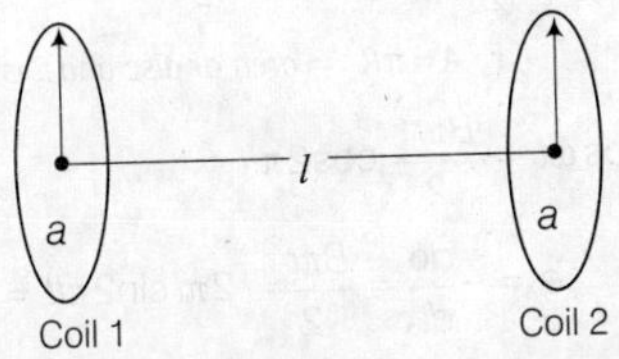

Flux induced in coil 2 due to coil 1,

$$\phi_{21} = Mi_1$$

As, we know flux induced on coil 2 due to coil 1,

$$\phi_{21} = B_1 A_2$$

$$\phi_{21} = \frac{\mu_0 i a^2}{2l^3}(a^2)$$

$$\frac{\mu_0 i a^4}{2l^3} = Mi$$

Mutual inductance of two loop system, $M = \frac{\mu_0 a^4}{2l^3}$.

18. (a) It is obvious that flux linkage in one ring due to current in other coaxial ring is maximum when $x = 0$ (as shown in figure) or the ring are also coplanar. Hence, under this condition, their mutual induction is maximum.

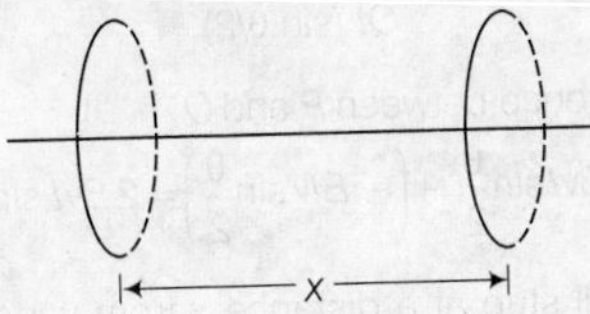

19. (b) If inductance of solenoid increases, then reactance of circuit also increases, then obviously current will decrease and lamp becomes dim.

20. (c)

(i) Just after switch S is closed, flux in M starts increasing in left direction, so in N also the flux starts increasing in left direction. This will induce current in N in a direction, so that the flux is in right direction. This is possible if induced current in N is from A to B.

(ii) In this case just reverse of (i) will happen, because after closing the switch, the flux in M starts decreasing in left direction.

(iii) After a long time a closing the switch, flux becomes constant. Hence, no current is induced.

(iv) Just after closing switch S, flux starts increasing, because M moves away, so due to this flux through N will decrease. But there will be a net increase in flux in N in left direction. This is the case similar to (i).

Hence, (i) → (p), (ii) → (q), (iii) → (r), (iv) → (s)

21. (c) From Faraday's law,

$$V = -L\frac{dI}{dt}$$

$$\Rightarrow \quad \frac{V_1}{V_2} = \frac{-L_1 \frac{d}{dt} I_1}{-L_2 \frac{d}{dt} I_2}$$

$$= \frac{L_1}{L_2} = \frac{8}{2} = \frac{4}{1}$$

Power given to both coils is same

$$\Rightarrow \quad V_1 I_1 = V_2 I_2$$

$$\Rightarrow \quad \frac{I_1}{I_2} = \frac{V_2}{V_1} = \frac{L_2}{L_1} = \frac{2}{8} = \frac{1}{4}$$

Energy stored in the coil is

$$U = \frac{1}{2}LI^2$$

$$\Rightarrow \quad \frac{U_1}{U_2} = \frac{\frac{1}{2}L_1 I_1^2}{\frac{1}{2}L_2 I_2^2} = \left(\frac{L_1}{L_2}\right)\left(\frac{I_1}{I_2}\right)^2 = \left(\frac{8}{2}\right)\left(\frac{2}{8}\right)^2 = \frac{4}{1} \times \frac{1}{16} = \frac{1}{4}$$

Hence, (i) → (q), (ii) → (p), (iii) → (q)

22. (a) Consider a strip at a distance x from the wire of thickness dx. Magnetic flux associated with this strip

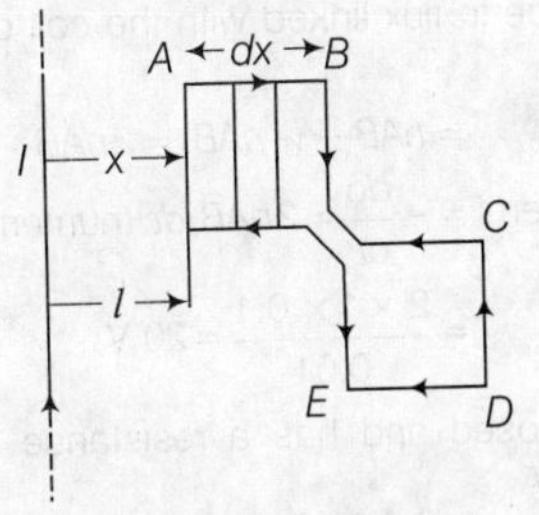

i.e. $$\phi = B(x)\, a dx = \frac{\mu_0 I a}{2\pi x} dx$$

$$\phi = \frac{\mu_0 I a}{2\pi}\left[\int_l^{a+l} \frac{dx}{x} + \int_{a+l}^{2a+l} \frac{dx}{x}\right]$$

$$= \frac{\mu_0 I a}{2\pi} \ln\left(\frac{2a + l}{l}\right)$$

Mutual inductance of the pair of current carrying wire,

$$M = \frac{\phi}{I}$$

$$\Rightarrow \quad M = \frac{\mu_0 a}{2\pi} \ln\left(\frac{2a + l}{l}\right)$$

23. (a) As, emf induced in the closed loop $ABCDEF$,

$$e = -M\frac{dI}{dt} = -M\frac{d}{dt}(I_0 \cdot t)$$

$$\Rightarrow \quad e = -MI_0 = -\frac{\mu_0 I_0 a}{2\pi} \ln\left(\frac{2a + l}{l}\right)$$

24. (b) Heat produced in the loop $ABCDEF$ in time t is

$$= \frac{\varepsilon^2}{R} t = \frac{\left[\frac{\mu_0 I_0}{2\pi} \ln\left(\frac{2a + l}{l}\right)\right]^2 at}{8\lambda}$$

25. (7) If I current flows through the circular loop, then magnetic flux at the location of square loop is

$$B = \frac{\mu_0 I R^2}{2(R^2 + Z^2)^{3/2}}$$

Substituting the value of $Z = \sqrt{3}R$, we have

$$B = \frac{\mu_0 I}{16R}$$

Now, total flux through the square loop is

$$\phi_T = NBS \cos\theta = (2)\left(\frac{\mu_0 I}{16R}\right) a^2 \cos 45°$$

Mutual inductance between the loops is given by

$$\therefore \quad M = \frac{\phi_T}{I} = \frac{\mu_0 a^2}{2^{7/2} R}$$

$$P = 7$$

26. (9) Induced emf = 9 mV

M = coefficient of mutual induction of B w.r.t. A

$$= \frac{N_B \phi_{BA}}{I_A} = \frac{700 \times 90 \times 10^{-6}}{3.5}$$

$$= 18 \times 10^{-3} \text{ H}$$

So, $$E_B = M\frac{dI_A}{dt} = 18 \times 10^{-3} \times 0.5 = 9 \text{ mV}$$

Revisal Problems (JEE Main)

1. (c) Given, area $= 10 \times 20\ cm^2 = 2 \times 10^{-2}\ m^2$

$B = 0.5T, \quad N = 60$

$\omega = 2\pi \times 1800/60$

$$\because \quad e = -\frac{d(N\phi)}{dt} = -N\frac{d}{dt}(BA\cos\omega t) = NBA\omega \sin\omega t$$

$$\therefore \quad e_{max} = NAB\omega = 60 \times 2 \times 10^{-2} \times 0.5 \times 2\pi \times 1800/60 = 113\ V$$

2. (a) The change in flux linked with the coil on rotating it through 180° is

$$= nAB - (-nAB) = 2nAB$$

$$\therefore \quad \text{Induced emf} = -\frac{d\phi}{dt} = 2nAB/dt \text{ (numerically)}$$

$$= \frac{2 \times 1 \times 0.1}{0.01} = 20\ V$$

The coil is closed and has a resistance of $2.0\ \Omega$. Therefore, $i = 20/2 = 10\ A$

3. (b) The flux linked with the coil, when the plane of the coil is perpendicular to the magnetic field is

$$\phi = nAB\cos\theta = nAB$$

The change in flux on rotating the coil by 180° is

$$d\phi = nAB - (-nAB) = 2nAB$$

$$\therefore \quad \text{Induced charge} = \frac{d\phi}{R} = \frac{2nAB}{R}$$

$$\therefore \quad \text{Induced charge} = \frac{2 \times 100 \times 0.001 \times 1}{10} = 0.02\ C$$

4. (a) The direction of current in the solenoid is clockwise. On displacing it towards the loop a current in the loop will be induced in clockwise direction, so as to oppose its approach. Therefore, the direction of induced current as observed by the observer will be anti-clockwise.

5. (b) When North pole of the magnet is moved away, then South pole is induced on the face of the loop in front of the magnet i.e. as seen from the magnet side, a clockwise induced current flows in the loop. This makes free electrons to move in opposite direction, to plate *a*. Thus, excess positive charge appear on plate *b*.

6. (b) $E = -L\dfrac{dI}{dt}$

$$= -100 \times 10^{-3}\, \frac{(0 - 100) \times 10^{-3}}{2 \times 10^{-3}} = 5.0\ V$$

7. (c) $E = -\dfrac{d\phi}{dt}$

or $\quad d\phi = -E\,dt = (0 - \phi)$

or $\quad \phi = 4 \times 10^{-3} \times 0.1 = 4 \times 10^{-4}\ Wb$

8. (d) $e = \dfrac{d\phi}{dt} = -\dfrac{d}{dt}[10t^2 + 5t + 1] \times 10^{-3}$

$$= -[10 \times 10^{-3}(2t) + 5 \times 10^{-3}]$$

At $t = 5s$,

$$e = -[10 \times 10^{-2} + 5 \times 10^{-3}]$$

$$\Rightarrow |e| = 0.105\ V$$

9. (a) For each spoke, the induced emf between the centre *O* and the rim will be the same

$$e = \frac{1}{2}B\omega L^2 = B\pi L^2 \qquad (\because \omega = 2\pi f)$$

Further for all spokes, centre *O* will be positive while rim will be negative. Thus all emf's are in parallel giving total emf

$$e = B\pi L^2 f$$

independent of the number of the spokes.

Substituting the values

$$e = 4 \times 10^{-5} \times 3.14 \times (0.5)^2 \times 2 = 6.28 \times 10^{-5}\ V$$

Note *If a copper disc of radius R is rotating about its own axis, with angular frequency ω in magnetic field B, which is perpendicular to the disc, then the induced emf between its centre and rim will be*

$$e = \frac{1}{2}B\omega R^2 \quad \text{or} \quad e = BAf = B\pi R^2 f$$

$(\because A = \pi R^2$ = *area of disc and f is frequency of rotation*)

10. (b) $\phi = BA\cos\omega t = \dfrac{B\pi r^2}{2}\cos 2\pi ft$

$$e = -\frac{d\phi}{dt} = \frac{B\pi r^2}{2} \cdot 2\pi f \sin 2\pi ft = B\pi^2 r^2 f \sin 2\pi ft$$

Peak value $= B\pi^2 r^2 f$

11. (b) $E = Blv = 4 \times 10^{-4} \times 50 \times \dfrac{360 \times 1000}{60 \times 60} = 2\ V$

12. (a) A positive charge will be induced at the *P* of rod *OP* and the end *O* becomes negative with respect to *P*.

The emf induced at the $P = Bl\,v\sin\dfrac{\theta}{2}$

Similarly, emf induced at end $Q = -Blv\sin\dfrac{\theta}{2}$

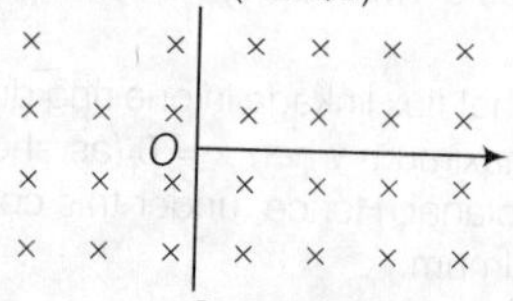

Potential difference between *P* and *Q*

$$= Bvl\sin\frac{\theta}{2} - \left(-Blv\sin\frac{\theta}{2}\right) = 2Bvl\sin\frac{\theta}{2}$$

13. (a) For a small strip at a distance *x* from vertex,

area of strip is $dA = \dfrac{b}{h}x \cdot dx$

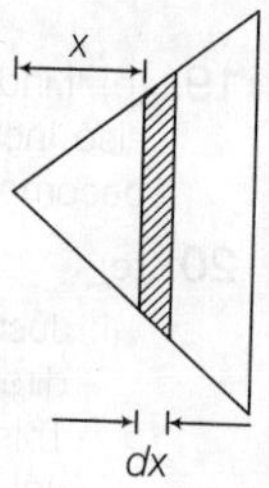

Field of wire at location of the strip $B = \dfrac{\mu_0 I}{2\pi(a + x)}$

Flux linked with strip is

$$d\phi = BdA = \frac{\mu_0 I}{2\pi(a + x)} \cdot \frac{b}{h} \cdot x \cdot dx = \frac{\mu_0 Ib}{2\pi h}\left(\frac{xdx}{a + x}\right)$$

Total flux linked with triangular coil,

$$\phi = \frac{\mu_0 Ib}{2\pi h}\int_a^h \frac{xdx}{(a + x)} = \frac{\mu_0 Ib}{2\pi h}\left[h - a\ln\left(\frac{a + h}{a}\right)\right]$$

$$\therefore \text{ Mutual inductance of the coil is } M = \frac{\phi}{I} = \frac{\mu_0 b}{2\pi h}\left(h - a\ln\left(\frac{a + h}{a}\right)\right)$$

$$\Rightarrow M = \frac{4\pi \times 10^{-7} \times 0.2}{2\pi \times 0.1}\left(0.1 - 0.1\ln\left(\frac{10 + 10}{10}\right)\right) = 1.22 \times 10^{-8}\ H$$

14. (d) When frame turned through angle θ, flux through the frame is

$$\phi = \mathbf{B} \cdot \mathbf{A} = BA\cos\theta$$

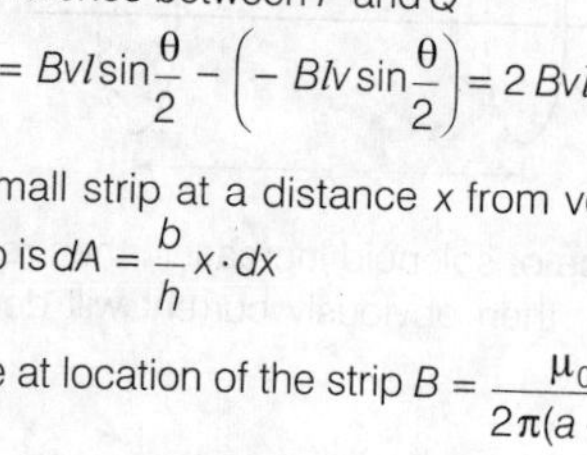

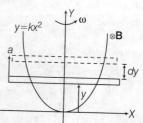

$$\text{Area } A = \int_0^y 2x\,dy = \int_0^y \frac{2}{\sqrt{k}} \cdot \sqrt{y} \cdot dy = \frac{4}{3\sqrt{k}} \cdot y^{3/2}$$

Also, $\quad y = \dfrac{1}{2}at^2$

$$\therefore \quad \phi = \frac{B}{3}\sqrt{\frac{2}{k}} \cdot a^{3/2} \cdot t^3 \cos\theta$$

15. (c), Given circuit is,

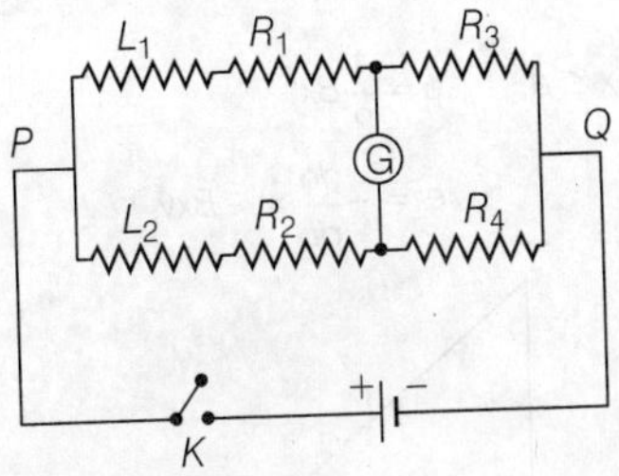

As galvanometer shows no deflection, $V_A = V_B$.

So, $V_{PA} = V_{PB}$

$$L_1 \frac{di}{dt} + i_1 R_1 = L_2 \frac{di_2}{dt} + i_2 R_2 \quad \text{...(i)}$$

and $V_{AQ} = V_{BQ}$

$$i_1 R_3 = i_2 R_4 \quad \text{...(ii)}$$

Substituting for i_1 in Eq. (i) from Eq. (ii), we get

$$L_1 \frac{R_4}{R_3} \cdot \frac{di_2}{dt} + i_2 \frac{R_1}{R_3} = L_2 \frac{di_2}{dt} + i_2 R_2$$

At $t = 0$, $i_2 = 0$.

So, $L_1 \frac{R_4}{R_3} - L_2 = 0 \Rightarrow \frac{L_1}{L_2} = \frac{R_3}{R_4}$

At $t = \infty$,

$$\frac{di_2}{dt} = 0 \Rightarrow i_2 = \frac{V}{R_2 + R_4}$$

$$\Rightarrow R_2 - \frac{R_4 R_1}{R_3} = 0$$

$$\Rightarrow \frac{R_1}{R_2} = \frac{R_3}{R_4}$$

So, combining both, we have

$$\frac{L_1}{L_2} = \frac{R_1}{R_2} = \frac{R_3}{R_4}$$

16. (b) Immediately after closing the switch, inductor acts as an open circuit.

17. (d) Flux through the loop

$$= \frac{\mu_0 I a}{4\pi} . \log 2$$

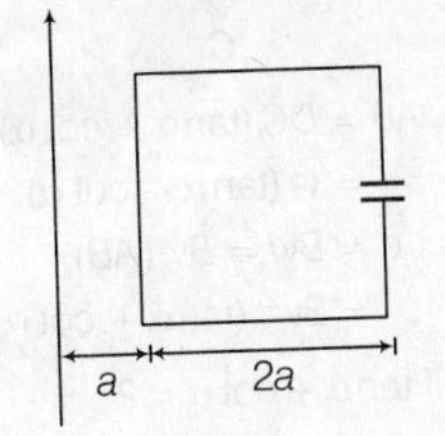

Induced emf in coil,

$$E = \frac{-d\phi}{dt} = \frac{-\mu_0}{\pi} a I_0 \omega \log 2 \cos \omega t$$

Charge on capacitor

$$Q = CE = \frac{C\mu_0}{\pi} a I_0 \omega \log 2 \cos \omega t$$

$\therefore$ Charge varies as

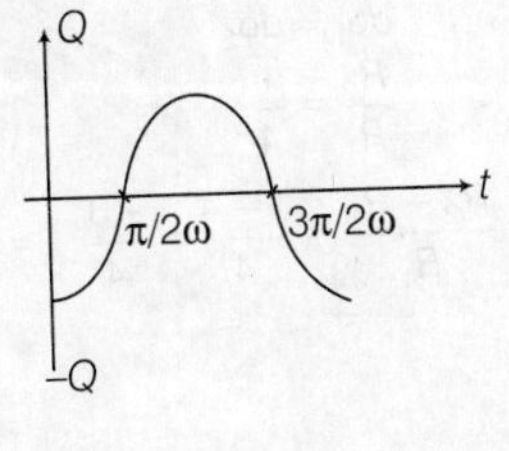

18. (d) $e = Blv$

at $t = 2$, $e = 3 \times 10^{-4}$ and at $t = 10$ s, loop is completely inside magnetic field. So, there is no flux change. Hence, $e = 0$V.

19. (a) $F = ilB$ and $i = (Blv)/R$

Hence, force required = F so,

$$F = \frac{B^2 l^2 v}{R}$$

$$F = \frac{(0.5)^2 \times (0.10)^2 \times (0.03)^2 \times 0.01}{1}$$

$$= 2.25 \times 10^{-8} \text{ N}$$

20. (a) $\phi = \mathbf{B} \cdot \mathbf{A}$

$$= (0.02\,\hat{\mathbf{i}}) \cdot (30\hat{\mathbf{i}} + 16\hat{\mathbf{j}} + 23\hat{\mathbf{k}}) \times 10^{-4}$$

$$= 0.6 \times 10^{-4} \text{ Wb} = 60\mu \text{ Wb}$$

21. (c) The induced emf,

$$e = -d\phi/dt = -\frac{d}{dt}(3t^2 + 2t + 3) \times 10^{-3}$$

Thus, $e = (-6t - 2) \times 10^{-3}$

At $t = 2$ s,

$$e = (-6 \times 2 - 2) \times 10^{-3} = -14 \text{ mV}$$

22. (c) Induced emf = Bv (effective length)

$$= Bv \cdot 2l \sin\theta/2$$

$$= 2Blv \sin\theta/2$$

23. (c) The area swept by radius OC in one half circle is $\pi r^2/2$. The flux change in time $T/2$ is thus $(\pi r^2 B/2)$. The induced emf is

$$e = \frac{\pi r^2 B/2}{T/2} = \frac{B\omega r^2}{2}$$

The induced current is

$$I = e/R = B\omega r^2/2R$$

24. (d) If we find of the small coil and then calculate flux through long solenoid, the problem becomes very difficult. So, we use the following fact about mutual inductance.

$$M_{21} = M_{12}, \frac{\phi_2}{I_1} = \frac{\phi_1}{I_2}$$

Thus, if I current flows in long solenoid, then flux ϕ through small coil is the same as the flux ϕ_2 that is obtained when I current flows through the small coil. Therefore,

$$\phi_2 = \phi_1 = \text{(field at small coil)} \times \text{(area)} \times \text{(turns)}$$

$$= \left(\mu_0 \frac{N_2}{l_2} I\right)(AN_1) = \frac{\mu_0 N_1 N_2 A I}{l_2}$$

25. (d) The induced current interacts with earth's magnetic field, causing repulsion which causes ring to jump but induced current will be soon zero (as current in the coil reaches its maximum value) and the ring falls back to base.

So, Statement I is not correct.

But, Statement II is correct.

26. (a) $Z = \sqrt{R^2 + X_C^2} = \sqrt{\left(R^2 + \frac{1}{(\omega C)^2}\right)}$

So, when capacitance is more (when dielectric is filled), X_C reduces and Z will be less and current will be large. Both Statements I and II are correct and Statement II explains Statement I.

27. (d) Multistrand wire used for connecting supply and Rheostat in potentiometer circuit have sufficient inductance to produce a back emf. But this causes an induced current in same direction which may delay the zero deflection of galvanometer. Reverse deflection is not possible. So, Statement I is incorrent but Statement II is correct.

28. (b) Statement I is incorrect due to word 'always' in it.

If the field is non-uniform from position point of view, no emf will be induced. If magnetic field is non-uniform from time point of view, an emf is induced. Also, Statement II is correct.

Revisal Problems (JEE Advanced)

1. (c) The induced emf between A and $B = e = Blv$

$$i = \frac{Blv}{R}$$

$$\text{Power} = i^2R = \frac{B^2l^2v^2}{R}$$

So, when v is doubled rate of heat dissipation becomes four times.

2. (b) Induced emf $= \frac{Mdi}{dt} = 0.5\,\frac{3-2}{0.01} = 50\,\text{V}$

3. (b) Let $AB = BC = l$

Velocity $= v$, $e_1 = Blv\sin\theta$

$e_2 = Blv\cos\theta$

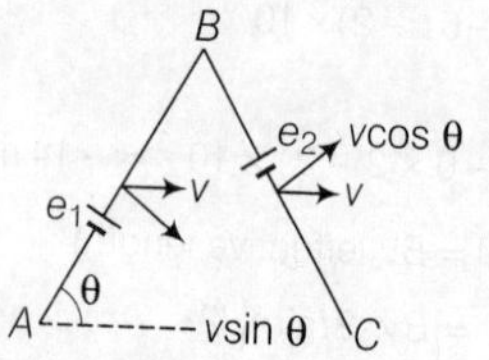

For induced emf at $A >$ induced emf at C

$e_2 > e_1$

Higher value of e_2 will ensure A to be at higher potential or $Blv(\cos\theta - \sin\theta) > 0$

$$\cos\theta > \sin\theta \quad \text{or} \quad \theta < 45^\circ$$

4. (d) The graph can be drawn qualitatively.

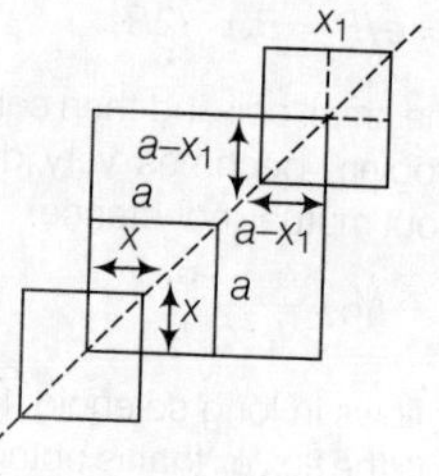

As shown for $0 \le x \le a$

$$\phi = Bx^2$$

$$e = -\frac{d\phi}{dt} = -2B\frac{xdx}{dt}$$

$$= -2Bxv$$

At $x = 0, e = 0$,

At $x = \frac{a}{2}, e = -Bav$, as $x \to a$, but $x \ne a$ $e \approx -2Bav$.

As, $\quad x = a, e = 0$ as $\phi =$ constant

Distance along diagonal $x_0 = x\sqrt{2}$, thus $x_0 = f(x)$

When the loop is coming out, x_1 should be selected carefully for which

$$\phi = B(a - x_1)^2$$

$$e = -\frac{d\phi}{dt} = -2B(a - x_1)(-v) = 2B(a - x_1)v$$

At $x_1 = 0, e = 2Bav$

At $x_1 = a, e = 0$

Thus, just before the loop enters the field completely inside it, $e = -2BaV$. When loop is completely inside the fluid, $e = 0$ (ϕ = constant). As, the loop starts coming out of the field ($e = 2Bav$) so, option (d) is correct.

5. (c) For $0 \le x \le a$, $\quad \phi = \frac{1}{2}Bx^2$

$$e = -\frac{d\phi}{dt} = -Bxv$$

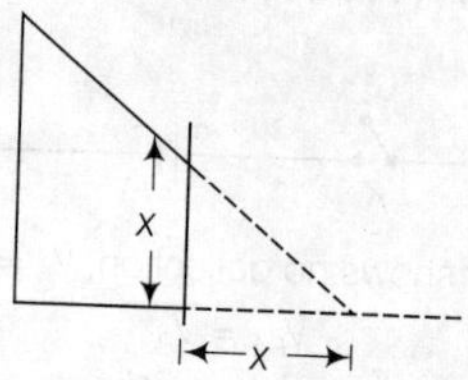

At $x = 0$, $\quad e = 0$

At $x = \frac{a}{2}$, $\quad e = -\frac{Bav}{2}$

At $x = a$, $\quad e \approx -Bav \qquad (x = \text{constant} \therefore e = 0)$

For $a < x \le 2a$,

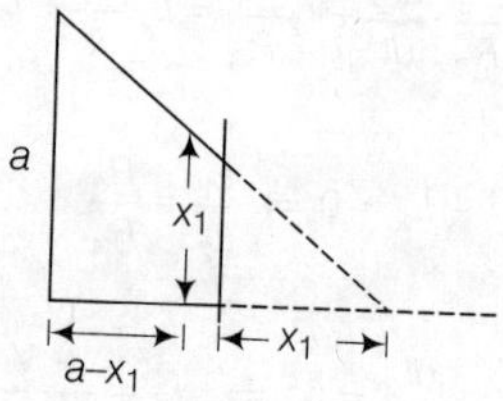

For $2a \le x \le 3a$,

$$\phi = B\left[\frac{a^2}{2} - \frac{x_1^2}{2}\right]$$

$$e = -\frac{d\phi}{dt} = Bx_1v$$

6. (b) $CD = vt$

$AD = DC\cot\alpha\, DB = DC\tan\alpha$

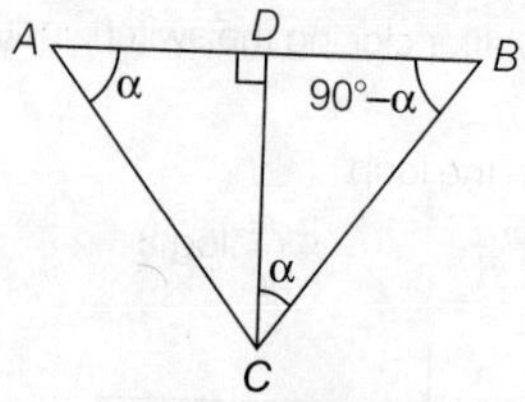

$$AB = DC(\tan\alpha + \cot\alpha)$$

$$= vt(\tan\alpha + \cot\alpha)$$

$$e = Bvl = Bv(AB)$$

$$= Bv^2t(\tan\alpha + \cot\alpha)$$

Given, $e = 2Bv^2t$, if $\tan\alpha + \cot\alpha = 2$

or $\tan\alpha = \cot\alpha = 1$, if $\alpha = 45^\circ$

7. (c) $\frac{d\phi}{R} = idt$

Case I $\quad \frac{d\phi_1}{R_1} = \frac{1}{2}it$

Case II $\quad \frac{d\phi_2}{R_2} = \frac{1}{2}4it = 2it$

But $\quad d\phi_1 = d\phi_2$

$$\frac{R_2}{R_1} = \frac{1}{4}$$

$$\frac{R_2 - R_1}{R_1} = \frac{1-4}{4} = \frac{-3}{4} \quad \Rightarrow 75\%\ \text{(decrease)}$$

8. (b) $i_L = \frac{V}{R}(1 - e^{-t/\tau_1}), i_c = \frac{V}{R}e^{-t/\tau_2}$

$$i_L = i_C \text{ at } t = CR\ln 2$$

$$1 - e^{-t/\tau_1} = e^{-t/\tau_2}$$

Solving, $\frac{L}{R} = CR$

$\Rightarrow$ $R = \sqrt{\frac{L}{C}}$

9. (d) Net magnetic flux across inductance, $\phi = LI$

where, $\phi = 100\,(5 \times 10^{-5})$

$\Rightarrow$ $\phi = 5 \times 10^{-3}$ Wb

$\Rightarrow$ $5 \times 10^{-3} = L(2)$

$\Rightarrow$ $L = 2.5 \times 10^{-3}$ H

So, magnetic energy associated with the coil is $U_m = \frac{1}{2}LI^2$

$\Rightarrow$ $U_m = \frac{1}{2}(2.5 \times 10^{-3})(4)$

$\Rightarrow$ $U_m = 5$ mJ $= 0.005$ J

10. (a) When the key is at position (2) for a long time then energy stored in the inductor is

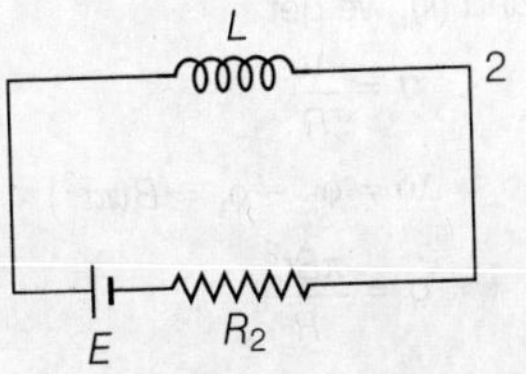

$$U_B = \frac{1}{2}Li_0^2 = \frac{1}{2}L\left(\frac{E}{R_2}\right)^2 = \frac{LE^2}{2R_2^2}$$

This whole energy will be dissipated in the form of heat when the inductor is connected to R_1 and no source is connected.

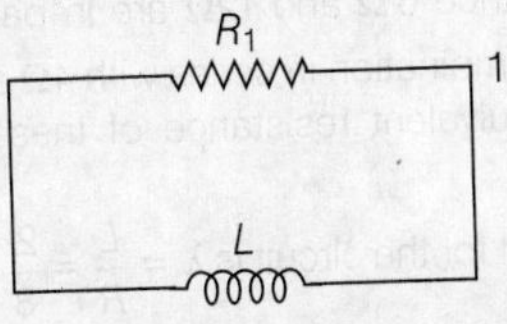

11. (a) When S_2 is closed then current across inductor $i = \frac{\varepsilon}{2R}$

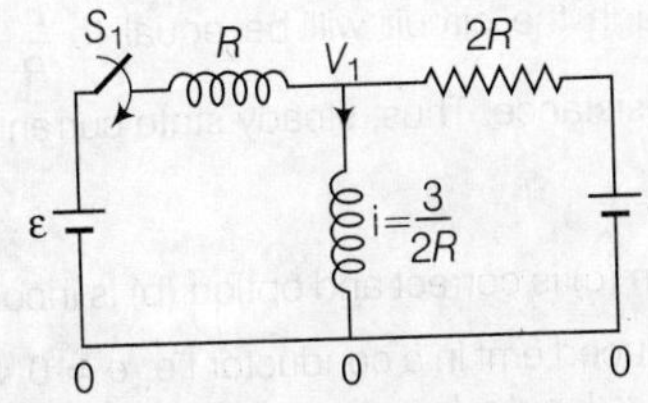

$\therefore$ $\frac{\varepsilon - V_1}{R} + \frac{\varepsilon - V_1}{2R} = \frac{\varepsilon}{2R}$ $\left(\because V_1 = \frac{2\varepsilon}{3}\right)$

$\therefore$ Potential difference $(V) = \varepsilon - \frac{2\,\varepsilon}{3} = \frac{\varepsilon}{3}$

12. (c) Net induced current in a circuit,

$$i = \frac{2\varepsilon_1 + 2\varepsilon_2}{R_1 + R_2} = \frac{\varepsilon}{R_1 + R_2}$$

where, $\varepsilon = \frac{d\phi}{dt}$ is the net emf in the circuit.

Then, potential difference across two voltmeters

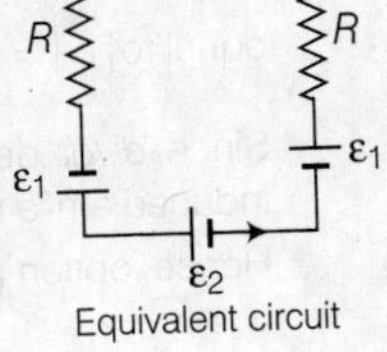

Equivalent circuit

$\therefore$ $V_1 - V_2 = (\varepsilon - iR_1) - (\varepsilon - iR_2)$

$$= \frac{\varepsilon(R_2 - R_1)}{R_1 + R_2} = \left|\frac{\varepsilon(R_1 - R_2)}{R_1 + R_2}\right|$$

13. (d) Initially the capacitor provides zero resistance. Therefore, initial currents is $E/2R$.

Finally, the capacitor provides infinite resistance, therefore, final current is $E/2(R + r)$, assuming r be the small resistance provide by the inductor.

Hence, after a long time interval current through battery will be less than current through it initially.

14. (b) Area of a triangular loop across a uniform magnetic field,

$$A = \frac{1}{2} \times 6 \times 4 - \frac{1}{2} \times 2\,vt\,\tan 37° \times vt$$

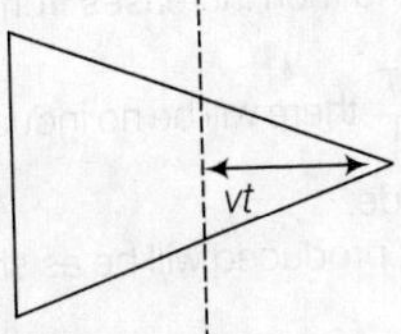

So, magnetic flux linked with loop,

$$\phi = BA$$

Thus, emf induced in a loop,

$$e = \frac{-d\phi}{dt}$$

$$= +\,Bv^2\,\frac{3}{4} \times 2t = iR'$$

$\Rightarrow$ $i \propto t$

$\Rightarrow$ $P \propto t^2$

Power generated by triangular loop

i.e. $P \propto t^2$ (parabolic variation).

15. (b) Induced emf, $\varepsilon = -\frac{dq}{dt}$

and induced current, $I = \frac{|\varepsilon|}{R}$

or $\frac{dq}{dt} = \frac{1}{R} = \frac{d\phi}{dt}$

or $dq = \frac{1}{R}d\phi$

$$\int dq = \frac{1}{R}\int d\phi$$

Charge flowing through the galvanometer,

$\Rightarrow$ $q = \frac{\phi}{R} = \frac{BA}{R}$

16. (a) Consider a free electron in the disc at point P distant x from centre of disc.

The magnetic force on free electron is

$eVB = e\omega\,xB$ (towards left)

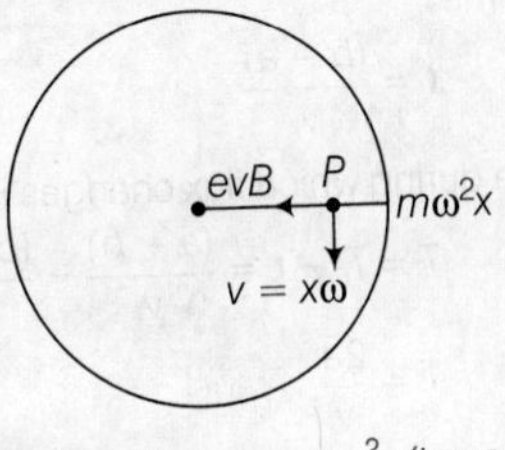

centrifugal force $= m\omega^2 x$ (towards right)

If the net force on the electron at P is zero

i.e. $e\omega x\,B = m\omega^2 x$

or $\omega = \frac{eB}{m}$

There shall be no flow of free electrons radially outwards and hence, no electric field shall develop within the disc.

17. (d) As, emf induced when a loop rotates in a magnetic field is ($e = NA\,B\,\omega \sin \omega t$), which is sinusoidal function. Therefore, induced current will also be sinusoidal as shown in the figure.

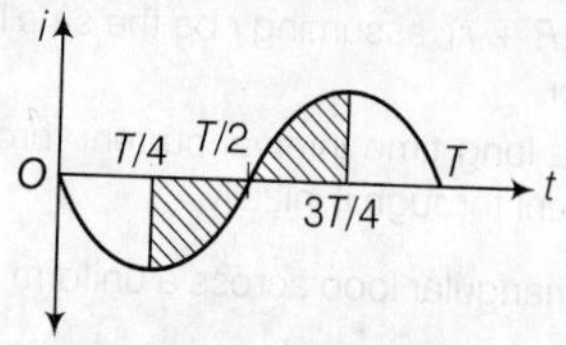

The curve will be negative, because initially the flux decreases from $t = 0$ to $T/4$ and then increases from $t = \frac{3T}{4}$ to $\frac{T}{4}$.

But from $t = \frac{T}{4}$ to $\frac{3T}{4}$, there will be no induced current as there is no field on the right side.

Hence, the current produced will be as shown in the figure.

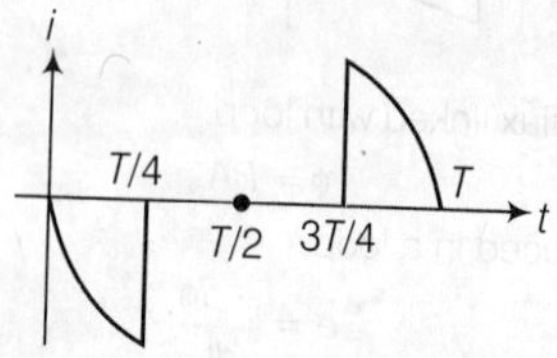

18. (b) The emf will induce when flux changes i.e. when the area changes. In the given problem, flux changes from situations (i) to (ii) (see figure).

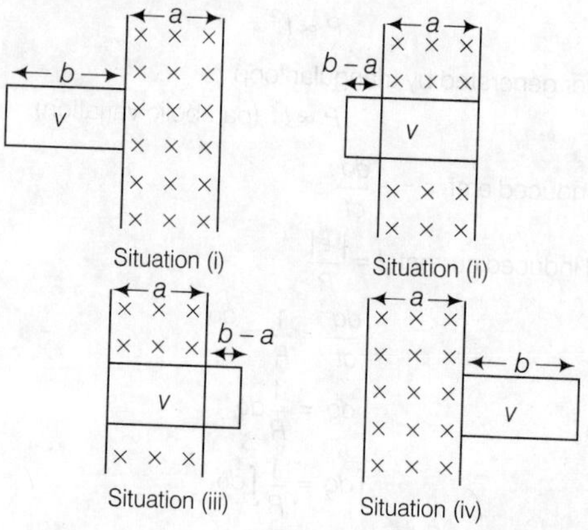

But no flux will change from situations (ii) to (iii) (as area inside the magnetic field remains unchanged). Again from situations (iii) to (iv), flux changes.

The total time take from situations (i) to (iv) is

$$T_0 = \frac{(a+b)}{v}$$

But during this time, flux does not change from situations (ii) to (iii) which takes a time

$$t = \frac{(b-a)}{v}$$

Hence, the time during which flux changes is

$$T = T_0 - t = \frac{(a+b)}{v} - \frac{(b-a)}{v}$$

$$T = \frac{2a}{v}$$

19. (b) In mutual Inductance, emf induced in a coil,

i.e. $\varepsilon = M\frac{dI}{dt} = M\frac{d}{dt}[I_0 \sin \omega t] = MI_0\, \omega \cos \omega t$

Maximum value of emf induced in the second coil,

i.e. $\varepsilon_{max} = MI_0\omega$ (maximum value of cos $\omega t = \omega$)

$= 0.005 \times 10 \times 100\,\pi = 5\pi V$

20. (b,c,d) By using Lenz's law.

21. (a,c,d) If loop rotates about Z-axis, variation of flux linkage is zero. So, no emf in induced. When it rotates about Y-axis, flux linkage changes, however in insulator there cannot be motional emf.

22. (a,b,c,d) If the equilibrium current is i_0. Then, current at any time in the transient state is given by

$$i = i_0(1 - e^{-t/\tau})$$

Putting $t = 0, i = 0$

The voltage drop across inductor is E that oppose the applied emf E.

$$E' = E \text{ (numerically)}$$

Since, current is zero, heat loss will be zero.

23. (a,d) Induced emf is given by

$$E = \frac{d\phi}{dt}$$

$$\Rightarrow \int_{\phi_1}^{\phi_2} d\phi = \Delta\phi = \int E dt \quad \text{...(i)}$$

Total charge flown in the loop, $q = \int i\,dt$

$$q = \int \frac{E}{R} dt \quad \text{...(ii)}$$

Using Eqs. (i) and (ii), we get

$$q = \frac{\Delta\phi}{R}$$

$$\Delta\phi = \phi_2 - \phi_1 = B(\pi r^2)$$

$$q = \frac{\pi B r^2}{R}$$

i.e. $q \propto B$

$q \propto r^2$

and $q \propto \frac{1}{R}$

24. (a,c) As resistance 6 Ω and 12Ω are in parallel with each other their parallel combination in series with 4Ω and the inductance of 2H. Hence, equivalent resistance of these three resistance is equal to 8 Ω.

∴ Time constant for the circuit is $\lambda = \frac{L}{R} = \frac{2}{8} = 0.255$

Hence, (a) is correct.

In steady state, no emf will be induced in the inductance. Hence, current through the circuit will be equal to $\frac{E}{R}$ where, R is the equivalent resistance. Thus, steady state current will be equal to $\frac{6}{8} = 0.75$ A.

Hence, option (c) is correct and option (b) is incorrect.

25. (a,c,d) As induced emf in a conductor i.e. $e = 0$, when conductor moves along its length. In options (c) and (d) conductor moves at right angle to its length and B is perpendicular to that conductor, $e = B\lambda v$.

26. (c,d) When switch is just closed in the circuit shown, at that moment current through the circuit is zero. Hence, emf induced across inductance L will be equal to emf E of the battery.

But as the current through the circuit increases, the induced emf in the solenoid decreases. But induced emf in the solenoid is equal to $|e| = L\frac{di}{dt}$

Since, di/dt decreases as time passes, therefore, the graph for induced emf e and time t will be as shown in option (c).

Hence, option (a) is incorrect and option (c) is correct.

The graph for current should be such that at initial moment current is zero and current increases with time in such a way that the rate of increase of current gradually decreases. Hence, slope of the current-time curve should decrease with time. Therefore, the graph between current and time will be as shown in option (d). Hence, option (b) is incorrect and option (d) is correct.

27. (c) Statement II is incorrect as there is no such rule. Self explanatory.

28. (c) **ds** is perpendicular to surface having magnitude ds and θ is the angle between **B** and ds at that element. So, $d\phi_B$ varies from element to element.

29. (c) When two coils of inductance L_1 and L_2 are joined together then, their mutual inductance is given by

i.e. $$M = K\sqrt{L_1 \cdot L_2}$$

From the relation, if self-inductance of primary and secondary coil are doubled the mutual inductance of the coils will be doubled. Also, Statement II is incorrect M and L_1L_2.

30. (a) (i) $\rightarrow$ (r), (ii) $\rightarrow$ (q), (iii) $\rightarrow$ (s), (iv) $\rightarrow$ (p)

31. (a) (i, ii) Current in inductor when switch is open

$$I_0 = \frac{E}{R}$$

Initially induced emf will be equal to E and finally it is zero. So, energy stored will be zero.

(iii, iv) Here current becomes zero suddenly.

so, $\frac{dI}{dt}$ is large.

Hence, induced emf, $L\frac{dI}{dt}$ will be large. Finally energy stored in inductor will be zero.

32. (b) Initial charge $= CV_0 = Q_0 = 10 \times 10^{-3} \times 5 = 50$ mC

When capacitor is connected at position I.

$$E - IR - \frac{q}{C} = 0$$

$$\int_0^t \frac{1}{RC}\,dt = \int_{\theta_0}^{q_0} \frac{dq}{EC - q}$$

$$\Rightarrow \quad q = 50[2 - e^{-t}]\,\text{mC}$$

Voltage across the capacitor at that time $t = 1$ s, we get

$$V = \frac{q}{C} = \frac{50\,[2 - 1/e]}{10 \times 10^{-3}}$$

$$= 5 \times 10^3[2 - (1/e)]\ \text{V}$$

33. (a) The maximum current flowing in the L-C circuit.

$$\frac{1}{2}LI^2 = \frac{1}{2}CV^2$$

$$\Rightarrow \quad I = \left[2 - \frac{1}{e}\right] \times 10^4\ \text{A}$$

34. (d) Frequency of L-C oscillation

i.e. $$\frac{1}{2\pi\sqrt{LC}} = \frac{1}{2\pi\sqrt{10 \times 10^{-3} \times 2.5 \times 10^{-3}}}$$

$$= \frac{100}{\pi}\ \text{Hz}$$

35. (c) Speed of waves in the wire is

$$v = \sqrt{\frac{T}{\mu}} = \sqrt{\frac{267}{3 \times 10^{-3}}} = 298\ \text{ms}^{-1}$$

In simplest mode of vibration, length of wire

= Distance between two nodes

$= \frac{\lambda}{2} = 0.64$ m

$\therefore$ λ = wavelength = 1.28 m

So, frequency of oscillation (fundamental frequency of wire),

$$f = \frac{v}{\lambda} = \frac{298}{1.28} = 233\ \text{Hz}$$

36. (c) Changing flux of magnetic field through the circuit containing the wire will drive to the left in the wire as it moves up and to the right as it moves down. So, the emf will have this same frequency 233 Hz.

37. (a) Vertical coordinate of centre of wire is given by

$$x = A\cos\omega t$$

$$= 1.5\cos(2\pi \times 233t)$$

$$\text{Velocity, } v = \frac{dx}{dt}$$

$$= -(1.5)(2\pi \times 233) \times \sin(2\pi \times 233t)$$

So, its maximum speed is

$$v_{max} = 1.5 \times 2\pi \times 233 = 22\ \text{m/s}$$

So, induced emf $= Blv$

$$= 4.5 \times 10^{-3} \times 0.02 \times 22$$

$$= 1.98 \times 10^{-3}\ \text{V}$$

38. (4) $\oint \mathbf{E} \cdot d\mathbf{l} = \left|\frac{d\phi}{dt}\right|$

$$E(4L) = L^2\frac{dB}{dt}$$

$$E = \frac{L}{4}(4t) = Lt$$

$$= 2(2) = 4\ \text{Vm}^{-1}$$

39. (8) $q = t^2 - 4$,

At $t = 3$ s, q is positive.

$$i = \frac{dq}{dt} = 2t$$

and $$\frac{di}{dt} = 2$$

At $t = 3$ s, $q = 5$C

and $i = 6$A

$$V_A - \frac{q}{C} - \frac{Ldi}{dt} - iR = V_B$$

$$V_A - V_B = \frac{q}{C} + \frac{Ldi}{dt} + iR$$

$$= \frac{5}{5} + (0.5)(2) + 6(1)$$

$$= 8\text{V}$$

1. (d) $\because$ Induced emf is rate of change of magnetic flux.

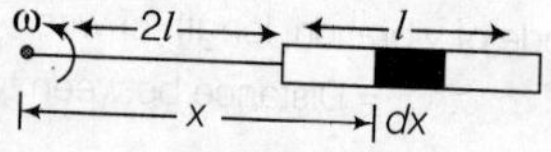

$$e = \int Bv\,dx \qquad (\because v = \omega x)$$

$$\Rightarrow \quad e = \int_{2l}^{3l} (\omega x) \cdot B \cdot dx$$

Using, $\int x^n dx = \dfrac{x^{n+1}}{n+1} \Rightarrow e = B\omega\left[\dfrac{x^2}{2}\right]_{2l}^{3l}$

$$e = B\omega\frac{[(3l)^2 - (2l)^2]}{2} = \frac{5Bl^2\omega}{2}$$

2. (a) The magnetic field due to the bigger circular loop is

$$B = \frac{\mu_0 I R^2}{2\,(x^2 + R^2)^{3/2}}$$

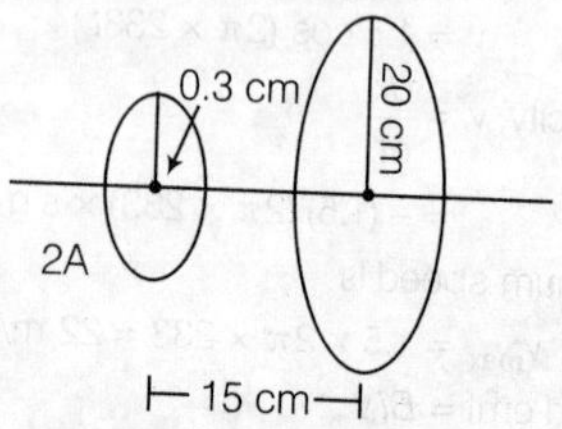

Given, $I = 2$ A, $\quad R = 20 \text{ cm} = 20 \times 10^{-2}$ m,

$$x = 15 \text{ cm} = 15 \times 10^{-2} \text{ m}$$

$$B = \frac{\mu_0 \times 2 \times (20 \times 10^{-2})^2}{[(0.2)^2 + (0.15)^2]^{3/2}}$$

Flux linked, $\phi = BA$

where, A is area of small circular loop.

$$A = \pi r^2 = \pi \times (0.3 \times 10^{-2})^2$$

$$\therefore \quad \phi = \frac{\mu_0 \times 2 \times (20 \times 10^{-2})^2}{2\,[(0.2)^2 + (0.15)^2]^{3/2}} \times \pi \times (0.3 \times 10^{-2})^2$$

$$\phi = 9.2 \times 10^{-11} \text{ Wb}$$

3. (d) According to Lenz's law, electromagnetic induction takes place in the aluminium plate due to which eddy current is developed which oppose the motion or vibrations of coil.

This causes loss in energy which results in damping of oscillatory motion of the coil.

4. (b) Induced emf, $e = B_H\, l v = 0.30 \times 10^{-4} \times 20 \times 5.0 = 3\,\text{mV}$

5. (c) Induced emf, $e = B_H\, l v = 5.0 \times 10^{-5} \times 2 \times 1.50$

$$= 0.15 \times 10^{-3} \text{ V} = 0.15 \text{ mV}$$

6. (d) $M = \dfrac{\mu_0 N_1 \times N_2 \times A}{l}$

where, $N_1 = 300$ turns, $N_2 = 400$ turns

$A = 10 \text{ cm}^2 = 10 \times 10^{-4} \text{m}^2$ and $l = 20 \text{ cm} = 20 \times 10^{-2}$m

Substituting the values in the given formula, we get

$$M = 2.4\,\pi \times 10^{-4} \text{ H}$$

7. (d) Rise of current in L-R circuit is given by

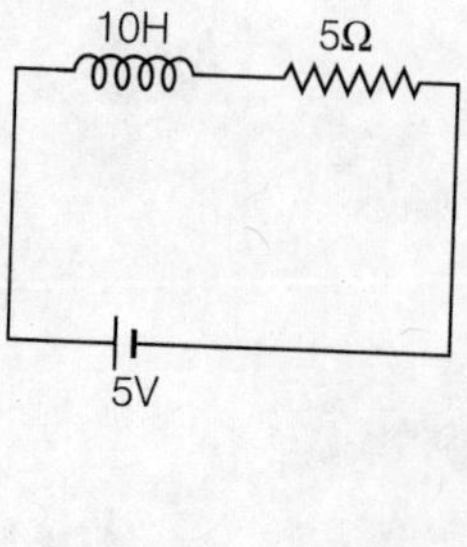

$$I = I_0(1 - e^{-t/\tau})$$

where, $\quad I_0 = \dfrac{E}{R} = \dfrac{5}{5} = 1$ A

Now, $\quad \tau = \dfrac{L}{R} = \dfrac{10}{5} = 2$ s

After 2s, i.e. at $t = 2$s,

Rise of current, $I = (1 - e^{-1})$A

8. (b) $\phi = 10t^2 - 50t + 250$

From Faraday's law of electromagnetic induction,

$$e = -\frac{d\phi}{dt}$$

$$\therefore \quad e = -[10 \times 2t - 50]$$

$$\therefore \quad e|_{t = 3\text{s}} = -[10 \times 6 - 50] = -10 \text{ V}$$

9. (b) The rate of change of flux or emf induced in the coil is $e = -n\dfrac{d\phi}{dt}$.

$\therefore$ Induced current, $I = \dfrac{e}{R'} = -\dfrac{n}{R'}\dfrac{d\phi}{dt}$ ...(i)

Given, $R' = R + 4R = 5R, \quad d\phi = W_2 - W_1, dt = t$

[here, W_1 and W_2 are flux associated with one turn]

Putting the given values in Eq. (i), we get

$$I = -\frac{n}{5R} \cdot \frac{(W_2 - W_1)}{t}$$

10. (b) The flux associated with coil of area A and magnetic induction B is

$$\phi = BA\cos\theta = \frac{1}{2}B\pi r^2 \cos\omega t \qquad \left(\because A = \frac{1}{2}\pi r^2\right)$$

$$\therefore e_{\text{induced}} = -\frac{d\phi}{dt} = -\frac{d}{dt}\left(\frac{1}{2}B\pi r^2 \cos\omega t\right) = \frac{1}{2}B\pi r^2 \omega \sin\omega t$$

$$\therefore \quad \text{Power, } P = \frac{e^2_{\text{induced}}}{R} = \frac{B^2\pi^2 r^4 \omega^2 \sin^2\omega t}{4R} \qquad (\because P = V^2/R)$$

$$\text{Hence, } P_{\text{mean}} = \langle P \rangle = \frac{B^2\pi^2 r^4 \omega^2}{4R} \cdot \frac{1}{2} = \frac{(B^2\pi r^2 \omega)^2}{8R}$$

$$\left(\because \langle \sin^2\omega t \rangle = \frac{1}{2}\right)$$

11. (b) The emf induced between ends of conductor,

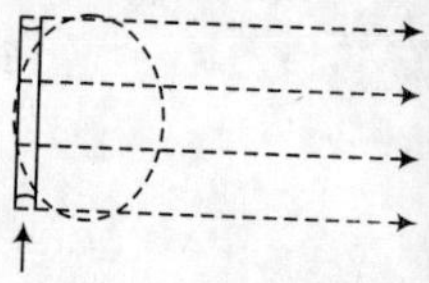

$$e = \frac{1}{2}B\omega L^2 = \frac{1}{2} \times 0.2 \times 10^{-4} \times 5 \times (1)^2$$

$$= 0.5 \times 10^{-4} \text{ V} = 5 \times 10^{-5} \text{ V} = 50\,\mu\text{V}$$

12. (b) Mutual inductance M between two coils is given by $M = \mu_0 n_1 n_2 \pi r_1^2 L$

where, n_1, n_2 are number of turns, r_1 is the radius of coil and L is the length. From the above formula, it is clear that mutual inductance depends on distance between the coils and geometry (πr^2 = area) of two coils.

13. (d) $e = -L\dfrac{dI}{dt} = -L\dfrac{(-2-2)}{0.05} \Rightarrow 8 = L\dfrac{(4)}{0.05}$

$$\therefore \quad L = \frac{8 \times 0.05}{4} = 0.1 \text{ H}$$

14. (d)

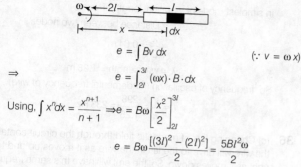

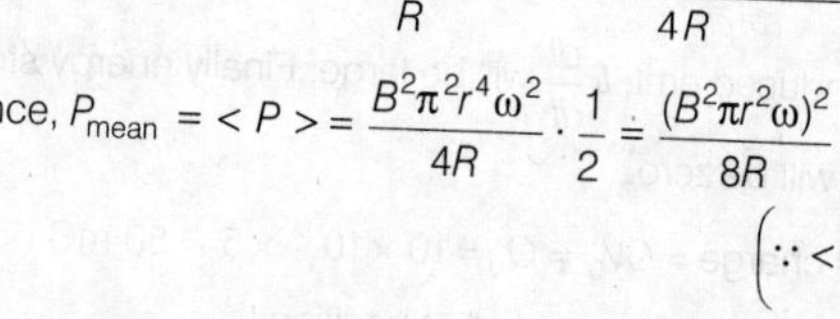

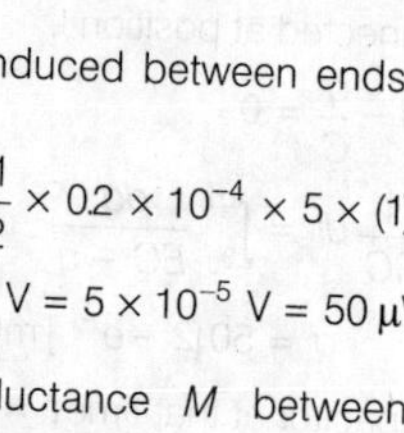

Here, inductors are in parallel.

$$\therefore \frac{1}{L} = \frac{1}{3} + \frac{1}{3} + \frac{1}{3} \quad \text{or} \quad L = 1 \text{ H}$$

JEE Advanced & IIT JEE Archive

1. (b) The induced electric field is given by,

$$\oint \mathbf{E} \cdot d\mathbf{l} = -\frac{d\phi}{dt}$$

or $$El = -s\left(\frac{dB}{dt}\right)$$

$\therefore$ $$E(2\pi R) = -(\pi R^2)(B)$$

or $$E = -\frac{BR}{2}$$

2. (b) $\frac{M}{L} = \frac{Q}{2m}$

$\therefore \quad M = \left(\frac{Q}{2m}\right)L$

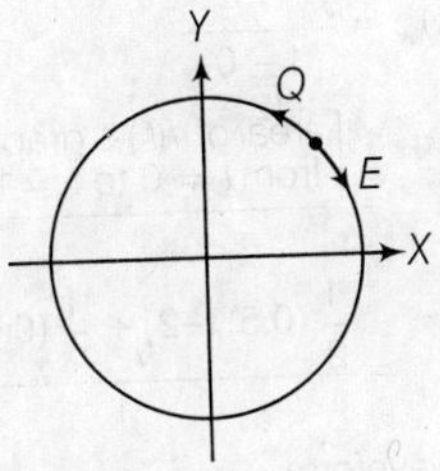

$\Rightarrow$ $$M \propto L,$$

where, $\gamma = \frac{Q}{2m} = \left(\frac{Q}{2m}\right)(I\omega)$

$$= \left(\frac{Q}{2m}\right)(mR^2\omega) = \frac{Q\omega R^2}{2}$$

Induced electric field is opposite. Therefore,

$$\omega' = \omega - \alpha t$$

$$\alpha = \frac{\tau}{I} = \frac{(QE)R}{mR^2} = \frac{(Q)\left(\frac{BR}{2}\right)R}{mR^2} = \frac{QB}{2m}$$

$\therefore$ $$\omega' = \omega - \frac{QB}{2m}.1 = \omega - \frac{QB}{2m}$$

$$M_f = \frac{Q\omega' R^2}{2} = Q\left(\omega - \frac{QB}{2m}\right)\frac{R^2}{2}$$

$\therefore$ $$\Delta M = M_f - M_i = -\frac{Q^2BR^2}{4m}$$

$$M = -\gamma\frac{QBR^2}{2} \qquad \left(\text{as, } \gamma = \frac{Q}{2m}\right)$$

3. (7) If I current flows through the circular loop, then magnetic flux at the location of square loop is

$$B = \frac{\mu_0 IR^2}{2(R^2 + Z^2)^{3/2}}$$

Substituting the value of Z $(= \sqrt{3}R)$,

we have

$$B = \frac{\mu_0 I}{16R}$$

Now, total flux through the square loop is

$$\phi_T = NBS\cos\theta$$

$$= (2)\left(\frac{\mu_0 T}{16R}\right)a^2\cos 45^\circ$$

Mutual inductance,

$$M = \frac{\phi_T}{I} = \frac{\mu_0 a^2}{2^{7/2}R}$$

$\therefore$ $$p = 7$$

4. (c) Due to the current in the straight wire, net magnetic flux from the circular loop is zero. Because in half of the circle magnetic field is inwards and in other half, magnetic field is outwards. Therefore, change in current will not cause any change in magnetic flux from the loop. Therefore, induced emf under all conditions through the circular loop is zero.

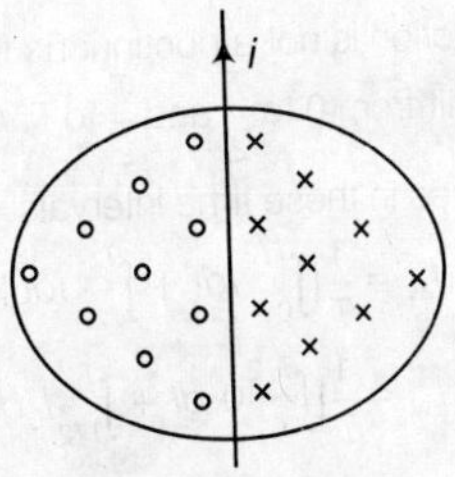

5. (c) $L = 2\pi R$

$\therefore \qquad R = L/2\pi, \qquad 2T\sin(d\theta) = F_m$

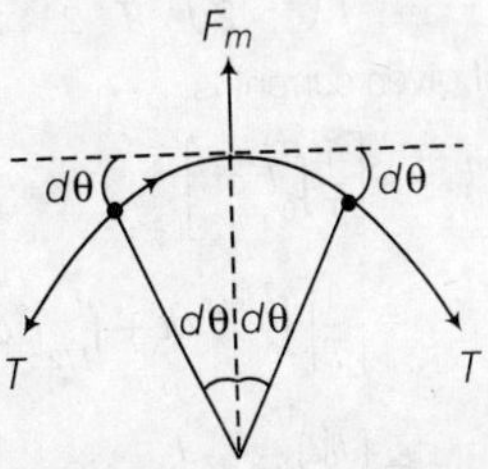

For small angles, $\sin(d\theta) \approx d\theta$

$\therefore \qquad 2T(d\theta) = I(dL)B\sin 90^\circ = I(2R \cdot d\theta) \cdot B$

$\therefore \qquad T = IRB = \frac{ILB}{2\pi}$

6. (c) Cross $\otimes$ magnetic field passing from the closed loop is increasing. Therefore, from Lenz's law induced current will produce dot $\odot$ magnetic field. Hence, induced current is anti-clockwise.

7. (b,d) Induced emf $e = -\frac{d\phi}{dt}$. For identical rings induced emf will be same. But currents will be different. Given $h_A > h_B$. Hence, $v_A > v_B$ as $\left(h = \frac{v^2}{2g}\right)$.

If $\rho_A > \rho_B$, then, $I_A < I_B$. In this case given condition can be fulfilled if $m_A < m_B$.

If $\rho_A < \rho_B$, then $I_A > I_B$. In this case given condition can be fulfilled if $m_A \le m_B$.

8. (a) The induced current in the ring will interact with horizontal component of magnetic field and both will repel each other.

This repulsion will balance the weight of ring.

Hence, option (a) is correct.

9. (a) As, there is no actual contact between train and tracks at high speeds.

$\therefore$ There is no friction resulting in an economical ride.

10. (d) As, the train slows down, the electromagnetic force decreases, it is not possible to keep train levitated also, an electromagnetic drag force is acting on the train.

11. (b) Due to motion of train current is induced in the coils of track which opposes the magnetic field of train, causing it to levitate.

12. (c) In uniform magnetic field, change in magnetic flux is zero. Therefore, induced current will be zero.

4. Alternating Current

DPP-1 Average, Peak and rms Value of AC

1. (i) Average value of a time varying current is given by

$$I_{av} = \frac{1}{T}\int_0^T I\,dt$$

As given function is not a continuous function, we have to break the limit from 0 to $\frac{T}{2}$ and $\frac{T}{2}$ to T. As, function has different values in these time intervals.

$$\therefore \quad I_{av} = \frac{1}{T}[\int_0^{T/2} I\,dt + \int_{T/2}^{T} I\,dt\,]$$

$$\Rightarrow \quad I_{av} = \frac{1}{T}[\int_0^{T/2} 0\cdot dt + \int_{T/2}^{T} I_0\,dt\,]$$

$$= \frac{1}{T}\cdot I_0 \int_{T/2}^{T} 1\cdot dt = \frac{I_0}{T}(t)_{T/2}^{T}$$

$$= \frac{I_0}{T}\left(T - \frac{T}{2}\right) = \frac{I_0}{T}\cdot\frac{T}{2} = \frac{I_0}{2}$$

rms value of given current is

$$I_{rms} = \left[\frac{1}{T}\int_0^T I^2 dt\right]^{1/2}$$

$$= \left[\frac{1}{T}\left[\int_0^{T/2} I^2 dt + \int_{T/2}^{T} I^2 dt\right]\right]^{1/2}$$

$$= \left(\frac{I_0^2}{2}\right)^{1/2} = \frac{I_0}{\sqrt{2}}$$

(ii) $$I_{av} = \frac{1}{\pi}\int_0^{\pi} I_0 \sin\omega t\, d(\omega t)$$

$$= \frac{I_0}{\pi}\int_0^{\pi}\sin\omega t\, d(\omega t) = \frac{I_0}{\pi}[-\cos\omega t\,]_0^{\pi}$$

$$= \frac{I_0}{\pi}\times 2 = \frac{2}{\pi}I_0 = 0.637\, I_0$$

and $$I_{rms} = \sqrt{\frac{1}{\pi}\int_0^{\pi} I_0^2 \sin^2\omega t\, d(\omega t)}$$

$$= \sqrt{\frac{I_0^2}{2\pi}\int_0^{\pi}(1-\cos 2\omega t)\, d(\omega t)}$$

$$= \frac{I_0}{\sqrt{2\pi}}\sqrt{\left[\omega t - \frac{\sin 2\omega t}{2}\right]_0^{\pi}} = \frac{I_0}{\sqrt{2}} = 0.707\, I_0$$

(iii) In this case, current function again changes its character, so we have to break the integration limits.

$$I_{av} = \frac{1}{2}\int_0^2 i\,dt = \frac{1}{2}\left[\int_0^1 i_1 dt + \int_1^2 i_2 dt\right]$$

$$= \frac{1}{2}\left[\int_0^1 500t\,dt + \int_1^2 500\,dt\right]$$

$$= \frac{1}{2}\times 500\times\left\{\left(\frac{t^2}{2}\right)_0^1 + (t)_1^2\right\}$$

$$= \frac{1}{2}\left(500\times\frac{1}{2} + 500\right) = \frac{750}{2} = 375\,\text{A}$$

Also, $$I_{rms} = \sqrt{\frac{1}{2}\left\{\int_0^1 i_1^2 dt + \int_1^2 i_2^2 dt\right\}}$$

$$= \sqrt{\frac{1}{2}\left(\int_0^1 (500t)^2 dt + \int_1^2 (500)^2 dt\right)}$$

$$= 500\sqrt{\frac{1}{2}\left[\left(\frac{t^3}{3}\right)_0^1 + (t)_1^2\right]} = 500\sqrt{\frac{1}{2}\left(\frac{1}{3}+1\right)}$$

$$= 500\times\frac{\sqrt{2}}{\sqrt{3}} = \frac{500\sqrt{2}}{\sqrt{3}}\,\text{A}$$

2.

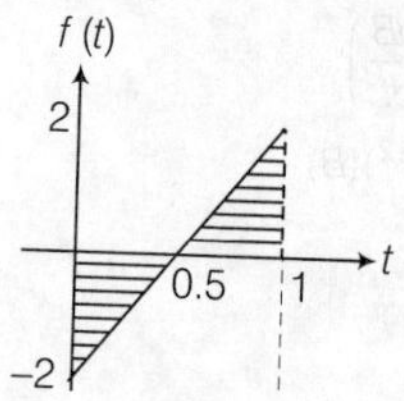

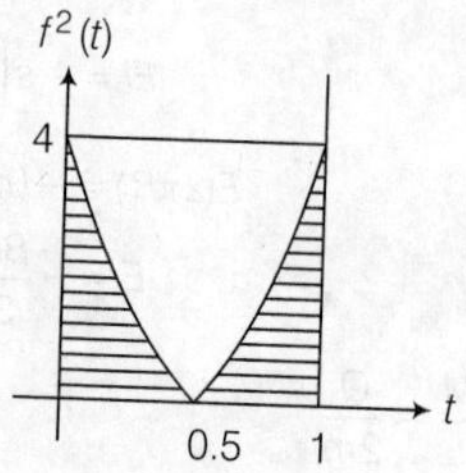

As we know that average value of waveform

i.e. $$f(t)_{av} = \frac{\int_0^1 f(t)\,dt}{1-0}$$

$$= \frac{\left[\begin{array}{l}\text{area of } f(t)\text{-}t \text{ graph}\\ \text{from } t = 0 \text{ to } t = 1\,\text{s}\end{array}\right]}{1}$$

$$= \frac{\frac{1}{2}(0.5)(-2) + \frac{1}{2}(0.5)\times 2}{1} = 0$$

So, rms value of waveform

i.e. $$f(t)_{rms} = \sqrt{\frac{\int_0^1 f^2(t)\,dt}{1-0}}$$

$$= \sqrt{\frac{\text{area under } f^2(t)\text{-}t \text{ graph from } t = 0 \text{ to } t = 1\,\text{s}}{1}}$$

$$= \sqrt{\frac{\frac{1}{3}\times 0.5\times 4 + \frac{1}{3}\times 0.5\times 4}{1}} = \sqrt{\frac{4}{3}}$$

3. (i) Given, $E_{rms} = 220$ V, $R = 40\,\Omega$

$\therefore$ The rms value of a current,

i.e. $$I_{rms} = \frac{E_{rms}}{R} = \frac{220}{40} = 5.5\,\text{A}$$

(ii) Maximum instantaneous current,

$$I_0 = \sqrt{2}\times I_{rms}$$
$$= 1.414\times 5.5\,\text{A}$$
$$= 7.8\,\text{A}$$

(iii) Suppose the alternating current is given by

$$I = I_0 \sin\omega t$$

Let the AC take its maximum and rms values at instants t_1 and t_2, respectively. Then,

$$I = I_0 \sin\omega t_1$$

$$\Rightarrow \quad I_0 = I_0 \sin\omega t_1$$

which implies $$\omega t_1 = \frac{\pi}{2}$$

and $$I_{rms} = \frac{I_0}{\sqrt{2}} = I_0 \sin\omega t_2$$

which implies $$\omega t_2 = \frac{3\pi}{4} = \frac{\pi}{2} + \frac{\pi}{4}$$

Time taken by the current to change from its maximum value to the rms value, i.e.

$$t_2 - t_1 = \frac{\pi}{4\omega} = \frac{\pi}{4\times 2\pi f}$$

$$= \frac{\pi}{4\times 2\pi\times 50}$$

$$= \frac{1}{400}\,\text{s} = 2.5\,\text{ms}$$

4. (i) Ordinary moving coil galvanometer is based on magnetic effect of current which in turn depends on the direction of current. So, it cannot be used to measure AC. During one half cycle of AC, its pointer moves in one direction and during next half cycle, it will move in the opposite direction.

Now, the average value of AC over a complete cycle is zero. Even if we measure an alternating current of low frequency, the pointer will appear to be stationary at the zero position due to persistence of vision.

So, we can measure AC by using a **hot-wire ammeter** which is based on heating effect of current and this effect is independent of the direction of current.

(ii) The mean square current for the rms current for the period $t = 0$ to $t = \tau$ is given by

$$i^2 = \frac{1}{\tau}\int_0^\tau i_0^2\left(\frac{t}{\tau}\right)^2 dt$$

$$= \frac{i_0^2}{\tau^3}\int_0^\tau t^{-2}\, dt = \frac{i_0^2}{\tau^3}\left[\frac{t^3}{3}\right]_0^\tau$$

$$= \frac{i_0^2}{\tau^3} \times \frac{\tau^3}{3} = \frac{i_0^2}{3}$$

So, the rms current,

i.e. $$i_{rms} = \sqrt{i^2} = \sqrt{\frac{i_0^2}{3}} = \frac{i_0}{\sqrt{3}}$$

5. Resultant current due to mixing of both current is

$$I = I_1 + I_2$$

$\Rightarrow$ $$I = 10 + 10\sin\omega t$$

So, $$I_{rms} = \sqrt{\frac{1}{2\pi}\int_0^{2\pi} I^2 d\theta}$$

$\Rightarrow$ $$I_{rms}^2 = \frac{1}{2\pi}\int_0^{2\pi} I^2 d\theta \qquad (\because \theta = \omega t)$$

$$= \frac{1}{2\pi}\int_0^{2\pi}(10 + 10\sin\omega t)^2 d\theta$$

$$= \frac{100}{2\pi}\int_0^{2\pi}(1 + \sin^2\theta + 2\sin\theta)\, d\theta$$

$$= \frac{100}{2\pi}[(2\pi - 0) + \pi + (0 - 0)] = 150$$

$$I_{rms} = \sqrt{150} = 12.247 \text{ A}$$

As, peak value of current,

$$I_{max} = \sqrt{2}\, I_{rms}$$

and $$I_{av} = \frac{2}{\pi} I_{max}$$

$\therefore$ $$I_{av} = \frac{2}{\pi} \times \sqrt{2} \times \sqrt{150} = 11.06 \text{ A}$$

6. (b) From i-t graph, area from $t = 0$ to $t = 2$ s, we get

$$= \frac{1}{2} \times 2 \times 10 = 10 \text{ A}$$

$\therefore$ $$\text{Average current } (i_{av}) = \frac{(\text{Area } i\text{-}t \text{ graph})}{\text{time interval}}$$

$$= \frac{10}{2} = 5 \text{ A}$$

7. (d) Equation of a current can be written as

$i = 2\sin 100\pi t + 2\sin(100\pi t + 120°)$

So, phase difference is 120°

$$I_{peak} = \sqrt{2^2 + 2^2 + 2(2)(2)\cos 120°}$$

$$I_{peak} = 2 \text{ A}$$

Effective value of current,

$$I_{rms} = \frac{I_{peak}}{\sqrt{2}}$$

$$= \frac{2}{\sqrt{2}} = \sqrt{2} \text{ A}$$

8. (a) The alternating current at any instant t is given by

$$I = I_0 \sin\omega t$$

Assuming that current remains constant for a small time dt, then the amount of charge that flows through the circuit in small time dt will be

$$dq = I dt = I_0 \sin\omega t\, dt$$

The total charge that flows through the circuit in small time dt will be

$$q = \int dq = \int_0^{2\pi/\omega} I\, dt$$

$$= \int_0^t I_0 \sin\omega t\, dt$$

$$= I_0\left[\frac{-\cos\omega t}{\omega}\right]_0^{2\pi/\omega}$$

$$= \frac{-I_0}{2\pi/T}[\cos 2\pi - \cos 0°]$$

$$= \frac{-I_0 T}{2\pi}[1 - 1] = 0$$

The average value of AC over one complete cycle of AC, we get

$$I_{av} = \frac{q}{T} = 0$$

9. (b)

I. The average of sinusoidal AC values any whole number of cycles is zero.

II. The rms value of current, $i_{rms} = 2.5$ A

$\therefore$ $$(i_{av}^2) = (i_{rms})^2$$

$\Rightarrow$ $$i_{av}^2 = (2.5)^2$$

$\Rightarrow$ $$i_{av} = 2.5 \text{ A}$$

III. As, rms value of current,

$$i_{rms} = \frac{i_0}{\sqrt{2}}$$

$\therefore$ Current amplitude, $i_0 = \sqrt{2} \times i_0 = \sqrt{2} \times (2.5) = 3.5$ A

IV. As we know, $V_{rms} = 220 \text{ V} = \dfrac{V_0}{\sqrt{2}}$

$\therefore$ Supply voltage amplitude,

$$V_m = \sqrt{2M}\,(V_{rms}) = \sqrt{2} \times 220 \text{ V} = 311 \text{ V}$$

10. (d) Given, $e = e_1 \sin\omega t + e_2 \cos\omega t$

rms value of alternating voltage, i.e.

$$e_{1\,rms} = \frac{e_1}{\sqrt{2}} \text{ and similarly } e_{2\,rms} = \frac{e_2}{\sqrt{2}}$$

using phasor diagram, we get

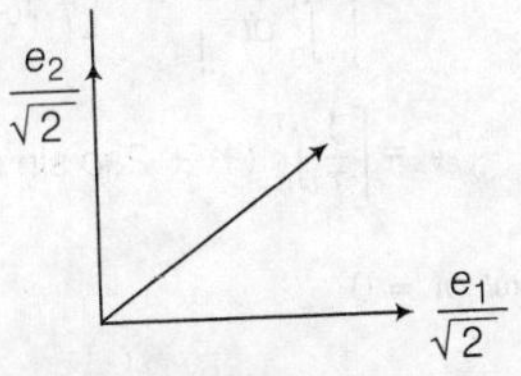

Net resultant root mean square of voltage is given by

$$e_{rms} = \sqrt{\frac{e_1^2 + e_2^2}{2}}$$

11. (b) The time period of sinusoidal voltage, $T = 0.4$ s

So, frequency $f = \dfrac{1}{T} = \dfrac{1}{0.4} = 2.5$ Hz

So, N lags M by 0.1 s which is equivalent to

$$\left(\frac{0.1}{0.4}\right) \times 2\pi \text{ or } \frac{\pi}{2} \text{ rad}$$

Thus, the lead of N over M is $-\dfrac{\pi}{2}$ rad.

12. (b) As, peak value of voltage, V_0

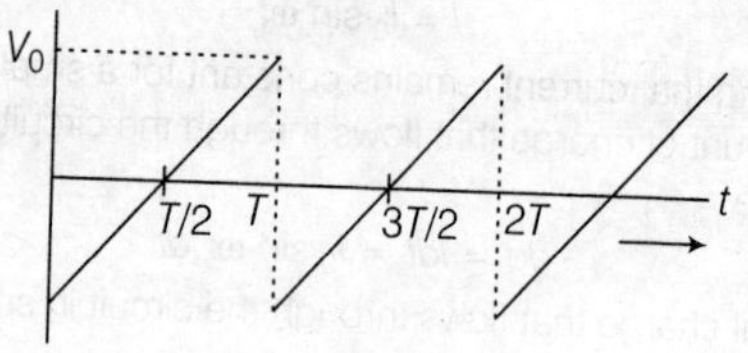

$$V(t) = \frac{2V_0}{T}t$$

rms value of the saw-tooth voltage of peak value V_0

$$V_{rms} = \sqrt{<V^2>} = \sqrt{\frac{\int_{-T/2}^{+T/2} \frac{4V_0^2}{T^2} t^2 dt}{\int_{-T/2}^{T/2} dt}} = \sqrt{\frac{V_0^2}{3}} = \frac{V_0}{\sqrt{3}}$$

13. (c) As, $E_1 = E_0 \sin \omega t$

$$E_2 = E_0 \sin [\omega t + (\pi/3)]$$

According to superposition principle, we get

$$E = E_1 + E_2 = E_0 \sin \omega t + E_0 \sin\left[\omega t + \left(\frac{\pi}{3}\right)\right]$$

$$= E_0\left[2 \sin\left\{\omega t + \left(\frac{\pi}{6}\right)\right\} \cos\left(\frac{\pi}{6}\right)\right]$$

$$= \sqrt{3} E_0 \sin\left[\omega t + \left(\frac{\pi}{6}\right)\right]$$

14. (a) Given, $I = I_0 + I_1 \sin \omega t$

So, root mean square value of current,

i.e. $$i_{rms} = \sqrt{<i^2>} = \sqrt{\frac{\int_0^T (I_0^2 + I_1^2 \sin^2 \omega t + 2I_0 I_1 \sin \omega t)\, dt}{\int_0^T dt}}$$

$$= \sqrt{\frac{\int_0^{T_0}\left(I_0^2 + \frac{I_1^2}{2}(1 - \cos 2\omega t) + 2I_0 I_1 \sin \omega t\right) dt}{\int_0^T dt}}$$

$$= \sqrt{\frac{(I_0^2 + 0.5I_1^2)\,T}{T}} = \sqrt{I_0^2 + 0.5 I_1^2}$$

15. (d) As, current at any instant in the circuit will be

$$I = I_{DC} + I_{AC} = a + b \sin \omega t$$

So, effective current,

$$I_{eff} = \left[\frac{\int_0^T I^2 dt}{\int_0^T dt}\right]^{1/2} = \left[\frac{1}{T}\int_0^T (a + b \sin \omega t)^2\, dt\right]^{1/2}$$

i.e. $$I_{eff} = \left[\frac{1}{T}\int_0^T (a^2 + 2ab \sin \omega t + b^2 \sin^2 \omega t)\, dt\right]^{1/2}$$

But $\frac{1}{T}\int_0^T \sin \omega t\, dt = 0$

and $\frac{I}{T}\int_0^T \sin^2 \omega t\, dt = \frac{1}{2}$

So, $$I_{eff} = \left[a^2 + \frac{b^2}{2}\right]^{1/2}$$

16. (d) Equation of line AB is given by

$$y = 10 + \left(\frac{10}{T}\right)t$$

$$y_{av} = \int_0^T \frac{1}{T} y\, dt = \frac{1}{T}\int_0^T \left(10 + \frac{10}{T}t\right) dt$$

$$= \frac{1}{T}\left[10t + \frac{5t^2}{T}\right]_0^T = 15$$

17. (a) Mean square value of the signal

i.e. $$\frac{1}{T}\int_0^T y^2\, dt = \int_0^T \left[10 + \frac{10}{T}t\right]^2 dt$$

$$= \frac{1}{T}\int_0^T \left(100 + \frac{100}{T^2}t^2 + \frac{200}{T}t\right) dt$$

$$= \frac{1}{T}\left[100t + \frac{100t^3}{3T^2} + \frac{100t^2}{T}\right]_0^T = \frac{700}{3}$$

So, rms value $= 10\sqrt{\frac{7}{3}}$

18 (d) Output waveform is

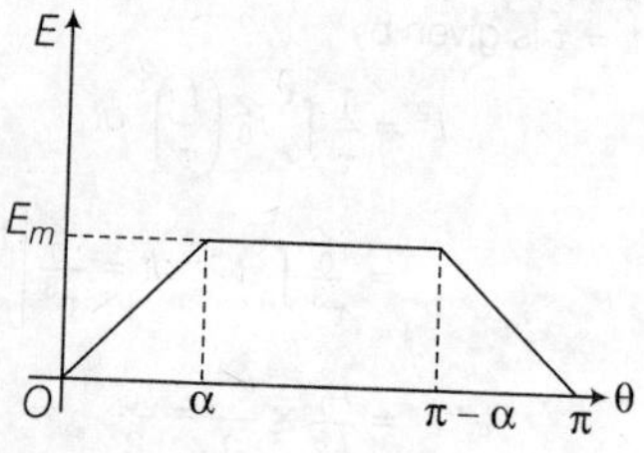

Area under current $= \frac{1}{2}(\pi + \pi - 2\alpha)E_m$

So, average value $= \frac{2(\pi - \alpha)}{2\pi}E_m$

$$E_{av} = \left(\frac{\pi - \alpha}{\pi}\right)E_m$$

When $\alpha = \frac{\pi}{6}$,

$$E_{av} = \left(\frac{\pi - \pi/6}{\pi}\right)E_m = \frac{5}{6}E_m$$

When $\alpha = \frac{\pi}{2}$,

$$E_{av} = \left(\frac{\pi - \frac{\pi}{2}}{\pi}\right)E_m = \frac{E_m}{2}$$

$\therefore$ $$\text{Ratio} = \frac{\frac{5}{6}E_m}{\frac{E_m}{2}} = \frac{5}{6} \times \frac{2}{1} = \frac{5}{3}$$

19. (c) $$E_{rms} = \sqrt{\left\{\frac{1}{\pi}\left[2\int_0^\alpha \frac{E_m^2}{\alpha^2}\theta^2 d\theta + \int_\alpha^{\pi-\alpha} E_m^2 d\theta\right]\right\}}$$

$$= \left[\frac{E_m^2}{\pi}\left(2\left(\frac{\theta^3}{3\alpha^2}\right)_0^\alpha + (\theta)_\alpha^{\pi-\alpha}\right)\right]^{1/2}$$

$$= \sqrt{\frac{E_m^2}{\pi}\left(\pi - \frac{4}{3}\alpha\right)} = E_m\sqrt{\left(\frac{\pi - \frac{4}{3}\alpha}{\pi}\right)}$$

For $\alpha = 0, E_{rms} = E_m$

20. (c) For $\alpha = \frac{\pi}{6}$,

$$E_{rms} = E_m\sqrt{\left(\frac{\pi - \frac{4}{3}\cdot\frac{\pi}{6}}{\pi}\right)} = \sqrt{\frac{7}{9}}E_m$$

For $\alpha = \frac{\pi}{2}$,

$$E_{rms} = \frac{E_m}{\sqrt{3}}$$

$\therefore$ $$\text{Ratio} = \sqrt{\left(\frac{7}{9} \times \frac{3}{1}\right)} = \sqrt{\frac{7}{3}}$$

21. (a,b,c) Various cases where average value of AC current in a half-time period, i.e. It may be positive, negative or zero depending on initial value of time from which average is taken.

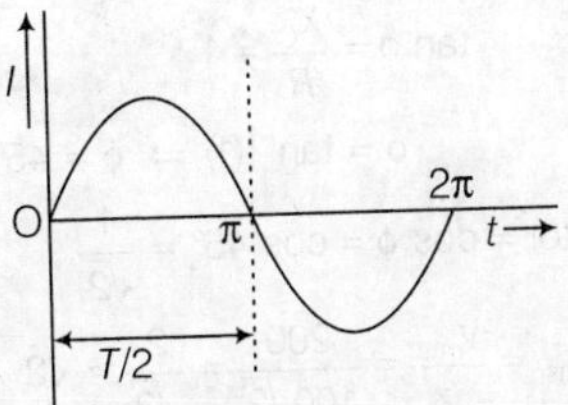

Positive average value

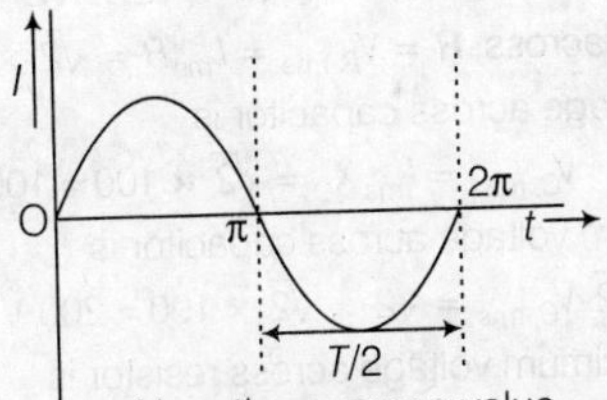

Negative average value

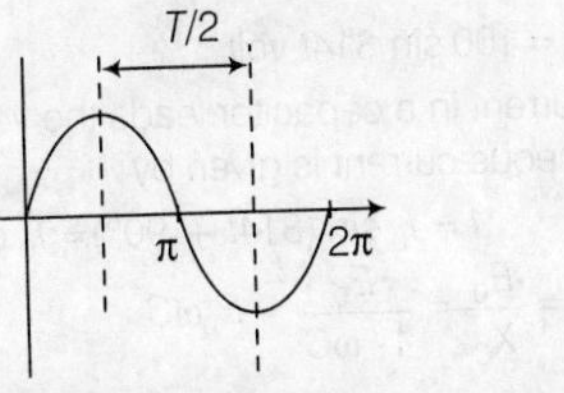

Zero average value

22. (a,b,c) As, heat produced, $H = \frac{I^2RT}{4.2}$

Thus, heat produced in a conductor is independent of the nature of current. If the current flowing in a conductor changes its direction, then also heat being produced in it will not change.

23. (c,d) $V = 100\sin 100\pi t \cos 100\pi t$

$$= 50 \times (2\sin 100\pi t \cos 100\pi t)$$
$$= 50\sin 200\pi t$$

∴ Source peak voltage is 50 V.

Also, $f = \frac{\omega}{2\pi}$

$$= \frac{200\pi}{2\pi} = 100 \text{ Hz}$$

24. (a,b,c,d) Slope of line representing voltage line, $\tan\theta = \frac{V_0}{T/2}$

Equation of line will be

$$V = \frac{V_0}{T/2}t - 1$$

$$\Rightarrow \quad V = V_0\left(\frac{2t}{T} - 1\right)$$

Also,
$$V_{av} = \frac{\int_0^{T/2} V\,dt}{\int_0^{T/2} dt}$$

$$= \frac{2}{T}V_0\left(\frac{T}{4} - \frac{T}{2}\right) = \frac{V_0}{2}$$

$$V_{rms} = \left(\frac{\int_0^T V^2\,dt}{\int_0^T dt}\right)^{1/2}$$

$$= \frac{V_0}{\sqrt{T}}\left[\int_0^T\left(\frac{4t^2}{T^2} - \frac{4t}{T} + 1\right)dt\right]^{1/2}$$

$$= \frac{V_0}{\sqrt{3}}$$

Also, $V_0 > V_{rms} > V_{av}$

25. (1)

(i) As, $i = 4\sin\omega t$

Case I rms value of current from $t = 0$ to $t = \frac{\pi}{\omega}$, we get

$$i_{rms} = \sqrt{\frac{\int_0^{\frac{\pi}{\omega}} I_m^2 \sin^2 \omega t\,dt}{\frac{\pi}{\omega}}} = \frac{I_m}{\sqrt{2}} \quad \text{...(i)}$$

(ii) ***Case II*** rms value of current from $t = \frac{\pi}{2\omega}$ to $\frac{3\pi}{2\omega}$, we get

$$i'_{rms} = \sqrt{\frac{\int_{\frac{\pi}{2\omega}}^{\frac{3\pi}{2\omega}} I_m^2 \sin^2 \omega t\,dt}{\frac{\pi}{\omega}}} = \frac{I_m}{\sqrt{2}} \quad \text{...(ii)}$$

Dividing Eq. (i) by Eq. (ii), we get

$$\frac{i_{rms}}{i'_{rms}} = \frac{1}{1}$$

Note *The rms values for one cycle and half-cycle (either positive or negative half-cycle) is same.*

26. (8) At $t = 5$ ms,

$$V_{AC} = \frac{1}{C}\int_0^{5k\times10^{-3}} i\,dt$$

$$= 10^6 \times 10^{-3}\left[\int_0^{3\times10^{-3}} 4dt - \int_{3\times10^{-3}}^{5\times10^{-3}} 2\,dt\right]$$

$$= 12 - 4 = 8 \text{ V}$$

Due to this charging effect, at $t = 5k$ ms, $V = 8k$ V.

With a constant current I_{DC}, capacitor voltage at $5k$ms is

$$V_{DC} = \frac{1}{C}\int_0^{5k\times10^{-3}} I_{DC}\,dt$$

$$= 10^6(I_{DC})(5k \times 10^{-3}) = 10^3(5k)I_{DC}$$

As, $V_{DC} = V_{AC}$ at $5k$ms

∴ $10^3\, 5k\, I_{DC} = 8k$

⇒ $10^3(5I_{DC}) = 8$

27. (3) $V_{rms} = 220$ V, $f = 50$ Hz

$$V_{peak} = V_m = \sqrt{2}, V_{rms} = 220\sqrt{2} \text{ V}$$

$$\omega = 2\pi f = 100\pi \text{ rad s}^{-1}$$

As, $V = V_m \sin\omega t$

⇒ $V = 220\sqrt{2}\sin(100\pi t)$

When $V = V_{rms}$, $220 = 220\sqrt{2}\sin(100\pi t_1)$

⇒ $100\pi t_1 = \frac{\pi}{4}$

⇒ $t_1 = \frac{\pi}{4 \times 100\pi} = \frac{1}{400}$

⇒ $t_1 = 2.5$ms

When $I = 0$,

$0 = 220\sqrt{2}\sin(100\pi t_2)$

⇒ $\sin(100\pi t_2) = 0$

⇒ $100\pi t_2 = \pi$

⇒ $t_2 = \frac{1}{100} = 10 \times 10^{-3}$

$t_2 = 10$ ms

∴ Time required $= 10 - 2.5 = 7.5$ ms $= 3 \times 2.5$ ms

28. (a) For triangular wave form (ii), $I_{rms} = \frac{I_0}{\sqrt{3}}$

Full cycle average current is zero for (i), (ii) and (iii).

For (iii), average current for positive half-cycle is i_0.

For (ii), average current for positive half-cycle is $\frac{i_0}{2}$.

Hence, (i) → (s), (ii) → (p,r,s), (iii) → (s), (iv) → (q)

DPP-2 Power Consumption in an AC Circuit

1. For given circuit,

$$L = 50\times 10^{-3}\text{ H}, R = 40\ \Omega$$

$$C = 50\times 10^{-6}\text{ F}, X_L = \omega L, X_C = \frac{1}{\omega C}$$

Impedance of circuit,

$$Z = \sqrt{R^3 + (X_L - X_C)^2} = \sqrt{40^2 + (50-20)^2}$$

$$= 50\ \Omega$$

(i) $I_{rms} = \dfrac{V_{rms}}{t} = \dfrac{100}{50} = 2\text{ A}$

(ii) Power supplied by the source $= V_{rms} I_{rms}\cos\phi$

$$= 100\times 2\times \frac{R}{Z}$$

$$= 100\times 2\times \frac{40}{50} = 160\text{ W}$$

(iii) Power used up by the resistor $= I_{rms}^2 R$

$$= (2)^2\times 40 = 160\text{ W}$$

2. (i) Instantaneous power delivered by source,

$$P = VI$$

$$= 10\sin\left(250\pi t + \frac{\pi}{3}\right)2\sin(250\pi t)$$

$$= 10\times 2\sin\left(250\pi t + \frac{\pi}{3}\right)\sin(250\pi t)$$

Using, $2\sin A\sin B = \cos(A-B) - \cos(A+B)$

$$\Rightarrow \quad P = 10\left(\cos\frac{\pi}{3} - \cos\left(500\pi t + \frac{\pi}{3}\right)\right)$$

At $t = \dfrac{2}{3}$ ms,

$$P = 10\left(\cos\frac{\pi}{3} - \cos\left(500\pi\times\frac{2}{3}\times 10^{-3} + \frac{\pi}{3}\right)\right)$$

$$= 10\left(\cos\frac{\pi}{3} - \cos\frac{2\pi}{3}\right) = 10\left(\cos\frac{\pi}{3} - \left(-\cos\frac{\pi}{3}\right)\right)$$

$$= 10\times 2\times\cos\frac{\pi}{3} = 20\times\frac{1}{2} = 10\text{W}$$

(ii) Average power consumed by circuit is

$$P_{av} = \frac{1}{T}\int_0^T VI\,dt = V_{rms} I_{rms}\cos\phi$$

$$\therefore \quad P_{av} = \frac{V_{max}}{\sqrt{2}}\cdot\frac{I_{max}}{\sqrt{2}}\cos\phi$$

$$= \frac{10}{\sqrt{2}}\cdot\frac{2}{\sqrt{2}}\cos\frac{\pi}{3}$$

$$= \frac{10}{\sqrt{2}}\cdot\frac{2}{\sqrt{2}}\cdot\frac{1}{2} = 5\text{ W}$$

3. As we know,

In R-C series circuit,

Capacitance, $C = \dfrac{100}{\pi}\ \mu\text{F}$

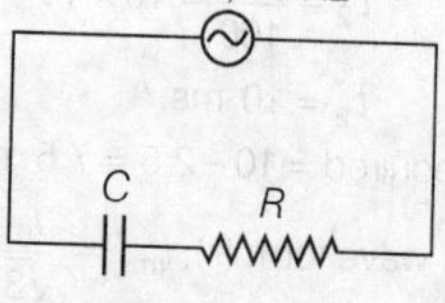

Net capacitive reactance,

$$X_C = \frac{1}{\omega C} = \frac{1}{2\pi f C} = \frac{1\times\pi\times 10^6}{2\pi\ (50)\times 100} = 100\ \Omega$$

(i) As, impedance across R-C series circuit

i.e. $Z = \sqrt{R^2 + X_C^2} = \sqrt{(100)^2 + (100)^2} = 100\sqrt{2}\ \Omega$

(ii) Power factor angle,

$$\tan\phi = \frac{X_C}{R} = 1$$

$$\therefore \quad \phi = \tan^{-1}(1) \Rightarrow \phi = 45°$$

(iii) Power factor $= \cos\phi = \cos 45° = \dfrac{1}{\sqrt{2}}$

(iv) Current, $I_{rms} = \dfrac{V_{rms}}{Z} = \dfrac{200}{100\sqrt{2}} = \dfrac{2}{\sqrt{2}} = \sqrt{2}$ A

(v) Maximum current, $I = I_{rms}\sqrt{2} = \sqrt{2}\times\sqrt{2} = 2$ A

(vi) Voltage across, $R = V_{R,\,rms} = I_{rms}R = \sqrt{2}\times 100 = 100\sqrt{2}$ V

and voltage across capacitor is

$$V_{C,\,rms} = I_{rms}X_C = \sqrt{2}\times 100 = 100\sqrt{2}\text{ V}$$

(vii) Maximum voltage across capacitor is

$$\sqrt{2}\ V_{C,\,rms} = \sqrt{2}\times\sqrt{2}\times 100 = 200\text{ V}$$

and maximum voltage across resistor is

$$\sqrt{2}\ V_{R,\,rms} = \sqrt{2}\times\sqrt{2}\times 100 = 200\text{ V}$$

4. (i) Given, $E = 100\sin 314t$ volt

As the current in a capacitor leads the voltage by 90°, so the instantaneous current is given by

$$I = I_0\sin(314t + 90°) = I_0\cos 314t$$

where, $I = \dfrac{E_0}{X_C} = \dfrac{E_0}{1/\omega C} = E_0\omega C$

But, $E_0 = 100$ V, $\omega = 314$ rad s^{-1}

$$C = 637\times 10^{-6}\text{ F}$$

$$I_0 = 100\times 314\times 637\times 10^{-6} = 20\text{ A}$$

Hence, instantaneous current,

$$I = 20\cos 314t\text{ A}$$

(ii) Instantaneous power,

$$P = EI = 100\sin 314t\times 20\cos 314t$$

$$= 1000\sin 628t\text{ watt}$$

(iii) Angular frequency of power,

$$\omega_P = 628\text{ rad s}^{-1}$$

$\therefore$ Frequency of power, $f_P = \dfrac{\omega_P}{2\pi} = \dfrac{628}{2\pi} = 100$ Hz

(iv) The maximum energy stored in the capacitor is

$$U_0 = \frac{1}{2}CE^2 = \frac{1}{2}\times 637\times 10^{-6}\times(100)^2 = 3.185\text{ J}$$

5. (a) Given, $V = 5\cos\omega t$

$$I = 2\sin\omega t$$

As, $V = 5\cos\omega t = 5\sin\left(\omega t + \dfrac{\pi}{2}\right)$ ...(i)

As we know,

$V = V_0\sin(\omega t + \phi)$ ...(ii)

Comparing Eqs. (i) and (ii), we get

$$\phi = \frac{\pi}{2}$$

So, power dissipated in the instrument

i.e. $P = I_{rms}V_{rms}\cos\phi$

$$= I_{rms}V_{rms}\cos\frac{\pi}{2} \qquad \left(\because \cos\frac{\pi}{2} = 0\right)$$

i.e. P = zero

6. (b) Current is maximum when, energy stored in the capacitor = energy stored in the inductor

$$\Rightarrow \quad \frac{1}{2}CV^2 = \frac{1}{2}LI^2$$

$$\therefore \quad L = \frac{CV^2}{I^2} = \frac{100\times 10^{-6}\times(50)^2}{5^2} = 0.01\text{ H}$$

7. (c) ***Case I*** In DC, heat produced in a resistance, i.e.

$$H_1 = I_0^2 Rt \quad \text{...(i)}$$

Case II In AC, heat produced, i.e.

$$H_2 = \frac{V_0 I_0}{2} \frac{R}{Z} t = \frac{(I_0 Z) I_0 Rt}{2Z} = \frac{I_0^2 Rt}{2} \quad \text{...(ii)}$$

Dividing Eq. (ii) by Eq. (i), we get

$$\frac{H_2}{H_1} = \frac{I_0^2 Rt}{2} \times \frac{1}{I_0^2 Rt}$$

$$\frac{H_2}{H_1} = \frac{1}{2}$$

$$H_1 : H_2 = 2 : 1$$

8. (a) As average power of an AC circuit, i.e.

$$P = V_{rms}\, I_{rms} \cos \phi = V_{rms}\, I_{rms} \times \frac{R}{Z}$$

$$\left(\text{Power factor, i.e. } \cos \phi = \frac{R}{Z}\right)$$

In *C-R* series circuit with an AC source,

Net reactive impedance, i.e.

$$Z = \sqrt{R^2 + \left(\frac{1}{\omega C}\right)^2}$$

So, increasing the value of ω, then impedance will decreases. Hence, power consumption across *C-R* series circuit will increases.

9. (c) Given, $V = 2 \sin (100t)$, $L = 0.1\,\text{H}, R = 10\,\Omega$

In *L-C-R* circuit, power factor, i.e.

$$\cos \phi = \frac{R}{Z}$$

$$\frac{1}{\sqrt{2}} = \frac{10}{\sqrt{10^2 + \left(100 \times 0.1 - \frac{1}{100C}\right)^2}}$$

$$\Rightarrow \sqrt{(10)^2 + \left(100 \times 0.1 - \frac{1}{100C}\right)^2} = 10\sqrt{2}$$

$$100 + \left(10 - \frac{1}{100C}\right)^2 = 100 \times 2$$

$$\Rightarrow \quad C = 500\,\mu\text{F}$$

10. (a) As, rms value of current,

$$I_{rms} = V_{rms}$$

So, net impedance across *L-C-R* circuit,

$$Z = \sqrt{R^2 + (X_L - X_C)^2} = \sqrt{(100)^2 + [\omega L - X_C]^2}$$

$$= \sqrt{(100)^2 + \left[\left(100 \times \pi \times \frac{1}{\pi}\right) - (X_C)\right]}$$

$$\left(\frac{200}{2.2}\right)^2 = (100)^2 + ((100) - (X_C))^2$$

$$\Rightarrow \quad X_C = 100\,\Omega$$

As, $\tan \phi = \dfrac{X_C}{R} = \dfrac{200}{200} = 1 \quad \Rightarrow \quad \phi = \tan^{-1}(1)$

$$\phi = 45°$$

Power factor, $\cos \phi = \dfrac{1}{\sqrt{2}}$

11. (b) In AC circuit, $I = I_0 \sin\left(\omega t - \dfrac{\pi}{2}\right)$

As, $\phi = \dfrac{\pi}{2}$

So, power consumption in the circuit

i.e. $P_{av} = E_{rms}\, I_{rms} \cos \phi = E_{rms}\, I_{rms} \cos \dfrac{\pi}{2} = 0$

12. (b) Wattless component of AC

$$= I_V \sin \phi = \frac{E_V}{Z} \cdot \frac{X_L}{Z} = \frac{E_V X_L}{Z^2} = \frac{200 \times \omega L}{(R^2 + \omega^2 L^2)}$$

As, $\omega L = 0.7 \times 2\pi \times 50 = 220\,\Omega$

Hence, wattless component of AC circuit

$$= \frac{200 \times 220}{(220)^2 + (220)^2} = 0.5\text{ A}$$

13. (b) As we know that root mean square current of the sinusoidal wave form, $I = \dfrac{I_0}{\sqrt{2}}$.

So, power output of the heater,

i.e. $P = (I_{rms})^2 R = \left(\dfrac{I_0}{\sqrt{2}}\right)^2 \times R = \dfrac{I_0^2 R}{2}$

14. (a) Since, current leads emf, therefore this is an *R-C* circuit.

$$\therefore \quad \tan \phi = \frac{X_C - X_L}{R} \quad (\because \phi = 45°)$$

$$\Rightarrow \quad X_C = R \quad (X_L = 0 \text{ as there is no inductor})$$

$$\frac{1}{\omega C} = R$$

$$\text{LHS} = \frac{1}{100 \times 10 \times 10^{-6}}$$

$$\Rightarrow \quad R = 10^3\,\Omega$$

15. (a) In *L-C-R* series circuit, as

$L = 20 \times 10^{-3}\,\text{H}, C = 100\,\mu\text{F} = 100 \times 10^{-6}\,\text{F}$ and $R = 50\,\Omega$

Net impedance, $Z = \sqrt{R^2 + (X_C - X_L)^2}$

$$Z = \sqrt{(50)^2 + (31.85 - 6.28)^2}$$

$$\Rightarrow \quad Z^2 = 3154\,\Omega$$

So, energy dissipated across *L-C-R* series circuit is

$$P = \frac{E_{rms}^2}{Z^2} \times R = \frac{(10/\sqrt{2})^2}{3154} \times 50 = 0.8\text{ W}$$

Heat produced in 20 min $= 0.8 \times (20 \times 60) = 960$ J

16. (a) As we know if resistance from the circuit is removed and the value of inductance is doubled, then

$$X_C - X_L = 31.85 - 2(6.28) = 19.29\,\Omega$$

So, variation of current,

$$I_m = \frac{10}{19.29} = 0.52\text{ A}$$

Hence, $I = 0.52 \sin\left(314t + \dfrac{\pi}{2}\right)$

$$= 0.52 \cos 314t$$

17. (c) For an *L-C* oscillatory circuit,

$$L\frac{d^2q}{dt^2} + \frac{q}{C} = 0$$

For an oscillating mass,

$$m\frac{d^2x}{dt^2} + kx = 0$$

Hence, mass corresponds to inductor and produces necessary inertia for oscillation.

$$\therefore \quad \text{Mass} = 125\text{ kg}$$

18. (a) Spring constant *k* corresponds to the reciprocal of the capacitance. Since, total energy is given by

$$U = \frac{Q^2}{2C}, \quad C = \frac{Q^2}{2U} = \frac{175 \times 10^{-6}}{2\,(5.7 \times 10^{-6})}$$

$$= 2.69 \times 10^{-3}\text{ F}$$

$$\therefore \quad k = \frac{1}{2.69 \times 10^{-3}} = 372\text{ Nm}^{-1}$$

19. (c) Speed corresponds to the current in the circuit.
So, maximum speed corresponds to maximum current.

$$i_{max} = Q_{max}\omega = \frac{Q_{max}}{\sqrt{LC}}$$

$$= \frac{175\times10^{-6}}{\sqrt{1.25\times2.69\times10^{-3}}} = 3.02\times10^{-3}\text{ A}$$

$$v_{max} = 3.02\times10^{-3}\text{ m/s} = 0.302\text{ cms}^{-1}$$

20. (a,c) Given, voltage drop,

$$V_S = 200\sqrt{2}\sin(\omega t + 15^\circ)$$

and current, $i = 2\sin\left(\omega t + \frac{\pi}{4}\right)$

Average power consumed across AC circuit, i.e.

$$P = \frac{V_0 I_0}{2}\cos\phi$$

Phase angle across power factor,

$$\phi = 45^\circ - 15^\circ = 30^\circ$$

$$\cos\phi = \cos 30^\circ = \frac{\sqrt{3}}{2}$$

Power consumed, i.e.

$$P = 200\sqrt{2}\times2\cos30^\circ$$
$$P = 200\sqrt{6}\text{ W}$$

21. (a,c,d) In R-L-C series circuit,

$$V_R^2 + V_L^2 = 100^2$$
$$|V_C - V_L| = 120$$

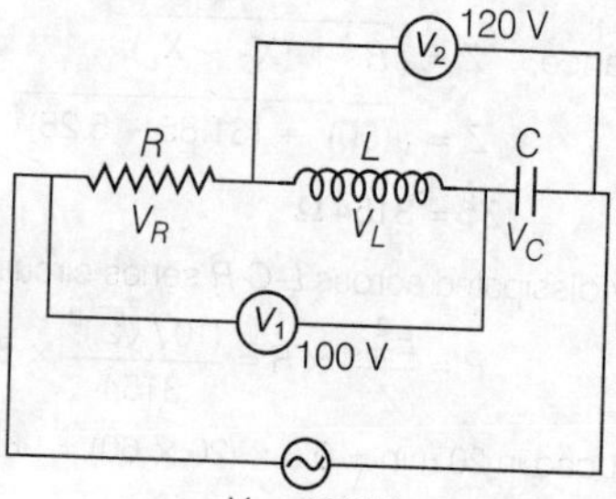

$$V = 130\text{ V}$$

$$130^2 = V_R^2 + (V_C - V_L)^2 \Rightarrow V_R = 50\text{ V}$$

$\Rightarrow$ Voltage across inductor, $V_L = \sqrt{100^2 - 50^2} = 50\sqrt{3}$ V

and voltage drop across capacitor,

$$V_C = 120 + 50\sqrt{3}\text{ V}$$

Power factor, $\cos\phi = \frac{V_R}{R} = \frac{50}{130} = \frac{5}{13}$

Since, $V_C < R_L$, so circuit is capacitive.

22. (a,b,c) Power transferred from the driving source to the driven oscillator is $P = I^2Z\cos\phi$. Therefore, power factor $\cos\phi \geq 0$ and $P \geq 0$. For wattless component the driving force shall give no energy to the oscillator ($P = 0$, when $\phi = 90^\circ$). Further, the driving force cannot syphon out energy of the oscillator, i.e. P cannot be negative.

23. (a) As, $X_L = \omega L = 2\pi\times50 = \frac{100}{\pi}\times10^{-3} = 10\,\Omega$

and $X_C = \frac{1}{\omega C} = \frac{1}{2\pi\times50\times\frac{100}{\pi}\times10^{-6}} = 100\,\Omega$

Impedance of the circuit

i.e. $Z = \sqrt{R^2 + (X_L - X_C)^2}$

$$= \sqrt{(100)^2 + (10-100)^2} = 134.5\,\Omega$$

rms value of the current through the circuit is

$$I_{rms} = \frac{220}{Z} = \frac{220}{Z} = \frac{220}{134.5} = 1.63\text{ A}$$

rms value of voltage drop across the capacitor is

$$V_{CV} = I_{rms}X_C = 1.63\times100 = 163\text{ V}$$

Average power dissipated across the resistor is

$$P_{av} = I_{rms}^2\times R = (1.63)^2\times100 = 265.7\text{ W}$$

Thus, average power dissipated in inductor and capacitor would be zero.

24. (b) $Z = \sqrt{(X_L - X_C)^2 + R^2}$

$\therefore\ Z = 0$ when $\omega = \frac{1}{\sqrt{LC}}$

Thus, **(i) $\rightarrow$ (s)**

$$I = \frac{E}{Z}$$

When $f \rightarrow f_r,\ Z \rightarrow R$

and I reaches maximum.

Thus, **(ii) $\rightarrow$ (r)**

$X_C = \frac{1}{\omega C}$, when $f \rightarrow \infty$, $X_C \rightarrow 0$

Thus, **(iii) $\rightarrow$ (q)**

and so I increases linearly.

As, $X_L = \omega L \Rightarrow \frac{1}{X_L} \propto \frac{1}{f}$

Thus, **(iv) $\rightarrow$ (p)**

Hence, (i) $\rightarrow$ (s), (ii) $\rightarrow$ (r), (iii) $\rightarrow$ (q), (iv) $\rightarrow$ (p)

25. (d) Heat produced $\propto I^2$

So, scale is non-linear.

26. (d) Average power consumed depends also over the phase difference between voltage and current,

$$P_{av} = V_{rms}I_{rms}\cos\phi$$

27. (4) $X_L = L\omega = 2\pi fL = 3.14\,\Omega$

$$Z = \sqrt{R^2 + X_L^2} = \sqrt{1^2 + (3.14)^2} = \sqrt{10.86} = 3.3\,\Omega$$

Also, $\text{lag} = \frac{\phi}{\omega}$

$$= \frac{\tan^{-1}\left(\frac{\omega L}{R}\right)}{\omega} = \frac{\tan^{-1}\left(\frac{Z}{R}\right)}{\omega} = \frac{\tan^{-1}(3.14)}{\omega}$$

$$= \frac{72^\circ}{\omega} = \frac{\frac{72\times\pi}{180}}{2\pi\times50} = \frac{1}{250}\text{ s}$$

$$4\times10^{-3}\text{ s} = k$$

So, $1000\,k = 4$

28. (5) When power is $\frac{1}{2}\times$ maximum power

$$I_{rms}^2 R = \frac{1}{2}(I_{rms}^2)_{max}R$$

or $\frac{V_{rms}^2}{Z^2}R = \frac{1}{2}\left(\frac{V_{rms}^2}{R^2}\right)R \quad \left(\text{as, } I_{max} = \frac{V_{rms}}{R}\right)$

$\Rightarrow$ $Z^2 = 2R^2$

or $R^2 + \left(\omega L - \frac{1}{\omega C}\right)^2 = 2R^2$

$\Rightarrow$ $\omega^4L^2C^2 - 2L\omega^2C - R^2\omega^2C + 1 = 0$

$\Rightarrow$ $L^2C^2\omega^4 - (2LC + R^2C^2)\omega^2 + 1 = 0$

$\Rightarrow$ $(2)^2(10\times10^{-6})^2\omega^4 - [2\times2\times(10\times10^{-6})]$
$+10^2\times(10\times10^{-6})]\omega^2 + 1 = 0$

On solving, we get

$$\omega^2 = 51130\text{ or }48894$$
$$\omega_1 = \sqrt{48894} = 221\text{ rad s}^{-1}$$
$$\omega_2 = \sqrt{51130} = 226\text{ rad s}^{-1}$$

Hence, $\omega_2 - \omega_1 = 226 - 221 = 5$

DPP-3 AC Source with *R-L-C* Connected in Series

1. Given, $R = 200\,\Omega$,

$$C = 15\,\mu F = 15 \times 10^{-6}\,F, V_{rms} = 220\,V$$

and $\quad V = 50\,Hz, \omega = 2\pi r = 2\pi \times 50 = 314\,rad/s$

$\therefore$ Current in the circuit, $I_{rms} = \dfrac{V_{rms}}{Z} = \dfrac{V_{rms}}{\sqrt{R^2 + \left(\dfrac{1}{\omega C}\right)^2}}$

$$= \frac{220}{\sqrt{(200)^2 + \left[\dfrac{1}{15 \times 10^{-6} \times 314}\right]^2}}$$

$$= \frac{220}{\sqrt{(200)^2 + (212.2)^2}} = 0.76\,A$$

rms voltage across the resistor,

$$V_R = I_{rms}\,R = 0.76 \times 200 = 152\,V$$

and rms voltage across the capacitor,

$$V_C = I_{rms}\,X_C = 0.76 \times 212.2 = 160\,V$$

Algebraic sum of V_R and V_C,

$$V_R + V_C = 152 + 160 = 312\,V$$

Obviously, it is more than the source voltage of 220 V.

As, V_R and V_C are not in same phase and hence cannot be added up algebraically. In fact, V_R and V_C differ in phase by $\dfrac{\pi}{2}$.

So, $\quad V_{R+C} = \sqrt{V_R^2 + V_C^2}$

$$= \sqrt{(152)^2 + (160)^2} = 220\,V$$

which is same as the supply voltage.

2. (i) In case of a parallel *L-C-R* circuit,

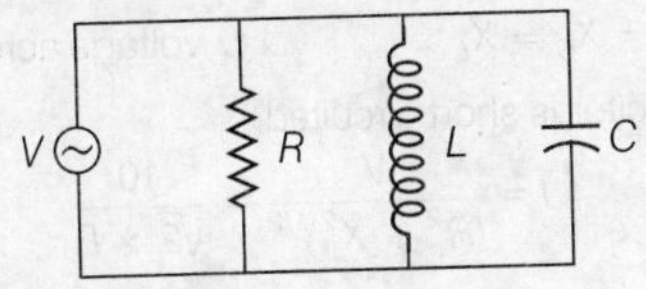

At any instant of time all elements have same voltage drop across them.

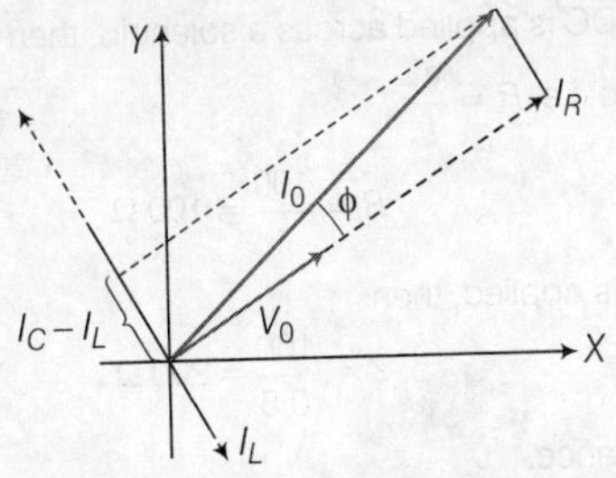

(ii) Current through resistor is in phase with V_0, $I_R = \dfrac{V_0}{R}\sin\omega t$

Current through inductor will lag behind voltage by $\dfrac{\pi}{2}$.

$$I_L = \frac{V_0}{X_L}\sin\left(\omega t - \frac{\pi}{2}\right)$$

Current through capacitor leads the voltage by $\dfrac{\pi}{2}$

$$I_C = \frac{V_0}{X_C}\sin\left(\omega t + \frac{\pi}{2}\right)$$

Total circuit current, $I_0 = \sqrt{I_R^2 + (I_C - I_L)^2}$

$$= \sqrt{\left(\frac{V_0}{R}\right)^2 + \left(\frac{V_0}{X_C} - \frac{V_0}{V_L}\right)^2}$$

$$= \frac{V_0}{R}\sqrt{\left(R\omega C - \frac{1}{R\omega L}\right)^2}$$

So, current as a function of time is given by

$$I(t) = \frac{V_0}{R}\sqrt{\left(1 + \left(R\omega C - \frac{R}{\omega L}\right)^2\right)}\sin(\omega t + \phi)$$

where phase difference between voltage and current is ϕ given by

$$\tan\phi = \left(\frac{\dfrac{V_0}{X_C} - \dfrac{V_0}{X_L}}{\dfrac{V_0}{R}}\right) = \frac{R}{X_C} - \frac{R}{X_L}$$

(iii) Circuit impedance is given by

$$Z = \frac{V_0}{I_0} = \frac{V_0}{\dfrac{V_0}{R}\sqrt{1 + \left(R\omega C - \dfrac{R}{\omega L}\right)^2}}$$

$$= \frac{R}{\sqrt{1 + \left(R\omega\,C - \dfrac{R}{\omega L}\right)^2}}$$

(iv) Power factor for the circuit is power dissipated divided by power supplied.

$$\therefore \quad \cos\phi = \frac{I_{Rrms}^2\,R}{V_{rms}I_{rms}}$$

$$= \frac{I_R^2 R}{V_0 I_0} = \frac{\left(\dfrac{V_0}{R}\right)^2 R}{V_0\dfrac{V_0}{R}\sqrt{1 + \left(R\omega\,C - \dfrac{R}{\omega L}\right)^2}}$$

$$= \frac{1}{\sqrt{1 + \left(R\omega C - \dfrac{R}{\omega L}\right)^2}}$$

3. (i) For a lightly damped *L-C-R* circuit,

$$T = \frac{2\pi}{\omega'} \approx \frac{2\pi}{\omega}$$

Initial energy, $U_{max} = \dfrac{Q_0^2\,e^{-\frac{R}{L}t}\cos^2(\omega t + \phi)}{2C}$

$$= \frac{Q_0^2\,e^{-\frac{R}{L}t}}{2C}$$

ΔU = Energy lost in 1 cycle

$$= Q_0^2\,e^{-\frac{R}{L}t} - Q_0^2\,e^{-\frac{R}{L}\left(t + \frac{2\pi}{\omega}\right)}$$

Fraction of energy lost in one complete cycle is given by

$$\frac{\Delta U}{U} = \frac{Q_0^2\,e^{-\frac{R}{L}t} - Q_0^2\,e^{-\frac{R}{L}\left(t + \frac{2\pi}{\omega}\right)}}{Q_0^2 e^{-\frac{R}{L}t}} = 1 - e^{-\frac{2\pi R}{\omega L}}$$

$$\therefore \quad \frac{\Delta U}{U} = 1 - e^{-\frac{2\pi R}{\omega L}} \approx 1 - \left(1 - \frac{2\pi R}{\omega L}\right)$$

(In above step we had taken $e^x = 1 + x + \ldots$)

So, $\quad \dfrac{\Delta U}{U} = \dfrac{2\pi R}{\omega L} = \dfrac{2\pi}{Q}$

where, Q = Q- factor = $\dfrac{\omega L}{R}$ of the circuit.

So, a high Q-value means smaller damping and less energy input required to maintain oscillations.

(ii) Equivalent values of each type of element in series

$$R_{eq} = R_1 + R_2$$

$$\frac{1}{C_{eq}} = \frac{1}{C_1} + \frac{1}{C_2}$$

$$L_{eq} = L_1 + L_2$$

So, impedance of circuit is given by

$$Z = \sqrt{R_{eq}^2 + \left(\omega L_{eq} - \frac{1}{\omega C_{eq}}\right)^2}$$

$$Z = \sqrt{(R_1 + R_2)^2 + \left(\omega L_1 + \omega L_2 - \frac{1}{\omega C_1} - \frac{1}{\omega C_2}\right)^2}$$

4. The inductive and capacitive reactances are

$$X_L = \omega L = (10000)\ \text{rad s}^{-1} \times 60\ \text{mH}$$

$$= 10000 \times 60 \times 10^{-3} = 600\ \Omega$$

$$X_C = \frac{1}{\omega C} = \frac{1}{10000 \times 0.50 \times 10^{-6}} = 200\ \Omega$$

Impedance Z of the circuit is given by

$$Z = \sqrt{R^2 + (X_L - X_C)^2}$$

$$= \sqrt{(300)^2 + (600 - 200)^2} = 500\ \Omega$$

With source voltage amplitude $E_0 = 50$ V, the current amplitude is given by

$$I_0 = \frac{E_0}{Z} = \frac{50\ \text{V}}{500\ \Omega} = 0.10\ \text{A}$$

The phase angle ϕ is given by

$$\phi = \tan^{-1}\frac{(X_C - X_L)}{R}$$

$$= \tan^{-1}\left[-\frac{400}{300}\right] = -53°$$

So, voltage amplitude V_{RO}, V_{LO} and V_{CO} across the resistor, inductor and capacitor, respectively

$$V_{RO} = I_0 R = 0.10 \times 300 = 30\ \text{V}$$

$$V_{LO} = I_0 X_L = 0.10 \times 600 = 60\ \text{V}$$

$$V_{CO} = I_0 X_C = 0.10 \times 200 = 20\ \text{V}$$

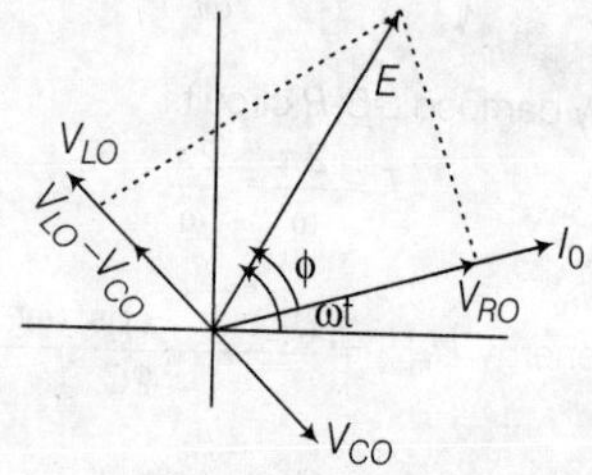

Note *$X_L > X_C$ and hence the voltage amplitude across the inductor is greater than across the capacitor and ϕ is negative. The value $\phi = -53°$ means that the voltage leads the current by 53°.*

5. (a) In *L-C-R* series circuit,

$$R = 50\ \Omega, V_L = 40\ \text{V}, V_C = 40\ \text{V}$$

and $V_{net} = 100$ V

$$V = \sqrt{V_R^2 + (V_L - V_C)^2}$$

$$100 = \sqrt{V_R^2 + (40 - 40)^2}$$

So, voltage across resistance,

$$V_R = 100\ \text{V}$$

As we know that rms value of a current,

$$I_{rms} = \frac{V_{rms}}{Z}$$

At resonating condition,

$\therefore$ $Z = R$

$$I_{rms} = \frac{100}{50} = 2\text{A}$$

6. (d) Given, $R = 3\ \Omega$, $X_L = 4\ \Omega$

In *L-R* circuit,

$$\tan\phi = \frac{X_L}{R} = \frac{4}{3} \Rightarrow \phi = 53°$$

Current across inductor,

$$I = I_0 \sin(\omega t - 53°)$$

As, $$I_0 = \frac{V_0}{Z} = \frac{10}{\sqrt{3^2 + 4^2}} = 2\ \text{A}$$

$$I = 2\sin(\omega t - 53°) \quad (\because \phi = 53°)$$

$\therefore$ Voltage across the inductor at $t = \frac{T}{2}$ is

$$(V_L)_0 = I_0 X_L = 2 \times 4$$

$$V_L = (V_L)_0 = \sin\left(\omega t - 53° + \frac{\pi}{2}\right)$$

At $t = \frac{T}{2}$,

$$V_L = 2 \times 4 \times \sin 37°$$

$$V_L = 4.8\ \text{V}$$

7. (c) Given, $f = 50$ Hz, $C = 100\ \mu$F, $I = 1.57$ A

Voltage drop across capacitance,

$$V_0 = I_0 X_C$$

$$V_0 = (1.57)\left(\frac{1}{2\pi f C}\right)$$

$$V_0 = \frac{1.57 \times 1}{2 \times 3.14 \times 50 \times 100 \times 10^{-6}}$$

$$V_0 = 50\ \text{V}$$

As, instantaneous current across capacitor,

$$I = I_0 \cos\omega t = I_0 \sin\left(\omega t + \frac{\pi}{2}\right)$$

As $t = 0$, $I = I_0$

So, instantaneous voltage across the capacitor,

$$V = V_0 \sin(\omega t)$$

$$E = 50\sin(100\pi t)$$

8. (b) Here, $R = X_C = X_L$ ($\because$ voltage across them is same)

When capacitor is short circuited,

$$I = \frac{V}{(R^2 + X_L^2)^{1/2}} = \frac{10}{\sqrt{2} \times R}$$

$\therefore$ Potential drop across inductance $= IX_L = IR = \frac{10}{\sqrt{2}}$ V

9. (d) When DC is applied across a solenoid, then

Resistance, i.e. $R = \frac{V}{I}$

$$R = \frac{100}{1} = 100\ \Omega$$

When AC is applied, then

$$Z = \frac{100}{0.5} = 200\ \Omega$$

Net impedance,

$$Z = \sqrt{R^2 + X_L^2}$$

$$200 = \sqrt{(100)^2 + X_L^2}$$

$$\Rightarrow (200)^2 = (100)^2 + X_L^2$$

$$\Rightarrow 30000 = X_L^2$$

$$\Rightarrow X_L = 173.205\ \Omega$$

Now, $X_L = \omega L$

$$\Rightarrow 173.205 = 2\pi f L$$

$$\Rightarrow \frac{173.205}{2\pi \times f} = L$$

$$\Rightarrow \frac{173.205}{2 \times 3.14 \times 50} = L$$

$$\Rightarrow 0.55\ \Omega = L$$

$$L = 0.55\ \Omega$$

10. (b) As the current I leads the emf by $\frac{\pi}{4}$ rad so that the circuit is an R-C circuit.

As, $\tan\phi = \frac{X_C}{R} = \frac{1}{\omega CR}$

or $\tan\frac{\pi}{4} = \frac{1}{\omega CR}$ or $\omega CR = 1$

$\Rightarrow$ $100 \times CR = 1$

$\Rightarrow$ $CR = \frac{1}{100}\ \text{s}^{-1}$

For option (b),

$$RC = 1\,\text{k}\Omega \times 10\,\mu\text{F}$$
$$= 10^3 \times 10^{-5} = \frac{1}{100}\ \text{s}^{-1}$$

11. (a,c) We have

$$X_L = \omega L = 10 \times 0.5 = 50\ \Omega$$

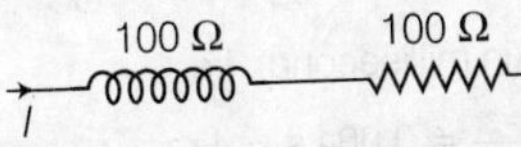

$$X_1 = \frac{1}{\omega C} = \frac{1}{100 \times 100 \times 10^{-6}} = 100\ \Omega$$

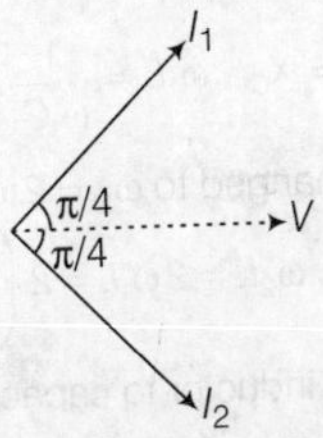

$$Z_1 = 100\sqrt{2}$$
$$I_1 = \frac{20}{100\sqrt{2}} = \frac{1}{5\sqrt{2}}$$

V across $100\ \Omega = \frac{1}{5\sqrt{2}} \times 100 = \frac{20}{\sqrt{2}} \times \frac{\sqrt{3}}{\sqrt{2}} = 10\sqrt{2}$

Phase difference between I_1 and V

$$\cos\phi_1 = \frac{R_1}{Z_1} = \frac{100}{100\sqrt{2}}$$
$$\phi = \frac{\pi}{4}$$

I_1 and V

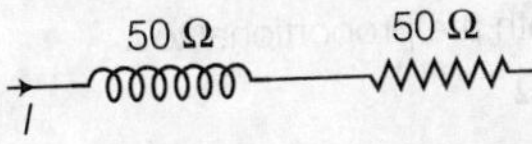

$$Z_2 = 50\sqrt{2},\ I_2 = \frac{20}{50\sqrt{2}} = \frac{2}{5\sqrt{2}}$$

V across $50\ \Omega = \frac{2}{5\sqrt{2}} \times 50$

$$= \frac{20}{\sqrt{2}} = 10\sqrt{2}$$

$$\phi_2 = \frac{\pi}{4}\ I_2 \log V \text{ by } \frac{\pi}{4}$$

$$I = I_1 = I_2$$
$$I_{net} = \sqrt{I_1^2 + I_2^2}$$
$$I = \sqrt{\frac{4}{25 \times 2} + \frac{1}{25 \times 2}}$$
$$= \sqrt{\frac{5}{50}} = \frac{1}{\sqrt{10}} = 0.316$$

12. (c,d) When AC voltage of 220 V is applied to a capacitor C, the charge on the plates in phase with the applied voltage. As current developed leads the applied voltage by a phase angle of 90°. Therefore, power delivered to the capacitor per cycle is

$$P = E_V I_V \cos 90° = \text{zero}$$

13. (a,d) The impedance of an L-C-R circuit is given by

$$Z = \sqrt{R^2 + (X_L - X_C)^2}$$
$$= \sqrt{R^2 + \left(2\pi\nu L - \frac{1}{2\pi\nu C}\right)^2}$$

The variation of Z with ν as shown in figure. As ν increases, Z decreases, current increases. At $\nu = \nu_r$, Z is minimum, current is maximum. Beyond $\nu = \nu_r$, Z increases and current decreases. Hence, the circuit elements likely in the circuit are L and C or R, L and C.

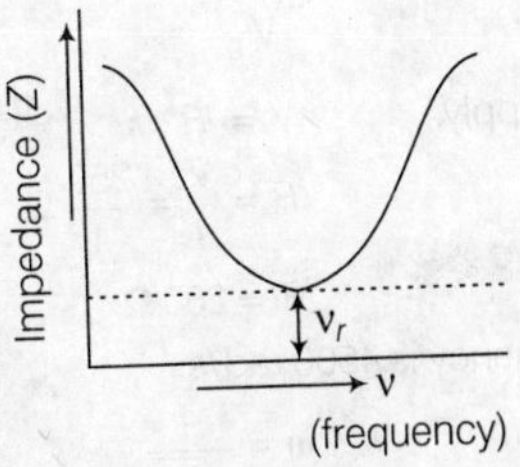

14. (b,c) With C alone or L alone, the phase difference between current and emf is $\frac{\pi}{2}$. In L-C circuit, the phase difference is π while R-C circuit, the phase difference is between 0 and $\frac{\pi}{2}$.

15. (b,c)

$$Z = \sqrt{R^2 + X_C^2}$$
$$Z = \sqrt{R^2 + \left(\frac{1}{\omega C}\right)^2}$$

In case (i), capacitance C will be more. Therefore, impedance Z will be less. Hence, current will be more. So, option (b) is correct.

Further, $V_C = \sqrt{V^2 - V_R^2} = \sqrt{V^2 - (IR)^2}$

In case (ii), since I is more, V_C will be less.

So, option (c) is correct.

16. (b) At resonance as $Z = 0$,

$$I = \frac{V}{R} = \frac{60}{120} = \frac{1}{2}\ \text{A}$$

As $V_L = IX_L = I\omega L \Rightarrow L = \frac{V_L}{I\omega}$

So, $L = \frac{40}{(1/2) \times 4 \times 10^5} = 0.2\ \text{mH}$

17. (a) As frequency, $\omega = \frac{1}{\sqrt{LC}}$

$\Rightarrow C = \frac{1}{L\omega^2}$

i.e. $C = \frac{1}{0.2 \times 10^{-3} \times (4 \times 10^5)^2} = \frac{1}{32}\ \mu\text{F}$

18. (c) Now in case of series L-C-R circuit,

$$\tan\phi = \frac{X_L - X_C}{R}$$

So, current will lag the applied voltage by 45° if

$$\tan 45° = \frac{\omega L - \frac{1}{\omega C}}{R}$$

$$1 \times 120 = \omega \times 2 \times 10^{-4} - \frac{1}{\omega\,(1/32) \times 10^{-6}}$$

$$\omega^2 - 6 \times 10^5\omega - 16 \times 10^{10} = 0$$

i.e. $\omega = \frac{6 \times 10^5 \pm \sqrt{(6 \times 10^5)^2 + 64 \times 10^{10}}}{2}$

i.e. frequency,

$$\omega = \frac{6 \times 10^5 + 10 \times 10^5}{2} = 8 \times 10^5\ \text{rad s}^{-1}$$

19. (c) Given, circuit can work on DC. It is not an L-C-R series circuit. Also, current increases with frequency and reaches ω maximum resonance is possible only when L and C are in series.
So, the given circuit must be

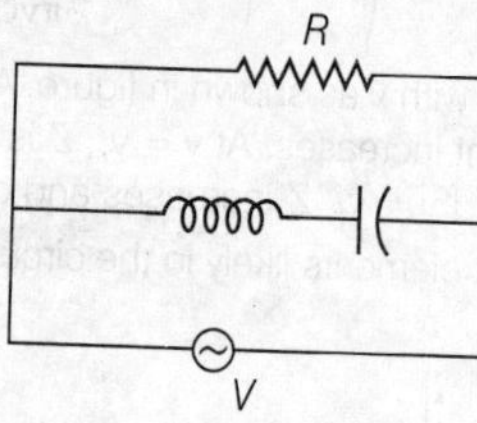

20. (a) With DC supply,

$$V = IR$$

$$R = \frac{V}{I} = \frac{250}{1}\,\Omega$$

$\Rightarrow$ $$R = 250\,\Omega$$

Resonant frequency is 4500 rad/s.

$\Rightarrow$ $$\omega = \frac{1}{\sqrt{LC}}$$

$\Rightarrow$ $$LC = \frac{1}{\omega_0^2} = \frac{1}{(4500)^2}$$

Also, with AC supply,

$$I_R = \frac{V}{R}\sin\omega t = \frac{250}{250}\sin\omega t$$

$\Rightarrow$ $$I_R = 1\sin\omega t$$

and $$I_x = \frac{V}{X}\cos\omega t$$

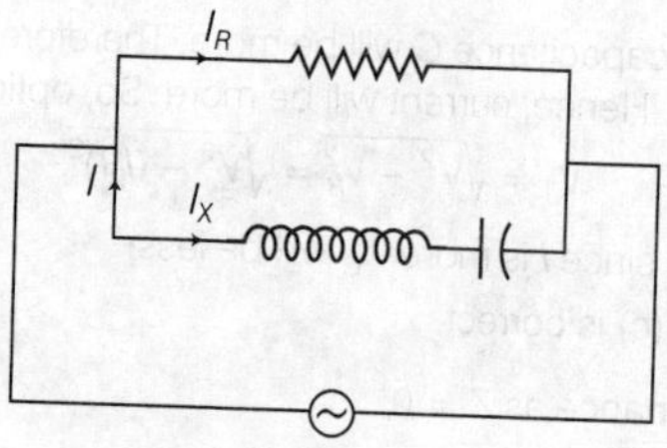

As, $$I = I_R + I_X$$

$$I = \sin\omega t + \frac{V}{X}\cos\omega t$$

Let $$I_0\cos\phi = 1 \text{ and } \frac{V}{X} = I_0\sin\phi$$

We have

$\Rightarrow$ $$I = I_0\cos\phi\,\sin\omega t + I_0\sin\phi\,\cos\omega t$$

$\Rightarrow$ $$I = I_0\sin(\omega t + \phi)$$

Now, $$I_0^2\cos^2\phi + I_0^2\sin^2\phi = 1 + \left(\frac{V}{X}\right)^2$$

$\Rightarrow$ $$I_0^2 = 1 + \left(\frac{V}{X}\right)^2$$

$\Rightarrow$ $$(1.25)^2 = 1 + \left(\frac{250}{X}\right)^2$$

$\Rightarrow$ $$X^2 = \frac{250 \times 250}{0.5625}$$

$$X = \frac{250}{0.75} = \frac{1000}{3}$$

$\Rightarrow$ $$X = 333.33\,\Omega$$

21. (d) As reactance, $X = \omega L - \dfrac{1}{\omega C}$

$\Rightarrow$ $$\frac{1000}{3} = 2250L - \frac{1}{2250\,C} \qquad \text{...(i)}$$

As $LC = \dfrac{1}{\omega_0^2}$, we have

$$L = \frac{1}{C\omega_0^2} = \frac{1}{C(4500)^2}$$

Substituting the values in Eq. (i), we have

$$\frac{1000}{3} = 2250\left(\frac{1}{C \times (4500)^2}\right) - \frac{1}{2250 \times C}$$

$\Rightarrow$ $$\frac{1}{C}\left\{\frac{2250}{4500 \times 4500} - \frac{1}{2250}\right\} = \frac{1000}{3}$$

$\Rightarrow$ $$\frac{1}{C}\left(\frac{1}{9000} - \frac{1}{2250}\right) = \frac{1000}{3}$$

$\Rightarrow$ $$\frac{1}{C} \times \frac{3}{9000} = \frac{1000}{3}$$

$\Rightarrow$ $$C = 10^{-6}\,\text{F} = 1\,\mu\text{F}$$

22. (4) Given, $\omega = 500\,\text{rad s}^{-1}$ and $Z = R\sqrt{1.25}$

As impedance, $Z = \sqrt{R^2 + \left(\dfrac{1}{\omega C}\right)^2} = R\sqrt{1.25}$

$$R^2(1.25) = R^2 + \left(\frac{1}{\omega C}\right)^2 \Rightarrow \frac{1}{\omega C} = 0.5R$$

or time constant (in millisecond), i.e.

$$RC = \frac{1}{0.5 \times 500\,\text{s}^{-1}} = 0.004\,\text{s} = 4\,\text{ms}$$

23. (4) At a certain frequency ω_1,

$$X_{L_1} = X_{C_1},\ \omega_1 L = \frac{1}{\omega_1 C}$$

Now, frequency is changed to $\omega_2 = 2\omega_1$

$$X_{L_2} = \omega_2 L = 2\omega_1 L = 2\left(\frac{1}{\omega_1 C}\right) = 2\left(\frac{2}{\omega_2 C}\right) = 4 \times C_2$$

Ratio of reactance of inductor to capacitor, i.e.

$$\frac{X_{L_2}}{X_{C_2}} = 4$$

24. (a)

(p) In this case, steady state current is zero ($I = 0$), because of the presence of capacitor. Entire potential will be across capacitor in steady state. Hence, $V_1 = 0$ and $V_2 = V$.

(q) In steady state, $V_1 = 0$, so $V_2 = 0$. Also, V_2 is proportional to I.

(r) $X_L = \omega L = 2\pi f L = 2\pi \times 50 \times 5 \times 10^{-3} = 1.885\,\Omega$

$R = 2\,\Omega$

Since, $R > X_L$, so $V_2 > V_1$

Here, $I \neq 0$

V_1 and V_2 both are proportional to I.

(s) $X_L = 1.885\,\Omega$

$$X_C = \frac{1}{\omega C} = \frac{1}{2\pi f C} = \frac{1}{2\pi \times 50 \times 3 \times 10^{-6}} = 1061\,\Omega$$

Hence, $X_C > X_L$, so $V_2 > V_1$

(t) $X_C = 1.061\,\text{k}\Omega$, $R = 1\,\text{k}\Omega$

$X_C > R$, so $V_2 > V_1$

25. (a) $X_2 = L\omega$ and $X_C = \dfrac{1}{C\omega}$

$\therefore$ $X_L \to 0$ as $\omega \to 0$ and $X_C \to \infty$ as $\omega \to 0$

Hence, an inductor is low pass filter and a capacitor blocks low frequency signals.

26. (a) At resonance

$$\omega_r = \frac{1}{\sqrt{LC}}$$

and $$X_L = L\omega = L\frac{1}{\sqrt{LC}} = \sqrt{\frac{L}{C}}$$

Potential drop across inductor,

$$V_L = IX_L, \text{ where } I = \frac{V}{Z}$$

$$\frac{V_{\text{rms}}}{Z}\sqrt{\frac{L}{C}} = QV_{\text{rms}}$$

DPP-4 Resonance

1. (i) Resonant frequency,

$$\omega_r = \frac{1}{\sqrt{LC}} \text{ rad s}^{-1}$$

or $$f_r = \frac{1}{2\pi\sqrt{LC}} \text{ Hz}$$

$$\Rightarrow \quad \omega_r = \frac{1}{\sqrt{4\times10^{-3}\times0.1\times10^{6}}} \text{ rad s}^{-1}$$

$$= 5\times10^4 \text{ rad s}^{-1}$$

$$= \frac{5}{2\pi}\times10^4 \text{ Hz}$$

(ii) Circuit current at resonance,

$$I = \frac{V}{R} = \frac{10}{5} = 2\text{A}$$

$\therefore$ V_4 = Potential drop across inductor

$$= IX_L = I\omega L$$

$$= 5\times10^4\times10^{-3}\times2$$

$$= 400\text{ V}$$

V_C = voltage drop across capacitor = IX_C

$$\therefore \quad V_C = I\times\frac{1}{\omega C}$$

$$= 2\times\frac{1}{5\times10^4\times0.1\times10^{-6}} = 400\text{ V}$$

(iii) Voltage drop across resistor,

$$V_R = IR = 2\times5 = 10\text{ V}$$

2. Resonant frequency,

$$\omega_r = \frac{1}{\sqrt{LC}} = \frac{1}{\sqrt{10^{-2}\times10^{-6}}} = 10^4 \text{ rad s}^{-1}$$

So, driving frequency = 10% less than ω_r

$$= \frac{9}{10}\omega_r = 9\times10^3 \text{ rad s}^{-1}$$

So, average power dissipated,

$$P_{av} = V_{rms}\, I_{rms}\cos\phi = I_{rms}^2 R$$

$$= \frac{1}{2}I_{max}^2 R = \frac{1}{2}\times(0.704)^2\times3$$

$$= 0.74\text{ W}$$

As, $$f = \frac{\omega}{2\pi}\cdot\frac{\text{cycles}}{\text{s}} = \frac{9\times10^3}{2\pi}$$

Hence, power dissipated per cycle $= \dfrac{P_{av}\text{ (per second)}}{\text{Number of cycles/s}}$

$$= \frac{0.74}{(9\times10^3/2\pi)}$$

$= 5.16\times10^{-4}$ joules per cycle

Reactances at this frequency are

$$X_L = L\omega = 9\times10^3\times10^{-2} = 90\,\Omega$$

$$X_C = \frac{1}{\omega C} = \frac{1}{9\times10^3\times10^{-6}} = 111.11\,\Omega$$

X = net reactance of circuit

$$= X_L - X_C = 90 - 111.11 = -21.11\,\Omega.$$

Impedence of the circuit $Z = \sqrt{R^2+X^2}$

$$\Rightarrow \quad Z = \sqrt{3^2+(-21.11)^2}$$

$$= \sqrt{9+445.63} = 21.32\,\Omega$$

Also, E_{max} = 15 V So, current amplitude

$$I_m = \frac{E_{max}}{Z}$$

$$\therefore \quad I_m = \frac{15}{21.32} = 0.704\text{ A}$$

3. (i) Current amplitude is maximum at resonance and source frequency at resonance is given by

$$f_0 = \frac{1}{2\pi\sqrt{LC}} = \frac{1}{2\times\pi\sqrt{0.12\times480\times10^{-9}}} = 663.14\text{ Hz}$$

and $$\omega_0 = 2\pi f_0 = 4166.66 \text{ rad s}^{-1}$$

The maximum current amplitude is given by

$$I_{max} = \frac{E_0}{R} = \frac{230\sqrt{2}}{23} = 14.14\text{ A}$$

(ii) Average power absorbed by the circuit is maximum at resonance for which frequency is calculated above as 663.14 Hz.

Maximum power absorbed is given by

$$P_{max} = \frac{E_V^2}{R} = \frac{(230)^2}{23} = 2300\text{ W}$$

(iii) Q-factor is given by

$$Q = \frac{1}{R}\sqrt{\frac{L}{C}} = \frac{1}{23}\sqrt{\frac{0.12}{480\times10^{-9}}} = 21.74$$

4. (i) Clearly, the resonant frequency ω_0 is same for all three graphs X, Y and Z.

As, $\omega_0 = \dfrac{1}{\sqrt{LC}}$ and $L_X = L_Y = L_Z$

So, $C_X = C_Y = C_Z$

(ii) The maximum value of current at resonance is

$$I_0 = \frac{E_0}{R}, \text{ i.e. } I_0 \propto \frac{1}{R}$$

But $$I_0^X > I_0^Y > I_0^Z$$

$$\therefore \quad R_X < R_Y < R_Z$$

Quality factor, $Q = \dfrac{\omega L}{R}$

As ω and L are same in all three cases, so

$$Q \propto \frac{1}{R}$$

Now, $R_X < R_Y < R_Z$

So, $Q_X < Q_Y > Q_Z$

i.e. Q is maximum in case X.

At resonance frequency ω_0, $X_L = X_C$, so

$$Z = \sqrt{R^2+(X_L-X_C)^2} = R$$

Thus, the impedance of the circuit is purely resistive in nature.

5. (a) In L-C-R series circuit, we have

$V_R = 60$ V, $$V_R = IR$$

$$I = \frac{60}{120} = \frac{1}{2}\text{ A}$$

Voltage drop across inductor, $V_L = IX_L$

$$40 = \frac{1}{2}X_L$$

Net reactive inductance, $X_L = 80\,\Omega$, i.e. $\omega L = 80\,\Omega$

$$L = \frac{80}{4\times10^3} = 20\text{ mH}$$

As we know that angular frequency, $\omega = \dfrac{1}{\sqrt{LC}}$

Squaring both sides, we get

$$\omega^2 = \frac{1}{LC}$$

$$(4\times10^3)^2 = \frac{1}{20\times10^{-3}\times C}$$

$$\Rightarrow \quad C = \frac{1}{16\times10^6\times20\times10^{-3}}$$

$$C = \frac{25}{8}\,\mu\text{F}$$

6. (a) As angular frequency, $\omega = \frac{1}{\sqrt{LC}}$ ...(i)

When capacitance is made one-fourth, then $\omega' = \omega$

$$\omega' = \frac{1}{\sqrt{L' \frac{C}{4}}} = \omega \quad \text{...(ii)}$$

So, $L' = 4L$

7. (c) In *L-C-R* series circuit,

$$I_{max} = 1.0 \text{ A}$$

Bandwidth corresponds to the frequencies at which

$$I_m = \frac{I_{max}}{\sqrt{2}} = \frac{1}{\sqrt{2}} = 0.7 \text{ A}$$

From the graph $\omega_1 = 0.8$ rad/s

$$\omega_2 = 1.2 \text{ rad/s}$$

$$\omega_2 - \omega_1 = (1.2 - 0.8) = 0.4 \text{ rad s}^{-1}$$

8. (c) At resonance,

$$Z = R$$

So, $(I_{rms})_{max} = \frac{V_{rms}}{R}$

$$6 = \frac{24}{R}$$

$$R = 4\,\Omega$$

When *R* connected to battery, then current flows through resistor, we get

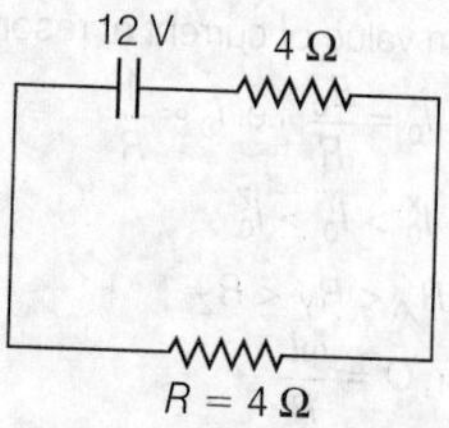

$$I = \frac{12}{4+4} = 1.5 \text{ A}$$

9. (d) At resonance, $V_L = V_C$, i.e. out of phase.

Voltage across *L-C* combination

$$= V_L - V_C = 0$$

10. (c) According to resonant frequency f_r,

i.e. $f_r = \frac{1}{2\pi\sqrt{LC}}$

When *C* is changed to 2*C*, *L* should be changed to *L*/2, so that f_r remains unchanged.

11. (a) $R = 1\,\text{k}\Omega$

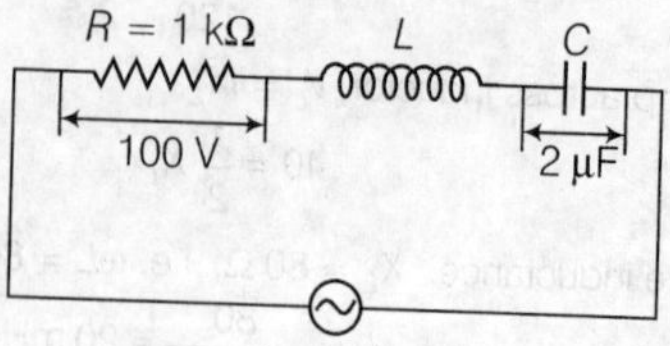

At resonant frequency,

$$\omega L = \frac{1}{\omega C}$$

$$\omega = \frac{1}{\sqrt{LC}}$$

$\Rightarrow$ $X_L = X_C$

So, $I_0 = \frac{V}{R}$

Then, voltage drop across inductor *L*, i.e.

$$V_L = I_0 X_L = I_0 X_C = \frac{100}{10^3} \times \frac{1}{200 \times 2 \times 10^{-6}} = \frac{1000}{4} = 250 \text{ V}$$

12. (b) We know that resonant frequency of a tuning circuit is given as

$$f_0 = \frac{1}{2\pi\sqrt{LC}} \quad \Rightarrow \quad C' = \frac{1}{4\pi^2 f_0^2 L}$$

We are given in the range of frequency and we have to find the range of variable capacitor.

$$C_1 = \frac{1}{4\pi^2 (500 \times 10^3)^2 \times 400 \times 10^{-6}} = 253 \text{ pF}$$

By putting $f_0 = 1.5$ MHz, we get

$$C_2 = \frac{1}{4\pi^2 (1.5 \times 10^6)^2 \times 400 \times 10^{-6}} = 28 \text{ pF}$$

So, range of variable capacitor 28 pF-253 pF.

13. (b)

I. The resonance frequency of a rejector (parallel) *L-C-R* circuit is given by

$$f = \frac{1}{2\pi}\sqrt{\frac{1}{LC} - \frac{R^2}{L^2}}$$

$$= \frac{1}{2\pi}\sqrt{\frac{1}{(1.6 \times 10^{-2})(250 \times 10^{-12})} - \frac{(20)^2}{(1.6 \times 10^{-2})^2}}$$

$$= 7.96 \times 10^4 \text{ Hz}$$

II. The circuit impedance at resonance is given by

$$Z = \frac{L}{CR} = \frac{1.6 \times 10^{-2}}{(250 \times 10^{-12}) \times (20)} = 3.2 \times 10^6\,\Omega$$

14. (b)

I. The resonance frequency is a rejector (parallel) *L-C-R* circuit is given by

$$f = \frac{1}{2\pi}\sqrt{\frac{1}{LC} - \frac{R^2}{L^2}}$$

$$= \frac{1}{2\pi}\sqrt{\frac{1}{(1.6 \times 10^{-2})(250 \times 10^{-12})} - \frac{(20)^2}{(1.6 \times 10^{-2})^2}}$$

$$= 7.96 \times 10^4 \text{ Hz}$$

II. The circuit impedance at resonance is given by

$$Z = \frac{L}{CR} = \frac{1.6 \times 10^{-2}}{(250 \times 10^{-2})(20)} = 3.2 \times 10^6\,\Omega$$

15. (c) The circuit will have inductive nature if

$$\omega > \frac{1}{\sqrt{LC}} \left(\omega L > \frac{1}{\omega C}\right)$$

Hence, (a) is false. Also if circuit has inductive nature the current will lag behind voltage. Hence, (d) is also false.

If $\omega = \frac{1}{\sqrt{LC}} \left[\omega L = \frac{1}{\omega C}\right]$, the circuit will have resistance nature.

Hence, (b) is false.

$\therefore$ Power factor, $\cos\phi = \frac{R}{\sqrt{R^2 + \left(\omega L - \frac{1}{\omega C}\right)^2}} = 1$ if $\omega L = \frac{1}{\omega C}$.

16. (b,d) At resonance, the series combination of *L* and *C* gives zero impedance and voltage across *L-C* combination is zero.

17. (a,b,c) At resonance, $Z = R$

$$I = \frac{V}{Z} = \frac{V}{R}$$

$$\omega = \frac{1}{\sqrt{LC}}$$

$\Rightarrow$ $f = \frac{1}{2\pi\sqrt{LC}}$

$\Rightarrow$ $f = 500$ Hz

$V_L = V_C$ and voltage across *C* is 180° out of phase with the voltage across *L*.

i.e. $V_{LC} = V_L - V_C$

18. (a,b,d) For an R-C high pass filter,

$$\frac{\Delta V_{out}}{\Delta V_{in}} = \frac{R}{\sqrt{R^2 + X_C^2}}$$

$$\Rightarrow \quad \frac{1}{2} = \frac{0.5}{\sqrt{(0.5)^2 + (X_C)^2}}$$

$$\Rightarrow \quad (0.5)^2 + (X_C)^2 = 1$$

$$\Rightarrow \quad X_C^2 = 1 - (0.5)^2 = 1 - 0.25$$

$$\Rightarrow \quad X_C^2 = 0.75 = \frac{75}{100} = \frac{3}{4}$$

$$\Rightarrow \quad X_C = \frac{\sqrt{3}}{2}$$

Also, as $\omega \to 0, \frac{1}{\omega C} \to \infty \Rightarrow \frac{\Delta V_{out}}{\Delta V_{in}} \to 0$

and as $\omega \to \infty, \frac{1}{\omega C} \to 0 \quad \Rightarrow \frac{\Delta V_{out}}{\Delta V_{in}} \to \frac{R}{R} = 1$

19. (a,c) For one half of a cycle, let left side of generator is positive, top diode conducts and the diode switches OFF. Circuit is equivalent to

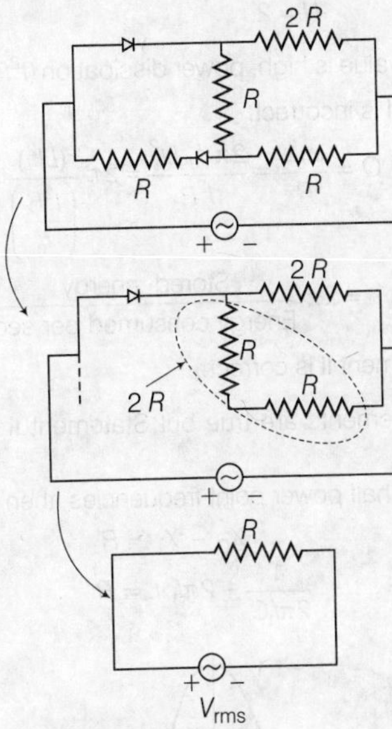

$$R_{eq} = R$$

So, power consumption is $P = \frac{V_{rms}^2}{R}$

In next half of cycle, right side of generator is positive, the upper diode is an open circuit and the lower diode has zero resistance. Equivalent circuit is

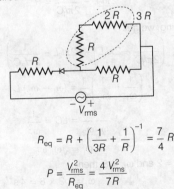

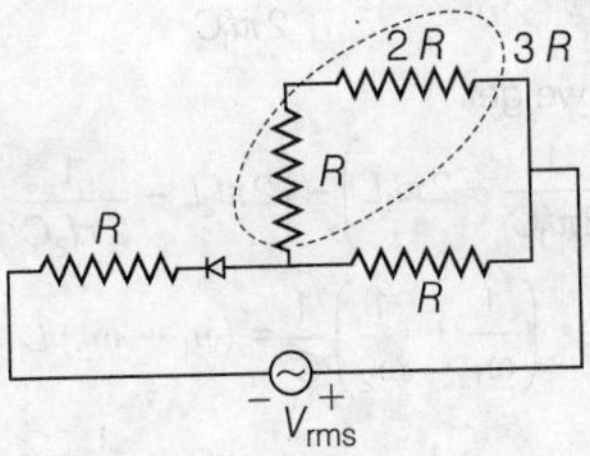

$$R_{eq} = R + \left(\frac{1}{3R} + \frac{1}{R}\right)^{-1} = \frac{7}{4}R$$

$$\Rightarrow \quad P = \frac{V_{rms}^2}{R_{eq}} = \frac{4\,V_{rms}^2}{7R}$$

Overall time average power is

$$\left[\frac{V_{rms}^2/R + \frac{4}{7}V_{rms}^2/R}{2}\right] = \frac{11}{14}\cdot\frac{V_{rms}^2}{R}$$

20. (b,d) $\omega = 1000\ \text{rad s}^{-1}, R = 400\ \Omega$

$$C = 5 \times 10^{-6}\ \text{F}, L = 0.5\ \text{H}$$

$$V_{max} = 100\ \text{V}$$

$$\omega L = 500\ \Omega, \omega C = 200\ \Omega$$

Circuit impedance,

$$Z = \sqrt{R^2 + \left(\omega L - \frac{1}{\omega C}\right)^2}$$

$$= \sqrt{(400)^2 + (300)^2} = 500\ \Omega$$

$$I_{max} = \frac{\Delta V_{max}}{Z} = \frac{100}{500} = 0.2\ \text{A}$$

Average power dissipated in the circuit is

$$P = I_{rms}^2\ R = \left(\frac{I_{max}^2}{2}\right)R = \frac{(0.2)}{2} \times 400 = 8\ \text{W}$$

21. (b) Given, $L = 2$ H, $C = 18\ \mu$F

Current is maximum at resonance,

$$\omega = \frac{1}{\sqrt{LC}} \quad \Rightarrow \quad f = \frac{1}{2\pi\sqrt{LC}}$$

$$\Rightarrow \quad \frac{1}{2\pi} \times \frac{1}{\sqrt{2 \times 18 \times 10^{-6}}} = \frac{250}{3\pi}\ \text{Hz}$$

22. (b) Current is maximum at resonance. So, maximum of current

$$(I_{rms})_{max} = \frac{V_{rms}}{Z} = \frac{V_{rms}}{R}$$

$$= \frac{20}{10\ \text{k}\Omega} = \frac{20}{10} \times 10^{-3} = 2\ \text{mA}$$

23. (b) With S_1 open and S_2 closed, inductor is in circuit and capacitor is shorted out.

So, reactance of circuit = $X_L = L\omega$

Resistance of circuit = R

Impedance of circuit,

$$Z = \sqrt{X_L^2 + R^2} = \sqrt{L^2\omega^2 + R^2}$$

Phase difference between current and voltage (current leads) is

$$\tan\phi = \frac{X_L}{R} = \frac{L\omega}{R}$$

$$\phi = \tan^{-1}\left(\frac{X_L}{R}\right) = \tan^{-1}\left(\frac{L\omega}{R}\right)$$

So, current, $i = i_{max}\cos(\omega t + \phi)$

where, $i_{max} = \frac{V_{max}}{Z} = \frac{V_{max}}{\sqrt{L^2\omega^2 + R^2}}$

$$\therefore \quad i = \frac{V_{max}}{\sqrt{L^2\omega^2 + R^2}}\cos\left(\omega t + \tan^{-1}\left(\frac{L\omega}{R}\right)\right)$$

24. (a) During oscillations, maximum energy stored in capacitor is

$$U = \frac{1}{2}C(\Delta V_C)^2 = \frac{1}{2}CI^2X_c^2$$

$$U_{max} = \frac{1}{2}C\,I_{max}^2\,X_C^2 \left.\right\}\text{when } I \to I_{max}\ Z \to R$$

$$= \frac{1}{2}C\,\frac{V_{max}^2}{R^2}\cdot\frac{1}{\omega^2C^2}$$

$$= \frac{1}{2}\cdot C\cdot\frac{V_{max}^2}{R^2}\cdot\frac{1}{\omega^2C^2}$$

Note *$I \to I_{max}$ at resonance*

and $X_L = X_C \Rightarrow C = \frac{1}{\omega^2 L}$

$$\therefore \quad U_{max} = \frac{V_{max}^2\,L}{2R^2}$$

25. (a) Maximum energy of inductor

$$= \frac{1}{2} L I_{max}^2 \quad \Big| \; I \to I_{max} \text{ at resonance.}$$
$$= \frac{1}{2} L \cdot \frac{V_{max}^2}{R^2}$$
$$= \frac{V_{max}^2 L}{2R^2}$$

26. (a)

(i) For L-R series circuit, the phasor figure is shown as below:

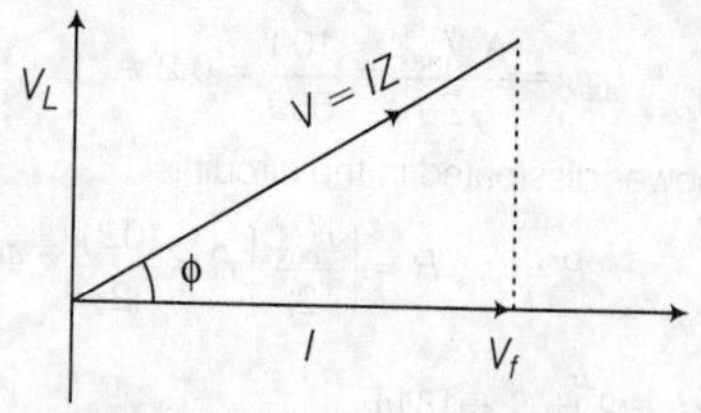

I lags voltage by an angle ϕ ($< \pi/2$)

I lags V_L by an angle $\pi/2$.

(ii) For R-C series AC circuit, I leads V by an angle less than $\pi/2$.

(iii) Depending on the value of L, C and R, circuit would be either capacitive, inductive or purely resistive. All possibilities can be there.

(iv) In purely resistive AC circuit, current and voltage are in same phase.

27. (c) When resistance is lowered $Z = \sqrt{R^2 + (X_L - X_C)^2}$ reduces and so I_{max} increases.

Thus, **(i) → (q)**

When capacitance is changed to C_1, resonance occurs at a lower frequency.

Thus, **(ii) → (p)**

When inductance is increased, resonance occurs at a higher frequency.

Thus, **(iii) → (r)**

Hence, (i) → (q), (ii) → (p), (iii) → (r)

28. (8) At resonance,

$$\Rightarrow \quad I = \frac{V}{R} = \frac{60}{120} = \frac{1}{2} \text{ A}$$

and as $\quad V_L = IX_L = I\omega L$

$$\Rightarrow \quad L = \frac{V_L}{I\omega} = \frac{40}{\frac{1}{2} \times 4 \times 10^5} = 0.2 \text{ mH}$$

Also, $\quad \omega_r = \frac{1}{\sqrt{LC}}, C = \frac{1}{L\omega_r^2}$

$$\Rightarrow \quad C = \frac{1}{0.2 \times 10^{-3} \times (4 \times 10^5)} = \frac{1}{32} \mu\text{F}$$

In case of series L-C-R circuit

$$\frac{X_L - X_C}{R} = \tan \phi$$

As, current lags behind voltage by $\frac{\pi}{4}$ rad.

$$X_L - X_C = R$$

$$\Rightarrow \quad L\omega - \frac{1}{C\omega} = R$$

$$\Rightarrow \quad \omega \times 2 \times 10^{-4} - \frac{1}{\omega \times \left(\frac{1}{32}\right) \times 10^{-6}} = 1 \times 120$$

$$\Rightarrow \quad \omega^2 - 6 \times 10^5 \omega - 16 \times 10^{10} = 0$$

$$\Rightarrow \quad \omega = \frac{6 \times 10^5 \pm \sqrt{(6 \times 10^5)^2 + (64 \times 10^{10})}}{2}$$
$$= 8 \times 10^5 \text{ rad s}^{-1}$$

$\therefore \quad k = 8$

29. (2) Reactance of coil $= X_L = L\omega$

$$= 2\pi \times 600 \times 10^3 \times 1.3 \times 10^{-3}$$
$$= 4900 \, \Omega$$

At resonance, $X_L = X_C$

$X_C = 4900 \, \Omega$

Input resistance $= Q \, X_C = 30 \times 4900 = 147000 \, \Omega$

Required Q for circuit is $\frac{f_0}{50} = \frac{600}{50} = 12$

Equivalent input resistance required $= Q' \, X_C$

$= 12 \times 4900 = 58800 \, \Omega$

Let shunt resistance is R'.

Then, $\quad 58800 = \frac{147000 \, R'}{147000 + R'}$

$\Rightarrow \quad R' = 98 \text{ k}\Omega = 2 \times 49 \text{ k}\Omega$

$\Rightarrow \quad N = 2$

30. (d) When Q-value is high, power dissipation (I^2R) is less.

∴ Statement I is incorrect.

Also, $\quad Q = \frac{\omega_0 L}{R} = \frac{2\pi f_0 LI^2}{I^2 R} = 2\pi \frac{(LI^2)}{\left(\frac{I^2 R}{f_0}\right)}$

$$= 2\pi \frac{\text{Stored energy}}{\text{Energy consumed per sec}}$$

Hence, Statement II is correct.

31. (b) Both Statements are true but Statement II does not explain Statement I.

If f_1 and f_2 are half power point frequencies, then

At f_1, $\quad X_C - X_L = R$

$$\Rightarrow \quad \frac{1}{2\pi f_1 C} - 2\pi f_1 L = R$$

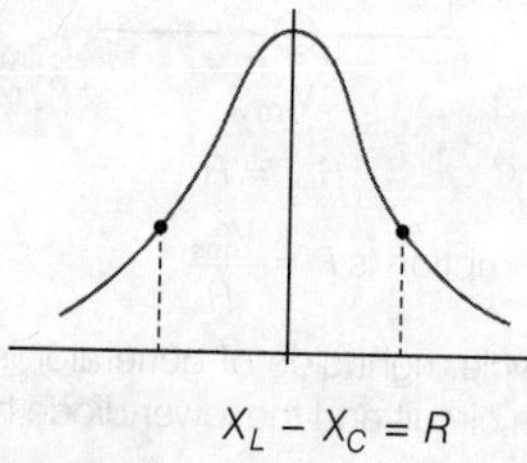

At f_2, $\quad X_L - X_C = R$

$$2\pi f_2 L - \frac{1}{2\pi f_1 C} = R$$

Subtracting, we get

$$\left(\frac{1}{2\pi f_1 C} - 2\pi f_1 L\right) - \left(2\pi f_2 L - \frac{1}{2\pi f_2 C}\right) = 0$$

$$\Rightarrow \quad \left(\frac{1}{\omega_1} + \frac{1}{\omega_2}\right)\frac{1}{C} = (\omega_1 + \omega_2) L_1$$

$$\Rightarrow \quad \frac{1}{\omega_1 \omega_2} = LC = \frac{1}{\omega_r^2}$$

$$\Rightarrow \quad \omega_r = \sqrt{\omega_1 \omega_2}$$

When $\omega_1 = 2$ and $\omega_2 = 8$, then

$$\omega_r = \sqrt{2 \times 8} = 4 \text{ s}^{-1}$$

DPP-5 Transformer

1. (i) Emf equation of transformer is $E_P = \sqrt{2} \times \pi f N_1 B_m A$

where, B_m = maximum flux density in the core. Substituting the values, we have

$$500 = \sqrt{2}\pi \times 50 \times 400 \times B_m \times 60 \times 10^{-4}$$

$$\Rightarrow \quad B_m = 0.938 \text{ Wb/m}^2$$

(ii) As, $\dfrac{E_S}{E_P} = \dfrac{V_S}{V_P} = \dfrac{N_S}{N_P}$

$$\Rightarrow \quad V_S = V_P \frac{N_S}{N_P} = 500 \times \frac{1000}{400} = 1250 \text{ V}$$

2. Here, $h = 300$ m, $V = 100 \text{ m}^3/\text{s}$, $\eta = 60\%$

Electric power = ?, $g = 9.8 \text{ m/s}^2$

As, hydroelectric power $= \dfrac{\text{Work done}}{\text{Time taken}} = \dfrac{\text{Force} \times \text{Distance}}{\text{Time taken}}$

$= \text{Force} \times \text{Velocity}$

$= (\text{Pressure} \times \text{Area}) \times \text{Velocity}$

But Area × Velocity = Volume/Second = V

$\therefore$ Hydroelectric power $= p \times V$

As efficiency is 60%, therefore

$$\text{Power available} = \frac{60}{100} pV = \frac{3}{5}(h\rho g)\, V$$

where, $\rho = 10^3 \text{ kg/m}^3$ for water

So, power available for plant, i.e.

$$P = \frac{3}{5} \times 300 \times 10^3 \times 9.8 \times 100$$

$$= 1.764 \times 10^8 \text{ W} = 176.4 \text{ MW} \quad (1 \text{ MW} = 10^6 \text{ W})$$

3. (i) Line resistance $= l x = 2 \times 15 \times 0.5 = 15\ \Omega$

The rms value of current in line $= \dfrac{P}{V} = \dfrac{800 \times 1000}{4000} = 200$ A

So, line power loss $= I_{rms}^2 R = (200)^2 \times 15 = 600$ kW.

(ii) Assuming zero leakage loss, power supplied by plant = line loss + used power = 600 + 800 = 1400 kW.

(iii) Line voltage drop $= IR = 200 \times 15 = 3000$ V

So, step-up transformer at plant is 440 - 7000 V.

4. (i) Here, $n_P = 2000$, $n_S = 50$, $E_P = 120$ V, $R_S = 0.6\ \Omega$

According to conservation of energy required that

$$\therefore \quad \frac{E_P}{E_S} = \frac{n_P}{n_S}$$

$$\Rightarrow \quad E_S = \frac{n_S}{n_P} \times E_P = \frac{50}{2000} \times 120 = 3 \text{ V}$$

$\therefore$ Voltage across the secondary coil, i.e. $E_S = 3$ V

(ii) So, current in the bulb, i.e. $I_S = \dfrac{E_S}{R_S} = \dfrac{3}{0.6} = 5$ A

(iii) As we know that $\dfrac{I_P}{I_S} = \dfrac{n_S}{n_P}$

So, current in primary coil,

$$I_P = \frac{n_S}{n_P} \times I_S = \frac{50}{2000} \times 5 = 0.125 \text{ A}$$

(iv) As power in primary and secondary coil

i.e. $E_P I_P = E_S I_S = 120 \times 0.125 = 15$ W

5. (d) Assuming that transformer to be ideal one so that there are no energy losses, then

$$\frac{V_1}{V_2} = \frac{N_1}{N_2}$$

$$V_2 = \frac{N_2}{N_1} \times V_1 = 8 \times 120$$

$$I_2 = \frac{8 \times 120}{10^4} = 96 \text{ mA}$$

6. (c) $P_{in} = 240 \times 0.7 = 168$ W, $P_{out} = 140$ W

Efficiency of the transformer, i.e.

$$\eta = \frac{P_{out}}{P_{in}} \times 100 = \frac{140}{168} \times 100 = 83.3\%$$

7. (b) Here, $\eta = 80\%$, $P_i = 4 \text{ kW} = 4000$ W

$E_P = 100$ V, $E_S = 240$ V, $I_P = ?$, $I_S = ?$

As, input power $(P_i) = E_P I_P$

$$\therefore \quad I_P = \frac{P_i}{E_P} = \frac{4000}{100} = 40 \text{ A}$$

As, $\eta = \dfrac{E_S I_S}{E_P I_P} \Rightarrow \dfrac{80}{100} = \dfrac{240\, I_S}{4000}$

$$\Rightarrow \quad I_S = \frac{80 \times 4000}{100 \times 240} = 13.3 \text{ A}$$

8. (c) Here, $V = 200$ V, $I = 5$ A, $\eta = ?$

$$n = \frac{3000}{60} = 50 \text{ rps}, R = 8.5\ \Omega$$

If E is the back emf, then $I = \dfrac{V - E}{R}$

or $V - E = IR$

$\Rightarrow \quad E = V - IR = 200 - 5 \times 8.5 = 157.5$ V

Input power, $P_{in} = VI = 200 \times 5 = 1000$ W

Output power $P_{out} = VI - I^2R = I(V - IR) = IE$

$= 5 \times 157.5 = 787.5$ W

$$\text{Efficiency, } \eta = \frac{\text{Output power } (P_o)}{\text{Input power } (P_i)}$$

$$= \frac{787.5}{1000} = 0.7875 = 78.75\%$$

9. (d) $P_{in} = P_{out} \Rightarrow V_{in} I_{in} = V_{out} I_{out}$

$220 \times 5 = 22000 \times I_{out}$

$I_{out} = 0.05$ A

10. (a) As, we know that long distance power transmission at low voltage and high current is neither efficient nor economical. If I is the current in the cable and R its resistance, the power wasted in the cable is I^2R. Since, $I = \dfrac{P}{V}$ for a given amount of power P, the power loss is less if I is less or V is high.

Thus, $P = VI$ = constant

At high V, I is less and loss $= I^2R$ will be less.

11. (a) In a step-up transformer, $N_S > N_P$, i.e. the turns ratio is greater than 1 and therefore, $V_S > V_P$. The output voltage is greater than the input voltage.

So, $i_P > i_S$

12. (b) For transporting DC, both the wires are equally suitable, but for transporting AC we prefer wire of multiple strands.

13. (b,d) As, we know that efficiency of a transformer is defined as the ratio of output power to the input power.

In an ideal transformer, where there is no power loss, i.e. $\eta = 1$, (i.e. 100%).

So, flux per turn in primary is equal to flux per turn in secondary. While power associated with primary coil at any moment equals to power associated with secondary coil.

14. (a,b,c,d) In the transmission of AC over long distances. The loss of power in the transmission line is I^2R, where I is strength of a current and R is the resistance of the wires. To reduce the power loss, AC is transmitted over long distances at extremely high voltages. This reduces I in the same ratio.

Therefore, I^2R becomes negligibly low. As I has been reduced sufficiently, I^2R remains negligible even when R is not very small. It means we can use even thin line wires of large resistance R instead of thick one. This saves us a lot of material (copper). Therefore, cost of transmission is reduced considerably.

15. (b) In transformer, $\frac{N_P}{N_S} = \frac{I_S}{I_P}$

Current in secondary coil

$$I_S = I_P \frac{N_P}{N_S} = 4 \times \frac{140}{280} = 2\text{ A}$$

16. (a,c) Power loss $= I_{rms}^2 R = \left(\frac{P}{\Delta V_{rms}^2}\right)^2 R = \left(\frac{P}{\Delta V_{rms}}\right)^2 \cdot (x\, l)$

$$= \left(\frac{5\times 10^6}{5\times 10^5}\right)^2 4.5\times 10^{-4} \times 6.44\times 10^5$$

$$= (10)^2 \times 290 = 29\text{ kW}$$

It is impossible to transmit so much power at such low voltage. Maximum power transfer occurs when load resistance equals to the line resistance of 290 Ω is

$$\frac{(4.5\times 10^3)^2}{2.2\times 290} = 17.2\text{ kW}$$

which is much below 5000 kW.

17. (b,c) Efficiency $= \frac{\text{Useful output}}{\text{Total input}} = \frac{2.7}{8} = 0.34 = 34\%$

Power loss = Output power − Input power

$$= 8 - 2.7 = 5.3\text{ W}$$

18. (a) During step-up transformer,

$$K = \frac{N_P}{N_S} = \frac{V_P}{V_S} \cdot \frac{1}{10} = \frac{4000}{V_S}$$

$\Rightarrow$ $V_S = 40000$ V

While in step-down transformer, we get

$$\frac{N_P}{N_S} = \frac{V_P}{V_S} = \frac{40000}{200} = \frac{200}{1}$$

19. (b) $P = VI$

$\Rightarrow$ $I = \frac{P}{V} = \frac{600\times 10^3}{4000} = 150$ A

Total resistance of cable,

$$R = 0.4 \times 20 = 8\,\Omega$$

Power loss during transmission $= I^2R = (150)^2\,(8) = 180000$ W

$= 180$ kW (1 kW = 1000 W)

Thus, power dissipated during transmission is 30% of 600 kW.

20. (c) Using $\frac{V_2}{V_1} = \frac{N_2}{N_1} = n$

We have $V_1 = 1000$, $n = \frac{1}{5}$

$\Rightarrow$ $V_2 = \frac{1}{5} V_1 = \frac{1}{5} \times 1000 = 200$ V

21. (c) Rated output voltage = 9000 W

Efficiency = 90%

So, $\eta = \frac{\text{Output}}{\text{Input}}$

$\Rightarrow$ Input $= \frac{100}{90} \times 9000 = 10000$ W

So, loss = Input − Output = 10000 − 9000 = 1000 W

As from loss = 700 W

So, copper loss (total) = 1000 − 700 = 300 W.

Now, for secondary,

$$P = V_2 I_2$$

$\Rightarrow$ $9000 = 200\, I_2$

$\Rightarrow$ $I_2 = \frac{90}{2} = 45$ A

and $\frac{I_1}{I_2} = \frac{V_2}{V_1} = n$

$\Rightarrow$ $I_1 = I_2 \times n = 45 \times \frac{1}{5} = 9$ A

Primary copper loses $= I^2R \times \frac{1}{n} = 9^2 \times 1 \times \frac{100}{90} \approx 100$ W

So, secondary copper loss is around 300 − 100 = 200 W.

22. (b) Secondary power output = VI

Secondary voltage = 200 V

Equating copper loss of secondary I^2R

We get resistance of secondary is 0.1 Ω.

23. (6) As input power, $P_{in} = 4$ kW

Output power, $P_{out} = 90\%$ of $P_{in} = \frac{90}{100} \times 4$ kW = 3.6 kW

So, output power of a transformer, $P = I^2R$

$$3.6\text{ kW} = I^2 Z$$

Impedance, $Z = \frac{3.6\times 10^3}{6\times 6} = 100\,\Omega = \frac{600}{6} = 100\,\Omega$

24. (8) Assuming that no energy is lost as conservation of energy, i.e. $\frac{N_2}{N_1} = \frac{V_2}{V_1}$

So, voltage induced in secondary coil, i.e.

$$V_2 = V_1 \frac{N_2}{N_1} = 200 \times 4 = 800\text{ V}$$

25. (6) Total flux in the core $= \phi_{11} + \phi_{12} = 0.2 + 0.4 = 0.6$ mWb.

Self-inductance of primary coil in $L_1 = \frac{N_1\phi_1}{I_1}$

$\Rightarrow$ $L_1 = \frac{500\times 0.6\times 10^{-3}}{5.0} = 60$ mH

M = coefficient of mutual induction $= \frac{N_2\phi_{12}}{I_1} = \frac{1500\times 0.4\times 10^{-3}}{5}$

$$= 120\times 10^{-3} = 120\text{ mH}$$

Coefficient of coupling, $k = \frac{\phi_{12}}{\phi_{11}} \times \frac{0.4\times 10^{-3}}{0.6\times 10^{-3}} = \frac{2}{3}$

and $M = k\sqrt{L_1 L_2}$

$\Rightarrow$ $120\times 10^{-3} = \frac{2}{3}\sqrt{60\times L_2} \times 10^{-3}$

$\Rightarrow$ $L_2 = 540\times 10^{-3} = 540$ mH

$\Rightarrow$ $L_2 = 90\times 6$ mH

26. (6) Primary current,

$$I_P = \frac{\text{Total power absorbed by load}}{V_P}$$

$\Rightarrow$ $V_P = \frac{8\times 100 + 200 + 15000}{6.67}$

$$= \frac{16000}{(20/3)} = 2400\text{ V}$$

and $\frac{N_P}{N_S} = \frac{V_P}{V_S}$

$\Rightarrow$ $N_P = N_S \times \frac{V_P}{V_S} = 72 \times \frac{2400}{240} = 720$

$\Rightarrow$ $\frac{N_P}{120} = 6$

27. (d) In AC generator, slip ring arrangement is used. In DC generator, split ring arrangement is used. Choke coil is the best way of reducing AC. A transformer work on the principle of mutual induction.

Revisal Problems (JEE Main)

1. (d) For instantaneous emf to be 25 V, we must have

$$25 = 50 \sin (314t)$$

or $\frac{1}{2} = \sin (314t)$ or $\frac{\pi}{6} = 314t$

or $t = \frac{\pi}{6 \times 314} = \frac{1}{600}$ s

2. (d) The time difference equivalent to phase difference $\Delta\phi$ is given by

$$\frac{\Delta\phi}{2\pi} = \frac{\Delta t}{T}$$

Thus, $\Delta t = \left(\frac{T}{2\pi} \times \Delta\phi\right) = \frac{1}{50} \times \frac{1}{2\pi} \times \frac{\pi}{4} = \frac{1}{400} = 2.5 \times 10^{-3}$ s

3. (b) The frequency of the AC = number of rotations of the coil per second × number of pairs of pole

$$= \frac{600}{60} \times 5 = 50$$

[10 poles means 5 pairs of poles (*N-S*)]

4. (c) From the first data, resistance of the solenoid is $R = 12/2 = 6\,\Omega$. From the second data, the impedance of the solenoid is

$$Z = \frac{12\text{ V}}{1\text{ A}} = 12\,\Omega$$

Since, $Z = \sqrt{R^2 + (\omega L)^2}$ or $Z^2 = R^2 + (\omega L)^2$

We have

$$12^2 = 6^2 + (2\pi \times 50L)^2$$

or $100\pi L = \sqrt{12^2 - 6^2} = \sqrt{108} = 10.4$

$$L = \frac{10.4}{314} = 33\text{ mH}$$

5. (b) $V_0 = \sqrt{V_R^2 + V_C^2}$

$$V_C^2 = V_0^2 - V_R^2 = 220^2 - 110^2 = 110^2(4 - 1)$$

or $V_C = 110\sqrt{3} = 190$ V

6. (b) Inductive reactance in the circuit is

$$X_L = \omega L = 2\pi fL$$

$$= 2 \times 3.14 \times (500) \times (8.1 \times 10^{-3})$$

$$= 25.4\,\Omega$$

Capacitive reactance in the circuit is

$$X_C = \frac{1}{\omega C} = \frac{1}{2\pi fC} = \frac{1}{2 \times 3.14 \times 500 \times (12.5 \times 10^{-6})} = 25.4\,\Omega$$

Since, $X_L = X_C$, then impedance is

$$Z = \sqrt{[R^2 + (X_L - X_C)^2]} = R = 10\,\Omega$$

$\therefore$ $i_{rms} = \frac{E_{rms}}{Z} = \frac{100\text{ V}}{10\,\Omega} = 10$ A

Hence, the potential difference across the resistance is

$$i_{rms} \times R = 10\text{ A} \times 10\,\Omega = 100\text{ V}$$

7. (b) Power factor,

$$\cos\phi = \frac{\text{Real power}}{V_{rms} I_{rms}}$$

or $\cos\phi = \frac{600}{5 \times 160} = 0.75$

8. (c) $P_{out} = V_S I_S$

Now, $V_S = \frac{N_S}{N_P} \times V_P = \frac{1}{20} \times 2500 = 125$ V

Thus, $P_{out} = 125 \times 80 = 10$ kW

(It is assumed that the load is resistive so that the power factor is unity).

9. (d) The reactance $X_L = \omega L = 100 \times 1$

This impedance is

$$Z = \sqrt{R^2 + X_L^2} = 100\sqrt{2}$$

The current (peak), $I_0 = \frac{E_0}{Z} = \frac{200}{100\sqrt{2}} = \sqrt{2}$ A

Thus, $P = E_{rms} I_{rms} \cos\phi = \frac{E_0 I_0}{2} \times \frac{R}{Z} = \frac{200 \times \sqrt{2}}{2} \times \frac{100}{100\sqrt{2}}$

$$= 100\text{ W}$$

10. (c) The capacitive reactance is

$$X_C = \frac{1}{\omega C} = \frac{1}{2\pi fC} = \frac{1}{2\pi \times 50 \times \left(\frac{25}{\pi} \times 10^{-6}\right)} = 4000\,\Omega$$

The impedance of the circuit is

$$Z = \sqrt{(R^2 + X_C^2)} = \sqrt{[(3000)^2 + (4000)^2]} = 5000\,\Omega$$

Power factor, $\cos\phi = \frac{R}{Z} = \frac{3000}{5000} = 0.6$

Power dissipation,

$$P = V_{rms} \times I_{rms} \times \cos\phi$$

$$= V_{rms} \times \frac{V_{rms}}{Z} \times \cos\phi$$

$$= 200 \times \frac{2000}{5000} \times 0.6 = 4.8\text{ W}$$

11. (b) Initially, the current lags behind the potential difference. Hence, the circuit contains resistance and inductance. The power of the circuit is

$$P = V_{rms} \times i_{rms} \times \cos\phi$$

But $i_{rms} = \frac{V_{rms}}{Z}$

where, $Z = \sqrt{[R^2 + (\omega L)^2]}$ is the impedance of the circuit.

$\therefore$ $P = V_{rms} \times \frac{V_{rms}}{Z} \times \cos\phi$

or $Z = \frac{(V_{rms})^2 \times \cos\phi}{P} = \frac{(220)^2 \times 0.8}{550} = 70.4\,\Omega$

Power factor, $\cos\phi = \frac{R}{Z}$

$\therefore$ $R = Z\cos\phi = 70.4 \times 0.8 = 56.32\,\Omega$

Now, $Z^2 = R^2 + (\omega L)^2$

$\therefore$ $(\omega L)^2 = Z^2 - R^2 = (70.4)^2 - (56.4)^2 = 1784$

$\therefore$ $\omega L = 42.2\,\Omega$

The impedance of the circuit after inserting the capacitance is given by

$$Z = \sqrt{\left[R^2 + \left(\omega L - \frac{1}{\omega C}\right)^2\right]}$$

Now, the power factor is given by

$$\cos\phi = \frac{R}{Z} = \frac{R}{\sqrt{\left[R^2 + \left(\omega L - \frac{1}{\omega C}\right)^2\right]}}$$

Clearly, for making power factor = 1, it must be that

$$\omega L = \frac{1}{\omega C}$$

or $C = \frac{1}{\omega(\omega L)}$

But, $\omega = 2\pi f = 2 \times 3.14 \times 50 = 314$

$\therefore$ $C = \frac{1}{314 \times 42.2} = 75 \times 10^{-6}\text{ F} = 75\,\mu\text{F}$

12. (a) Resonant frequency, $\omega_R = \frac{1}{\sqrt{(LC)}}$

$$= \frac{1}{\sqrt{(10 \times 10^{-3}\text{ H}) \times (1 \times 10^{-6}\text{ F})}} = 10^4 \text{ per second}$$

The frequency 10% lower than this is

$$\omega = 10^4 - 10^4 \times \frac{10}{100} = 9 \times 10^3 \text{ per second}$$

At this frequency, we have

$$X_L = \omega L = 9 \times 10^3 \times (10 \times 10^{-3}) = 90\,\Omega$$

and $X_C = \frac{1}{\omega C} = \frac{1}{9 \times 10^3 \times (1 \times 10^{-6})} = 111.11\,\Omega$

$\therefore$ Impedance, $Z = \sqrt{R^2 + (X_L - X_C)^2} = \sqrt{(3)^2 + (90 - 111.11)^2}$

$$= 21.32\,\Omega$$

$\therefore$ Current amplitude is

$$i_0 = \frac{E_0}{Z} = \frac{15}{21.32} = 0.704\text{ A}$$

Average power dissipated is

$$P = \frac{1}{2} E_0 i_0 \cos\phi$$

Here, $\cos\phi = \frac{R}{Z} = \frac{3}{21.32} = 0.141$

$\therefore$ $P = \frac{1}{2} \times 15 \times 0.704 \times 0.141 = 0.744\text{ W}$

13. (c) Inductive reactance in the circuit is

$$\omega L = 2 \times 3.14 \times 750 \times 0.1803 = 850\,\Omega$$

Capacitive reactance

$$\frac{1}{\omega C} = \frac{1}{2 \times 3.14 \times 750 \times (10 \times 10^{-6})} = 21.2\,\Omega$$

Impedance of the circuit,

$$Z = \left[R^2 + \left(\omega L - \frac{1}{\omega C}\right)^2\right]^{\frac{1}{2}}$$

$$= [(100)^2 + (850 - 21.2)^2]^{1/2} = 835\,\Omega$$

Power dissipated, $P = V_{rms}\left[R^2 + \left(\omega L - \frac{1}{\omega C}\right)^2\right]^{\frac{1}{2}} i_{rms} \times \cos\phi$

$$= V_{rms} \times \frac{V_{rms}}{Z} \times \frac{R}{Z} = 20 \times \frac{20}{835} \times \frac{100}{835}$$

$$= 0.0574\text{ W}$$

Heat produced in the resistance = 2 J/°C × 10° C = 10°C = 20 J

Let this heat be produced in t second. Then,

$$Pt = 20\text{ J}$$

or $t = \frac{20\text{ J}}{0.0574\text{ W}} = 348\text{ s}$

14. (a) $\tan 60° = \frac{\omega L}{R}$, $\tan 60° = \frac{1/\omega C}{R}$

$\therefore$ $\omega L = \frac{1}{\omega C}$

$\therefore$ Impedance of the circuit, $Z = \left[R^2 + \left(\omega L - \frac{1}{\omega C}\right)^2\right]^{\frac{1}{2}} = R$

Current in the circuit, $i_0 = \frac{V_0}{Z} = \frac{V_0}{R} = \frac{200}{100} = 2\text{ A}$

Average power, $\overline{P} = \frac{1}{2} V_0 i_0 \cos\phi$

But, $\tan\phi = \frac{\omega L - \frac{1}{\omega C}}{R} = 0$

$\therefore$ $\cos\phi = 1$

$\therefore$ $\overline{P} = \frac{1}{2} \times 200 \times 2 \times 1 = 200\text{ W}$

15. (d) $\omega_r = \frac{1}{\sqrt{LC'}}$, $\omega'_r = \frac{1}{\sqrt{L'C'}}$

For $\omega_r = \omega'_r$, $\frac{1}{\sqrt{LC}} = \frac{1}{\sqrt{L'(4C)}}$

or $LC = L'(4C)$ or $L' = \frac{L}{4}$

16. (c) Suppose the frequency in the alternating circuit is f. If the inductive reactance (ωL) is equal to the capacitive reactance ($1/\omega C$) in the circuit, then there is resonance in the circuit. Thus,

$$\omega L = \frac{1}{\omega C} \text{ or } 2\pi f L = \frac{1}{2\pi f C}$$

or $f = \frac{1}{2\pi}\sqrt{\left(\frac{1}{LC}\right)}$

Here, $L = 0.16$ H and $C = 0.81\,\mu\text{F} = 0.81 \times 10^{-6}$ F

$\therefore$ $f = \frac{1}{2 \times 3.14}\sqrt{\left(\frac{1}{0.16 \times (0.81 \times 10^{-6})}\right)} = 442$ cycles/s

17. (a) If C be the capacity of the condenser connected in the circuit and L be the self-inductance of the coil, then the resonant frequency of the circuit is given by

$$f = \frac{1}{2\pi}\sqrt{\left(\frac{1}{LC}\right)} \text{ or } C = \frac{1}{L \times (2\pi f)^2}$$

Here, $L = 10\text{ mH} = 10 \times 10^{-3}$ H,

$f = 1$ megacycle/s $= 1 \times 10^6$ cycles/s

$\therefore$ $C = \frac{1}{(10 \times 10^{-3}) \times (2 \times 3.14 \times 1 \times 10^6)^2}$

$$= 2.5 \times 10^{-12}\text{ F} = 2.5\text{ pF}$$

18. (c) For an AC ammeter, $Q \propto I^2$

$$90 \propto 25^2 \text{ and } 180 \propto I^2$$

$\Rightarrow$ $2 = \left(\frac{I}{25}\right)^2$

$\Rightarrow$ $I = 25\sqrt{2}$ A

19. (a) Reactance of coil,

$$X_C = L\omega = 2\pi \times 600 \times 10^3 \times 1.3 \times 10^{-3} = 4900\,\Omega$$

At resonance $X_L = X_C = 4900\,\Omega$

Input resistance $= QX_C = 30 \times 4900 = 147000\,\Omega$

Bandwidth requires $= 50 \times 10^3$ Hz

So, required Q-value of circuit

$$Q = \frac{f_0}{50} = \frac{600}{50} = 12$$

Equivalent input resistance $= QX_C = 12 \times 4900 = 58800$

Let resistance used are shunt $= R'$

Then, $58800 = \frac{147000\,R'}{147000 + R'}$

$\Rightarrow$ $R' = \frac{147000 \times 58800}{88200} = 98\text{ k}\Omega$

20. (b) Both Statements I and II are correct but Statement II is not the correct explanation of Statement I.

At resonance,

$$V_L = QV_R$$

$\Rightarrow$ $\frac{V_L}{V_R} = Q > 1$

21. (d) Statement I is incorrect and Statement II is correct.

$i_1 + i_2$ does not have an amplitude of 60 A.

22. (b) Both Statements I and II are correct but Statement II is not the correct explanation of Statement I.

Impedance is not a sinusoidally varying quantity.

$\therefore$ It is not a phasor.

23. (d) Statement I is incorrect. Charge on capacitor of an *L-C-R* has an amplitude.

$$Q_0 = CV_0 = CI_0X_C = C\frac{V_0}{Z}X_C = \frac{CV_0}{\omega C\sqrt{R^2 + \left(\omega L - \frac{1}{\omega C}\right)^2}}$$

$$= \frac{V_0}{\sqrt{\omega^2R^2 + \omega^2(L_1 - 1/C)^2}}$$

$\frac{dQ_0}{d\omega} = 0$ when Q_0 is minimum.

$$\Rightarrow \quad \omega = \sqrt{\frac{1}{LC} - \frac{R^2}{2L^2}}$$

Also, Statement II is correct.

24. (a) For input power = Output power,

$$I_P^2Z_P = I_S^2Z_S$$

$$\Rightarrow \quad \left(\frac{N_P}{N_S}\right)^2 = \left(\frac{I_S}{I_P}\right)^2, \quad \frac{N_P}{N_S} = \sqrt{\frac{Z_P}{Z_S}} = \sqrt{\frac{45\times10^3\ \Omega}{8\ \Omega}} = 75$$

25. (a) $U = \frac{1}{2}LI^2e^{-2(R/L)t}$

$$\Rightarrow \quad U = U_0e^{-2(R/L)\,t}$$

When $t = 2.3\,\tau = 2.3\frac{L}{R}$

$$U = U_0e^{-2(2.3)} = U_0\,e^{-4.6}$$

$$= 0.01/U_0 = \frac{1}{100}U_0$$

Revisal Problems (JEE Advanced)

1. (a) As we know, $X_L = X_C$

So, it is condition of resonance in *L-C-R* circuit.

where, V_L lags current by 90° and V_C lags current by 90°.

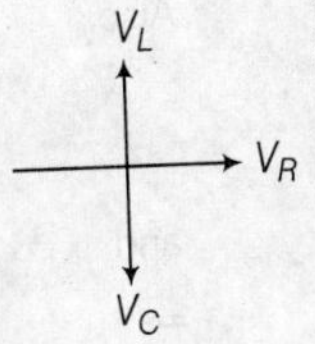

Thus, *I* is in phase with applied voltage.

So, the V_R is Leads V_C by 90°.

2. (a) As heat produced in an *L-C-R* circuit, i.e. Heat = $(i_{rms})^2\,Rt$

The rms value of current $i_{rms} = \frac{i_0}{\sqrt{2}} = \frac{25}{Z\sqrt{2}}$

where, $\quad Z = \sqrt{R^2 + (X_L - X_C)^2} = 5\,\Omega$

$$\text{Heat} = \left(\frac{25}{5\sqrt{2}}\right)^2 4\times80 = 4000\ \text{J}$$

Amplitude of wattless current is $i_0 \sin\phi$

where, $\quad \phi = \tan^{-1}\left(\frac{X_L - X_C}{R}\right) \Rightarrow \phi = 37°$

Amplitude of wattless current = $i_0 \sin 37° = \frac{25}{5}\times\frac{3}{5} = 3$ A

3. (c) For given parallel *L-C-R* circuit, we get

$$V_S = V_S \sin\omega t$$

$$I_1 = I_{01}\sin\left(\omega t - \frac{\pi}{2}\right)$$

$$I_2 = I_{02}\sin(\omega t + \theta)$$

Then, $\quad \tan\theta = \frac{X_C}{R}$

So, phase difference $= \theta + \frac{\pi}{2} = \tan^{-1}\left(\frac{X_C}{R}\right) + \frac{\pi}{2}$

4. (d) As, voltage drop across inductance, i.e. $V = -L(dI/dt)$, *V* is proportional to the slope of the *I-t* graph, which is constant and negative for the first half (0 to *T*/2) and positive and constant for the second half (*T*/2 to *T*).

Note $|V| = L\frac{di}{dt}$ *in this case.*

5. (d) As, for potential across capacitor in discharging *R-C* circuit $V = V_0e^{-t/\tau}$, when

Time constant τ of this circuit, $t = \tau$

$$V = V_0e^{-1} = \frac{V_0}{e} = \frac{25}{2.718} = 9.2\ \text{V}$$

Corresponding to $V = 9.2$ V, *t* lies between 100 s and 150 s.

6. (b) Charging current, $I = \frac{E}{R}e^{\frac{-t}{RC}}$

Taking log on both sides, we get $\log I = \log\left(\frac{E}{R}\right) - \frac{t}{RC}$

When *R* is doubled, then slope of curve increases. Also, at $t = 0$, the current will be less.

Graph *Q* represents the best solid curve.

7. (a,c,d) At resonating condition, we have

$$V_1 = V_2 \Rightarrow X_L = X_C$$

So, resonating frequency,

$$f = \frac{1}{2\pi\sqrt{LC}} = 125\ \text{Hz}, \quad I_0 = \frac{V_0}{R} = \frac{200}{100} = 2\ \text{A}$$

$$V_1 = V_2 = IX_L = I(\omega L) = IL\frac{1}{\sqrt{LC}} = 1000\ \text{V}$$

8. (a,d) The line that draws power supply to our home from street supplies alternating current, whose average value/mean value is zero over a cycle. As the line has some resistance ($R \neq 0$), therefore voltage and current differ in phase ϕ such that $|\phi| < \frac{\pi}{2}$.

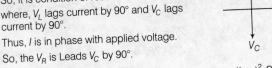

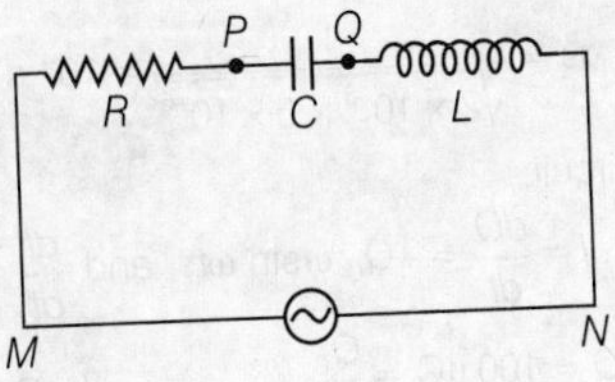

9. (a,b) As we know

$$(f = 900\ \text{Hz}) < (f_{res} = 1000\ \text{Hz})$$

So, $X_C > X_L$ circuit becomes capacitive and current leads emf.

At resistance, impedance is minimum.

Not only at resonance, but also at all conditions voltage across *L* and *C* differ in phase by 180° when connected in series.

10. (a,c) In an *L-C-R* circuit, we have

Resonant frequency,

$$f_{res} = \frac{1}{2\pi\sqrt{LC}} = \frac{1}{2\pi\sqrt{\frac{1}{\pi}\times\frac{1}{\pi}\times10^{-6}}}$$

But at resonance, $Z = R$, we get

So, $\quad$ Current, $I = \frac{V}{Z} = \frac{V}{R}$

When *L* and *C* are in series, then voltage across capacitor and inductor is 180° out of phase.

11. (c) $\tan\phi = \left|\frac{X_L - X_C}{R}\right| = 1$, so $\phi = \frac{\pi}{4}$

So, current flowing in *L-C-R* circuit,

$$I = I_0\sin\left(\omega t + \frac{\pi}{4}\right)$$

As, $\quad X_C > X_L$

So, *I* is leading voltage by $\frac{\pi}{4}$.

$$\text{Current, } I_0 = \frac{V_0}{Z} = \frac{V_0}{\sqrt{2}\,R}$$

12. (d) In an *L-C-R* circuit, phase angle, i.e.

$$\tan\phi = \left|\frac{X_C - X_L}{R}\right| = \left|\frac{\frac{1}{\omega C} - \omega L}{R}\right| = 1, \text{ i.e. } \phi = \frac{\pi}{4}$$

Net reactive impedance, i.e.

$$Z = \sqrt{R^2 + (X_C - X_L)^2} = \sqrt{R^2 + R^2} = \sqrt{2}\,R$$

The rms value of a current,

$$I_{rms} = \frac{V_{rms}}{Z} = \frac{V_0}{\sqrt{2}\,(\sqrt{2}\,R)} = \frac{V_0}{2R}$$

$$I_0 = \sqrt{2}\,I_{rms} \Rightarrow I_0 = \frac{V_0}{\sqrt{2}R}$$

So, potential drop in capacitance is

$$(V_{rms})_C = (I_{rms})\,X_C = \frac{V_0}{2R} \times \frac{1}{\omega C} = \frac{V_0}{2R\omega C}$$

13. (a) Potential across inductance,

$$(V_L)_0 = I_0 X_L = \frac{V_0 \omega L}{\sqrt{2}\,R}$$

$$V_L = (V_L)_0 \sin\left(\omega t + \frac{\pi}{4} + \frac{\pi}{2}\right)$$

as V_L is leading I by $\frac{\pi}{2}$.

$$V_L = \frac{V_0 \omega L}{\sqrt{2}\,R}\sin\left(\omega t + \frac{3\pi}{4}\right)$$

14. (a) Charge stored in the capacitor oscillates $Q = Q_0 \cos\omega t$

where, $Q_0 = 200\,\mu C = 2 \times 10^{-4}$ C

Resonance frequency or oscillation frequency, $\omega = \frac{1}{\sqrt{LC}}$

$$\Rightarrow \quad \omega = \frac{1}{\sqrt{2 \times 10^{-3} \times 5 \times 10^{-6}}} = 10^4\ s^{-1}$$

Current in circuit,

$$I = \frac{dQ}{dt} = -Q_0\,\omega \sin\omega t \quad \text{and} \quad \frac{dI}{dt} = -Q_0\omega^2 \cos\omega t$$

Now, when $Q = 100\,\mu C = \frac{Q_0}{2}$

Then, $Q = Q_0 \cos\omega t$

or $\frac{Q_0}{2} = Q_0 \cos\omega t \Rightarrow \cos\omega t = \frac{1}{2}$ or $\omega t = \frac{\pi}{3}$

Taking $\cos\omega t = \frac{1}{2}$, $\frac{dI}{dt} = -Q_0\omega^2 \cos\omega t$

$$\Rightarrow \quad \left|\frac{dI}{dt}\right| = 2 \times 10^{-4} \times (10^4) \times \frac{1}{2} = 10^4\ As^{-1}$$

15. (c) $I = I_0 \sin\omega t = \frac{dQ}{dt}$

$\Rightarrow \quad I = -Q_0\omega \sin\omega t$

$\therefore \quad |I_{max}| = Q_0\omega = 2 \times 10^{-4} \times 10^4 = 2$ A

16. (b) From energy conservation, $\frac{1}{2}LI^2_{max} = \frac{1}{2}LI^2 + \frac{1}{2}\frac{Q^2}{C}$

or $\quad Q = \sqrt{LC(I^2_{max} - I^2)}$

$$I = \frac{1}{2} I_{max}$$

$$\Rightarrow \quad Q = \sqrt{2 \times 10^{-3} \times 5 \times 10^{-6} \times (2^2 - 1^2)} = \sqrt{3} \times 10^{-4}\ C$$

17. (2) $Q = Q_{max} e^{-Rt/2L} \cos(\omega dt)$

$$I_{max} \propto e^{-Rt/2L}$$

$$\frac{1}{K} = r^{-Rt/2L} \quad \Rightarrow \quad -\log K = -\frac{Rt}{2L}$$

$$\Rightarrow \quad -\log K = \frac{-R}{2L} \times \log 2 \times \frac{2L}{R} \quad \Rightarrow \quad K = 2$$

18. (9) Charge on capacitor is $Q = Q_{max}\cos\omega t$, $\omega = \frac{1}{\sqrt{LC}}$

Energy stored in the capacitor,

$$U = \frac{Q^2}{2C} = \frac{Q^2_{max}\cos^2\omega t}{2C} = U_0 \cos^2\omega t$$

When $U = \frac{1}{4}U_0$

$$U_0\cos^2\omega t = \frac{1}{4}U_0$$

$$\Rightarrow \quad \cos^2\omega t = \frac{1}{4} \Rightarrow \omega t = \frac{\pi}{3}$$

$$\therefore \quad \frac{t}{\sqrt{LC}} = \frac{\pi}{3} \text{ or } \frac{t^2}{LC} = \frac{\pi^2}{9}$$

Inductance, $L = \frac{9t^2}{\pi^2 C}$

19. (d) Phase difference, $\phi = \frac{\pi}{4}$

Also, $\quad \tan\phi = \frac{X_L}{R} \quad \Rightarrow X_L = R$ as $\phi = \frac{\pi}{4}$

$$I_{max} = \frac{V_{max}}{Z} \quad \Rightarrow \quad Z = \frac{V_{max}}{I_{max}} = \frac{100}{10} = 10$$

and $\quad Z^2 = R^2 + X_L^2$

$\Rightarrow \quad Z^2 = 2R^2 \qquad (\because X_L = R)$

$\Rightarrow \quad R^2 = \frac{Z^2}{2} = \frac{100}{2} = 50$

$\Rightarrow \quad R = 5\sqrt{2}$

So, $\quad X_L = 5\sqrt{2}$

and $\quad X_L = L\omega$

$\Rightarrow \quad L = \frac{X_L}{\omega} = \frac{1}{10\sqrt{2}}$

Hence, (i) → (q), (ii) → (q), (iii) → (p), (iv) → (s)

20. (c) Charge on capacitor is

$$q = q_0 \cos\omega t$$

where, q_0 = maximum charge

So, charge is maximum when $\cos\omega t = \pm 1$

or $\quad \omega t = 0, \pi, 2\pi, 3\pi, \ldots\ldots$

Also, $\quad \omega = \frac{1}{\sqrt{LC}}$

∴ Charge is maximum when

$$t = 0, \pi\sqrt{LC}, 2\pi\sqrt{LC}, 3\pi\sqrt{LC}, \ldots\ldots \text{ etc.}$$

Thus, **(i) → (q,s)**

Current in circuit I is

$$I = \frac{dq}{dt} = -q_0 \sin\omega t$$

Hence, current is maximum when $\sin\omega t = \pm 1$

$$\Rightarrow \quad \omega t = \frac{\pi}{2}, \frac{3\pi}{2}, \frac{5\pi}{2}, \frac{7\pi}{2}, \ldots\ldots$$

So, $t = \frac{\pi}{2}\sqrt{LC}, \frac{3\pi}{2}\sqrt{LC}, \frac{5\pi}{2}\sqrt{LC}, \ldots\ldots$ etc.

Thus, **(ii) → (p,r)**

Electrical energy oscillates as

$$U_E = \frac{q_0^2}{2C}\cdot\cos^2(\omega t + \phi)$$

∴ At $t = 0, \pi\sqrt{LC}, 2\pi\sqrt{LC}, 3\pi\sqrt{LC}, \ldots\ldots$ etc.

Thus, **(iii) → (q,s)**

Magnetic energy oscillates as

$$U_B = \frac{1}{2}L\omega^2 Q^2 \sin^2(\omega t + \phi)$$

∴ At $t = \frac{\pi}{2}\sqrt{LC}, \frac{3\pi}{2}\sqrt{LC}, \ldots\ldots$ etc.

Thus, **(iv) → (p,r)**

Hence, (i) → (q,s), (ii) → (p,r), (iii) → (q,s), (iv) → (p,r)

JEE Main & AIEEE Archive

1. (b) With B connected to C, given circuit is a discharging L-R circuit with

$$i = i_0 e^{-t/\tau}, \text{ where } \tau = \text{time constant} = \frac{L}{R}$$

Also, $i = i_0$ at $t = 0$

So, $$i = i_0 e^{-t/\tau}$$

and $$\frac{di}{dt} = \frac{-i_0}{\tau} \cdot e^{-t/\tau}$$

$$V_R = iR = Ri_0 e^{-t/\tau}$$

$$V_L = -L\frac{di}{dt} = \frac{Li_0}{\tau} \cdot e^{-t/\tau} = Ri_0 e^{-t/\tau}$$

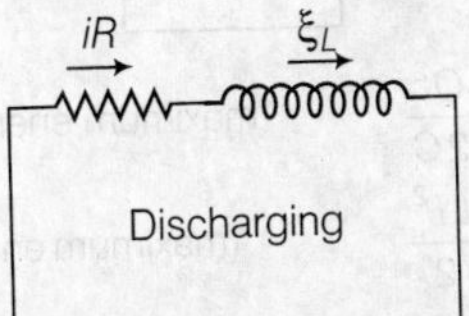

Hence, $\frac{V_R}{V_L} = 1$

2. (a) For charging capacitor, q is given as

$$q = q_0(1 - e^{-t/\tau}) = CV(1 - e^{-t/\tau})$$

At $t = 2\tau$,

$$q = CV\,(1 - e^{\frac{-2\tau}{\tau}})$$

$\Rightarrow$ $$q = CV\,(1 - e^{-2})$$

3. (a) Time constant τ is the duration when the value of potential drops by 37% of its initial maximum value (i.e. V_0/e). Here, 37% of 25 V = 9.25 V which lies between 150 to 200 s in the graph.

4. (d) The given circuit is under resonance as $X_L = X_C$.

($\because$ same phase change in L-R and C-R circuits)

Hence, power dissipated in the circuit is

$$P = \frac{V^2}{R} = 242 \text{ W}$$

5. (d) $I_1 = \frac{E}{R_1} = \frac{12}{4} = 3$ A

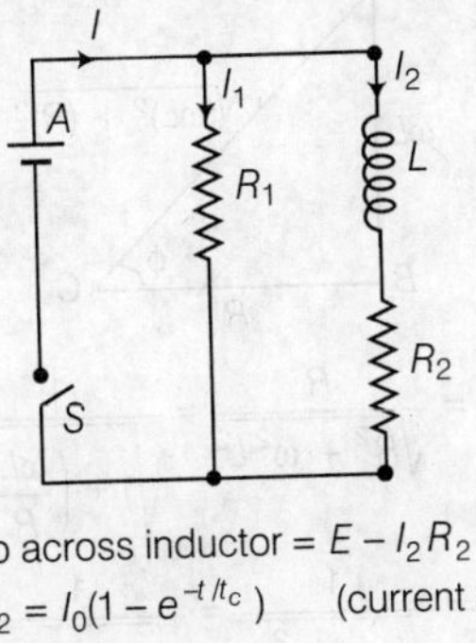

$\because$ Potential drop across inductor $= E - I_2R_2$

$$I_2 = I_0(1 - e^{-t/t_c}) \quad \text{(current as a function of time)}$$

$\Rightarrow$ $$I_0 = \frac{E}{R_2} = \frac{12}{2} = 6 \text{ A}$$

and $$t_c = \frac{L}{R_2} = \frac{400 \times 10^{-3}}{2} = 0.2$$

$$I_2 = 6(1 - e^{-5t})$$

Potential drop across inductor,

$$L = E - R_2I_2$$
$$= 12 - 2 \times 6(1 - e^{-5t})$$
$$= 12e^{-5t} \text{ V}$$

6. (b) For the given circuit, current is lagging the voltage by $\frac{\pi}{2}$, so circuit is purely inductive and there is no power consumption in the circuit. The work done by battery is stored as magnetic energy in the inductor.

7. (a) This is a combined example of discharging L-R circuit.

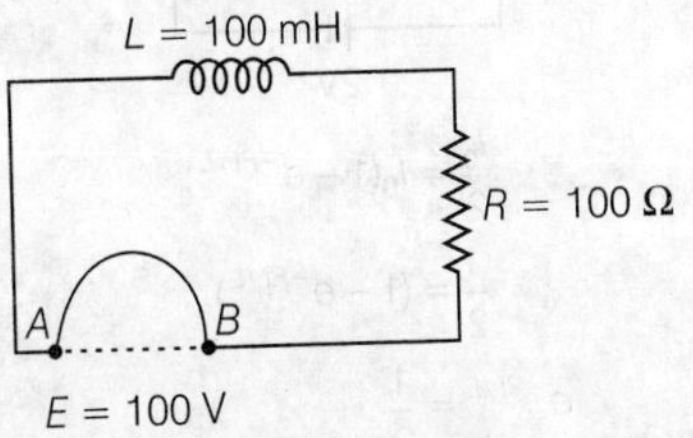

The current through circuit just before shorting the battery,

$$I_0 = \frac{E}{R} = \frac{100}{100} = 1 \text{ A}$$

(as inductor would be shorted in steady state)

After this decay of current starts in the circuit, according to the equation,

$$I = I_0 e^{-t/\tau}$$

where, $\tau = L/R$

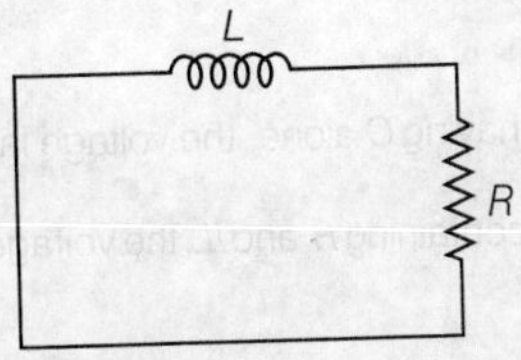

So, current after 1ms will be

$$I = 1 \times e^{-(1 \times 10^{-3})/(100 \times 10^{-3}/100)} = \frac{1}{e} \text{ A}$$

($\because t = 1$ millisecond $= 1 \times 10^{-3}$ s and $L = 100 \times 10^{-3}$ H)

8. (d) The emf generated would be maximum when flux (cutting) would be maximum, i.e. angle between area vector of coil and magnetic field is 0°. The emf generated is given by (as a function of time)

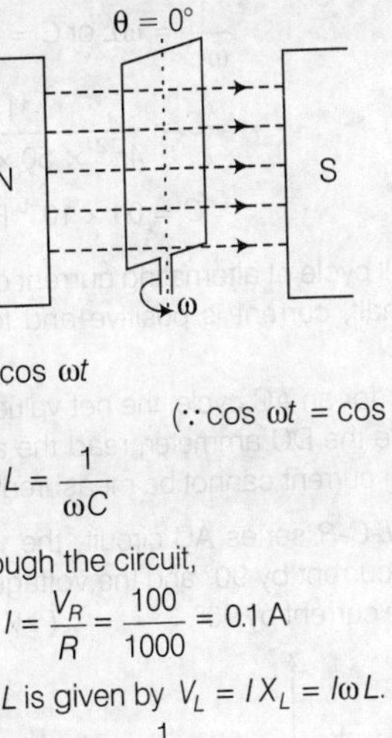

$$e = NBA\omega \cos \omega t$$

$\Rightarrow$ $$e_{max} = NBA\omega \quad (\because \cos \omega t = \cos \theta = 1 \Rightarrow \theta = 0°)$$

9. (c) At resonance, $\omega L = \frac{1}{\omega C}$

Current flowing through the circuit,

$$I = \frac{V_R}{R} = \frac{100}{1000} = 0.1 \text{ A}$$

So, voltage across L is given by $V_L = IX_L = I\omega L$.

But, $$\omega L = \frac{1}{\omega C}$$

$\therefore$ $$V_L = \frac{I}{\omega C} = V_C$$

$$= \frac{0.1}{200 \times 2 \times 10^{-6}} = 250 \text{ V}$$

10. (b) The current at any instant is given by

$$I = I_0(1 - e^{-Rt/L})$$

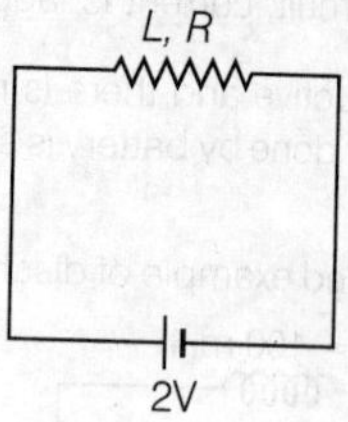

$\Rightarrow \quad \frac{I_0}{2} = I_0(1 - e^{-Rt/L})$

or $\quad \frac{1}{2} = (1 - e^{-Rt/L})$

or $\quad e^{-Rt/L} = \frac{1}{2}$

or $\quad \frac{Rt}{L} = \ln 2$

$\therefore \quad t = \frac{L}{R} \ln 2 = \frac{300 \times 10^{-3}}{2} \times 0.693$

$= 150 \times 0.693 \times 10^{-3}$

$= 0.10395 \text{ s} = 0.1 \text{ s}$

11. (a) Power factor $= \cos\phi = \frac{R}{Z} = \frac{12}{15} = \frac{4}{5} = 0.8$

12. (c)

(a) In a circuit having C alone, the voltage lags the current by $\frac{\pi}{2}$.

(b) In a circuit containing R and L, the voltage leads the current by $\frac{\pi}{2}$.

(c) In an L-C circuit, the phase difference between current and voltage can have any value between 0 to $\frac{\pi}{2}$ depending on the values of L and C.

(d) In a circuit containing L alone, the voltage leads the current by $\frac{\pi}{2}$.

13. (c) Given, $\quad L = 10 \text{ H}, f = 50 \text{ Hz}$

For maximum power, $X_C = X_L$ $\quad$ ($\because$ resonance condition)

or $\quad \frac{1}{\omega C} = \omega L$ or $C = \frac{1}{\omega^2 L}$

$\therefore \quad C = \frac{1}{4\pi^2 \times 50 \times 50 \times 10}$

or $\quad C = 0.1 \times 10^{-5} \text{ F} = 1\,\mu\text{F}$

14. (c) The full cycle of alternating current consists of two half cycles. For one half, current is positive and for second half, current is negative.

Therefore, for an AC cycle, the net value of current average out to zero. While the DC ammeter, read the average value. Hence, the alternating current cannot be measured by DC ammeter.

15. (d) In an L-C-R series AC circuit, the voltage across inductor L leads the current by 90° and the voltage across capacitor C lags behind the current by 90°. $\quad (\because V = V_L \sim V_C = 50 - 50 = 0)$

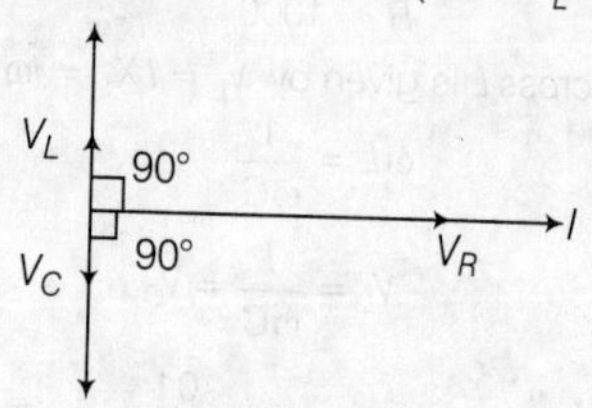

Hence, the voltage across L-C combination will be zero.

16. (c) In the condition of resonance,

$$X_L = X_C \text{ or } \omega L = \frac{1}{\omega C} \quad \ldots(i)$$

Since, resonance frequency remains unchanged, so

$\sqrt{LC}$ = constant or LC = constant

$\therefore \quad L_1 C_1 = L_2 C_2$

or $\quad L \times C = L_2 \times 2C$ or $L_2 = \frac{L}{2}$

17. (c) In an L-C circuit, the energy oscillates between inductor (in the magnetic field) and capacitor (in the electric field).

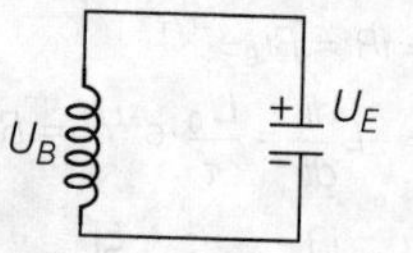

$U_{E_{max}} = \frac{Q^2}{2C}$ (maximum energy stored in capacitor)

$U_{E_{max}} = \frac{Li^2}{2}$ (maximum energy stored in inductor)

where, I is the current at this time.

For the given instant, $U_E = U_B$

i.e. $\quad \frac{q^2}{2C} = \frac{LI^2}{2} \quad \ldots(i)$

From energy conservation,

$U_E + U_B = U_{E_{max}} = U_{B_{max}}$

$\frac{q^2}{2C} + \frac{1}{2}LI^2 = \frac{Q^2}{2C}$

$\Rightarrow \quad \frac{2q^2}{2C} = \frac{Q^2}{2C}$ [from Eq. (i)]

or $\quad q = \frac{Q}{\sqrt{2}}$

18. (a) The core of a transformer is laminated to reduce energy loss due to eddy currents.

19. (b) From the relation, $\tan\phi = \frac{\omega L}{R}$

Power factor, $\cos\phi = \frac{1}{\sqrt{1 + \tan^2\phi}}$

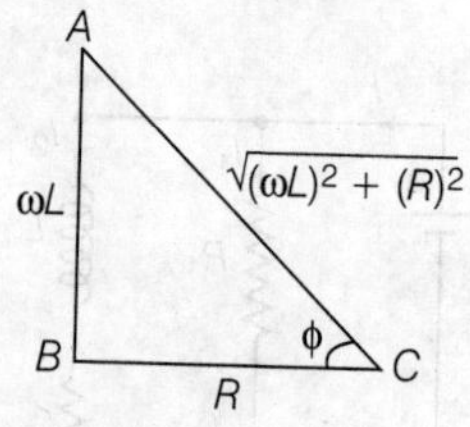

$\Rightarrow \quad \cos\phi = \frac{R}{\sqrt{R^2 + \omega^2 L^2}} = \frac{1}{\sqrt{1 + \left(\frac{\omega L}{R}\right)^2}}$

$= \frac{1}{\sqrt{1 + \tan^2\phi}} = \frac{1}{\sqrt{1 + \left(\frac{\omega L}{R}\right)^2}} = \frac{R}{\sqrt{R^2 + \omega^2 L^2}}$

20. (b) Given, $\quad I_P = 4 \text{ A}, N_P = 140$

and $N_S = 280$

From the formula, $\frac{I_P}{I_S} = \frac{N_S}{N_P}$

or $\quad \frac{4}{I_S} = \frac{280}{140}$

So, $\quad I_S = 2 \text{ A}$

1. (b) $P = Vi$

$\therefore \quad i = \frac{P}{V} = \frac{600 \times 10^3}{4000} = 150$ A

Total resistance of cable,

$R = 0.4 \times 20 = 8\ \Omega$

$\therefore$ Power loss in cable $= i^2R = (150)^2(8)$

$= 180000$ W $= 180$ kW

This loss is 30% of 600 kW.

2. (a) During step-up transformer,

$$\frac{N_P}{N_S} = \frac{V_P}{V_S}$$

or $\quad \frac{1}{10} = \frac{4000}{V_S}$

or $\quad V_S = 40000$ V

In step, down transformer,

$$\frac{N_P}{N_S} = \frac{V_P}{V_S} = \frac{40000}{200} = \frac{200}{1}$$

3. (a,c,d) $\quad \frac{dQ}{dt} = I$

$\Rightarrow \quad Q = \int I\,dt = \int (I_0 \cos \omega t)\,dt$

$\therefore \quad Q_{max} = \frac{I_0}{\omega} = \frac{1}{500} = 2 \times 10^{-3}$ C

Just after switching,

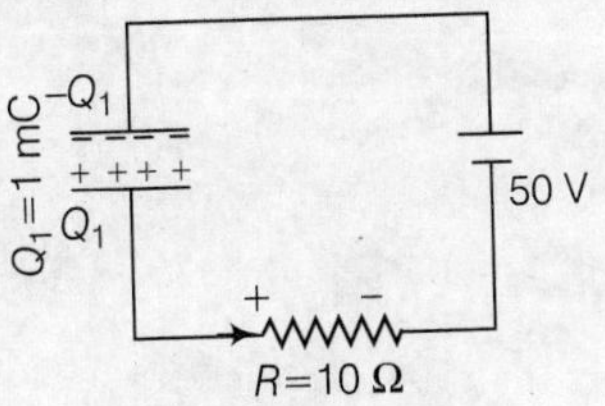

In steady state,

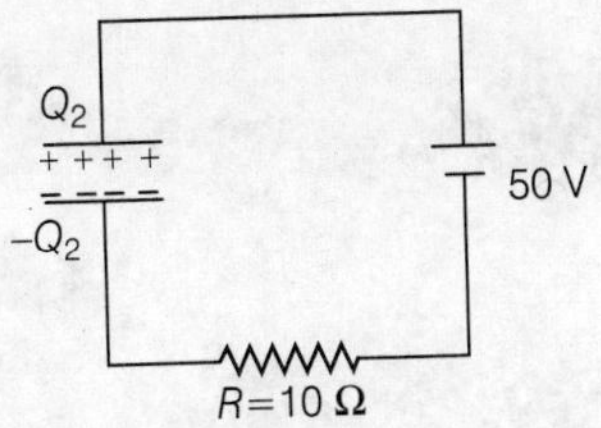

At $\quad t = \frac{7\pi}{6\omega}$ or $\omega t = \frac{7\pi}{6}$

Current comes out to be negative from the given expression. So, current is anti-clockwise.

Charge supplied by source from $t = 0$ to $t = \frac{7\pi}{6\omega}$

$$Q = \int_0^{\frac{7\pi}{6\omega}} \cos (500t)\,dt$$

$$= \left[\frac{\sin 500t}{500}\right]_0^{\frac{7\pi}{6\omega}} = \frac{\sin \frac{7\pi}{6}}{500} = 1 \text{ mC}$$

$\therefore$ Option (a) is correct.

Apply Kirchhoff's loop law just after changing the switch to position D

$$50 + \frac{Q_1}{C} - IR = 0$$

Substituting the values of Q_1, C and R we get

$$I = 10 \text{ A}$$

$\therefore$ Option (c) is correct.

In steady state, $Q_2 = CV = 1$ mC

$\therefore$ Net charge flown from battery = 2 mC.

Option (d) is correct.

4. (b,c) $Z = \sqrt{R^2 + X_C^2} = \sqrt{R^2 + \left(\frac{1}{\omega C}\right)^2}$

In case (b), capacitance C will be more. Therefore, impedance Z will be less. Hence, current will be more.

$\therefore$ Option (b) is correct.

Further,

$$V_C = \sqrt{V^2 - V_R^2} = \sqrt{V^2 - (IR)^2}$$

In case (b), since current I is more.

Therefore, V_C will be less. So, option (c) is correct.

$\therefore$ Correct options are (b) and (c).

5. (4) $\quad Z = \sqrt{R^2 + X_C^2} = R\sqrt{1.25}$

$\therefore \quad R^2 + X_C^2 = 1.25\,R^2$

or $\quad X_C = \frac{R}{2}$

or $\quad \frac{1}{\omega C} = \frac{R}{2}$

$\therefore$ Time constant $= CR = \frac{2}{\omega} = \frac{2}{500}$ s $= 4$ ms

6. (b) $Z = \sqrt{R^2 + X_C^2}$ $I_{rms} = \frac{V_{rms}}{Z}$, $P = I_{rms}^2 R$, where $X_C = \frac{1}{\omega C}$

As ω is increased, X_C will decrease or Z will decrease. Hence, I_{rms} or P will increase. Therefore, bulb glows brighter.

7. (a) In circuit (p), I cannot be non-zero in steady state.

In circuit (q),

$$V_1 = 0 \text{ and } V_2 = 2I = V \text{ (also)}$$

In circuit (r), $V_1 = X_L I = (2\pi f L) I = (2\pi \times 50 \times 6 \times 10^{-3}) I = 1.88\,I$

$$V_2 = 2I$$

In circuit (s), $V_1 = X_L I = 1.88I$

$$V_2 = X_C I = \left(\frac{1}{2\pi f C}\right) I = \left(\frac{1}{2\pi \times 50 \times 3 \times 10^{-6}}\right) I$$

$$= (1061)\,I$$

In circuit (t)

$$V_1 = IR = (1000)\,I$$

$$V_2 = X_C\,I = (1061)\,I$$

Therefore, the correct options are as under

(i) $\rightarrow$ (r,s,t), (ii) $\rightarrow$ (q,r,s,t), (iii) $\rightarrow$ (q,p,q), (iv) $\rightarrow$ (q,r,s,t)

8. (b) Charge on capacitor at time t is

$$q = q_0\,(1 - e^{-t/\tau})$$

Here, $\quad q_0 = CV$ and $t = 2\tau$

$\therefore \quad q = CV\,(1 - e^{-2\tau/\tau}) = CV\,(1 - e^{-2})$

9. (d) From conservation of energy,

$$\frac{1}{2} LI_{max}^2 = \frac{1}{2} CV^2$$

$\therefore \quad I_{max} = V\sqrt{\frac{C}{L}}$

10. (c) Comparing the L-C oscillations with normal SHM, we get

$$\frac{d^2Q}{dt^2} = -\omega^2 Q$$

Here, $\quad \omega^2 = \frac{1}{LC}$

$\therefore \quad Q = -LC\frac{d^2Q}{dt^2}$

Last 3 Years' Questions
JEE Main & Advanced
(2016-2018)

JEE Main

2016

1. Two identical wires A and B, each of length l, carry the same current I. Wire A is bent into a circle of radius R and wire B is bent to form a square of side a. If B_A and B_B are the values of magnetic field at the centres of the circle and square respectively, then the ratio $\frac{B_A}{B_B}$ is

[Magnetic Effect of Current]

(a) $\frac{\pi^2}{8}$ (b) $\frac{\pi^2}{16\sqrt{2}}$ (c) $\frac{\pi^2}{16}$ (d) $\frac{\pi^2}{8\sqrt{2}}$

2. Hysteresis loops for two magnetic materials A and B are as given below:

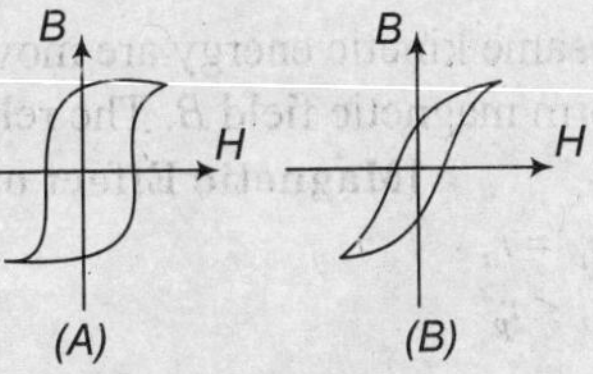

These materials are used to make magnets for electric generators, transformer core and electromagnet core. Then, it is proper to use

[Magnetism]

(a) A for electric generators and transformers (b) A for electromagnets and B for electric generators
(c) A for transformers and B for electric generators (d) B for electromagnets and transformers

3. An arc lamp requires a direct current of 10 A at 80 V to function. If it is connected to a 220 V (rms), 50 Hz AC supply, the series inductor needed for it to work is close to

[Alternating Current]

(a) 80 H (b) 0.08 H (c) 0.044 H (d) 0.065 H

4. A galvanometer having a coil resistance of 100 Ω gives a full scale deflection when a current of 1 mA is passed through it. The value of the resistance which can convert this galvanometer into ammeter giving a full scale deflection for a current of 10 A, is

[Magnetic Effect of Current]

(a) 0.01 Ω (b) 2 Ω (c) 0.1 Ω (d) 3 Ω

2017

5. In a coil of resistance 100 Ω, a current is induced by changing the magnetic flux through it as shown in the figure. The magnitude of change in flux through the coil is

[Electromagnetic Induction]

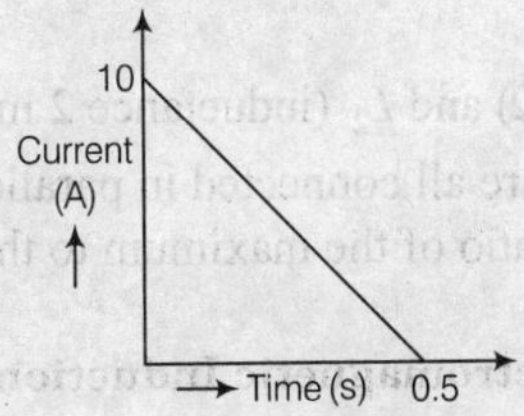

(a) 225 Wb (b) 250 Wb (c) 275 Wb (d) 200 Wb

6. A magnetic needle of magnetic moment 6.7×10^{-2} Am^2 and moment of inertia 7.5×10^{-6} kg m^2 is performing simple harmonic oscillations in a magnetic field of 0.01 T. Time taken for 10 complete oscillations is **[Magnetism]**

(a) 8.89 s (b) 6.98 s

(c) 8.76 s (d) 6.65 s

7. When a current of 5 mA is passed through a galvanometer having a coil of resistance 15 Ω, it shows full scale deflection. The value of the resistance to be put in series with the galvanometer to convert it into a voltmeter of range 0-10 V is **[Magnetic Effect of Current]**

(a) 2.045×10^3 Ω (b) 2.535×10^3 Ω

(c) 4.005×10^3 Ω (d) 1.985×10^3 Ω

2018

8. In an AC circuit, the instantaneous emf and current are given by

$$e = 100 \sin 30t,\ i = 20 \sin\left(30t - \frac{\pi}{4}\right)$$

In one cycle of AC, the average power consumed by the circuit and the wattless current are, respectively

[Alternating Current]

(a) 50 , 10 (b) $\frac{1000}{\sqrt{2}}$, 10

(c) $\frac{50}{\sqrt{2}}$, 0 (d) 50 , 0

9. An electron, a proton and an alpha particle having the same kinetic energy are moving in circular orbits of radii r_e, r_p, r_α respectively, in a uniform magnetic field B. The relation between r_e, r_p, r_α is **[Magnetic Effect of Current]**

(a) $r_e > r_p = r_\alpha$ (b) $r_e < r_p = r_\alpha$

(c) $r_e < r_p < r_\alpha$ (d) $r_e < r_\alpha < r_p$

10. The dipole moment of a circular loop carrying a current I, is m and the magnetic field at the centre of the loop is B_1. When the dipole moment is doubled by keeping the current constant, the magnetic field at the centre of the loop is B_2. The ratio $\frac{B_1}{B_2}$ is **[Magnetism]**

(a) 2 (b) $\sqrt{3}$

(c) $\sqrt{2}$ (d) $\frac{1}{\sqrt{2}}$

11. For an *R-L-C* circuit driven with voltage of amplitude v_m and frequency $\omega_0 = \frac{1}{\sqrt{LC}}$, the current exhibits resonance. The quality factor, Q is given by **[Alternating Current]**

(a) $\frac{\omega_0 L}{R}$ (b) $\frac{\omega_0 R}{L}$

(c) $\frac{R}{\omega_0 C}$ (d) $\frac{CR}{\omega_0}$

JEE Advanced

2016

1. Two inductors L_1 (inductance 1mH, internal resistance 3 Ω) and L_2 (inductance 2 mH, internal resistance 4 Ω), and a resistor R (resistance 12 Ω) are all connected in parallel across a 5V battery. The circuit is switched on at time $t = 0$. The ratio of the maximum to the minimum current (I_{max}/I_{min}) drawn from the battery is

[Electromagnetic Induction, Type 3]

Note *Type-1 Only one correct option,* ***Type-2*** *More then one correct option,* ***Type-3*** *Single digit Answer,* ***Type-4*** *Numerical value based*

2. A rigid wire loop of square shape having side of length L and resistance R is moving along the X-axis with a constant velocity v_0 in the plane of the paper. At $t = 0$, the right edge of the loop enters a region of length $3L$ where there is a uniform magnetic field B_0 into the plane of the paper, as shown in the figure. For sufficiently large v_0, the loop eventually crosses the region. Let x be the location of the right edge of the loop. Let $v(x)$, $I(x)$ and $F(x)$ represent the velocity of the loop, current in the loop, and force on the loop, respectively, as a function of x. Counter-clockwise current is taken as positive.

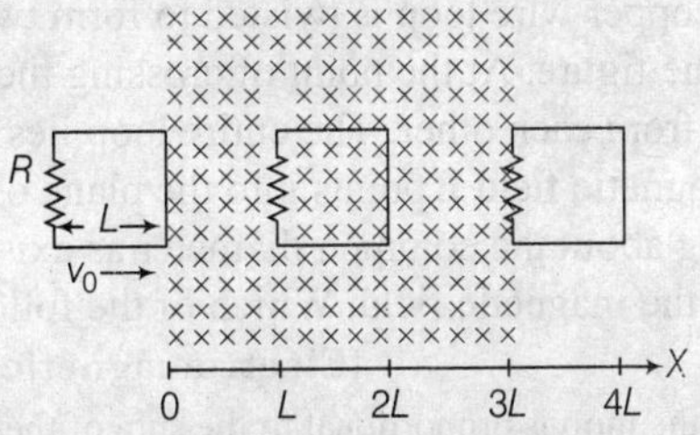

Which of the following schematic plot(s) is (are) correct? (Ignore gravity)

[Electromagnetic Induction, Type 2]

(a)

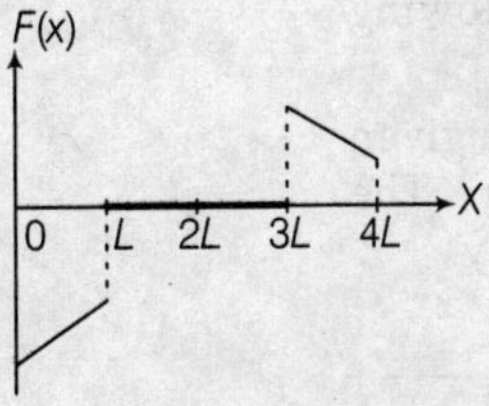

(b)

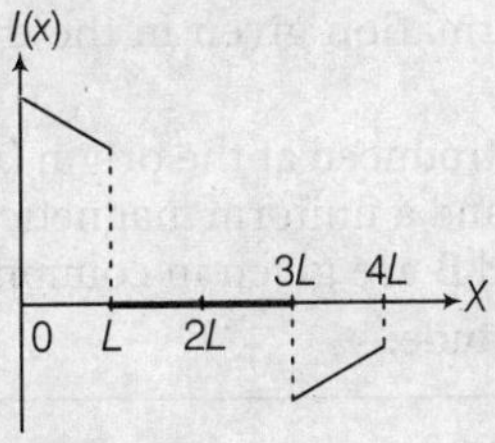

(c)

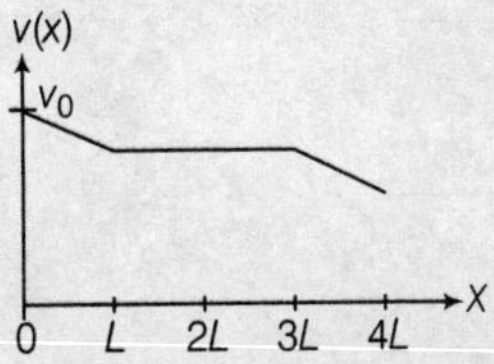

(d)

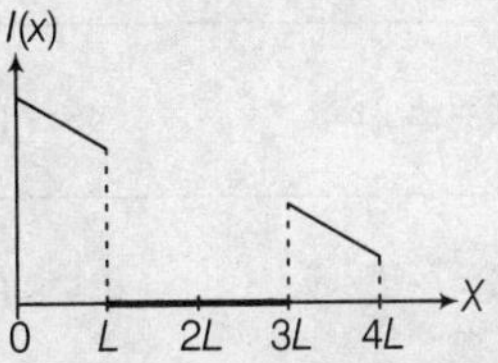

3. A conducting loop in the shape of a right angled isosceles triangle of height 10 cm is kept such that the 90° vertex is very close to an infinitely long conducting wire (see the figure). The wire is electrically insulated from the loop. The hypotenuse of the triangle is parallel to the wire. The current in the triangular loop is in counterclockwise direction and increased at a constant rate of $10\,\mathrm{As}^{-1}$. Which of the following statement(s) is (are) true?

[Electromagnetic Induction, Type 2]

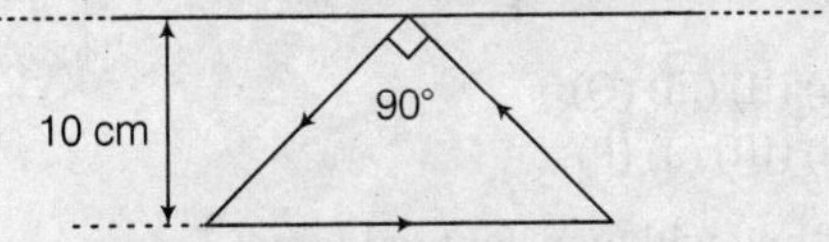

(a) There is a repulsive force between the wire and the loop

(b) If the loop is rotated at a constant angular speed about the wire, an additional emf of $\left(\frac{\mu_0}{\pi}\right)$ volt is induced in the wire

(c) The magnitude of induced emf in the wire is $\left(\frac{\mu_0}{\pi}\right)$ volt

(d) The induced current in the wire is in opposite direction to the current along the hypotenuse

2017

4. In the circuit shown, $L = 1\,\mu\mathrm{H}$, $C = 1\,\mu\mathrm{F}$ and $R = 1\,\mathrm{k}\Omega$. They are connected in series with an AC source $V = V_0 \sin \omega t$ as shown. Which of the following options is/are correct?

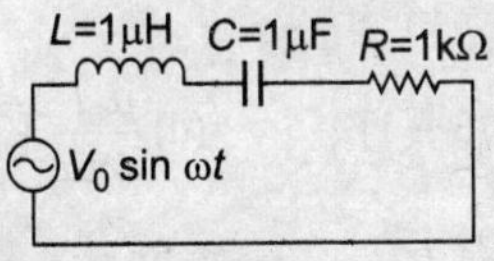

[Alternating Current, Type 2]

(a) At $\omega \sim 0$ the current flowing through the circuit becomes nearly zero

(b) The frequency at which the current will be in phase with the voltage is independent of R

(c) The current will be in phase with the voltage if $\omega = 10^4\,\mathrm{rads}^{-1}$

(d) At $\omega >> 10^6\,\mathrm{rads}^{-1}$, the circuit behaves like a capacitor

5. A circular insulated copper wire loop is twisted to form two loops of area A and $2A$ as shown in the figure. At the point of crossing the wires remain electrically insulated from each other. The entire loop lies in the plane (of the paper). A uniform magnetic field **B** points into the plane of the paper. At $t = 0$, the loop starts rotating about the common diameter as axis with a constant angular velocity ω in the magnetic field. Which of the following options is/are correct? **[Electromagnetic Induction, Type 2]**

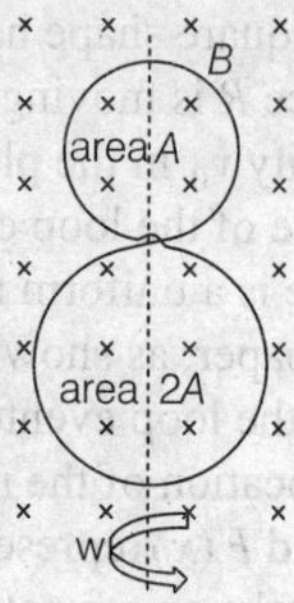

(a) the emf induced in the loop is proportional to the sum of the areas of the two loops
(b) The rate of change of the flux is maximum when the plane of the loops is perpendicular to plane of the paper
(c) The net emf induced due to both the loops is proportional to cos ωt
(d) The amplitude of the maximum net emf induced due to both the loops is equal to the amplitude of maximum emf induced in the smaller loop alone

Directions (Q.Nos. 6-8) Matching the information given in the three columns of the following table.

A charged particle (electron or proton) is introduced at the origin ($x = 0, y = 0, z = 0$) with a given initial velocity **v**. A uniform electric field **E** and a uniform magnetic field **B** exist everywhere. The velocity **v**, electric field **E** and magnetic field **B** are given in columns 1, 2 and 3, respectively. The quantities E_0, B_0 are positive in magnitude.

Column 1	Column 2	Column 3
(I) Electron with $\mathbf{v} = 2\frac{E_0}{B_0}\hat{x}$	(i) $\mathbf{E} = E_0\hat{z}$	(P) $\mathbf{B} = -B_0\hat{x}$
(II) Election with $\mathbf{v} = \frac{E_0}{B_0}\hat{y}$	(ii) $\mathbf{E} = -E_0\hat{y}$	(Q) $\mathbf{B} = B_0\hat{x}$
(III) Proton with $\mathbf{v} = 0$	(iii) $\mathbf{E} = -E_0\hat{x}$	(R) $\mathbf{B} = B_0\hat{y}$
(IV) Proton with $\mathbf{v} = 2\frac{E_0}{B_0}\hat{x}$	(iv) $\mathbf{E} = E_0\hat{x}$	(S) $\mathbf{B} = B_0\hat{z}$

[Magnetic Effect of Current]

6. In which case would the particle move in a straight line along the negative direction of Y-axis (i.e. move along $-\hat{y}$)?
(a) (IV) (ii) (S)
(b) (II) (iii) (Q)
(c) (III), (ii) (R)
(d) (III) (ii) (P)

7. In which case will the particle move in a straight line with constant velocity?
(a) (II) (iii) (S)
(b) (III) (iii) (P)
(c) (IV) (i) (S)
(d) (III) (ii) (R)

8. In which case will the particle describe a helical path with axis along the positive z-direction?
(a) (II) (ii) (R)
(b) (III) (iii) (P)
(c) (IV) (i) (S)
(d) (IV) (ii) (R)

9. A symmetric star shaped conducting wire loop is carrying a steady state current I as shown in the figure. The distance between the diametrically opposite vertices of the star is $4a$. The magnitude of the magnetic field at the center of the loop is **[Magnetic Effect of Current, Type 1]**

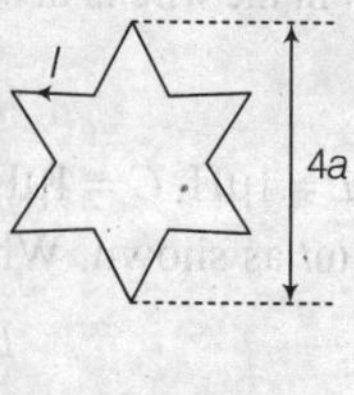

(a) $\frac{\mu_0 I}{4\pi a}6[\sqrt{3} - 1]$
(b) $\frac{\mu_0 I}{4\pi a}6[\sqrt{3} + 1]$
(c) $\frac{\mu_0 I}{4\pi a}3[\sqrt{3} - 1]$
(d) $\frac{\mu_0 I}{4\pi a}3[2 - \sqrt{3}]$

10. A source of constant voltage V is connected to a resistance R and two ideal inductors L_1 and L_2 through a switch S as shown. There is no mutual inductance between the two inductors. The switch S is initially open. At $t = 0$, the switch is closed and current begins to flow. Which of the following options is/are correct? **[Electromagnetic Induction, Type 2]**

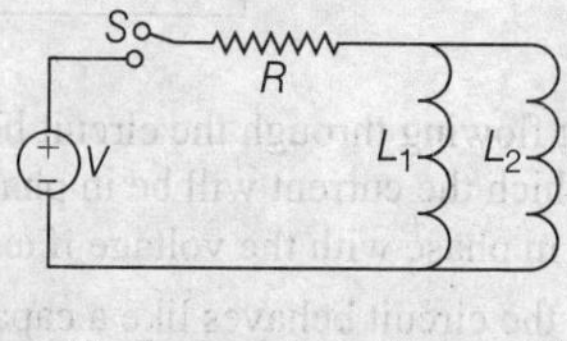

(a) After a long time, the current through L_1 will be $\frac{V}{R}\frac{L_2}{L_1 + L_2}$

(b) After a long time, the current through L_2 will be $\frac{V}{R}\frac{L_1}{L_1 + L_2}$

(c) The ratio of the currents through L_1 and L_2 is fixed at all times ($t > 0$)

(d) At $t = 0$, the current through the resistance R is $\frac{V}{R}$

11. A uniform magnetic field B exists in the region between $x = 0$ and $x = \frac{3R}{2}$ (region 2 in the figure) pointing normally into the plane of the paper. A particle with charge $+Q$ and momentum p directed along X-axis enters region 2 from region 1 at point P_1 $(y = -R)$.

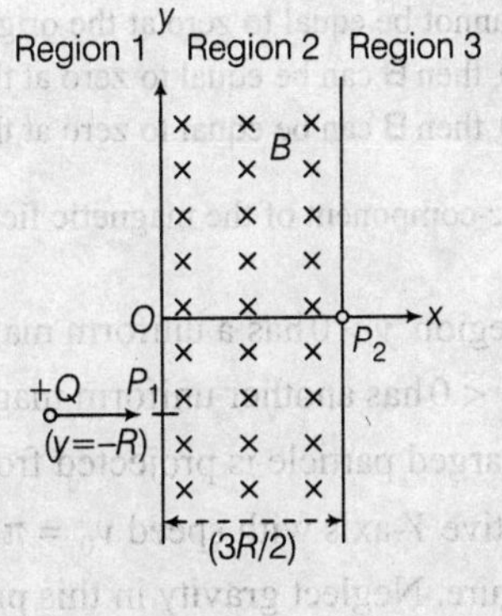

Which of the following option(s) is/are correct?

[Magnetic Effect of Current, Type 2]

(a) When the particle re-enters region 1 through the longest possible path in region 2, the magnitude of the change in its linear momentum between point P_1 and the farthest point from Y-axis is $\frac{p}{\sqrt{2}}$

(b) For $B = \frac{8}{13}\frac{p}{QR}$, the particle will enter region 3 through the point P_2 on X-axis

(c) For $B > \frac{2}{3}\frac{p}{QR}$, the particle will re-enter region 1

(d) For a fixed B, particles of same charge Q and same velocity v, the distance between the point P_1 and the point of re-entry into region 1 is inversely proportional to the mass of the particle

12. The instantaneous voltages at three terminals marked X, Y and Z are given by $V_X = V_0 \sin \omega t$,

$V_Y = V_0 \sin\left(\omega t + \frac{2\pi}{3}\right)$ and $V_Z = V_0 \sin\left(\omega t + \frac{4\pi}{3}\right)$.

An ideal voltmeter is configured to read rms value of the potential difference between its terminals. It is connected between points X and Y and then between Y and Z. The reading(s) of the voltmeter will be **[Alternating Current, Type 2]**

(a) $V_{YZ}^{rms} = V_0\sqrt{\frac{1}{2}}$

(b) $V_{XY}^{rms} = V_0\sqrt{\frac{3}{2}}$

(c) independent of the choice of the two terminals

(d) $V_{XY}^{rms} = V_0$

2018

13. In the figure below, the switches S_1 and S_2 are closed simultaneously at $t = 0$ and a current starts to flow in the circuit. Both the batteries have the same magnitude of the electromotive force (emf) and the polarities are as indicated in the figure. Ignore mutual inductance between the inductors. The current I in the middle wire reaches its maximum magnitude I_{max} at time $t = \tau$. Which of the following statements is (are) true ?

[Electromagnetic Induction, Type 2]

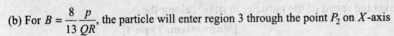

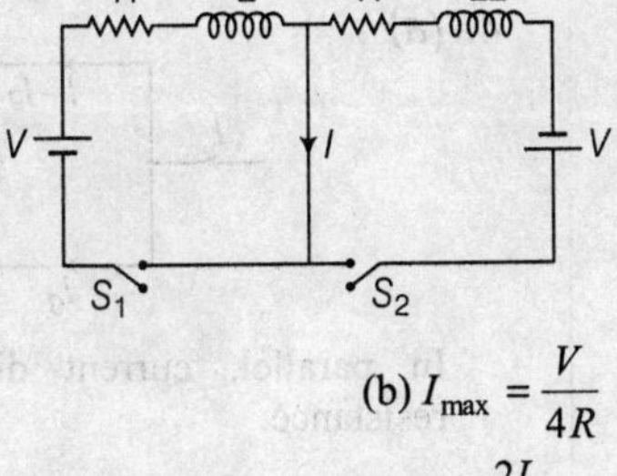

(a) $I_{max} = \frac{V}{2R}$

(b) $I_{max} = \frac{V}{4R}$

(c) $\tau = \frac{L}{R}\ln 2$

(d) $\tau = \frac{2L}{R}\ln 2$

14. Two infinitely long straight wires lie in the xy-plane along the lines $x = \pm R$. The wire located at $x = +R$ carries a constant current I_1 and the wire located at $x = -R$ carries a constant current I_2. A circular loop of radius R is suspended with its centre at $(0, 0, \sqrt{3}R)$ and in a plane parallel to the xy-plane. This loop carries a constant current I in the clockwise direction as seen from above the loop. The current in the wire is taken to be positive, if it is in the $+\hat{\mathbf{j}}$-direction. Which of the following statements regarding the magnetic field **B** is (are) true?

[Magnetic Effect of Current, Type 2]

(a) If $I_1 = I_2$, then **B** cannot be equal to zero at the origin (0,0,0)
(b) If $I_1 > 0$ and $I_2 < 0$, then **B** can be equal to zero at the origin (0,0,0)
(c) If $I_1 < 0$ and $I_2 > 0$, then **B** can be equal to zero at the origin (0,0,0)
(d) If $I_1 = I_2$, then the z-component of the magnetic field at the centre of the loop is $\left(-\dfrac{\mu_0 I}{2R}\right)$

15. In the xy-plane, the region $y > 0$ has a uniform magnetic field $B_1\hat{\mathbf{k}}$ and the region $y < 0$ has another uniform magnetic field $B_2\hat{\mathbf{k}}$. A positively charged particle is projected from the origin along the positive Y-axis with speed $v_0 = \pi \text{ ms}^{-1}$ at $t = 0$, as shown in figure. Neglect gravity in this problem. Let $t = T$ be the time when the particle crosses the X-axis from below for the first time. If $B_2 = 4B_1$, the average speed of the particle, in ms^{-1}, along the X-axis in the time interval T is............ .

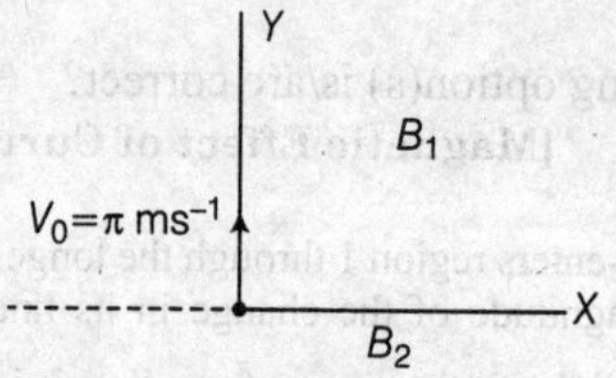

[Magnetism, Type 4]

16. A moving coil galvanometer has 50 turns and each turn has an area $2 \times 10^{-4} \text{ m}^2$. The magnetic field produced by the magnet inside the galvanometer is 0.02 T. The torsional constant of the suspension wire is 10^{-4} N-m rad^{-1}. When a current flows through the galvanometer, a full scale deflection occurs, if the coil rotates by 0.2 rad. The resistance of the coil of the galvanometer is 50 Ω. This galvanometer is to be converted into an ammeter capable of measuring current in the range 0 – 1.0 A. For this purpose, a shunt resistance is to be added in parallel to the galvanometer. The value of this shunt resistance in ohms, is ....

[Magnetic Effect of Current, Type 4]

Answer *with* Explanations

JEE Main

1. *(d)* B at centre of a circle $= \dfrac{\mu_0 I}{2R}$

B at centre of a square $= 4 \times \dfrac{\mu I}{4\pi \cdot \dfrac{l}{2}}[\sin 45° + \sin 45°]$

$= 4\sqrt{2}\dfrac{\mu_0 I}{2\pi l}$

Now, $R = \dfrac{L}{2\pi}$ and $l = \dfrac{L}{4}$ (as $L = 2\pi R = 4l$)

where, L = length of wire.

$\therefore \quad B_A = \dfrac{\mu_0 I}{2 \cdot \dfrac{L}{2\pi}} = \dfrac{\pi \mu_0 I}{L} = \pi\left[\dfrac{\mu_0 I}{L}\right]$

$B_B = 4\sqrt{2}\dfrac{\mu_0 I}{2\pi\left(\dfrac{L}{4}\right)} = \dfrac{8\sqrt{2}\mu_0 I}{\pi L} = \dfrac{8\sqrt{2}}{\pi}\left[\dfrac{\mu_0 I}{L}\right]$

$\therefore \quad \dfrac{B_A}{B_B} = \pi^2 : 8\sqrt{2}$

2. *(d)* We need high retentivity and high coercivity for electromagnets and small area of hysteresis loop for transformers.

3. *(d)* $V^2 = V_R^2 + V_L^2 \Rightarrow 220^2 = 80^2 + V_L^2$

Solving we get,

$V_L \approx 205$ V

$X_L = \dfrac{V_L}{I} = \dfrac{205}{10} = 20.5\ \Omega = \omega L$

$L = \dfrac{20.5}{2\pi \times 50} = 0.065$ H

4. *(a)*

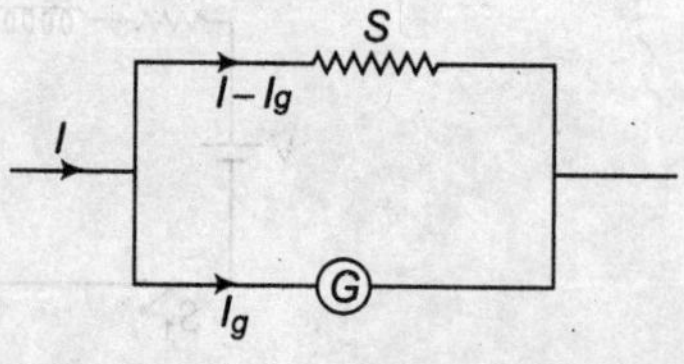

In parallel, current distributes in inverse ratio of resistance.

Hence, $\dfrac{I - I_g}{I_g} = \dfrac{G}{S} \Rightarrow S = \dfrac{GI_g}{I - I_g}$

As I_g is very small, hence $S = \dfrac{GI_g}{I}$

$$b = \frac{(100)(1 \times 10^{-3})}{10} = 0.01\,\Omega$$

5. *(b)* Induced constant, $I = \dfrac{e}{R}$

Here, e = induced emf

$$= \frac{d\phi}{dt}$$

$$I = \frac{e}{R} = \left(\frac{d\phi}{dt}\right) \cdot \frac{1}{R}$$

$$d\phi = IRdt$$

$$\phi = \int IRdt$$

$\therefore$ Here, R is constant

$\therefore$ $\phi = R\int Idt$

$\int I \cdot dt$ = Area under $I-t$ graph

$$= \frac{1}{2} \times 10 \times 0.5 = 2.5$$

$\therefore$ $\phi = R \times 2.5 = 100 \times 2.5 = 250$ Wb.

6. *(d)* Time period of oscillation is

$$T = 2\pi\sqrt{\frac{I}{MB}}$$

$$\Rightarrow \quad T = 2\pi\sqrt{\frac{7.5 \times 10^{-6}}{6.7 \times 10^{-2} \times 0.01}} = 0.665\text{ s}$$

Hence, time for 10 oscillations is $t = 6.65$ s.

7. *(d)* Suppose a resistance R_s is connected in series with galvanometer to convert it into voltmeter.

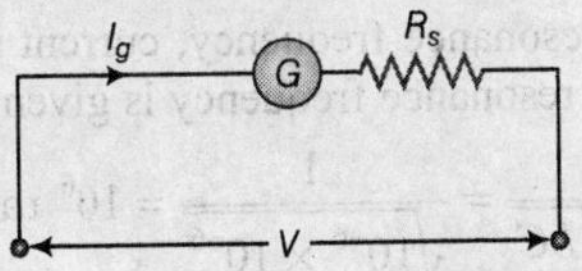

$$I_g(G + R_s) = V \Rightarrow R = \frac{V}{I_g} - G$$

$\Rightarrow$ $R = 1985 = 1.985\text{ k}\Omega$ or $R = 1.985 \times 10^3\ \Omega$

8. *(b)* Given, $e = 100\sin 30t$

and $\quad i = 20\sin\left(30t - \dfrac{\pi}{4}\right)$

$\therefore$ Average power ,

$$P_{av} = V_{rms}I_{rms}\cos\phi = \frac{100}{\sqrt{2}} \times \frac{20}{\sqrt{2}} \times \cos\frac{\pi}{4} = \frac{1000}{\sqrt{2}}\text{ watt}$$

Wattless current is,

$$I = I_{rms}\sin\phi = \frac{20}{\sqrt{2}} \times \sin\frac{\pi}{4} = \frac{20}{2} = 10\text{A}$$

$\therefore$ $P_{av} = \dfrac{1000}{\sqrt{2}}$ watt and $I_{wattless} = 10$A

9. *(b)* From $Bqv = \dfrac{mv^2}{r}$, we have

$$r = \frac{mv}{Bq} = \frac{\sqrt{2mK}}{Bq}$$

where, K is the kinetic energy.

As, kinetic energies of particles are same;

$$r \propto \frac{\sqrt{m}}{q} \Rightarrow r_e : r_p : r_\alpha = \frac{\sqrt{m_e}}{e} : \frac{\sqrt{m_p}}{e} : \frac{\sqrt{4m_p}}{2e}$$

Clearly, $r_p = r_\alpha$ and r_e is least $\quad [\because m_e < m_p]$

So, $\quad r_p = r_\alpha > r_e$

10. *(c)* **Key Idea** *As $m = IA$, so to change dipole moment (current is kept constant), we have to change radius of loop.*

Initially, $\quad m = I\pi R^2$ and $B_1 = \dfrac{\mu_0 I}{2R_1}$

Finally, $\quad m' = 2m = I\pi R_2^2$

$\Rightarrow$ $\quad 2I\pi R_1^2 = I\pi R_2^2$ or $R_2 = \sqrt{2}R_1$

So, $\quad B_2 = \dfrac{\mu_0 I}{2(R_2)} = \dfrac{\mu_0 I}{2\sqrt{2}R_1}$

Hence, ratio $\dfrac{B_1}{B_2} = \dfrac{\left(\dfrac{\mu_0 I}{2R_1}\right)}{\left(\dfrac{\mu_0 I}{2\sqrt{2}R_1}\right)} = \sqrt{2}$

$\therefore$ $\quad$ Ratio $\dfrac{B_1}{B_2} = \sqrt{2}$

11. *(a)* Sharpness of resonance of a resonant L-C-R circuit is determined by the ratio of resonant frequency with the selectivity of circuit. This ratio is also called ''Quality Factor'' or Q-factor.

$$Q\text{-factor} = \frac{\omega_0}{2\Delta\omega} = \frac{\omega_0 L}{R} = \frac{1}{\omega_0 CR}$$

JEE Advanced

1. **(8)**

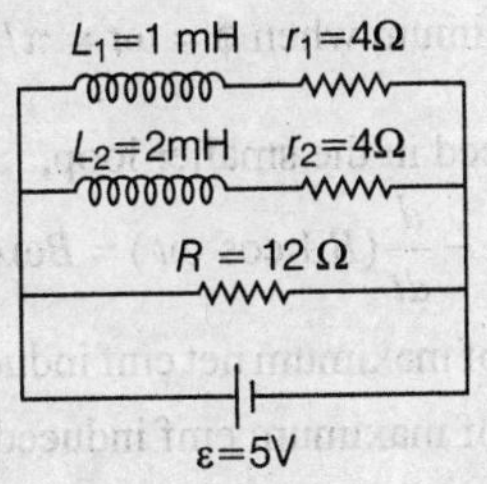

$I_{max} = \dfrac{\varepsilon}{R} = \dfrac{5}{12}$A $\quad$ (Initially at $t = 0$)

$I_{min} = \dfrac{\varepsilon}{R_{eq}} = \varepsilon\left(\dfrac{1}{r_1} + \dfrac{1}{r_2} + \dfrac{1}{R}\right)$ $\quad$ (finally in steady state)

$$= 5\left(\frac{1}{3} + \frac{1}{4} + \frac{1}{12}\right) = \frac{10}{3}\text{A}$$

$$\frac{I_{max}}{I_{min}} = 8$$

2. ***(b,c)***

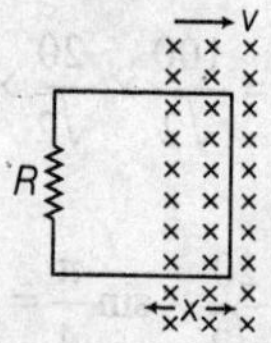

When loop was entering ($x < L$)

$$\phi = BLx \Rightarrow \quad e = -\frac{d\phi}{dt} = -BL\frac{dx}{dt}$$

$$|e| = BLv$$

$$i = \frac{e}{R} = \frac{BLv}{R} \text{ (anti- clockwise)}$$

$$F = ilB \qquad \text{(Left direction)}$$

$$= \frac{B^2L^2v}{R} \qquad \text{(in left direction)}$$

$$\Rightarrow \quad a = \frac{F}{m} = -\frac{B^2L^2v}{mR} \Rightarrow a = v\frac{dv}{dx}$$

$$v\frac{dv}{dx} = -\frac{B^2L^2v}{mR}$$

$$\Rightarrow \quad \int_{v_0}^{v} dv = -\frac{B^2L^2}{mR}\int_0^x dx$$

$$\Rightarrow \quad v = v_0 - \frac{B^2L^2v}{mR}x$$

(straight line of negative slope for $x < L$)

$$I = \frac{BL}{R}v$$

$\Rightarrow$ (I *vs* x will also be straight line of negative slope for $x < L$)

$$L \le x \le 3L$$

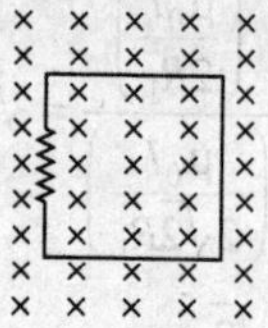

$$\frac{d\phi}{dt} = 0; \; e = 0, i = 0; F = 0$$

$$x > 4L \quad e = Blv$$

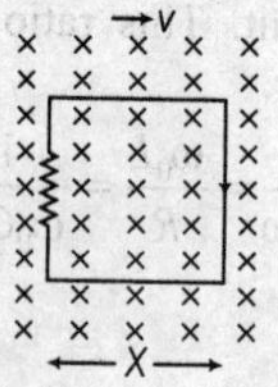

Force also will be in left direction.

$$i = \frac{BLv}{R} \qquad \text{(clockwise)}$$

$$a = -\frac{B^2L^2v}{mR} = v\frac{dv}{dx} \Rightarrow \quad F = \frac{B^2L^2v}{R}$$

$$\int_L^x -\frac{B^2L^2}{mR}dx = \int_{v_i}^{v_f} dv$$

$$\Rightarrow \quad -\frac{B^2L^2}{mR}(x - L) = v_f - v_i$$

$$v_f = v_i - \frac{B^2L^2}{mR}(x - L) \text{ (straight line of negative slope)}$$

$$I = \frac{BLv}{R} \rightarrow \text{ (clockwise)}$$

(straight line of negative slope)

3. ***(a,c)*** By reciprocity theorem of mutual induction, it can be assumed that current in infinite wire is varying at 10A/s and EMF is induced in triangular loop.

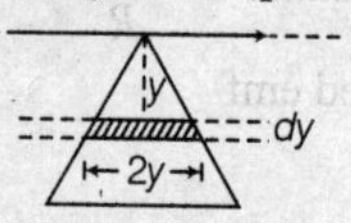

Flux of magnetic field through triangle loop, if current in infinite wire is ϕ, can be calculated as follows:

$$d\phi = \frac{\mu_0 i}{2\pi y}\cdot 2y\,dy$$

$$d\phi = \frac{\mu_0 i}{\pi}dy \Rightarrow \phi = \frac{\mu_0 i}{\pi}\left(\frac{l}{\sqrt{2}}\right)$$

$$\Rightarrow \quad \text{EMF} = \left|\frac{d\phi}{dt}\right| = \frac{\mu_0}{\pi}\left(\frac{l}{\sqrt{2}}\right)\cdot\frac{di}{dt}$$

$$= \frac{\mu_0}{\pi}(10\text{cm})\left(10\frac{\text{A}}{\text{s}}\right) = \frac{\mu_0}{\pi}\text{volt}$$

If we assume the current in the wire towards right then as the flux in the loop increases we know that the induced current in the wire is counter clockwise. Hence, the current in the wire is towards right. Field due to triangular loop at the location of infinite wire is into the paper. Hence, force on infinite wire is away from the loop. By cylindrical symmetry about infinite wire, rotation of triangular loop will not cause any additional EMF.

4. ***(a,b)*** At $\omega \approx 0$, $X_C = \frac{1}{\omega C} = \infty$. Therefore, current is nearly zero.

Further at resonance frequency, current and voltage are in phase. This resonance frequency is given by,

$$\omega_r = \frac{1}{\sqrt{LC}} = \frac{1}{\sqrt{10^{-6} \times 10^{-6}}} = 10^6 \text{ rad/s}$$

We can see that this frequency is independent of R.

Further, $X_L = \omega L$, $X_C = \frac{1}{\omega C}$

At, $\omega = \omega_r = 10^6$ rad/s, $X_L = X_C$.

For $\omega > \omega_r$, $X_L > X_C$. So, circuit is inductive.

5. ***(b,d)*** The net magnetic flux through the loops at time t is

$$\phi = B(2A - A)\cos\omega t = BA\cos\omega t$$

$$\text{so, } \left|\frac{d\phi}{dt}\right| = B\omega A\sin\omega t$$

$$\therefore \left|\frac{d\phi}{dt}\right| \text{ is maximum when } \phi = \omega t = \pi/2$$

The emf induced in the smaller loop,

$$\varepsilon_{\text{smaller}} = -\frac{d}{dt}(BA\cos\omega t) = B\omega A\sin\omega t$$

$\therefore$ Amplitude of maximum net emf induced in both the loops = Amplitude of maximum emf induced in the smaller loop alone.

6. *(c)* For particle to move in negative y-direction, either its velocity must be in negative y-direction (if initial velocity $\neq 0$) and force should be parallel to velocity or it must experience a net force in negative y-direction only (if initial velocity = 0)

7. *(a)* $\mathbf{F}_{net} = \mathbf{F}_e + \mathbf{F}_B = q\mathbf{E} + q\mathbf{v} \times \mathbf{B}$

For particle to move in straight line with constant velocity, $\mathbf{F}_{net} = 0$

$$\therefore \qquad q\mathbf{E} + q\mathbf{v} \times \mathbf{B} = 0$$

8. *(c)* For path to be helix with axis along positive z-direction, particle should experience a centripetal acceleration in xy-plane.

For the given set of options only option (c) satisfy the condition. Path is helical with increasing pitch.

9. *(a)*

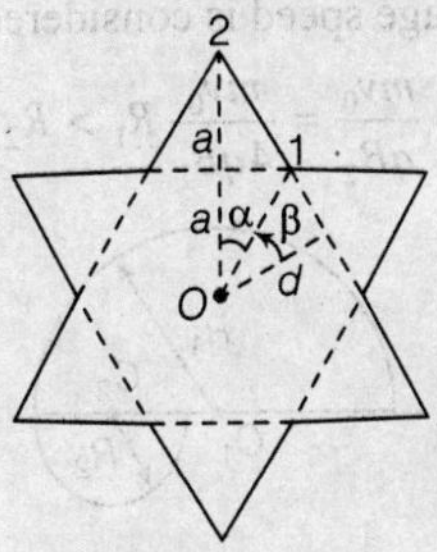

$$B_{12} = \frac{\mu_0 I}{4\pi d}[\sin\alpha + \sin\beta]$$

$$\alpha = 60° \text{ and } \beta = -30°$$

$$= \frac{\mu_0 I}{4\pi d}\left[\frac{\sqrt{3}}{2} - \frac{1}{2}\right]$$

$$B_{12} = \frac{\mu_0 I}{4\pi d}\left[\frac{\sqrt{3}-1}{2}\right]$$

$$d = a$$

$$B_0 = 12\,B_{12} = 12 \times \frac{\mu_0 I}{4\pi d}\left[\frac{\sqrt{3}-1}{2}\right]$$

$$= \frac{\mu_0 I}{4\pi a} 6[\sqrt{3}-1]$$

10. *(a,b,c)*

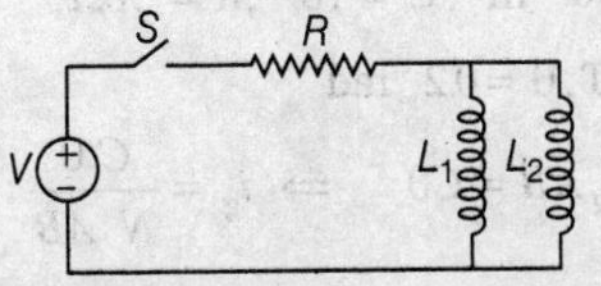

Since inductors are connected in parallel

$$V_{L_1} = V_{L_2}$$

$$L_1 \frac{dI_1}{dt} = L_2 \frac{dI_2}{dt}$$

$$L_1 I_1 = L_2 I_2$$

$$\frac{I_1}{I_2} = \frac{L_2}{L_1}$$

Current through resistor at any time t is given by

$$I = \frac{V}{R}(1 - e^{-\frac{RT}{L}}),$$

where $L = \dfrac{L_1 L_2}{L_1 + L_2}$

After long time $I = \dfrac{V}{R}$

$$I_1 + I_2 = I \qquad \text{...(i)}$$

$$L_1 I_1 = L_2 I_2 \qquad \text{...(ii)}$$

From Eqs. (i) and (ii), we get

$$I_1 = \frac{V}{R}\frac{L_2}{L_1 + L_2}, \quad I_2 = \frac{V}{R}\frac{L_1}{L_1 + L_2}$$

(d) Value of current is zero at $t = 0$

Value of current is V/R at $t = \infty$

Hence option (d) is incorrect.

11. *(b,c)* (a)

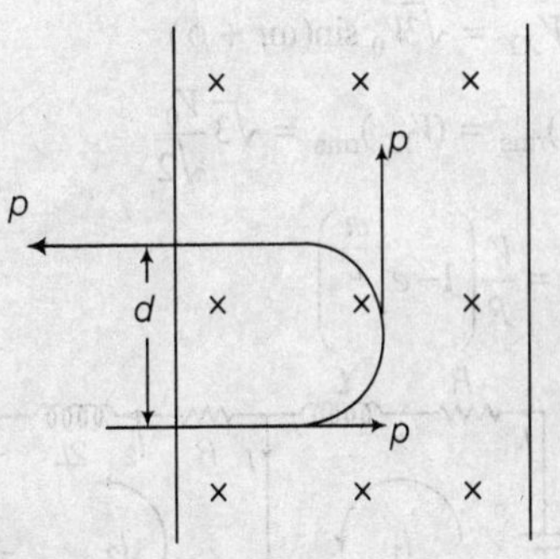

$$|\Delta\mathbf{P}| = \sqrt{2}p$$

(b) $r(1 - \cos\theta) = R$

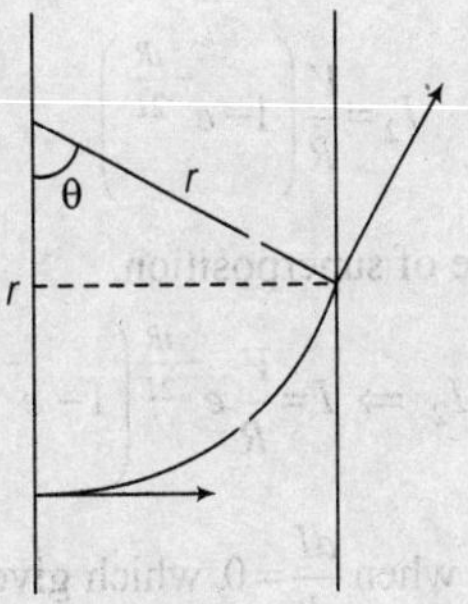

$$r\sin\theta = \frac{3R}{2} \Rightarrow \frac{\sin\theta}{1-\cos\theta} = \frac{3}{2}$$

$$\frac{2\sin\frac{\theta}{2}\cos\frac{\theta}{2}}{2\sin^2\frac{\theta}{2}} = \frac{3}{2} \quad \cot\frac{\theta}{2} = \frac{3}{2}$$

$$\Rightarrow \qquad \tan\frac{\theta}{2} = \frac{2}{3} \Rightarrow \tan\theta = \frac{2\left(\frac{2}{3}\right)}{1 - \frac{4}{9}} = \frac{\frac{4}{3}}{\frac{5}{9}} = \frac{12}{5}$$

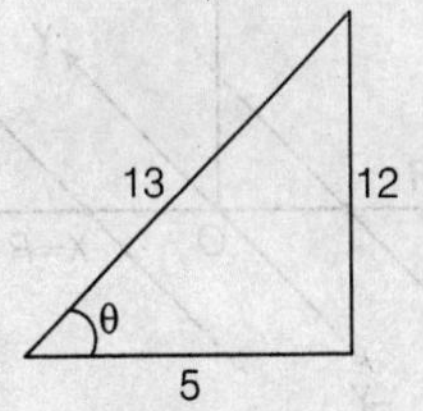

$$\sin\theta = \frac{12}{13}$$

$$r\left(\frac{12}{13}\right) = \frac{3R}{2}; \quad r = \frac{13R}{8} = \frac{P}{QB}; \quad B = \frac{8P}{13QR}$$

(c) $\dfrac{P}{QB} < \dfrac{3R}{2}, B > \dfrac{2P}{3QR}$

(d) $r = \frac{mv}{QB}$, $d = 2r = \frac{2mv}{QB} \Rightarrow d \propto m$

12. *(b,c)* $V_{XY} = V_0 \sin\left(\omega t + \frac{2\pi}{3}\right) - V_0 \sin \omega t$

$$= V_0 \sin\left(\omega t + \frac{2\pi}{3}\right) + V_0 \sin(\omega t + \pi)$$

$$\Rightarrow \quad \phi = \pi - \frac{2\pi}{3} = \frac{\pi}{3}$$

$$\Rightarrow \quad V_0' = 2V_0 \cos\left(\frac{\pi}{6}\right) = \sqrt{3}\, V_0$$

$$\Rightarrow \quad V_{XY} = \sqrt{3}V_0 \sin(\omega t + \phi)$$

$$\Rightarrow (V_{XY})_{\text{rms}} = (V_{YZ})_{\text{rms}} = \sqrt{3}\frac{V_0}{\sqrt{2}}$$

13. *(b, d)* $I_1 = \frac{V}{R}\left(1 - e^{-\frac{tR}{L}}\right)$

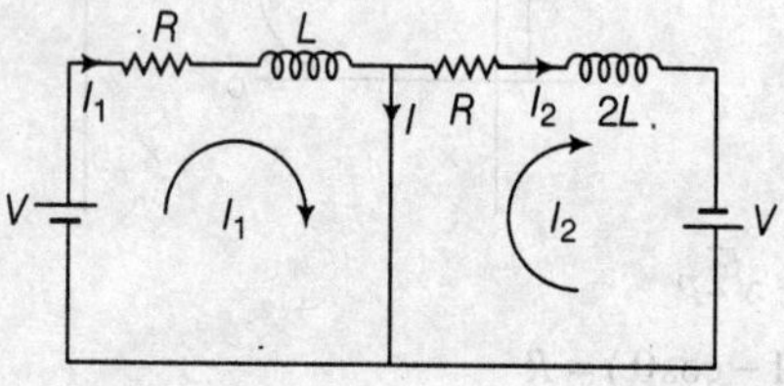

$$I_2 = \frac{V}{R}\left(1 - e^{-\frac{tR}{2L}}\right)$$

From principle of superposition,

$$I = I_1 - I_2 \Rightarrow I = \frac{V}{R} e^{-\frac{tR}{2L}}\left(1 - e^{-\frac{tR}{2L}}\right) \quad \ldots\text{(i)}$$

I is maximum when $\frac{dI}{dt} = 0$, which gives

$$e^{-\frac{tR}{2L}} = \frac{1}{2} \text{ or } t = \frac{2L}{R}\ln 2$$

Substituting this time in Eq. (i), we get

$$I_{\max} = \frac{V}{4R}$$

14. *(a,b,d)*

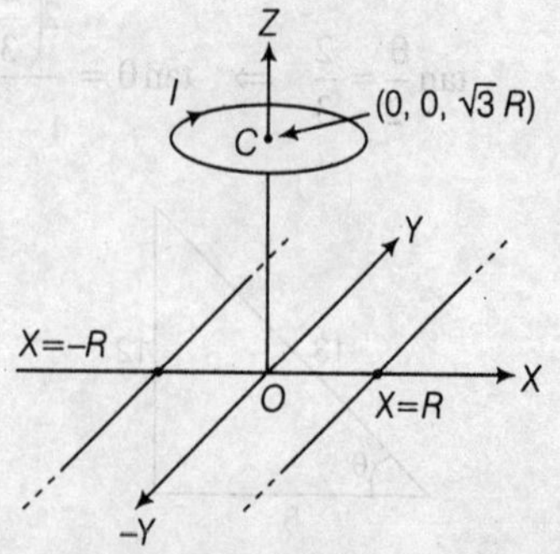

(a) At origin, $\mathbf{B} = 0$ due to two wires if $I_1 = I_2$, hence $(\mathbf{B}_{\text{net}})$ at origin is equal to **B** due to ring. which is non-zero.

(b) If $I_1 > 0$ and $I_2 < 0$, **B** at origin due to wires will be along $+\hat{\mathbf{k}}$. Direction of **B** due to ring is along $-\hat{\mathbf{k}}$ direction and hence **B** can be zero at origin.

(c) If $I_1 < 0$ and $I_2 > 0$, **B** at origin due to wires is along $-\hat{\mathbf{k}}$ and also along $-\hat{\mathbf{k}}$ due to ring, hence **B** cannot be zero.

(d)

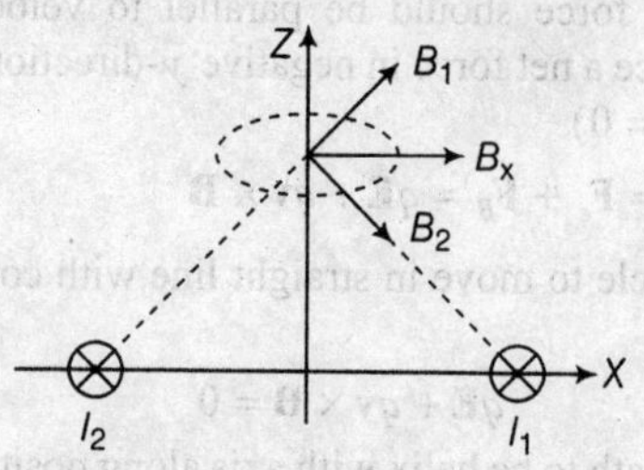

At centre of ring, **B** due to wires is along x-axis. Hence z-component is only because of ring which $B = \frac{\mu_0 i}{2R}(-\hat{\mathbf{k}})$.

15. *(2 m/s)* If average speed is considered along x-axis,

$$R_1 = \frac{mv_0}{qB_1}, R_2 = \frac{mv_0}{qB_2} = \frac{mv_0}{4qB_1} \; R_1 > R_2$$

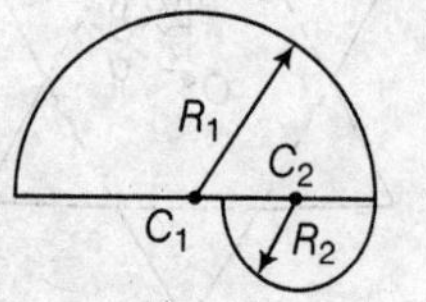

Distance travelled along x-axis, $\Delta x = 2(R_1 + R_2) = \frac{5mv_0}{2qB_1}$

$$\text{Total time} = \frac{T_1}{2} + \frac{T_2}{2} = \frac{\pi m}{qB_1} + \frac{\pi m}{qB_2}$$

$$= \frac{\pi m}{qB_1} + \frac{\pi m}{4qB_1}$$

$$= \frac{5\pi m}{4qB_1}$$

$$\text{Magnitude of average speed} = \frac{\frac{5mv_0}{2qB_1}}{\frac{5\pi m}{4qB_1}} = 2 \text{ m/s}$$

16. *(5.55)* Given, $N = 50$,

$A = 2 \times 10^{-4} \text{ m}^2, C = 10^{-4}, R = 50\Omega,$

$B = 0.02 \text{ T}, \theta = 0.2 \text{ rad}$

$$\therefore \quad N i_g AB = C\theta \quad \Rightarrow i_g = \frac{C\theta}{N\,AB}$$

$$= \frac{10^{-4} \times 0.2}{50 \times 2 \times 10^{-4} \times 0.02} = 0.1\text{A}$$

$$\therefore \quad V_{ab} = i_g \times G = (i - i_g)S$$

$$0.1 \times 50 = (1 - 0.1) \times S$$

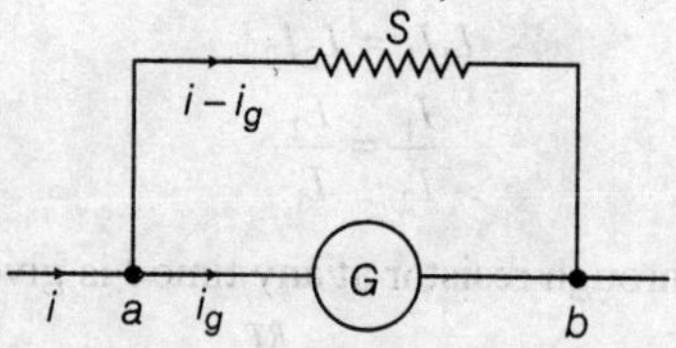

$$5 = 0.9 \times S$$

$$\therefore \quad S = \frac{50}{9}\,\Omega = 5.55\,\Omega$$